BARRON'S

HOW TO PREPARE FOR THE

ENVIRONMENTAL SCIENCE

ADVANCED PLACEMENT EXAMINATION

Gary S. Thorpe, M.S.
Beverly Hills High School
Beverly Hills, CA

Acknowledgments
I would like to thank my wife, Patti, and my two daughters, Krissi and Erin, for their patience and understanding while I was writing this book. Special thanks also goes to Dr. Jerry Bobrow of Bobrow Test Preparation Services for coordinating this project and, most of all, to the many professional and dedicated high school APES teachers who have contributed their work and talent to make this book possible.

About the Author
Gary Thorpe has taught science at the secondary and college level for 30 years. He lives in Los Angeles with his wife and two daughters, and also works as a patrol officer for the Los Angeles Police Department.

All inquiries should be addressed to:
Barron's Educational Series, Inc.
250 Wireless Boulevard
Hauppauge, New York 11788
http://www.barronseduc.com

International Standard Book No. 0-7641-2161-8

Library of Congress Catalog Card No. 2002066720

Library of Congress Cataloging-in-Publication Data
Thorpe, Gary S.
Barron's how to prepare for the AP : advanced placement test in environmental science / Gary S. Thorpe.
p. cm.
Includes index.
ISBN 0-7641-2161-8
1. Environmental sciences—Examinations, questions, etc. 2. Advanced placement programs (Education)—Examinations—Study guides. I. Title: How to prepare for the AP. II. Title.
GE76 .T48 2002
363.7′0076—dc21 2002066720

Printed in the United States of America
9 8 7 6 5 4 3 2 1

Contents

Part I: Introduction and Strategies

Part II: Specific Topics

UNIT I: INTERDEPENDENCE OF EARTH'S SYSTEMS—FUNDAMENTAL CONCEPTS

PART I: Introduction and Strategies

Preface

The AP Environmental Science Exam is coming up! Your thorough understanding of months and months of reading, college-level lectures, tests, labs, lab write-ups, field studies, and notes are to be evaluated in a 3-hour examination. It's just you and the AP exam. In preparing to do the very best possible, you have two options:

1. Read your entire textbook again.
2. Use Barron's How To Prepare for the AP Environmental Science Exam.

I'm glad you chose option 2. I've taught Environmental Science at the college-level for years, and I've put together in a reasonable number of pages what I feel are the essential facts and issues that you will need to review for the exam. Concepts are separated, broken down and explained, much like you would do when you take notes or make note-cards. With other AP exams to study for and other time commitments, you need a quick and neat book that you can finish in a few weeks and that covers just about everything you might expect to find on the exam. You have that book in your hands.

This guide is divided into three parts.

Part I: Introduction

Part I contains the following sections: Questions Commonly Asked About the AP Environmental Science Exam, Strategies for Taking the AP Environmental Science Exam, and Methods for Writing the Essays.

Part II: Specific Topics

Each chapter lists key vocabulary words, key concepts, legislation, and important case studies. Each chapter also includes fifteen multiple-choice questions, and their solutions and explanation. Finally, each chapter includes a sample free-response essay question including a sample rubric that the College Board might use to evaluate your response. Don't expect your answer to match the one given in this book. It won't. The essays in the book were contributed by Advanced Placement Environmental Science (APES) teachers who had months to write and perfect their essays. No other preparation book contains such a wide variety of input from known experts in environmental science. You will see what actual APES instructors consider to be superb essays.

Self-contained chapters cover the flow of energy, the cycling of matter, the Earth, the atmosphere, the biosphere, population dynamics, water, minerals and soil, biological dynamics, solid wastes, pollution and its impact on human health, global changes, consequences of higher-order interactions, economic consequences, cultural and aesthetic considerations, environmental ethics, environmental laws and regulations, and conservation. *Every topic tested* on the AP Environmental Science Exam, as published by the College Board, is covered in this book.

Part III: AP Environmental Science Practice Exams

This book contains two complete AP Environmental Science Exams consisting of both multiple-choice and free-response questions. All solutions are fully worked out and explained. Please note that on the actual APES exam, there are no 1/2-point credits given on free-response questions. In this book, however, I have used them to spell out as clearly as possible what can be valuable in your responses.

This book is not a textbook. It was never meant to be one. The last thing you need right now is another environmental science textbook. In fact, most of the material presented is in the form you would use if you had to outline your own textbook. Furthermore, many of the sample essays in this book have come from APES teachers. You will get a broad range of writing styles and philosophies as you go through the essays. If you have forgotten concepts or if something is new to you, use this preparation guide *with your textbook* to prepare for the AP Environmental Science Exam. Now turn to the Study Guide Checklist on page 6 and check each item as you complete the task. When you have checked all the items, you will be ready for the AP Environmental Science Exam.

Introduction

Purpose

The Advanced Placement Environmental Science Exam is offered once a year, in May. In May 2001, 18,880 students nationwide took the test, and each year the number gets larger. It is one of the fastest-growing AP programs. This book offers a comprehensive review of the key concepts taught in a college Environmental Science course, with many examples, case studies, and diagrams included that you can use for the exam. In addition, this book contains:

- Overviews of each topic.
- Lists of key terms and definitions for each chapter. Essential vocabulary words have been identified for each chapter and have been placed in each chapter respectively.
- Outlines of only the material that you will be tested on as described by the College Board. This book "teaches the test." Only information that you will be tested on is put into the book.
- Organizational charts where information is presented in logical, easy-to-remember, and comparative formats.
- Multiple-choice questions for each chapter covering the material and complete explanations for each question.
- Key legislation including summaries of important environmental laws.
- Case studies wherein important current and past environmental issues are summarized for you to incorporate into your essays.
- A comprehensive alphabetical index at the back of the book, directing you quickly to the page where you can find more information on that topic.
- Free-response questions (essays) written by the author *and* APES instructors and college professors who teach the subject. These experts have been assembled from throughout the United States to bring to you the very best examples of writing essays dealing with current topics in environmental science.
- Two complete AP practice exams with explanations and answers to the multiple-choice questions and complete rubrics (grading standards) for the free-response section. The rubrics are those that would be used to evaluate the essay and writing samples from both experts and students showing what *(and what not)* to do in answering the essays.

Suggested Uses for This Book

- I recommend using this book with your textbook throughout your high school course. Read the appropriate chapter before you take the exam in class. Answer the multiple-choice questions at the end of the chapter you are covering and review the questions that you miss. Review the key terms. You cannot do well on essays unless you know the vocabulary. Analyze the free-response question. You will save many,

many hours of "cramming" if you use this book throughout your class. Before your final exam, take the two practice exams at the end of this book.

or

- For last-minute "crammers" (and I confess, I am one of you) who have just a few weeks before the AP Environmental Science Exam, this book is perfect. Being a crammer myself, I know from experience what information you will need to use before an exam and how it needs to be presented in a form that can be quickly used. You will not need to go back to your textbook. Everything you will be tested on is in this book. There is no need to review notes—it's all been done for you. Go through one chapter per day. The weekend before the exam, take the two practice exams.

A Note for the Teacher

Use this book in your class from the very first week. You may want to have your students take a practice exam in the first few days as a pretest to get a baseline of what information they bring to the class. Some of them may have already had AP Biology or Honors Biology and know some of this material already. Others may be new to the Advanced Placement program or may have only had a regular high school biology class and are starting from a different point. Nevertheless, they are all eager to do well.

You may want to make overheads of the charts in this book. You will find that you can cover twice the amount of information in the same time if they have this book with them as you are going through diagrams or charts on the screen. Try making overheads of the essays too. As you go through the essays using the overhead, they can make notes in their books. And, the notes will always be there. You won't have "I can't find it" or "I didn't bring it" types of excuses. It helps if each student has his or her own copy of this book, that way, as a teacher, you can assign vocabulary terms, assign certain charts to memorize, etc., and everyone will be "on the same page." *Encourage* your students to write in this book.

Another idea that works for me is to give a "brainstorm" quiz at the end of each chapter. Have the students take out a piece of paper and give them 5 minutes to write down as many of the key terms from the chapter as they can think of. Don't test for definitions. This is the same thinking process that they will use when they are faced with the free-response section of the AP exam. Set a certain minimum number of terms that they have to think of and give "extra credit" for any above that number. Somehow they generally know the definitions; what they have trouble with is thinking of the key terms (which lead to concepts) in the first place to put into their essays. Once they come up with a list of key terms for a particular concept, interconnecting them and then writing about them is a much easier task.

Just before the AP exam, give the same practice exam that you gave at the beginning of the class (or the other version) under actual test conditions. You can then evaluate the students on how much progress they have made in your course and how prepared they are for the AP exam.

Finally, do you or your class have ideas on how to make this book even better? Would you like to see one of your essays that you use or one of your student's best essays be used in a future edition of this book? A book like this is never finished. It can always be made better. I encourage you and your students to send me material that you would like to see in the next edition. You might want to have a "contest" for the best essay. All material that is incorporated will be referenced to the author and his or her school. Please submit material to Barron's Educational Series, Inc., c/o AP Environmental Science, 250 Wireless Boulevard, Hauppauge, New York 11788. This book only gets better with your and your students' input and suggestions.

Wishing you only 5's.

Gary S. Thorpe

Study Guide Checklist

❑ 1. Read the Advanced Placement Course Description—Environmental Science (also commonly known as the Acorn Book) produced by the Educational Testing Service (ETS) and available from your AP Environmental Science teacher, testing office, counseling center, or The College Board.

❑ 2. Read the Preface on page 1 of this book.

❑ 3. Read "Questions Commonly Asked About the AP Environmental Science Exam" on page 12.

❑ 4. Read "Strategies for Taking the AP Environmental Science Exam" on page 15.

❑ 5. Go through each chapter, carefully reviewing each key term. You should be familiar with the meanings of these words and with the principles underlying them. They can serve as the foundations for your essays. The more of these key terms that you can use effectively in your explanations (provided that they are relevant), the better your free-response answers will be. Next, review each key concept. Key concepts are broken down into charts, diagrams, and bulleted lists that will help you remember. Take the multiple-choice quiz found in each chapter and go over any mistakes that you may have made by referring to the explanations in the answer key. Then, carefully read the essay found in each chapter. Carefully study the rubric that is included with each essay. Then, without referring to the essay, take the question and write your own answer and compare it to the one given in the book. Don't expect your answer to match the one in this book. It won't. Many of the essays in this book were contributed by the world's top APES teachers. They had months to write and perfect their essays. They also had access to reference material as they wrote. You won't. However, if over time, you find that your essays are getting better in terms of content and organization, then you are on your way to getting a 5!

❑ Chapter 1: Energy

❑ Chapter 2: The Cycling of Matter

❑ Chapter 3: The Solid Earth

❑ Chapter 4: The Atmosphere

❑ Chapter 5: The Biosphere

❑ Chapter 6: Human History and Global Distribution

❑ Chapter 7: Water

❑ Chapter 8: Food and Other Agricultural Products

❑ Chapter 9: Land

❑ Chapter 10: Air, Water, and Soil Pollution

- ❑ Chapter 11: Solid Waste
- ❑ Chapter 12: Human Health
- ❑ Chapter 13: Cause and Effect Relationships
- ❑ Chapter 14: Economic Forces
- ❑ Chapter 15: Environmental Ethics
- ❑ Chapter 16: Environmental History, Laws, and Regulations: Regional, National, and International

- ❑ 6. Take AP Environmental Science Practice Exam 1. Review your errors.
- ❑ 7. Take AP Environmental Science Practice Exam 2. Review your errors.

Format of the AP Environmental Science Exam*

The AP Environmental Science Exam is 3 hours long and is divided equally in time between a multiple-choice section and a free-response section. The multiple-choice section, which constitutes 60% of the final grade, consists of 100 multiple-choice questions designed to cover the breadth of the students' knowledge and understanding of environmental science. Thought-provoking problems and questions based on fundamental ideas from environmental science are included along with questions based on the recall of basic facts and major concepts. The number of multiple-choice questions taken from each major topic area is reflected in the percentage of the course as designated in the outline of topics (see page 9).

The free-response section emphasizes the application of principles in greater depth. In this section, students must organize answers to broad questions, thereby demonstrating reasoning and analytical skills as well as the ability to synthesize material from several sources into cogent and coherent essays. Four free-response questions are included in this section, which constitutes 40% of the final grade: one data-set question, one document-based question, and two synthesis and evaluation questions.

To provide maximum information about differences in students' achievements in environmental science, the examination is designed to yield average scores of about 50% of the maximum possible scores for both the multiple-choice and free-response sections. Thus, students should be aware that they might find the AP Environmental Science Exam more difficult than most classroom examinations. However, it is possible for students who have studied most, but not all topics, to obtain an acceptable grade.

Section I: Multiple-Choice Questions

90 minutes
100 questions: 60% of total grade

Section II: Free-Response Questions

90 minutes: 40% of total grade
1 data-set question, 1 document-based question, and 2 synthesis and evaluation questions

*Advanced Placement Program Course Description—Environmental Science, May 2002, AP College Board.

Topics Covered by the AP Environmental Science Exam*

The percentage after each major topic indicates the approximate proportion of questions on the examination that pertain to the specific topic. The examination is constructed using the percentages as guideline for question distribution.

I. Interdependence of Earth's Systems: Fundamental Principles and Concepts (25%)

A. Energy
1. Forms and quality of energy
2. Energy units and measurements
3. Sources and sinks, conversions
4. Conventional and alternative sources

B. The Cycling of Matter
1. Water
2. Carbon
3. Major nutrients
    a. Nitrogen
    b. Phosphorus
    c. Sulfur
4. Differences between cycling of major and trace elements

C. The Solid Earth
1. Earth history and the geologic time scale
2. Earth dynamics—earthquakes, plate tectonics, the rock cycle, soil formation, volcanism

D. The Atmosphere
1. Atmospheric history: origin, evolution, composition, structure
2. Atmospheric dynamics: weather and climate

E. The Biosphere
1. Natural Areas
2. Organisms: adaptations to their environments
3. Populations and communities: exponential growth and carrying capacity
4. Ecosystems and change: biomass, energy transfer, succession
5. Evolution of life: natural selection, genetic diversity, extinction

*Advanced Placement Program Course Description—Environmental Science, May 2001, AP College Board.

II. Human Population Dynamics (10%)

A. Human History and Global Distribution
1. Numbers
2. Demographics—birth and death rates
3. Patterns of resource utilization
4. Carrying capacity—local, regional, and global
5. Cultural, aesthetic, and economic influences

III. Renewable and Nonrenewable Resources: Distribution, Ownership, Use, and Degradation (15%)

A. Water
1. Fresh: agricultural, industrial, and domestic
2. Oceans: fisheries and industrial

B. Minerals

C. Soils
1. Soil types
2. Erosion and conservation

D. Biological
1. Natural areas
2. Genetic diversity
3. Food and other agricultural products

E. Energy
1. Conventional sources
2. Alternative sources

F. Land
1. Residential and commercial
2. Agricultural and forestry
3. Recreational and wilderness

IV. Environmental Quality (20–25%)

A. Air, Water, and Soil Pollution
1. Major pollutants
   a. Types, such as SO_2, NO_x, and pesticides
   b. Thermal pollution
   c. Measurement and units of measure (ppm, pH, micrograms)
   d. Point and nonpoint sources (domestic, industrial, agricultural)
2. Effects of pollutants
   a. Aquatic systems
   b. Vegetation
   c. Natural features, buildings, and structures
   d. Wildlife
3. Pollution reduction, remediation, and control

B. Solid Waste
1. Types, sources, and amounts
2. Current disposal methods and limitations
3. Alternative practices in solid waste management

C. Human Health
1. Agents: chemical and biological
2. Effects: acute and chronic, dose-response relationships
3. Relative risks: evaluation and response

V. Global Changes and Their Consequences (15–20%)

A. First-order effects (changes)
1. Atmosphere: CO_2, CH_4, stratospheric ozone
2. Oceans: surface temperatures, currents
3. Biota: habitat destruction, introduced exotics, overharvesting

B. Higher-order interactions (consequences)
1. Atmosphere: global warming, increasing ultraviolet radiation
2. Oceans: increasing sea level, long-term climate change, impact on El Niño
3. Biota: loss of biodiversity

VI. Environment and Society: Trade-Offs and Decision Making (10%)

A. Economic Forces
1. Cost-benefit analysis
2. Marginal costs
3. Ownership and externalized costs

B. Cultural and Aesthetic Considerations

C. Environmental Ethics

D. Environmental History, Laws, and Regulations (international, national, and regional)

E. Issues and options (conservation, preservation, restoration, remediation, sustainability, mitigation)

Questions Commonly Asked About the AP Environmental Science Exam

Q: What is the AP Environmental Science Exam?

A: The AP Environmental Science Exam is given once a year to high school students and tests their knowledge of concepts in a one-semester college-level course on environmental science. The student who passes the AP exam may receive college credit for taking AP Environmental Science in high school. Passing is generally considered to be achieving a score of 3, 4, or 5. The test is administered each May. It has two sections:

- Section I, worth 60% of the total score, is 90 minutes long and consists of 100 multiple-choice questions. The total score for Section I is the number of correct answers minus ¼ for each wrong answer. If you leave a question unanswered, it does not count at all. A student generally needs to answer from 50 to 60% of the multiple-choice questions correctly to obtain a 3 on the exam. The multiple-choice questions are based on fundamental ideas from environmental science and include questions based on the recall of basic facts and major concepts.
- Section II, worth 40% of the total score, is 90 minutes long and consists of four parts: one data-set question where calculations are required, one document-based question, and two synthesis and evaluation questions.

Q: What are the advantages of taking AP Environmental Science?

A: Students who pass the exam may, at the discretion of the college in which the student enrolls, be given full college credit for taking the class in high school.

- Taking the exam improves your chance of getting into the college of your choice. Studies show that students who successfully participate in AP programs in high school stand a much better chance of being accepted by selective colleges than students who do not.
- Taking the exam reduces the cost of a college education. In the many private colleges that charge upwards of $700 a unit, a one-semester environmental science course could cost as much as $3,500! Taking the course during high school saves money.
- Taking the exam may reduce the number of years needed to earn a college degree.
- If you take the course and the exam while still in high school, you will not be faced with the college course being closed or overcrowded.
- For those of you who are not going on in a science career, passing the AP Environmental Science exam may fulfill the science requirement at the college, thus making more time available for you to take other courses.
- Taking AP Environmental Science greatly improves your chances of doing well in the college course. You will have already covered most of the topics during your high school AP Environmental Science program, and you will find yourself setting the curve in college!

Q: Do all colleges accept AP exam grades for college credit?

A: Almost all of the colleges and universities in the United States and Canada, and many in Europe, take part in the AP program. The vast majority of the 2,900 U.S. colleges and universities that receive AP grades grant credit and/or advanced placement. Even colleges that receive only a few AP candidates and may not have specific AP policies are often willing to accommodate AP students who inquire about advanced-placement work.

To find out about a specific policy for the AP exam(s) you plan to take, write to the college's Director of Admissions. You should receive a written reply telling you how much credit and/or advanced placement you will receive for a given grade on an AP exam, including any courses you will be allowed to enter.

The best source of specific and up-to-date information about an individual institution's policy is its catalog or website. Other sources of information include The College Handbook with College Explorer CD-ROM and College Search. For more information on these and other products, log on to the College Board's online store at http://store.collegeboard.com/.

Q: How is the AP exam graded and what do the scores mean?

A: The AP exam is graded on a five-point scale:

5: Extremely well qualified. About 9% of the students who take the exam earn this grade.

4: Well qualified. Roughly 25% earn this grade.

3: Qualified. Generally 21% earn this grade.

2: Possibly qualified. Generally considered "not passing." About 17% of the students who take the exam earn this grade.

1: Not qualified. About 28% earn this grade.

Of the 18,880 students from 1,198 high schools who took the AP Environmental Science Exam in 2001, the average grade was 2.71 with a standard deviation of 1.35. Eleven hundred and thirty-eight colleges received AP scores from students who passed the AP Environmental Science exam.

Section I, the multiple-choice section, is machine graded. Each question has five answers to choose from. Remember, there is a penalty for guessing: one-quarter of a point is taken off for each wrong answer. A student generally needs to answer 50 to 60% of the multiple-choice questions correctly to obtain a 3 on the exam. Each answer in Section II, the free-response section, is read several times by different environmental science instructors who pay great attention to consistency in grading.

Q: Are there old exams out there that I could look at?

A: Yes! Questions (and answers) from previous exams are available from The College Board. Request an order form by contacting: AP Services, P.O. Box 6671, Princeton, NJ 08541; (609) 771-7300 or (888) 225-5427; Fax (609) 530-0482; TTY: (609) 882-4118; or email: apexams@ets.org. You can also log on to www.collegeboard.com/ap/ and follow the links to find current questions from past exams on-line.

Q: What materials should I take to the exam?

A: Be sure to take your admission ticket, some form of photo and signature identification, your social security number, several sharpened No. 2 pencils, a good eraser, and a watch.

Q: When will I get my score?

A: The exam itself is generally given in the second or third week of May. The scores are usually available during the second or third week of July.

Q: Should I guess on the test?

A: Except in certain special cases explained later in this book, you should not guess. There is a penalty for guessing on the multiple-choice section of the exam. As for the free-response section, it simply comes down to whether you know the material or not.

Q: Suppose I do terribly on the exam. *May I cancel the test and/or the scores?*

A: You may cancel an AP grade permanently only if the request is received by June 15 of the year in which the exam was taken. There is no fee for this service, but a signature is required to process the cancellation. Once a grade is cancelled, it is permanently deleted from the records.

You may also request that one or more of your AP grades are not included in the report sent to colleges. There is a $5 fee for each score not included on the report.

Q: May I write on the test?

A: Yes. Because scratch paper is not provided, you'll need to write in the test booklet. Make your notes in the booklet near the questions so that if you have time at the end, you can go back to your notes to try to answer the question.

Q: How do I register or get more information?

A: For further information contact: AP Services, P.O. Box 6671, Princeton, NJ 08541-6671; (609) 771-7300 or (888) 225-5427; Fax (609) 530-0482; TTY: (609) 882-4118; or e-mail: apexams@ets.org.

Strategies for Taking the AP Environmental Science Exam

The Educational Testing Service will send you your AP Environmental Science score in July. Depending upon your choice, the scores are also sent to colleges and universities. The scores are reported on the following scale:

- 5—Extremely well qualified
- 4—Well qualified
- 3—Qualified
- 2—Possibly qualified
- 1—No recommendation

Most colleges and universities accept a score of 4 or 5 for credit and placement. Many colleges and universities may accept a score of 3 for credit and/or placement. Scores of 1 or 2 are not accepted by colleges and universities for either credit or placement.

The rule of thumb in determining how well you will probably do on the AP Environmental Science Exam is to look at the number of multiple-choice questions you answer correctly on the practice exams in this book. If you consistently get between 50 and 60% correct, you should be able to score a minimum of a 3. If you score consistently between 65 and 75% of the multiple-choice questions correct, you should be able to achieve a 4, and if you score 80 to 100 percent correctly, you should be looking at a 5. This assumes, of course, that you adequately answer the questions in the free-response section.

Section I: The Multiple-Choice Section

- Read the entire question. Underline key words in the question such as
  all of the following EXCEPT
  which of the following
  increases
  decreases
  are commonly used
  is responsible
  principles
  most accurately compares
  is recognized as
  best describes
  is correct
  results
  reflects
  is most likely
  is true
  all the following are true EXCEPT

least likely
which best DEFINES

- Look for and underline key vocabulary words in the questions and answer choices.
- Read the entire answer. Using the process of elimination can usually increase your ability to find the correct answer.
- In questions where there is an "All of the above" or more than one correct choice, make sure you look for multiple answers.
- Be aware of "negative" questions such as "all of the following EXCEPT."
- Don't guess wildly. A one-quarter-point penalty is assessed for each incorrect answer. If you can eliminate two or more choices, you should attempt to answer the question, circle it and return to it after you finish the rest of the questions. I have provided you two strategies below for answering multiple-choice questions—The Plus-Minus System and The Elimination Strategy.
- Go with your first instinct. Usually, your first response is the correct one. Only change answers if you are absolutely certain.
- Be aware of the time limitation. Unlike many other AP exams, you have 100 multiple-choice questions to answer in 90 minutes.

The "Plus-Minus" System

Many students who take the AP Environmental Science Exam do not get their best possible score on Section I (multiple-choice) because they spend too much time on difficult questions and fail to leave themselves enough time to answer the easy ones. Don't let this happen to you. Because every question is worth the same amount, consider the following guidelines.

1. Note in your test booklet the starting time of Section I. Remember that you have just under a minute per question.
2. Go through the entire test and answer all the easy questions first. Generally, the questions become more difficult as you move through Section I.
3. When you come to a question that seems impossible to answer, mark a large minus sign (–) next to it in your test booklet. You are penalized for wrong answers, so do not guess at this point. Move on to the next question.
4. When you come to a question that seems solvable but appears too time-consuming, mark a large plus sign (+) next to that question in your test booklet. Do not guess; move on to the next question.
5. Your time allotment is just under 1 minute per question, so a "time-consuming" question is one that you estimate will take you more than a minute to answer. Don't waste time deciding whether a question gets a plus or a minus. Act quickly. The intent of this strategy is to save you valuable time. After you have worked all the easy questions, your booklet should look something like this:

 1.
+2.
 3.
–4.
 5.

and so on.

6. After doing all the problems you can do immediately (the easy ones), go back and work on your + problems.
7. If you finish working your + problems and still have time left, you can do either of two things:
   a. Attempt the – problems, but remember not to guess under any circumstances.
   b. Forget the – problems and go back over your completed work to be sure you didn't make any careless mistakes on the questions you thought were easy to answer.

   You do not have to erase the pluses and minuses you made in your question booklet.

The Elimination Strategy

Take advantage of being able to mark in your test booklet. As you go through the + questions, eliminate choices from consideration by marking them out in your question booklet. Mark with question marks any choices you wish to consider as possible answers. See the following example:

~~A.~~
B. ?
~~C.~~
~~D.~~
E. ?

This technique will help you avoid reconsidering those choices that you have already eliminated and will thus save you time. It will also help you narrow down your possible answers.

If you are able to eliminate all but two possible answers, such as B and E in the previous example, you may want to guess. Under these conditions, you stand a better chance of raising your score by guessing than by leaving the answer sheet blank.

Types of Multiple-Choice Questions

The AP Environmental Science Exam relies on a variety of multiple-choice questions including identification and analysis. Those kinds of questions can be further identified as generalizations, comparing and contrasting concepts and events, sequencing a series of related ideas or events, cause-and-effect relationships, definitions, solutions to a problem, hypothetical situations, chronological problems, multiple correct answers, and negative questions. More than 75% of the multiple-choice questions fit into these categories.

In addition to identification and analysis questions, there are stimulus-based questions that rely on your interpretation and understanding of maps, graphs, charts, tables, pictures, flowcharts, photographs or sketches, cartoons, short narrative passages, surveys and poll data, quotations that come from primary source documents, and passages from environmental legislation and statutes, as well as questions that relate directly to laboratory and field investigations that you were *supposed* to have done.

Identification and Analysis Questions

Question Type	Example
Definitional	Any factor that influences a natural process under study is a(n) (A) independent variable (B) dependent variable (C) control (D) placebo (E) experimental value
Cause-and-effect relationships	__________ contributes to the formation of __________ and thereby compounds the problem of __________. (A) ozone, carbon dioxide, acid rain (B) carbon dioxide, carbon monoxide, ozone depletion (C) sulfur dioxide, acid deposition, global warming (D) nitrous oxide, ozone, industrial smog (E) nitric oxide, ozone, photochemical smog
Sequencing a series of related ideas or events	Which of the following statements describe the process of how environmental legislation would pass through Congress? I. Reports the bill out of the appropriate committee II. Debates the bill on the floor of the respective houses III. Rejects or accepts amendments to the bill IV. Resolves any differences in a conference committee (A) I only (B) I and II only (C) I, II, and III only (D) I, III, and IV only (E) I, II, III, and IV
Generalization	What is generally considered to be the most significant factor in terms of being a causative agent for cancer? (A) smoking (B) diet (C) stress (D) heredity (E) pollution

Solution to a problem

A field biologist had been keeping density counts of the number of coastal side-blotched lizards (*Uta stansburiana hesperis* Richardson) that had recently been introduced into a large section of California desert. The yearly counts per acre were: 1998: 4; 1999: 8; 2000: 16. Assuming that the carrying capacity had not been reached, which of the following would be true?

(A) In 2001, there should be 20 lizards due to arithmetic growth.
(B) In 2001, there should be 20 lizards due to exponential growth.
(C) In 2001, there should be 32 lizards due to exponential growth.
(D) In 2001, there should be around 16 lizards due to population stability.
(E) It is impossible to predict how many lizards there would be in 2001.

Hypothetical situation

Converting to a solar-hydrogen energy source could theoretically be achieved by

(A) attracting private investors
(B) passing legislation that would fund "seed money" for entrepreneurs
(C) passing legislation that would discontinue government subsidies of fossil fuels
(D) educating the public as to the environmental benefits of solar-hydrogen fuel sources
(E) all of the above

Chronological problem

The following events are related to major case studies of environmental pollution. Place the events in chronological order.

I. Bhopal, India
II. *Exxon Valdez*
III. Donora, Pennsylvania
IV. Three Mile Island
V. Chernobyl, Ukraine

(A) I, II, III, IV, V
(B) V, IV, III, II, I
(C) I, III, V, II, IV
(D) I, V, II, IV, III
(E) III, IV, V, I, II

Comparing and contrasting concepts and events

Which of the following acts or treaties accurately compares the political, environmental, and economic goals of the participating nations to the goal of reducing global greenhouse gas emissions?

(A) Clean Air Act of 1955
(B) National Environmental Policy Act (NEPA) 1969
(C) Comprehensive Environmental Response, Compensation, and Liability (Superfund) Act (CERCLA) 1980
(D) Pollution Prevention Act of 1990
(E) Kyoto Protocol of 1997

Multiple correct answers

Nitrogen is assimilated in plants in what form?

(A) NO_2^-
(B) NH_3
(C) NH_4^+
(D) NO_3^-
(E) Choices B, C, and D

Negative questions

An effective method to decrease the amount of pesticide use would include all of the following EXCEPT

(A) using monoculture techniques
(B) rotating crops
(C) using pheromones
(D) using polyculture techniques
(E) using insect-resistant crops

A word of caution: If you can reasonably attack the question through a process of elimination or if you have factual knowledge regarding the content of the question, then you should attempt to answer it. You may be able to identify the type of question, but if you don't understand the issue or vocabulary that is used, you are not going to be able to find the correct choice. In that case, skip the question and avoid the possibility of a "guessing penalty."

Stimulus-Based Multiple-Choice Questions

Question Type	Example
Short narrative passage	"Human beings are adaptable. They have survived and conquered when other species have become extinct. Granted, that we as humans have a long way to go in order to live in harmony with nature, nevertheless, the future looks bright for man to conquer all environmental problems. There is no limit to what problems humans can solve. If we all work together, we can solve any problem. The future is bright." The quote above would be most closely associated with (A) Malthusian principles. (B) neo-Luddites. (C) cornucopian fallacy. (D) nihilists. (E) utilitarianism.
Short quotation	As the troposphere warms, the stratosphere begins to cool and causes the formation of ice-clouds above the South Pole. The ice-clouds tend to accelerate the decrease of ozone above the South Pole by providing a surface for chemical reactions to occur. This is an example of (A) serendipity. (B) a positive feedback loop. (C) a negative feedback loop. (D) synergy. (E) chaos.
Environmental case law	Which act established the first clear water purity standards? (A) Water Resources Planning Act (B) Water Resources Development Act (C) Safe Drinking Water Act (D) Surface Water Treatment Rule (E) Water Quality Act
Sketch or diagram	The following diagram indicates:

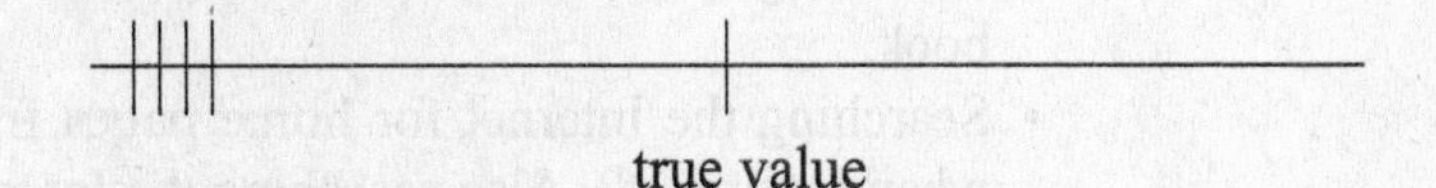

(A) a measurement that is accurate and precise.
(B) a measurement that is accurate but not precise.
(C) a measurement that is precise but not accurate.
(D) a measurement that is neither accurate nor precise.
(E) a conclusion cannot be made regarding this distribution.

Graph or table interpretation

Two varieties of the same species of voles (meadow mice), albino and red-backed, were used in an experiment. Both varieties were subjected to the predation of a hawk, under controlled laboratory conditions. During the experiment, the floor of the test room was covered on alternate days with white ground cover that matched the albino voles and red-brown cover that matched the red-backed voles. The results of fifty trials are shown in the following tabulation.

NUMBER OF VOLES CAPTURED

Variety	White Cover	Red-Brown Cover	Total
Albino	35	57	92
Red-backed	60	40	100
Total	95	97	192

Which result would have been most likely if only red-brown floor covering had been used?

(A) Ninety-seven red-backed voles would have survived.
(B) Ninety-five albino voles would have survived.
(C) The survival rate of the albino voles would have decreased markedly.
(D) A greater number of red-backed voles would not have survived.
(E) There would have been no change in the results.

A Final Word About the Multiple-Choice Questions

Throughout the APES course, your teacher will be giving you a variety of sample multiple-choice questions with five possible answers. Other methods include:

- Building up your own database of questions by using questions from this book.
- Sharing and collecting questions that other students come up with (the Internet is wonderful for this).
- Creating a class database where everyone contributes ten questions per chapter in all question formats.
- Obtaining questions from the Educational Testing Service as found in their Acorn book.
- Searching the Internet for home pages from other APES classes (I counted over 60 when I searched). Also searching the Internet for home pages from students who have already taken the AP Environmental Science Exam and can share with you their thoughts, tips, and experiences.
- Making your own class web page that includes multiple-choice questions and links for more information.

- Creating reciprocal agreements with other schools for material ("we'll give you 300 of our questions if you give us 300 of yours").
- Visiting the many college-text publishers that have websites where you can get more multiple-choice questions and working through these questions.
- Debriefing. Right after you finish the AP exam come immediately back and let your teacher know the type of questions you found easy and what type you found more difficult; let them know how well-prepared you were and where more attention could have been placed. This will help your teacher and the class next year to know where to place more emphasis. After a few years of student feedback, your teacher will have a pretty good idea of what to expect.

Section II: The Free-Response (Essay) Section

There are three types of free-response questions:

1. data-analysis
2. document-based
3. synthesis and evaluation

The data-analysis, or data-set question, presents data in tabular or graphical form as stimulus material. It measures the student's ability to interpret and analyze data. The document-based question or DBQ, presents stimulus material in the form of real-life documents such as newspaper articles or product advertisements. The DBQ measures the student's ability to apply knowledge of environmental science to topics that are timely, relevant, and authentic. The two synthesis and evaluation questions are in-depth essay questions, often with multiple parts. They measure the ability to synthesize and evaluate ideas by using concepts of environmental science.

In scoring free-response essays, points are awarded only for arguments that are supported by scientific facts and principles. Each question is scored on a scale of 1 to 10, with a maximum score of 10 points, although the sum of points available to separate parts of a question may total more than 10 points. Parts of questions may also have maximum parts scores to ensure that students earning a score of 10 have addressed all parts of the question.

All four essay questions must be answered in essay form and within 90 minutes. This means that if evenly divided, each question should take about 23 minutes. Lists, outlines, and unlabeled drawings, which are meant to substitute for college-level essays, are not acceptable and will receive no credit.

There are different approaches that you can take in answering the four required essay questions. Depending on the specifics of the question, you may have to write an essay on:

- A formal thesis ("assess the accuracy"; "to what extent", etc.)
- An introductory statement that lists the tasks that you are going to answer (give examples that illustrate"; "discuss the nature of", etc.)
- Data (graphs, charts, or quotes)
- Calculations ("calculate how much heat is lost", etc.)

The following chart provides general criteria for scoring the free-response essays:

Excellent thesis or introductory statement (5)	Good explanation of data; responds to every section of the question using specific and accurate examples to support the answer.
A thesis statement is evident in form and relates to the question (3–4)	An introductory statement presents *some* examples in a logical manner to support the data; may present weaker or factually incorrect arguments or fewer examples to support the answer.
Presents a weak thesis or introductory statement (2–3)	Shows some relationship of examples or data to thesis or statement; may leave out pertinent information that the question is asking or may present incorrect examples to support answer.
Has no organized thesis or introductory statement (1–2)	Uses narrative form with few examples of data; may contain incomplete or factually incorrect information and may not answer the entire question.
Thesis or introductory statement that cannot be proven or is not related to question (1)	Scope of answer may be acceptable, but examples either do not support thesis or data and are factually incorrect.
Thesis or introductory argument is not referred to in the body of the essay (1)	Few, if any, factual example or case studies to support the statement; includes factually incorrect, irrelevant, or inaccurate information.
Makes an effort to answer the question by providing some discussion of the issue(s) raised (1)	Fails to provide relevant details or specific examples to answer the question.
No effort to answer the question (1)	Any examples are factually wrong.

After examining the chart, it is clear that there are many more ways to answer an essay question incorrectly than there are ways to answer it correctly. There is only one way to answer an essay question at the highest level (5): answer it correctly and in exemplary style. Cramming in a few short weeks prior to the exam will not prepare you to write your essays at the highest level. You need to practice your writing style throughout the course and have your teacher provide you, in a timely manner, constructive criticism on how to improve your essay for next time. Writing essays without feedback from your teacher, where each sentence and each paragraph that you write is evaluated and careful attention is paid to rubrics that are noted and referenced on your essay, will not prepare you. You are simply "spinning your wheels" or doing advanced "busy work" and not moving forward. Repeating errors and poor writing styles only reinforces them.

It is important to remember that when answering an essay, the goal is to give a solid introductory statement. You must then give specific examples to support your introductory statement(s). The four essays are worth 40% of the examination. Each essay must be completed within a reasonable amount of time. Taking too long on one essay because you know the information and want to impress the grader with your knowledge will only result in your having less time for those questions that you do not know as well and that may require even more time to develop a reasonable answer. All essay questions are weighted equally. Do not overkill an essay—there is no extra credit. As you can see, the development of a thesis, strong introductory statement, or data analysis and numerous supporting examples are essential to getting the best possible score on the AP Environmental Science Exam.

Examples of Key Terms Used in Free-Response Essays

Key Term	Meaning	Example
Analyze the effects . . .	Evaluate the impact	Analyze the effects of major global wind patterns in determining the distribution of biomes.
Assess the accuracy . . .	Determine the truth of the statement	Assess the accuracy of the statement, "The reason that there are so many people on the Earth is that more people are having more children."
Compare the strengths and weaknesses . . .	Show differences	Compare the strengths and weaknesses of three government intervention programs that have been adopted into law to combat pollution.
Critically evaluate evidence that both supports and refutes . . .	Give examples that agree and disagree	Critically evaluate evidence that both supports and refutes the claim that greenhouse gases are responsible for the hole in the ozone layer over the Antarctic.
Define and evaluate the contention . . .	Give a definition and analyze the point of view	Define ecofeminism and evaluate the contention that male-dominated societies are responsible for environmental imbalance and social oppression.
Discuss . . .	Give examples that illustrate	Discuss the rapid increase of African "killer bees" in the United States as an example of the processes of natural selection.
Evaluate the claim . . .	Determine the validity	Evaluate the claim that increasing the standard of living and improving the role of women will result in a decrease in natality.

Key Term	Meaning	Example
Explain . . .	Offer the meaning, cause, effect, influence	Explain three different theories of moral responsibility to the environment. Include in your discussion the major proponents of each theory and their specific idea(s) or "school."
To what extent . . .	Explain the relationship and role	To what extent does moisture determine a soil's characteristic? Provide specific soil types, overview of that soil's horizon, and its categorical geographic distribution.

Notice how the key phrases encourage the development of a thesis statement. Additional sample and practice free-response essay questions appear at the end of each chapter and in the two practice exams.

The Restatement

In Section II, you should begin all questions by numbering your answer. You do not need to work the questions in order. However, the graders must be able to identify quickly which question you are answering. You may wish to underline any key words or key concepts in your answer. Do not underline too much, however, because doing so may obscure your reasons for underlining. In free-response questions that require specific calculations, you may also want to underline or draw a box around your final answer(s).

Before you begin to write, brainstorm. That is, make a list in the test booklet of all of the concepts and ideas that are pertinent to the question. You should have practiced this skill in your class. After you have finished your list, number the items in the order in which you wish to discuss them. A few minutes spent in organizing the essay before you begin writing will result in a much more coherent and complete answer.

Strategies for Answering the Free-Response Essay Questions

1. Read the entire question. Underline key terms (given for each chapter in this book) that will help you understand what the question is looking for.
2. Base your decision on your overall familiarity with the subject matter of the questions and the specific areas related to the general question asked.
3. Read all the questions, looking for potential thesis development, appropriate opening statement or reference to data, and supporting examples to illustrate your arguments.
4. On the exam, jot down a possible thesis, appropriate opening statement, or reference to data, and specific examples that relate to the question.
5. Outline your argument with your thesis, appropriate opening statement, or reference to data, making sure that you provide a specific example for every statement made. At this point, try organizing your essay going from general to specific in your outline.

6. Use factual data, both historical and current, to support your answer. If you are not sure of a specific name, date, or term, don't use it. However, if you can describe an example filling in most of the details, do so.
7. Make sure that your supporting evidence and data support your thesis and your opening statement or references to data.
8. Start writing your essay beginning with your thesis statement, appropriate opening statement, or references to data, and main arguments.
9. Using your outline, use the specific factual data and examples to support your thesis. Make sure that your answer refers to the key term(s) used in the question.
10. If the question gives you specific data areas to comment on, make sure that your examples and answer refer to the areas in the question.
11. Reiterate your thesis statement in your summary.
12. Reread your essay. Revise it or add to it if necessary. You may use a single, straight line to cross out material that does not belong or is in error. Be aware of your time limit, but don't let that stop you from adding, changing, or crossing out material in your essay if you need to.

How to Write the Free-Response Essay

Once you follow the strategy advice, the actual writing of the free-response essay is an easier task (however, it is never easy). Many students have a phobia about free-response questions. View your essay as the letter T. The top of the T represents the thesis statement, and it is broader than the evidence below it, which supports the thesis statement.

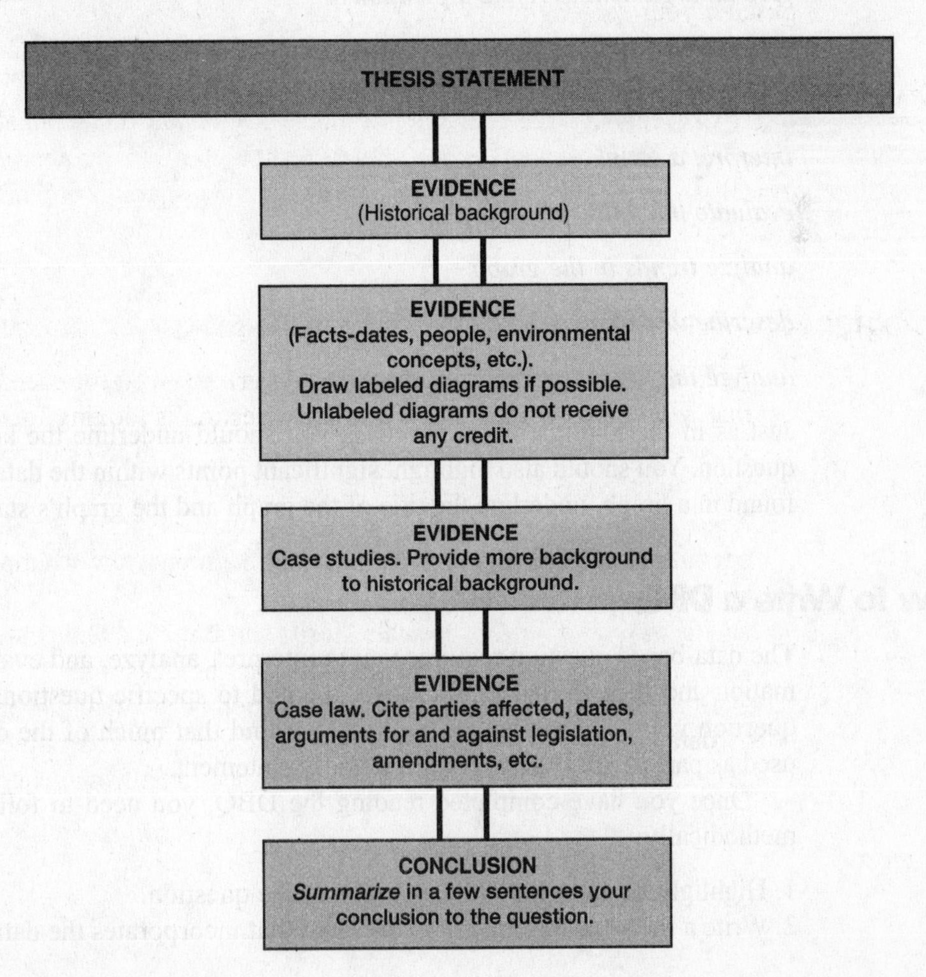

Data-Based Questions (DBQ)

One of the questions that you will need to answer is a data-based question. DBQs consist of charts, graphs, documents, and newspaper articles of environmental issues, among others. DBQs may be historical or contemporary in nature. You will be expected to analyze, evaluate, and interpret material from a variety of sources. Conclusions are reached, inferences are made, and the actual response is written using a format similar to the more traditional free-response essay. Although a formal thesis is not necessary, the evaluator will look for evidence that you understand the data, can explain the data in relation to the question, and can give specific examples to answer the question as it relates to the data.

Examples of Key Terms Used in DBQs

Directions are very clear for you to analyze and integrate an analysis with a general understanding of environmental science. Be especially aware of the following phrases:

using data, identify

explain

using data, describe

using quotation, identify and explain

give an argument to refute a position

give general similarities and differences

identify trends

interpret a graph

evaluate the following chart

analyze trends in the graph

describe and analyze a chart

analyze the following environmental law

Just as in the multiple-choice section, you should underline the key terms used in the question. You should also highlight significant points within the data itself. If the data are found in a graph, underline the title of the graph and the graph's statistical parameters.

How to Write a DBQ

The data-based question requires you to interpret, analyze, and evaluate a base of information and then to make conclusions related to specific questions. In looking for the question you want to respond to, keep in mind that much of the data provided can be used as part of your thesis or introductory statement.

Once you have completed reading the DBQ, you need to follow a series of steps methodically:

1. Highlight or underline the key terms in the question.
2. Write a thesis or introductory statement that incorporates the data into your response.

3. After you finish your thesis or introductory statement, provide specific examples that illustrate your understanding of the data.

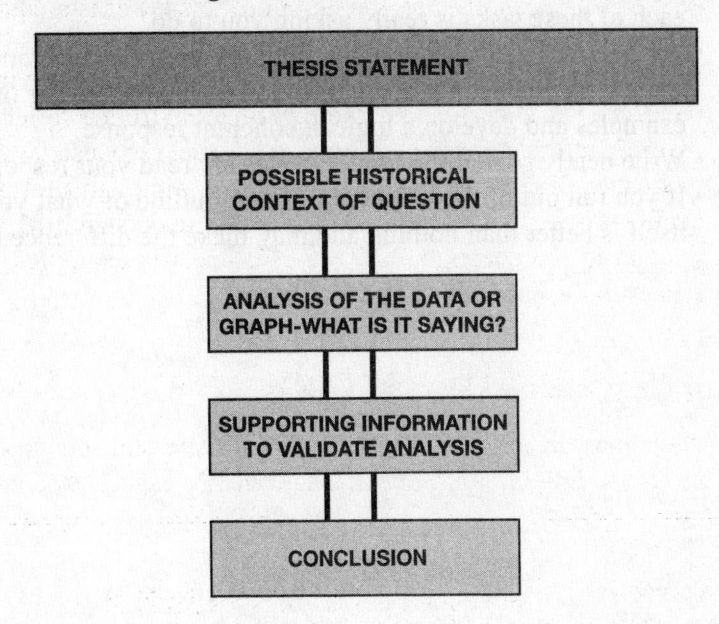

4. Like the non-data-based free-response essay, you need to develop a thesis, give supporting examples, and then provide a historical or contemporary description, which reinforces your thesis or opening statement. The major difference in developing a DBQ versus a traditional free-response essay is that in a DBQ the data are provided for you to analyze, and in the traditional free-response, you must supply the data to support your thesis.

Summary

- Four mandatory essays are required and are worth 40% of the examination score.
- Brainstorm the question first. Write down as many key terms as you can remember regarding the topic or case laws, case studies, and the like. Map out the question in 2 to 3 minutes before you even begin to write your answer. After you have your thoughts of what you are going to say down on paper, spend a minute and numerically prioritize them in the way you want to discuss them. Believe me, the 5 minutes you spend in this activity will make the difference between an outstanding essay and a mediocre one.
- Each essay should be written within 23 minutes. Take a watch.
- Non-data-based essays may require either a formal thesis or introductory statement that relates to the tasks that you are writing about.
- DBQs should have an opening statement that makes references to the data.
- All essays should give supporting evidence in the form of specific historical or contemporary environmental examples (case studies).
- Read the question carefully and perform all tasks listed.
- Some questions may allow you to interrelate material from other courses. AP Biology, AP Chemistry, AP Physics, AP Statistics, and AP U.S. Government and Politics are especially helpful because many of the concepts in AP Environmental Science can be found in these areas. Don't be afraid to include this material.

- Tasks will vary from essay to essay. One essay may ask you to explain while another essay may ask you to illustrate, describe, or discuss. Make sure that you know what each of these tasks is really asking you to do.
- All free-response questions will require you to understand the nature of the question, writing a response that demonstrates an understanding of the question using specific examples and develop a logical, coherent response.
- Write neatly so that the grader can easily read your response.
- If you run out of time, provide a basic outline of what you were going to say. This in itself is better than nothing and may make the difference between a 4 and a 5.

PART II: Specific Topics

UNIT I: INTERDEPENDENCE OF EARTH'S SYSTEMS—FUNDAMENTAL CONCEPTS

CHAPTER 1

Energy

Areas That You Will Be Tested On

A: Forms and Quality of Energy
B: Energy Units and Measurement
C: Sources and Sinks, Conversions (see Appendix I)
D: Conventional and Alternative Sources

Key Terms

Note: Definitions for all key terms listed below can be found in the glossary starting on page 648. For additional definitions of relevant terms from this chapter, refer to *www.barronseduc.com/0764121618.html.*

active solar heating
anthracite
alpha particle
atomic mass number
atomic number
biogas
biomass
bituminous coal
breeder nuclear fission reactor
British Thermal Unit (BTU)
calorie
cogeneration
energy efficiency
first-law efficiency
first law of energy or thermodynamics
fission
fusion
geothermal energy
high-quality energy
high-throughput society
joule
kilocalorie
kilowatt
law of conservation of energy
lignite
low-throughput society
methane
passive solar heating system
second-law efficiency
second law of energy or thermodynamics
third law of thermodynamics
watt

Key Concepts

Nonrenewable Energy Sources

- Fossil fuel resources will be totally depleted in 500 years.
- Fossil fuels (coal, natural gas, crude oil) produce 90% of energy used.
- Fossil fuels are responsible for air pollution, acid rain, and global warming.
- Costs associated with fossil fuels include extraction of fossil fuel, transportation, processing, power plant, and transmission.

Coal

Coal is produced by decomposition of ancient organic matter during Carboniferous period (286 million years ago) under high temperature and pressure. Sulfur from decomposition and H_2S produced by anaerobic bacteria became trapped in coal. There are three types of coal: lignite (brown coal, softest, low heat content), bituminous (soft, high sulfur content, 50% of U.S. reserve), and anthracite (hard, high heat content and low sulfur, 2% of U.S. reserve). Peat is pre-coal and is used in some countries for heat but has low heat content. Coal supplies 25% of the world's energy—China and the United States use the most coal (52% of U.S. electricity is generated from coal). Other coal users include former Soviet Union, Germany, India, Australia, Poland, and the United Kingdom. In the United States, 87% of coal is used by power plants. Low-sulfur coal in the United States is primarily low-energy lignite and sub-bituminous located west of Mississippi. High-sulfur coal in the United States is found primarily in the Eastern United States. Legislation in the United States (Clean Air Amendments) requires up to 90% reduction in sulfur released in burning of coal, which can be achieved by (1) cleaning coal prior to burning, (2) redesigning boilers, and (3) scrubbing, or adding limestone or lime into effluent.

PROS	CONS
Abundant, known world reserves (1000 billion tons) to last more than 200 years at current rate of consumption (4 billion metric tons per year); 65 years if 2% increase per year occurs.	Most extraction in the United States is done through strip mining (60%). 40% of coal mining is done through underground mining and is dangerous (fires and collapse).
Unidentified world reserves estimated to last 1000 years at current rate, 150 years with 2% per year increase in consumption.	Unhealthy (black lung disease); up to 20% of coal becomes fly ash, boiler slag, and sludge. Releases mercury and radioactive particles into air.
In the United States, reserves estimated to last 300 years at current rate.	Harms land, subsidence, acidic water pollution associated with extraction, restoring land is expensive and decreases net energy yield.

High net energy yield.	Expensive to process and transport; cannot be used efficiently for transportation needs.
U.S. government subsidies keep prices low.	Releases carbon monoxide (CO), carbon dioxide (35% of all CO_2) sulfur dioxide (70% of all SO_2) and nitrogen oxides (30% of all NO and NO_2), which kill hundreds of thousands of people each year; produces fly ash, toxic metals, and radioactive particles (more than a nuclear power plant would).
	Pollution causes global warming. Scrubbers and other antipollution control devices are expensive.

Solid Coal to Gas and Liquid Fuels

Coal can be converted to SNG (synthetic natural gas), methanol (CH_3OH), or synthetic gasoline through coal liquefaction.

PROS	CONS
Transport through pipelines.	Low net energy yield.
Produces less air pollution.	Plants expensive to build.
Supply is large.	Would accelerate depletion of coal due to inefficiency in process.
	Land disruption due to mining, large amounts of water required.
	Releases large amounts of CO_2.
	More expensive than coal.

Natural Gas (Methane—CH_4)

Natural gas is produced by decomposition of ancient organic matter under high temperatures and pressures. Conventional sources found associated with oil. Unconventional sources include: coal beds, Devonian shale rock, tight sands, dissolved in deep zones of hot water, and gas hydrates. LNG (liquefied natural gas) makes worldwide distribution possible. LPG (liquefied petroleum gas) consists of propane (C_3H_8) and butane (C_4H_{10}) and is removed from methane. LPG is used in rural areas without natural gas pipelines. Natural gas supplies 10% of electricity generated in the United States.

Russia and Kazakhstan have approximately 40% of world reserves of natural gas, the United States has 3%, and the Middle East has approximately 25%.

PROS	CONS
Pipelines in place in the United States and other developed countries.	H_2S and SO_2 are released during processing.
Relatively inexpensive with it being cheaper than oil.	LNG processing is expensive and dangerous and results in a lower net energy.
United States reserves estimated to be 75 years at current rate of consumption; world reserves, about 125 years.	Leakage of pipes and tanks; leakage of CH_4 has a greater impact on global warming than does CO_2.
High net energy yield; burns hotter than any other fossil fuel.	Environmental damage produced when drilling platforms are constructed.
Produces less air pollution than any other fossil fuel (~16% of all CO_2 emission).	Extraction releases contaminated waste-water and brine.
Extraction is not as harmful to environment as that of coal or uranium.	Land subsidence (sinking).
Easily processed; inexpensive to transport.	Disruption to wildlife and habitats (controversy of Arctic Wildlife Refuge in Alaska).
Can be used in fuel cells.	
Can be used in combine-cycle natural gas systems.	
Can be concentrated and liquefied (LNG) for transportation over rail or ship.	
Viewed by many as a transitionary fossil fuel as the world switches to alternative sources (solar, wind, nuclear, etc.).	

Oil

Oil is a fossil fuel produced by decomposition of deeply buried organic material such as plants under high temperatures and pressures for millions of years. Of all known oil, 65% is found in 1% of all fields—most in Middle East. Oil supplies 3% of electricity produced in the United States. It is used for transportation (500,000,000 cars alone in the world), industry, and agriculture. The United States uses 300 million gallons of gasoline per day. Petrochemicals are used in the manufacture of fertilizer, paints, pesticides, plastics, synthetic material, and medicines.

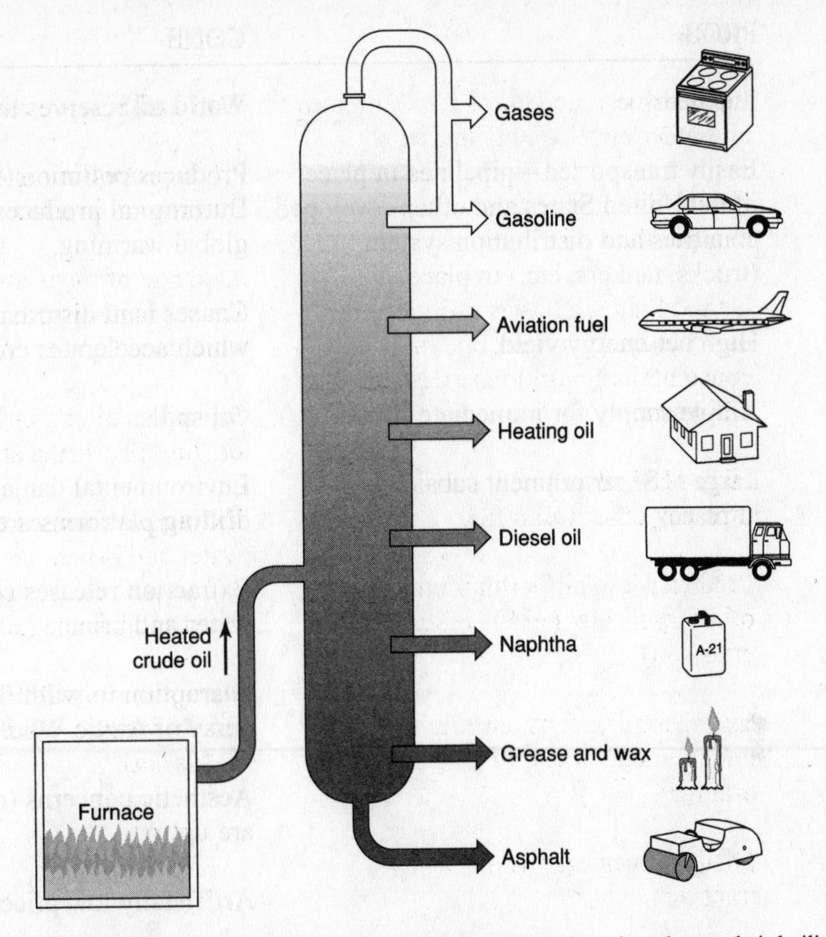

Figure 1.1. Cracking tower. Components of crude oil are separated out based upon their boiling points.

World Oil Supplies OPEC (Organization of Petroleum Exporting Countries) owns two thirds of the world's oil reserves and produces approximately half of the world's oil. It includes Algeria, Ecuador, Gabon, Indonesia, Iran, Iraq, Kuwait, Libya, Nigeria, Qatar, Saudi Arabia, United Arab Emirates, and Venezuela. The United States owns approximately 2–4% of the world's oil reserves but uses almost one third of the world's oil, with the majority of that being used for transportation. The United States imports almost half of the oil it uses. Most of the oil in the Middle East goes to Europe, Japan, and Southeast Asia. Supply–demand theory indicates major economic issue for United States.

How long will world oil supplies last? Most of the world's energy is supplied by burning oil. At *current* rate of consumption, it will last about 45 years. The United States' oil reserves will be depleted in approximately 25 years. Higher prices for oil may stimulate oil discovery or make other sources of oil (shale oil and tar sands) more economical.

PROS	CONS
Inexpensive.	World oil reserves limited and declining.
Easily transported—pipelines in place in the United States and other developed countries and distribution system (trucks, tankers, etc.) in place.	Produces pollution (SO_2, NO, NO_2). Burning oil produces CO_2 that leads to global warming.
High net-energy yield.	Causes land disturbances in drilling process, which accelerates erosion.
Ample supply for immediate future.	Oil spills.
Large U.S. government subsidies in place.	Environmental damage produced when drilling platforms are constructed.
	Extraction releases contaminated waste-water and brine.
	Disruption to wildlife and habitats (controversy of Arctic Wildlife Refuge in Alaska).
	Aesthetic concerns (oil wells and refineries are ugly).
	Artificially low prices encourage waste.

Heavy Oil Produced from Oil Shale

Oil shale contains a waxy mixture of hydrocarbons called kerogens. Oil shale is heated and the vapor condensed. Shale oil is a synfuel—fuels derived from solid fossil fuel (oil from kerogen in oil shale or oil and gas from coal).

PROS	CONS
Vast amounts in the western United States—Green River formation in Colorado, Utah, and Wyoming. Could supply U.S. oil needs for ≅50 years. 2/3 of world supply is in the United States.	Lower net energy yield than oil.
Global supply (3 trillion barrels of oil) is estimated to be 200 times greater than oil reserves.	Requires lots of water to process.
	Tears up land, tailings can leach harmful salts, carcinogens, and toxic metals into ground water. Massive amounts of tailings need to be removed.
	Expensive to produce.
	SO_2 pollution.
	90% recovery possible with surface mining (open-pit or strip mining), but 60% available by underground mining.

Tar Sand or Oil Sand

Tar sand or oil sand is a mixture of clay, sand, water, and bitumen (high-sulfur heavy oil). It is heated until oil floats to top and then refined. Total resource is estimated to be 1,000 billion barrels. Major deposits in Canada (70% of all reserves), Soviet Union, Venezuela and Colombia.

PROS	CONS
Could supply about 2 years of world's oil supply.	Net energy is low.
	Requires large amounts of water with resulting polluted water.
	Releases large amounts of air pollution.

Renewable Alternative Energy Sources

Biomass

Burning biomass (plants and wood) accounts for 15% of world's current energy source, 4% for the United States. One billion people use wood as their primary source of fuel. Biomass can be converted into liquid fuels (biofuels) for storage and transportation.

PROS	CONS
Renewable energy source as long as rate of consumption does not exceed rate of replenishment.	Current use of biomass is unregulated causing rate of consumption to exceed rate of replenishment.
Rate of use balanced, with rate of renewal does not disrupt CO_2 levels.	Requires adequate water and fertilizer-sources which are declining. Use of fertilizer and pesticides harm wildlife, pollute water, and would reduce biodiversity.
Less SO_2, NO, and NO_2 produced than coal.	Deforestation.
Can be sustainable if issues of deforestation, soil erosion, and inefficiency of burning are controlled.	Soil erosion.
Could supply half of world's electrical needs.	Loss of wildlife habitats.
Biomass plantations (fast-growing species—cottonwoods, poplars, sycamores, shrubs, and etc.) can be located in less desirable locations and can reduce soil erosion and restore degraded land. May be up to 200 million acres in the U.S. that could support biomass plantations.	Inefficient methods of burning causing air pollution.
Reduces impact on landfills.	Expensive to transport.
	When converted to electricity, 70% of energy derived from burning biomass is lost.

Heat Stored in Water

Large differences in temperature between deep cold water and surface warm water create a temperature gradient. Ocean thermal energy conversion (OTEC) plants would channel this difference in energy into usable electricity. Other sources include saline and freshwater solar ponds.

PROS	CONS
No pollution concerns.	Technology in research stage.
Unlimited source.	Ocean plants would be expensive to build and maintain.
Moderate net-energy yield.	Cannot compete in price with conventional sources.
Land-based plants for freshwater solar ponds would be moderate in costs and maintenance.	

Hydroelectric Power

Dams are built to trap water, water is released and channeled through turbines, which turn electrical generators. Hydroelectric power supplies 9% of electricity generated in the United States, 3% worldwide.

PROS	CONS
Dams control flooding.	Dams create large flooded areas, destroy wildlife habitats, uproot people, expensive to build, decreases natural fertilization of prime agricultural land in river valleys below dam, traps sediment, keep fish from migrating.
Low operating and maintenance costs.	
No polluting waste products.	
Long life spans.	Falling water may pick up nitrogen gas and kill fish.
Moderate-to-high net useful energy yield.	People love the beauty of a wild river and 'white water' for recreation.

Hydrogen

Burning hydrogen gas produces heat which is then converted into electricity. Hydrogen could be mixed with oxygen gas in fuel cells. Methods of producing hydrogen include (1) reforming, (2) electrolysis of water, (3) photoelectrolysis, (4) coal gasification, (5) biomass gasification, (6) thermolysis, and (7) biological production. It can be stored as: (1) compressed gas, (2) liquid hydrogen, (3) solid metal hydrides, (4) absorption on activated charcoal, and (5) encapsulated into glass spheres.

PROS	CONS
Waste product is pure water.	Takes energy to produce hydrogen.
Unlimited supply of water to derive hydrogen from.	Changing from a fossil-fuel system in place to a hydrogen-system would involve huge financial costs.
Does not destroy wildlife habitats; minimal environmental impact.	Hydrogen gas is explosive.
Energy to produce hydrogen could come from sun.	Using solar technology to produce hydrogen is currently too expensive.
Easily transported through pipelines.	Major storage issue major for individual transportation needs (cars).
Electricity produced from hydrogen could be used for mass transportation needs (rail systems).	
Not explosive when stored in compounds.	
Could be used in fuel cells in which hydrogen and oxygen gases are mixed with no environmental waste products.	

Solar

Solar energy consists of trapping and harnessing radiant energy from the sun to provide heat or electricity. Electrical power and/or heat can be generated or trapped at home sites through photovoltaic cells and solar collectors, or at a central solar-thermal plant location. For more information, go to http://www.solarpaces.org/resources/technologies.html.

PROS	CONS
Supply of solar energy is limitless and free.	Inefficient in areas where sunlight is limited or seasonal, home and centralized systems are expensive.
Reduces reliance on foreign oil imports.	Maintenance costs are high.
Pollution is nonexistent.	Systems deteriorate and must be replaced.
Creates jobs and industries.	Visual pollution.
No environmental impact.	Current efficiency is between 10 and 25%, and is not expected to increase.
Can store energy during day and release it at night; good for remote areas.	Needs backup and storage systems.

Tides and Wave Source

The natural movement of tides and waves spin turbines that generate electricity. Only a few plants are operating worldwide (Bay of Fundy in Canada and United States and along the north coast of France).

PROS	CONS
No pollution.	Construction expensive.
Minimal environmental impact.	Few suitable sites.
Net-energy yield is moderate.	Equipment damaged by corrosion or storms.

Waste

Burning agricultural and other waste products in large incinerators is an energy source.

PROS	CONS
Crop residues are currently available.	Energy required to dry and transport material to centralized facilities is prohibitive.
Ash can be collected and recycled.	
Reduces impact on landfills.	Severe air pollution if not burned in centralized facilities.

Wind

Wind turns giant turbine blades, which run electrical generator. Turbines can be grouped in clusters called wind farms.

PROS	CONS
Unlimited source, all electrical needs of the United States could be met by wind in North Dakota, South Dakota, and Texas.	Steady wind is required to make it economical, backup systems needed when wind is not blowing.
	Visual pollution.
Wind farms can be quickly built.	
Maintenance is low and is automated.	Interfere with flight patterns of birds and birds of prey (hawks and falcons).
Moderate to high net-energy yield, and no pollution. Production of wind turbines would be a boost to economy.	Noise pollution.
	May interfere with communications—radio, TV, microwave.
Land underneath turbines can be used for agriculture.	

Nonrenewable Alternative Energy Sources

Geothermal

Heat contained in underground rocks and fluids from molten rock (magma), hot dry-rock zones and warm-rock reservoirs produces pockets of underground dry steam, wet steam, and hot water. Geothermal energy supplies less than 1% of energy needs in the United States. It is considered nonrenewable when use exceeds replacement. It is currently being utilized in Hawaii, Iceland, Japan, Mexico, New Zealand, Russia, and California (The Geysers—2000-MW facility, largest facility in the world). Areas of known geothermal resources tend to follow tectonic plate boundaries, both convergent (mountain and island arc forming areas) and divergent (oceanic ridge systems).

PROS	CONS
Moderate net-energy yield.	Reservoir sites are scarce.
Limitless and reliable source if managed.	Source can be depleted if not managed.
Little air pollution.	Noise.
Competitive cost.	Odor.
	May cause land to sink.
	Local climatic changes.
	Land damage involved for pipes and roads.
	Can degrade ecosystems due to hot water wastes and corrosive or saline water.

Nuclear

Nuclear energy supplies 20% of electricity in the United States and 7% worldwide. Most are light-water reactors (LWRs), as shown in Figure 1.2. Fuel rods within the core are composed of 97% U-238 and 3% U-235 and create a sustainable fission reaction. Control rods are moved in and out of the core to control the reaction. The heat that is produced is used to create steam, which is then used to turn power generators. Breeder reactors generate more nuclear fuel than they consume by converting U-238 into Pu-239. Nuclear power plants in the United States began to appear in the early 1960s due to (1) prospects of cheap power, (2) government subsidies, (3) liability protection afforded by the U.S. Government (Price-Anderson Act). Worldwide, nuclear power has leveled off, and there is little prospect for future plants except in China, where there are plans to build 50 new plants by 2020. France gets 75% of its energy needs from nuclear power but has stopped new construction. Reasons for decline of nuclear power include (1) cost overruns, (2) safety issues, (3) higher than expected operating costs as compared to other methods, (4) malfunction, (5) perception as a risky investment, and (6) mismanagement by local operators and the Nuclear Regulatory Commission (NRC). Worst accidents so far have been Chernobyl and Three Mile Island (see case studies that follow).

Department of Energy (DOE) estimates up to 50,000 radioactive contaminated sites in the United States with clean-up costs projected to be $1 trillion dollars. Contaminated sites in the former Soviet Union is many times higher than in the United States.

ESTIMATED HEALTH RISKS PER YEAR IN THE UNITED STATES

	Nuclear	Coal
Premature death	6,000	65,000
Genetic defects/damage	4,000	200,000

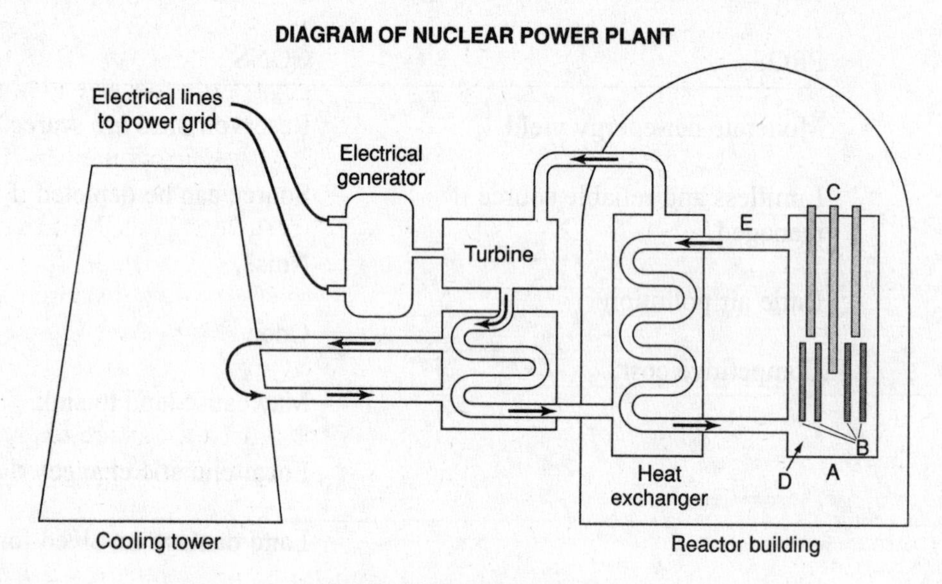

Figure 1.2. Nuclear power plant.

A. Core contains up to 50,000 fuel rods. Each pellet has the equivalent of 1 ton of coal.

B. Uranium oxide fuel: 97% U-238, 3% U-235.

C. Control rods move in and out of core to absorb neutrons and slow down reaction.

D. Moderator can be water, graphite (can produce Pu-239 for weapons), or deuterium oxide (heavy water).

E. Coolant removes heat and produces steam to generate electricity.

PROS	CONS
No air pollutants.	Safety—explosion, terrorists. One in three countries in the world have nuclear weapons or ability to build them. Black market exists for rogue nations or terrorists to buy nuclear weapons.
Releases only about one sixth the CO_2 as fossil-fuel plants, thus reducing global warming effect.	
Water pollution low.	Government subsidies.
Disruption of land is low to moderate.	Nuclear waste products.
	Current plants have a lifetime of only 15 to 40 years.
	Plants are expensive to build and run.
	High malfunction rate.
	Public fear (Three Mile Island and Chernobyl).
	Limited natural resource with extraction and refinement having major environmental impact.
	Breeder reactors are costly and potentially dangerous in design (production of Pu- 239).
	Low net-energy yield—energy needed for mining uranium, processing ore, transportation, building and operating plant, dismantling plants, storing wastes.

Economics of Energy Alternatives

1. No attempt to control the price of an energy resource, allowing price to fluctuate in an open market. Consequence: true cost of energy dictates use and conservation of resource, stimulates research and development of more efficient sources.
2. Keep energy prices artificially low through government subsidies and tax breaks. Consequence: perpetuates the status quo and discourages research and development of alternative sources.
3. Keeping energy prices artificially high. Consequence: by withdrawing subsidies and tax breaks and/or adding additional taxes, government income increases at the expense of the consumer, encourages improvement in conservation, may reduce dependence on imports, decreases use of energy sources with limited future supply, dampens economic growth, and puts an economic burden on those who can least afford it. However, it stimulates employment by requiring labor for increasing efficiency.

Politics of Sustainable Energy

1. Utilize a combination of energy resources taking into account available local energy resources, technology, economic issues, pollution risks, etc.
2. Future energy resources will have high development costs with low-to-moderate net-energy yields.
3. Energy projects must have community input with priorities established due to limited financial resources.
4. Energy projects must include financial incentives of improving efficiency and conservation.
5. Governments can increase efficiency by
   - Increasing fuel-efficiency standards;
   - Increasing energy-efficiency standards for appliances and industries;
   - Sponsoring research and development;
   - Giving tax credits or rebates for using more efficient vehicles, buildings, and appliances;
   - Using free bates—taxing inefficient products and giving the income to those who purchase efficient products;
   - Taxing energy;
   - Limiting government subsidies granted to inefficient energy resources, thus allowing an open market and competition in the energy industry.

COMPARATIVE WORLDWIDE ELECTRICITY CONSUMPTION

Country	Population Estimate in 2000	Per Capita Electricity Consumption (kW-h)
Canada	31,000,000	17,770
United States	276,000,000	12,190
Japan	127,000,000	7,290
Worldwide Average	6,081,000,000	2,030
China	1,262,000,000	800
India	1,014,000,000	410

MAJOR U.S. RESOURCE CONSERVATION AND ENVIRONMENTAL LEGISLATION

Legislation	Description
Energy Policy and Conservation Act, 1975	Authorizes the president's authority to draw from the strategic petroleum reserve as well as establishes a permanent home-heating oil reserve in the Northeast. Clarifies when the president can draw from these reserves, expands the Department of Energy Weatherization Assistance Program, and requires the secretaries of the interior and energy to undertake a national inventory of onshore oil and natural gas reserves.
Energy Policy Act, 1992	Provides for improved energy efficiency. It includes provisions to allow for greater competition in energy sales.
National Energy Security Act, 1978 Energy Tax Act, 1978	The United States first experienced the economic impact of international oil disruptions in 1973 when an Arab Oil Embargo caused long lines at gas stations, lost productivity, declines in the stock market and economic recession. The first congressional response to the petroleum crisis was the Energy Security Act and the Energy Tax Act of 1978. The Acts created favorable tax legislation and research and development commitments to expand the use of fuel ethanol in the United States to stimulate domestic production of ethanol.
National Appliance Energy Conservation Act, 1987	Sets minimum efficiency standards for numerous categories of appliances, including residential comfort equipment.
Price-Anderson Act, 1957	Limits the liability of the nuclear industry in the event of a nuclear accident in the United States. It covers incidents that occur through operation of nuclear plants as well as transportation and storage of nuclear fuel and radioactive wastes.
U.S. Public Utility Regulatory Act (PURPA), 1978	Requires utilities to interconnect with small-scale independent power producers and pay fair market price for the electricity produced.

Factor-Label Method for Calculations

Every AP Environmental Science Exam has one calculation problem in the free-response section. It is very important for you to know how to set up these types of problems. The method that I will show you here is called the Factor-Label Method. Let's start with a stupid problem—no, you won't be required on the AP Environmental Science Exam to figure out how much money a farmer made on selling canaries, but the steps are the same when we do another problem in the free-response section later on in this chapter. We'll just change the word "canary" to "joule," we'll change "pigs" to "kilowatts," and so on.

PROBLEM: A farmer started with 5 goats. He traded all of his goats for sheep at an exchange rate of 3 sheep for 1 goat. He then traded his sheep for pigs at a rate of 1 sheep for 2 pigs. Next, he traded his pigs for canaries. For every 3 pigs, he received 27 canaries. He then sold all of the canaries for a rate of $3.25 per canary. How much money did the farmer make on his canaries?

Step 1: Write an equal sign.

$$=$$

Step 2. Ask yourself, what units do I want to end up in? Pigs? Canaries? Sheep? Nope, I want to end up in dollars. So put the dollar sign to the right of the equal sign. It will look like this now:

$$= \$$$

Step 3: Now ask yourself, what did the problem start me off with? Did the farmer start with canaries? Pigs? Nope, he started with 5 goats. So, way to the left of the equal sign, put 5 goats over 1. It will look like this now:

$$\frac{5 \text{ goats}}{1} = \$$$

Step 4: See the word "goats" above? You have to get rid of that word because you want to be in "dollars" when you finish. So put the word "goats" in a fraction so that it will cancel with the goats in the first factor. All factors get multiplied. It will look like this now:

$$\frac{5 \text{ goats}}{1} \times \frac{\quad}{\text{goats}} = \$$$

Step 5: Now ask yourself, "What do I know about goats?" Do I know something about goats and canaries? Do I know something about goats and sheep? Nope. I do know that the farmer traded his goats for sheep, at a rate of 3 sheep for 1 goat. At this point, put a 1 in front of goat and put "3 sheep" as the numerator. Notice the word "goats" cancel. It will look like this:

$$\frac{5\ \cancel{\text{goats}}}{1} \times \frac{3\ \text{sheep}}{1\ \cancel{\text{goats}}} = \$$$

Step 6: Now you are in "sheep." You need to get rid of "sheep." Where will you put it in the next factor? It should now look like this:

$$\frac{5\ \cancel{\text{goats}}}{1} \times \frac{3\ \text{sheep}}{1\ \cancel{\text{goats}}} \times \frac{}{\text{sheep}} = \$$$

Step 7: What do you know about his trading sheep? It should look like this:

$$\frac{5\ \cancel{\text{goats}}}{1} \times \frac{3\ \cancel{\text{sheep}}}{1\ \cancel{\text{goats}}} \times \frac{2\ \text{pigs}}{1\ \cancel{\text{sheep}}} = \$$$

Step 8: Now get rid of "pigs"

$$\frac{5\ \cancel{\text{goats}}}{1} \times \frac{3\ \cancel{\text{sheep}}}{1\ \cancel{\text{goats}}} \times \frac{2\ \cancel{\text{pigs}}}{1\ \cancel{\text{sheep}}} \times \frac{27\ \text{canaries}}{3\ \cancel{\text{pigs}}} = \$$$

Step 9: You're almost done. Now you can get rid of the canaries. Remember from the problem that the farmer received \$3.25 for each canary. Remember to put "canary" in the denominator to cancel, leaving only one unit left—dollars.

$$\frac{5\ \cancel{\text{goats}}}{1} \times \frac{3\ \cancel{\text{sheep}}}{1\ \cancel{\text{goats}}} \times \frac{2\ \cancel{\text{pigs}}}{1\ \cancel{\text{sheep}}} \times \frac{27\ \cancel{\text{canaries}}}{3\ \cancel{\text{pigs}}} \times \frac{\$3.25}{1\ \cancel{\text{canary}}} = \$$$

Step 10: You are now in \$. That is what you want to be in. Now, stop and do the math.

= \$ 877.50

Case Studies

Chernobyl, Ukraine—1986. Explosion in a then-Russian nuclear power plant sent highly radioactive debris throughout northern Europe. Estimates run as high as 32,000 deaths and 62,000 square miles remain contaminated. About 500,000 people were exposed to dangerous levels of radiation. Cost estimates run as high as \$400 billion. The cause was determined to be both design and human error.

Luz International. During the 1980s, Luz produced large amounts of electricity in the Mojave Desert in California using photovoltaic cells. The price for their electricity was becoming competitive with other sources. Luz supplemented solar-derived electricity with electricity generated from natural gas. Filed for bankruptcy in 1991 due to lack of coherent U.S. energy policy: initial tax incentives were diminished. Luz became too large too fast.

Three-Mile Island, Pennsylvania—1979. The reactor lost its coolant, and the core melted. About 100,000 people were forced to leave their homes. Radiation that was leaked was relatively low; however, the estimated cost was $1 billion. The cause was mechanical and human failure.

Multiple-Choice Questions

1. Which of the following forms of energy is a renewable resource?

   (A) Synthetic oil
   (B) Breeder fission
   (C) Biomass
   (D) Oil shale
   (E) Synthetic natural gas

2. Which of the following forms of energy has a low long-term (next 50 years) estimated availability?

   (A) Low-temperature heating from solar energy
   (B) Synthetic oil and alcohols from coal
   (C) Photovoltaic production of electricity
   (D) Coal
   (E) Petroleum

3. Which of the following forms of energy is characteristic of having high net useful energy?

   (A) Tar sands
   (B) Wind energy
   (C) Fission
   (D) Synthetic natural gas
   (E) Geothermal energy

4. Which of the following alternatives would not lead to a sustainable energy future?

   (A) Phase out nuclear power subsidies.
   (B) Create policies to encourage governments to purchase renewable energy devices.
   (C) Assess penalties or taxes on continued use of coal and oil.
   (D) Decrease fuel-efficiency standards for cars, appliances, and HVAC systems.
   (E) Create tax incentives for independent power producers.

5. At today's rate of consumption, known U.S. oil reserves will be depleted

(A) in approximately 100 years.
(B) in approximately 50 years.
(C) in approximately 25 years.
(D) in approximately 10 years.
(E) in approximately 3 years.

6. Which country currently ranks number one in both coal reserves and use of coal as an energy source?

(A) Russia
(B) United States
(C) China
(D) India
(E) Brazil

7. The lowest average generating cost (cents per kilowatt-hour) comes from what energy source?

(A) Large hydroelectric facilities
(B) Geothermal
(C) Nuclear
(D) Solar photovoltaics
(E) Coal

8. The fastest-growing renewable energy resource today is

(A) nuclear energy.
(B) coal.
(C) wind.
(D) large scale hydroelectric plants.
(E) geothermal energy.

9. The least efficient energy conversion device listed is the

(A) steam turbine
(B) fuel cell
(C) fluorescent light
(D) incandescent light
(E) internal combustion engine

10. Which is NOT an advantage of using nuclear fusion?

(A) Abundant fuel supply
(B) No generation of weapons-grade material
(C) No air pollution
(D) No high-level nuclear waste or generation of weapons material
(E) All are advantages.

11. Given the following choices, which one has the greatest ability to perform useful work (high-quality energy)?

    (A) Coal
    (B) Oil
    (C) Solar
    (D) Wind
    (E) Electricity

12. Only about 10% of the potential energy of gasoline is used in powering an automobile. The remaining energy is lost into space as low-quality heat. This is an example of the

    (A) First Law of Thermodynamics.
    (B) Second Law of Thermodynamics.
    (C) Law of Conservation of Energy.
    (D) first-law efficiency.
    (E) second-law efficiency.

13. The Law of Conservation of Mass and Energy states that matter can neither be created nor destroyed and that the total energy of an isolated system is constant despite internal changes. Which society offers the best long-term solution to the constraints of this law?

    (A) Low-throughput society
    (B) High-throughput society
    (C) Matter-recycling society
    (D) Free-market society
    (E) Global market society

14. Which of the following methods CANNOT be used to produce hydrogen gas?

    (A) Reforming
    (B) Thermolysis
    (C) Producing it from plants
    (D) Coal gasification
    (E) All are methods of producing hydrogen gas

15. Energy derived from fossil fuels supplies what percent of the world's energy needs?

    (A) 10%
    (B) 33%
    (C) 50%
    (D) 85%
    (E) 97%

Answers to Multiple-Choice Questions

1. **C**	6. **C**	11. **E**
2. **E**	7. **A**	12. **B**
3. **B**	8. **C**	13. **A**
4. **D**	9. **D**	14. **E**
5. **C**	10. **E**	15. **D**

Explanations for Multiple-Choice Questions

1. **(C)** Renewable resources are those resources that theoretically will last indefinitely either because they are replaced naturally at a higher rate than they are consumed or because their source is essentially inexhaustible. Biomass can either be the burning of wood and agricultural wastes or urban wastes that can be incinerated.

2. **(E)** At the current rate of consumption, global oil reserves are only expected to last another 45 to 50 years. With projections of increased consumption in the near future, this figure will be even lower.

3. **(B)** Net useful energy is defined as the total amount of useful energy available from an energy resource over its lifetime minus the amount of energy used (first law of thermodynamics), wasted (second law of thermodynamics), and used in processing and transporting it to the end-user. Wind energy is essentially an unlimited resource in favorable sites.

4. **(D)** The key word in the question is "not." To foster a sustainable energy future, one would have to increase fuel-efficiency standards for cars, appliances, and HVAC (heating, ventilation and air-conditioning) systems.

5. **(C)** World oil demand is increasing at a rate of about 2% per year. *Known* U.S. oil reserves are projected to last about another 25 years. Potential reserves (Alaskan Arctic National Wildlife Refuge for example) might extend the estimate another 25 years.

6. **(C)** China gets approximately 75% of its energy from coal. Coal supplies over half of the fuel source to generate electricity in the United States. Utilities are the largest users of coal in the United States.

7. **(A)** Nonrenewable resources of energy (natural gas, oil, etc.) have had recent and dramatic price increases which have resulted in major increases in the cost of electricity. In the United States, this began in 2001 in California. Renewable resources of energy (hydroelectric and wind) provide the least expensive energy source for producing electricity. With increases in technology and production, wind energy is expected to be competitive in price with energy supplied by hydroelectric power plants.

8. **(C)** During the 1990s, wind power experienced a growth of 22% per year. Wind power supplies less than 2% of the energy used in the United States, primarily because it is a new industry. The country that is the largest user of wind power is Denmark, with 8% of its electricity being generated by the wind. Germany, Spain, and India are also large wind power users. China has enough potential wind power sites (especially in Inner Mongolia) to provide all that country's electricity needs.

9. **(D)** An incandescent light bulb is only 5% efficient, as compared to a fluorescent light at around 22% efficiency. A hydrogen fuel cell is approximately 60% efficient. The average current internal combustion engine, fueled by gasoline, is around 10% efficient. The United States wastes as much energy each day as two thirds of the world consumes.

10. **(E)** The major fuel used in fusion reactors, deuterium, could be readily extracted from ordinary water, which is available to all nations. The surface waters of the Earth contain more than 10 million million tons of deuterium, an essentially inexhaustible supply. The tritium required would be produced from lithium, which is available from land deposits or from seawater that contains thousands of years' supply. The worldwide availability of these materials would thus eliminate many current international tensions

caused by an imbalance in fuel supply. The amounts of deuterium and tritium in the fusion reaction zone would be so small that a large uncontrolled release of energy would be impossible. In the event of a malfunction, the plasma would strike the walls of its containment vessel and cool. Since no fossil fuels are used, there would be no release of chemical combustion products because they would not be produced. Similarly, there would be no fission products formed to present a handling and disposal problem. Radioactivity would be produced by neutrons interacting with the reactor structure, but careful materials selection would be expected to minimize the handling and ultimate disposal of activated materials. And finally, the materials and by-products of fusion are not suitable for use in the production of nuclear weapons.

11. **(E)** High-quality energy is defined as energy that is intense, concentrated and capable of performing useful work. Low-quality energy is diffused, dispersed, and low in temperature. Coal and oil have waste products, which means they are not totally concentrated. Solar and wind are diffuse, and the energy produced from them at this time is relatively low power. Electricity is pure, concentrated energy.

12. **(B)** The Second Law of Energy or Thermodynamics states that when energy is changed from one form to another, some of the useful energy is always degraded to lower-quality, more-dispersed (higher-entropy), and less-useful form of energy.

13. **(A)** A low-throughput society also known as a low-waste society or Earth-Wisdom society focuses on matter and energy efficiency. The society accomplishes this by

- Reusing and recycling nonrenewable matter resources,
- Using potentially renewable resources no faster than they are replenished,
- Using matter and energy resources efficiently,
- Reducing unnecessary consumption,
- Emphasizing pollution prevention and waste reduction,
- Controlling population growth.

14. **(E)** Reforming is a chemical process of splitting water molecules. Thermolysis uses extremely high temperatures to break water molecules apart. Hydrogen gas can be produced from algae by depriving the algae of oxygen and sulfur. Coal gasification is the conversion of coal into synthetic natural gas (SNG). The SNG can then be converted into hydrogen gas.

15. **(D)** Oil supplies approximately 36%, coal around 26%, and natural gas around 23%. The remaining 15% of the world's energy comes from (a) nuclear, solar, wind and hydropower at a total of around 9%; and (b) wood, peat, charcoal, and biomass at around a total of 6%.

Free-Response Question

Possible issues to address in energy free-response questions:

☐ What are the future energy supplies for this source?

☐ What is the net-energy yield from this source?

☐ What are the economic factors in developing, phasing in, and using this source?

☐ What are environmental factors associated with extracting, transporting and using the resource?

☐ What are the long-term benefits and risks?

A large, natural gas-fired electrical power facility produces 15 million kilowatt-hours of electricity each day it operates. The power plant requires an input of 13,000 BTUs of heat to produce 1 kilowatt-hour of electricity. One cubic foot of natural gas supplies 1,000 BTUs of heat energy.

(a) Showing all steps in your calculations, determine the
  (i) BTUs of heat needed to generate the electricity produced by the power plant in 24 hours.
  (ii) Cubic feet of natural gas consumed by the power plant each hour.
  (iii) Cubic feet of carbon dioxide gas released by the power plant each day. Assume that methane combusts with oxygen to produce only carbon dioxide and water vapor and that the pressures and temperatures are kept constant.
  (iv) Gross profit per year for the power company. The power company is able to sell electricity at \$50 per 500 kW-h. They pay a wholesale price of \$5.00 per 1000 cubic feet of natural gas.

(b) What environmental effect might the production of electricity through the burning of natural gas pose?

(c) Describe two other methods of producing electricity providing technological, economic and environmental pros and cons for each method discussed.

Answer to Free-Response Question

Note: Begin this problem by stating the information that is provided to you in a restatement. This will get rid of a lot of the words and allow you to just focus on the facts needed to solve the problem.

Restatement
- Gas-fired electrical power plant
- 15,000,000 kW-h of electricity every 24 hours (1 day)
- Plant requires 13,000 BTUs of heat to produce 1 kW-h of electricity
- 1 ft^3 of natural gas supplies 1000 BTUs of heat energy

(a) (i). Calculate BTUs of heat needed to generate electricity produced in 24 hours.

Start with an equal sign

$$=$$

Next, write down the units you want to be in.

$$= \text{BTUs}$$

Next, what did they give you to start with, or what were you limited to? The answer to this question is 24 hours.

$$\frac{24 \text{ hours}}{1} = \text{BTUs}$$

Next, what do you know about hours so that you can cancel the term? The answer is that 15,000,000 kW-h are produced every 24 hours.

$$\frac{24 \cancel{\text{hours}}}{1} \times \frac{1.5 \times 10^7 \text{ kW-h}}{24 \cancel{\text{hours}}} = \text{BTUs}$$

Note: We are now in kW-h and we have to end up in BTUs. Was there any information regarding kW-h and BTUs? The answer is yes. 13,000 BTUs are required for every 1 kW-h.

$$\frac{24 \text{ hours}}{1} \times \frac{1.5 \times 10^7 \text{ kW-h}}{24 \text{ hours}} \times \frac{13{,}000 \text{ BTUs}}{1 \text{ kW-h}} = \text{BTUs}$$

At this point you are in BTUs. That is what you want to be in. So stop and do the calculation.

= 2.0 × 10^{11} BTUs per day

I rounded my answer to two significant figures following significant figure rules. If you have not had significant figures yet, that is OK, and 1.95 × 10^{11} BTUs if perfectly OK.

(a)(ii) Calculate the number of cubic feet of natural gas consumed by the power plant each hour.

This time, I am just going to set up the problem without walking you through each step.

$$\frac{2.0 \times 10^{11} \text{ BTUs}}{1 \text{ day}} \times \frac{1 \text{ ft}^3}{1{,}000 \text{ BTUs}} \times \frac{1 \text{ day}}{24 \text{ hours}} = \text{ft}^3 \text{ natural gas / hour}$$

= 8.3 × 10^6 ft^3 natural gas · $hour^{-1}$

per hour or /hour is the same thing as $hour^{-1}$.

(a)(iii) Cubic feet of carbon dioxide gas released by the power plant each day.

The first thing we need to do is write a balanced equation for the reaction.

Natural gas is CH_4.
Oxygen gas is O_2.
Carbon dioxide is CO_2.
Water vapor is H_2O.

Putting them in an unbalanced equation results in

$CH_4 (g) + O_2 (g) \rightarrow CO_2 (g) + H_2O (g)$

The next thing we need to do is balance the equation. Check to make sure that you have the same number of atoms on each side of the equation. Remember, oxygen gas is diatomic.

$CH_4 (g) + 2O_2 (g) \rightarrow CO_2 (g) + 2H_2O (g)$

Remember from chemistry that if everything is a gas, and if the temperature and pressure are kept constant, then the coefficients can stand for volume. Therefore, we could say that 1 cubic foot of methane combines with 2 cubic feet of oxygen gas to give 1 cubic foot of carbon dioxide gas and 2 cubic feet of water vapor.

Now, look at our previous answer. We know that the power plant is consuming 8.3 × 10^6 ft^3 of CH_4 per hour. Therefore, let's start with that and convert to carbon dioxide and then make the problem represent the amount produced in 1 day.

$$\frac{9.3 \times 10^6 \text{ ft}^3 \text{ CH}_4}{\text{hour}} \times \frac{1 \text{ ft}^3 \text{ CO}_2}{1 \text{ ft}^3 \text{ CH}_4} \times \frac{24 \text{ hours}}{1 \text{ day}} = \text{ft}^3 \text{ CO}_2 \cdot \text{day}^{-1}$$

$$= \mathbf{2.2 \times 10^8 \text{ ft}^3 \text{ CO}_2 \cdot \text{day}^{-1}}$$

(a)(iv) Gross profit per year for the power company.

Restatement
- Customer pays \$50 for every 500 kW-h.
- Power company pays \$5.00 per 1,000 cubic feet of natural gas.
- Calculate the gross profit per year for the power company.

From the problem, we know that the power company produces 15 million kW-h of electricity each day it operates. If we assume that the power company runs 365 days per year and sells all the power that it produces (to produce power and not sell it would not be profitable), then the power company receives the following amount of income per year:

$$\frac{1.5 \times 10^7 \text{ kW-h}}{\text{day}} \times \frac{365 \text{ days}}{\text{year}} = 5.48 \times 10^9 \text{ kW-h} \cdot \text{yr}^{-1}$$

If they charge \$50 for every 500 kW-h, then they receive the following income:

$$\frac{5.48 \times 10^9 \text{ kW-h}}{\text{year}} \times \frac{\$50}{500 \text{ kW-h}} = \$5.48 \times 10^8 \text{ (\$548 million)}$$

Next, we need to figure out how much natural gas was used during the year to generate the electricity. From (a)(ii), we figured out that 8.36×10^6 ft³ of natural gas were used each hour. Let's convert that to cubic feet of natural gas used in one year:

$$\frac{8.36 \times 10^6 \text{ ft}^3 \text{ natural gas}}{\text{hour}} \times \frac{24 \text{ hours}}{1 \text{ day}} \times \frac{365 \text{ days}}{1 \text{ year}} = \frac{7.32 \times 10^{10} \text{ ft}^3 \text{ natural gas}}{\text{year}}$$

Now, we need to figure out how much the power company spends in one year for the natural gas that they burn to produce the electricity:

$$\frac{7.32 \times 10^{10} \text{ ft}^3 \text{ natural gas}}{\text{year}} \times \frac{\$5.00}{1000 \text{ ft}^3 \text{ natural gas}} = \$3.66 \times 10^8 \cdot \text{year}^{-1}$$

The last step is to figure out the gross profit (this is how much money the power company made without paying for other costs such as salaries, maintenance, new equipment, etc.).

5.48×10^8 came in from customers
– 3.66×10^8 paid out to the gas company

1.82×10^8 left (\$182 million gross profit)

$1.82 × 10^8

(b) An environmental effect that results from the burning of natural gas to produce electricity is the production of carbon dioxide gas, which is a greenhouse gas. As a greenhouse gas, the carbon dioxide gas molecule absorbs radiant energy reflected from the Earth's surface and reradiates the energy as long-wave infrared radiation back to the Earth. This "trapping" of the Earth's heat has serious environmental consequences in terms of its effect on global weather patterns, local climate conditions, and the glaciers and polar ice caps. Other environmental effects produced by burning of natural gas include the production of hydrogen sulfide gas (H_2S) and sulfur dioxide gas (SO_2) as a by-product of the production process. Sulfur dioxide gas when combined with atmospheric water vapor produces what is known as "acid rain." Acid rain has serious environmental effects primarily on aquatic organisms in lower trophic levels and on the reproduction biology of developing aquatic organisms. Hydrogen sulfide has serious environmental effects as a toxic pollutant and respiratory irritant. Leakage of CH_4 during processing adds methane to the atmosphere, which has a more deleterious effect as a greenhouse gas than does carbon dioxide. Drilling platforms to extract natural gas, the production of contaminated wastewater and brine, the possibility of land subsidence (sinking), and the disruption of wildlife and habitat refuges (i.e., Arctic Wildlife Refuge in Alaska) are other deleterious side effects of using natural gas. However, these negative side effects of natural gas have less environmental impact than using other forms of fossil fuels such as oil or coal.

(c) Two other methods of producing electricity include burning coal and using flowing water (hydroelectric power). The advantages of using coal are that it is plentiful and, according to the Law of Supply and Demand, inexpensive to use. Currently the world uses 4 billion metric tons of coal per year. At this rate, the world has enough coal to last for the next 200 years, and perhaps through projected unidentified world reserves, enough coal to last 1,000 years at the current rate of consumption. So not only is coal inexpensive to use, but it is plentiful too. Drawbacks to using coal include that current extraction methods are both environmentally damaging to the land and natural habitats (strip mining) and potentially dangerous to miners (cave-ins and black lung disease). Burning coal also releases into the atmosphere radioactive particles and mercury. Carbon dioxide produced from the burning of coal adds to the problem of global warming. Burning coal releases sulfur dioxide, which is directly related to acid rain production. And finally, the nitrogen oxides released (NO and NO_2), also add to acid rain.

Producing electricity through the flow of running water generally requires dams. Dams create large flooded areas, which destroy natural habitats and disrupt the lives of the people who have inhabited the area. The collection of sediment by the dam, the loss of this valuable sediment downstream making the land less productive, the obstruction that dams cause to migrating fish populations, and the aesthetic loss of a beautiful, white-water river are detractants of hydroelectric power. On the positive side, hydroelectric power does not significantly add to global pollution, as does the burning fossil fuels.

Now, take out a couple of sheets of paper and go back to the question. Let's not worry about time right now. We'll worry about time later. The important thing is to be able to do all parts of this question correctly. If you have to "peek" back at the answers, that's OK for now. Remember, the answers to the essays go far beyond what most students can do under limited-time conditions. What you want to work for is getting most of the material that I provide you, but don't stress if you don't get it all.

CHAPTER 2

The Cycling of Matter

Areas That You Will Be Tested On

A: Water
B: Carbon
C: Major Nutrients
  1. Nitrogen
  2. Phosphorus
  3. Sulfur
D: Difference Between Cycling of Major and Trace Elements

Key Terms

Note: Definitions for all key terms listed below can be found in the glossary starting on page 648. For additional definitions of relevant terms from this chapter, refer to *www.barronseduc.com/0764121618.html.*

abiotic	**assimilation**	**denitrification**
ammonia	**carbon dioxide**	**legumes**
ammonification	**cyanobacteria**	**nitrification**
ammonium	**decomposer**	**sink**

Key Concepts

Carbon Cycle

The symbol for carbon is C; the Lewis dot diagram is $\cdot\dot{\underset{\cdot}{C}}\cdot$. Carbon is the basic building block of life and the fundamental element found in carbohydrates, fats, proteins, nucleic acids (DNA and RNA). It is essential for life. Although carbon is found in rocks, it is a minor constituent when compared to oxygen atoms (45%) and silicon (30%) by weight. Carbon is found in the gas, carbon dioxide (CO_2). The atmosphere is composed of 0.036% CO_2 by volume.

Up until the last few hundred years, the rate of removal of carbon dioxide from the atmosphere exceeded the rate of its addition. In the late 18th century, at the beginning of the Industrial Revolution, the level of carbon dioxide in the atmosphere was 275 parts per million (ppm). Today, 200 years later, the amount of carbon dioxide in the atmosphere is approximately 365 ppm, more than a 30% increase. At current projected rates of increase, it is estimated that, by the year 2100, the concentration of CO_2 in the atmosphere could reach 600 ppm or higher. Reasons for this increase include the burning of fossil fuel (estimated to be 65% of the causative factor) and the modification and destruction of natural habitats, particularly forests, grasslands, and woodlands. Natural ecosystems store up to 100 times more carbon than converted land devoted to agriculture. Removing CO_2 from the atmosphere results in cooling temperatures; adding CO_2 to the atmosphere increases global warming.

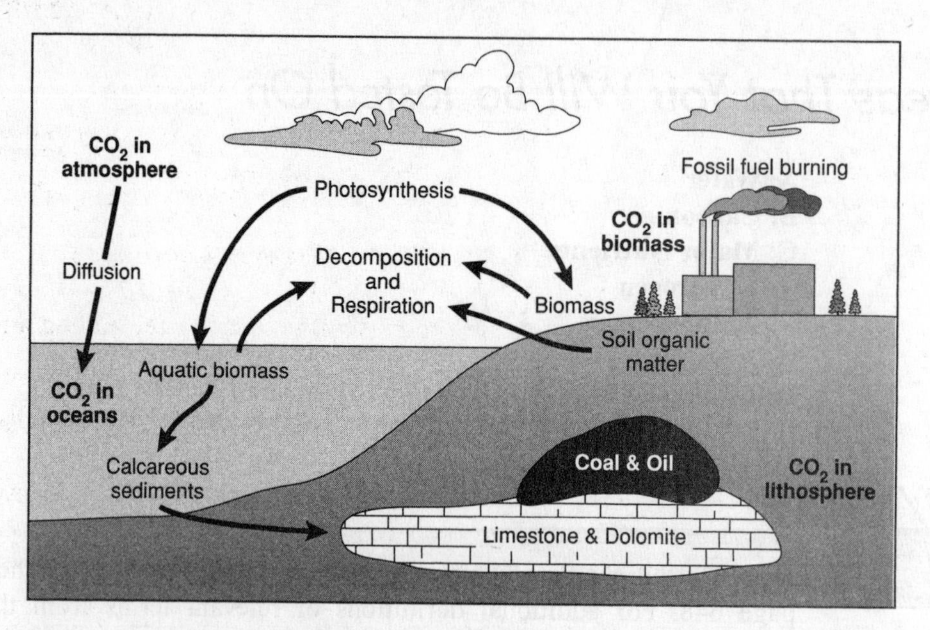

Figure 2.1. The carbon cycle viewed as chemical processes.

As shown in Figure 2.1, terrestrial producers (land plants) remove CO_2 from the atmosphere, while aquatic producers (aquatic plants) remove CO_2 from water, which raises the pH making the water more basic. Approximately 15% of the carbon in the atmosphere in the form of atmospheric carbon dioxide is taken up each year through photosynthesis.

CO_2 As a Reactant in Photosynthesis:

$$6CO_2\ (g) + 12H_2O\ (l) + \text{energy} \rightarrow C_6H_{12}O_6\ (aq) + 6CO_2 + 6H_2O$$

Note: For those of you who have had AP Biology, the Calvin-Benson cycle and the citric acid cycle below are presented as thematic tie-ins to material that you may have had in that course and are appropriate to mention here since they show carbon dioxide as a reactant and as a product in cycles essential to life. These two cycles are *NOT* on the AP Environmental Science Exam.

The Calvin cycle fixes atmospheric carbon dioxide to a three-carbon compound. This cycle then takes the NADPH and ATP formed in the light reactions of photosynthesis and utilizes them in the biosynthesis of complex carbon structures as shown in Figure 2.2.

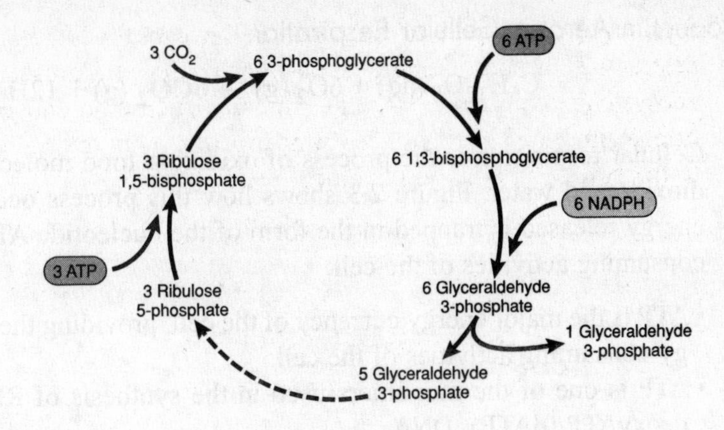

Figure 2.2. The Calvin-Benson cycle.

The Calvin-Benson cycle runs six times, each time adding a new carbon. Running through the Calvin cycle six times generates

6(5-carbon sugars: ribulose) + 6(carbon dioxides)

The six carbon dioxides are then reduced to glucose by the conversion of NADPH to NADP+. Glucose can now serve as a building block to make polysaccharides, other monosaccharides, fats, amino acids, nucleotides, and all the other molecules that living organisms require.

Some anaerobic bacteria found in restricted environments such as anaerobic hot spring in Yellowstone Park are able to produce glucose through reactions involving hydrogen sulfide and carbon dioxide:

$$6\ CO_2 + 6\ H_2S + \text{Energy} \rightarrow C_6H_{12}O_6 + 6\ S \text{ (no free oxygen)}$$

Photosynthetic autotrophs (blue-green algae also known as aerobic cyanobacteria) have also adapted to using carbon dioxide for life processes. Along with water and sunlight, they convert carbon dioxide into sugar. Further reactions within these types of organisms convert the sugars to amino acids, which are then used to build proteins and cellulose (a form of carbohydrate).

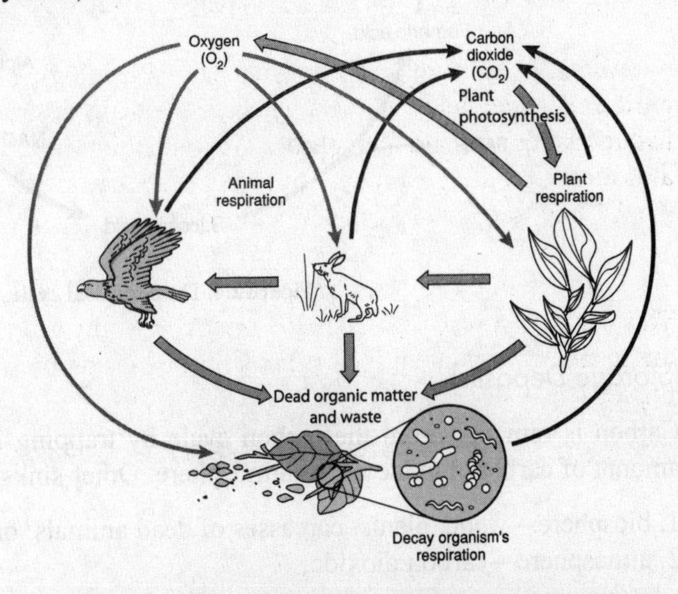

Figure 2.3. The carbon cycle viewed as an ecosystem cycle.

CO_2 As a Product in Aerobic Cellular Respiration

$$C_6H_{12}O_6\,(aq) + 6O_2\,(g) \rightarrow \mathbf{6CO_2\,(g)} + 12H_2O\,(l) + \text{energy}$$

Cellular respiration is the process of oxidizing food molecules, like glucose, to carbon dioxide and water. Figure 2.3 shows how this process occurs within a food web. The energy released is trapped in the form of the nucleotide ATP, for use by all the energy-consuming activities of the cell:

- ATP is the major energy currency of the cell, providing the energy for most of the energy-consuming activities of the cell.
- ATP is one of the monomers used in the synthesis of RNA and, after conversion to deoxyATP (dATP), DNA.
- ATP regulates many biochemical pathways.

Cellular respiration occurs in two phases:

1. Glycolysis—the breakdown of glucose to pyruvic acid, and
2. The complete oxidation of pyruvic acid to carbon dioxide and water through what is known as the citric acid cycle (see Figure 2.4).

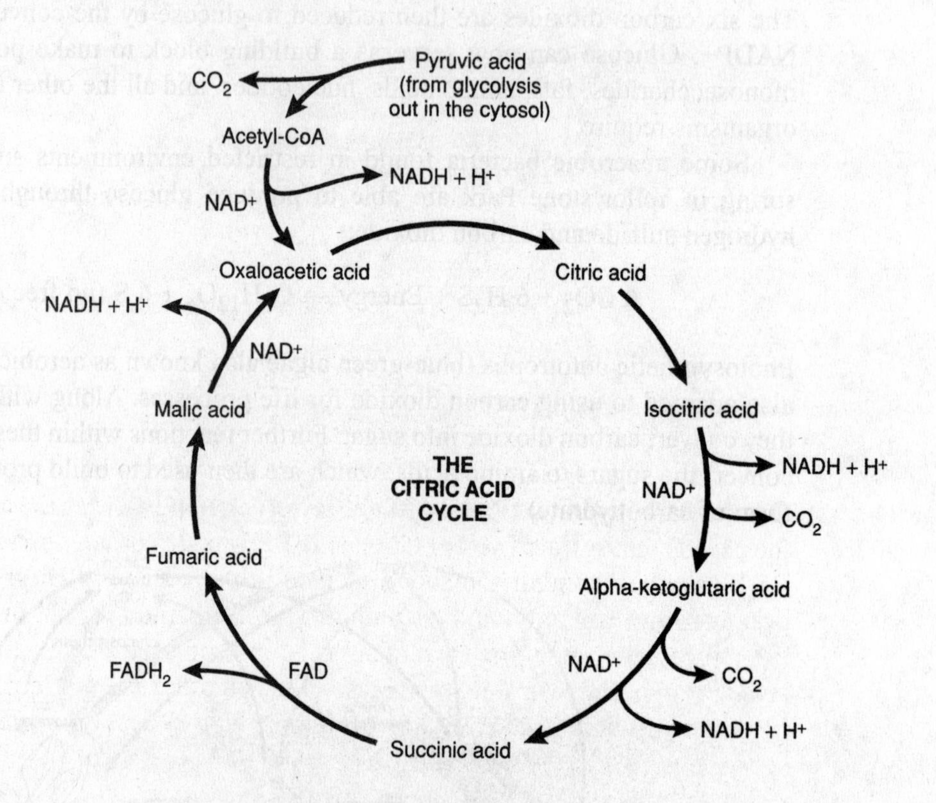

Figure 2.4. The citric acid cycle.

Carbon Sinks (Storage Deposits)

Carbon is removed from the carbon cycle by trapping it in sinks, which reduces the amount of carbon dioxide in the atmosphere. Other sinks include

1. biosphere—wood, plants, carcasses of dead animals, organic molecules;
2. atmosphere—carbon dioxide;

3. lithosphere—oil, coal, sedimentary rock such as limestone ($CaCO_3$), natural gas, oil shale, litter, organic matter;
4. hydrosphere—ocean floor sediments including shells and coral, dissolved carbon dioxide.

CARBON SINK SOURCES AND ABUNDANCE

Sink	Amount in Billions of Metric Tons
Marine sediments and sedimentary rocks	66,000,000 to 100,000,000
Ocean	38,000 to 40,000
Fossil fuel deposits	4000
Soil organic matter	1500 to 1600
Atmosphere	578 (as of 1700) to 766 (as of 1999)
Terrestrial plants	540 to 610

Carbon is released back into the environment by

- Fires,
- Burning fossil fuels,
- Geologic processes that bring carbon-containing sediments to the surface,
- Acid rain,
- Decomposition and respiration,
- Volcanoes through either volcanic rock, which then undergoes weatherization, or through release of carbon dioxide gas.

Carbon Balance Sheet

Oceans. 92 Gigatons/yr (1 Gt = 1×10^{12} kg) of carbon leaves the atmosphere and enters the oceans through acid precipitation and diffusion of carbon dioxide. The carbon is used for biosynthesis and/or remains dissolved in the ocean. 90 Gt/yr of carbon leaves the oceans from respiration and decomposition of organic material in the ocean and enters the atmosphere, resulting in increased acidity of ocean water (increased $H^+(aq)$ and $HCO_3^-(aq)$). As ocean waters increase in temperature due to global warming, their ability to absorb carbon dioxide gas decreases. Net: *Oceans are gaining 2 Gt of carbon every year.*

Land. 100 Gt/yr of carbon leaves the atmosphere and enters the lithosphere. The main avenue is through acid precipitation. The carbon is used for biosynthesis. 107 Gt/yr of carbon goes from the land to the atmosphere (100 Gt/yr comes from soil-based respiration and decomposition, 5 Gt/yr from burning of fossil fuels and 2 Gt/yr from deforestation). Net: *The atmosphere is gaining 7 Gt of carbon every year.*

Atmosphere. The final result is that the atmosphere is gaining 5 Gt of carbon per year (7 Gt entering – 2 Gt leaving = 5 Gt remaining). Nearly all the 5 Gt of carbon ends up as greenhouse gases.

The Carbonate-Silicate Cycle

The carbonate-silicate cycle is a very slow cycle that takes place over hundreds of millions of years. Carbonic acid that is formed from atmospheric carbon dioxide dissolving in water erodes rocks containing various silicates (remember, silicon is the second most abundant element in the Earth's crust—30% by weight). This acidic weathering also releases calcium ions ($Ca^{2+}(aq)$), carbonate ions ($CO_3^{2-}(aq)$), and bicarbonate ions ($HCO_3^-(aq)$), which eventually end up in the oceans. Marine organisms utilize the $Ca^{2+}(aq)$ to construct shells and other structures. When these marine organisms die, the shells and body parts sink and become concentrated. Over a very long time period, these carbon-containing deposits are transformed into sedimentary rock. Marine sediments and sedimentary rocks is the single largest carbon sink. Sedimentary material that ends up entering oceanic subduction zones is melted into magma, with the resulting formation and release of carbon dioxide gas back into the atmosphere through volcanic venting.

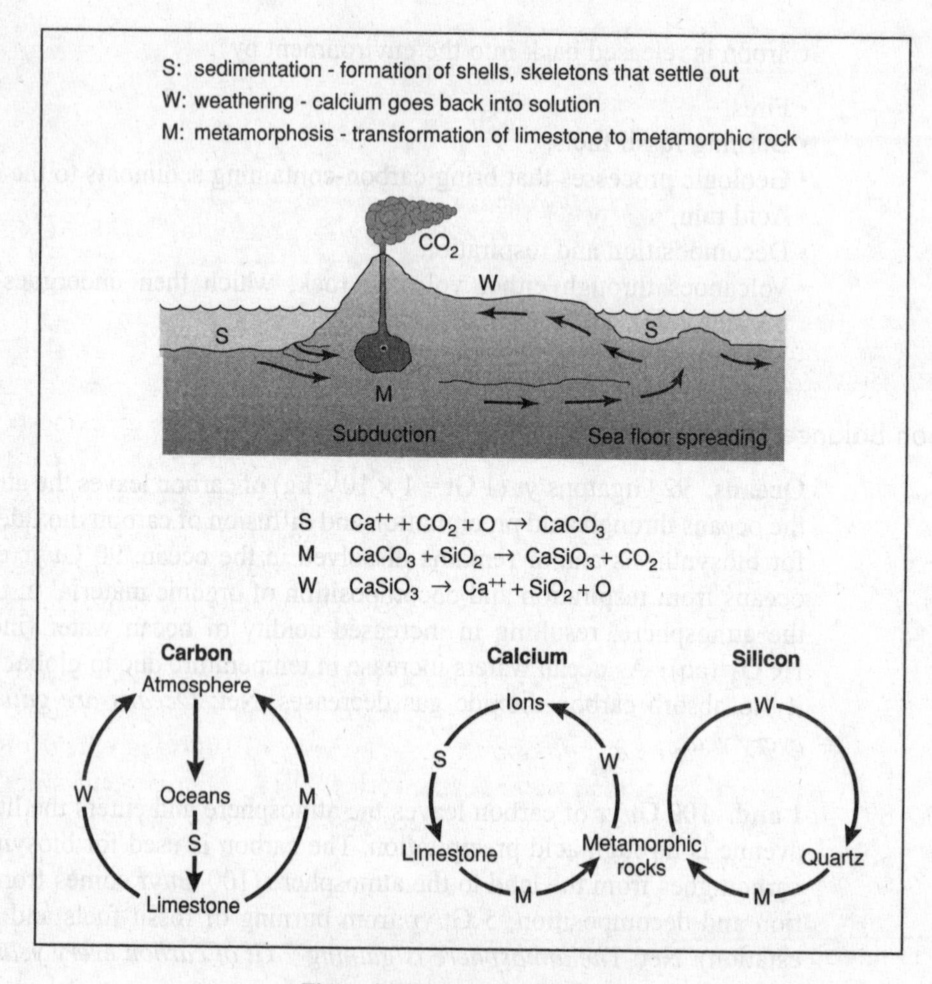

Figure 2.5. Inorganic carbon cycle.

Human Impact on the Carbon Cycle

1. Humans add carbon and carbon dioxide to the environment by burning fossil fuels and releasing carbon from sinks. Humans contribute one fourth as much CO_2 to the troposphere as nature with the potential of altering global climate.
2. Humans add carbon and carbon dioxide to the environment by burning fuel from wood and other organic material (e.g., slash and burn in the tropics).

Hydrologic (Water) Cycle

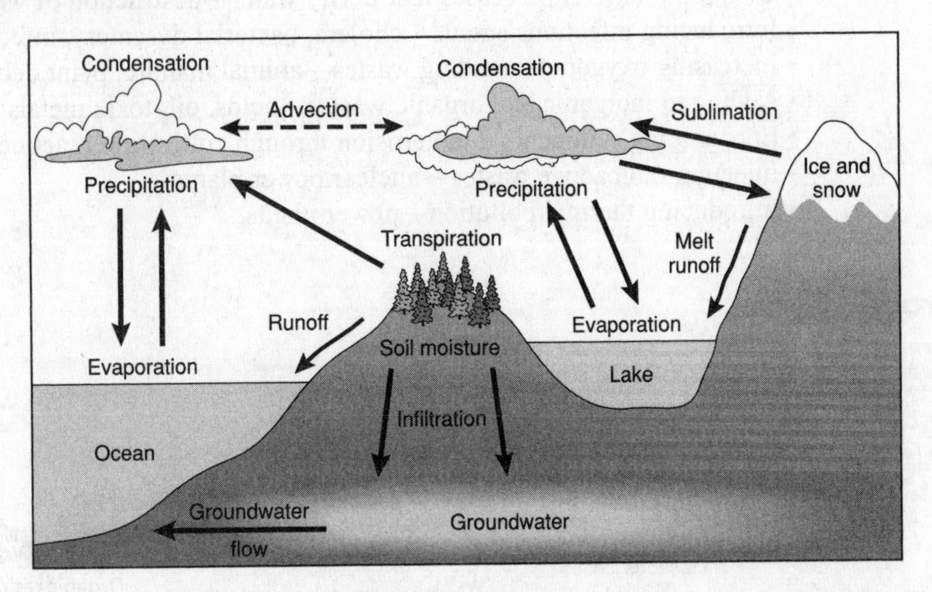

Figure 2.6. The water cycle.

The water cycle (Figure 2.6) is powered by energy from the sun and gravity. Solar energy evaporates water from oceans, lakes, rivers, streams, soil, and vegetation. Approximately 85% of the water vapor in the air initially comes from oceans. 97% of the water on Earth is found in the oceans. 2% is found in ice caps and glaciers. The remaining 1% is found in rivers, lakes, atmosphere, ground water, soil, and living organisms. The system is in dynamic equilibrium: the rate of evaporation equals the rate of condensation. Warm air can hold more water vapor than cooler air. There are five processes at work in the hydrologic cycle: (1) condensation, (2) precipitation, (3) infiltration, (4) runoff, and (5) evapotranspiration.

Water in the atmosphere is recycled on the average about every 10 days. In rivers, it is recycled about every 20 days; in glaciers, about every 50 years; in lakes, about 100 years; and in deep ground water, about every 10,000 years. The implication of contaminated deep ground water is noteworthy. The fact that ground water is being removed at a rate faster than it can be replaced makes ground water, for all practical purposes, a nonrenewable resource.

Human Impact on the Hydrologic Cycle

1. Withdrawing large amounts of fresh water from lakes, underground aquifers, rivers, etc. This is often accompanied by saltwater intrusion and ground water depletion.

2. Clearing of land for agriculture or habitation thereby

- Increasing runoff,
- Decreasing infiltration,
- Increasing flood risks,
- Accelerating soil erosion,
- Increasing potential for landslides.

3. Augmenting pollution by

- Adding nutrients through agricultural runoff (nitrates, phosphates, ammonium, etc.);
- Disturbing natural processes that purify water—destruction of wetlands;
- Introducing infectious agents—cholera, bacterial dysentery, infectious hepatitis, etc.;
- Increasing oxygen-demanding wastes—animal manure, plant debris, sewage;
- Adding to inorganic and organic wastes—acids, oil, toxic metals;
- Encouraging sediment—land erosion through cultivation practices and urbanization;
- Injecting radioactive wastes—nuclear power plants;
- Introducing thermal pollution—power plants.

Nitrogen Cycle

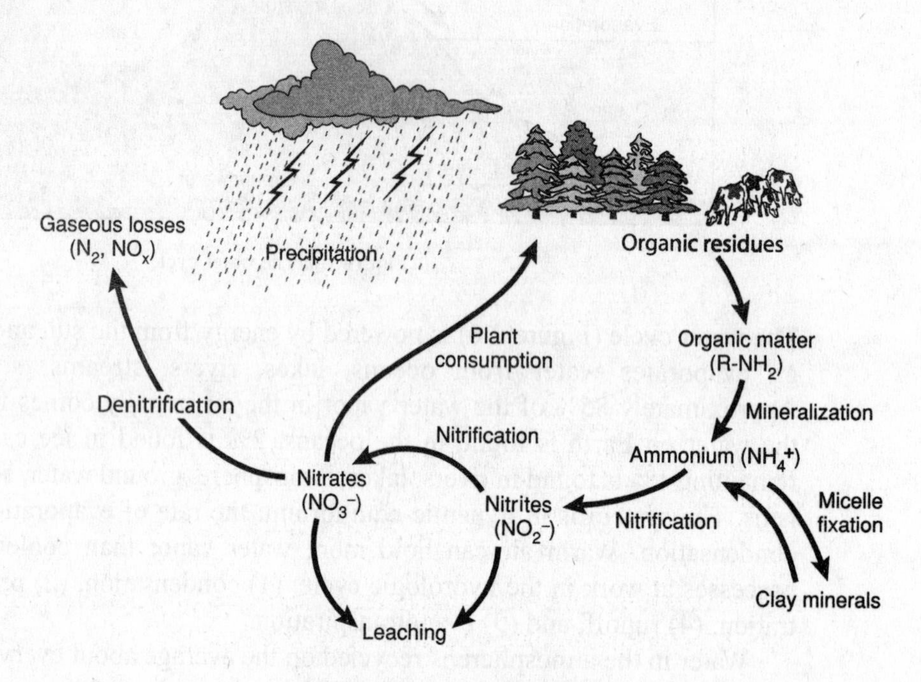

Figure 2.7. The nitrogen cycle.

The symbol for nitrogen is N, and its Lewis dot-diagram is •N̤̈•. Nitrogen is an essential element needed to make amino acids, proteins, DNA, and RNA. N_2 makes up 78% of the atmosphere (one million times more nitrogen is found in the atmosphere than is contained in the biosphere). Other nitrogen stores include organic matter in soil and oceans. Plants absorb nitrogen mostly through nitrate ions ($NO_3^-(aq)$) and to a lesser extent through ammonium ions ($NH_4^+(aq)$) due to its toxicity (Figure 2.7). Animals absorb the required amounts of nitrogen by consuming organic matter containing nitrogen. The organic matter can either be plants, live or dead animals, or products of decomposition.

Decomposers, especially bacteria, fungi, and actinomycetes convert ammonia (NH_3) to ammonium (NH_4^+) compounds through a process called mineralization. The ammonium ions are then attracted to small clay particles through a process called micelle fixation. The ammonium ions are then converted through a process called nitrification into nitrite ions (NO_2^-) by autotrophic chemosynthetic bacteria belonging to the genus *nitrosomonas* and into nitrate ions (NO_3^-) by autotrophic chemosynthetic bacteria belonging to the genus *nitrobacter*. The chemical process that both of these bacteria use is called nitrification, and functions through the chemical process of oxidation. The nitrate ion (NO_3^-) produced by *nitrobacter* is very soluble (from chemistry, "all nitrates are soluble") and is washed away from the soil by a process called leaching. This leached nitrate ion then eventually ends up in the oceans. From there, the nitrate ion can return to the atmosphere through a process called denitrification, which converts the nitrate ions into molecular nitrogen (N_2) and nitrous oxide (N_2O). Nitrous oxide is a greenhouse gas and also enters the atmosphere through the burning of fossil fuels, the burning of biomass, and the use of inorganic fertilizers. Denitrification can also occur through the action of heterotrophic bacteria (bacteria that absorb their nutrients rather than manufacture them) by chemically reducing (or reduction) NO_3^- into N_2 or N_2O. This process supplies the heterotrophic bacteria with oxygen required for metabolic mechanisms.

Through the Haber process ($N_2 + 3H_2 \rightarrow 2NH_3$), the ammonia produced can be converted to ammonium salts and used as inorganic fertilizer.

The Nitrogen Cycle involves the following processes:

1. **Nitrogen fixation.** Specialized bacteria (cyanobacteria and actinomycetes in soil and water and *Rhizobium* bacteria in root systems of legumes) convert N_2 to ammonia, NH_3. The legumes (bean family with over 13,000 species) form a mutualistic relationship with *Rhizobium;* the legumes receive nitrogen in a form they can utilize and the *Rhizobium* receive carbohydrates from the plant. The root nodules, which were produced as a response to the invasion of the root hairs and cortical cells by the *Rhizobium* bacteria, also supply a moist environment for the *Rhizobium* to live. It is estimated that nitrogen fixation adds 140 million metric tons of usable nitrogen to the biosphere each year. Nonlegumes (e.g., birch and alder trees) can also form nodules, which support nitrogen fixation. Nitrogen fixation can also occur without nodules by reactions involving cyanobacteria (blue-green algae):

$$N_2 + 3H_2 \rightarrow 2NH_3$$

2. **Nitrification.** Ammonia produced is converted by aerobic bacteria to nitrite ions (NO_2^-), which are then converted to nitrate ions (NO_3^-):

$$NH_3 \xrightarrow{\text{aerobic bacteria}} NO_2^- \xrightarrow{\text{bacteria}} NO_3^-$$

3. **Assimilation.** Plant roots absorb ammonia, ammonium ions, and nitrate ions to manufacture DNA, amino acids, and proteins:

$$\text{plant roots} + NH_3 + NH_4^+ + NO_3^- \xrightarrow{\text{bacteria}} \text{DNA} + \text{Amino Acids} + \text{Proteins}$$

4. **Ammonification.** Specialized bacteria break down dead organic material containing nitrogen and convert the nitrogen to ammonia (NH_3) and water-soluble salts containing ammonium ion (NH_4^+):

$$\text{Organic Material} \xrightarrow{\text{bacteria}} NH_3 + NH_4^+$$

5. **Denitrification.** Specialized bacteria convert the NH_3 and NH_4^+ into NO_2^- and NO_3^- and then into N_2 and N_2O (nitrous oxide), which are then released into the atmosphere:

$$NH_3 + NH_4^+ \xrightarrow{\text{bacteria}} NO_2^- + NO_3^- \xrightarrow{\text{bacteria}} N_2 + N_2O$$

Human Impact on the Nitrogen Cycle

1. Burning fuels

   a. Nitrogen and oxygen in the air combine to form nitrous oxide:

   $$N_2 + O_2 \rightarrow 2NO$$

   b. Nitrous oxide (NO) combines with atmospheric oxygen to produce nitrogen dioxide:

   $$2NO + O_2 \rightarrow 2NO_2$$

   c. Nitrogen dioxide combines with water vapor in the air to form nitric acid (HNO_3):

   $$NO_2 + H_2O \rightarrow HNO_3$$

   d. Nitric acid falls to Earth (acid deposition or acid rain).

2. Nitrous oxide (N_2O) is released into atmosphere by anaerobic bacteria acting on livestock wastes and through inorganic fertilizers applied to soil.
3. Nitrogen is removed from the soil through mining of ammonium nitrate (NH_4NO_3) used for fertilizer.
4. Depleting nitrogen from topsoil by overplanting nitrogen-rich crops.
5. Leaching of water-soluble nitrate ions (NO_3^-) from soil through irrigation and into groundwater.
6. Agricultural runoff adds nitrogen compounds to aquatic ecosystems, which in turn stimulates photosynthesizing algae and other plants. Aerobic decomposers break down excess aquatic bloom thereby depleting water of dissolved oxygen, a process called eutrophication.
7. Discharge of municipal sewage.
8. Man releases three times more nitrogen oxides (NO, NO_2, and N_2O) and ammonia (NH_3) into the nitrogen cycle than does nature.

Phosphorous Cycle

The symbol for phosphorus is P, and its Lewis structure is $\cdot\ddot{\underset{\cdot}{P}}\cdot$. The phosphorus cycle is different from the water, carbon, and nitrogen cycles because phosphorus is found in sedimentary rock, not in the atmosphere. Generally, phosphorus is found in the form of the phosphate ion (PO_4^{3-}) or the hydrogen phosphate ion (HPO_4^{2-}), which is found in terrestrial rock formations and ocean sediments. It is essential in the production of DNA, fats in cell membranes of both plants and animals, bones, teeth, and shells. Phosphorus production is not dependent on bacterial action and is not found to any significant amount in the atmosphere. Phosphate salts are only slightly soluble. Phosphorus is released from terrestrial rock formations by weathering and action of acid rain and is a relatively slow process. Phosphorus becomes dissolved in soil water and is then taken up by plant roots; however, it is often a limiting factor for both terrestrial and aquatic plant growth due to limited concentrations in soil, rock, and water. Phosphorus is also found in guano, an excrement of fish-eating birds. Erosion accelerates terrestrial phosphate to ocean sources. Phosphorus for commercial purposes (detergents and fertilizers) is mined in Bone Valley, Florida (near Tampa). U.S. reserves of phosphorus is estimated to be 8 billion metric tons. An inorganic fertilizer rated as 6-24-26 would be 6% nitrogen, 24% phosphorus, and 26% potassium.

Human Impact on the Phosphorus Cycle

1. Mining of large quantities of rocks containing phosphates used for production of inorganic fertilizer and detergents. Mining scars land. Production is expected to increase in the near future.
2. Clear cutting tropical areas for agriculture, decreasing available phosphorus. Most phosphorus is found in vegetation in tropical zones; little is found in the soil. Upon decomposition of plant material by decomposers, phosphorus is washed away into streams.
3. Allowing runoff of animal wastes from feedlots, runoff of fertilizers, and discharge of municipal sewage plants.
4. Allowing stream runoff, which causes phosphorus to accumulate in lakes and ponds. This accumulation causes an increase in growth of cyanobacteria (blue-green algae), green algae, and aquatic plants. This increase causes sunlight to be blocked and for aerobic decomposers to decrease oxygen content in water, thereby killing aquatic animals and plants.

Sulfur Cycle

The symbol for sulfur is S, and its Lewis structure is $\cdot\ddot{\underset{\cdot\cdot}{S}}\cdot$. The sulfur cycle is primarily a gaseous cycle. Most sulfur is found in underground rocks (FeS_2 or "Fool's Gold" and $CaSO_4$ or gypsum) and deep oceanic deposits. Natural release of sulfur into the atmosphere comes from weathering, gases released from sea floor vents, and volcanic eruptions and venting. Most sulfur enters the atmosphere through human activity, particularly through the burning of high-sulfur fossil fuels such as coal and oil. Plants assimilate sulfur in the form of sulfate salts.

When inorganic sulfur compounds are reduced by organisms as a source of nutrition, they are said to be assimilated in a process known as assimilative metabolism. Normally this assimilative metabolism produces a –SH (sulfhydryl group). The reduced sulfur is converted into cell material in the form of macromolecules such as sulfur-containing proteins. Bacteria, Archea, fungi, algae, and higher plants have the ability to use assimilative metabolism. When the sulfur source is used as an electron acceptor for energy metabolism, it is referred to as dissimilative metabolism. In this form of metabolism, large amounts of sulfur are reduced and excreted to the environment.

Hydrogen sulfide (H_2S) and sulfur dioxide (SO_2) gases enter the sulfur cycle through volcanoes and geysers. Hydrogen sulfide is also released on breakdown of organic material in swamps, bogs, and tidal flats (presence of anaerobic decomposers). Some species of marine phytoplankton also produce fairly large quantities of dimethyl sulfide or DMS (CH_3SCH_3).

Sulfur dioxide in the atmosphere reacts with oxygen gas to produce sulfur trioxide (SO_3) gas. Sulfur trioxide then reacts with water droplets to form sulfuric acid (H_2SO_4), which is a major contributor to acid rain. Ammonia (NH_3) and sulfur dioxide (SO_2) react in the atmosphere to produce tiny, solid particles that fall to Earth in a process known as acid deposition. Sulfur dioxide and sulfate aerosols create human health problems and damage crops. These gases also damage buildings, reduce visibility, and absorb ultraviolet radiation. The cloud cover that they create increases the Earth's albedo (reflectivity) and tends to produce a cooling effect, which may offset a minor portion of the heating caused by carbon dioxide and other greenhouse gases. This negative feedback mechanism may help substantiate the Gaia hypothesis.

Human Impact on the Sulfur Cycle

1. Burning of coal and oil (containing sulfur) and petroleum refining. This releases twice as much sulfur dioxide (SO_2) into the troposphere as nature.
2. Smelting of metal ores that contain sulfur.
3. Utilizing sulfuric acid in industrial processes and manufacture.
4. Refining of petroleum.

Nutrient Cycling: Macro and Micro Nutrients

The cycling of nutrients through the biosphere involves inorganic chemical reactions and specific organic metabolic reactions. How essential nutrients are cycled through the biosphere, that is, their specific biochemical pathways in living organisms, their role in geological mechanisms, the rate at which they are cycled, and the changes that they undergo through chemical reactions in the environment are known as biogeochemical cycles. These reactions occur in the atmosphere, hydrosphere, and lithosphere.

Macronutrients are elements that are required by living organisms in relatively large amounts, typically making up more than 0.2% of the dry weight of the organism. This weight is primarily made up in the form of carbohydrates (including cellulose, sugars, and starches), which are in turn made up of carbon, hydrogen, and oxygen; support structures such as bone or shell made primarily from calcium, oxygen, and phosphorus; and proteins and fats, which are made up of many different elements.

Micronutrients are elements that are required by living organisms in trace amounts and often constitute less than 0.2% of the dry weight of the organism. Micronutrients are important primarily in enzymatic pathways. See the following tables for a list of essential macro- and micronutrients required by living organisms and their role.

Nutrients can *enter* the ecosystem by: (1) weathering of rocks (aluminum, calcium, iron, magnesium, phosphorus, potassium, silicon, and sodium), (2) lightning (NO produced from N_2 and O_2), (3) precipitation (calcium, chlorine, nitrogen, sodium, and sulfur), (4) photosynthesis (oxygen), and (5) respiration (carbon).

Nutrients can *leave* ecosystems through: (1) soil erosion caused by poor cultivation techniques, overgrazing, and clear cutting; (2) leaching, in which water carries nutrients down into the soil and possibly into the ground water; (3) a gas phase, especially apparent of soil nitrogen; (4) emigration of animals; and (5) harvesting of crops. Some nutrients (especially nitrogen) can also leave an ecosystem in gaseous form. This occurs when soil conditions are wet and low in oxygen content (anaerobic), whereby nitrogen compounds in the soil are chemically reduced to a gaseous form and escape into the atmosphere. Up to 80% of the nitrogen that is applied to crops through the use of fertilizer in the form of ammonia or ammonium compounds may be lost this way in a process called denitrification.

Nutrient cycling that occurs on land occurs primarily in the upper layers of the soil. This area is subject to weathering, erosion, leaching, precipitation, application of fertilizer and other soil additives, atmospheric deposition, input of wastes from other organisms, and the action of decomposers. Decomposition begins with dead organic matter being broken into smaller pieces by soil organisms including worms, arthropods (ants, beetles, termites), and gastropods (slugs and snails). The material is broken down further by heterotrophic bacteria and fungi. The rate of decomposition depends on the amount of moisture in the soil, temperature, and nature of the soil.

MACRONUTRIENTS (MAKE UP MORE THAN 0.2% OF DRY WEIGHT)

Nutrient	**Role**
Calcium	Principal skeletal mineral in bones and teeth, muscle contraction and relaxation, nerve function, intracellular regulation, extracellular enzyme cofactor, blood clotting, and blood pressure. Involved in formation of cell walls. Participates in translocation of sugars. In plants, it improves fruit and nut formation. Root and leaf development. Involved in uptake of other nutrients. Improves postharvest quality of fruits and vegetables. Aids in the control of certain fungal and bacterial diseases.
Carbon	Forms carbohydrates with oxygen and hydrogen. Carbon is a major component of organic molecules, which are the building blocks of all organisms.
Hydrogen	Hydrogen is a major component of organic molecules, which are the building blocks of all organisms.
Nitrogen	Involved in making proteins.
Oxygen	Oxygen is a major component of organic molecules, which are the building blocks of all organisms.
Phosphorus	Acid-base balance, DNA/RNA structure, energy, and enzyme cofactor. Found in every cell.
Potassium	Protein synthesis, fluid balance, muscle contraction, and nerve transmission.
Sodium	Acid-base balance and fluid retention. Involved in nerve impulse transmission.
Sulfur	Component of biotin, thiamin, insulin, and some amino acids. Involved in formation of nodules and chlorophyll synthesis; a structural component of amino acids and enzymes. In plants, it improves cold resistance and disease resistance. Assists decomposition of crop residue. Involved in protein formation and uptake of other nutrients.

MICRONUTRIENTS OR TRACE ELEMENTS (MAKE UP LESS THAN 0.2% OF DRY WEIGHT)

Nutrient	Role
Aluminum	A micronutrient required for proper development and growth in plants. Requirements in human nutrition have not been established.
Boron	Involved in formation of cell walls, terminal buds, and pollen tubes. Participates in regulation of starch production and translocation of sugars and starches. In plants, it improves quality and disease resistance. Involved in seed, flower, and fruit formation.
Chlorine	Fluid balance, aids digestion in stomach.
Chromium	Energy release, sugar and fat metabolism, potentiates the action of insulin.
Cobalt	As a component of vitamin B_{12}, aids in nerve function and blood formation.
Copper	Absorption of iron, part of many enzymes. Involved in photosynthetic and respiration systems. Assists chlorophyll synthesis and used as reaction catalyst. In plants, it improves nitrogen utilization. Involved in protein formation and root metabolism
Fluorine	Bone and teeth formation; decreases dental caries.
Iodine	Component of the hormone thyroxin, which aids in metabolism regulation and fetal development.
Iron	Hemoglobin formation in red blood cells, myoglobin formation in muscle, oxygen carrier, energy utilization. Involved in respiration and chlorophyll synthesis. In plants, it improves plant appearance. Required for vigorous growth.
Magnesium	Bone mineralization, protein synthesis, enzymatic reactions, muscular contraction, nerve transmission. Involved in photosynthetic and respiration system. Active in uptake of phosphate and translocation of phosphate and starches. In plants, it improves seed production and formation of seed oil and fat. Involved in uptake of other nutrients.
Manganese	Involved in regulation of enzymes and growth hormones. Assists in photosynthesis and respiration. In plants, it improves germination and hastens maturity. Involved in uptake of carbon, magnesium, and phosphorous. Manganese is important in resistance development to both root and foliar diseases caused by fungi.
Molybdenum	Component of a several enzymes and involved in enzymatic reduction of nitrates to ammonia. Assists in conversion of inorganic phosphate to organic form. In plants, it improves nodule formation and fixation of nitrogen. Assists protein formation. Required for the synthesis and activity of the enzyme nitrate reductase (reduces nitrates to ammonium in the plant).
Selenium	Protects against oxidation.

MICRONUTRIENTS OR TRACE ELEMENTS
(MAKE UP LESS THAN 0.2% OF DRY WEIGHT) (continued)

Nutrient	Role
Silicon	Promotes the synthesis of collagen and formation of bone.
Vanadium	Involved in enzyme activities in the body.
Zinc	Transport of vitamin A, taste, wound healing, sperm production, fetal development. Plays a part in many enzymes, hormones (insulin), genetic material, and proteins. Involved in production of growth hormones and chlorophyll. Active in respiration and carbohydrate synthesis. In plants, it improves plant appearance, seed production and absorption of water. Involved in protein and carbohydrate formation.

Case Study

Hubbard Brook Experimental Forest. Study centered on how deforestation affects nutrient cycles. The forest consisted of several watersheds each drained by a single creek. Impervious bedrock was close to the surface, which prevented seepage of water from one forested hillside, valley, and creek ecosystem to another.

Conclusions:

- An undisturbed mature forest ecosystem is in dynamic equilibrium with respect to chemical nutrients. Nutrients leaving ecosystem are balanced by nutrients entering the ecosystem.
- Inflow and outflow of nutrients was low compared to levels of nutrients being recycled within the ecosystem.
- When deforestation occurred, water runoff increased. Consequently, soil erosion increased which caused a large increase in outflow of nutrients from the ecosystem. Increase in outflow of nutrients causes water pollution.
- Nutrient loss could be reduced by clearing trees and vegetation in horizontal strips. Remaining vegetation reduced soil erosion.

Multiple-Choice Questions

1. The Hubbard Brook experiment demonstrated that
   (A) nutrient loss could be increased by clearing trees and vegetation in horizontal strips. The remaining vegetation reduced soil erosion.
   (B) inflow and outflow of nutrients was high compared to levels of nutrients being recycled within the ecosystem.
   (C) when deforestation occurred, water runoff increased. Consequently, soil erosion decreased, which caused large increase in outflow of nutrients from the ecosystem.
   (D) an undisturbed mature forest ecosystem is in dynamic equilibrium with respect to chemical nutrients. Nutrients leaving the ecosystem are balanced by nutrients entering ecosystem.
   (E) all of the above

2. Which of the following sinks is NOT a primary depository for the element listed?

   (A) Carbon—coal
   (B) Nitrogen—nitrogen gas in the atmosphere
   (C) Phosphorus—marble and limestone
   (D) Sulfur—deep ocean deposits
   (E) All are correct

3. Burning of fossil fuels coupled with deforestation increases the amount of ______ in the atmosphere.

   (A) NO_2
   (B) CO_2
   (C) SO_2
   (D) O_3
   (E) all of the above are correct

4. In the nitrogen fixation cycle, cyanobacteria in soil and water and *Rhizobium* bacteria in root systems of legumes are responsible for converting

   (A) organic material to NH_3 and NH_4^+.
   (B) NH_3, NH_4^+, and NO_3^- to DNA, amino acids, and proteins.
   (C) NH_3 and NO_2^- to NO_3^-.
   (D) N_2 and H_2 to NH_3.
   (E) NH_3 to NO_2^- and then to NO_3^-.

5. An industrial method used to manufacture nitrogen-rich fertilizer is known as

   (A) nitrogen-fixation process
   (B) Haber process
   (C) ammonium conversion
   (D) cracking
   (E) nitrogen-enrichment

6. Nitrogen is assimilated in plants in what form?

   (A) NO_2^-
   (B) NH_3
   (C) NH_4^+
   (D) NO_3^-
   (E) Choices B, C, and D

7. Plants primarily assimilate sulfur in what form?

   (A) Sulfates
   (B) Sulfites
   (C) Hydrogen sulfide
   (D) Sulfur dioxide
   (E) Elemental sulfur

8. Man increases sulfur into the atmosphere and thereby increases acid deposition by all of the following activities EXCEPT

   (A) industrial processing.
   (B) processing (smelting) ores to produce metals.
   (C) burning coal.
   (D) petroleum refining.
   (E) clear-cutting.

9. Phosphorus is being added to the environment by all of the following activities EXCEPT

   (A) runoff from feedlots.
   (B) clear cutting in tropical areas.
   (C) stream runoff.
   (D) burning coal and petroleum.
   (E) mining to produce inorganic fertilizer.

10. Carbon dioxide is a reactant in

    (A) photosynthesis.
    (B) cellular respiration.
    (C) Haber process.
    (D) nitrogen-fixation.
    (E) none of the above.

11. Human activity adds significant amounts of carbon dioxide to the atmosphere by all of the following EXCEPT

    (A) brush clearing.
    (B) burning wood.
    (C) burning petroleum.
    (D) clear cutting.
    (E) agricultural runoff.

12. Clearing of land for either habitation or agriculture does all of the following EXCEPT

    (A) increases runoff.
    (B) increases flood risks.
    (C) increases potential for landslides.
    (D) increases infiltration.
    (E) accelerates soil erosion.

13. All the following have an impact on the nitrogen cycle EXCEPT

    (A) the application of inorganic fertilizers applied to the soil.
    (B) the action of aerobic bacteria acting on livestock wastes.
    (C) the overplanting of nitrogen-rich crops.
    (D) the discharge of municipal sewage.
    (E) the burning of most fuels.

For Questions 14 and 15, refer to the following choices:

(A) Sulfur
(B) Nitrogen
(C) Iron
(D) Cobalt
(E) Molybdenum

14. A macronutrient essential to the formation of proteins.

15. A micronutrient that is a component of several enzymes and involved in enzymatic reduction of nitrates to ammonia. It assists in conversion of inorganic phosphate to organic form. In plants, it improves nodule formation and fixation of nitrogen. It assists protein formation and is required for the synthesis and activity of the enzyme nitrate reductase (reduces nitrates to ammonium in the plant.)

Answers to Multiple-Choice Questions

1. **D**	6. **E**	11. **E**
2. **C**	7. **A**	12. **D**
3. **B**	8. **E**	13. **B**
4. **D**	9. **D**	14. **B**
5. **B**	10. **A**	15. **E**

Explanations for Multiple-Choice Questions

1. **(D)** In choice A, horizontal strips would help to *prevent* nutrient loss through erosion. In choice B, inflow and outflow of nutrients was stable when compared to the level of nutrients being recycled. In choice C, soil erosion would *increase* due to deforestation. Choice D is valid only in an undisturbed forest ecosystem, which is the premise

of the choice. The Hubbard Brook Experimental Forest in New Hampshire, the longest-running forest-ecosystem experiment in North America, did show that in a *disturbed* forest ecosystem, one that had an input of acid rain, forest soils became depleted of natural buffering cations such as Ca^{2+} and Mg^{2+} over the years. Those cations were replaced by hydrogen and aluminum cations, which resulted in tree loss.

2. **(C)** The primary sink for phosphorus is ocean sediment and certain islands off South America and the Pacific island nation of Nauru that have or had high amounts of bird guano.

3. **(B)** Burning of fossil fuels releases sulfur oxides (SO_x), carbon oxides (CO_x—carbon dioxide on complete combustion, carbon monoxide on incomplete combustion), and nitrogen oxides (NO_x). Ozone is a photochemical oxidant that is produced in a secondary atmospheric reaction involving the formation of atomic oxygen through splitting nitrogen dioxide (NO_2) and is not produced directly by burning fossil fuels. Deforestation, or the removal and burning of trees and other vegetation on a large scale (slash and burn) to expand agricultural or grazing lands, releases primarily carbon dioxide. Since the question said "coupled," the gas that is common to both processes is carbon dioxide.

4. **(D)** This is the first step in the nitrogen cycle and is called nitrogen fixation. Refer to Step 1 in the nitrogen cycle for details on nitrogen fixation.

5. **(B)** The industrial production of ammonia through the Haber process is the same chemical reaction as nitrogen fixation: $N_2 + 3H_2 \rightarrow 2NH_3$. The differences between the bacterial process and a process that occurs in factories are the pressures and temperatures required in the industrial manufacturing. The production of ammonia through the Haber process ranks as one of the most produced chemicals in the world, the reason being the world's need for fertilizer.

6. **(E)** The nitrite ion (NO_2^-) is toxic to plants. In the nitrogen cycle, at a step called assimilation, plant roots can absorb ammonia (NH_3), ammonium ion (NH_4^+) and nitrate ion (NO_3^-).

7. **(A)** Hydrogen sulfide (H_2S) and sulfur dioxide (SO_2) are toxic to living organisms. Most of the sulfur in the world is stored either in the elemental form and is extracted using hot steam through a method called the Frasch process or as sulfate compounds. Some sulfate compounds are water soluble, allowing the sulfate anion (SO_4^{2-}) to be absorbed by the plant. Free sulfur is not soluble in water and is not able to be absorbed. Plants cannot effectively absorb sulfite ions (SO_3^{2-}).

8. **(E)** Clear cutting produces carbon dioxide (CO_2), not SO_x.

9. **(D)** Animal manure and bird guano are rich in phosphate. In the tropics, most of the nutrients are contained within the trees and vegetation, with little being retained in the soil. Phosphorus therefore would be released back into the environment by clear cutting. It would then be subject to runoff. Mining phosphates for fertilizer and industrial products takes phosphorus out of sinks and puts it into the environment for cycling. Burning coal and petroleum does not add appreciable phosphorus to the environment.

10. **(A)** Photosynthesis is written as

$$6CO_2\ (g) + 12H_2O\ (1) + \text{Energy} \rightarrow C_6H_{12}O_6\ (aq) + 6CO_2\ (g) + 6H_2O$$

Cellular respiration is the reverse reaction

$$C_6H_{12}O_6\ (aq) + 6O_2\ (g) + 6H_2O \rightarrow 6CO_2 + 12H_2O\ (1) + \text{Energy}$$

11. **(E)** Agricultural runoff, primarily from fertilizers and feedlots, adds nitrates and phosphates to streams. All other choices involve combustion, which produces carbon dioxide.

12. **(D)** Infiltration is the movement of water into the soil. Removing vegetation decreases infiltration by not allowing water to percolate slowly through the soil. Removing vegetation will increase runoff since water cannot be absorbed by the soil fast enough. Since runoff is increased, the potential for floods increase. Since floods and runoff increase, the soil can become saturated and lose its integrity and result in a landslide. Runoff carries with it topsoil and nutrients thus accelerating soil erosion.

13. **(B)** The bacteria that "digest" livestock wastes are *anaerobes* and operate only in anaerobic environments (no free oxygen). They produce nitrous oxide (N_2O). Other bacteria can be used to generate methane gas from animal manure in what is known as a "digester." In this process, anaerobic digestion is a two-part process and each part is performed by a specific group of organisms. The first part is the breakdown of complex organic matter (manure) into simple organic compounds by acid-forming bacteria. The second group of microorganisms, the methane formers, break down the acids into methane and carbon dioxide. This methane gas can then be used for heating. Inorganic fertilizers lose up to 80% of their nitrogen to the atmosphere through denitrification caused by bacterial action. Municipal sewage and nitrogen-rich crops release nitrogen into the environment. Burning most fuels assumes that fossil fuels, which produce (NO_x), are included.

14. **(B)** Refer to the macronutrient chart.

15. **(E)** Refer to the micronutrient chart.

Free-Response Question

by Sarah E. Utley
Environmental Content Specialist, Center for Digital Innovation,
AP Environmental Science Division, University of California at Los Angeles

APES Fertilizer Company creates a soil additive that contains 10% (w/v) of nitrogen, carbonate, and phosphorous. Using this information, answer the following questions.

(a) Choose either carbon or nitrogen and clearly explain the biogeochemical cycle of that nutrient. Include all the important steps in the cycle and indicate movement from the atmosphere, water, and soil where necessary.

(b) Agricultural runoff from overapplication of fertilizers is a major source of water pollution. Chose either nitrogen or phosphorous and fully explain the cause and effect of this excess on the aquatic ecosystem.

(c) Rather than rely on the application of fertilizers, current agricultural research is encouraging the practice of more conservation-minded farming. Name and explain one such environmental practice that can decrease dependence on human-applied nutrients without causing a significant decrease in crop yield.

Free-Response Answer

Rubric

Part (a): 5 points
Part (b): 3 points
Part (c): 2 points

Ten points possible for full credit.

RUBRIC	ESSAY
1 pt. Movement through the Earth's life support systems (atmosphere, lithosphere, biomass).	(a) The Earth is a closed system to matter. A finite amount of matter (such as life-supporting nutrients) is found in the biosphere, and these matter or nutrients cycle throughout the Earth's varied support systems in different forms. Nitrogen, an essential nutrient for life, is *found primarily in the atmosphere* (making up 78%), but it also *cycles through the lithosphere* and the *biomass of organisms*. Unfortunately, the producers (mostly plants) cannot absorb nitrogen in its most commonly found form: atmospheric N_2.
½ pt. Introduction to nitrogen cycle.	In order for this gaseous element to be used by living organisms, it has to be *converted in a series of steps (called the nitrogen cycle)*.
½ pt. Definition of nitrogen fixation. ½ pt. Sources of fixation.	The first step in the nitrogen cycle is called *nitrogen fixation*. This is the process by which *atmospheric nitrogen (N_2) is converted to ammonia (NH_3)*. There are some natural sources of *nitrogen fixing* such as lightning and volcanoes, but most of the *biological fixing* is done by nitrogen-fixing bacteria that live underwater or in the roots of plants (especially legumes). The N_2 is converted to NH_3 using the enzyme *nitrogenase*.
½ pt. Next step in cycle: nitrification.	After the nitrogen is fixed into ammonia, it goes through a process called *nitrification*. This is a two-step process where special soil bacteria first
½ pt. Definition of nitrification (step 1) ½ pt. Definition of nitrification (step 2)	convert the *ammonia to nitrite (NO_2^-)*. Then the nitrite is converted to *nitrate (NO_3^-)*, the form that plants can most easily absorb by their roots. Plants can also take up a limited amount of ammonia.
½ pt. Next step in cycle: assimilation.	The process by which the *roots absorb the nitrate (and the ammonia) is called assimilation*. During this step in the cycle, the plants incorporate the nitrogen into their tissues. This absorbed nitrogen is used by the plant to form nitrogen-containing molecules such as proteins and nucleic acids. Since animals cannot directly absorb nitrogen, they assimilate it into their body tissues when they eat the plants or other plant-eating animals. After the plants and animals absorb the nitrogen into their tissues, it is excreted as urea or uric acid. In addition, when an organism dies, the nitrogen in the tissues is released back into the cycle.
½ pt. Next step in cycle: ammonification.	These *nitrogen-containing substances are decomposed by bacteria into ammonia in a process called ammonification*.
½ pt. Next step in cycle: denitrification.	Then, during *denitrification, other anaerobic bacteria convert the ammonia found in the soil into nitrate and nitrite*. The *nitrite and nitrate are further converted into gaseous N_2* and *released back into the atmosphere*.
½ pt. Final step in cycle: conversion back into gaseous N_2.	The gaseous N_2 is then available to be refixed and used by another organism.
	(b) When fertilizers such as phosphorous are over-applied, the *rain or the slope* of the land tends to cause the *excess to run off into local waterways*. Since phosphorous is often a

RUBRIC

1 pt. Cause of the aquatic pollution.
½ pt. Initial effect of excessive fertilizer.

½ pt. Effect after initial algal bloom.

½ pt. Result of overapplication of fertilizer.

½ pt. Final result of excessive application.

1 pt. Name of the conservation method.
½ pt. Initial description of the method.

½ pt. Further explanation of the conservation-minded agricultural method.
½ pt. Reason why the crop yield will not decrease.

½ pt. Further explanation of the yield maintenance.

1 pt. Movement through the Earth's life support systems (atmosphere, lithosphere, biomass).

½ pt. Introduction to the carbon cycle. Capture of atmospheric carbon into plant tissue.

½ pt. Incorporation of carbon into animal tissue.

½ pt. Release of carbon back into the atmosphere.

½ pt. Alternate source of carbon into cycle.

ESSAY

limiting factor in aquatic ecosystems, this *excess amount* of phosphorous from the fertilizers results in a *proliferation of algae*. When these algal populations die, their *bodies are decomposed by large amounts of bacteria that consume much of the dissolved oxygen* found in the water. This can lead to a *shortage of dissolved oxygen*, which causes *the aquatic organisms, such as fishes and plants, to die from suffocation*. The result is a sterile lake or stream that lacks the nutrients necessary to support life.

(c) Conservation-minded agriculture can lead to a decreased dependence on human-applied nutrients. One of the most common, environmentally conscious agricultural practices is *conservation or no-till farming*. In this practice, *the fields are not plowed under at the end of growing season*. The old stalks and vegetation are left on the fields over the winter. Then in the spring, *special machinery is used to plant the new crop on the untilled field*. By leaving the previous season's biomass on the land, the *natural decomposition process* can occur. The nutrients that would normally have been removed and lost are *allowed to decompose back into the soil*. This leads to a decreased need for artificial fertilizers without a loss of crop yield.

Form Two

(a) Carbon, in the form of proteins, carbohydrates, lipids, and nucleic acids, are essential to life. This element is *found in almost all the life support* systems of the world: the *atmosphere is made of 0.036% carbon (in the form of CO_2); the hydrosphere (oceans) contains dissolved CO_2; and the limestone rocks of the lithosphere are made of carbon*. Much of the world's carbon supply is also stored in the biomass of living and decaying organisms. During photosynthesis, *autotrophs* such as plants and other photosynthetic organisms *use the sun's energy to convert the atmospheric carbon dioxide and water into the carbohydrate glucose*. When heterotrophs or *consumers eat* these autotrophs, they *incorporate the carbon into their tissues*. During the process of *cellular respiration, this glucose energy is burned, and the carbon is returned to the atmosphere in the form of carbon dioxide*. Because the autotrophs require a high amount of carbon dioxide during photosynthesis, most atmospheric carbon moves through the cycle rapidly. Much of the world's carbon, however, is not an active part of the carbon cycle. This carbon is temporarily not available to living organisms. The *biomass of plants and animals contains carbon that will not be released until the decomposition of*

RUBRIC	ESSAY
½ pt. Alternate source of carbon into cycle.	*the organism. Limestone rock (either on land or under the oceans) is made of calcium carbonate ($CaCO_3$).* The calcium carbonate was formed when aquatic organisms chemically combined carbon dioxide and calcium in ocean water. This compound is used to build the skeletons and shells of the organisms. When the organisms died, the shells fell to the bottom of the ocean and were buried for millions of years. Over time, the shells were converted to limestone.
½ pt. Release of carbon (limestone) into the active carbon cycle.	Presently, *the limestone is slowly dissolved by water erosion. The dissolved carbon enters the oceans or atmosphere as carbon dioxide.* Also, much of the world's *carbon is*
½ pt. Alternate source of carbon (sink) into the cycle.	*stored in sinks*, compartmentalized underground deposits that contain sealed amounts of a nutrient. Millions of years ago much of the world's carbon was stored in the form of
½ pt. Conversion of carbon (biomass) into carbon-based fossil fuels.	fossil fuels. These *fossilized plant and animal remains were converted into oil, coal, and natural gas under incredible pressure and time*. These carbon reserves were not an active part of the cycle for millions of years, but with the increased
½ pt. Cause of increase of carbon in the cycle.	*burning of fossil fuels, the amount of carbon in the atmosphere has been increasing.*
	(b) Nitrogen is an important part of the body systems of all organisms. Organisms require nitrogen in their proteins and nucleic acids, molecules that are essential for life. Unfortunately, because most of the world's nitrogen is gaseous and not easily incorporated into body tissues (of
1 pt. Cause of the aquatic pollution.	both terrestrial and aquatic organisms), nitrogen is frequently a limiting factor to population growth. To increase the amount of nitrogen available for plant growth, farmers fre-
½ pt. Initial effect of excessive fertilizer.	quently add nitrogen as a fertilizer. *When fertilizers such as nitrogen are overapplied, the rain or the slope of the land tends to cause the excess to run off into local waterways.*
½ pt. Effect after initial algal bloom.	This excess amount of nitrogen from the fertilizers results in a *proliferation of algae*. When these algal populations die,
½ pt. Result of overapplication of fertilizer.	their *bodies are decomposed by large amounts of aerobic bacteria* that consume much of the dissolved oxygen found in the water. This excess bacterial growth can lead to a *short-*
½ pt. Final result of excessive application.	*age of dissolved oxygen*. The lack of oxygen causes other *aquatic organisms, such as fishes and plants, to die from suffocation*. The result is a sterile lake or stream that lacks the nutrients necessary to support life.
	(c) There are many agricultural practices that can decrease
½ pt. Name of the conservation method.	reliance on human-applied nutrients. One of the most common conservation-minded farming techniques is the *rotation of crops*. Rather than plant the same type of crop on the same field for several years in a row, farmers are now *rotating the*

RUBRIC	ESSAY
½ pt. Initial description of the method.	*type of crop that is planted on a field.* For example, rather than plant corn on a field for several years in a row, the farmer will plant corn one season, soybeans the second sea-
½ pt. Further explanation of method.	son, and let the field lie fallow during the third season. Since
½ pt. Reason why crop yield will not decrease.	each *type of plant requires different nutrient levels*, the *varied planting allows certain nutrients to replenish in the soil.*
½ pt. Further explanations of the yield.	This procedure *decreases the need for human-applied fertilizers*. Most farms plant at least two different crops so the rotation is not inconvenient or far from the typical crop situation.

CHAPTER 3

The Solid Earth

Areas That You Will Be Tested On

A: Earth History and the Geologic Time Scale
B: Earth Dynamics—Earthquakes, Plate Tectonics, the Rock Cycle, Soil Formation, Volcanism

Key Terms

Note: Definitions for all key terms listed below can be found in the glossary starting on page 648. For additional definitions of relevant terms from this chapter, refer to www.barronseduc.com/0764121618.html.

A horizon	**epoch**
abyssal floor	**era**
B horizon	**fault**
C horizon	**glaciers**
continental crust	**magma**
continental drift	**oceanic plate**
continental margin	**plate tectonics**
continental shelf	**P wave**
continental slope	**Richter scale**
convergent plate boundary	**sedimentary rock**
core	**shield volcano**
crust	**subduction zone**
epicenter	**S wave**

Key Concepts

Earth History and Geologic Time Scale

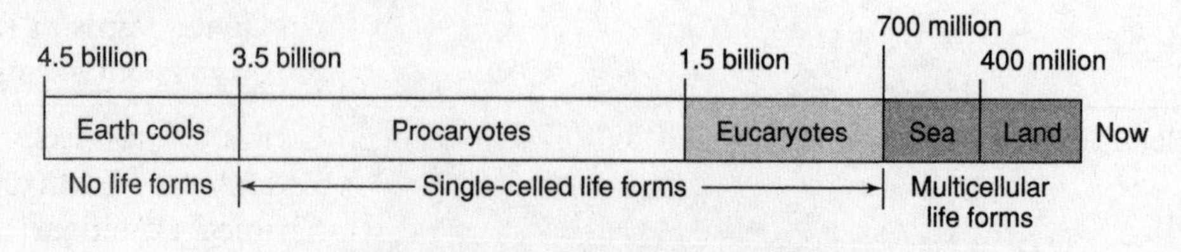

Figure 3.1. Timeline of life development.

THE GEOLOGIC TIME SCALE

Era	When period began (millions of years ago)	Period	Animal life	Plant life	Major geologic events
Cenozoic	2	Quarternary	Rise of civilizations	Increase in number of herbs and grasses	Ice Age
	65	Tertiary	Appearance of first men; dominance on land of mammals, birds, and insects	Dominance of land by flowering plants	
Mesozoic	135	Cretaceous		Dominance of land by conifers; first flowering plants appear	Building of the Rocky Mountains
	180	Jurassic	Age of dinosaurs		
	225	Triassic	First birds		
Paleozoic	275	Permian	Expansion of reptiles		Building of the Appalachian Mountains
	350	Carboniferous	Age of amphibians	Formation of great coal swamps	
	413	Devonian	Age of fishes		
	430	Silurian	Invasion of land by invertebrates	Invasion of land by primitive plants	
	500	Ordovician	Appearance of first vertebrates (fish)	Abundant marine algae	
	570	Cambrian	Abundant marine invertebrates	Appearance of primitive marine algae	
Precambrian		—	Primitive marine life		

Figure 3.2. Geologic time scale.

EONS

Name	Time	Description
Phanerozoic	Present to 570 million years ago. Includes the Cenozoic, Mesozoic, and Paleozoic eras.	Begins when invertebrates were common. Marked by an abundance of fossil evidence of life, especially higher forms, in the corresponding rocks. Multicellular organisms. Most animal phyla present, diverse algae; explosive evolution of higher life forms.
Proterozoic	570 million to 2.5 billion years ago. Includes the Precambrian period.	Major land masses and shallow seas and the buildup of oxygen and the appearance of the first multicellular eukaryotic life forms. Eukaryotic (one-cell) organisms predominant.
Archean	2.5 billion to 3.8 billion years ago. Includes the Precambrian period.	An atmosphere with little free oxygen, the formation of the first igneous rocks and oceans, and the development of the first living forms of life—single-celled prokaryotic organisms.
Hadean	3.8 billion to 4.6 billion years ago.	Earth's oldest rocks. No life present. Formation of Earth and continents, chemical evolution.

ERAS

Name	Time	Description
Cenozoic	Present to 65 million years ago. Includes the Quaternary and Tertiary periods.	Formation of modern continents, glaciation, and the diversification of mammals, birds, and plants.
Mesozoic	65 to 245 million years ago. Includes the Cretaceous, Jurassic, and Triassic periods.	Development of flying reptiles, birds, and flowering plants. Appearance and extinction of dinosaurs.
Paleozoic	245 to 570 million years ago. Includes the Permian, Pennsylvanian, Mississippian, Devonian, Silurian, Ordovician, and Cambrian periods.	Appearance of marine invertebrates, primitive fishes, land plants, and primitive reptiles.

PERIODS

Name	Time	Description
Quaternary	Present to 1.6 million years ago. Includes Holocene and Pleistocene epochs.	Appearance and development of humans and the Pleistocene Ice Age.
Tertiary	1.6 to 65 million years ago. Includes Pliocene, Miocene, Oligocene, Eocene, and Paleocene epochs.	Appearance of modern flora, apes, and other large mammals.
Cretaceous	65 million to 144 million years ago.	Development of flowering plants. Large diversity in dinosaurs but ending with their sudden extinction approximately 65 million years ago. Formation of the Andes Mountains. African and South American plates begin to separate. Climate cooling. Shallow seas are prominent.
Jurassic	144 million to 208 million years ago.	Dinosaurs continued to be the dominant land animal and the earliest birds and mammals appear. Sierra-Nevada Mountains form. Shallow seas prominent and expanding. Stable, warm climate.
Triassic	208 million to 245 million years ago.	Diversification of land life, the rise of dinosaurs, and the appearance of the earliest mammals. Large deserts. Warm climate. Shallow seas are limited.

PERIODS (continued)

Name	Time	Description
Permian	245 million to 286 million years ago.	Formation of the supercontinent Pangaea, the rise of conifers, and the diversification of reptiles. Ending with the largest known mass extinction of life forms. Climate begins to warm. Limited shallow seas.
Pennsylvanian	286 million to 320 million years ago.	Development of the amniotic egg, the diversification of amphibians, and widespread swamp forests. First reptiles appear along with winged insects. Some glaciers forming in Southern Hemisphere.
Mississippian	320 million to 360 million years ago.	Submergence of extensive land areas under shallow seas. Primitive ferns and insects evolve. Widespread forests. Mountain building causes arid regions in the interior of continents.
Devonian	360 million to 408 million years ago	Development of lobe-finned fishes, the appearance of amphibians and insects, and the first trees appear. Appalachian Mountains form. Primitive vascular plants become extinct. Rise of landmasses. Cooling climate.
Silurian	408 million to 438 million years ago.	Development of jawed, bony fishes, early invertebrate land animals, insects, and land plants. Colonization of land. Continental areas are generally flat. Major extinction occurs. Beginning of mountain building.
Ordovician	438 million to 505 million years ago.	Appearance of primitive (jawless) fishes and fungi. Shallow seas are extensive. Climate warming. Animal diversification.
Cambrian	505 million to 570 million years ago.	Desert land areas, shallow warm seas near equator, and rapid early diversification of marine life (especially invertebrates). Warm climate. Most animal phyla present, diverse algae.
Precambrian	570 million to 4.6 billion years ago.	Comprising most of the Earth's history and marked by the appearance of primitive forms of life. During most of this time the Earth was not suitable for life due to composition of atmosphere and temperatures.

EPOCHS

Name	Time	Description
Holocene	Present to 10,000 years ago	Development of human civilizations. Pleistocene ice age.
Pleistocene	10,000 to 1.6 million years ago	Alternate appearance and recession of northern glaciers, the appearance and worldwide spread of hominids, and the extinction of numerous land mammals, such as the mammoths, mastodons, and saber-toothed tigers.
Pliocene	1.6 million to 5.3 million years ago	Appearance of distinctly modern animals, development of humans walking upright, formation of Cascade Mountains. Continued cooling of climate.
Miocene	5.3 million to 24 million years ago	Development of grasses and grazing mammals. Chimpanzee and human lines evolve. Extensive glaciers in southern hemisphere. Cooling of climate.
Oligocene	24 million to 37 million years ago	Further development of modern mammalian fauna, especially browsers, including the rise of the true carnivores. Formation of the Alps and continued development of the Himalayan mountain chains. Volcanoes active in Rocky Mountains.
Eocene	37 million to 58 million years ago	Warm climates and the rise of most modern mammalian families. Primitive monkeys and the beginning of the Himalayan mountain area. Australian plate separates from Antarctica. Indian plate collides with Asia.
Paleocene	58 million to 65 million years ago	Appearance of placental mammals and rodents. Continental collisions leading to the formation of the Rocky Mountains and the Himalayas. Shallow continental seas becoming less prominent.

Earthquakes

Faulting or abrupt movement on an existing fault, along tectonic plate boundary zones, or along midoceanic ridges. Massive amounts of stored or potential energy, held in place by friction, are released into kinetic energy in a very short period of time. The area where the energy is released from is called the earthquake focus. From this area, seismic waves travel in all directions outward. Directly above the focus, on the Earth's surface is the epicenter, as shown in Figure 3.3.

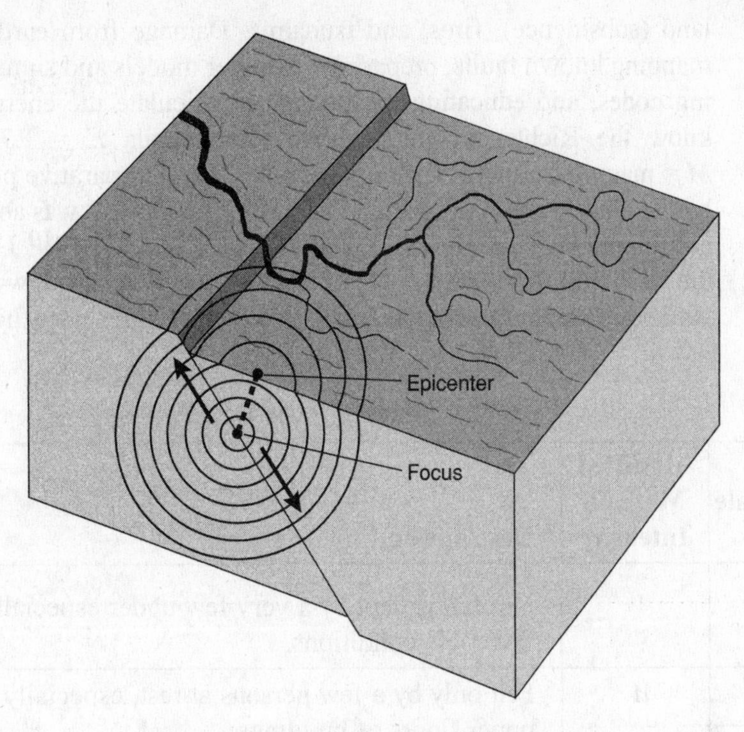

Figure 3.3. Relationship of an epicenter to a focus.

WAVES

Body waves—waves that travel through the Earth	**P waves**—primary waves. Can travel through solid, liquid, or gas. Caused by expansion and contraction of bedrock. Speed is between 5 and 8 kilometers per second (~13,000 to 21,000 mph). Waves can be heard. **S waves**—secondary waves. Slower waves that can only travel through solids. Produced when material moves either vertically or horizontally through action of shear stress.
Surface waves—waves that occur and travel near the Earth's surface.	Produce rolling and/or swaying motion. Speed slower than body waves. Cause considerable ground motion and damage.

The severity of an earthquake depends upon the amount of potential energy that had been stored, the distance the rock mass moved when the energy was converted to kinetic motion, and the makeup of the rock material in terms of transmitting this energy through seismic waves. Remember from physics that energy is carried either through waves and/or particles (wave–particle duality of nature). Over 400 earthquakes strong enough to be felt occur each day worldwide. Primary effects are shaking and either vertical or horizontal ground displacement. People, buildings, pipelines, bridges, and other infrastructures are affected. Secondary effects include rockslides, flooding due to the sinking of land (subsidence), fires, and tsunamis. Damage from earthquakes can be reduced by mapping known faults, preparing computer models and simulations, strengthening building codes, and educating the public. To calculate the energy of an earthquake, if you know the Richter magnitude, use the formula $E = 1.74 \times 10^{(5 + 1.44M)}$, where M = magnitude and E = energy in joules. For comparative purposes, the energy released by the atomic bomb dropped on Bikini Island in 1946 was about 10^{12} J and each year the consumption of energy in the United States is about 10^{19} J. For more information, visit the National Earthquake Information Center at http://www.neic.cr.usgs.gov/ or the National Geophysical Data Center (NOAA) at http://www.ngdc.noaa.gov/ngdc.html.

EARTHQUAKE SCALES

Richter Scale Magnitude	Modified Mercalli Intensity	Description	~Energy Released (joules)
1.0–3.0	I	Not felt except by a very few under especially favorable conditions.	6×10^7
3.0–3.9	II	Felt only by a few persons at rest, especially on upper floors of buildings.	
3.0–3.9	III	Felt quite noticeably by persons indoors, especially on upper floors of buildings. Many people do not recognize it as an earthquake. Standing motorcars may rock slightly. Vibrations similar to the passing of a truck. Duration estimated.	
4.0–4.9	IV	Felt indoors by many, outdoors by few during the day. At night, some awakened. Dishes, windows, doors disturbed; walls make cracking sound. Sensation like heavy truck striking building. Standing motorcars rock noticeably.	
4.0–4.9	V	Felt by nearly everyone; many awakened. Some dishes and windows broken. Unstable objects overturned. Pendulum clocks may stop.	
5.0–5.9	VI	Felt by all, many frightened. Some heavy furniture moved; a few instances of fallen plaster. Damage slight.	Energy equivalent to Hiroshima atomic bomb
6.0–6.9	VII	Damage negligible in buildings of good design and construction; slight to moderate damage in well-built ordinary structures; considerable damage in poorly built or badly designed structures; some chimneys broken.	6×10^{13} to 1×10^{15} J 1994 Northridge, CA, quake
6.0–6.9	VIII	Damage slight in specially designed structures; considerable damage in ordinary substantial buildings with partial collapse. Damage great in poorly built structures. Fall of chimneys, factory stacks, columns, monuments, and walls. Heavy furniture overturned.	6×10^{13} to 1.5×10^{15} J 1994 Northridge, CA, quake

EARTHQUAKE SCALES (continued)

Richter Scale Magnitude	Modified Mercalli Intensity	Description	~Energy Released (joules)
6.0–6.9	IX	Damage considerable in specially designed structures; well-designed frame structures thrown out of plumb. Damage great in substantial buildings, with partial collapse. Buildings shifted off foundations.	6×10^{13} to 1.5×10^{15} J
7.0 +	X	Some well-built wooden structures destroyed; most masonry and frame structures destroyed with foundations. Rails bent.	2×10^{15} + J 1999 Turkey (12,000 killed)
7.0 +	XI	Few, if any (masonry) structures remain standing. Bridges destroyed. Rails bent greatly.	2×10^{15} +
7.0 +	XII	Damage total. Lines of sight and level are distorted. Objects thrown into the air.	2×10^{15} + 1976 Tangshan, China (250,000 killed) 1906 San Francisco 1960 Southern Chile (5,700 killed)

Plate Tectonics

The Continental Drift Theory. In 1915, German meteorologist and geophysicist, Alfred Wegener, published his theory of continental drift in the book, *The Origin of Continents and Oceans.* He proposed that all the present-day continents originally formed one landmass, which he called Pangaea (meaning "all lands" in Greek). Wegener believed that this supercontinent began to break into smaller continents around 200 million years ago. He supported his theories by scientific evidence, such as (1) the discovery of fossilized tropical plants beneath the Greenland icecap, (2) the fact that there were glaciated landscapes in the tropics of Africa and South America, (3) tropical regions on some continents had polar climates in the past based on paleoclimatic data, (4) the continents fit together like pieces of a puzzle, and (5) similarities existed in rocks between the East coasts of North and South America and the West coasts of Africa and Europe. Continental drift gained acceptance in the 1960s when the theory of plate tectonics provided a mechanism that would account for the movement of the continents.

The Sea-Floor Spreading Theory. During the 1960s, alternating patterns of magnetic (polarity) properties were discovered in rocks found on the sea floor. Similar patterns were found on either side of midoceanic ridges found near the center of the oceanic basins. Dating of the rocks indicated that as one moved away from the ridge, the rocks became older. This evidence led to the theory of sea-floor spreading, which suggested that new crust was being created at these volcanic rift zones.

Although new crust is being formed at midoceanic ridges, older oceanic crust is being destroyed by being subducted under lighter continental crust in deep oceanic trenches. Further research indicates that the crust is not one continuous surface—that oceanic crust and continental crust exists in very large sections known as oceanic plates and continental plates that essentially "float" on the asthenosphere.

Types of Converging Plates

Two Oceanic Plates Converging

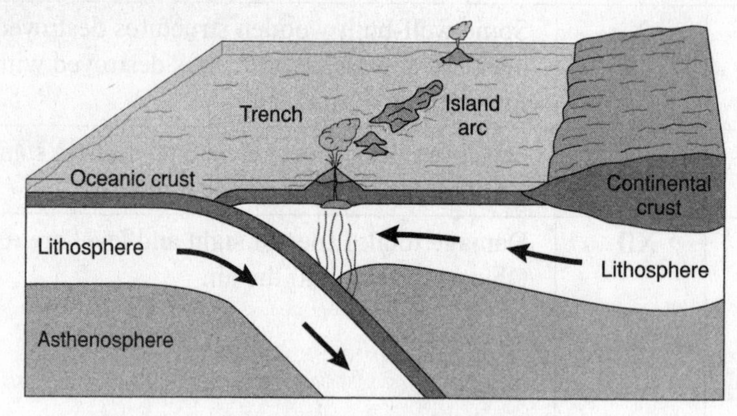

Figure 3.4. Collision of two oceanic plates.

One oceanic plate is subducted, or sinks, below the other, creating a deep trench, as shown in Figure 3.4. The Marianas trench is formed by the fast-moving Pacific Plate colliding with the slower-moving Philippine Plate. This type of collision produces a jerky motion, earthquakes, and a large amount of friction, which results in chains of volcanic islands called island arcs.

Oceanic Plate Converging with a Continental Plate

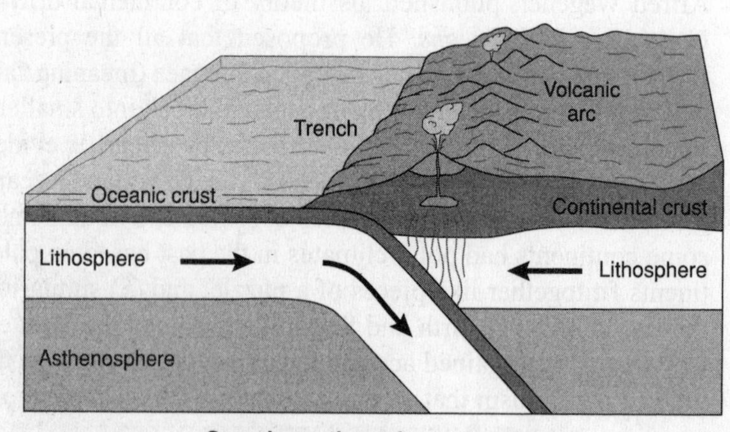

Figure 3.5. An oceanic plate colliding with a continental plate.

The denser oceanic plate subducts under the less-dense continental plate, as shown in Figure 3.5. Friction produced by the colliding plates produces heat, which in turn produces continental volcanoes, which then become mountains. Some volcanic mountains in California were produced in this manner.

Two Continental Plates Colliding

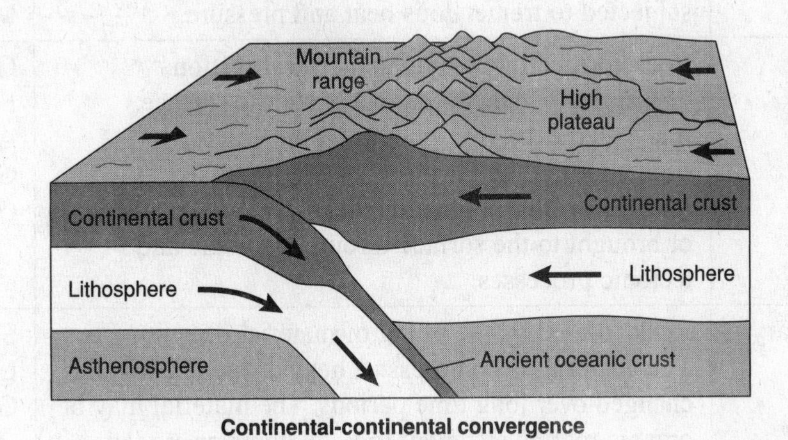

Figure 3.6. Two continental plates colliding.

Mountain ranges are produced on the surface where two continental plates collide, as shown in Figure 3.6. Rocks in the mountain range may be of oceanic origin, which is evidence that the continental crust was originally formed at midoceanic ridges.

The Rock Cycle

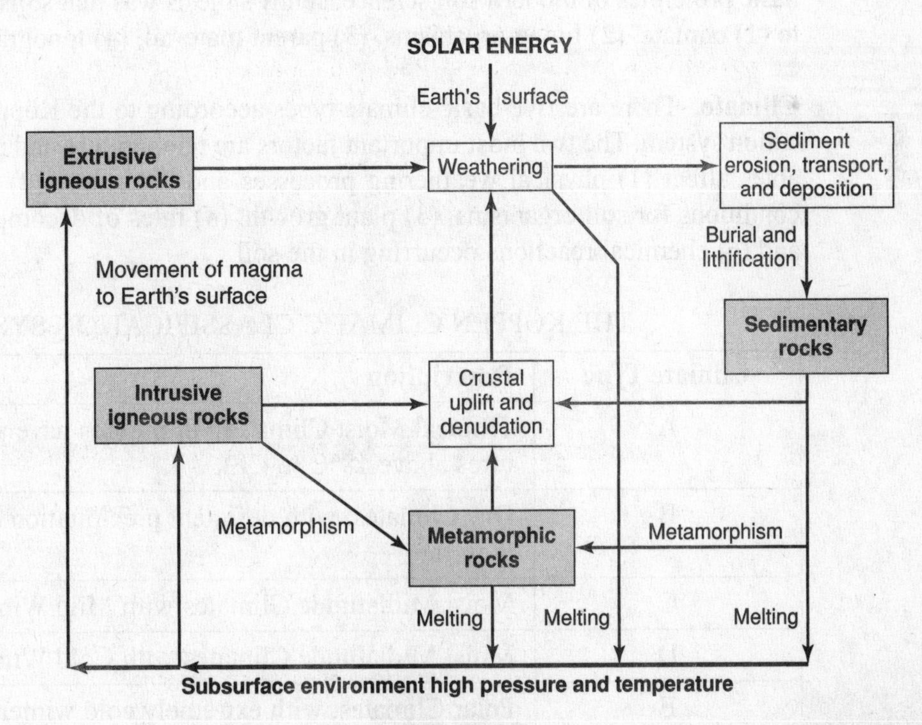

Figure 3.7.

There are three main categories of rocks—metamorphic, igneous, and sedimentary (see Figure 3.7).

Type of Rock	Description	Example
Metamorphic	Igneous or sedimentary rock that has been subjected to tremendous heat and pressure.	Slate, Schist Marble, Gneiss, Quartzite
Igneous	Rock formed by cooling and crystallization of magma. If the cooling occurs at the surface, it is called extrusive igneous. If the cooling occurs within the Earth, it is called intrusive igneous. Intrusive igneous rocks can be exposed or brought to the surface through geologic and tectonic processes.	Granite, Basalt, Quartz, Feldspars, Rhyolite, Andesite,Gabbro, Granodiorite, Dolerite, Obsidian
Sedimentary	Rock formed by the piling of material over time. This material is compressed, heated, and chemically changed over long time periods. The material may be organic marine sediment (e.g., diatoms or eroded rock debris) or chemical precipitates.	Sandstone, Shale, Limestone, Dolomite, Gypsum, Chalk

Soil Formation

In 1941, Hans Jenny, Swiss doctor, artist, and researcher, wrote *Factors of Soil Formation.* This was a system of quantitative pedology (the scientific study of soils, including their origins, characteristics, and use), which summarized and illustrated the basic principles of modern soil science. Jenny's thesis was that soils develop in response to (1) climate, (2) living organisms, (3) parent material, (4) topography, and (5) time.

Climate. There are five basic climate types according to the Köppen climatic classification system. The two most important factors are temperature and moisture. These variables affect (1) physical weathering processes and determine (2) microenvironmental conditions for soil organisms, (3) plant growth, (4) rates of decomposition, (5) soil pH, and (6) chemical reactions occurring in the soil.

THE KÖPPEN CLIMATIC CLASSIFICATION SYSTEM

Climate Type	Description
A	Tropical Moist Climates: all months have average temperatures above 18°C (64°F).
B	Dry Climates: with deficient precipitation during most of the year.
C	Moist Midlatitude Climates with Mild Winters
D	Moist Midlatitude Climates with Cold Winters
E	Polar Climates: with extremely cold winters and summers

Living Organisms. Living organisms are intricately associated with determining soil profile characteristics. Nutrient cycling as seen through the nitrogen cycle and the role of *Rhizobium sp*., fungi, actinomycetes, autotrophic chemosynthetic bacteria (*nitrobacter*), autotrophic chemosynthetic bacteria (*nitrosomonas*), cyanobacteria, arthropods, worms, and gastropods who help to reduce the size of litter and facilitate decomposition are only a few examples of this relationship.

Parent Material. Parent material refers to the rock and minerals from which the soil derives. The nature of the parent rock, which can either be native to the area or transported to the area by wind, water, or glacier has a direct effect on the ultimate soil texture, chemistry, and cycling pathways.

Topography. Topography refers to the physical characteristic of the location where the soil is formed. Topography can be viewed on a macro-scale (soil type in a certain valley) or it can be viewed on a micro-scale (soil type in a field). Topographic factors that affect soil characteristics include drainage, slope direction, elevation, and wind exposure.

Time. With sufficient time, a mature soil profile reaches a state of equilibrium. Feedback mechanisms involving both abiotic and biotic factors work to preserve the mature profile.

PRINCIPAL SOIL PROCESSES

Process	**Description**
Calcification	Common in grasslands. Occurs when the rate of evapotranspiration exceeds the rate of precipitation. This results in the upward movement of alkaline salts (calcium carbonate) from groundwater into the B horizon. Caliche may form.
Gleization	Common in areas of cold climate (which limits percolation) and poor drainage—bogs. Organic matter accumulates in the upper layers. Clay layers limit the porosity of the soil, causing plant litter and animal waste to be accumulated on the surface. Soil depth is limited and humus collects along the surface.
Laterization	Common in tropical and subtropical soils. High temperatures and moisture cause rapid weatherization of rock. Rainfall causes the leaching of soil nutrients, except for iron and aluminum compounds. Soils are typically acidic due to the loss of basic cations (Ca^{2+}, Mg^{2+}, K^+, and Na^+). Eluviation (movement of nutrients downward through soil profile) is high. Nutrients are stored in vegetation. Nutrient level in soil is low.
Podzolization	Common in cold, midlatitude areas. Supports coniferous plant life (evergreens, needles, scalelike leaves). Soil is acidic due to decomposition of coniferous litter. Leaching removes basic cations from soil (Ca^{2+}, Mg^{2+}, K^+, and Na^+). Aluminum and iron compounds found in A horizon.
Salinization	Occurs in dry climates. Similar to calcification except it lacks rainwater and its downward action to keep the alkaline salts from reaching the A horizon. In this case, the alkaline salts occur near the surface.

Soil Profiles

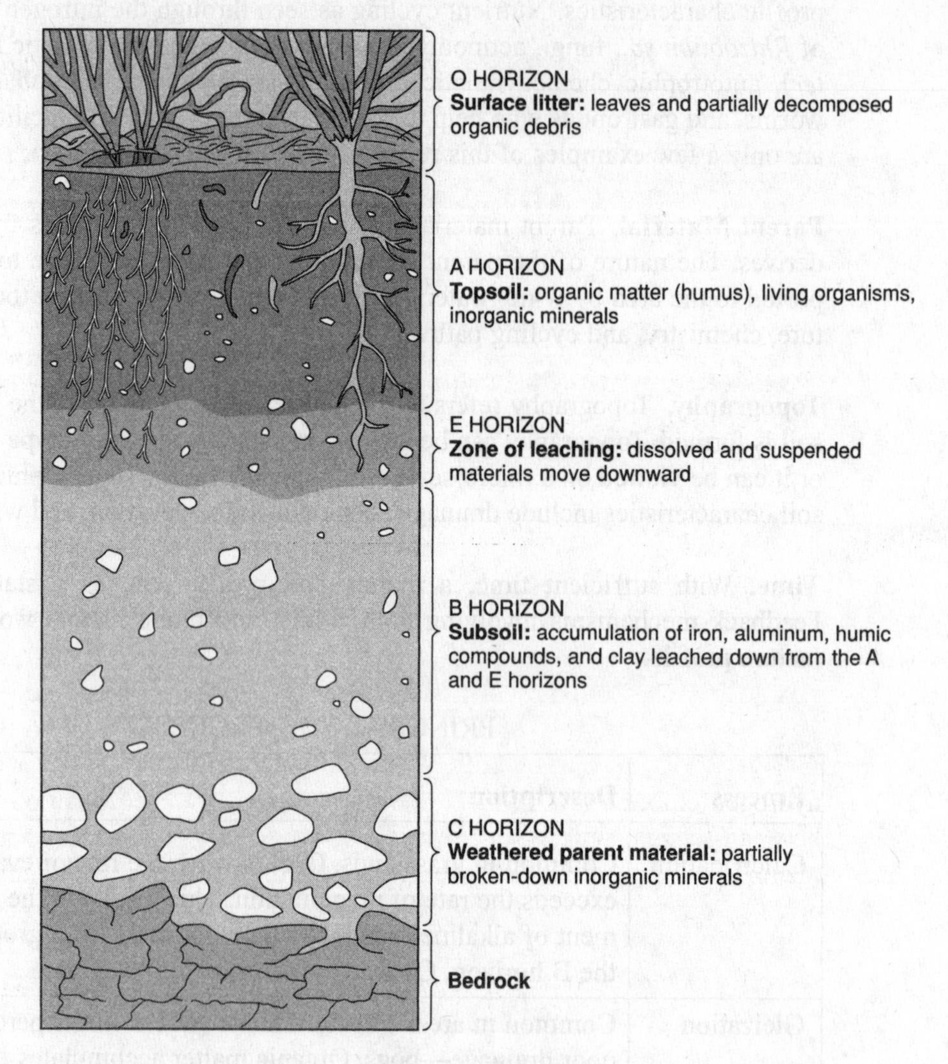

Figure 3.8. Soil profile.

SOIL FORMATION AND PROFILES

O horizon—top surface litter layer. Decomposed leaves and organic matter. Normally brown or black. Rich in bacteria, fungi, insects, and earthworms.
A horizon—topsoil layer. Humus and minerals. Roots are in this area. Also rich in living organisms. If dark brown or black (rich in nitrogen and organic material), good for crops. If gray, yellow, or red (low in organic matter), poor for crops.
B horizon—subsoil layer. Mostly inorganic (minerals). Clay particles present. Receives material from A horizon through illuviation. May be colored by iron oxides (red), aluminum oxides (yellow) or white due to calcium carbonate.
C horizon—weathered parent material. Consists of broken fragments of parent rock.
R horizon—unweathered bedrock.

TYPES OF SOIL

Clay—very fine particles. Compacts easily. Forms large, dense clumps when wet. Low permeability to water therefore upper layers become waterlogged. Holds positively charged ions. Less than 0.002 mm in size.
Gravel—coarse. An unconsolidated mixture of rock fragments or pebbles.
Loams—a mixture of clay, sand, silt, and humus. Best soils for crops. Holds water but does create waterlogging.
Sand—a sedimentary material coarser than silt. Water flows through them too fast for most crops. Good for crops requiring low amounts of water. 0.06–2.0 mm.
Silt—a sedimentary material consisting of very fine particles intermediate in size between sand and clay. 0.002–0.06 mm.

Volcanism

Active volcanoes produce magma (molten rock) at the surface. Other types of volcanoes are classified as intermittent, dormant, or extinct. Of volcanoes, 95% occur at subduction zones and midoceanic ridges. The remaining 5% occur at hot spots—areas where plumes of magma come close to the surface. Volcanoes may produce ejecta (lava rock and/or ash), molten lava, and/or toxic gases. The most common gases released by magma are steam (H_2O), followed by carbon dioxide (CO_2), sulfur dioxide (SO_2), and hydrogen chloride (HCl). Gases may contribute to acid deposition. Correlation exists between seismic activity and volcanic activity. Eruptions occur when pressure within a magma chamber forces molten magma up through a conduit and out a vent. The type of eruption depends on the gases, the amount of silica in the magma (which determines viscosity), and how free the pipe is (whether the volcano flows or explodes). Benefits include producing new landforms and nutrients produced from erosion of lava rock. Methods of dealing with volcanoes include (1) modeling and data analysis for better prediction, (2) better evacuation plans, (3) study of precursors such as changes in the cone, (4) changes in temperature and gas composition, (5) thermal and magnetic changes, and (6) changes in seismic activity. See Figure 3.9.

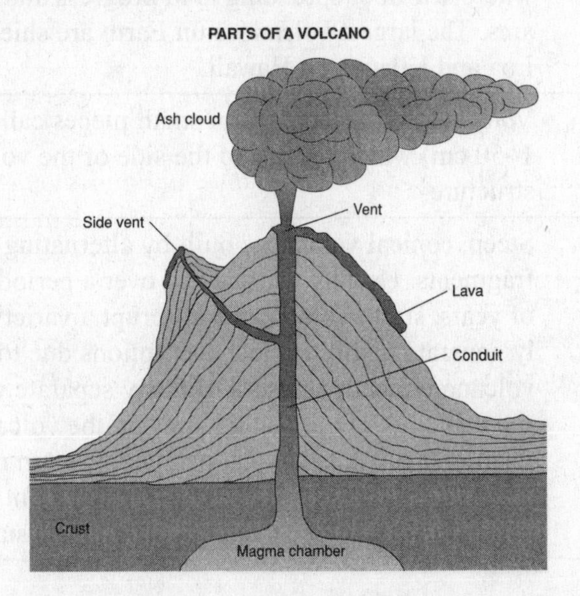

Figure 3.9. Parts of a volcano.

TYPES OF VOLCANOES

Basalt plateau	Characterized by having magma derived from basalt that tends to flow horizontally. Flat, gentle sloping volcanoes, and non-explosive. Basalt plateaus are found near the Columbia River in Washington, western India, Australia, Iceland, Antarctica, Brazil, and Argentina.
Cinder cones	One of the most common types of volcanoes. A steep, conical hill of volcanic fragments called cinders that accumulates around a vent. Formed from Stombolian eruptions. The rock fragments, often called cinders or scoria, are glassy and contain numerous gas bubbles "frozen" into place as magma exploded into the air and then cooled quickly. Cinder cones range in size from tens to hundreds of meters tall and usually occur in groups, often on the flanks of strato and shield volcanoes. Example: Paricutín, Mexico, 1943.
Compound or complex	Volcanic system that consists of two or more vents. Example: Caldera complexes. Calderas produce terrific explosions due to a high water content in the granitic magma. The magma builds a dome that allows pressure to build. Example: Krakatau.
Mud volcanoes	Small volcano-shaped cone of mud and clay, usually less than 1–2 m tall. These small mud volcanoes are built by a mixture of hot water and fine sediment (mud and clay) that either (1) pours gently from a vent in the ground like a fluid lava flow or (2) is ejected into the air like a lava fountain by escaping volcanic gas and boiling water. The fine mud and clay typically originates from solid rock—volcanic gases and heat escaping from magma deep below turn ground water into a hot acidic mixture that chemically changes the rock into mud- and clay-sized fragments.
Shield volcanoes	Volcanoes that have broad, gentle slopes and were built by the eruption of runny, fluid basalt lava. Basalt lava tends to build enormous, low-angle, gently sloping cones because it flows across the ground easily and can form lava tubes that enable lava to flow tens of kilometers from an erupting vent with very little cooling. The fluid nature of the lava prevents it from piling into steep mounds. Shield volcanoes also occur along the midoceanic ridge, where sea-floor spreading is in progress and along subduction related volcanic arcs. The largest volcanoes on Earth are shield volcanoes. Examples: Mauna Loa and Kilauea on Hawaii.
Spatter cones	Volcanoes that eject lava in small pieces called spatter (0.4–20 inches, or 1–50 cm) which returns to the side of the volcano to build a steep-sided structure.
Strato-volcanoes or composite	Steep, conical volcanoes built by alternating layers of lava and rock fragments. Usually constructed over a period of tens to hundreds of thousands of years, strato-volcanoes may erupt a variety of magma types. They commonly generate highly explosive eruptions due to clogs in the craterpipe. A strato-volcano typically consists of many separate vents, some of which may have erupted cinder cones and domes on the volcano's flanks. They frequently form impressive, snow-capped peaks, often more than 1.5 miles (2,500 meters). Examples: Vesuvius in Italy, Mount Rainier in Washington, Mt. Fuji in Japan, Mount Hood in Oregon, and Shasta in California.

TYPES OF VOLCANIC ERUPTIONS

Hawaiian or Lava Fountains	Jets of lava spray into the air by the rapid formation and expansion of gas bubbles in the molten rock Lava fountains typically range from about 10 to 100 m (33 to 328 feet) in height, but they occasionally reach more than 500 m (1,640 feet). Lava fountains erupt from isolated vents, along fissures, within active lava lakes, and from a lava tube when water gains access to the tube in a confined space.
Pelean	Also known as a "Nuee Ardent" (glowing cloud) eruption. A large quantity of gas, dust, ash, and incandescent lava fragments are blown out of a central crater, fall back, and form tongue-like, glowing avalanches that move down-slope at velocities as great as 100 miles per hour (160 km per hour). Such eruptive activity can cause great destruction and loss of life if it occurs in populated areas, as demonstrated by the devastation of St. Pierre during the 1902 eruption of Mount Pelee on Martinique, West Indies or the Mayan Volcano in the Philippines in 1968.
Phreatomagmatic	Steam-driven explosions that occur when water beneath the ground or on the surface is heated by magma, lava, hot rocks, or new volcanic deposits (e.g., tephra and pyroclastic-flow deposits). The intense heat of such material (as high as 1,170°C or 2,140°F for basaltic lava) may cause water to boil and flash to steam, thereby generating an explosion of steam, water, ash, blocks, and bombs. The eruption ends when the water supply is exhausted.
Plinian	Large explosive events caused by viscous magma containing high amounts of gas exploding within the volcano. Forms enormous dark columns of volcanic rock fragments (tephra) and gas high into the stratosphere (>11 km or 7 miles) forming massive ashfalls and ashflows. Starts suddenly after long dormant periods. Example: Mt. Pinatubo.
Strombolian	Characterized by the intermittent explosion or fountaining of basaltic lava from a single vent or crater. Each episode is caused by the release of volcanic gases, and they typically occur every few minutes or so, sometimes rhythmically and sometimes irregularly. The lava fragments generally consist of partially molten volcanic bombs that become rounded as they fly through the air.
Submarine	The most common type of volcanic eruption on Earth, and yet it is not always seen. Violent, steam-blast eruptions take place when seawater pours into active shallow submarine vents. Lava, erupting onto a shallow sea floor or flowing into the sea from land, (called pillow lava) may cool so rapidly that it shatters into sand and rubble. The result is the production of huge amounts of fragmental volcanic debris. The famous "black sand" beaches of Hawaii were created virtually instantaneously by the violent interaction between hot lava and seawater. During an explosive submarine eruption in the shallow open ocean, enormous piles of debris are built up around the active volcanic vent. Ocean currents rework the debris in shallow water, while other debris slumps from the upper part of the cone and flows into deep water along the sea floor. Fine debris and ash in the eruptive plume are scattered over a wide area in airborne clouds. Coarse debris in the same eruptive plume rains into the sea and settles on the flanks of the cone. Pumice from the eruption floats on the water and drifts with the ocean currents over a large area. It is hypothesized that one million submarine volcanoes exist. Kavachi (Solomon Islands), Metis Shoal (Tonga), Kick-'em-Jenny (Caribbean), Loihi Seamount (Hawaii), Axial Seamount (Offshore Oregon, USA), and Bayonnaise Rocks (Japan). All Taiwan's volcanoes are submarine.

TYPES OF VOLCANIC ERUPTIONS (continued)

Surtseyan	Also known as hydrovolcanic eruptions, Surtseyan eruptions take place mainly in shallow seas and lakes. Generated by the interaction of magma with either groundwater or surface water—water pressure is less than in submarine eruptions. Considered to be the "wet" equivalents of Strombolian-type eruptions, although they are much more explosive. This high explosivity is a hallmark of hydrovolcanic Surtseyan activity as are the thick, dark-pointed fragments, often accompanied by "bombs." As the water is heated, it flashes to steam and expands explosively, thus fragmenting the magma into exceptionally fine-grained ash, called tuffs, that accumulate around the vent. Examples: Surtsey, Iceland, 1963; Ukinrek, Alaska, 1977; Capelinhos, Azores (1957).
Vulcanian	Ejects new lava fragments that do not take on a rounded shape during their flight through the air. This may be because the lava is too viscous or already solidified. These moderate-sized explosive eruptions commonly eject a large proportion of volcanic ash and also breadcrust bombs and blocks. Gives off a lot of gas, ash, cinders, pumice, and noise. May lay dormant for thousands of years.

TYPES OF LAVA

a`a	Fast-flowing flows that have a rough, spiny surface composed of broken lava blocks called clinkers. Produced by fire-fountaining volcanoes. It cools faster than Pahoehoe flows. It is also thicker, meaning this type of lava has a higher viscosity Discontinuous surface.
Block	A highly fluid, slow-moving lava flow. The surface is relatively smooth, cubic in appearance, and discontinuous.
Pahoehoe	Thin, ropey, billowy-smooth-flowing layers of hot, fluid basalt that travel through lava tubes. Characterized by a glassy, discontinuous, smooth plastic skin that cools very slowly, remaining hot until it reaches the ocean.
Sheet	Fast-flowing, very fluid, and tends to fill-in and pond in low-lying areas. Continuous surface.

Case Studies

Dust Bowl. The Dust Bowl occurred during 1930s in Oklahoma, Texas, and Kansas. It was caused by plowing the prairies and the subsequent loss of natural grasses that rooted the soil. Drought and winds that occurred blew most of the topsoil away causing people to leave area. As a result, in 1935, the Soil Erosion Act was passed which established the Soil Conservation Service.

Mt. Saint Helens. Located in Washington, Mt. Saint Helens erupted in 1980. The three zones of damage were (1) tree—removal, (2) tree—down, and (3) seared. The earthquake removed trees, increased soil erosion in the area, destroyed wildlife and their habitat, and polluted air with gases and ash. Ash extended high enough in the atmosphere to be carried around the Earth. It also produced mudflows, heat melted glacial ice and snow clogged rivers with debris causing flooding. Fifty-seven people were killed, and homes and property were destroyed.

Multiple-Choice Questions

1. The majority of the rocks in the earth's crust are:

   (A) igneous
   (B) metamorphic
   (C) sedimentary
   (D) basaltic
   (E) volcanic

2. Which of the following is an example of igneous rock?

   (A) marble
   (B) slate
   (C) limestone
   (D) granite
   (E) sandstone

3. The smallest particle of soil is known as:

   (A) clay
   (B) sand
   (C) silt
   (D) gravel
   (E) humus

4. "Acid rain," also known as acid deposition, affects soils by:

   (A) decreasing soil porosity
   (B) decreasing the pH
   (C) decreasing soil aeration
   (D) lowering nutrient capacity
   (E) all of the above

5. The higher the amount of ________ in the soil, the better its nutrient-holding capacity.

   (A) clay
   (B) silt
   (C) sand
   (D) gravel
   (E) none of the above

6. A volcano with broad, gentle slopes and built by the eruption of runny, fluid-type basalt lava would be:

   (A) Mt. Saint Helens
   (B) Krakatau
   (C) Mauna Loa or Kilauea
   (D) Vesuvius
   (E) Mount Rainier

7. Which of the following is at a convergent boundary where two plates carrying continental lithosphere are presently colliding?

(A) Appalachian Mountains
(B) Himalayan Mountains
(C) Andes Mountains
(D) Rocky Mountains
(E) None of the above

8. The "dust bowl" of the 1930's resulted in the passage of what legislation?

(A) Endangered American Wilderness Act
(B) Soil and Water Conservation Act
(C) Federal Land Management Act
(D) Public Rangelands and Improvement Act
(E) Soil Erosion (Conservation) Act

9. Poor nutrient-holding capacity, good water infiltration capacity, and good aeration properties are unique characteristics of

(A) clay
(B) silt
(C) sand
(D) loam
(E) humus

10. An alkaline dark soil, rich in humus, found in a semiarid climate would be most characteristic of:

(A) deserts
(B) grasslands
(C) tropical rain forests
(D) deciduous forests
(E) coniferous forests

11. Which period in geological time describes the following: "Development of flowering plants. Large diversity in dinosaurs but ending with their sudden extinction approximately 65 million years ago. Formation of the Andes mountains. African and South American plates begin to separate. Climate cooling. Shallow seas are prominent."

(A) Tertiary
(B) Quaternary
(C) Jurassic
(D) Cretaceous
(E) Permian

12. The process of weathering produces what type of rock?

    (A) igneous
    (B) metamorphic
    (C) sedimentary
    (D) volcanic
    (E) none of the above

13. A rock that would most likely contain a fossil would be:

    (A) igneous
    (B) metamorphic
    (C) sedimentary
    (D) volcanic
    (E) all of the above

14. The most common element found in the earth's crust is:

    (A) oxygen
    (B) hydrogen
    (C) iron
    (D) silicon
    (E) aluminum

15. A horizon of soil also known as the topsoil layer and which contains humus, minerals, roots, and is also rich in living organisms is known as the

    (A) A layer
    (B) B layer
    (C) C layer
    (D) D layer
    (E) O layer

Answers to Multiple-Choice Questions

1. **A**	6. **C**	11. **D**
2. **D**	7. **B**	12. **E**
3. **A**	8. **E**	13. **C**
4. **E**	9. **C**	14. **A**
5. **A**	10. **B**	15. **A**

Explanations for Multiple-Choice Questions

1. **(A)** Igneous rocks are solidified from magma. Magma that reaches the surface and which cools quickly, forms fine-grained rocks. These rocks can be categorized as basalt, rhyolite, andesite, etc. If the magma cools slowly, the rocks are coarser in nature and include granite, gabbro, etc. Had the question been worded, "the majority of the rocks on the surface of the Earth," the answer would have been sedimentary.

2. **(D)** Igneous rocks are formed by cooling and crystallization of magma. If the cooling occurs at the surface, it is called extrusive igneous. If the cooling occurs within the Earth, it is called intrusive igneous. Intrusive igneous rocks can be exposed or brought to the surface through geologic and tectonic processes. Other examples of igneous rock include basalt and quartz.

3. **(A)** Clay consists of very fine particles. It easily compacts. Forms large, dense clumps when wet. Low permeability to water therefore upper layers become waterlogged. Holds positively charged ions. Less than 0.002 mm in size.

4. **(E)** Acid rain, also known as acid deposition has several effects on soils. The effects include: (1) calcium and magnesium compounds are leached from the soil—which reduces the soil's buffering capacity. The effect is to reduce soil pH. (2) Aluminum ions that are normally part of insoluble soil compounds and not free as aluminum ions (Al^{3+}), are released by acids and reduce the plant's ability to utilize soil nutrients and water. (3) Acids also release other cations from insoluble compounds, which are toxic to plants. These include mercury, lead, cadmium, and other heavy metals. (4) Acidic soils also promote the growth of acid-loving mosses that tend to retain water in the soil, thereby decreasing soil aeration; i.e. the soil becomes waterlogged. (5) The mosses also create an environment in which mycorrhizal fungi counts are decreased which reduces the ability of the roots to absorb nutrients. (6) Finally, acid deposition and the effects listed, decrease resistance of plants and trees making them more susceptible to disease, insects, drought, etc.

5. **(A)** Clay soils are not the best soils for growing crops. The small size of clay particles makes them ideal for water retention, but poor for water permeability. Clay particles become compacted when wet and form dense clumps. The soil particle that is best for water permeability is sand. The clay's small particle size does not allow adequate air spaces either. Again, sand particles are the best when it comes to the type of particle best for allowing air spaces or pores. However, the question was not asking what is the best soil to grow crops; the answer to that would be loam. The question was asking about nutrient-holding capacity of soil types. In that limited sense, clay particles have the highest capacity to hold and retain nutrients (compounds, ions, etc.) due to their size and electrochemical nature (bonding ability to polar compounds). Sand particles are the poorest for their nutrient-holding capacity. The bottom line to the best soil for growing crops is a balanced mixture—one that contains clay, silt, sand, loam, humus, etc. Each soil type has unique properties, and a mixture allows the best properties of all to be present.

6. **(C)** Mauna Loa and Kilauea on Hawaii are shield volcanoes. Basalt lava tends to build enormous, low-angle, gently sloping cones because it flows across the ground easily and can form lava tubes that enable lava to flow tens of kilometers from an erupting vent with very little cooling. The fluid nature of the lava prevents it from piling into steep mounds. Shields volcanoes also occur along the mid-oceanic ridge, where seafloor spreading is in progress and along subduction related volcanic arcs. The largest volcanoes on Earth are shield volcanoes.

7. **(B)** The Appalachian and Rocky Mountains were formed at ancient convergent plate boundaries, but neither lies at a plate boundary today. The Andes lie at a convergent boundary where oceanic lithosphere is being subducted under the South American continent.

8. **(B)** The Soil Conservation Act of 1935 established the Soil Conservation Service that deals with soil erosion problems, carries out numerous soil surveys, and does research on soil salinity. It also provides computer databases for scientific research, such as pesticides.

9. **(C)** Sand is a sedimentary material coarser than silt. Water flows through it too fast for most crops. Good for crops requiring low amounts of water. 0.06–2.0 mm

10. **(B)** Soil typically found in semi-arid grassland areas is called mollisol—it is rich in organic matter.

11. **(D)** Refer to Periods chart above.

12. **(E)** You might call this a trick question. Weathering does not produce a unique type of rock. Weathering is the erosion of rock material through physical (mechanical) and/or chemical methods. Weathering breaks down rock material. The sediment that is produced from weathering and erosion of any type of rock can be transported and buried to produce sedimentary rock, but weathering does not produce the sedimentary rock—it is formed by the piling of material over time. This material is compressed, heated, and chemically changed over long time periods. Igneous rock is formed by cooling and crystallization of magma. Metamorphic rock can be either igneous or sedimentary rock that has been subjected to tremendous heat and pressure. Volcanic rock is produced by the solidification of magma.

13. **(C)** Since sedimentary rock is formed by the piling of material over time, if conditions are right, organisms that die may become buried by eroded material and become fossilized. When you look at a fossil, you are not looking at the actual shell or bone—you are looking at the impression that they made in the material. What you are looking at is mineral material that filled the space where the bone or shell was originally.

14. **(A)** The crust is the top layer of the earth. Thickness varies from 6 to 40 miles (10 to 65 km). Eight elements make up 99% of the weight of the crust (in decreasing amounts—O, Si, Al, Fe, Ca, Na, K, Mg). Divided into continental (30%) and oceanic (70%) sections.

15. **(A)** The A layer is called the topsoil layer. Humus and minerals. Roots are in this area. Also rich in living organisms. If dark brown or black—rich in nitrogen and organic material—good for crops. If gray, yellow or red—low in organic matter and poor for crops.

Free-Response Question

by Dr. Ian Kelleher
Brooks School
North Andover, MA
B.S. Univ. of Manchester, England
Ph.D. University of Cambridge, England
Chemistry

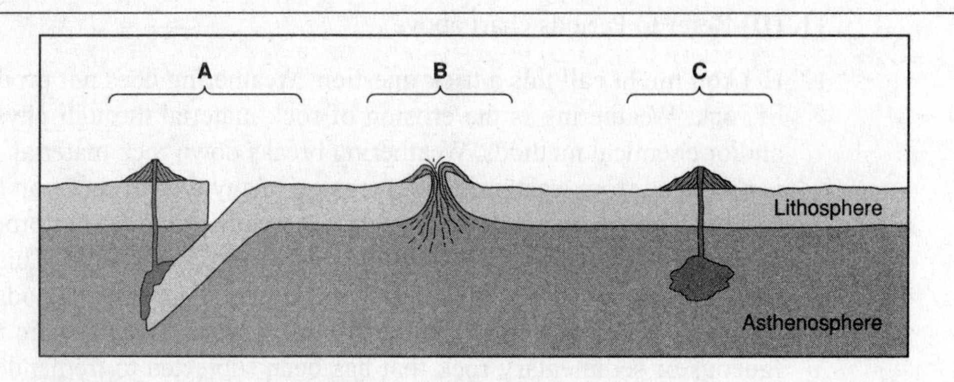

(a) Geological features A–C in Figure 1 above are formed as a consequence of plate tectonics. Choose **two** of these features and:

(i) describe what is happening there

(ii) give an actual geographic example

(b) Charles Darwin was the geologist, botanist and zoologist on the research vessel *Beagle* when he made observations that lead to his book, *The Origin of Species,* in which the theory of evolution was first introduced. A century later, scientists developed the theory of plate tectonics, describing how the solid earth formed. Describe two ways in which evolution may occur as a consequence of plate tectonics.

(c) Mount Pinatubo in the Philippines erupted in 1991. Examine the temperature graph below and answer the following questions.

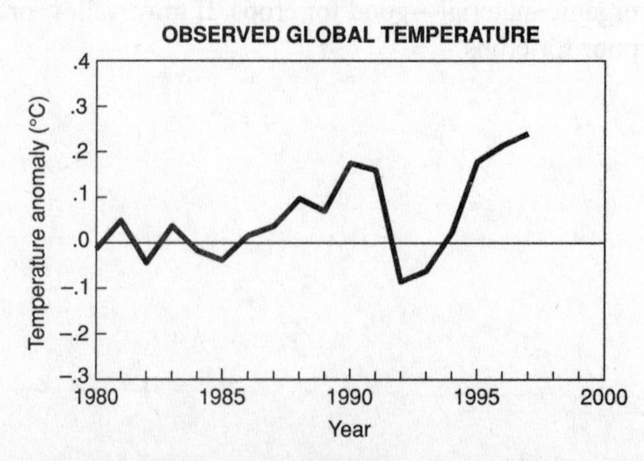

(i) Compare the Earth's climate before and after the eruption of Mount Pinatubo.

(ii) Explain how the eruption of Mount Pinatubo might affect short-term and long-term climate change.

Free-Response Answer

(A) Two of the following three:

1. (a) Feature A is a subduction zone. One lithospheric plate is subducting (sinking) below another, largely due to differences in density (the more dense plate sinks). This is an example of a destructive or convergent plate boundary. As the subducted plate sinks to greater depths, the temperature increases to the point where it begins to melt. This molten magma is less dense than the solid rock around it, so it rises up and forms a chain of volcanic mountains parallel to the plate boundary. When oceanic crust subducts under continental crust, these are known as island arcs.
   (b) Japan is an example of an island arc. The Cascade Mountains in Washington state, USA are an example of a volcanic arc.
2. (a) Feature B is a constructive or divergent plate boundary. Lithospheric plates are moving apart, with the space created between them being filled by hot molten magma coming up from the asthenosphere which cools and adds to the crust. In oceanic crust they are known as mid-ocean ridges. When they form on continental crust they are known as rift valleys.
   (b) The Mid-Atlantic Ridge is an example of a mid-ocean ridge. The African Rift Valley is an example of a rift valley.
3. (a) Feature C is a hot spot. This is a place in the asthenosphere where the temperature is higher than average such that localised melting occurs. This molten rock, being less dense due to its temperature and state of matter, rises up where it forms a volcano on the Earth's surface. Over geological time the location of the hot spot remains constant, whereas the lithospheric plate moves over it. This causes a chain of volcanoes to form over time from a single hot spot.
   (b) The Hawaiian Island chain is an example of the consequences of a hot spot.

(B) Two of the following:
(parts a and b of one category may count as two examples)

Geographic Separation
Evolution may occur as a consequence of geographic separation of one population of a species into two or more populations. Plate tectonics may cause this separation by:

(a) A divergent (constructive) plate boundary could cause one land mass to be divided into two or more distinct parts, perhaps even separated by an ocean. For example, identical fossils can be found on the east coast of South America and the west coast of Africa, indicating that this was once the same, connected land mass. After these two continents diverged, different species would evolve from this common ancestor as a reaction to the different environments on the different land masses.

(b) Faulting occurring as a consequence of plate tectonics may cause a river to be diverted. The new path of the river could divide a population into two and serve as a barrier between these two new populations. Alternatively, the fault itself could act as a geographic barrier. Different conditions in geographically separated regions would eventually lead to the evolution of different species as each population adapted to its environment in different ways.

Climate Change

Plate tectonics may result in climate change in one of the following ways:

(a) The Earth's atmosphere has changed considerably throughout geological history, largely as a consequence of gases emitted through volcanic activity caused by plate tectonics. These changes in atmosphere have caused climate change. For example, there is evidence that the Earth was much hotter hundreds of millions of years ago.

(b) Lithospheric plates move over the surface of the Earth at speeds of a few centimeters a year. Individual plates have moved thousands of miles over geological history. As the latitude of a plate in particular changes, so will its climate. For example, some rocks in Alaska indicate that they were originally deposited at a time when the plate was had a tropical climate, and so must have been close to the equator.

Species evolve in reaction to these climate changes. As examples, animals will adapt to shifting food sources as different plants grow in different climates. A species of animal may develop a fur coat over time as temperatures drop, or begin to lose one as temperatures increase.

As lithospheric plates move to latitudes farther from the equator, climates will have greater seasonal variations. This could lead to evolutionary adaptions such as plants shedding their leaves or animals hibernating during winter months to conserve energy.

Adaptive Radiation Following Mass Extinction

Evidence suggests that some mass extinctions in Earth's history may have been caused by large scale volcanic activity, such as flood basalts, which occur as a consequence of plate tectonics. These mass extinctions tend to be followed by periods with high rates of evolution and increase in species diversity, known as adaptive radiation, as ecological niches are filled.

(C)

(a) After the eruption of Mount Pinatubo the Earth's temperature was approximately 0.3°C lower for the next two years. The Earth's temperature rose by approximately 0.1°C during the third year. By the fourth year after the eruption the Earth's temperature had returned to the pre-eruption level.

(b) In the short term, dust and other particulates released into the atmosphere from the eruption would block the sun's rays. This decrease in energy reaching the Earth would result in lower global temperatures.

In the long term, gases such as carbon dioxide released during the eruption would accumulate in the stratosphere. There they would contribute to the greenhouse effect; they would absorb energy radiated back from the Earth, leading to an increase in global temperature. The degree of temperature change from the Mount Pinatubo eruption via this mechanism would be much less than that by dust blocking the sun, but the effect would last much longer.

CHAPTER 4
The Atmosphere

a frozen Continent... beat with perpetual storms... the parching Air Burns frore, and cold performs th' effect of Fire.

Milton, *Book II—Paradise Lost*

Areas That You Will Be Tested On

A: Atmospheric History: Origin, Evolution, Composition, and Structure
B: Atmospheric Dynamics: Weather and Climate

Key Terms

Note: Definitions for all key terms listed below can be found in the glossary starting on page 648. For additional definitions of relevant terms from this chapter, refer to *www.barronseduc.com/0764121618.html*.

atmospheric pressure	**jet stream**
climate	**La Niña**
cold front	**midlatitude cyclone**
convective lifting	**monsoon**
El Niño–Southern Oscillation	**ozone layer**
ferrel cell	**polar jet stream**
greenhouse effect	**rain shadow effect**
Hadley cell	**temperature inversion**
hurricanes	**weather**

Key Concepts

Origin of the Atmosphere

First Atmosphere

The Earth is about 4.6 billion years old. The first atmosphere was created within the first few hundred million years.

- Composition was probably H_2 and He.
- These gases are relatively rare on Earth compared to other places in the universe and were probably lost to space early in Earth's history because:

1. Earth's gravity is not strong enough to hold lighter gases
2. Earth still did not have a differentiated core (solid inner/liquid outer core), which creates Earth's magnetic field (magnetosphere = Van Allen Belt) that deflects solar winds.

- After the core differentiated, the heavier gases could be retained

Second Atmosphere

- Second atmosphere was produced by volcanic outgassing.
- Gases produced were probably similar to those created by modern volcanoes (H_2O, CO_2, SO_2, CO, S_2, Cl_2, N_2, H_2), NH_3 (ammonia), and CH_4 (methane).
- There was no free O_2 at this time (not found in volcanic gases).
- As the Earth cooled, H_2O produced by outgassing could exist as liquid in the Early Archean (3.8 billion to 2.5 billion years ago) allowing oceans to form.
- Evidence comes from pillow basalts and deep marine sediments in greenstone belts.

Oxygen Production

Oxygen was produced about 2 billion to 3 billion years ago when first life appeared.

- Photochemical dissociation is the breakup of water molecules by ultraviolet rays and produced O_2 levels approximately 1 to 2% of current levels. At these levels, O_3 (ozone) can form to shield the Earth's surface from UV rays.
- Photosynthesis ($CO_2 + H_2O$ + sunlight $\rightarrow$ organic compounds + O_2). Atmospheric oxygen was first produced by cyanobacteria, but eventually by higher plants, which supplied the rest of O_2 to atmosphere. The current level is ~21%.
- Oxygen consumers include animal respiration (much later), burning of fossil fuels (much, much later), and chemical weathering through oxidation of surface materials.
- Throughout the Archean (3.8 billion to 2.5 billion years ago), there was little to no free oxygen in the atmosphere (<1% of present levels). What little was produced by cyanobacteria, was probably consumed by the weathering process. After rocks at the surface were sufficiently oxidized, more oxygen could remain free in the atmosphere.
- During the Proterozoic (2.5 billion to 544 million years ago), the amount of free O_2 in the atmosphere rose from 1 to 10 %. Most of this was released by cyanobacteria, which increased in abundance in the fossil record. Present levels of O_2 were probably not achieved until ~400 million years ago.

Evidence from the Rock Record

- Iron (Fe) is extremely reactive with oxygen. If we look at the oxidation state of Fe in the rock record, we can infer a great deal about atmospheric evolution.
- In the Archean, we find minerals that only form in nonoxidizing environments—pyrite (fool's gold; FeS_2) and uraninite (UO_2). These minerals are easily dissolved out of rocks under present atmospheric conditions.
- Banded iron formations (BIFs) are deep-water deposits in which layers of iron-rich minerals alternate with iron-poor layers, primarily chert. They are common in rocks 2.0 billion to 2.8 billion years old, but they do not form today.
- Red beds (continental siliciclastic deposits) are never found in rocks older than 2.3 billion years, but they are common during the Phanerozoic. Conclusion: the amount of O_2 in the atmosphere has increased with time.

BIOLOGICAL EVIDENCE

- Chemical building blocks of life could not have formed in the presence of atmospheric oxygen. Chemical reactions that yield amino acids are inhibited by the presence of very small amounts of oxygen.
- Oxygen prevents growth of the most primitive living bacteria such as photosynthetic bacteria, methane-producing bacteria, and bacteria that derive energy from fermentation. Conclusion: because today's most primitive life forms are anaerobic, the first forms of cellular life probably had similar metabolisms.
- Today these anaerobic life forms are restricted to anoxic (low-oxygen) habitats such as swamps, ponds, and lagoons.

Climatic History

Climatic history can be developed through the following evidence:

1. Data from instruments (temperature, precipitation, wind speed, wind direction, and pressure). Problems: data are only available for the last few hundred years, and world data (especially in less-populated areas and oceans) are incomplete.

2. Written accounts, which are subjective in nature.

3. Fossil data ("proxy data") including
   - Tree ring data,
   - Fossilized plants,
   - Insect and pollen samples,
   - Gas bubbles trapped in glaciers,
   - Physical and chemical properties of deep ice,
   - Lake sediments,
   - Stalactites and stalagmites,
   - Marine fossils (including coral analysis),
   - Sediments (clay, ice rafted debris, and dust analysis), and
   - Isotope ratios (oxygen isotope ratios in fossilized remains).

Most *past* climatic change has been due to

1. Variation in the Earth's orbit—Milankovitch theory states that three characteristics can affect climatic change: (1) *eccentricity*—shape of the Earth's orbit around the sun; (2) *precession of the equinox*—a 23,000 year cycle that involves the "wobble" of the Earth's axis; (3) *obliquity*—changes in the tilt of the Earth's axis. The current angle is 23.5°. A smaller angle results in less climatic variation. A greater angle results in more climatic variation.
2. Changes in CO_2 concentration in the atmosphere—see Chapter 13.
3. Volcanic eruptions and the influence of gases and ash—most dust and ash returns to the Earth's surface within six months and is not the primary cause for cooling of global temperatures. The primary cause of cooling temperatures due to volcanoes is the sulfur dioxide gas that can remain in atmosphere for up to three years and that results in haze that reduces solar input.
4. Changes in solar output—changes in solar output of only 1% per 100 years would change the Earth's temperature by up to 1°F. Times of sunspot activity (every 11, 90, and 180 years) corresponds to decreases in solar radiation reaching the Earth.

Furthermore, every 22 years, the Sun's magnetic field reverses, which may have correlation with drought cycles on Earth. See Figure 4.1.

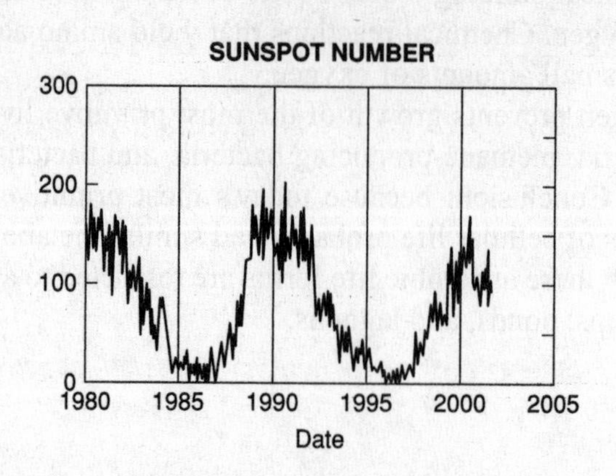

Figure 4.1. Sunspot cycles.

MAJOR CLIMATE PERIODS*

Time Period	Description
0 to 12,000 B.C.	Pleistocene ice age. Glacial ice sheets (periodically grew and receded) covered North America, Europe, and Asia. Temperatures averaged 4 to 5° cooler than present.
5,000 to 3,000 B.C.	Known as climatic optimum. World temperatures averaged 1 to 2° warmer than present. World civilizations began.
3,000 to 2,000 B.C.	Cooling trend. Drop in sea level. Emergence of islands and coastal areas.
2,000 to 1,500 B.C.	Warming trend followed by cooling trend.
1500 to 750 B.C.	Cooler temperatures. Growth in size of glaciers. Sea level dropped 6 to 9 feet from today's level.
150 B.C. to 800 A.D.	Warming trend.
A.D. 150 to 900	Cooling trend. Nile River and Black Sea froze.
A.D. 900 to 1200	Known as little climatic optimum. Warming trend. Vikings in Greenland and Iceland. Droughts and floods.
A.D. 1550 to 1850	Little ice age. Temperatures about 1° cooler than present.
A.D. 1850 to Present	Warming trend. Major droughts in the U.S. in the 1930's (Dust Bowl—see case study in Chapter 3). 1998—warmest year in last 1200 years. 1999—La Niña.

*many anomalies occurred within the years shown

COMPOSITION OF THE ATMOSPHERE

Element	Symbol	Percent of Atmosphere	Description
Nitrogen*	N_2	78.08%	Fundamental nutrient for living organisms. Deposits on Earth through nitrogen fixation and reactions involving lightning and subsequent precipitation. Returns to atmosphere through combustion of biomass and denitrification.
Oxygen*	O_2	20.95%	Oxygen molecules are produced through photosynthesis and utilized in cellular respiration.
Water Vapor*	H_2O	0–4%	Largest amounts occur near equator, over oceans, and in tropical regions. Areas where atmospheric water vapor can be low are polar areas and subtropical desert areas.
Argon	Ar	0.93%	Last atmospheric component discovered in 1894 by Ramsey and Raleigh in England.
Carbon Dioxide*	CO_2	0.036%	First discovered as atmospheric component in 1752. Volume has increased ~25% in last 300 years due to burning fossil fuels and deforestation. Produced during cellular respiration and decay of organic matter and is a reactant in photosynthesis. A greenhouse gas. Humans responsible for ~5500 million tons per year into atmosphere. Average time of a CO_2 molecule in the atmosphere is ~100 years.
Neon	Ne	0.002%	Inert gas.
Helium	He	0.0005%	Inert gas.
Methane*	CH_4	0.0002%	Contributes to greenhouse effect. Since 1750, atmospheric concentration of methane has increased ~150% due to use of fossil fuels, coal mining, landfills, grazers, and rice fields (anaerobic flooding of fields produces methane). Human activity is responsible for ~400 million tons per year as compared to ~200 million tons per year by nature. Average cycle of a methane molecule in the atmosphere is 10 years.
Hydrogen	H_2	0.00005%	Early atmosphere had high concentration of hydrogen.
Nitrous Oxide*	N_2O	0.00003%	Atmospheric concentration increasing at about 0.3% per year. Sources include burning of fossil fuels, widespread use of fertilizers, burning of biomass, soil fertilization, changes in land-use (deforestation), and conversion to agricultural land. Humans responsible for ~6 million tons per year as compared to nature at ~20 million tons per year. Contributor to greenhouse effect. Average time of a N_2O molecule in atmosphere is ~170 years.
Ozone*	O_3	0.000005%	97% of O_3 is found in stratosphere (ozone layer), 9–35 miles (15–55 km) above Earth's surface. Absorbs UV radiation. Ozone is produced in production of photochemical smog. O_3 is decreasing, especially over Antarctica. Chlorofluorocarbons (CFCs) are primary cause of degradation of O_3.

*important for life processes on Earth.

Atmospheric Structure

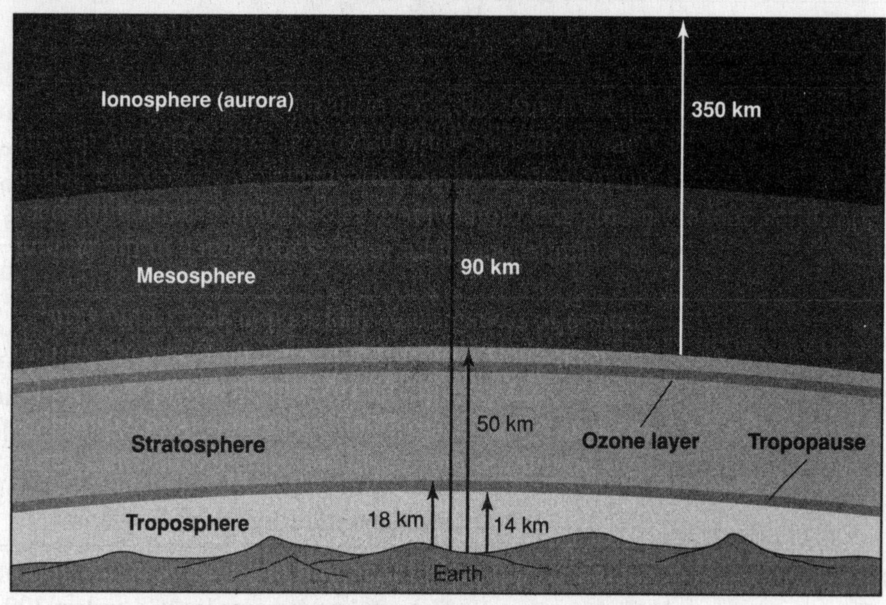

Figure 4.2. Layers of the atmosphere.

LAYERS OF THE ATMOSPHERE

Layer	Description
Troposphere (0–7 miles above surface)	Closest to Earth. 75% of atmosphere's mass is in troposphere. (See Figure 4.2.) Temperature decreases with altitude, reaching –60°C (–76°F) near the top of the trophosphere. Weather occurs in this zone. Maximum air temperatures occur in this zone near the surface. Air temperature decreases with height (~7°C every 0.6 mile) and is known as the environmental lapse rate.
Tropopause (7–13 miles above surface)	Temperature remains stable (isothermal). Boundary between troposphere and stratosphere. Jet streams occur in the tropopause.
Stratosphere (13–30 miles above surface)	Temperature increases with altitude in this zone due to absorption of heat by ozone. O_3 is produced by UV radiation and lightning. This layer contains the ozone layer, which protects organisms from too much UV radiation. The stratosphere is almost completely free of clouds or other forms of weather. The stratosphere provides some advantages for long-distance flight because it is above stormy weather and has strong, steady, horizontal winds.
Stratopause (30–31 miles above surface)	Isothermal layer separating stratosphere and mesosphere.

LAYERS OF THE ATMOSPHERE (continued)

Layer	Description
Mesosphere (31–50 miles above surface)	Extends from stratosphere upwards to about 50 miles. Temperature decreases with altitude in this zone. Temperature is around –130°F (–90°C) at top of mesosphere. The mesosphere is the coldest of the atmospheric layers (colder than Antarctica's lowest recorded temperature). It is cold enough to freeze water vapor into ice clouds. You can see these clouds if sunlight hits them after sunset. They are called noctilucent clouds (NLC). NLCs are most readily visible when the Sun is from 4° to 16° below the horizon. The mesosphere is also the layer in which meteors burn up while entering the Earth's atmosphere. From the Earth, they are seen as shooting stars.
Mesopause (50–52 miles above surface)	Isothermal layer separating mesosphere from thermosphere. Extremely cold temperatures.
Thermosphere (52–300 miles above surface)	Highest zone. Temperature increases rapidly in this zone due to gamma rays, X-rays and UV radiation bombarding molecules and the energy being converted to heat. Molecules are converted to positive ions with ejection of free electrons. Area is also known as ionosphere. Temperatures increase with height reaching maximums above 3000°F (1800°C). Air temperature, however, is a measure of the kinetic energy of air molecules, not of the total energy stored by the air. Therefore, because the air is so thin within the thermosphere, such temperature values are not comparable to those of the troposphere or stratosphere. Although the measured temperature is very hot, the thermosphere would actually feel very cold because the total energy of only a few air molecules residing there would not be enough to transfer any appreciable heat to the skin.
Exosphere (300–6200 miles above surface)	The atmosphere here merges into space in the extremely thin air. Air atoms and molecules are constantly escaping to space from the exosphere. In this region of the atmosphere, hydrogen and helium are the prime components and are only present at extremely low densities. This is the area where many satellites orbit the Earth.

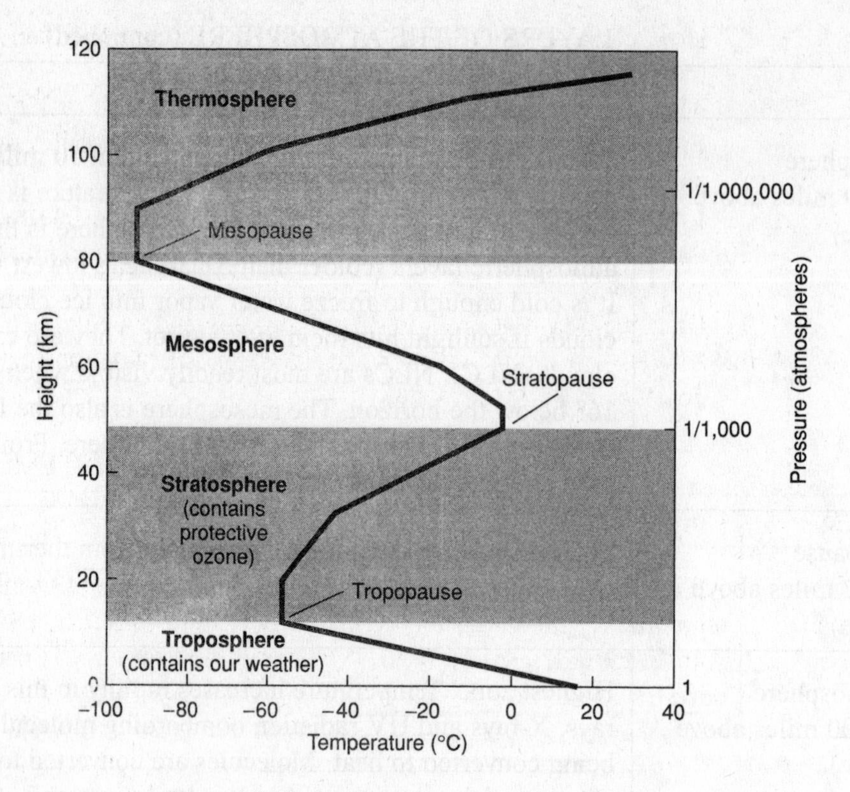

Figure 4.3. Changes in temperature in the atmosphere.

Weather and Climate

Weather is caused by the movement or transfer of heat energy (see Figure 4.3). Energy is transferred wherever there is a temperature difference between two objects. Energy can be transferred through (1) radiation (flow of electromagnetic radiation and the method that the Earth receives solar energy), (2) conduction (energy is transferred by the collisions that take place between heat-carrying molecules), and/or (3) convection. Convection is the primary way energy is transferred from hotter to colder regions in the Earth's atmosphere and is the primary determinant of weather patterns. Convection involves the movement of the more energetic molecules in air. Convection takes place both vertically (when air near the ground becomes warmer, and therefore less dense, than air above it as shown in the troposphere, Figure 4.3) and horizontally (pressure differences, which develop because of temperature differences, generate wind). Regions nearer the equator receive much more energy than regions nearer the poles and are consequently much warmer. These latitudinal differences in surface temperature create global-scale flows of energy within the atmosphere, giving rise to the major weather patterns of the world. Without convection and the transfer of energy, the equator would be ~15° warmer and the Arctic would be ~25° colder.

Climate can be influenced by:

1. Latitude (the higher the latitudes the less solar radiation);
2. Air masses (characteristic temperatures and humidities for a region);
3. Location of high- and low-pressure zones;
4. Heat exchange between air over oceans and air over land;
5. Presence or absence of mountain ranges;
6. Wind patterns;

7. How close an area is to the ocean (oceans are thermally more stable than land masses);
8. Altitude—for every 1000 feet rise in elevation, there is 3°F drop in temperature, and for every 300 feet in altitude, it is equivalent to 62 miles north in latitude in biome similarity;
9. Large-scale land changes—urbanization and deforestation;
10. Pollution and its influence on the greenhouse effect, including inputs from volcanoes;
11. Distance between Earth and the Sun (annual change);
12. Albedo—the reflectivity of a surface to solar radiation (e.g., snow has a high albedo, dark soil a low albedo).

KÖPPEN CLIMATE CLASSIFICATION SYSTEM

Climate	Characteristics	Subclimates
Tropical Moist (A)	All months have average temperatures above 32°F (0°C). Annual rainfall > 60 in. (152 cm). Extend 15° to 25° north and south from equator.	Af : No dry season. Rainfall year round. Monthly temperature variation < 3°. Cumulus and cumulonimbus clouds. Daily average temperature: high: 90°F (32°C); low: 72°F (22°C). Am: Tropical monsoonal—short dry season; heavy monsoonal rains in other months. Aw : Tropical savanna—winter dry season. Wet season precipitation < 39 in. (99 cm).
Dry Climates (B)	Low amounts of precipitation most months of the year. Exist 20° to 35° north and south of equator. Often surrounded by mountains. Deserts cover 12% of Earth and contain xerophytic vegetation. Steppes are grassland that cover 14% of surface of Earth.	BWh : Subtropical desert—low-latitude desert. BSh : Subtropical steppe—low-latitude dry. BWk : Midlatitude desert. BSk : Midlatitude steppe: midlatitude dry.
Moist Midlatitude with Mild Winters (C)	Extend from 30° to 50° latitude, primarily on east and west borders of continents. Midlatitude cyclones are common during winter. Thunderstorms common during summer.	Csa : Mediterranean—mild with dry, hot summer. Example: inland California. Csb : Mediterranean—mild with dry, warm summer. Example: west coast of southern California. Cfa : Humid subtropical—mild with no dry season hot, muggy summer with frequent thunderstorms. Example: southeastern United States. Cwa : Humid subtropical—mild with dry winter, hot summer. Cfb : Marine west coast—humid climate with short dry summers. Heavy precipitation during mild winter due to midlatitude cyclones. Cfc: Marine west coast—mild with no dry season, cool summer.

KÖPPEN CLIMATE CLASSIFICATION SYSTEM (continued)

Climate	Characteristics	Subclimates
Moist Midlatitude with Cold Winters (D)	Warm to cool summers and cold winters. Exist above 50° north and south of equator. Coldest months less than –22°F (30°C), warmest months > 50°F (10°C). Winters have snowstorms, strong winds, and cold air from Arctic air masses.	Dfa: Humid continental—humid with severe winter, no dry season, hot summer. Dfb: Humid continental—humid with severe winter, no dry season, warm summer. Dwa: Humid continental—humid with severe, dry winter, hot summer. Dwb: Humid continental—humid with severe, dry winter, warm summer. Dfc: Subarctic—severe winter, no dry season, cool summer. Dfd: Subarctic—severe, very cold winter, no dry season, cool summer. Dwc: Subarctic—severe, dry winter, cool summer. Dwd: Subarctic—severe, very cold and dry winter, cool summer.
Polar Climates (E)	Year-round cold temperatures. Warmest months < 10°F (12°C). Examples: Greenland, Antarctica, northern Canada and Russia.	ET: Tundra—polar tundra, no true summer, permafrost. Lichens, mosses, dwarf trees and sparse woody shrubs. EF: Ice Cap—perennial ice.

Temperature (Troposphere)

- Heat energy (as measured through temperature, which is a function of the speed of the molecular motion of a substance) flows from a warmer substance to a cooler substance until thermal equilibrium is reached (temperatures are equal).
- Temperature decreases with altitude.
- °C = (°F – 32.0) / 1.80 °F = (°C × 1.80) + 32.0
- Daily temperature cycles are primarily due to the Earth's rotation (once every 24 hours) on its axis. At night, infrared radiation emitted from the Earth's surface exceeds solar input through isolation. Conduction and convection transfer heat from the Earth's surface to the atmosphere. Therefore, daily minimum temperatures occur just before sunrise.
- Annual temperature cycles are determined by amount of net radiation reaching Earth as determined by the Earth's rotation around the Sun and the tilt of the Earth on its axis. During the winter, the Earth is closer to the sun than in summer, but due to the tilt (23.5°), the Earth receives less net radiation.
- Due to the tilt of the Earth on its axis, latitudes near the equator region receives more net solar radiation than the polar regions, which in turn determines day-length.
- The specific heat of water is five times greater than land; therefore, oceans heat and cool more slowly than land.
- Albedo (how much heat is reflected off of a surface) affects temperature.

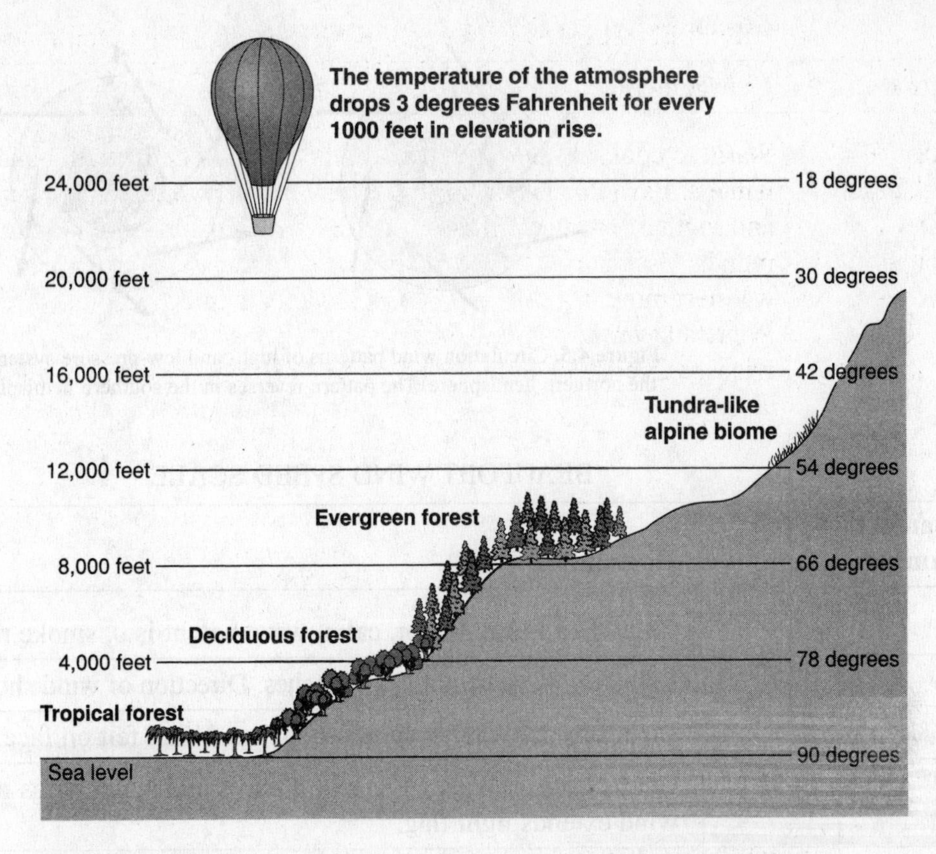

Figure 4.4. Change in temperature in response to change in altitude. Every 300 feet in altitude is equivalent to 62 miles north in latitude in biome similarity.

Wind

Due to the rotation of the Earth on its axis, its rotation around the sun, and the tilt of the Earth's axis, the sun heats the atmosphere unevenly. Air closer to the Earth's surface is the warmest and rises. Air at higher elevations is cooler and denser and, therefore, sinks. This process of convection is the primary determinant for wind.

- Horizontal wind moves from areas of high pressure to areas of low pressure. Vertically, winds move from areas of low pressure to areas of high pressure. Speed is determined by pressure differences (pressure gradient), which are linear. Example: tripling the pressure difference, would triple the wind speed and is derived from Newton's second law of motion ($F = ma$).
- Earth's rotation on its axis causes winds not to blow straight. This is called the Coriolis force which only influences wind direction and causes prevailing winds to spiral out from high-pressure areas and spiral in toward low-pressure areas in the northern hemisphere. The Coriolis force deflects wind to the right causing surface winds to blow counterclockwise and inward into a surface low, and clockwise and out of a surface high. In the southern hemisphere, it deflects winds to the left, causing surface winds to blow clockwise and inward around surface lows, and counterclockwise and outward around surface highs. The Coriolis effect increases as one gets closer to the poles. See Figure 4.5.
- Wind speed is measured with an anemometer and direction by a wind vane. Wind speed can be communicated nonquantitatively by the Beaufort Scale.
- Wind direction is based upon where the wind is coming from. Example: a wind coming from the east and moving toward the west is called an easterly.

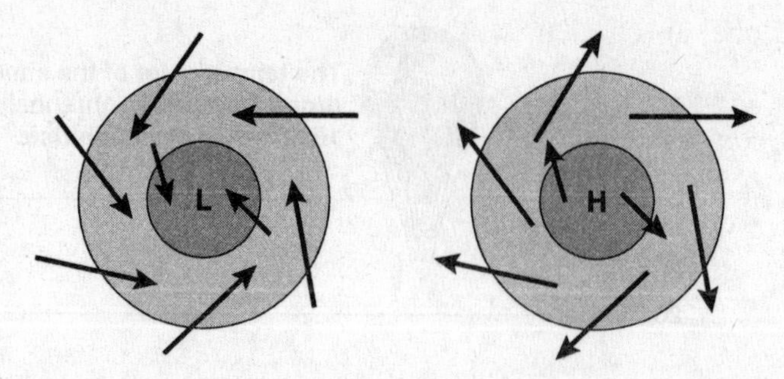

Figure 4.5. Circulation wind patterns of high- and low-pressure systems in the northern hemisphere. The pattern reverses in the southern hemisphere.

BEAUFORT WIND SPEED SCALE

Beaufort Number	Speed (mph)	Description
0	<1	Calm. Sea like a mirror, calm, wave height is 0, smoke rises vertically.
1	1–3	Light air. Wave height is 2–4 inches. Direction of wind shown by smoke drift.
2	4–7	Light breeze. Wave height 4–6 inches. Wind felt on face; leaves rustle.
3	8–12	Gentle breeze. Waves 1–2 feet. Leaves and small twigs in constant motion; wind extends light flag.
4	13–18	Moderate breeze. 3½-foot small waves. Raises dust and loose paper; small branches are moved.
5	19–24	Fresh breeze. Moderate waves taking a more pronounced long form, 6- to 7-foot waves. Small trees in leaf begin to sway; crested wavelets form on inland waters.
6	25–31	Strong breeze. Large waves begin to form, 9–10 feet. Large branches in motion; umbrellas used with difficulty.
7	32–38	Near to moderate gale. White foam from breaking waves begins to be blown in streaks along the direction of the wind , 13- to 14-foot waves. Whole trees in motion; inconvenience felt when walking against wind.
8	39–46	Gale. Moderately high waves of greater length, 18-foot-high waves. Breaks twigs off trees.
9	47–54	Strong gale. High waves, 23 foot. Slight structural damage occurs. Tree branches break off.
10	55–63	Storm. Very high waves with long overhanging crests, 29-foot waves. Seldom-experienced inland; trees uprooted; considerable structural damage occurs. Trees uprooted.
11	64–75	Violent storm. Exceptionally high waves (small and medium-size ships might be for a time lost from view behind waves); sea is completely covered with long white patches of foam lying along the direction of wind, 37-foot waves. Widespread damage.
12	> 75	Hurricane. The air is filled with foam and spray, wave height > 37 feet. Extensive, severe, widespread damage.

Global air circulation is affected by

1. Uneven heating of the Earth's surface,
2. Seasons,
3. Coriolis effect deflects winds clockwise in northern hemisphere and counterclockwise in southern hemisphere,
4. Amount of solar radiation reaching the Earth over long period of time,
5. Convection cells created by warm ocean waters, and
6. Ocean currents, which are caused by differences in water density, winds, and Earth's rotation.

Pressure

The force of gravity pulls the layers of air molecules down and close to the Earth's surface. In fact, 99% of the total mass of the atmosphere is within 20 miles of the Earth's surface. This force translates itself into what we call air pressure. Pressure is a force, or weight, exerted on a surface per unit area and is measured in the United States in inches of mercury (in. Hg). In most countries, air pressure is measured in hectopascals (hPa). A hectopascal is the pressure equal to 100 Newtons per square meter and is often referred to as a millibar (mb). A millibar is a force of 100 Newtons acting on a square meter. At sea level, commonly observed values of air pressure range between 970 and 1040 mb, or 76 cm (30 in.) of mercury.

$$1 \text{ mb} = 0.02953 \text{ in. Hg}$$
$$1 \text{ mb} = 1 \text{ hPa}$$

- Air pressure decreases with height.
- Rule of thumb for air less than 3000 feet: pressure drops about 1 in. Hg for each 1000 foot altitude gain or 1 mb for each 8 m of altitude gain. See Figure 4.6

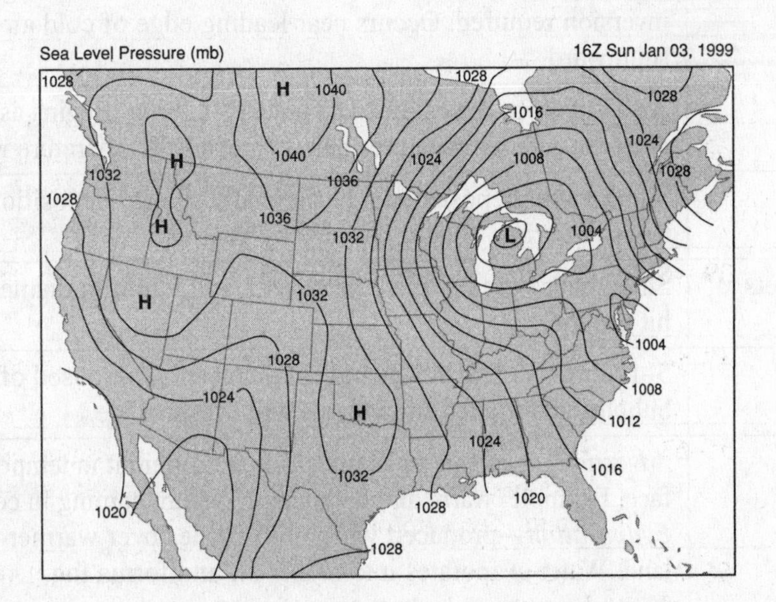

Figure 4.6. The solid white contours represent pressure contours (isobars) in millibars. The isobars have an interval of 4 mb. Wind speed is directly related to the distance between the isobars. The closer they are together, the stronger the wind.

Moisture

Moisture is one of life's most important requirements. It determines whether land will be a desert, with few living organisms, or whether it will be a tropical rain forest, teeming with life. It not only provides the supply for clouds and rainfall but also plays a vital role in energy exchanges within the atmosphere, which affects the Earth's energy balance. Water vapor in the atmosphere also behaves like a greenhouse gas, trapping heat that is trying to escape from the Earth to space. The presence of water vapor in the atmosphere significantly strengthens the Earth's natural greenhouse effect.

Moisture enters the atmosphere by evaporation from all surface bodies of water and includes puddles, ponds, streams, rivers, lakes, and oceans. Moisture also enters the atmosphere by transpiration from the leaves of plants and trees.

When air containing water vapor cools in the atmosphere (e.g., when it rises), the water vapor condenses to form tiny droplets of liquid water (or ice) in clouds. Eventually the water stored within clouds is returned to the Earth's surface in precipitation—rain, hail, sleet, or snow—where it is returned to the soil for uptake by vegetation or to surface streams, rivers, and lakes and ultimately the sea. Water vapor in the atmosphere also behaves like a greenhouse gas, trapping heat that is trying to escape from the Earth to space. The presence of water vapor in the atmosphere significantly strengthens the Earth's natural greenhouse effect. This cycle of evaporation, condensation, and precipitation is called the water cycle.

TYPES OF PRECIPITATION

Precipitation	Description
Rain	Liquid, falls to Earth, diameter > 0.5 mm.
Freezing rain	Occurs when water drop comes in contact with freezing surface. Temperature inversion required. Occurs near leading edge of cold air from the North moving southward.
Sleet	Also known as ice pellets. Diameter < 0.5 mm. Begins as rain; as drops fall, they enter air with temperatures below freezing. Temperature inversion usually required.
Snow	Water vapor deposits on a hexagonally shaped deposition nuclei at temperatures below freezing. Liquid phase is by-passed.
Snow pellets	Small dense grains of ice between 2 and 5 mm in diameter. Bounce when they hit ground.
Hail	5–190 mm (0.2 to 7.5 inches) in diameter. Composed of concentric rings of air bubbles and melted snowflakes with rings of ice.
Fog	*Advection*—generated by winds that are different in temperature than the Earth's surface. Example: warm air containing moisture coming in contact with colder land. *Evaporation*—produced when cold air lies over warmer water or warm, moist land. Water evaporates into colder air and forms fog. *Frontal*—produced when warm fronts pass through an area of colder temperatures. Rain falling in colder air evaporates when it meets warmer air forming fog near the ground. *Radiation*—ground fog. Develops when land cools by losing heat at night. *Upslope*—produced when air rises and cools.

GLOBAL DISTRIBUTION OF MOISTURE

Area	Type of Moisture
Deserts (subtropical)	Low precipitation. Dominated by subsiding air. No mechanism for lifting air masses. Remember that deserts are defined by a lack of moisture, not temperature extremes.
Continents	Distance from oceans and mountain ranges determines amount of precipitation moving inland.
Polar areas	Dry because cold air does not hold as much moisture.
Equator	High in rainfall due to constant solar heating that causes convection. Also contributing to high precipitation is converging of northern and southern air masses and frontal lifting.
Midlatitudes	Typical high cyclonic activity and frontal lifting due to polar and subtropical air masses meeting at a polar fronts determines precipitation. Air generally moves from west to east.
Mountain ranges	Mountain ranges near edges of continents may experience high rainfall due to orographic uplift. Leeward side of mountains may be dry due to rainshadow effect.

RAINFALL AMOUNTS IN BIOMES

Biome	Rainfall (in.)
Deserts	< 10
Tundra	< 10
Grasslands	10–30
Coniferous Forests	12–33
Deciduous Forests	30–60
Rain Forests	> 60

Clouds

Clouds are aggregates or collections of tiny water droplets and/or ice crystals suspended in the atmosphere. Clouds exist in different sizes and shapes and have unique characteristics. Clouds form by rising air, with warm air rising above cold air as seen at the edges of the fronts. As warmer air rises, it expands and consequently drops in temperature. As the temperature falls, the air is not able to hold as much water vapor; therefore, humidity increases, approaching 100% with the air becoming saturated with water molecules. Excess water vapor begins to coalesce forming tiny water droplets, hence a cloud. High-level clouds (prefixes with "cirr") form above 20,000 feet, and due to the cold temperatures at this level, clouds are primarily ice crystals. High-level clouds are thin, white, or can have a variety of colors close to sunrise or sunset. Mid-level clouds (prefix "alto") occur between 6000 and 20,000 feet in altitude. Warmer temperatures in this area of the atmosphere cause these clouds to be composed primarily of water droplets. Low-level clouds ("strat"), which occur below 6000 feet, are primarily composed of water droplets, although, if temperatures are cold enough, they can also contain ice particles and snow.

CLOUD TYPES

Name	Description
Altocumulus	Middle level layered cloud (6500–20,000 ft), rippled elements, generally white with some shading. May produce light showers.
Altostratus	Middle level (6500–20,000 ft), gray sheet, thinner layer allows sun to appear as through ground glass. Precipitation: rain or snow.
Cirrocumulus	High level (20,000 ft +), small rippled elements; ice crystals. No precipitation.
Cirrostratus	High level (20,000 ft +), transparent sheet or veil, halo phenomena; ice crystals. No precipitation.
Cirrus	High level (20,000 ft +), white tufts or filaments; made up of ice crystals. No precipitation.
Cumulonimbus	Low level, very large cauliflower-shaped towers, often "anvil tops." Phenomena: thunderstorms, lightning, squalls. Precipitation: showers of rain or snow.
Cumulus	Low level, individual cells, vertical rolls or towers, flat base. Precipitation: showers of rain or snow.
Nimbostratus	Thicker, darker, and lower based sheet. Precipitation: heavier intensity rain or snow.
Stratocumulus	Low-level layered cloud (ground–6500 ft), series of rounded rolls, generally white. Precipitation: drizzle.
Stratus	Low-level layer (ground–6500 ft), or mass, gray, uniform base; if ragged, referred to as "fractostratus." Precipitation: drizzle.

Fronts and Global Circulation Patterns

Air Masses. An air mass is a large body of air that has similar temperature and moisture properties throughout. At any given time, an estimated fifty distinct air masses are scattered across the face of the planet (see Figure 4.7). Air masses can be categorized based upon where they were formed, which have unique temperature characteristics and include (1) equatorial (E); (2) tropical (T); (3) polar (P); and (4) arctic (A). They can be classified based upon their moisture content and include the categories: (1) continental (c) and (2) maritime (m). Maritime air masses are generally fairly moist, containing considerable amounts of water vapor, which is ultimately condensed and released as rain or snow. By contrast, continental air is usually a lot drier. If temperature and moisture characteristics of air masses are combined, the following *common* combinations are possible.

CHARACTERISTICS OF AIR MASSES

Air Mass	Characteristics
Continental arctic, cA	Extremely cold temperatures and very little moisture. These usually originate north of the Arctic Circle, where days of 24-hour darkness allow the air to cool to sometimes record-breaking low temperatures. Such air masses often plunge southward across Canada and the United States during winter, but very rarely form during the summer because the sun warms the Arctic.
Continental polar, cP	Cold and dry, but not as cold as Arctic air masses. They form over the northernmost portions of North America, Europe, and Asia and often dominate the weather picture across the United States during winter. Continental polar masses do form during the summer but usually influence only the northern United States. These air masses are the ones responsible for bringing clear and pleasant weather during the summer to the north.
Maritime polar, mP	Cool and moist. They usually bring cloudy, damp weather to the United States. Maritime polar air masses form over the northern Atlantic and the northern Pacific Oceans. They most often influence the Pacific Northwest and the Northeast. Maritime polar air masses can form any time of the year and are usually not as cold as continental polar air masses.
Maritime tropical, mT	Warm temperatures with copious moisture. Maritime tropical air masses are most common across the eastern United States and originate over the warm waters of the southern Atlantic Ocean and the Gulf of Mexico. These air masses can form year round, but they are most prevalent across the United States during summer. Maritime tropical air masses are responsible for the hot, humid days of summer across the south and the east.
Continental tropical, cT	Hot and very dry. They usually form over the Desert Southwest and northern Mexico during summer. They can bring record heat to the Plains and the Mississippi Valley during summer, but they usually do not make it to the east and the southeast. As they move eastward, moisture evaporates into the air, making the air mass more like a maritime tropical air mass. Continental tropical air masses very rarely form during winter, but they usually keep the Desert Southwest scorching above 100°F during summer.

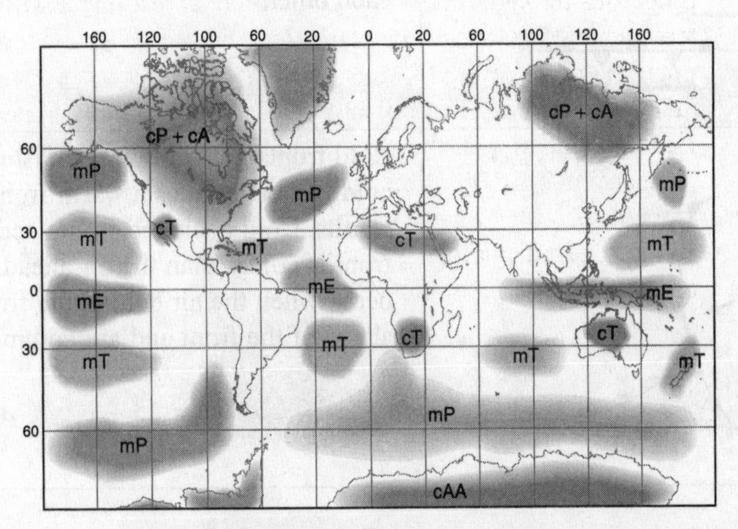

Figure 4.7. Typical air masses of the world.

Fronts. When two *different* air masses meet, the boundary between them forms a front. For a front to be identified, there must be a difference in the temperature, dew point, or wind direction between the air on each side of the boundary.

TYPES OF FRONTS

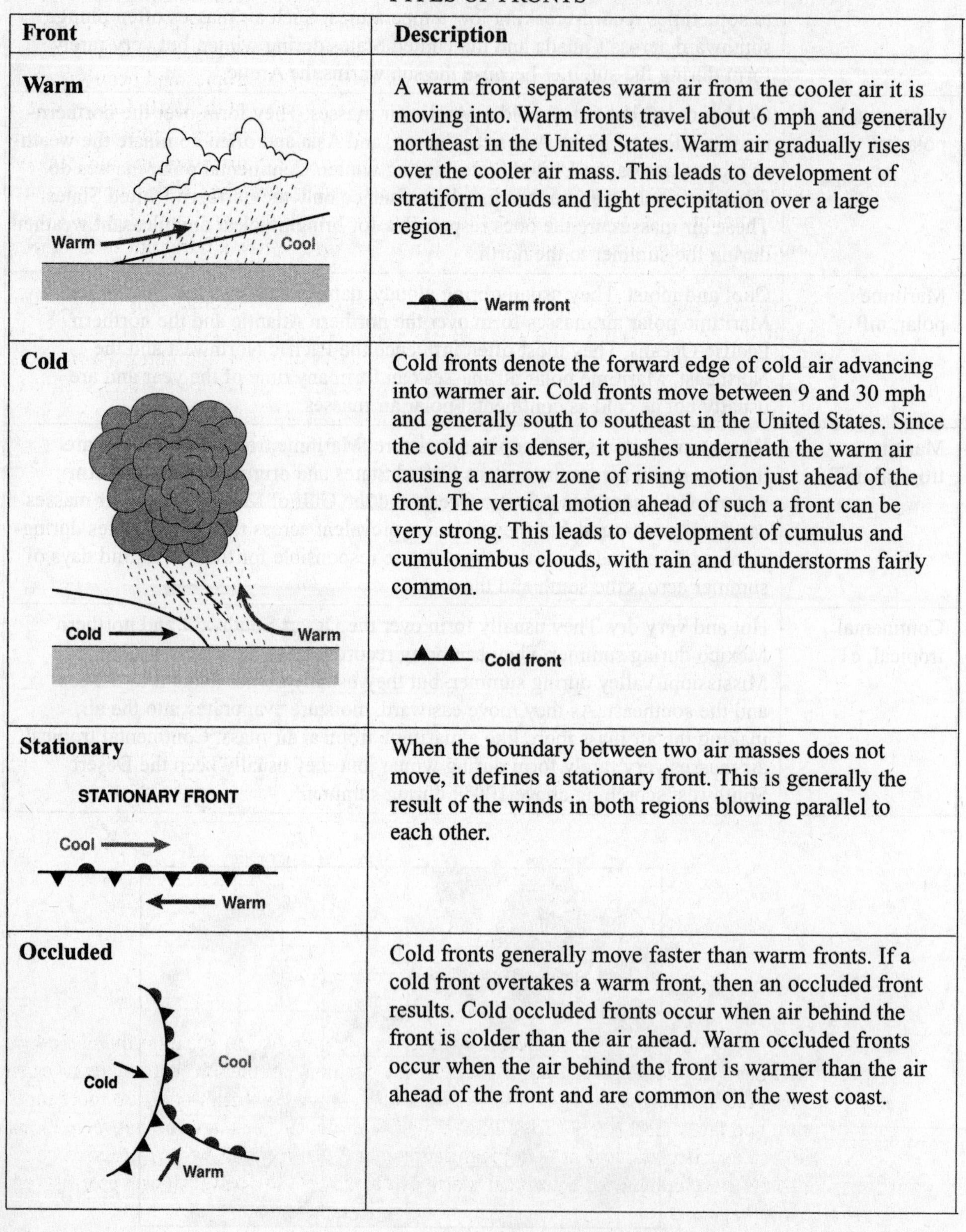

Front	Description
Warm	A warm front separates warm air from the cooler air it is moving into. Warm fronts travel about 6 mph and generally northeast in the United States. Warm air gradually rises over the cooler air mass. This leads to development of stratiform clouds and light precipitation over a large region.
Cold	Cold fronts denote the forward edge of cold air advancing into warmer air. Cold fronts move between 9 and 30 mph and generally south to southeast in the United States. Since the cold air is denser, it pushes underneath the warm air causing a narrow zone of rising motion just ahead of the front. The vertical motion ahead of such a front can be very strong. This leads to development of cumulus and cumulonimbus clouds, with rain and thunderstorms fairly common.
Stationary	When the boundary between two air masses does not move, it defines a stationary front. This is generally the result of the winds in both regions blowing parallel to each other.
Occluded	Cold fronts generally move faster than warm fronts. If a cold front overtakes a warm front, then an occluded front results. Cold occluded fronts occur when air behind the front is colder than the air ahead. Warm occluded fronts occur when the air behind the front is warmer than the air ahead of the front and are common on the west coast.

Hurricanes

Hurricanes are the most severe weather phenomena on the planet, causing billions in property loss and thousands of lost lives each year. In 1992, hurricane Andrew cost over $26 billion in the Dade County area of Florida. In 1998, hurricane Mitch was responsible for 11,000 deaths in Central America. Hurricanes, by international agreement, are more properly termed tropical cyclones. Other names that hurricanes go by are cyclones, typhoons, and baguio. Hurricanes, which must have wind speeds greater than 75 miles per hour at the storm center, begin over warm oceans of the tropics and occur between the Tropic of Cancer and the Tropic of Capricorn. In this area of the Earth (the intertropical convergence zone that occurs between 10° and 23.5° north and south latitude, see Figure 4.8), the sun is directly overhead for part of the year, providing the highest possible solar insolation (ocean temperature must be > 80°F), which provides the energy for huge amounts of evaporation from the oceans and the formation of cumulus and cumulonimbus clouds through atmospheric lifting. This also causes a fairly uniform temperature pattern in the tropics. This area also is a location where the northeast and southeast trade winds converge, adding more moisture to the air and enhancing the development of convective rain clouds. A subtropical high-pressure zone, characterized by descending airflow creates hot daytime temperatures and low humidities (which aids evaporation) and further adds to the build-up of conditions necessary to create hurricanes. Finally, the convergence of easterly winds (trade winds) that occurs in the tropics adds to conditions that create hurricanes. The Coriolis force is required to initiate cyclonic flow that is necessary for the development of hurricanes; therefore, hurricanes cannot develop 5° north or south of the equator. Temperature inversions prevent the development of hurricanes.

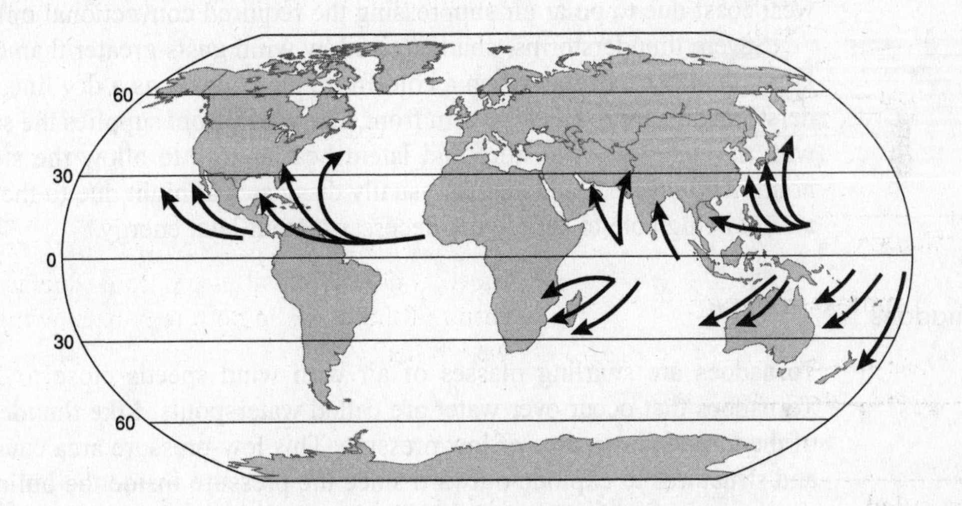

Figure 4.8. Areas where hurricanes develop.

Stages of building a hurricane include (1) the presence of separate thunderstorms that have developed over tropical oceans; (2) cyclonic circulation that begins to cause these thunderstorms to move in a circular motion, allowing them to pick up more moisture and latent heat energy from the ocean; (3) if wind speeds develop between 23 and 40 mph, it is classified as a tropical depression (characterized by low-pressure cells); (4) the development of a tropical storm is characterized by even lower pressure and consequently high wind speeds (40–75 mph); and (5) the final development of the hurricane (wind speeds greater than 75 mph).

In the center of the hurricane is the "eye"—an area that is characterized by low air pressure and descending air. Going outward from the eye are strong winds and precipitation stemming from a vertical wall of thunderstorm cloud.

The energy of a hurricane dissipates when the hurricane travels over land or moves over cooler bodies of water.

Rainfall amounts can exceed 24 inches in 24 hours. This amount of rainfall can cause flooding over land. Storm surges, or the increase in the height of the ocean near the eye of a hurricane, also cause extensive flooding when the hurricane hits land (Hurricane Camille in 1969 had a storm surge of 23 feet).

Thunderstorms

Thunderstorms occur from the equator northward as far as Alaska, but more commonly over tropical land areas. Thunderstorms may have associated with them hail, strong winds, lightning, thunder, rain, and tornadoes. Thunderstorms occur when moist air rises into the atmosphere due to either unequal warming of the Earth's surface, orographic lifting, or frontal zone lifting. As the air rises it becomes cooler and loses its latent heat. Since the air is cooler it cannot hold as much moisture (dew point) and cumulus clouds begin to form accompanied by updrafts. At about 42,000 feet, strong downdrafts begin to occur along with precipitation. Cumulus clouds then develop into cumulonimbus clouds, which can reach as high as 65,000 feet. Thunderstorms are short-lived, lasting about an hour and occur frequently in the southeastern United States and southern Arizona. They are more common in the eastern United States but are not common on the west coast due to polar air suppressing the required convectional uplift.

Severe thunderstorms, characterized by wind gusts greater than 60 mph, and/or hail larger than ¾ in. occurs when a cold front, also known as a dry line, precedes the thunderstorm and approaches a warm front. The warm front supplies the severe thunderstorm with the necessary moisture and latent heat energy to allow the storm to last several hours. The severe thunderstorm usually dissipates at night due to the presence of cooler air not being able to supply the necessary latent-heat energy.

Tornadoes

Tornadoes are swirling masses of air with wind speeds close to 300 mph or more. Tornadoes that occur over water are called waterspouts. Like thunderstorms, the center of the tornado is an area of low pressure. This low-pressure area causes many buildings and structures to explode outward since the pressure inside the building is greater than the pressure outside. The destruction of a tornado usually extends no more than 0.5 miles in width and 15 miles in length. In the United States, tornadoes are frequent from April through July and occur from central Texas and Oklahoma to Illinois and Indiana in a belt known as Tornado Alley (see Figure 4.9). Other areas where tornadoes are frequent include South Africa, Australia, northern India, and Argentina.

In the last 50 years, the total damage due to tornadoes has been estimated to be $25 billion. In the 1930s, close to 2000 people died from tornadoes. Due to advances in weather forecasting and warning systems, the deaths from tornadoes has decreased despite increases in population in these areas (e.g., from 1986 to 1995, just over 400 people died from tornadoes in the United States).

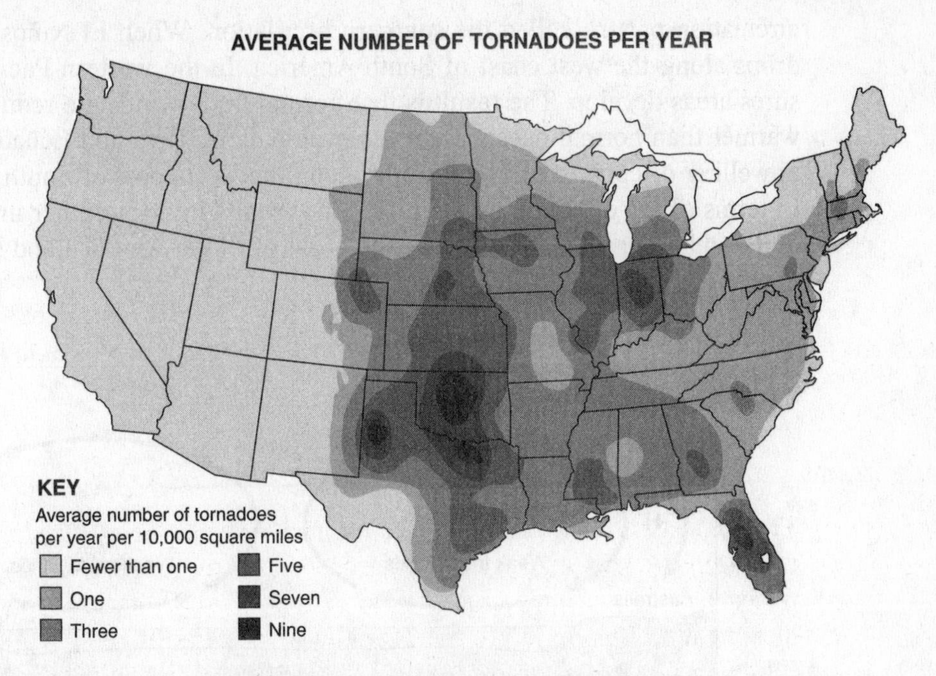

Figure 4.9. Average number of tornadoes per year.

Urban Climate

Climatic patterns can be influenced by human activity. Deforestation, urbanization, conversion of land for agriculture, release of pollutants including greenhouse gases, and the burning of fossil fuels and its influence on acid precipitation formation are just a few examples of how humans have and continue to alter world climatic patterns.

Urban areas store more heat than native land due to the specific heat of the materials used in cities—concrete and asphalt. Urban areas also have lower albedos than native lands, that is, they do not reflect light to the same extent. Compounding the heat-island effect is heat added to the urban environment through industrial, commercial, and residential processes including transportation, lighting, space heating, and cooling. The burning of fossil fuels alone in New York City during the winter is almost three times greater than the heat absorbed by the sun. Native lands also are able to reduce their latent heat through transpiration and evaporation, processes that are severely limited in urban areas. Buildings restrict wind flow, thereby reducing the amount of heat that can be carried away. All this adds up to make temperatures in cities as high as 6° warmer than comparable native land.

Increased pollution and particulates, combined with an increase in convectional uplift in urban areas, tends to increase the amount of rainfall urban areas receive—up to 10% greater than comparable native land.

El Niño and La Niña

El Niño occurs generally during late December along the coast of Ecuador and Peru and typically lasts from a few weeks to a few months. However, an El Niño that developed in 1991 lasted until 1995. Warm surface waters develop with the reduction of nutrient-rich cold-water upwelling (see Figure 4.10). El Niños are linked to a Pacific Ocean circulation pattern called the southern oscillation. When El Niños appear, air pressure drops along the west coast of South America. In the western Pacific, weak high-pressures areas develop. The result is that normal trade winds are reduced, which results in warmer than normal ocean water to develop along Peru and Ecuador. The result is that upwelling of cold, deep ocean water along the west coast of South America is reduced. In terms of the effect on weather, El Niño results in dry weather in the western Pacific, higher than normal rainfall along the coast of South America and storms in the central Pacific.

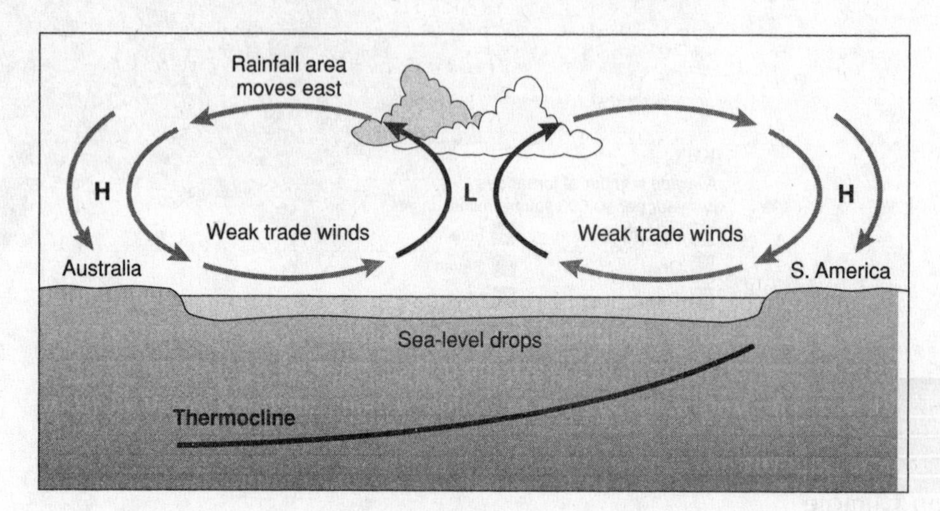

Figure 4.10. Development of El Niño.

La Niña La Niña is characterized by unusually cold ocean temperatures in the eastern equatorial Pacific. La Niña tends to bring nearly opposite effects of El Niño to the United States—wetter than normal conditions across the Pacific Northwest and dryer and warmer than normal conditions across much of the southern tier. The impacts of El Niño and La Niña at these latitudes are most clearly seen in wintertime. In the continental U.S., during El Niño years, temperatures in the winter are warmer than normal in the North Central States, and cooler than normal in the Southeast and the Southwest. During a La Niña year, winter temperatures are warmer than normal in the Southeast and cooler than normal in the Northwest. The increased temperatures in the Southeast United States during La Niña years correlates with the substantial increase in hurricanes that occur during this same time period. La Niña is also responsible for heavier than normal monsoons in India and Southeast Asia, and wetter and cooler winter weather in southeastern Africa.

Midlatitude Cyclones

A midlatitude cyclone is a large weather system (up to 1200 miles in diameter) that can travel up to 750 miles per day (generally to the east) and includes a well-defined surface low-pressure area and associated warm, cold, and occluded fronts. The path of midlatitude cyclones is determined to a great extent by the polar jet stream and by the winds accompanying the cold front. They are common during the winter months in the United States and are the predominant weather patterns. They are less destructive than hurricanes, which form over oceans and involve greater releases of potential energy.

Factors that lead to the lowering of the pressure at the surface are:

- Diverging airflow at high altitudes that reduces the mass of the air over the surface low;
- Inflow of warm, moist air at low and mid levels; and
- Latent heat release caused by convection in the warm air mass sector of the growing storm system.

Midlatitude cyclones (see Figure 4.11) develop through the following stages (cyclogenesis):

1. Cold air from northern polar regions comes in contact with warmer air from the south.
2. The two air masses coming together cause orographic uplift whereby the warmer air is uplifted into the higher areas of the upper atmosphere. Precipitation, which occurs at the center of the low, in the form of rain, hail (spring and summer only), sleet, or snow is common where the warmer air is being uplifted to higher altitudes, and thus not able to hold as much water vapor.
3. As the warm air is rising higher, a swirling motion begins (counterclockwise in the northern hemisphere) and then intensifies with the center of the vortex having lower pressure than the outside.

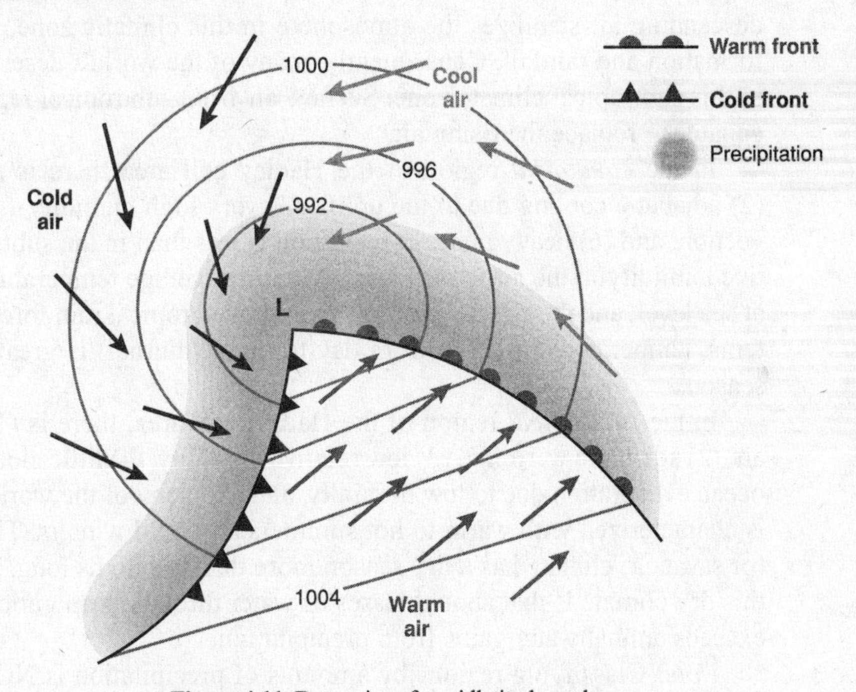

Figure 4.11. Dynamics of a midlatitude cyclone.

Circulation Patterns

The worldwide system of winds, which transports warm air from the equator where solar heating is greatest toward the higher latitudes where solar heating is diminished, gives rise to the Earth's climatic zones. Three types of air circulation cells associated with latitude exist: Hadley cell (from the equator to 30° north and south latitude; Ferrel cell (between 30° and 60° north and south latitude), and the polar cell (from 60° north and south latitudes to the poles), as shown in Figure 4.12. The general circulation of air serves to transport heat energy from warm equatorial regions to colder temperate and polar regions.

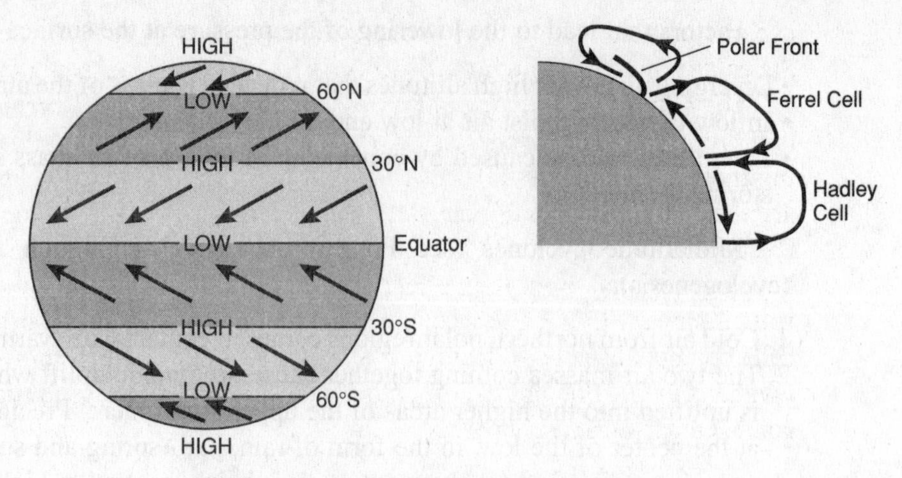

Figure 4.12. The Hadley, Ferrel, and polar cells.

Hadley Cell Air heated near the equator rises and spreads out north and south. After cooling in the upper atmosphere, the air sinks back to the Earth's surface within the subtropical climate zone between 25° and 40° north and south latitudes. This cooler descending air stabilizes the atmosphere in this climatic zone, preventing much cloud formation and rainfall. Consequently, many of the world's desert climates can be found in this subtropical climate zone. Surface air from subtropical regions returns toward the equator to replace the rising air.

In the *equatorial* region of the Hadley cell area there is (1) upward air motion, (2) adiabatic cooling due to the uplift, (3) very high humidity, (4) high clouds, (5) convection, and (6) heavy rains. Evaporation is less than in the subtropics because the relative humidity in the air is very high. Monthly average temperatures can be around 90°F at sea level, and there is no winter. Vegetation is tropical rain forest. In these tropical systems, temperature variation from day to night (diurnal) is greater than from season to season.

In the *subtropical* region of the Hadley cell area, there is (1) downward motion of air, (2) adiabatic warming, (3) low relative humidity, (4) little cloud formation, (5) strong ocean evaporation due to low humidity, and (6) many of the world's deserts. The climate is characterized with warm to hot summers and mild winters. The tropical wet-and-dry (or savanna) climate has a dry season more than 2 months long. The essential feature of this dry climate is that annual losses of water through evaporation at the Earth's surface exceeds annual water gains from precipitation.

Note: Classifying regions by amounts of precipitation is NOT something you want to do on the AP Environmental Science Exam! Evaporation is a function of temperature, and temperature is a function of altitude. Altitudes can vary considerably even at the same latitudes (e.g., there is snow on mountains near the equator). Also, 36 inches of annual rainfall might produce a humid climate in Seattle but would result in semiarid conditions in the tropics that receive between 100 to 400 inches of rain annually.

Ferrel Cell Ferrel cells develop between the tropics and the polar circles, between 30° and 60° north and south latitude. The descending winds of the Hadley cell diverge as moist tropical air moves toward the poles in winds known as the Westerlies. As the Westerlies meet the cold Polar air, they are forced to rise resulting in midlatitude depressions. Midlatitude climates can have severe winters and cool summers due to midlatitude cyclone patterns. The western United States is drier in summer than the eastern United States due to oceanic high pressure that brings cool *dry* air down from the north.

The climate of this area is known as Humid Temperate Domain and is governed by both tropical and polar air masses. Much of the precipitation in this belt comes from rising moist air along fronts as described above in midlatitude cyclones. Defined seasons are the rule, with strong annual cycles of temperature and precipitation. The seasonal fluctuation of temperature is greater than the change in temperature occurring in a 24-hour cycle. Climates of the middle latitudes have a distinctive winter season, while tropical climates within Hadley cell areas do not. This area of the Earth controlled by Ferrel cells contains broadleaf deciduous and coniferous evergreen forests. There are six subdivisions in this domain: warm continental, hot continental, subtropical, marine, prairie, and Mediterranean.

Polar Cell The Polar cells originate as icy cold, dry air, which is heavy and descends from the troposphere to the ground. The air moves south as a wedge of heavy cold air. This air meets with the warm tropical air from the midlatitudes that rises over it forming the midlatitude depressions. The air then returns to the poles, cooling, sinking, and repeating the cycle. Sinking air suppresses precipitation; thus the polar regions are deserts. (*Remember that deserts are not defined by temperature; they are defined by the amount of* available *liquid water.*) Very little available water exists in this area because it is tied up in the frozen state as ice. Furthermore, the amount of snowfall per year is relatively small, and what snow that does fall just adds to the existing amounts. Climates in this area known as the Polar Domain are controlled chiefly by polar and arctic air masses. In general, climates in the Polar Domain are characterized by low temperatures, severe winters, and small amounts of precipitation, most of which falls in summer. The annual fluctuation of temperature is greater than the change in temperature occurring in a 24-hour cycle. In this area where summers are short and temperatures are generally low throughout the year, temperature rather than precipitation is the critical factor in plant distribution and soil development. Two major divisions exist—the tundra and the subarctic taiga.

Multiple-Choice Questions

1. The zone of the atmosphere in which weather occurs is known as the

   (A) ionosphere.
   (B) mesosphere.
   (C) troposphere.
   (D) thermosphere.
   (E) stratosphere.

2. Ninety-nine percent of the volume of gases in the lower atmosphere, listed in descending order of volume, consist of

   (A) oxygen, nitrogen, carbon dioxide, water vapor.
   (B) water vapor, nitrogen, oxygen, carbon dioxide.
   (C) nitrogen, carbon dioxide, oxygen, water vapor.
   (D) oxygen, carbon dioxide, nitrogen, water vapor.
   (E) nitrogen, oxygen, water vapor, argon, carbon dioxide.

3. Regional climates are most influenced by

(A) latitude and altitude.
(B) prevailing winds and latitude.
(C) altitude and longitude.
(D) latitude and longitude.
(E) Coriolis effect and trade winds.

4. A low-pressure air mass is generally associated with

(A) hot, humid weather.
(B) fair weather.
(C) tornadoes.
(D) cloudy or stormy weather.
(E) hurricanes.

5. La Niña would produce the following effects except

(A) more rain in Southeast Asia.
(B) wetter winters in the Pacific Northwest region of the United States.
(C) warmer winters in Canada and northeast United States.
(D) warmer and drier winters in the southwest and southeast United States.
(E) more Atlantic hurricanes.

6. On the leeward side of a mountain range, one would expect

(A) more clouds and rain than on the windward side.
(B) more clouds but less rain than on the windward side.
(C) colder temperatures.
(D) less clouds and less rain than on the windward side.
(E) no significant difference in climate compared to the windward side.

7. The ozone layer exists primarily in what section of the atmosphere?

(A) Troposphere
(B) Stratosphere
(C) Mesosphere
(D) Thermosphere
(E) Ionosphere

8. Along the equator

(A) warm, moist air rises.
(B) warm, moist air descends.
(C) warm, dry air descends.
(D) cool, dry air descends.
(E) cool, moist air descends.

9. The gas that is responsible for trapping most of the heat in the lower atmosphere is

   (A) water vapor
   (B) ozone
   (C) carbon dioxide
   (D) oxygen
   (E) nitrogen

10. Characteristics or requirements of a monsoon include all of the following EXCEPT

   (A) a seasonal reversal of wind patterns.
   (B) large land areas cut off from continental air masses by mountain ranges and surrounded by large bodies of water.
   (C) different heating and cooling rates between the ocean and the continent.
   (D) extremely heavy rainfall.
   (E) heating and cooling rates between the oceans and the continents are equivalent.

11. An atmospheric condition in which the air temperature rises with increasing altitude, holding surface air down and preventing dispersion of pollutants, is known as (a)

   (A) temperature inversion.
   (B) cold front.
   (C) warm front.
   (D) global warming.
   (E) upwelling.

12. Earth's earliest atmosphere probably consisted primarily of

   (A) H_2 and He.
   (B) CO_2 and water vapor.
   (C) ozone, sulfur dioxide, and methane.
   (D) nitrogen, oxygen, and carbon dioxide.
   (E) greenhouse gases.

13. The surface with the lowest albedo would be

   (A) snow.
   (B) ocean.
   (C) forest.
   (D) desert.
   (E) black topsoil.

14. Jet streams travel primarily

   (A) north to south.
   (B) south to north.
   (C) east to west.
   (D) west to east.
   (E) in many directions depending on a multitude of factors.

15. The correct arrangement of atmospheric layers, arranged in order from the most distance from the Earth's surface to the one closest to the Earth's surface, is

    (A) Troposphere–Tropopause–Stratosphere–Stratopause–Mesosphere–Mesopause–Thermosphere–Exosphere.
    (B) Exosphere–Thermosphere–Mesopause–Mesosphere–Stratopause–Stratosphere–Tropopause–Troposphere.
    (C) Exosphere–Stratosphere–Stratopause–Tropopause–Troposphere–Mesosphere–Mesopause–Thermosphere.
    (D) Thermosphere–Mesopause–Mesosphere–Troposphere–Tropopause–Stratopause–Stratosphere–Exosphere.
    (E) None of the above is correct.

Answers to Multiple-Choice Questions

1. **C**	6. **D**	11. **A**
2. **E**	7. **B**	12. **A**
3. **A**	8. **A**	13. **E**
4. **D**	9. **A**	14. **D**
5. **C**	10. **E**	15. **B**

Explanations for Multiple-Choice Questions

1. **(C)** The troposphere is the innermost layer (closest to Earth) of the atmosphere, which extends from about 11 miles above the Earth at the equator to about 5 miles above the Earth at the poles. Composition: 78% nitrogen (N_2); 21% oxygen (O_2); 0.0036% carbon dioxide (CO_2). Temperature declines the higher in altitude one goes in the troposphere until the lowest temperature in the troposphere is reached (~–57°C or –71°F).

2. **(E)** Nitrogen (78%); oxygen (21%); water vapor (0–4%); argon (1%); CO_2 (0.04%).

3. **(A)** Latitude is a measurement that expresses how far north (or south) from the equator a location is. The equator is at 0° latitude while the poles are at 90°. At sea level, the climate at the equator is typically hot and humid. At the poles, the weather is typically cold with variable humidity. Between these areas are the mid-latitudes. Altitude has a direct influence on climate. So much so, that even at the equator, at high altitudes, the temperature can be below freezing. A general rule of thumb for altitude is that for every 1,000 feet in altitude, there is a 3°F drop in temperature. As one ascends a mountain, they will pass through many climatic zones and biomes.

4. **(D)** A low-pressure air mass, also known as a low occurs when warm air, which is less dense, spirals inward toward the center of a low-pressure area. Since the center of the low-pressure area is of even less density and pressure, the air in this section rises and the warm air cools as it expands. The temperature begins to fall and may go below the dew point—the point at which air condenses into droplets. These droplets make up clouds. If the droplets begin to coalesce on condensation nuclei, rain follows.

5. **(C)** La Niña is characterized by unusually cold ocean temperatures in the eastern equatorial Pacific. La Niña tends to bring nearly opposite effects of El Niño to the United States—wetter than normal conditions across the Pacific Northwest and dryer and warmer than normal conditions across much of the southern tier. Winter temperatures are warmer than normal in the Southeast and cooler than normal in the Northwest. The increased temperatures in the Southeast United States during La Niña years correlates with the substantial increase in hurricanes that occur during this same time period. La Niña is also responsible for heavier than normal monsoons in India and Southeast Asia, and wetter and cooler winter weather in southeastern Africa.
6. **(D)** The rain-shadow effect occurs on leeward side of mountain—the side away from ocean. Moist air from ocean rises when it hits mountains, as air rises it cools and loses its moisture as rain and snow on the windward side. On the leeward side, air is dry and semiarid to arid conditions exist. Examples: eastern side of Sierra Nevada mountain range in California, Himalayas, and Karakorum ranges of south Asia, and Mount Waialeale on the island of Kauai.
7. **(B)** Ninety-seven percent of ozone (O_3) is found in the stratosphere, 9–35 miles (15–55 km) above the Earth's surface. Ozone absorbs UV radiation. Ozone is produced in production of photochemical smog and by lightning. O_3 is decreasing, especially over Antarctica. Chlorofluorocarbons are primary cause of degradation of O_3. Temperature increases with altitude in the stratosphere due to absorption of heat by ozone. The stratosphere is almost completely free of clouds or other forms of weather. The stratosphere provides some advantages for long-distance flight because it is above stormy weather and has strong, steady, horizontal winds.
8. **(A)** In the equatorial region of the Hadley cell area (0° to 25° north and south latitude) there is (1) upward air motion, (2) adiabatic cooling due to the uplift, (3) very high humidity, (4) high clouds, (5) convection, and (6) heavy rains. Evaporation is less than in the subtropics because the relative humidity in the air is very high.
9. **(A)** Water vapor in the lower atmosphere ranges from 0 to 4% by volume. Clouds and fog occur in the lower atmosphere. Water is constantly being evaporated from oceans and fresh water sources and entering the atmosphere through transpiration. Largest amounts of water vapor occur near the equator, over oceans, and in tropical regions. Areas where atmospheric water vapor can be low are polar areas and subtropical desert areas. Ozone occurs primarily in the upper atmosphere (stratosphere). CO_2 content of the troposphere is 0.036%.
10. **(E)** Monsoons are caused by winds from the southwest or south that brings heavy rainfall to southern Asia in the summer. They are created by temperature gradients that exist between ocean and land surfaces. They occur over very large areas, and they are seasonal. In summer, humid wind blows from cooler ocean areas (higher pressure) to warmer land masses (lower pressure). As air rises over land masses, it cools and is unable to retain water, producing great amounts of rain. In winter, the ocean is now warmer and the cycle reverses and the drier air travels from the continent out to the ocean. Monsoon winds exist in Australia, Africa, and North and South America.

11. **(A)** Temperature inversions are an atmospheric condition in which the air temperature rises with increasing altitude, holding surface air down and preventing dispersion of pollutants. There are two types of inversions: (1) radiation temperature inversion, which generally occurs at night. When a layer of warm air lies on top of a layer of cooler air, but as the sun warms the surface of the Earth, the inversion generally dissipates during the day, dispersing pollutants; and (2) subsidence temperature inversion, a large mass of warm air at high altitude moves in and traps colder air near the ground. Prevents mixing of air and dispersion of pollutants.

12. **(A)** The Earth is about 4.6 billion years old. The first atmosphere was created within the first few hundred million years and was composed of hydrogen and helium. These gases are relatively rare on Earth today compared to other places in the universe and were probably lost to space early in Earth's history because (1) Earth's gravity was not strong enough to hold lighter gases; and (2) Earth still did not have a differentiated core (solid inner/liquid outer core) which creates Earth's magnetic field (magnetosphere = Van Allen Belt) that deflects solar winds. After the core differentiated, heavier gases could be retained

13. **(E)** Albedo is a measure of reflection of sunlight from a surface. Dark objects absorb energy, light surfaces reflect sunlight. Dark topsoil would absorb the most energy and, therefore, would reflect the least resulting in the lowest albedo.

14. **(D)** Jet streams are large-scale upper air flows that travel from west to east. They are located between 3½ and 7½ miles above the Earth's surface and can be up to 50 km wide (30 miles) and 5 km deep (3 miles). They travel up to 125 mph (200 km/hr) and at times travel up to 250 mph. They are produced by large temperature differences. In the northern hemisphere, the two major jet streams are the (1) subtropical jet stream and (2) the northern jet stream. The subtropical jet stream travels near the southern edge of the continental United States (around 30° north latitude) and the northern jet stream moves like a snake across the northern border of the United States and into Canada following the edge of a cold air mass known as the circumpolar vortex. When warm, moist air from the Pacific Ocean and the Gulf of Mexico meets the frigid, drier air from the Arctic moving southward, conditions exist for rain, storms, and wind. During winter months, when the angle of the Earth above the equator produces less solar insolation, the Arctic air masses move farther south, bringing cooler temperatures to North America. During the summer months in the northern hemisphere, the added energy in the subtropical air masses push the Arctic air masses farther north, creating milder climatic conditions over North America.

15. **(B)** Refer to the chart on page 114.

Free-Response Question

Characteristics of air masses and their worldwide circulation patterns influence the spatial distribution of biomes and the organisms that inhabit them.

(a) Describe Hadley, Ferrel, and Polar cells in terms of what they are, how they develop, and their locations in reference to the equator.
(b) Describe the characteristics of the air mass that would occur within each type of cell. Describe climate conditions in terms of temperature, relative humidity, prevailing winds, and solar insolation.
(c) Describe ONE biome that would exist at sea level within the specific latitudes of a cell. Give examples of both plants and animals that would exist in that biome.

Free-Response Answer

Let's do this essay together, using it as a teaching tool, rather than just providing an answer and rubric. We'll do that throughout the book, switching from essays that are models that you can compare your answers to with complete rubrics, to other essays where we will break the essay into pieces and learn how to put it together in a coherent, organized piece of writing.

What is the first step that you should do? Correct, BRAINSTORM! Let's write a list of key words that would apply to the questions that we are asked to write about. Remember that order is not important; we'll do that later. Try to get at least ten key terms. Examine the question again to see if you missed anything.

Hadley
Ferrel
Polar
Temperature
Solar insolation
Humidity
Biomes
Plants
Animals

Now, that we have around ten key terms, let's expand the list by adding detail—things that we will talk about in our essay within each category. We can also begin to map out the order in which we want to discuss these items.

Hadley 0°–30°
- deserts
- equatorial regions
- tropical rain forests
- subtropical areas
- savannas

Ferrel 30°–60°
- Westerlies
- midlatitude depressions

- midlatitude cyclones
- humid temperate domain
- seasons
- forests
- prairies
- chaparral

Polar 60°–90°

- dry air
- desert
- polar domain
- low temperature
- tundra
- taiga

location

temperature

- heat moves from equator to colder areas

relative humidity

prevailing winds

solar insolation

biomes

animals

plants

Remember that the clock is ticking. You have less than 23 minutes to spend on this essay, so don't spend more than 5 minutes total in organization. But also remember that, just like in painting a house, preparation is essential for a good job. Now, look over the detailed list—are there other things YOU would like to add? If so, add them now.

We're going to start the essay now. What is the first step when we actually begin writing? Correct, the restatement. The restatement is part of your thesis statement. Also remember that, as we write, go from general concepts to specific examples. Let's start:

The world's biomes are primarily determined by climatic conditions. Deserts are characterized as areas of little precipitation, while tropical rain forests are characterized as areas receiving great amounts of precipitation.

This is my thesis statement. Looking at the question, it seemed to me that the question was looking for a tie-in or correlation between climate and its effect on biome distribution. My first sentence did that. I then followed the general thesis statement with specific references to illustrate that thought. Now I can begin describing what determines climatic conditions that in turn affect the type of life present within that zone.

Solar insolation, that is the amount of sunlight received on the Earth, is greatest at the equator and diminishes toward the poles. Since heat flows from warmer regions to cooler regions, the warmer air produced at the equator moves through major worldwide wind patterns and distributes this energy worldwide. As one moves from the equator (0° latitude) to either pole (90°), there are three major air circulation cells known as Hadley, Ferrel, and polar. (At this point, a brief labeled sketch is worth a thousand words!)

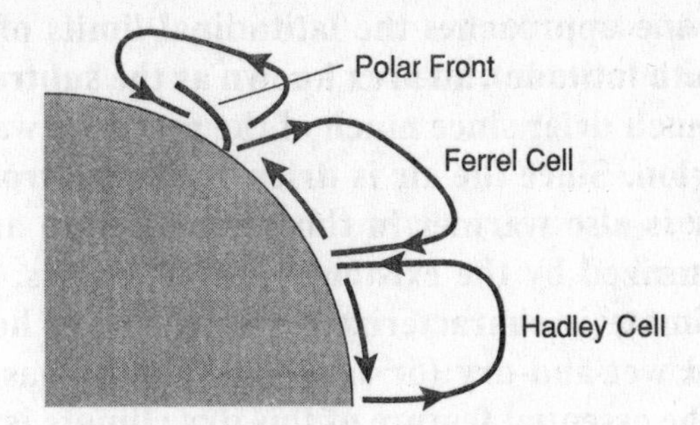

From the equator to 30° north and south latitude exists the Hadley cell. Since this area of the Earth receives the greatest solar radiation due to the Earth's tilt, this area of the Earth is the warmest. Near the equator, this warm, moist air rises. As the warm air rises, it begins to cool and become denser. Since cooler air cannot hold as much water vapor as warmer air, the humidity of the air increases to the point where high clouds are produced. This in turn causes great amounts of rain (200–225 cm or 79–89 inches per year). Monthly average temperatures can be around 90°F at sea level and there is no winter. Vegetation is tropical rain forest. In these tropical systems, temperature variations from day to night (diurnal) are greater than from season to season. Tropical rain forests, which generally occur between the Tropic of Cancer (23.5° N latitude) and the Tropic of Capricorn (23.5° S latitude) and extend about 1500 miles north and south of the equator, are found in South and Central America, West Africa, southern India, northeastern Australia, Indonesia, Philippines, Hawaii, and parts of Malaysia. There are three types of climatic weather patterns in the tropical areas according to the Köppen Climate Classification System: (1) rain throughout the year (Af); (2) monsoonal (Am), which consists of a short dry season followed by heavy rains in other months; and (3) tropical savanna (Aw), with characteristic wet and dry seasons.

Tropical rain forests have characteristically high-species diversity for both plants and animals (up to 100 different tree species per square kilometer as opposed to 3–5 species in temperate zones). Vegetation is dense. Lianas, bromeliads, and epiphytic orchids, ferns, and palms are present. Leaves are large in an effort to absorb sunlight, and there is little need to conserve water lost through transpiration. Soils are characteristically low in nutrients with the nutrients being stored in vegetation. Soil is characteristically acidic. Decomposition of organic material is very high due to temperature and moisture. Leaching is high. Abundant insects and animal biodiversity (up to 50% of all animal species) exists in the tropical rain forests. Examples of some animals that would inhabit the tropical rainforest would include:

***insects:* numerous species of butterflies (i.e., Monarch), ants, mosquitoes, dragonfly, millipedes, and moths**

***mammals:* jaguars, ocelots, bats, monkeys, opossums, bandicoots, echidnas, 3-toed sloths, capybaras, marmosets, peccaries, tarsiers, orangutans, tigers, gorillas, hippopotamus, and tapirs**

***birds:* macaws, hummingbirds, toucans, cassowaries, kookaburras, tree swifts, wood swallows, and parrots**

***reptiles:* anacondas, pythons, lizards, geckos, alligators, and caimans**

***amphibians:* poison arrow frogs, red-eyed tree frogs, and toads**

As one approaches the latitudinal limits of the Hadley cell (around 30° north and south latitude), an area known as the subtropics, the air begins to descend. This air is much drier since much of the moisture was returned to the Earth in the tropical region. Since the air is drier in this subtropical region, evaporation is greater. The air is also warmer in this region due to adiabatic warming. This area is also characterized by the existence of hot deserts, and there is little cloud formation. The climate is characterized with warm to hot summers and mild winters. The tropical wet-and-dry (or savanna) climate has a dry season more than 2 months long. The essential feature of this dry climate is that annual losses of water through evaporation at the Earth's surface exceed annual water gains from precipitation.

At this point, I have said what I want to say about Hadley cells. I am asking myself, "Have I answered all of the questions required?" Let's review to make sure before we finish the essay describing Ferrel and polar cells. The questions asked were:

1. to describe the cell. ✔
2. how it develops. ✔
3. location of the cell. ✔
4. characteristics of the air mass. ✔
   - temperature ✔
   - relative humidity ✔
   - prevailing winds ✔
   - solar insolation ✔
5. Pick one biome and list plants and animals that would exist there. ✔

I am satisfied that I have answered all the questions required for the Hadley cell. Since I picked the tropical rain forest as the biome existing within this cell as my choice, I do not need to write about any other biomes. I could mention what biomes exist in the remaining cells, but NOT describe the biomes in detail. Remember that the clock is running. Also, please remember this, and I will keep saying it over and over, these essays go FAR beyond what you might be able to do in 25 minutes. Don't look at these essays as what you need to do to get a 5. Use them as models to compare your essays with in terms of asking yourself, "Did I get *MOST or SOME* of the ideas mentioned."

I am satisfied with the pattern I developed in describing the Hadley cell. I am now going to follow that same pattern when I describe the Ferrel and polar cells.

Ferrel cells begin where the Hadley cells end. They exist between 30° and 60° N and S latitudes. The descending winds of the Hadley cell near 30° N latitude diverge north as moist tropical air moves toward the Arctic in winds known as the Westerlies. As these warmer Westerlies meet the colder polar air, they are forced to rise (orographic lifting) resulting in midlatitude depressions. Midlatitude climates can have severe winters and cool summers due to midlatitude cyclone patterns. The climate of this area is known as Humid Temperate Domain and is governed by both tropical and polar air masses. Much of the precipitation in this zone comes from rising moist air along fronts and results in midlatitude cyclone weather patterns. Defined seasons are the rule in biomes within the Ferrel cells, with strong annual cycles of temperature and precipitation. The seasonal fluctuation of temperature is greater than the change in temperature occurring in a 24-hour cycle. Climates of the middle latitudes have a distinctive winter season. This area of the Earth controlled by Ferrel cells contains broadleaf deciduous and coniferous evergreen

forests. There are six subdivisions in this domain: warm continental, hot continental, subtropical, marine, prairie, and Mediterranean.

I ask myself again, "Did I answer ALL questions necessary for the Ferrel cell?"
1. describe the cell. ✔
2. how it develops. ✔
3. location of the cell. ✔
4. characteristics of the air mass. ✔
   – temperature ✔
   – relative humidity or precipitation patterns ✔
   – prevailing winds ✔
   – solar insolation ✔
5. brief mention of biomes within that cell ✔

Now, check your watch. If you are like me, you are running out of time. Remember that we've got 3 more essays to write if this were the actual exam. Remember, no extra-credit for writing and writing. So, let's finish the essay up with the Polar cell and we should be done.

Polar cells, located between 60° N and S latitudes and the poles, originate as icy cold, dry air, which is heavy and descends from the troposphere to the ground. The air is cold due to the tilt of the Earth's axis, which results in this area receiving very little direct solar radiation. In the northern hemisphere, the air moves south as a wedge of heavy cold air. This air meets with the warm tropical air from the midlatitudes (within the Ferrel cell) that rises over it forming the midlatitude depressions. The air then returns to the poles, cooling, sinking, and repeating the cycle. Sinking air suppresses precipitation; thus, the polar regions are deserts. Very little available water exists in this area since it is tied up in the frozen state as ice. Furthermore, the amount of snowfall per year is relatively small, and what snow that does fall just adds to the existing amounts. Climates in this area, known as the Polar Domain, are controlled chiefly by polar and Arctic air masses. In general, climates in the Polar Domain are characterized by low temperatures, severe winters, and small amounts of precipitation, most of which falls in summer. The annual fluctuation of temperature is greater than the change in temperature occurring in a 24-hour cycle. In this area where summers are short and temperatures are generally low throughout the year, temperature rather than precipitation is the critical factor in plant distribution and soil development. Two major divisions exist—the tundra and the subarctic taiga.

Total Word Count: 1122

Now we come to the stage that matters most. *Practice.* (A man hailed a taxicab in New York City and asked the cabby, "What's the best way to get to Carnegie Hall?" The cabby thought about the question for a while and replied, "Practice a lot!"). The same is true here. Go back to the question and redo it in your words. But don't pick the tropical rain forest this time. Instead, let's pick the temperate deciduous forest within the Ferrel cell.

Characteristics of the biome include distinct seasons, temperatures average 75°F (but the extremes run from below 0°F to above 100°F), precipitation averages 2 to 5 feet per year including snowfall, and high humidity (60–80%).

Trees might include: maples, oaks, hickory, birch, chestnut, elm, basswood, linden, walnut or sweetgum (make sure the trees are deciduous, that is, shed their leaves in autumn). Provide a brief reason for the advantage of shedding leaves (the tree conserves energy during a period of high stress by entering a dormant state). Also, during the winter there are less hours of sunlight. With less sunlight, photosynthesis cannot produce as much sugar to keep up with the nutritional requirements of the tree; therefore, the tree becomes dormant and loses the leaves. The sunlight ties into the tilt of the Earth and the angle of incidence. You might also want to mention that species diversity is reduced, and that large stands of a single species are common.

Plants might include: rhododendrons, azaleas, mountain laurels, huckleberries, lichens, club mosses, true mosses

Animals might include:

rodents: beaver, muskrat, squirrel, mice, and rats

mammals: bears, deer, fox, porcupine, rabbit, raccoon, skunks, weasel

birds: cardinals, ducks, hawks, turkey

amphibians: frogs, salamanders

CHAPTER 5

The Biosphere

Nature encourages no looseness, pardons no errors

—Ralph Waldo Emerson

Areas That You Will Be Tested On

A: Natural Areas
B: Organisms: Adaptations to Their Environments
C: Populations and Communities: Exponential Growth and Carrying Capacity
D: Ecosystems and Change: Biomass, Energy Transfer, and Succession
E: Evolution of Life: Natural Selection, Genetic Diversity, and Extinction

Key Terms

Note: Definitions for all key terms listed below can be found in the glossary starting on page 648. For additional definitions of relevant terms from this chapter, refer to *www.barronseduc.com/0764121618.html*.

adaptation
adaptive radiation
allelopathy
allogenic succession
allopatric speciation
autogenic succession
autotroph
biome
carrying capacity
chaparral
climax community
convergent evolution
divergent evolution
ecological niche
food chain
K strategists
mutualism
natural selection
r-strategists
secondary consumer
succession
symbiosis

Key Concepts

Natural Areas

Biomes

Biomes are a major regional or global biotic community characterized chiefly by the dominant forms of plant life and the prevailing climate. Temperature and precipitation are the most important determinants.

- Biomes are classified by type of dominant plant (e.g., grassland, tropical deciduous forest, etc.).
- Species diversity within a biome is directly related to (1) net productivity, (2) availability of moisture, and (3) ambient temperature.

BIOME CHARACTERISTICS

Antarctic. Area surrounding the South Pole. Waters are abundant with life. Most land is frozen desert. Rainfall is less than 2 inches a year. 70% of all fresh water is tied up as ice in the Antarctic. Animals have adapted to living in this biome by (1) developing extra layers of fat, (2) developing chemicals in the blood to prevent it from freezing, (3) developing compact bodies to conserve heat, (4) developing thick skin and fur; (5) developing waterproof feathers above downy insulating feathers, and (6) migrating during the coldest months. Animals include amphipods, cod, copepods, cormorants, crabs, dolphin, gull, hagfish, krill (euphausids), limpets, marine snails, midge, mites, nudibranchs, octopus, penguin, petrels, sea anemones, sea squirts, sea urchins, shrimp, skates, sponges, squid, seals, skua, sheathbills, terns, and whales.

Benthos. Bottom portion of oceans. No sunlight, therefore no plants. Primary input of food is dead organic matter from above. Animals include nematodes, starfish, and occasional fish.

Coastal zones. Ocean depth less than 200 meters (656 feet). Includes estuaries, tidal wetlands, and coral reefs. Very high counts of animal and plant species (grasses and sedges) due to nutrients brought in from runoff from land. Coral reefs (see later) are extremely diverse communities of animal species.

Coral reefs. Warm, clear shallow ocean habitats near land and generally in the tropics (water temperature is favorable, 70°–85°F [21°–29°C]). Soft corals (sea fingers and sea whips) do not build reefs. Hard corals (brain and elkhorn) build reefs that are formed from the exoskeletons of coral polyps. Coral reefs exist off the coast of Brazil, the Caribbean, Eastern Africa, southern coast of India, northeast (Great Barrier Reef, 1300 miles long) and northwest coasts of Australia, Polynesia, off the coast of Florida, and the Red Sea. Three types of coral reefs: (1) fringing reefs—grow on continental shelf near coastline; (2) barrier reefs—parallel to shoreline but farther from coastline; (3) coral atolls—rings of coral that grow on top of sunken oceanic volcanoes. Coral reefs are disappearing at a fast rate due to (1) pollution, (2) dredging, and (3) sedimentation. Animals that live in coral reefs include fish (angelshark, blowfish, clown fish, fishtiger sharks, Galapagos shark, groupers, John Dory, lemon shark, nurse shark, parrot fish, puffer fish, rays, sand sharks, sea horses, snappers, and zebra bullhead sharks), crustaceans (crabs, lobsters, and shrimp), brittle stars, clams, conch, jellyfish, krill, octopus, oysters, sea anemones, sea urchins, scallops, various species of sponges, sea turtles, slugs, snails, squid, starfish, tusk shells, whelks, and zooplankton.

Deserts. Occur between 15° and 25° north and south latitude and generally in the interior of continents—away from ocean masses. Occupy about 20% of land area. Major deserts include southwest United States; Atacama and Patagonian in South America; Sahara in northwest Africa; Namib and Kalahari in southern Africa; Arabian, Turkestan, Thar, and Takla Makan-Gobi in Eurasia; and the Australian in Australia. Rainfall less than 50 cm/yr (20 in./yr). Air currents are descending, which generally diminishes the formation of rain, which requires ascending air (orographic lifting). Temperature is not a determinant, precipitation is, with most deserts < 25 mm (1 inch) of precipitation annually.

Cold and hot deserts. Cold deserts (Gobi and Antarctica) have very few frost-free days, while hot deserts (Sahara) have over 300 frost-free days per year. Soils often have abundant nutrients but lack organic matter. Relatively few large mammals. Dominant animals of warm deserts are snakes and reptiles. Many animals are nocturnal to conserve on water. Succulent plants (cactus) and short-lived annuals that depend on a short rainy season. Plants are spaced apart due to limiting factors and soil has very little humus. Desert animals include Australian (bilby, dingo, kangaroo, marsupial mole, quokka, rabbit-eared bandicoot); Arabian (camel, chameleon, civet, cobra, dromedary, Egyptian vulture, flamingo, fox, gazelle, hare, hedgehog, horse, hyena, ibex, jackal, jerboa, lizards, locusts, oryx, peregrine falcon, porcupine, scorpion, and vipers); southwest United States and Mexico (bats, bobcats, chuckwallas, coati, collard peccary, coyotes, desert iguana, desert tortoise, dragonfly, elf owl, Gila monster, jack rabbits, kangaroo rats, mule deer, pupfish, rattlesnakes, roadrunner, scorpion, tarantula, turkey vulture, and wild burro); southwest Africa (gazelle, gerbil, ground squirrel, hyena, jackal, meerkat, and springbok). South America (armadillo, cavy, eagles, foxes, guanaco, hawks, jaguarondi, llama, mara, puma, rhea, tinamou, and tuco-tuco); north Africa (addax antelope, barn owls, cape hare, dama deer, desert hedgehog, dorcas gazelle, fantailed raven, Fennec fox, gerbil, horned viper, jackal, jerboa, mongoose, Nubian buzzard, ostrich, sand fox, shrew, and spotted hyena); China (Bactrian camel, beetles, blue hill pigeon, gazelle, gerbil, jerboa, Pallas cat, Pallas sand grouse, snow leopard, wild mountain sheep, and wolf); and Antarctic (brown skua, mites, penguins, springtails, and worms).

Freshwater. Freshwater wetlands include freshwater swamps, marshes, bogs, prairie potholes (important stopping grounds for migrating birds), ponds, and peat bogs. Ground is saturated with freestanding water. Plants with specialized roots include black spruces, mangroves, and mosses. Soil does not contain oxygen. Important breeding areas, rich in insects, amphibians, and reptiles. Freshwater wetlands naturally filter water of pollutants, sediments, and nutrients. Prime areas for human development thereby resulting in large amounts of habitat destruction. Major recreational areas. Critical for fresh water supplies. Easily polluted. Dominant plants are phytoplankton (floating algae). Rooted flowering plants occur along shoreline (Example: water lilies). Animal life abundant. Zooplankton as well as finfish and shellfish. Estuaries are rich in nutrients and are breeding grounds for fish. Input includes runoff, groundwater flow, and streams. Eutrophic lakes are rich in nitrogen and phosphorus. Indicative of eutrophic lakes are high densities of plankton and zooplankton, which support large and diverse populations of fish. Water lilies, rushes, and insect larvae are present. During warm weather, oxygen content of water may decrease resulting in die-offs and naturally polluted water. Oligotrophic lakes are nutrient poor and are characterized by very clear water with low animal and plant biomass. Cultural eutrophication, that is human-introduced nutrients from agricultural runoff such as nitrates and phosphates affect this biome. Animals common in freshwater wetlands include amoeba, ants, beaver, black swan, Canadian goose, capybara, copepods, crayfish, dragonfly, ducks, earthworm, egret, many species of fish, flamingo, flies, frogs, Great Blue Heron, mosquitoes, muskrats, newts, otters, shrimp, snails, swans, toads, turtles, water striders, white-tailed deer, and zooplankton.

Grasslands. Found in areas too dry for forests and too wet for deserts. Rainfall is seasonal. Temperatures are moderate. Temperate grasslands are located north of the Tropic of Cancer and south of the Tropic of Capricorn. (Tropical grasslands are located *between* these two lines of latitude). Grasslands occupy approximately 25% of all land area.

Generally between 120 and 200 frost-free days per year. Few trees and shrubs (less than 1 tree per acre). Prairies in north central United States, tall grass prairies in eastern United States, steppes of Eurasia (primarily Mongolia and Russia), pampas in Argentina, campos of Uruguay and Brazil, veldts in East Africa and Madagascar, and rangelands in Australia and New Zealand. Dominant plants are grasses and perennials with extensively developed roots. Soils rich in organic matter (chernozemic). Drier prairies may have salinization. Due to fertility of soils, grasslands have been extensively used by man for agriculture, particularly grains. Animals common to grasslands include: Africa (aardvark, African elephant, African wild cat, antelopes, baboon, buffalo, cheetah, giraffe, gnu, hartebeest, hippopotamus, hyena, impala, jackals, kudu, leopard, lion, meerkat, mongoose, oryx, ostrich, rhinoceros, vulture, wildebeest, zebra); Australia (dingo, emu, kangaroo, wallaby, wombat, and introduced species such as camel, donkey, goat, horse, rabbit, and sheep); North America (badger, bees, black-footed ferret, bison, burrowing owl, California Condor, carrion beetle, common snipe, coyote, deer, dragonfly, eagles, elk, fox snake, golden owl, gopher snake, grasshopper, ground squirrels, jack rabbit, killdeer, lady beetle, larks, long-billed curlew, meadow vole, monarch butterfly, prairie chicken, prairie dog, pronghorn antelope, red fox, red-tailed hawk, shrew, skunk, tiger beetle, western meadow lark, and western tiger swallowtail); South America (armadillo, fox, jaguar, llama, opossum, puma, rhea, and tapir); Eurasia (golden pheasant, leopard gecko, snow leopard, and vole). Habitat destruction and poaching have seen numbers drastically decrease. Burrowing mammals (adaptation for protection since trees are not present and land is flat) include, jackrabbits, gophers, ground squirrels, and prairie dogs.

Hydrothermal vents. Occurs in deep ocean where hot water vents rich in sulfur compounds are found. Sulfur compounds provide energy for chemosynthetic bacteria, which are food for clams and worms. High water pressure requires specific adaptations of life forms.

Intertidal. The area of the shoreline exposed to water during high tide and air during low tide. Water movement brings in nutrients and removes waste products. Extremely rich in life and biodiversity, including algae, attached shellfish, birds, fish, kelp, sea anemones, and sea urchins. Specialized adaptations required for being partially submerged during the day—exoskeletons of crabs, barnacles, snails, etc. protect organisms from predators and dehydration; salt-removing mechanisms; lower metabolic rates due to cooler ambient temperatures; less availability of sunlight for photosynthesis; and oxygen availability. Phosphorus and iron are limiting nutrients. Susceptible and sensitive to pollution from land and sea (oil spills).

Ocean. Pelagic region—90% of total ocean is open ocean environment, but it is an area with few animal and plant species. Oceans occupy 75% of Earth's surface. Oceans contain 97% of Earth's water supply. Four oceans: Pacific, Atlantic, Indian, and Arctic. Seas are smaller and partly enclosed by land. Large seas include South China Sea, Caribbean, and the Mediterranean. An area of low diversity and low productivity. Low in nitrogen and phosphorus, which limits plants and the smaller organisms that feed on them. Large animals are found but at low densities. Animals include brittle star, clams, coral, copepods, crabs, crustaceans, dolphins, echinoderms, eels, jellyfish, krill, octopus, porpoise, rays, salmon, sea anemones, seals, sea urchins, sand dollars, scallops, sea cucumbers, sea turtles, sharks, shrimp, snails, squids, tuna, and whales.

Savannas. Warm year-round. Prolonged dry seasons. Scattered trees. Environment is intermediate between grassland and forest. Extended dry season followed by a rainy season. Australia; central, eastern, and south Africa; India; Madagascar; central South America; Southeast Asia; and Thailand. Consist of grasslands with stands of deciduous shrubs and trees that do not grow more than 30 meters (100 feet) high. Trees and shrubs generally shed leaves during dry season, which reduces need for water. Food is limited during dry season so that many animals migrate during this season. Soils are rich in nutrients. Contain large herds of grazing animals and browsing animals that provide resources for predators. Animals include Africa (aardvark, antelope, cheetah, elephant, gazelles, giraffe, gnu, hippopotamus, hornbills, hyena, impala, jackal, leopard, lion, meerkat, oryx, ostrich, pigeons, raptors, rhinoceros, serval, vulture, waterbuck, wildebeests, and zebras); Australia (finch, kangaroo, pigeon, wallaby and wombat); South America (capybara, deer, and rhea); and India (Asian elephant, rhinoceros, tiger, and water buffalo).

Taiga or boreal forests. Generally found between 45° and 60° north latitude. Occupies approximately 17% of land area (50 million acres or 20 million hectares or 78,000 square miles). Forests of cold climates of high latitudes and high altitudes. Two types of taigas: (1) open woodland and (2) dense forest. More precipitation than tundra and generally occurs during summer months due to midlatitude cyclones. There are between 90 and 120 frost-free days per year. Soils are generally poor in nutrients because of large amounts of leaching caused by rainfall and the soils are acidic due to decomposition of needles. They have a deep layer of litter and decomposition is slow because of temperature. Dominated by conifers (spruces, firs, pines). Flowering trees include aspens and birches. Dense stands of small trees cause understory to be low in light. Understory plants and trees (blueberry, cranberry) in deciduous forests have adapted to grow primarily in the fall and winter months to take advantage of light that can come through when dominant trees have lost their leaves. Low biodiversity due to harshness of environment. Source of lumber and paper pulp. Animals include Arctic fox, Arctic hare, badger, bald eagle, bats, bears, beaver, Canadian goose, caribou, chipmunks, dall sheep, deer, earthworms, ermine, fox, gray wolf, great horned owl, lemming, lynx, marten, moose, musk ox, muskrat, red-tailed hawk, reindeer, scorpion, short-tailed weasel, snow goose, snowshoe hares, snowy owl, shrews, squirrels, voles, wolves, and wolverines. Reptiles are not present due to temperatures. Disturbances include fires, storms, and insect infestation.

Temperate forests. Most of the eastern United States, central Europe, Korea, and China. Forests found in milder temperatures than boreal forests. There are between 120 and 250 frost-free days per year. Rapid decomposition due to mild temperatures and precipitation, which results in small litter layer. Because of mild climate, this biome has been greatly exploited by humans for agriculture and urban development. Soil is generally poor in nutrients. Tall deciduous trees (oaks, maples, beeches, hickory, basswood, cottonwood, elm, and willow). Hardwood trees are used to make furniture. Rich and diverse understory. Low density of large mammals due to shade that prevents much ground vegetation. Animals include bald eagle, beaver, bears, birds, deer, ducks, earthworm, fox, frog, insects, mice, muskrat, newt, opossum, porcupine, rabbit, raccoon, scorpion, skunk, squirrels, turkey, and weasel.

Temperate rain forests. Moderate temperatures and rainfall exceeding 250 cm/yr (100 inches/yr). Redwood forests, cedars, and Douglas firs. Low biodiversity because of high shade, which limits food for herbivores. Major resource for timber. Found in Canada, northern California, Oregon, and Washington state.

Temperate shrub lands (chaparral or sclerophyll forests). Located between 30° and 40° north and south latitude and on west coasts. Found in areas of Mediterranean climate—California coast, Cape region of South Africa, Chile, coastal Australia, and the Mediterranean region. Hot, dry summers with mild, cool, rainy winters. During fall, summer, and spring, a subtropical high-pressure zone exists over the area. Rain falls during winter due to midlatitude cyclones. Average rainfall is between 300–750 mm/yr (15–40 in./yr). Chaparral characterized by dense shrub growth that does not exceed but a few meters in height. Leaves do not fall during dry season due to stress of replacing leaves without adequate water. Decomposition is slow during dry months. Plants include acacia, arbutus, eucalyptus, oaks, and olive. Thorns are common for protection. Low rainfall concentrated in cool season. Few large mammals. Small mammals and reptiles present. Animals include southern California (butterflies, California quail, chipmunk, coyote, deer, fox, kangaroo rats, lizards, lynx, mountain lion, rabbit, rattlesnake, spiders, wood rats, and wren); Mediterranean (deer, dormouse, hare, hedgehog, lynx, mongoose, stork, vulture, weasel, wild boar, and wild sheep); coastal Australia (brush turkey, kangaroo, pademelon, scrub-birds, thornbill, and wallaby). Because of climate, areas are being utilized by humans for settlement. Vegetation is adapted to fires. Fires are common due to high oil content in brush. Erosion is common after fires.

Temperate woodlands. Drier climate than deciduous forests. Dominated by small trees—pinion pine and evergreen oaks. Stands of trees are open allowing light to reach ground. Fires are common. Range from New England to Georgia and the Caribbean islands.

Tropical rain forests. High and constant temperature. No frost. High rainfall (2000–2250 mm/yr or 75–100 in./yr). Found in Hawaii, Indonesia, northeastern Australia, parts of Malaysia, South and Central America, and the Philippines—in areas near the equator. Under Hadley cells. High species diversity for both plants and animals (up to 100 different tree species per square kilometer as opposed to three to five in temperate zones). Vegetation is dense. Lianas, bromeliads, and epiphytic orchids, ferns and palms are present. Leaves are large to try to absorb sunlight and there is no need to conserve water through losses due to transpiration. Soils are low in nutrients, and nutrients are stored in vegetation. Soil is acidic. Decomposition of organic material is very high due to temperature and moisture. Leaching is high. Abundant insects and animal biodiversity (up to 50% of all animal species). Animals include South America (anaconda, bats, butterflies, capybara, flycatchers, frogs, hummingbirds, iguanas, jaguar, lizards, macaw, marmosets, monkeys, ocelot, opossum peccaries, piranha, sloths, and tapirs); Australia (bandicoot, butterflies, duck-billed platypus, echidna, frogs, kookaburra, opossums, rats, and sugar gliders); Southeast Asia (butterflies, monkeys, orangutans, tarsiers, and tigers); West Africa (chimpanzee, gorilla, hippopotamus, monkeys, and parrot). Man is clearing tropical rain forests for agriculture and cattle raising, but because of poor nutrient levels in soil, these activities last only for a short time.

Tropical seasonal forests. Tropical seasonal forests occur in areas of seasonal, high rainfall (monsoon) followed by long, dry season. Warm temperatures all year. Africa, Central and South America, India, and Southeast Asia. Contain mixture of deciduous and drought-tolerant evergreen trees.

Tundra. 60° north latitude and above. Influenced by the Polar cell. Alpine tundra is located in mountainous areas, above the tree line with well-drained soil and dominant animals are small rodents and insects. Arctic tundra is frozen treeless plains, low rainfall, low average temperatures (summers average <10°C or 50°F) and contain many bogs and ponds. Frozen ground prevents drainage. Growing season lasts 50–60 days. Tundra is found in Alaska, Canada, Europe, Greenland, and Russia. Dominant vegetation includes flowering dwarf shrubs, grasses, lichens, mosses, and sedges. Soil has few nutrients because of low vegetation and little decomposition. There are between 60 and 100 frost-free days per year. Arctic tundra is higher latitude than alpine tundra. Herbivores include Arctic hare, caribou, lemmings, musk ox, and voles. Large mammals. Carnivores include Arctic fox, polar bear, snow owl, and wolves. Birds tend to migrate south during winter. Birds include Arctic terns, falcons, gulls, ptarmigans, ravens, sandpipers, snow geese, and swans. Fish include cod, flatfish, salmon, and trout. Insects include bees, flies, grasshoppers, mosquitoes, and moths. Reptiles and amphibians are not present due to low temperatures. Permafrost—permanently frozen ground and a barrier for roots. Little precipitation, but the ground is often waterlogged due to low rates of evaporation from the surface.

Upwellings. Deep, cold ocean water rich in nutrients rise to the surface and provide food for higher-order species. Occur along coasts of Arctic and Antarctic ice sheets, North and South America, and West Africa.

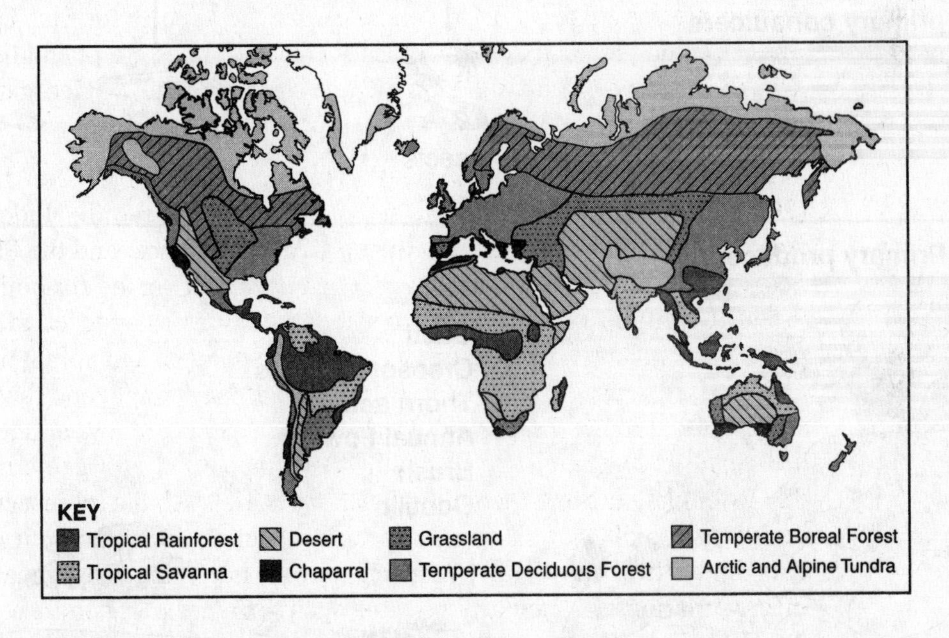

Figure 5.1. Major biomes of the world. (Adapted from H. J. de Blij and P. O. Miller. 1996. *Physical Geography of the Global Environment,* p. 290, John Wiley, New York.

Food Webs

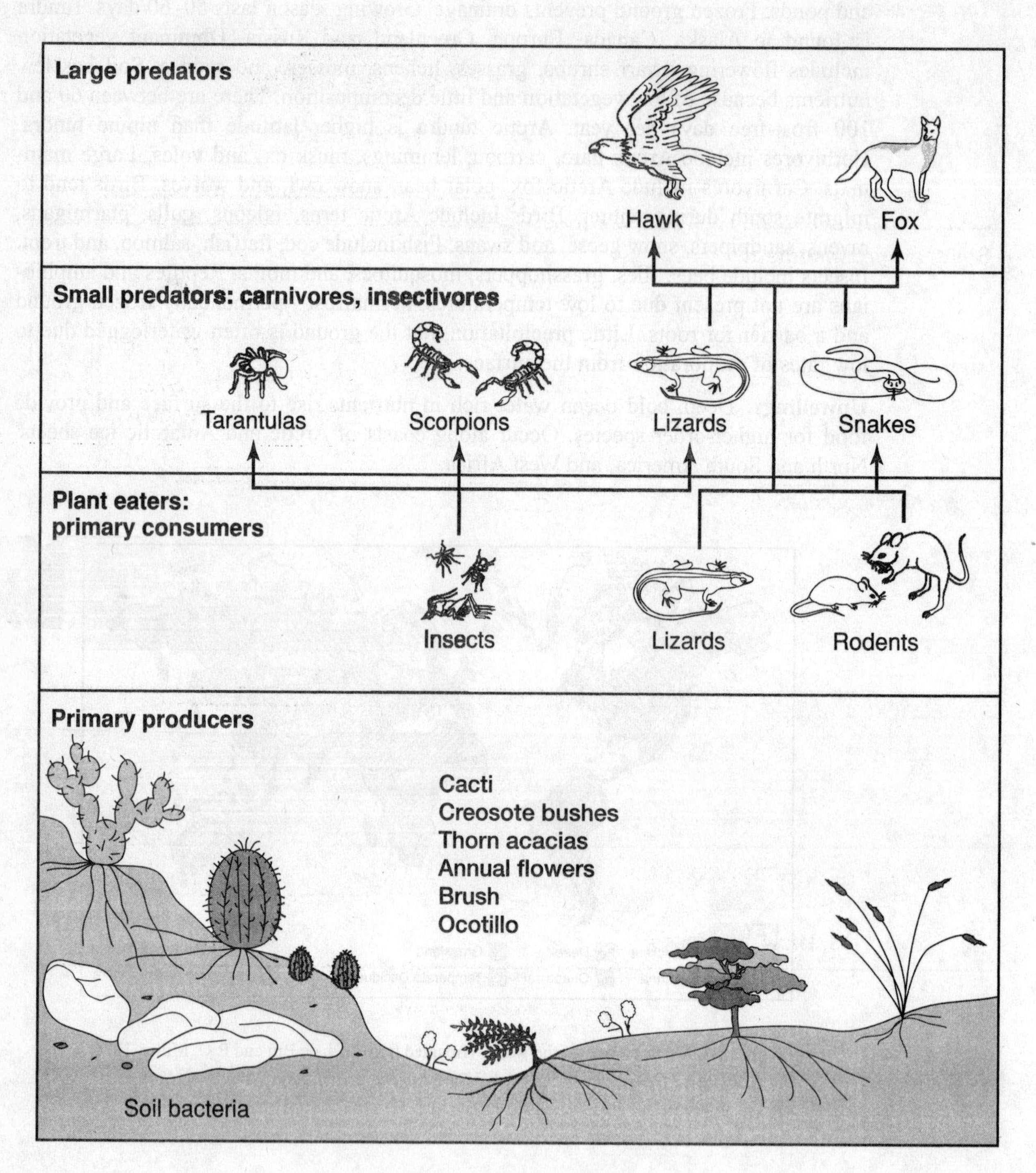

Figure 5.2. Desert food web.

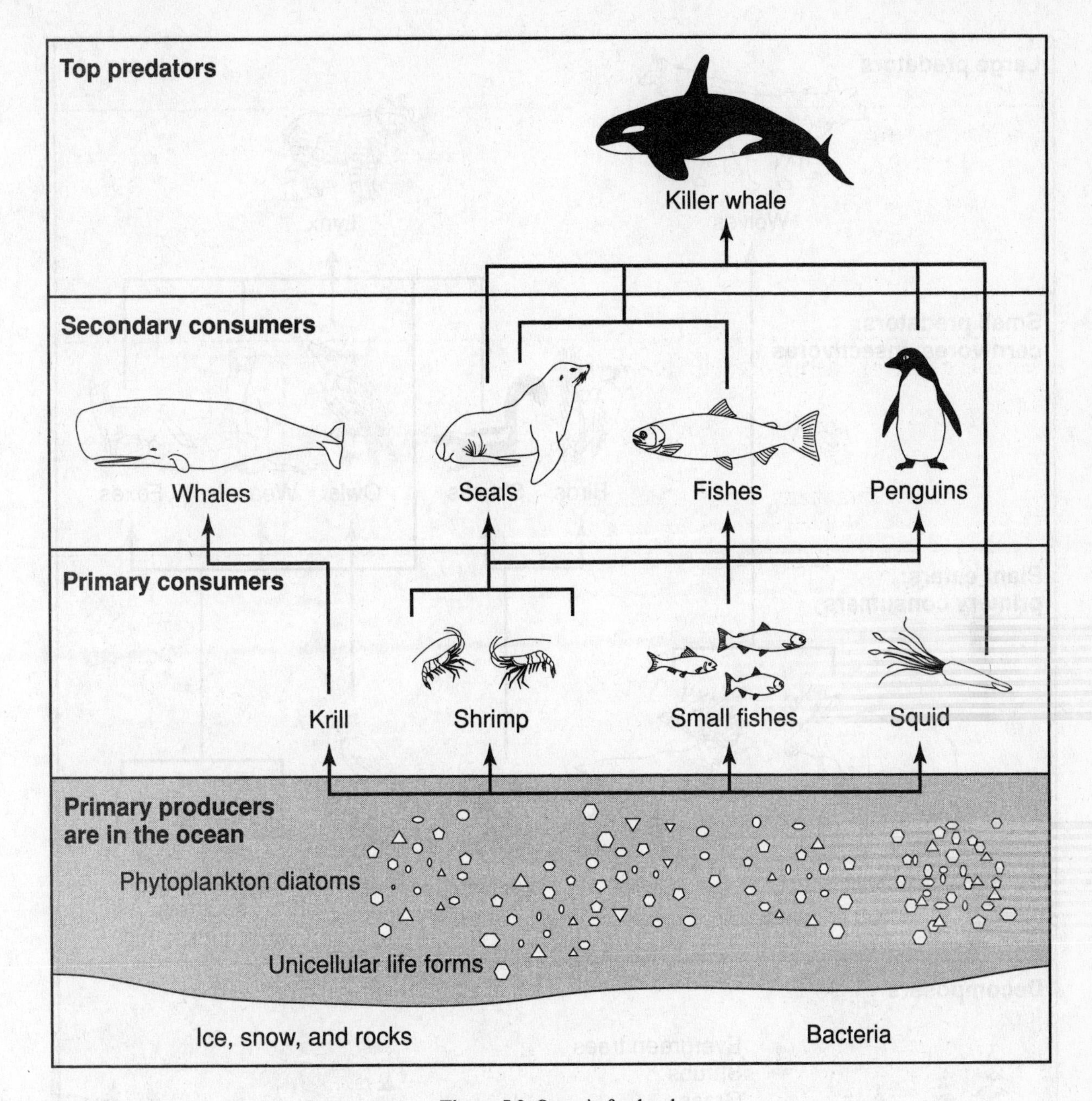

Figure 5.3. Oceanic food web.

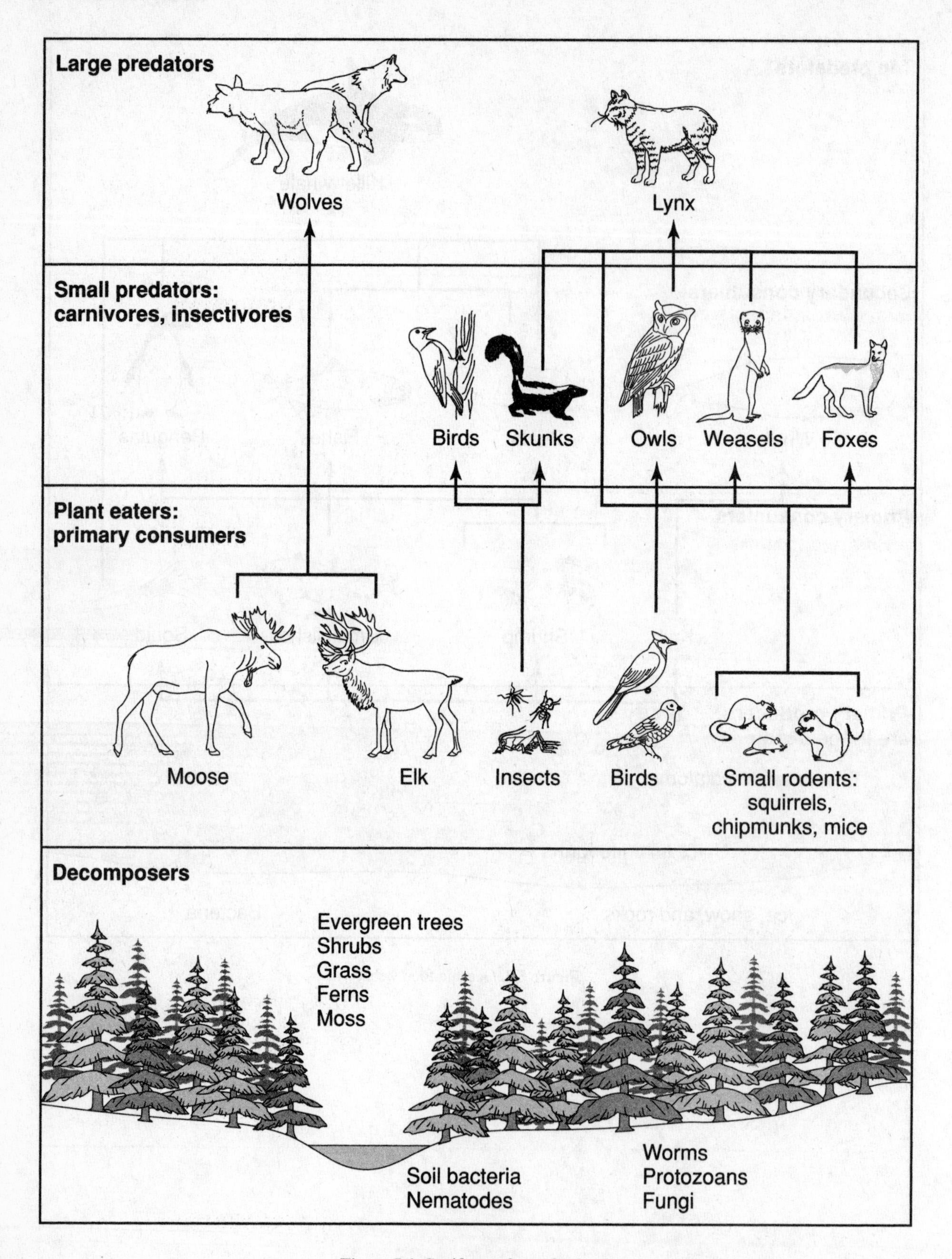

Figure 5.4. Coniferous forest food web.

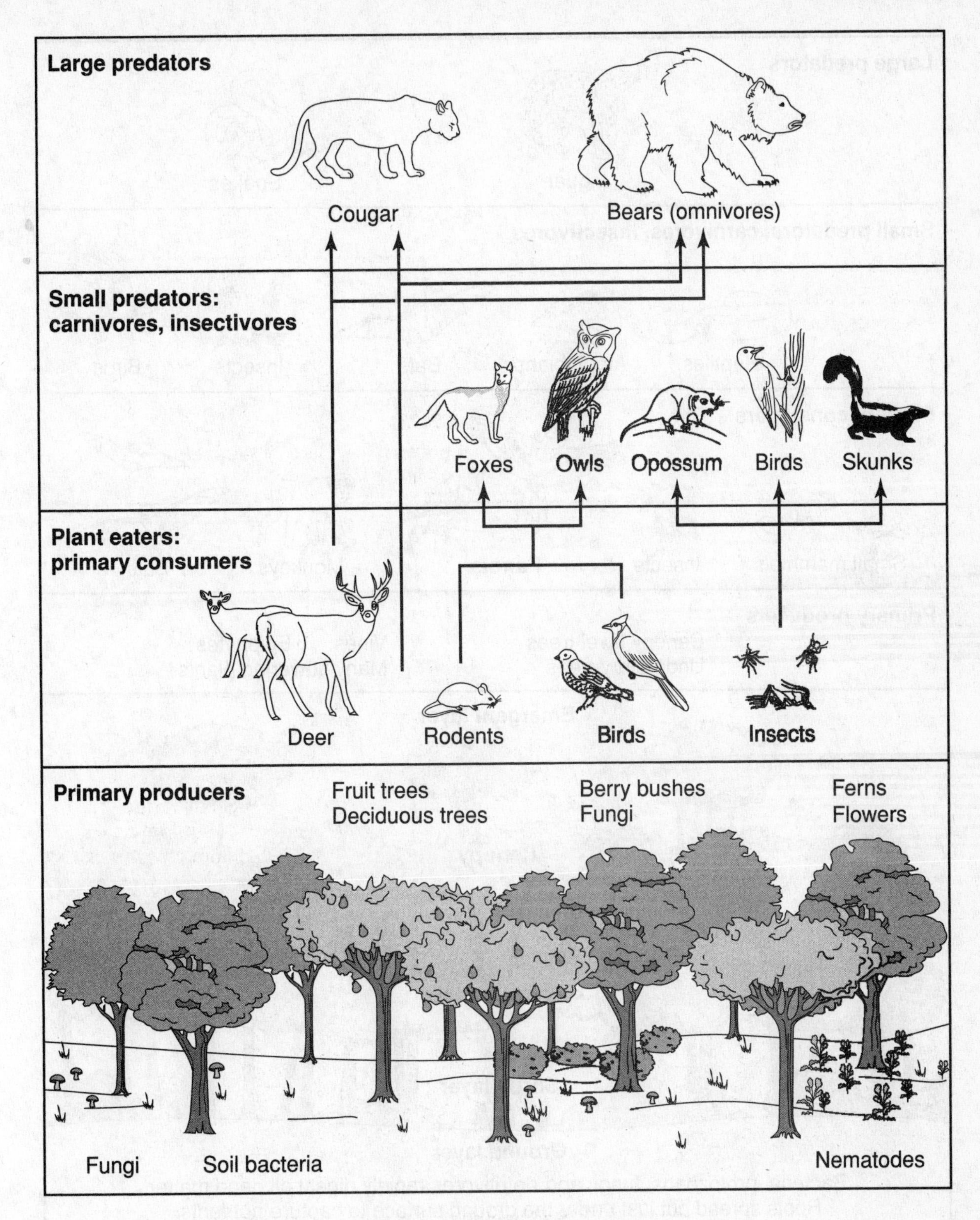

Figure 5.5. Deciduous forest food web.

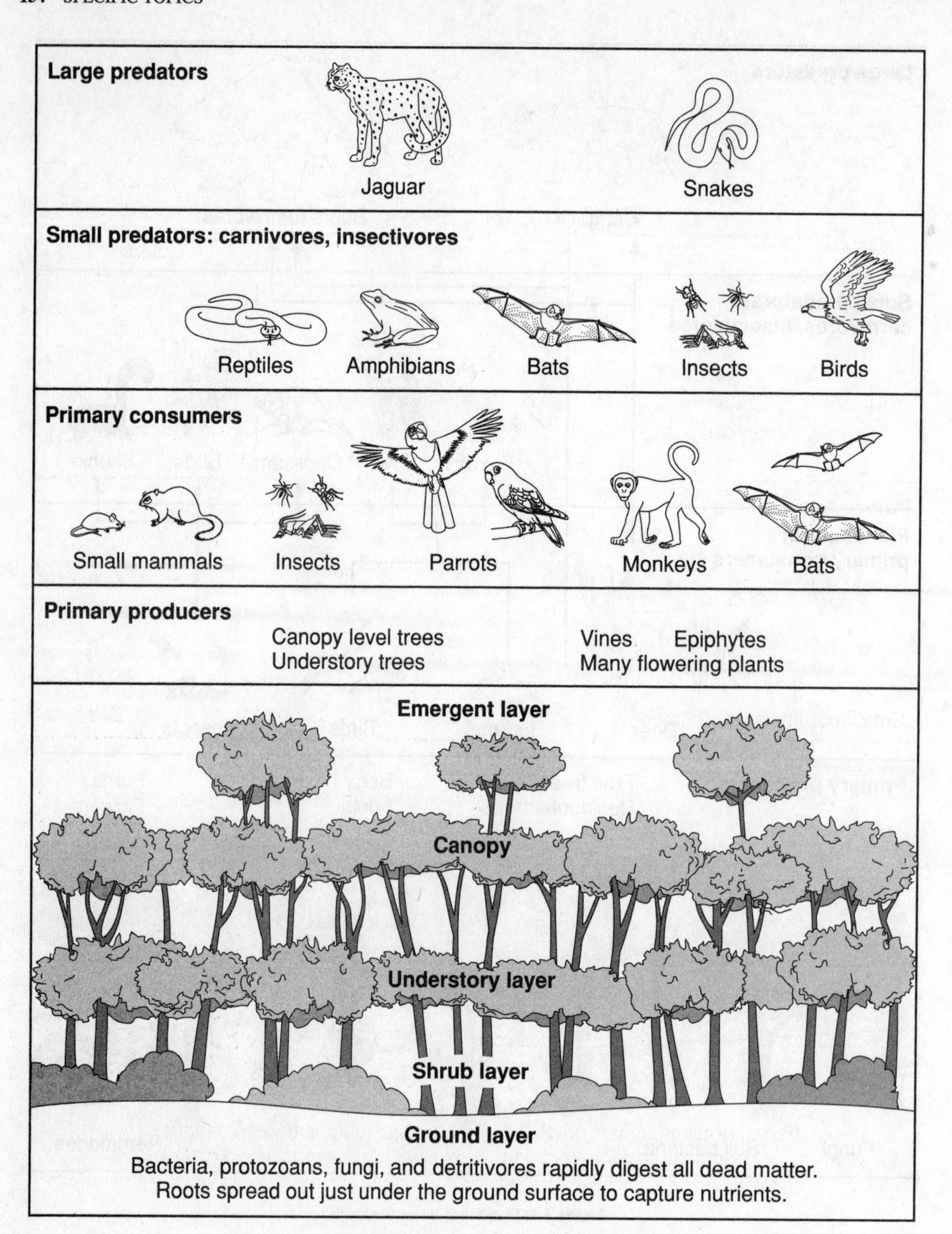

Figure 5.6. Tropical rain forest food web.

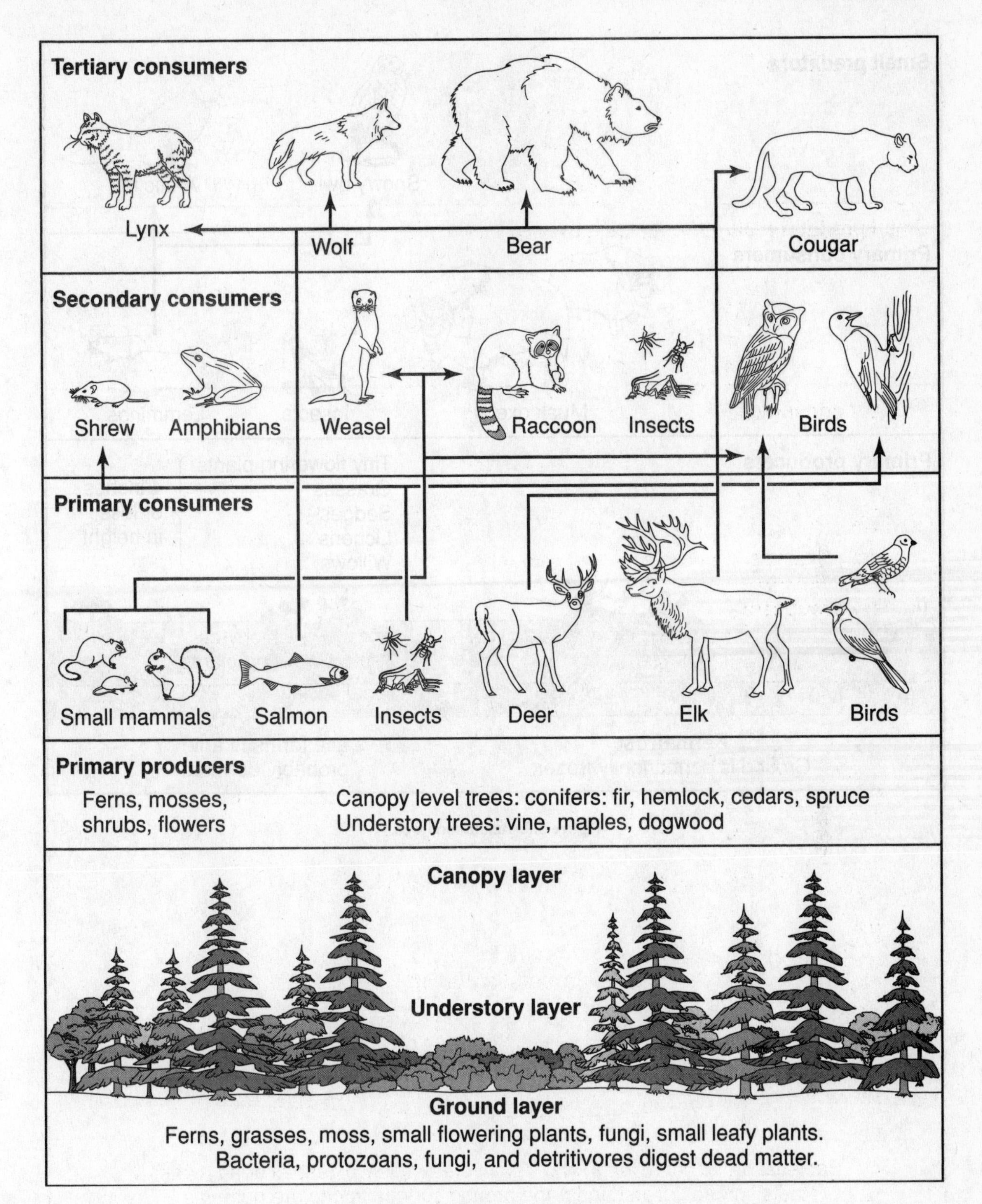

Figure 5.7. Temperate rain forest food web.

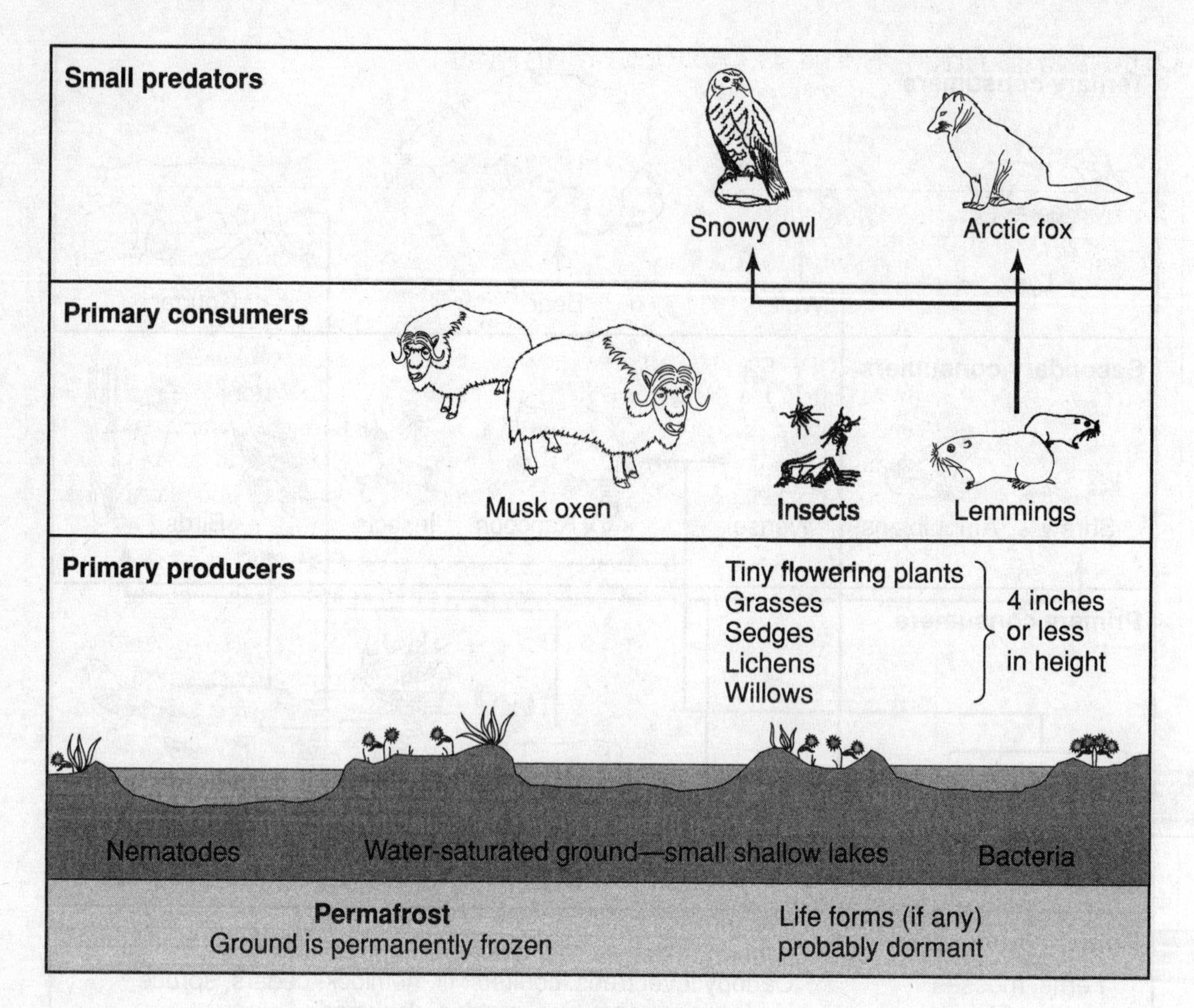

Figure 5.8. Tundra food web.

Organisms: Adaptations to Their Environments

SPECIFIC ADAPTATIONS

Environment	Adaptations
Aquatic	Water provides buoyancy—reduces need for support structures (legs, trunks), allows transportation with little or no expenditure of energy for reproduction and/or food gathering (e.g., floating organisms such as jellyfish). Water has high thermal capacity so most organisms do not spend energy on temperature regulation. Many organisms can obtain nutrients directly from water thereby reducing energy spent on searching for food (filter feeders—barnacles, clams, oysters, sponges, baleen whales). Water allows dispersal of gametes and larvae to new areas. Methods of conserving water as found in land animals and plants are not necessary. Water screens out UV radiation. Intertidal zone: organisms have evolved methods of not being swept away by waves (barnacles, mussels, digging under sand by clams); preventing water loss during low tide through shells.
Desert	Succulents—no leaves, store water, small surface area exposed to sunlight, vertical orientation to minimize exposure to sun, stomata open at night. Waxy leaves to minimize transpiration. Deep roots to tap groundwater (mesquite and creosote). Shallow roots to collect water after short rainfall (prickly pear, saguaro cactus). Sharp spines—reflects sunlight, discourage herbivores, provide insulation, and provide shade. Secrete toxins into soil to prevent interspecific competition. Store biomass in seeds. Some have short life spans and are dependent on water for germination (wildflowers). Others remain dormant between rainfalls (mosses and lichens). Animals are small and have small surface area, spend time underground in burrows, nocturnal. Dry feces and concentrated urine. Aestivation. Able to metabolize dry seeds. Kangaroo rat produces its own water. Insects and reptiles have thick outer coverings to minimize water loss.
Grasslands, prairies, steppes	Grasses grow out from the bottom so they can grow again after being nibbled on by grazing animals. Grass species are drought-resistant. Deciduous trees and shrubs in grasslands shed leaves during dry season to conserve water. Grazers and browsing animals eat vegetation at different heights so as to not compete (giraffes, tops; elephants, lower down; zebras, longer grasses and stems; gazelles, short grass). Some animals migrate to find water (wildebeests), some animals become dormant, and others survive on seeds during dry season. Some animals live in burrows to hide and escape predators in the open and their fur color matches the color of the ground.
Forests	*Tropic*: Some animals live in canopy where shelter and available food supplies (leaves, flowers, fruits) are abundant to escape predators. Epiphytes (orchids and bromeliads) live on trunks and branches of trees and catch organic matter falling from canopy. Some plants have very large leaves (philodendrons) to capture scarce light. Roots of trees are shallow and spread out to capture nutrients in poor soil. To compensate for little support by shallow roots, trees have buttresses. Flowers have elaborate devices to attract pollinators since wind is minimal in dense growth.

SPECIFIC ADAPTATIONS (Continued)

Environment	Adaptations
Forests (continued)	*Temperate Deciduous*: Broadleaf deciduous trees (oak, hickory, maple) drop leaves in winter and become dormant to conserve energy. *Evergreen Coniferous*: Small, waxy-coated needles are able to withstand cold and drought of winter. Low surface area. Decomposed needles make soil acidic, preventing many competing species to survive in soil environment. Some animals hibernate (bears) to conserve energy during winter when food supplies are scarce.
Temperate shrub lands	Chaparral plants have small, waxy-coated leaves to reduce transpiration. Produce toxins that leach into soil to prevent interspecific competition. Vegetation becomes dormant during dry season. Periodic fires reduce interspecific competition and allow seeds to germinate. Rodents are common and store seeds in underground burrows to conserve energy during droughts.
Tropics	Broadleaf evergreens found in tropics where moisture is not a problem. Large surface area to collect scarce light under forest canopy and radiate heat.
Tundra (polar grasslands)	Tundra (polar grasslands) plants are adapted to low sunlight, low amount of free water, high winds, low temperatures. Tundra plants primarily grow in summer. Leaves on plants have waxy outer coating. Other tundra plants survive winter as roots, stems, bulbs, and tubers. Lichens dehydrate during winter to avoid frost damage. Animals have thick coats of fur (arctic wolf, arctic fox, musk oxen), feathers to provide insulation (snowy owl), or compact bodies to prevent exposure, or they live underground (lemming).

BIOTIC RELATIONSHIPS

Interaction	Description
Amensalism (0 –)	The interaction between two species whereby one species suffers and the other species is not affected. *Alleopathy*—occurs when one species releases a chemical substance to inhibit the growth of another species. Example—sagebrush, mint, black walnut.
Commensalism (+ 0)	One species benefits from the relationship but the other is neither harmed nor benefitted. Example: certain mites carry out most of their life activities within the flower of a particular plant species. Hummingbirds that pollinate these flowers transport the mites from old to new flowers.
Competition (– –)	1. *Interspecific*—organisms from different species compete for the same resource. Example: in certain areas of the Southwest, the population of Gila woodpeckers is decreasing, and the population of starlings is increasing. Both birds nest in the cavities of giant cacti. 2. *Intraspecific*—organisms from the same species compete for the same resource. Example: see the essay on Africanized honeybees and European honeybees.

BIOTIC RELATIONSHIPS (Continued)

Interaction	Description
	3. *Exploitation*—competition that occurs when indirect effects reduce resource(s). Example: One species obtains the resource first such as when there's a big fruiting tree with orangutans, macaques, and colobines all eating at once. There's no interaction or aggression between them, but they're still reducing the amount available to others. 4. *Interference*—competition that occurs when an organism prevents physical establishment of another organism in the habitat. Example: hummingbirds exclude other hummingbirds (as well as bees and moths) from flowering plants; encrusting sponges use poisonous chemicals to overcome other species of sponges; shrubs release toxic chemicals that depress the growth of competitors; bacteria and fungi also release poisonous chemicals in competitive interactions with other microbes.
Mutualism (+ +)	The interaction between two species whereby both species mutually benefit. Example: mycorrhizae fungus helps plant to absorb soil nutrients—roots of plants supply carbohydrates to mycorrhizae. *Symbiotic*—physical interaction but either species cannot live without the other. Example: fungal–algal symbiosis in lichens. *Nonsymbiotic*—no physical interaction but either species cannot live without the other. Example: insect pollinators and flowers.
Neutralism (0 0)	An interspecific interaction whereby neither population really affects the other. Example: Grizzly bear and butterfly.
Parasitism (+ –)	Species that obtain food at the expense of their host. Example: certain young clams attach themselves to the gills of a fish. In a short time, each clam becomes surrounded by a capsule formed by cells of the fish. The clam feeds and grows by absorbing nutrients from the fish's body.
Predation (+ –)	Species that obtain food at the expense of their prey. Many feedback mechanisms have evolved to keep numbers stable. Example: lions prey upon zebras for food.
Saprophytism (+ 0)	Species, especially fungus or bacteria, that grow on and derive their nourishment from dead or decaying organic matter. Include microbes, some insects, grubs, snails, slugs, beetles, and ants. They aid in recycling valuable nutrients from dead organic matter that is then released back into the soil to be reabsorbed rapidly by plants and trees. Decayed matter contains essential nutrients like iron, calcium, potassium, and phosphorous, all of which are necessary to promote healthy plant growth.

LIMITS TO ADAPTATION

- To survive in changing environments, adaptive traits *must* be present in gene pool.
- Ability to survive changing environments is directly related to a species' reproductive capacity. Example: bacteria reproduce rapidly. A change in their environment or the introduction of a new antibiotic may reduce the population substantially. However, those that survive and possess a phenotypic adaptive trait that allowed them to survive, can reproduce quickly to replace the original population. This would not be true in the case of elephants. Elephants produce few offspring, the range of possible adaptive traits is limited in their gene pool due to smaller populations, and elephants that do survive a changing environment because they have a phenotypic adaptive trait probably cannot reproduce fast enough to prevent a population crash and eventual extinction.

Populations and Communities: Exponential Growth and Carrying Capacity

- A population is a group of organisms of the same species occupying the same general area.
- Members of a population can be dispersed in an area three ways:

1. Clumped (attraction)—some areas within the habitat are dense with organisms, while other areas contain few members.
2. Uniform (repulsion)—there is fairly uniform spacing between individuals.
3. Random (minimum interaction)—there is little interaction between members of the population leading to random spacing.

- Age structure and sex ratio of a population are two important determinants that can influence population size. Three factors determine the age structure of a population:
1. fecundity (birth rate)—The greater the fecundity, the faster the population increases in size.
2. Generation time (average span between birth of an individual and their offspring)—Shorter generation times result in faster population growth.
3. Death rate (rate at which individuals of a certain age die within the population)—Generally death rates are highest in the very young and very old. Population growth occurs when fecundity exceeds death rates.

- In terms of sex ratios, the fewer females, the slower the population growth becomes. This is limited when males are involved in raising offspring or males are monogamous.

Survivorship Curves

Survivorship curves show age distribution characteristics of species, reproductive strategies, and life history. Reproductive success is measured by how many organisms are able to reproduce, not by how many organisms are produced. Life history of species is a result of natural selection. Each survivorship curve represents a balance between natural resource limitations and inter- and intra-specific competition. For example, humans could not survive in a Type III mode where human females would produce thousands of offspring with only a few surviving predation. Likewise, ants could not survive in a Type I mode, where each queen ant produced only a few eggs during her lifetime and she spent her time and energy in raising the offspring. The three types of survivorship curves are shown in Figure 5.9.

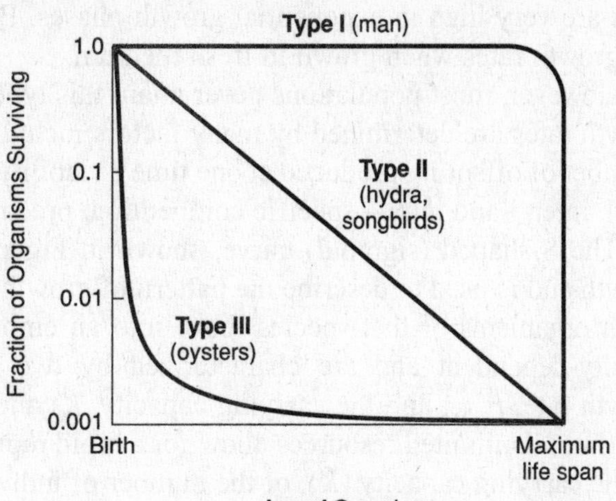

Figure 5.9. Survivorship curve.

Survivorship Curve Type	Description
I	Reproduction occurs fairly early in life. Most deaths occur at the limit of biological lifespan. Low mortality at birth; high probability of surviving to "old age". Frequency of death is high at "old age." Advances in prenatal care, nutrition, disease prevention, and cure have meant that the lifespan for humans has increased. Examples include humans, rotifers, annual plants, and sheep.
II	Individuals in all age categories have fairly uniform death rates. Predation affecting all age categories is primary means of death. Typical of organisms that reach adult stages very quickly—insects, birds, mud turtles, rodents, and perennial plants.
III	Typical of species that have great numbers of offspring and reproduce for most of their lifetime. Death is prevalent for younger members of the species (environmental loss and predation) and declines with age. Examples include sea turtles, trees, internal parasites, fish, invertebrates, perennial plants, oysters, and other marine invertebrates.

Population Size and Carrying Capacity

The intrinsic rate of population increase occurs when resources are unlimited and is characteristic of the species, not the environmental conditions. If a population in a community is left unchecked, the maximum population growth rate, r_{max},can increase exponentially (1, 2, 4, 8, 16, 32, 64, 128, 256, . . .) or (1, 3, 9, 27, 81, 243, . . .); and shows a characteristic J-shaped curve or plots as a straight line on semilogarithmic graph paper. Exponential growth occurs initially when populations are introduced into new or unfilled niches or when their numbers have been seriously reduced due to a catastrophic event and the population is rebounding. The difference between birth rates and death rates are very high in exponential growth phases. Bacteria often show initial exponential growth rates when grown in fresh medium.

However, most populations never attain this type of increase. Maximum population growth rates are determined by many factors including age of reproduction, clutch size (number of offspring produced at one time), viability of offspring (how many young survive), inter—and intra—specific competition, predation, and limitations of resources.

The S-shaped (sigmoid) curve, shown in Figure 5.10, is characteristic of logistic growth and is used to describe the pattern of growth over extended time of a population when organisms of that species move into an empty niche. Logistic growth rates are density-dependent and are characterized by two parameters (maximum population growth rate, r_{max}, and the carrying capacity, K) and two variables (population size, N, and time). Unlimited resources allow for a rapid reproductive phase. Eventually, however, the carrying capacity (K), or the number of individuals that can be supported by the available resources, is reached. In effect, each organism has less share of resources; consequently, the larger the population (N) becomes, the slower the growth rate becomes. The population size may then stabilize or fluctuate around the carrying capacity (Figure 5.11). At this point, individuals experience either higher death rates and/or lower fecundity. When the birth rate equals the death rate, a state of equilibrium is reached.

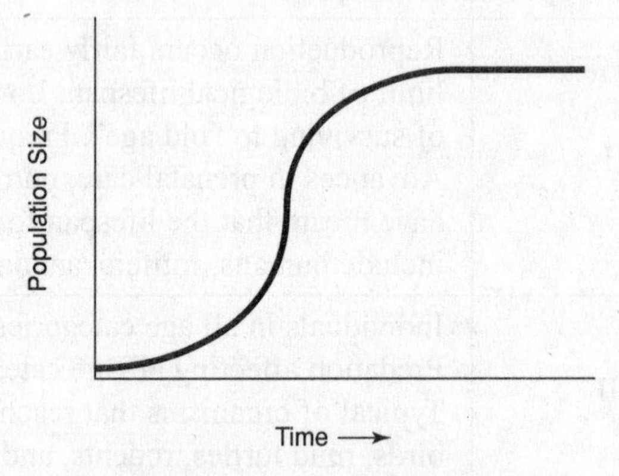

Figure 5.10. S-shaped curve.

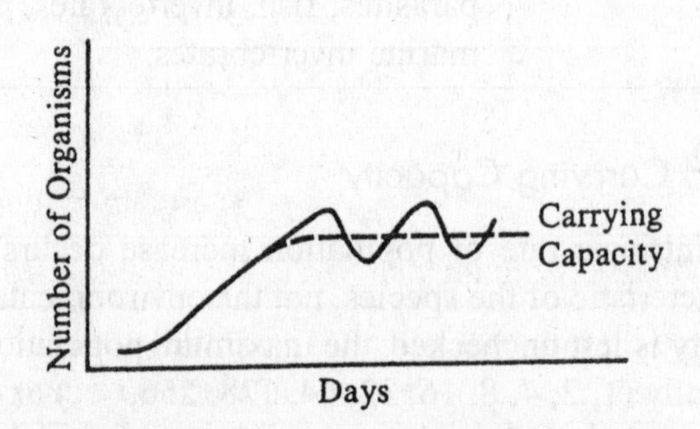

Figure 5.11. Fluctuations around the carrying capacity.

Stress due to overcrowding produces psychological and behavioral changes in a population. Difficulty in mating, high infant mortality, and inability to nurture the young may reduce the population growth rate. Parasitism may also increase as resistance is weakened by stress. When the death rate exceeds the birth rate, the density (size) of the population is reduced.

Predation not only removes the very old, the very young, and the sick from the population but also reduces the population of the prey. If the predators do not keep the prey population in balance, the carrying capacity is exceeded and the prey may starve (Figure 5.12). Predator and prey are closely interdependent.

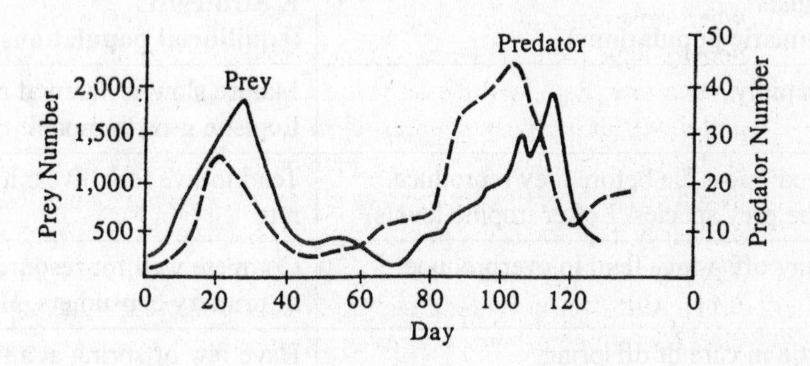

Figure 5.12. Predator-prey relationship.

Populations of algae, annual plants, and insects with short life spans are controlled by seasonal and nutritional environmental changes. A J-shaped growth curve may show a plunge in population as the reproductive potential declines because of environmental changes. The following year there may be an exponential increase in the population.

FACTORS THAT AFFECT POPULATION SIZE

Increase (+)	**Decrease (–)**
Favorable environmental conditions: light, temperature, and nutrients.	Unfavorable environmental conditions: insufficient light, temperature extremes and wide fluctuations, poor supply of nutrients
High natality	Low natality
Generalized niche	Specialized niche
Sufficient food supply	Deficient food supply.
Habitat is satisfactory	Habitat not satisfactory or has been seriously impacted
Few competitors	Too many competitors
Suitable predatory defense mechanisms	Unsuitable predatory defense mechanisms
Adequate resistance to disease and parasites	Little or no suitable defense mechanisms against disease or parasites
Abile to migrate	Unable to migrate
Flexible—able to adapt	Inflexible—unable to adapt

Differences Between r Strategists and K Strategists

In line with concepts of natural selection, species have adapted either to maximize their growth rates in environments that lack limits (r_{max}) (e.g., weeds) or to maintain their population size (*N*) at close to the environmental carrying capacity in stable environments (*K*) (e.g., lions).

r Strategists (opportunistic populations)	**K Strategists (equilibrial populations)**
Mature rapidly.	Mature slowly. Parental care may be involved. Logistic growth model.
Short-lived: most die before they reproduce. Tend to be prey species. Lower trophic levels.	Tend to live long lives: low juvenile mortality rate.
Have many offspring. Tend to overproduce.	Compete well for resources. Tend to be predators or primary consumers. Higher trophic levels.
Invest little in care of offspring.	Have few offspring at a time.
Most pests are r strategists.	Most endangered species are K strategists.
Wide fluctuations in population density: boom and bust population figures.	Population stabilizes near carrying capacity.
Opportunistic—invade new areas. Adapted to unstable environments. Colonizers, pioneers. Niche generalists.	Maintain numbers in stable ecosystems. Niche specialists.
Population size limited by density-independent limiting factors. Density independent factors include climate, weather, natural disasters, and requirements for growth.	Density-dependent limiting factors to population growth stem from intraspecific competition. Density-dependent factors include competition, migration, predation, and parasitism.
Organisms are generally small.	Organisms are generally larger.
Energy requirements to produce an individual are small.	Energy requirements to produce an individual are large.
Type III survivorship curve.	Type I or II survivorship curve.

Ecosystems and Change: Biomass, Energy Transfer, and Succession

Biomass

NET PRIMARY PRODUCTIVITY OF BIOMES

Biome	Relative Mean Dry Matter Productivity (Highest = 100)	Relative Mean Standing Dry Biomass (Highest = 100)
Swamps and Marshes	100	33
Tropical Rain Forests	73	100
Tropical Dry Forests	53	78
Temperate Evergreen Forests	43	78
Temperate Deciduous Forests	40	67
Savanna	30	9
Boreal Forests	27	44
Woodland-Shrubland	23	13
Cultivated Land	22	2
Grassland	20	4
Lakes and Streams	13	0.04
Alpine Tundra	5	1
Desert Scrubland	3	2
Extreme Deserts	0.1	0.04

- Swamps and marshes are the most productive biomes, producing the most biomass per year while tropical rain forests have the most standing biomass. Extreme deserts are the least productive biomes. Lakes and streams tie with extreme deserts for having the least standing biomass.
- Photosynthesis depends upon (1) sunlight; (2) CO_2 (in general, photosynthesis becomes more efficient in plants at higher concentrations of carbon dioxide); (3) water availability (in dry regions, there is a linear increase in net primary productivity with increased water availability); (4) nutrient availability (nitrogen and phosphorus limit the amount of plant productivity that can occur); and (5) temperature (ranges between 15–25°C [60–75°F] are necessary for optimum rates of plant metabolism). For photosynthesis to be an indicator of biomass productivity, *all* five variables must be within optimal range. If not, they become limiting factors for plant biomass. Tropical rain forests, swamps, and marshes are biomes that tend to have all five variables within optimal ranges, thereby being some of the most productive biomes. For example, in extreme deserts, sunlight may be optimal; however, water availability, temperature extremes, and nutrient availability are factors that reduce rates of photosynthesis thereby reducing net productivity as measured through biomass.

- Human activity influences biomass productivity. Approximately 15% of current land surfaces are either used for agriculture or human settlement, and 8% is dedicated for grazing. This land conversion has affected biome productivity compared to what the land was before human modification. Biogeochemical cycles that have been affected by man also disturb rates of photosynthesis, which consequently affect biomass. Increases in CO_2 through burning of fossil fuels increase the Earth's temperatures through greenhouse effect and thereby favor increases in photosynthesis; increases in CO_2 also favor higher rates of photosynthesis since CO_2 is a reactant in the chemical process (Le Chatelier's principle). Humans have increased soil nitrogen through (1) using fertilizers; (2) planting legumes; and (3) burning fossil fuels thereby releasing NO_x into the environment which cycles back to the ground and eventually becomes available to the plants. Plant productivity decreases if too much nitrogen is added (nitrogen saturation).

Energy Transfer

- Sun's energy is the driving force for energy required for all biologic processes. All energy used in biological systems can be traced back to sunlight used in photosynthesis to convert inorganic compounds into organic compounds.
- Less than 3% of all sunlight that reaches the Earth is used (or fixed) for photosynthesis. This energy is released in plants (autotrophs) and animals (heterotrophs—herbivores, carnivores, and detritivores) through cellular respiration. Exceptions are the chemosynthetic bacteria that are able to derive their energy needs ultimately from the nuclear decay of radioactive elements within the Earth's core, which provides energy to the mantle to create hydrothermal vents where these chemotrophs convert dissolved hydrogen sulfide and carbon dioxide into organic nutrient molecules. The energy that is trapped in these molecules is then absorbed by organisms that feed upon chemotrophs—tubeworms, some species of clams, crabs, mussels, and barnacles.
- Natural energy inputs into an ecosystem include solar radiation, input of seeds, animal migration into the ecosystem, and erosional processes. Natural energy outputs from an ecosystem include reflection of infrared into space, conduction of infrared energy into the environment (heat), seed dispersal out from the ecosystem, migration of animals away from the ecosystem, and leaching of soil nutrients into the ground. See Figure 5.13.

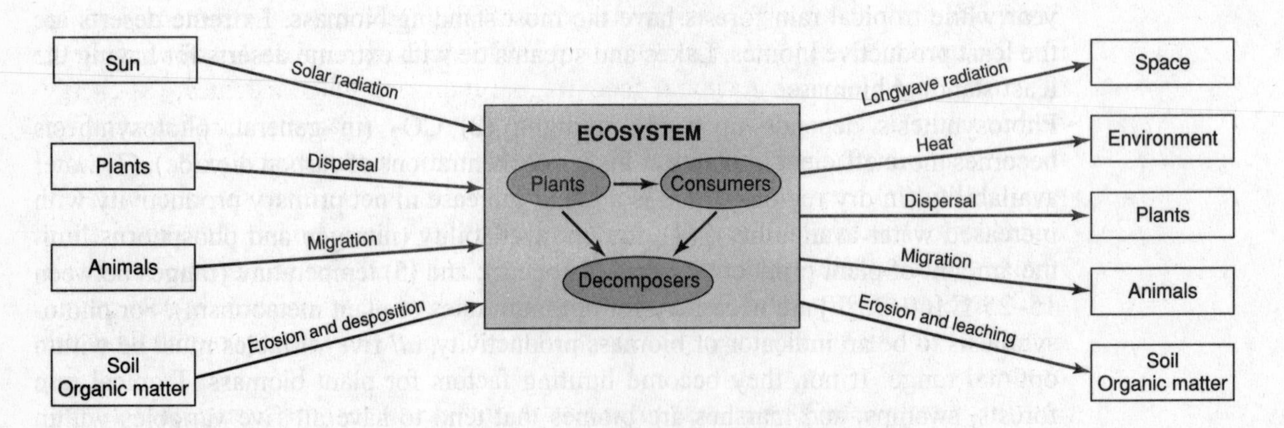

Figure 5.13. Inputs and outputs of energy into an ecosystem.

- Autotrophs (primary producers) are plants. They convert solar energy into stored chemical energy. Many plants have evolved mechanisms to help protect them from being consumed (e.g., thorns, poisons and organic compounds that make them bitter).
- Primary consumers (herbivores) are plant eaters. They get their energy by consuming primary producers. Losses of energy from plant material include digestive inefficiency and cellular respiration requirements—much of the plant tissue is unavailable for energy conversion. For example, elephants only digest about 40% of what they eat. They need tremendous amounts of vegetation (approximately 5% of their body weight, or up to 700 pounds of grass per day) and about 30 to 50 gallons of water. They also produce great quantities of waste products. The energy present in these waste products can be used by humans through burning of the biomass. Assimilation efficiencies for primary consumers ranges from 20 to 60%.
- Secondary consumers (primary carnivores) are meat eaters. They get their energy by consuming primary consumers. Primary consumers have developed defense mechanisms against predation including speed, flight, quills, tough hides, camouflage, horns, and antlers. Because energy is more concentrated in meat than in plant, the assimilation of energy by secondary consumers ranges from 50 to 90%. However, only a small amount ends up as secondary consumer biomass, because most of it is spent for growth, motion, heat, reproductive activities, and other metabolic requirements.
- Tertiary consumers (secondary carnivores) are meat eaters. They get their energy by consuming secondary consumers. All the processes described for primary carnivores (secondary consumers) are true for tertiary consumers. The amount of biomass at this (and rarely higher) trophic level is much smaller because less energy is available to organisms at this level.

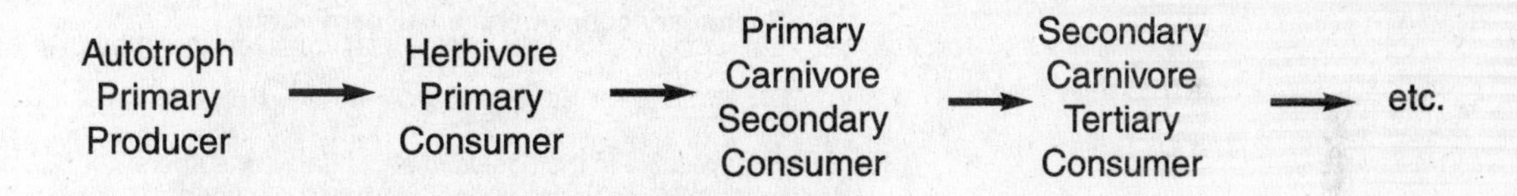

- Energy that is obtained by heterotrophs eventually is returned to the system (Law of Conservation of Energy) through (1) energy used for cellular respiration and waste heat; (2) energy expended for obtaining food, reproduction, and other activities; (3) energy released through decay of waste products and decay of the organism; and (4) transfer of the energy to an organism higher in the food chain through consumption.
- Amount of *net* energy that is transferred from one trophic level to the next is determined by inefficiencies (heat lost), energy expended to obtain food, cellular respiration required to maintain the organism, reproductive activities, and so on. The limit to a food chain and the number of trophic levels is determined when there is not enough energy left for higher trophic organisms to obtain the necessary energy required for life processes.
- Detritus food chains (organisms that consume organic wastes) are significantly different: (1) size of organisms are much smaller, (2) organisms exist in environments rich in nutrients so energy is not needed to obtain food, (3) organisms are generally not motile (cannot move on their own), (4) trophic levels are more complex and interrelated (included in detritus food chains are many species of algae, bacteria, fungi, protozoa,

slime molds, insects, arthropods, crustaceans, and worms), and (5) organisms are always active and release much of the energy trapped in organic molecules as heat.

- Rate of decomposition and numbers of decomposers is generally in equilibrium with amount of material available. For example, leaf litter amounts are fairly uniform within each biome. Rates are determined by temperature, oxygen, and available moisture. If oxygen is limited, decomposition may proceed through slower anaerobic mechanisms with release of energy (less when compared to aerobic mechanisms), carbon dioxide, water, organic sediments, and other organic molecules including simple alcohols, organic acids, and methane.
- The concept of linear food chains allows understanding of energy-relationships and trophic-level hierarchies. However, life is not so simple. The interrelationships among living organisms and codependences of all organisms to function together in balance creates the much more complex food webs.
- Figures 5.14 through 5.18 show biomass pyramids for various biomes. Notice that each higher trophic level contains less biomass (as measured by dry weight per unit area) and, consequently, less available energy at that level.

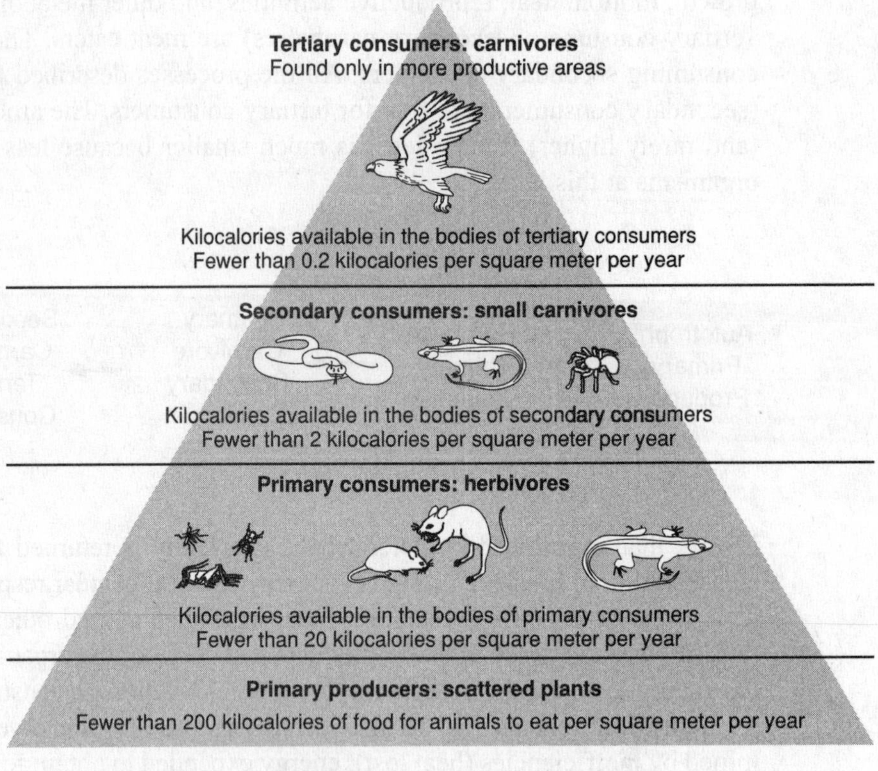

Figure 5.14. Desert biomass-energy pyramid.

Tertiary consumers: eat smaller predators

Kilocalories available in the bodies of tertiary consumers
8 kilocalories per square meter per year

Secondary consumers: eat smaller animals

Kilocalories available in the bodies of secondary consumers
80 kilocalories per square meter per year

Primary consumers: eat phytoplankton

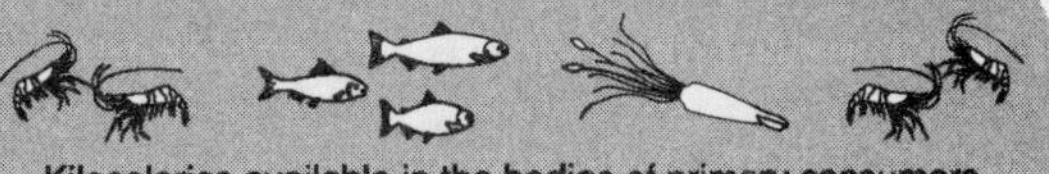

Kilocalories available in the bodies of primary consumers
800 kilocalories per square meter per year

Primary producers: phytoplankton
Wide seasonal variation in yield
Assumed spring productivity: rate of 8000 kilocalories
per square meter per year in food for animals to eat.

Figure 5.15. Marine biomass-energy pyramid.

Tertiary consumers: predators

Kilocalories available in the bodies of tertiary consumers
6 kilocalories per square meter per year

Secondary consumers: predators

Kilocalories available in the bodies of secondary consumers
60 kilocalories per square meter per year

Primary consumers: herbivores

Kilocalories available in the bodies of primary consumers
600 kilocalories per square meter per year

Primary producers: plants: deciduous trees, shrubs, grasses
6000 kilocalories per square meter per year

Figure 5.16. Deciduous forest biomass-energy pyramid.

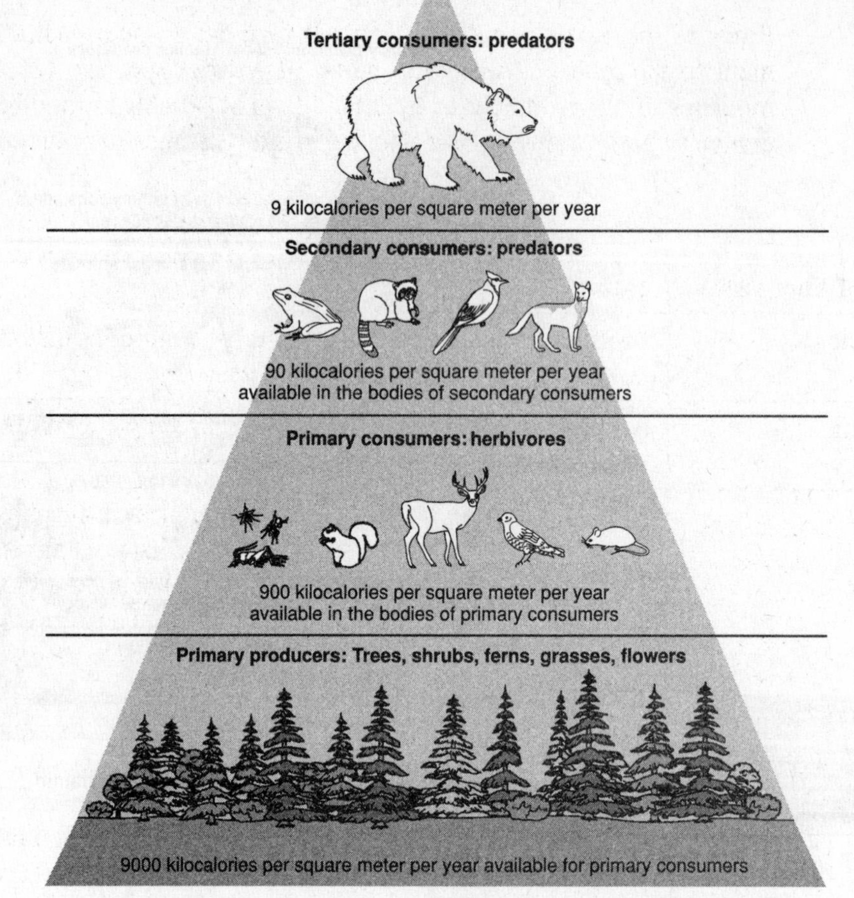

Figure 5.17. Temperate rain forest biomass-energy pyramid.

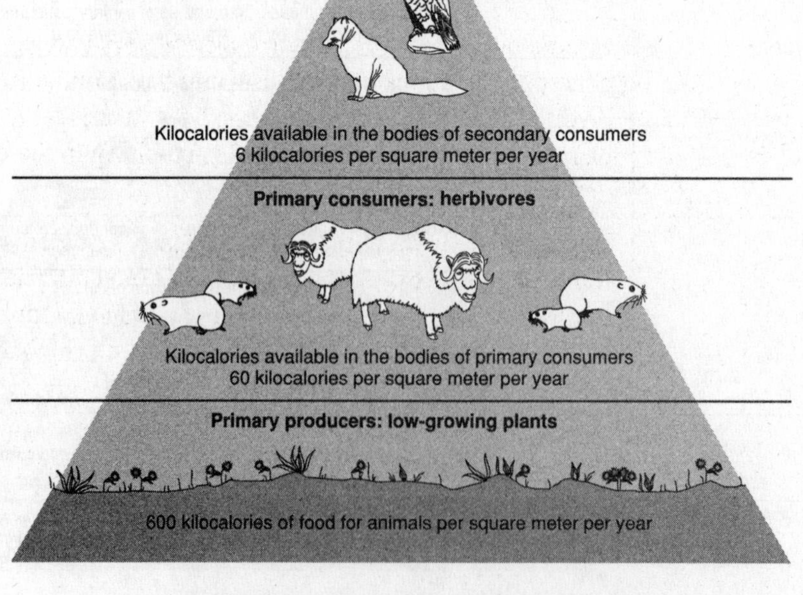

Figure 5.18. Tundra biomass-energy pyramid.

Succession

Rates of succession are affected by (1) facilitation (one species modifying an environment to the extent it meets the needs of another species), (2) inhibition (one species modifies the environment to an extent it is not suitable for another species), and (3) tolerance (when species are not affected by the presence of other species).

TYPES OF SUCCESSION

Types of Succession	Description
Allogenic	Caused by changes in environmental conditions that create conditions conducive to new plant communities.
Autogenic	Succession caused by changes both in the plant community and the environment.
Primary	The colonization and establishment of pioneer plant species on bare ground or water. Lichens are a symbiotic mutualistic relationship between algae that photosynthesize and produce nutrients and fungi that allows the lichen to attach to bare rock and retain water. Lichens provide a niche for organisms that secrete acids that can weather rock, which eventually becomes soil. Small organisms such as bacteria, protozoa, fungi, and worms begin to process soil into a suitable growth medium. Small annual plants eventually replace lichen. Shrubs begin to replace the annual plants, and finally a complex climax community of trees develops. Aquatic primary succession begins with input of sediment. Aquatic plants with primitive roots then become established. More sediment is accumulated, allowing a nutrient base for more complex aquatic plants. Over time, the sediment may build up and the area develops into a wet meadow. Example: pioneer plant species that became established on land that was destroyed by the volcano, Mt. Saint Helens.
Progressive	Occurs over time whereby the communities become more complex by having a higher species diversity and greater biomass.
Retrogressive	Occurs over time when environmental variables deteriorate and thereby changing the status quo and causing the communities to become less diverse and contain less biomass. May be allogenic such as the introduction of grazing animals into grasslands resulting in the destruction of the grasses.
Secondary	Begins in area where natural community or organisms has been disturbed, removed, or destroyed, but bottom soil or sediment remains. Examples: burned forests, logged forests, climatic change, disease, pest infestation, heavily polluted streams, land that has been flooded, and conversion of natural ecosystem to agriculture.

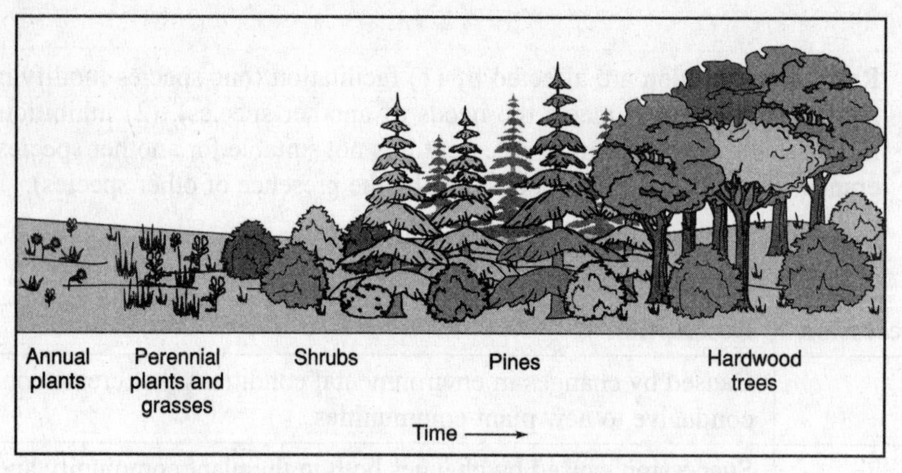

Figure 5.19. Stages of succession in a temperate deciduous forest. The time span from annual plants to hardwood trees is over 100 years.

CHARACTERISTICS OF SUCCESSION WITHIN PLANT COMMUNITIES

Characteristic	**Early Stage of Succession**	**Late Stage of Succession**
Biomass	Limited.	Depends upon biome and resource availability—high in tropics and marshes, limited in deserts; nevertheless, greater than in early stages.
Consumption of soil nutrients	Because biomass is limited and few nutrients exist within the plant structure, nutrients are quickly absorbed by these simpler plants.	Because biomass is greater and more nutrients are contained within plant structures, nutrient cycling between the plant and soil tends to be slower.
Impact of macroenvironment	These early plants depend primarily on conditions created by macroenvironmental change (e.g., fires, floods); therefore macroenvironmental conditions are significant.	These plant species appear only after pioneer plant communities have adequately prepared the soil for their existence. Adaptive mechanisms to the impact of macroenvironmental change are more advanced so that macroenvironmental changes are not as significant.
Leaf canopy	Close to ground initially. Since plants are spread out, light is available for plants at different heights creating a multiple leaf zone.	In established communities, light may be limited in lower zones, creating a less stratified leaf canopy.
Life span of plant	Tends to be short.	Tends to be longer.
Life span of seed	Long—seeds may become dormant and able to withstand wide environmental fluctuations.	Short—seeds are generally not able to withstand wide environmental fluctuations.

CHARACTERISTICS OF SUCCESSION WITHIN PLANT COMMUNITIES (Continued)

Characteristic	Early Stage of Succession	Late Stage of Succession
Life strategy	r strategists: mature rapidly, short-lived, species number is greater, species diversity is smaller, niche generalists.	K strategists: mature slowly, long-lived, species number is smaller, species diversity is greater, niche specialists.
Location of nutrients	Primarily in the soil and in leaf litter.	Primarily within the plant and top layers of soil.
Net primary productivity	High	Low.
Nutrient cycling by decomposers	Limited because humus has not developed to any great extent.	Complex because of established humus layer.
Nutrient cycling through biogeochemical cycles	Because nutrient sinks have not fully developed, the nutrients are available to cycle through established biogeochemical cycles fairly rapidly.	Because of nutrient sinks, nutrients may not be readily available to cycle through biogeochemical cycles.
Photosynthesis efficiency	Low	High.
Plant structure complexity	Tends to be simple.	Tends to be more complex.
Recovery rate of plants from environmental stress	Quickly and easily come back since these species have adapted to exist in areas where environmental conditions are limited.	These plants depend upon environmental conditions created by pioneer plants; therefore, recovery is slow and may take years.
Seed dispersal	Widespread	Limited in range.
Species diversity	Diversity is limited.	Diversity is high.
Stability of ecosystem	Because plant diversity is limited, the ecosystem is subject to instability if only one stressor is present; adjustment and equilibrium mechanisms are limited.	Because of a wide diversity of plant species and accompanying adaptation capacity within the gene pool, there is a greater capacity to recover from environmental stress.

Evolution of Life: Natural Selection, Genetic Diversity, and Extinction

I have called this principle, by which each slight variation, if useful, is preserved, by the term Natural Selection.

Charles Darwin, *The Origin of Species*

Theory of Evolution

- The world has changed over time. Organisms today are different from those many years ago. Many organisms have become extinct. Evidence = fossil record.
- The concept of "common ancestor" states that similarities among species (genetic and less reliable morphology) can be traced back through branches of a phylogenetic tree, as shown in Figure 5.20.
- Change generally takes a very long time (millions of years), and is supported by the fossil record. Exception: see punctuated equilibrium later in this chapter.
- Evolution occurs through the process of natural selection.

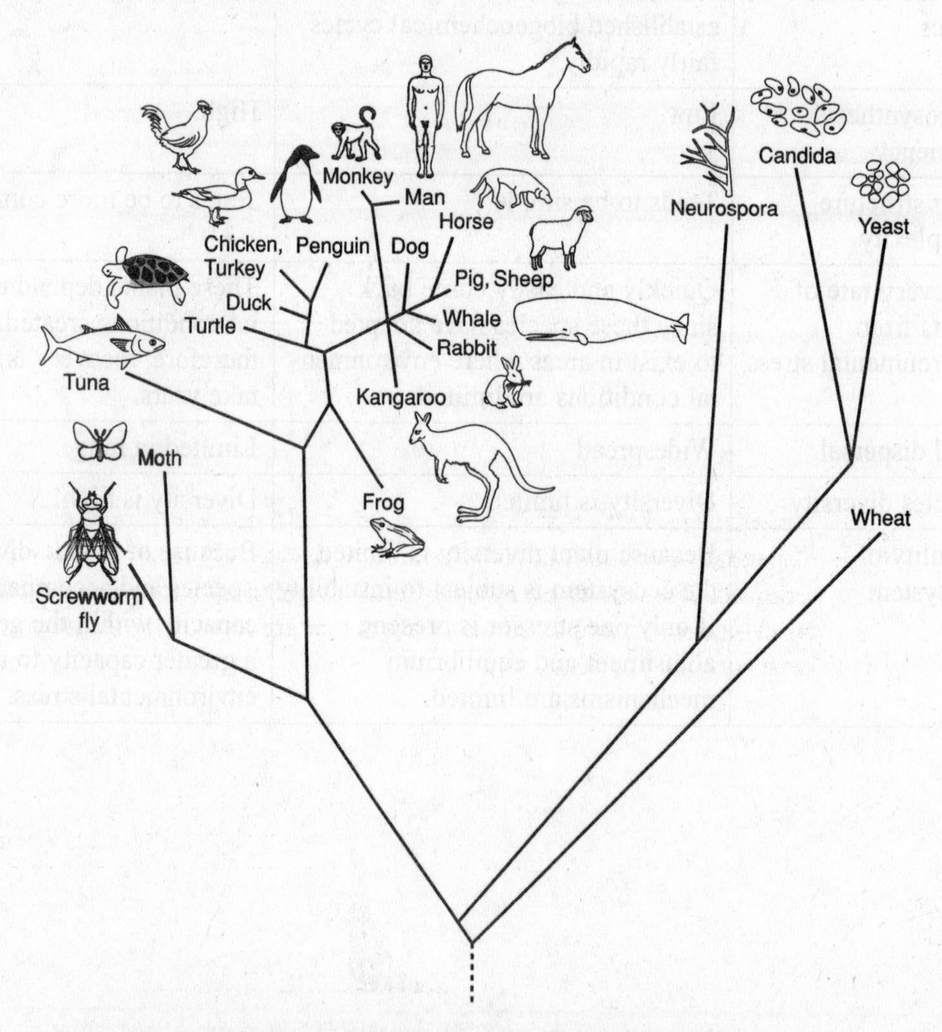

Figure 5.20. Phylogenetic tree traced through analysis of cytochrome c.

Natural Selection

- In 1859, Charles Darwin published *The Origin of Species* after years of observation on *H.M.S. Beagle.* Natural selection is the mechanism of how organisms evolve. Natural selection works on the *individual* level by determining which organisms have adaptations that allow them to survive and reproduce and to be able to pass those adaptive traits on to their offspring. Natural selection occurs over successive generations. Evolution works on the *species* level by describing how the species attains the genetic adaptations that allow them to survive in a changing environment. Without a changing environment, neither evolution nor natural selection would exist.
- The range of genetic variation within a species' gene pool determines whether or not the species, not the individual, has the capacity to adapt and survive to changes in the environment. The expression of that variation in phenotypic and behavioral expressions determines whether the individual survives, and is commonly referred to as "survival of the fittest." New genes enter the gene pool through mutation and combined with change in gene frequency resulting in evolution at the species level.
- Natural selection operates in three ways. It may be (1) stabilizing, (2) directional, or (3) disruptive.

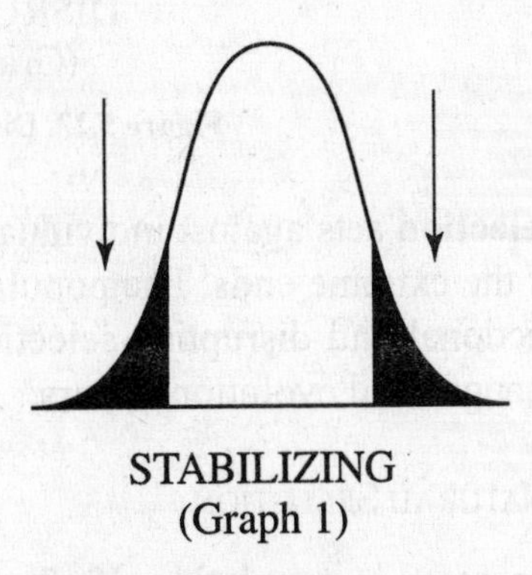

STABILIZING
(Graph 1)

Figure 5.21. Stabilizing selection.

Stabilizing selection affects the extremes of a population and is the most common form of natural selection. The individuals that deviate too far from the average conditions are removed. The results are a decrease in diversity, maintenance of a stable gene pool, and no evolution. Example: human babies that are too low in weight or too high in weight have survival problems.

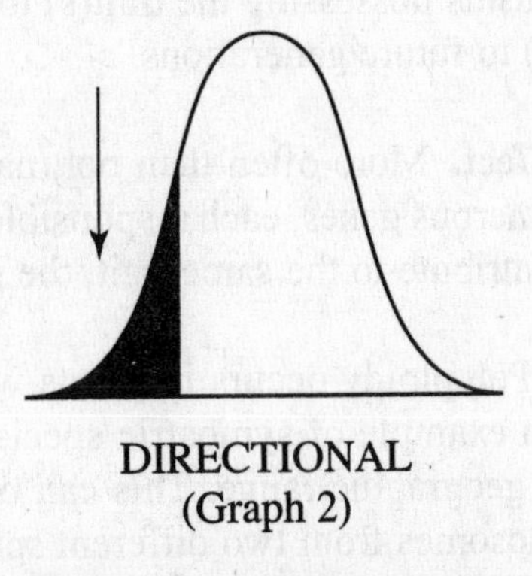

DIRECTIONAL
(Graph 2)

Figure 5.22. Directional selection.

Directional selection affects the extremes of a population. Individuals toward one end of the distribution may do especially well, resulting in a frequency distribution toward this advantage in subsequent generations. An example demonstrating directional selection is the case of industrial melanism of the peppered moth. Another example of directional selection involves the evolution of the horse. The early horse, *Hyracotherium*, present during the early Eocene was a small-bodied creature that moved well through heavy brush and woodlands. Today's horse, *Equus*, looks nothing like its ancestor with its long legs (designed for speed in open grassland) and changes in dental and toe structure.

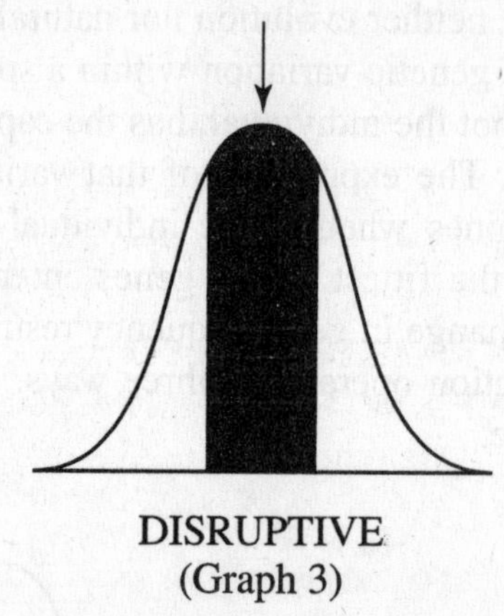

DISRUPTIVE
(Graph 3)

Figure 5.23. Disruptive selection.

Disruptive selection acts against individuals that have the average condition and favors individuals at the extreme ends. The population is essentially split into two.

Both directional and disruptive selection affects the gene pool. In both cases, the population changes and evolution occurs.

Process of Natural Selection

- Geometric increase in population: If all offspring survived, there would be astronomical numbers of individuals.
- Limited resources: The Earth has finite resources.
- Competition: There is a "struggle to survive" and competition exists for limited resources.
- Individual variations: There is variation in offspring. For natural selection, the variations must be gene-expressed and capable of being inherited.
- Disproportionate increase and persistence in phenotypic adaptation in successive populations: Variations that are advantageous to the individual in terms of survival allow more organisms possessing the trait(s) to survive and reproduce and pass on the characteristic(s) to future generations.

Polygenic effect. More often than not, natural selection is based upon the cumulative effects of numerous genes, each responsible for slight changes. When genes at more than one locus contribute to the same trait, the result is called a polygenic effect.

Polyploidy. Polyploidy occurs in plants when the entire set of chromosomes is multiplied. It is an example of sympatric speciation in which species arise within the same, overlapping, geographic range. This can occur through the process of hybridization in which chromosomes from two different species are artificially combined to form a new species (hybrid) or when chromosomes naturally fail to segregate at meiosis, producing diploid gametes. If the hybrid has adaptive traits to survive in the new environment, a new species of plant has been produced. Although the plants may not be able to reproduce sexually, they may be able to reproduce through vegetative means. Examples of polyploidy include cotton, tobacco, sugar cane, bananas, potatoes and wheat used for bread. More than half of all known species of plants today (260,000 species) may have originated through polyploidy. Occasionally, polyploidy occurs in certain species of earthworms, primitive fish, and frogs.

Genetic Diversity

Biodiversity attempts to describe diversity at three levels:

1. **Genetic diversity**—the range of all genetic traits, both expressed and recessive, that makes up the gene pool for a particular species.
2. **Species diversity**—the number of different species that inhabit a specific area. Tropical rain forests have a higher species diversity than extreme deserts.
3. **Ecosystem diversity**—the range of habitats that can be found in a defined area. Ecosystems are composed of both biotic and abiotic components.

BIOLOGICAL DIVERSITY

Diversity Increasers	Diversity Decreasers
Diverse habitats.	Environmental stress.
Disturbance in the habitat—fires, storms.	Extreme environments.
Environmental conditions (temperature, rainfall, food supply, etc) with low variation.	Extreme limitations in the supply of a fundamental resource.
Trophic levels with high diversity.	Extreme amounts of disturbance.
Middle stages of succession.	Introduction of species from other areas.
Evolution.	Geographic isolation.

Evolution and speciation. The direct result of evolution is speciation, resulting in genetically isolated, interbreeding organisms. When segments of a population are so isolated that gene flow ceases, the gene pool of each group is separated and subjected to particular selective pressures. Each group develops its own seasonal breeding pattern, courtship and mating behavior, subtle reproductive changes (genital and gametic), and habitat preference. Even if the isolating barrier is removed, gene flow between the two groups may not be possible. New species may have been formed.

Speciation is best exemplified by a group of birds, known as Darwin's finches, on the Galapagos Islands. The ancestral finch probably came from Ecuador, on the mainland of South America, traveling in a flock that was blown off course. The small population experienced different selective pressures on each of the Galapagos Islands. Consequently, the finches of each island evolved their own gene pool. Isolation and genetic drift fostered adaptive radiation and thus divergent evolution.

Groups of finches from a given population spread out and occupied the many available habitats on each island. Selective pressures of adaptation to changed environments ultimately resulted in genetic isolation and speciation that prevented hybridization. Adaptive radiation is the evolutionary pathway that accounts for the existence of various species of finch on the same island.

Very often populations that are completely unrelated resemble each other—a condition known as convergent evolution. Unrelated species subjected to the same environmental factors may adapt in a similar fashion. For example, both the whale and the fish have streamline bodies with fins, although they do not share a common ancestry.

Genetic drift. Genetic drift refers to random changes in the gene pool of a population. If a large and stable population were suddenly destroyed and only a few survivors remained, the population would be rebuilt by these survivors through reproductive means. However, the gene pool in this smaller and newer population might differ from that of the former population. As a result, mutations that were rare might become more concentrated and harmful genetic defects might show up with greater frequency.

There is an example in present day Afrikaaners. Afrikaaners are descendants from 30 families, a rather small population. A metabolic disorder called *porphyria variegata* is rare among most populations but occurs with great frequency among the South African descendants of Dutch settlers. This disease is characterized by excessive amounts of iron porphyrins in the blood. The urine turns red, and the afflicted person is very sensitive to light and ultimately suffers liver damage. The abnormal genes that cause this condition have been passed along to succeeding generations of the original settlers.

Gradualism versus punctuated equilibrium. Gradualism views evolution as a slow, stepwise development of a species over long periods of time (millions of years). Fossil evidence supports gradualism in many cases by showing a continuum from the ancestral species to the new species through transitional fossil intermediates.

In 1971, Niles Eldredge and Stephen Jay Gould proposed the theory of punctuated equilibrium. This theory proposes that some species arose suddenly in a short period of time (thousands of years) after long periods of stability or stasis. Thus, evolution may also occur in "bursts." These bursts of rapid evolution are thought to be triggered by changes in the physical or biological environment—perhaps a period of drought or the appearance of a new, more challenging predator. The abrupt appearance of flowering plants without a fossil record of their origin is an example.

Reproductive barriers. Gene flow between two closely related species occupying the same area is prevented by reproductive barriers. Reproductive barriers isolate gene pools.

Prezygotic barriers prevent or deter fertilization. An example of ecological isolation occurred between lions and tigers. Both animals had overlapping habitats in India with the tiger preferring the forest and the lion preferring the grassland resulting in isolation of the gene pools. In rare circumstances, a "liger" may be produced in a zoo. Examples of breeding time differences (temporal isolation) are found in North American frogs of the genus *Rana*. Each of five different species of *Rana* have different breeding times, which isolates their gene pools. Other prezygotic barriers include genitalia (mechanical isolation) and failure of gametic fusion (gametic isolation).

Hybrid sterility (e.g., mule), hybrid infertility in later generations (e.g., fruit flies raised under laboratory conditions), and aborted hybrids due to genetic incompatibility (e.g., zygote produced from a sheep and a goat) are postzygotic barriers. Postzygotic barriers prevent the development of viable or fertile offspring.

Extinction

- Extinction is a normal process and an opportunity for adaptive radiation (e.g., rise of mammals). The majority of all animals and plants that existed on Earth are now extinct. Species on average last 2 million to 10 million years before becoming extinct. *Natural* background rate of extinction is 1 to 2 species per year. Causes can be linked to either changes in the physical environment and/or evolution of better-adapted life forms in response to changes in the environment through natural selection.
- Five mass extinctions have occurred in Earth's history. A sixth recent extinction occurred during the end of the Pleistocene. During the Permian extinction, 95% of all terrestrial and marine species became extinct. This extinction coincides with coalescing of the continents into Pangaea (cooling in interior since farther from ocean heat source) and possible glaciation. The extinction during the Cretaceous period (K-T extinction) marks the end of the age of reptiles (dinosaurs) and begins the radiation of birds and mammals.

Era	Period	Number of Years Ago
Paleozoic	Ordovician	438 million to 505 million
Paleozoic	Devonian	360 million to 408 million
Paleozoic	Permian	245 million to 286 million
Mesozoic	Triassic	208 million to 245 million
Mesozoic	Cretaceous	65 million to 144 million
Cenozoic	Pleistocene	10,000 to 1.6 million

- Iridium is found in rocks of the K-T extinction period. It is common in meteorites, but not common on Earth. Luis Alvarez proposed that a large meteorite (10 km in diameter) traveling at 72,000 km/hr (45,000 mph) hit the Earth during this period producing a large dust cloud that cooled the Earth, reducing photosynthesis, and creating acid rains. Fires, tidal waves, and volcanic eruptions may have also contributed to extinction. Evidence of large craters in Earth surface (Chicxulub, Yucatan Peninsula, Mexico) and evidence in melted quarts found in layers of the Earth corresponding to Tertiary times, adds to the hypothesis.
- Pleistocene extinction (the most recent one) coincides with retreat of glaciers in North America. It also coincides with the migration of humans arriving in North America from Eurasia over the Bering Land Bridge and may have been due to early hunting.
- Other theories for extinction include:

1. **Glaciation** (cooling). Species cannot either adapt or migrate to prevent extinction. Evidence from glacial sediments under the sea floor of the North Sea and extinction of species adapted to warm climates. It links to extinctions occurring in the Ordovician, late Devonian, and end of Permian. An increase in precipitation may have been responsible for extinction during late Triassic.
2. **Volcanic activity.** Probably during Cretaceous and Permian times since this was a time of great volcanic activity. Dust entering upper atmospheres would have impacted glaciation by reducing solar insolation. Iridium D is also expelled from Earth's mantle during volcanic eruptions. Evidence in quartz crystals supports volcanic activity.

3. **Sea-level variation.** Related to glaciation. Changes in oxygen, salinity, or thermal heat may have contributed. Also contributing are losses to habitats. Linked to possible extinctions during Cambrian, Ordovician, and Permian.
4. **Cosmic radiation and cancer.** Possible explosions of supernovas releasing large amounts of radiation or neutrinos, which may have caused cancer.
5. **Nickel poisoning.** Nickel was common in asteroids and also prevents photosynthesis. Plants would have died as well as animals that ate them.
6. **Egg-eating mammals.** Small, early mammals depleted dinosaur eggs. Fails to explain concurrent marine extinctions.

Case Studies—Natural Selection, Allopatric Speciation, and the Environment

Peppered moths. Before 1800, most peppered moths, *Biston betularia,* in London were light-colored. Dark moths were rare. During the Industrial Revolution through the use of coal, trees became darker due to soot, and the light moths stood out. Because they stood out, they were subject to greater predation by birds than the dark moths that blended with the darker trees. By 1886, the dark moths had become more predominant. The dark moths had an adaptive trait that was an advantage for survival, thereby more of them were able to reproduce.

Galapagos finches. Fourteen different species existed on different islands 600 miles off the coast of Ecuador. Each of the finch species was separate and distinct and could not interbreed with any other species on neighboring islands. Each species had adapted to unique conditions on different islands (large seed eaters, small seed eaters, fruit eaters, insect eaters, ground dwellers, tree dwellers, etc.), as shown in Figure 5.24. Each population of finch had filled a unique niche. Adaptive radiation coupled with geographic isolation was responsible for diversification. The finches were examples of allopatric populations living in different geographical areas that became genetically different from one another. Allopatric speciation occurs when a population that inhabits a particular area is divided into two or more geographically separated populations. Allopatry is caused not only by migration of species but also by separation of population (geographic and reproductive isolation) and is due to major environmental changes such as glaciation and movement of land masses.

Figure 5.24. Examples of beaks from Galapagos finches.

Multiple-Choice Questions

For questions 1–3, one of the following lettered items refers to a numbered statement. Select the one lettered choice that best fits each statement. A choice may be used once, more than once, or not at all.

(A) Tropical rain forest
(B) Temperate rain forest
(C) Savanna
(D) Taiga
(E) Tundra

1. Forests of cold climates of high latitudes and high altitudes.

2. Warm year-round. Prolonged dry seasons. Scattered trees.

3. Low biodiversity due to high shade, which limits food for herbivores. Major resource for timber.

4. The annual productivity of any ecosystem is greater than the annual increase in biomass of the herbivores in the ecosystem because

   (A) plants convert energy input into biomass more efficiently than animals.
   (B) there are always more animals than plants in any ecosystem.
   (C) plants have a greater longevity than animals.
   (D) during each energy transformation, some energy is lost.
   (E) animals convert energy input into biomass more efficiently than plants do.

5. A type II survivorship curve would apply to

   (A) humans.
   (B) redwoods.
   (C) bacteria.
   (D) flies.
   (E) tapeworms.

6. All the following are factors that increase population size EXCEPT:

   (A) ability to adapt.
   (B) specialized niche.
   (C) few competitors.
   (D) generalized niche.
   (E) high natality.

7. A specialist faces _____ competition for resources, and has _____ ability to adapt to environmental changes. A generalist faces _____ competition for resources, and has _____ ability to adapt to environmental changes.

   (A) less, greater, greater, less
   (B) greater, less, less, greater
   (C) less, less, greater, greater
   (D) greater, greater, less, less
   (E) none of the above

8. Whether a land area supports a deciduous forest or grassland depends primarily on

   (A) changes in temperature.
   (B) latitude north or south of the equator.
   (C) consistency of rainfall from year to year.
   (D) changes in length of growing season.
   (E) none of the above.

9. The main difference between primary and secondary succession is that

   (A) primary succession occurs in the year before secondary succession.
   (B) primary succession occurs in barren, rocky areas and secondary succession does not.
   (C) secondary succession ends in a climax species and primary succession ends in a pioneer species.
   (D) secondary succession occurs in barren, rocky areas and primary succession does not.
   (E) all of the above statements are true.

10. The natural, background rate of extinction is ______________ species per year.

   (A) 0
   (B) 1–10
   (C) 10–20
   (D) 100–300
   (E) 1000–2000

11. Darwin noted that the Patagonian hare was similar in appearance and ecological niche to the European rabbit. However, the Patagonian hare is not a rabbit. It is a rodent related to the guinea pig. This example illustrates the principle known as

   (A) allopatric speciation.
   (B) adaptive radiation.
   (C) divergent evolution.
   (D) coevolution.
   (E) convergent evolution.

For Questions 12 through 15, choose the letter of the item that is MOST closely related to the numbered statement.

   (A) Adaptive radiation
   (B) Isolation
   (C) Natural selection
   (D) Stable gene pool
   (E) Convergent evolution

12. Members of the same species of moths are prevented from interbreeding because they live on opposite sides of a mountain range.

13. Darwin's finches are a good example of this biological principle.

14. Members of a large population mate at random.

15. In the evolutionary history of the horse, eohippus (the dawn horse) was replaced by the modern one-toed horse.

Answers to Multiple-Choice Questions

1. **D**	6. **B**	11. **E**
2. **C**	7. **C**	12. **B**
3. **B**	8. **C**	13. **A**
4. **D**	9. **B**	14. **D**
5. **D**	10. **B**	15. **C**

Explanations for Multiple-Choice Questions

1. **(D)** Taigas are generally found between 45° and 60° north latitude. They occupy approximately 17% of all land area (50 million acres worldwide, or 20 million hectares). Taigas are composed of forests of cold climates of high latitudes and high altitudes. There are two types of taigas: (1) open woodland and (2) dense forest. Precipitation generally occurs during summer months due to midlatitude cyclones. There are between 90 and 120 frost-free days per year. Soils are generally poor in nutrients because large amounts of leaching caused by rainfall, and the soils are acidic because of the decomposition of needles. Taigas have a deep layer of litter and decomposition is slow due to cold temperature. Taigas are dominated by conifers (spruces, firs, pines). Flowering trees include aspens and birches. Dense stands of small trees causes the understory to be low in light. Understory plants and trees (blueberry, cranberry) in deciduous forests have adapted to grow primarily in the fall and winter months to take advantage of light that can come through when dominant trees have lost their leaves. Taigas have low biodiversity due to harshness of environment.

2. **(C)** Savannas are warm year-round and have prolonged dry seasons with scattered trees. The environment is intermediate between grassland and forest. They have an extended dry season followed by a rainy season and are found in Australia; central, eastern, and south Africa; India; Madagascar; central South America; Southeast Asia; and Thailand. Savannas consist of grasslands with stands of deciduous shrubs and trees that do not grow more than 30 meters (100 feet) high. Trees and shrubs generally shed leaves during the dry season, which reduces the need for water. Food is limited during the dry season so many animals migrate during this season. Soils are rich in nutrients. Savannas contain large herds of grazing animals and browsing animals that provide resources for predators.

3. **(B)** Temperate rain forests have moderate temperatures and rainfall exceeding 250 cm (8 feet)/yr. Trees consist of redwoods, cedars, and Douglas firs. There is low biodiversity because of high shade, which limits food for herbivores. They are a major resource for timber and are found in Canada, northern California, Oregon, and Washington State.

4. **(D)** Less energy is available at each trophic level because energy is lost by organisms through respiration and incomplete digestion of food sources. Therefore, fewer herbivores can be supported by the vegetative material.

5. **(D)** Type II survivorship curves are characteristic for organisms in all age categories that have fairly uniform death rates. Predation affecting all age categories is the primary means of death. Type II survivorship curves are typical of organisms that reach adult stages very quickly such as flies.

6. **(B)** If you did not know the answer to this question, you can use a trick. Notice that two of the answers end with the same word, "niche". More than not, the answer is one of these choices. In most cases if you do not know the answer, the ¼-point penalty prevents you from taking a wild guess, since your guessing the right answer is only ⅕, or 20%. But in cases where both choices end with the same word, I recommend picking one (and usually the first one). Specialized niches are more susceptible to environmental changes that have a direct effect on the stability of populations; hence, their population size. Other factors that tend to have an effect on decreasing population size are unfavorable environmental conditions, limitations of food resources, large number of competitors, inability to migrate, and unsuitable adaptation mechanisms.

7. **(C)** Specialist species are adapted to a narrow range of habitats and conditions, while generalist species are able to live in a variety of environments. The specialist species *Australopithecus robustus* adapted to life on the African savanna and became adept at digesting savanna plants. However, *A. robustus* died out when these plants were killed in an ice age. *Homo erectus*, a generalist species, was able to eat meat and therefore could live in a wider variety of environments. *H. erectus* evolved a smaller gut, since the new dietary habits didn't require such long intestines, and a larger brain, due to increased protein intake. Larger brain size ultimately improved the intellectual capacity of the species.

8. **(C)** A number of climatic factors interact in the creation and maintenance of a biome. Where precipitation is moderately abundant (40 inches or more per year) and distributed fairly evenly throughout the year, the major determinant is temperature. Temperature is the major determinant for tropical rain forest, temperate deciduous forest, taiga, and tundra. It is not simply a matter of average temperature, but includes such limiting factors as whether it ever freezes or length of the growing season. The amount and seasonal distribution of rainfall is a primary determinant in establishing temperate rain forest, grassland, desert, or chaparral. The question of determining deciduous forest or grassland is more dependent on yearly patterns of rainfall because both biomes can exist over similar temperature ranges.

9. **(B)** Primary succession occurs on bare areas not previously supporting vegetation. Examples include areas of water, sand, or rock. Primary succession begins with soil building; developing from primitive plants (colonizers) reacting with the rock over long periods of time to eventually provide bits of soil that, in time, will support larger vegetation. With the accumulation of soil, new plants germinate, grow, and reproduce to begin the stages of a new succession. Secondary succession occurs in areas in which vegetation does grow but which have been altered by such external forces as fire, logging, and land clearing.

10. **(B)** The background rate of extinction is the number of extinctions that would be occurring naturally in the absence of human influence. Estimates range from one to ten species per year for the past 600 million years. It is difficult to estimate this rate, in part because the number of species in existence is not known. The background rate of extinctions establishes a baseline from which the severity of the current extinctions crisis can be measured. The current rate of extinction appears to be hun-

dreds, or perhaps even thousands, of times higher than the background rate. It is difficult to be precise because most of the disappearing species today have never been identified by scientists.

11. **(E)** Convergent evolution is the development of totally unrelated organisms in the same manner because of adaptation to similar environments. In many instances, animals that live in similar habitats resemble each other in outward appearance. These similar looking animals may, however, have quite different evolutionary origins. For example, swifts, swallows and martins all hunt for insects while they fly. They have streamlined bodies with long wings. Hummingbirds and sunbirds feed on nectar from flowers. They have long bills to reach the nectar at the base of flowers. Based on appearance only one might conclude that sunbirds are related to hummingbirds and that swifts are related to swallows and martins. In reality though, genetic techniques have shown that swifts are related to hummingbirds, while sunbirds are related to swallows and martins.

12. **(B)** "Species" is defined as a group of organisms that look similar and have the ability to interbreed and produce fertile offspring in the natural environment. For a new species to arise, either interbreeding or the production of fertile offspring must somehow cease among members of a formerly successful breeding population. For this to occur, populations or segments of a population must somehow become isolated. Two forms of isolation prevent interbreeding or cause infertility among members of the same species. These forms of isolation are geographic isolation and reproductive isolation. For example, about 100 years ago when the London subways were being built, mosquitoes (*Culex pipiens*) settled into the tunnels. Now after 100 years and hundreds of generations later, the mosquitoes have genetically mutated to such a degree that they can be considered a separate species. These new mosquitoes (*Culex molestus*) now feed on the inhabitants of the tunnels, including rats and mice in comparison to the species aboveground, *Culex pipiens* that feeds primarily on birds. Additionally, it has been reported that there are genetic variations among *Culex molestus* in the various London subway tunnels.

13. **(A)** Adaptive radiation is the development of many species derived from a single ancestral population. The Hawaiian silversword `ohana (family) is an example of adaptive radiation among plants in the world. Over the course of millions of years, the descendants of the pioneer plant evolved into 28 distinct species in three genera, occupying many different habitats.

14. **(D)** Most large populations have a stable gene pool—variations are tolerated in the absence of severe "pressure." This recognition was advanced by Hardy and Weinberg in 1908 and is known as the Hardy-Weinberg theorem. The allele frequency in a stable population can be represented quantitatively and assumes:

    1. Large population—to insure no sampling error from one generation to the next;
    2. Random mating—no assortive mating or mating by genotype;
    3. No mutations—even new mutations have little effect on allele frequencies from one generation to the next;
    4. No migration between populations;
    5. No selection—all genotypes reproduce with equal success.

15. **(C)** Natural selection is the process in nature by which, according to Darwin's theory of evolution, only the organisms best adapted to their environment tends to survive and transmit their genetic characteristics in increasing numbers to succeeding generations while those less adapted tend to be eliminated.

Free-Response Question

There are over 25,000 species of bees in the world today. Africanized honey bees (AHBs) (*Apis mellifera L. scutellata* (Lepetelier), sometimes referred to as killer bees, are the same species as European honey bees (EHB), but a different subspecies. Only by careful examination under a microscope, or through DNA tests, can AHBs be distinguished from EHBs. AHBs are called "Africanized" honeybees as a result of interbreeding experiments. The thought was that AHBs might be better suited for the tropical climates of South America than EHBs. In the 1950s, AHBs were inadvertently released in Brazil. AHBs began migrating (about 300 miles per year), arriving in southern Texas in 1990, Arizona and New Mexico in 1993, and southern California in 1994. Since 1994, however, migration has seemed to slow down significantly. Although the two subspecies are virtually indistinguishable, it is the behavior of the two subspecies that sets them apart. AHBs defend their colonies much more vigorously than do EHBs. When AHBs sting, more of them participate in extremely aggressive behavior. AHBs nest where EHBs do not, such as small, confined spaces near the ground such as water meter boxes, flowerpots, abandoned tires, and cracks in foundations.

(a) Discuss the influx of AHBs in the southwest United States in terms of the processes of natural selection. In what areas do the processes of natural selection apply to the colonization of AHBs and in what areas do they not? What issues would be warranted for further study?
(b) Provide possible explanations of why the expansion of AHBs may have stopped.
(c) What adaptations have honeybees in general developed to be able to survive?
(d) Provide some possible methods to control AHBs.
(e) What are the possible economic implications of AHBs to agriculture?

Free-Response Answer

Rubric

Part (a): 2 points
Part (b): 2 points
Part (c): 2 points
Part (d): 2 points
Part (e): 2 points

Ten points possible for full credit.

RUBRIC	ESSAY
½ pt. Recognition of genetic variation as a determinant in the process of natural selection.	(a) The *range of genetic variation within a species' gene pool determines whether or not the species, not the individual, has the capacity to adapt and survive to changes in the environment*. The colonization of the American southwest by Africanized honey bees (AHBs) differs from classical theories of natural selection in that AHBs and European honey bees (EHBs) belong to the same species, and therefore are presumably *able to interbreed and produce fertile offspring.*
½ pt. Recognition that sub-species are able to interbreed.	The two subspecies seem to differ only in behavioral patterns. As to whether the behavioral patterns of the two subspecies are so radically different as to cause a *prezygotic reproductive barrier* or not is worth investigation.
½ pt. Recognition that behavior may be a reproductive barrier.	Another issue that differs from classical natural selection theory is that, in this case, *the environment, for all practical purposes, is not changing;*
½ pt. Recognition that the environment is remaining constant and not acting to select one subspecies over another.	in effect, it is not a variable involved in survival rates. Both subspecies can and do survive in the environment, nor is the environment affecting disproportionate natality rates between the two subspecies based upon one subspecies having a survival advantage over the other.
½ pt. Recognition that the two sub-species are competing for the same limited natural resource(s)—a basic tenant of natural selection.	In terms of following classic natural selection theory, what exists in the scenario is *competition between the two subspecies for limited resources*. Both subspecies are competing for space and food supplies.
½ pt. Recognition that more study is needed in the advantage of aggressive behavior and its effect on survivability of the colony.	As to whether or not the *more aggressive behavior of protecting the nest, allows more AHBs to disproportionately survive*, thereby increasing the ratio of AHBs to EHBs, is another factor that could be studied.
	(b) It appears that since 1994, the expansion of AHBs into the southwestern United States has slowed down. Possible reasons might include:
½ pt. Specific adaptations of AHBs to tropical climates.	(1) AHBs *are adapted to tropical climates* (Africa and Brazil). Their ability to survive may be *restricted to climatic zones that have fairly warm temperatures* during most of the year (American southwest). Extremely cold tempera-

RUBRIC	ESSAY
½ pt. Temperature tolerances may be narrower in AHBs.	tures (common in the Midwest and northeast United States) may influence the viability of offspring and consequently their limited expansion into these areas.
½ pt. Geographic barriers. ½ pt. Mountain ranges. ½ pt. Low humidity in deserts. ½ pt. Lack of sufficient food resources in deserts. ½ pt. Variation in seasonal photoperiod. ½ pt. Timing of forage availability (seasons) as opposed to no seasons in the tropics.	(2) Geographic *barriers* may be another reason for the decline in the expansion of AHBs. *Mountain ranges, dry deserts with low humidity, and the lack of adequate food resources* may also be slowing the expansion. Finally, *variation in seasonal photoperiod, timing of forage availability, and/or the existence of parasites such as mites, pathogenic bacteria, or viruses,* may be having an effect on the newly introduced subspecies.
½ pt. Parasites. ½ pt. Pathogenic organisms. ½ pt. Clustering behavior. ½ pt. Predetermined roles. ½ pt. Temperature regulation of hive to create stable environment for young. ½ pt. Internal stability of nest temperature allows colonization into more diverse areas. ½ pt. Ability to communicate.	(c) Three adaptations of honeybees which have been essential to their evolution and biology are their (1) *clustering behavior*, that is working as a social unit with specific predetermined behavioral patterns and duties of each class in the hierarchy. (2) The ability to cool the nest through the process of evaporation of water collected outside the nest, *ensuring a stable internal temperature of the nest*. The ability to ensure temperature *stability within the nest allows honeybees to colonize a wide variety of environments,* as opposed to bees (*Meliponinae*) that lack this trait and are therefore restricted to thermally stable environments (tropics). And (3), the *ability to communicate information about food sources, direction, and distance through "dance behavior."* The ability to communicate such a wide variety of complex information is unparalleled in the animal kingdom (except for humans).
½ pt. Recognition that wide use of pesticides would not be effective. ½ pt. EHBs are more suitable. ½ pt. Bacterial agents or viruses would affect both subspecies. ½ pt. Only way to control AHBs would be to ensure competitive population numbers of EHBs. ½ pt. Reduce carrying capacity of environment.	(d) Since AHBs and EHBs are coexisting within the same areas, any attempt to control the AHBs through *pesticides or other extermination technique would probably have an effect on the EHBs as well.* For the most part, *EHBs and their less-aggressive behavior are more suited to coexist with man* for agricultural pollination purposes than are AHBs. Introduction of any bacterial agent, parasite, or other biological controlling mechanism would also probably affect the EHBs. One method that might work to control the domination of an area by AHBs would be to *ensure that there are sufficient EHBs in the area to* (1) *provide competition* for resources between the two subspecies, thereby *reducing the carrying capacity (K) of the area which in turn would limit expansion of the AHB populations*; and (2) by providing sufficiently high enough EHBs in an area, to *ensure interbreeding of AHBs queens with EHB males*. The goal of interbreeding would be to possibly dampen the highly aggressive behavioral characteristics of the AHBs. This

RUBRIC

½ pt. Interbreeding and its role to diminish aggressive behavior.

½ pt. Introduction of sterile males.

½ pt. Bees are necessary for agricultural production.

½ pt. People are closer today to agricultural areas and are thereby more impacted.

½ pt. Public fear and hysteria of AHBs.

½ pt. Availability of suitable sites for EHBs.
½ pt. Public pressure.
½ pt. Liability issues (insurance).
½ pt. Legislative mandate.
½ pt. Fewer beekeepers.
½ pt. Cost of maintaining EHBs would increase.
½ pt. As cost of bees goes up, so would cost of agricultural products.

ESSAY

would be an area in need of more research. Finally, *introduction of large numbers of sterile male AHBs into hives* might result in lower fertile AHB offspring. Physically disrupting hives would only enhance recolonization.

(e) Agriculture *as practiced today, requires the use of honeybees for pollination purposes*. Honeybees are rented and placed into fields for this purpose by beekeepers *Urban expansion, population increases, and sophisticated transportation systems have all made rural areas more accessible* to more people. AHBs strike terror in many people, and yet, documented cases of deaths due to AHBs are rather low, and at least are exaggerated. *Public pressure to eradicate "killer bees" will certainly involve economic effects*. The cost of renting EHBs would certainly increase due to issues of *suitability and availability of sites, public concern, legislative mandates, and destroying EHBs in an attempt to control AHBs*. As costs of renting EHBs increase, *profit margins could decrease*, forcing many *beekeepers out of business*. The result would be *less bees for the needs of agriculture*. This in turn could result in *less agricultural production and consequently higher prices*.

Word Count: 861

It is now your turn. Take the same question above and write your own answer. Remember, don't worry if you don't get all the points. Strive for half for right now. Your chances of getting a question on "killer bees" on the AP Environmental Science Exam are practically zero. That is not the point. The point is to practice writing—from initial brainstorming, organization, topic sentences, paragraph development, evidence to support your statements, and finally conclusions.

UNIT II: HUMAN POPULATION DYNAMICS

CHAPTER 6

Human History and Global Distribution

Areas That You Will Be Tested On

A: Numbers
B: Demographics—Birth and Death Rates
C: Patterns of Resource Utilization
D: Carrying Capacity—Local, Regional, and Global
E: Cultural, Aesthetic, and Economic Influences

Key Terms

Note: Definitions for all key terms listed below can be found in the glossary starting on page 648. For additional definitions of relevant terms from this chapter, refer to www.barronseduc.com/0764121618.html.

age structure	**exponential growth phase**
demographic transition	**immigration**
density-dependent factor	**limiting factor**
density-independent factor	**logistic growth curve**
developing country	**maximum sustained yield**
emigration	**total fertility rate**

Key Concepts

Numbers

Population Growth

- Human population has had three surges in growth due to: (1) the use of tools, fire and the cultural revolution that *Homo sapiens* brought through the ice ages; (2) the agricultural revolution about 10,000 years ago; and (3) the expoential rise (J-shape) in human population due to the industrial and medical revolution a mere 200 years ago (Figure 6.1). Each surge increased the population more than tenfold.

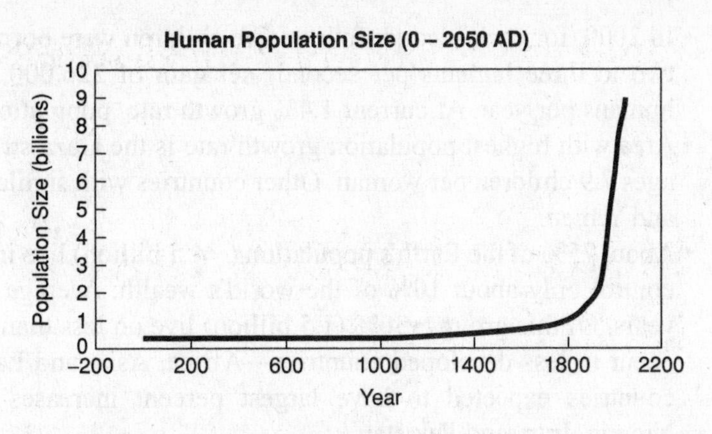

Figure 6.1. Estimated human population growth from 200 B.C. to A.D. 2050.

HUMAN POPULATION GROWTH OVER TIME

Time Period	Description
Before Agricultural Revolution ?–8000 B.C.	~1 million to 3 million humans on Earth. Subsisted by hunting animals and gathering natural plant material (nuts, berries, fruits, etc.), "hunter-gatherers."
8000–5000 B.C.	~50 million humans on Earth. Increases due to advances in agriculture, domestication of animals, and the end of a nomadic type of existence.
5000 B.C.–A.D. 0	~200 million humans on Earth. Rate of population growth during this period was about 0.03–0.05%, compared to about a worldwide average today of about 1.3%.
A.D. 0–A.D. 1300	~500 million. Population rate increased during the middle ages because new habitats were discovered. Factors that tended to decrease the rate included famines, wars, and disease (all caused by population density factors).
A.D. 1300–A.D. 1650	~600 million. The reason that there was only about a 100 million increase in the world population estimates over a span of 400 years is primarily due to plagues that affected Europe. Plague (bubonic) may have been responsible for up to a 25% mortality rate at its peak in 1650.
A.D. 1650–present	~6 billion currently. Annual percentage rate of natural population growth went from ~0.1% during the end of the plagues to about 1.3% today. Factors responsible for such dramatic increases include a large decrease in mortality rates due to more advanced and better health care, health insurance, drugs (especially antibiotics), improvements in hygiene and sanitation, advances in agriculture and techniques for producing and distributing more food, and education. Also included is the fact that people living in rural environments are more isolated from technological advances available to urban dwellers. At the beginning of the 20th century, about 15% of people lived in cities. At the end of the 20th century, this figure had climbed to over 50%.
Present–A.D. 2050	8 billion to 15 billion. This represents a 50 to 250% increase compared to current figures. Disparity in figures reflects uncertainty in trends for less-developed countries (birth control, AIDS, health care, famine, etc.). Some countries will experience declines in population figures—Russia, many countries in western Europe, South Korea, and Japan due to lower natality and higher mortality.

- In 2000, for each second: four to five children were born, two people died; net gain of two to three humans per second; net gain of 226,000 births per day or 78 million humans per year. At current 1.4% growth rate, population will double in ~50 years.
- Area with highest population growth rate is the Gaza strip in Israel (4%), which averages 7.9 children per woman. Other countries with similar growth rates are Iraq, Syria, and Yemen.
- About 75% of the Earth's population (~ 4.5 billion) live in less-developed countries but control only about 10% of the world's wealth. Average lifespan is approximately 50 years. Of this group, ~30% (1.5 billion) live on less than $1 per day. Most growth will occur in less-developed countries—Africa, Asia, and Latin America. Less-developed countries expected to have largest percent increases in population are Ethiopia, Nigeria, Iran, and Pakistan.

PERCENT ANNUAL GROWTH RATE AND DENSITY FIGURES
FOR MAJOR AREAS OF THE EARTH

Region	Inhabitants Per km² Agricultural Land	% Annual Growth Rate
Asia	423	1.8
Europe	213	0.2
Africa	80	3.0
Former Soviet Union	69	0.7
Latin America	58	1.9
North America	55	0.7
Oceania	15	1.4

- Countries expected to have declining population due to decreases in standard of living and/or AIDS: Russia, Zimbabwe, South Africa, Botswana, Zambia, and Namibia.
- Approximately 31 million people are HIV positive or have contracted AIDS worldwide, 90% of whom live in less-developed countries, particularly in Sub-Saharan Africa.
- Countries experiencing declining population due to higher education, career opportunities, and other competing factors include Italy, Austria, Germany, and Greece. In 1950, North America and Europe made up 22% of world population. By 2030, it is expected to only make up 9%, which shifts world political power and influence.
- In the United States, the population is expected to rise from 276 million in 2000 to 404 million to 507 million in 2050—a 46 to 86% increase. More than half of this increase is projected to be due to increases in immigration and children of those immigrants. This represents about a 1% per year growth rate. China also has about a 1% per year growth rate. Together, China (1.3 billion) and the United States (275 million) contribute about 40% of the world's population.
- ~21% of the Earth's population (~1.5 billion) live in developed countries and control about 90% of the world's wealth. Average lifespan is approximately 75 years. The per capita Gross Domestic Product in the United States in 2000 was ~$90 per day ($34,000 per year). This compares to figures of less than $2 per day for Rwanda and similarly less-developed countries.

- Growth rate for urban areas is four times greater than growth in rural areas, with about 5 billion people living in urban areas by 2025.
- Between 1965 and 2000, in developed countries, infant mortality rates dropped from 20 per 1000 live births to 8 per 1000. In developing countries, the rates dropped from 118 per 1000 live births to 63 per 1000 live births or almost 22,000 worldwide per day. Mortality rates have also decreased dramatically in the last 100 years due to increased food supplies, better and more efficient methods of food distribution, better overall nutrition, advances in health care, health care becoming more available to more people, immunization and vaccination programs, antibiotics, improvements in public sanitation, improvements in personal hygiene, and safer water supplies that have lessened the impact of waterborne diseases.
- In the United States each year, almost 1 million teenage girls become pregnant; almost ¾ are unplanned and ¼ of those pregnancies end up in abortions.
- Approximately 30 countries (mostly in Europe and including Japan) have either achieved a stable population (growth rate below 0.3%) or a declining population (see Figure 6.2). Reasons include family planning, housing availability and associated land prices, late marriages, and high costs of education. Japan leads the world in declining population. In 2000 the population of Japan reached 127 million. By 2100, it is estimated to decline in half to almost 67 million.

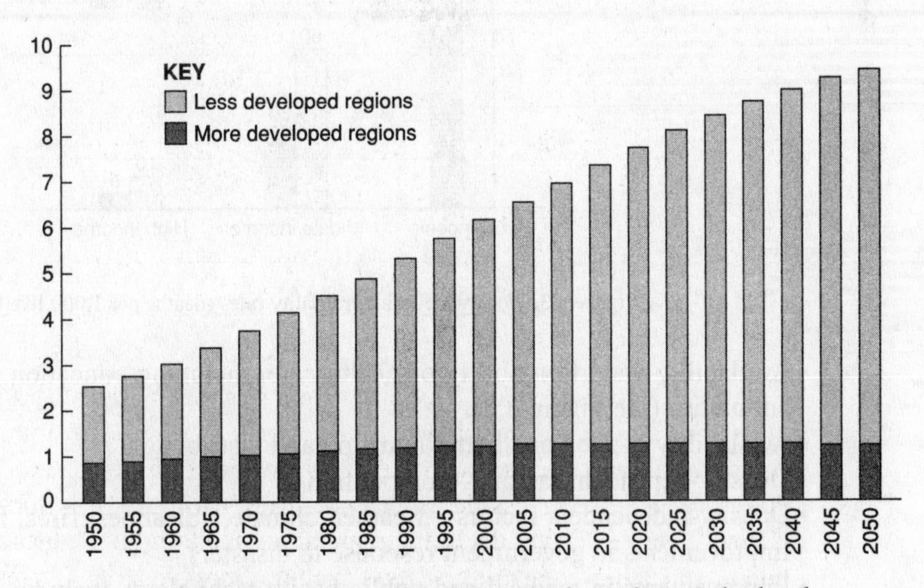

Figure 6.2. Comparison of population growth rates between developed countries and less-developed countries. (Source: United Nations).

Factors That Affect Birth and Fertility Rates

- Age of woman at time of marriage
- Availability and affordability of birth control—most governments spend less than 1% of their national budget on birth control and family planning services
- Availability of legal abortion
- Children needed in workforce
- Costs of raising and educating children
- Culture, religion, and tradition

- Density-dependent factors—competition, predation, parasitism, and so on.
- Density-independent factors—weather, climate, volcanoes, fires, floods, and so on.
- Drug and alcohol dependency of women
- Educational level and employment opportunities for women
- Government programs that reward (or coerce) parents through financial means to either limit or produce children (China levies penalties for having more than two children; the United States has tax credits for having more children. More than 90% of countries in the world have some form of family-planning programs.)
- Infant mortality
- Pension and retirement plans
- Rural versus urban lifestyle and the centralization of information and government services.

FACTORS THAT AFFECT DEATH RATES

Remember that the rapid growth of the world's population over the past 100 years was not caused by a rise in birth rates—it was due to a *decrease in the death rate!*

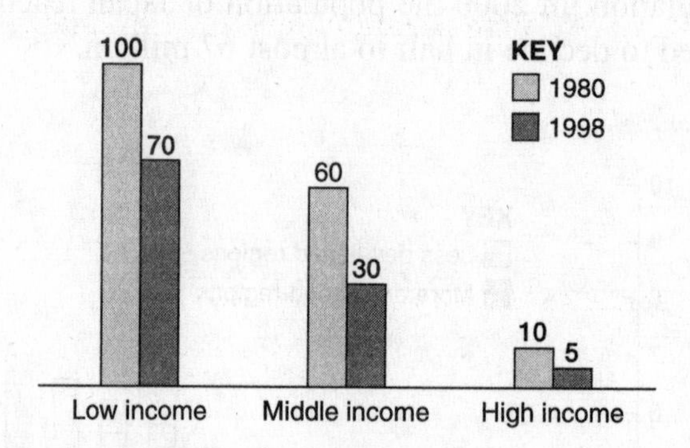

Figure 6.3. Worldwide infant mortality rates (deaths per 1,000 live births).

- Availability and affordability of health care, including education, immunization, and antibiotics (see Figure 6.3).
- Availability of food and better nutrition and distribution
- Density-dependent factors—competition
- Density-independent factors—weather, climate, volcanoes, fires, floods, and so on.
- Improvements in government response to disasters
- Improvements in medical and public health technology, including antibiotics, immunization, and insecticides
- Improvements in sanitation and pollution control
- Safer water supplies

Life Expectancy

- Life expectancy worldwide has risen on average by 4 months each year since 1970.
- Infant mortality rates fell from 80 per 1000 live births in 1980, to 54 per 1000 in 1998 (see Figure 6.4).
- Women tend to outlive men by 5 to 8 years in the countries with the highest life expectancies, but by only 0 to 3 years in countries where life expectancy is low.

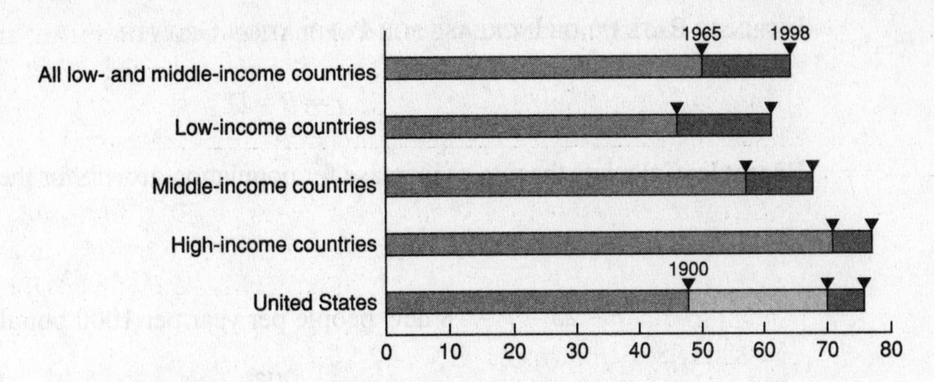

Figure 6.4. Life expectancy at birth. (Source: U.S. Bureau of the Census, *Historical Statistics of the United States.)*

- Between 1980 and 1998, the world's average life expectancy at birth rose from 61 to 67 years, with the most dramatic increases occurring in the low- and middle-income countries. Factors that affect life expectancy include (1) increased access to nutritious food, (2) access to health care, (3) safe water, (4) sanitation, (5) antibiotics and other medicines, (6) immunizations, and (7) education.

Calculating Growth

SIMPLE GROWTH RATE OF A POPULATION

$$N_1 = N_0 + B - D + I - E$$

where B = birth rate (natality), D = death rate (mortality), I = immigration (movement into the population), E = emigration (movement out of the population), N_0 = population size at an initial time, and N_1 = population size at a later time.

Example: In 1950, the population of a small suburb in Los Angeles, California, was 20,000. The birth rate was measured at 25 per 1000 population per year, while the death rate was measured at 7 per 1000 population per year. Immigration was measured at 600 per year while emigration was measured at 200 per year. Calculate the population size in 1951.

$$N_1 = N_0 + B - D + I - E$$

$$N_1 = 20{,}000 + 25\,(20) - 7\,(20) + 600 - 200$$

$$N_1 = 20{,}760$$

The population had grown by 760 people in one year.

INTRINSIC RATE (r) OF INCREASE FOR POPULATION GROWTH

$$r = B - D$$

Example: Calculate the rate of increase for population growth for the preceding problem:

$$r = B - D$$

$r = 25 - 7 = 18$ new people per year per 1000 population

RATE OF CHANGE OF POPULATION SIZE

$$\frac{dN}{dt} = rN$$

Example: Calculate the rate of change for the population described in the first example.

$$\frac{20{,}760 - 20{,}000}{1 \text{ year}} = 760 \text{ people per year}$$

NET GROWTH RATE OF A POPULATION (R_0)

$$R_0 = \frac{N_1}{N_0}$$

Example: Calculate the net growth rate of the population described in the first example. A population that was not growing would have a net growth rate of 1. A population that was doubling would have a net growth rate of 2, and so on.

$$R_0 = \frac{20{,}760}{20{,}000} = 1.038$$

DOUBLING TIME FOR A POPULATION

$$D_t = 70/R_0$$

Example: A population had a growth rate of 1.7 in 1995. In what year would the population be double its current value? Assume all other factors are constant.

The population in a region with a 1.7 rate of increase will double in 41 years (70/1.7 = 41 years).

Adding 41 years to 1995 gives the answer of 2036.

Note: The number 70 is the approximate equivalent of the natural log of 2 times 100.

ANNUAL % RATE OF NATURAL POPULATION CHANGE

$$\% = \frac{(B - D)}{1000} \times 100\%$$

Example 1: In 2000, the world birth rate was 22 births per 1000 people. The world death rate was 9 deaths per 1000 people. Calculate the percent world growth rate.

$$\% = \frac{22 - 9}{1000} \times 100\% = 1.3$$

Example 2: Using the table in the section entitled "Demographics—Birth and Death Rates," calculate the percent population growth rate for Uganda. According to the table, Uganda has a crude birth rate of 48.0 and a crude death rate of 18.4. Therefore,

$$\% = \frac{48 - 18.4}{1000} \times 100\% = 2.96\%$$

This agrees with the figure in the table and is almost three times the average world population growth rate.

LOGISTICAL GROWTH RATE
Population size is limited by the carrying capacity, K.

$$\frac{dN}{dt} = rN\frac{(K - N)}{K}$$

Note: You will not be tested on *this* equation. Relax!

Curbing Population Growth

1. Provide economic incentives for having less children.
2. Empower and educate women.
3. Extra income allows a higher educational level that postpones the age when couples have children.
4. Government family-planning services have been successful in many countries (Brazil, China, Colombia, Cuba, Costa Rica, Indonesia, Mexico, Thailand, Singapore, South Korea, Taiwan, and Venezuela) and unsuccessful in others (India, Pakistan, and Sub-Sahara African countries) due to religious and cultural factors. Family planning in China has been a success story due to repression, the ability to reach far out into the country and economic incentives such as salary bonuses, larger retirement benefits, better housing, better medical care and free tuition for one child. The family-planning program in India has been a disaster due to poor administration, traditional and cultural beliefs, and poor government support.
5. Higher economic standards usually result in fewer children because
   - Mothers and fathers are working and bringing home income,
   - Extra income allows families to focus on what is important for their children rather than focusing on meeting daily survival needs, and
   - two incomes provide for increased retirements so that the need for children to provide for aging parents is reduced.

6. Improve health care to reduce infant mortality—women would not need to have as many children if the children survived.
7. Increasing death rates is taboo, so the only option available is to reduce birth rates.
8. Organizations actively involved in curbing population growth worldwide include International Planned Parenthood Federation, Planned Parenthood Federation, United Nations Fund for Population Activities, U.S. Agency for International Development, Ford Foundation, and the World Bank.
9. Population growth stabilizes when death rate equals birth rate.
10. Reduce poverty and economic development.
11. Use universal family-planning services (90%+ of all countries have some form of family-planning services).

Demographics

Birth and Death Rates

CRUDE BIRTH RATE, CRUDE DEATH RATE, AND ANNUAL RATES OF NATURAL POPULATION CHANGE FOR SOME SELECTED COUNTRIES LISTED IN ORDER OF DECREASING PERCENT NATURAL POPULATION CHANGE (DATA SOURCE: *CIA—THE WORLD FACTBOOK 2000*).

Country	Crude Birth Rate	Crude Death Rate	Annual Rate of Natural Population Change (%)
Uganda	48.0	18.4	2.96
Ethiopia	45.1	17.6	2.75
Pakistan	32.1	9.5	2.26
Mexico	23.2	5.1	1.81
India	24.8	8.9	1.59
China	16.1	6.7	0.94
South Korea	15.1	5.9	0.93
United States	14.2	8.7	0.55
Canada	11.4	7.4	0.40
Japan	10.0	8.2	0.18
United Kingdom	11.8	10.4	0.14
Germany	9.4	10.5	–0.11

Emigration/Immigration

From 1820 until about 1875, most immigrants came from northern and western Europe. From 1875 until about 1920, immigrants from southern and eastern Europe dominated. In the last decades of the 20th century, immigrants from Latin America, the

Caribbean, and Asia were the most numerous. In 1998, the Immigration and Naturalization Service admitted a total of 660,477 immigrants (see Figure 6.5).

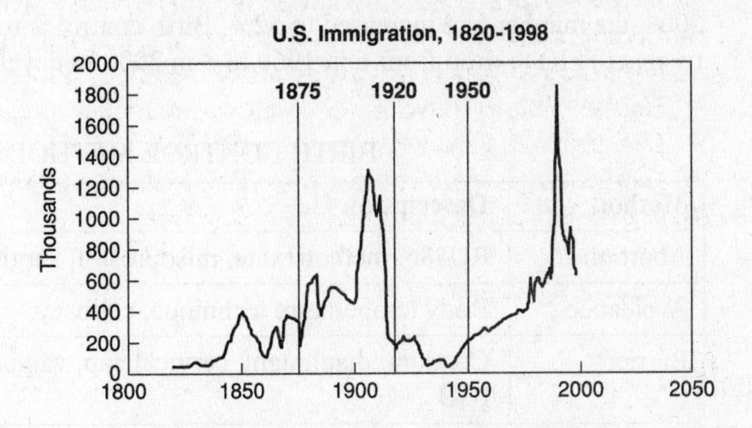

Figure 6.5. U.S. immigration from 1820 to 1998.

FACTORS THAT AFFECT IMMIGRATION AND EMIGRATION:

1. Depletion of or competition for natural resources.
2. Droughts.
3. Economic opportunity.
4. Politics (countries that accept largest numbers of immigrants include United States, Canada, and Australia).
5. Religious, political, or cultural persecution.
6. War.

HIV-AIDS

- Estimates place the number of HIV infections in the world today over 35 million, most in Sub-Sahara Africa, with infections in some countries close to 40%.
- In Asia, Cambodia, India, and Thailand will be most affected.
- In Latin America and the Caribbean, Brazil and Haiti will be most affected.
- In the United States, the rate of increase of new HIV infections has levelled off since 1992. As of 1999, the rate of HIV infection in the United States was 0.61%. Furthermore, in 1999, 850,000 people in the United States were living with HIV/AIDS with a mortality of approximately 20,000 for that year (See Figure 6.6).

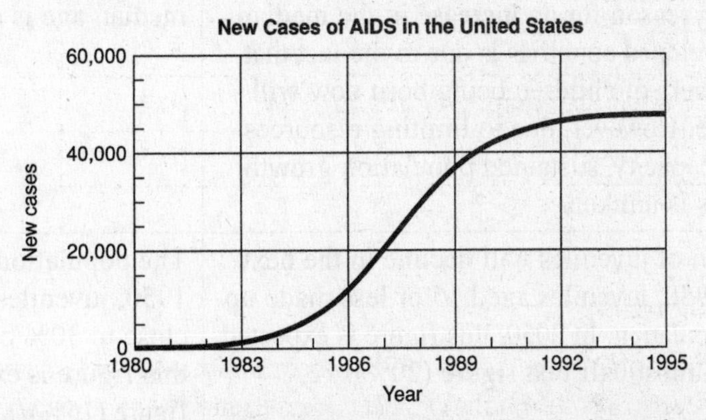

Figure 6.6. Growth curve for new cases of AIDS in the United States between 1980 and 1995. Notice how it follows the classic logistic growth model.

Family Planning

In 1960, 10% of women in developing countries used some form of birth control. In 2000, the number had increased to 52%. Birth control is responsible for the total fertility rate (TFR) to drop from 6 in 1960 to 3 in 2000 in developing countries.

BIRTH CONTROL METHODS

Method	Description
Abortion	RU486, methotrexate, misoprostol, surgical abortion
Avoidance	Body temperature technique, celibacy
Barrier	Condom, diaphragm, cervical cap, vaginal sponge, spermicide, IUD
Chemical	"The pill" (estrogen + progesterone), gossypol, Norplant, Depo-Provera, gonadotropin releasing-hormone agonist
Surgical	Tubal ligation, vasectomy, Filshie clip

Aging Trends

From 1950 to 2000, median age worldwide went from 23.5 to 26.0 years. In 2050, it is estimated that the median age will climb to 38 years.

COMPARISON OF LESS-DEVELOPED COUNTRIES TO MORE-DEVELOPED COUNTRIES IN TERMS OF AGING CHARACTERISTICS

Less-Developed Countries	More-Developed Countries
By the year 2050, it is projected that people aged 60+ may reach 21% or more of the population. This compares to less than one third that figure (6%) in 1950.	By the year 2050, it is projected that people aged 60+ may reach 30% or more of the population. This compares to less than half that figure (12%) in 1950.
Median age in 2000 was 24 (half the population was above 24 and half the population was below 24). By 2050, the median age is expected to reach 37. The surprising reason for an increase in the median age in less-developed countries is due to the fact that the large numbers of children being born now will advance in age. However, due to limiting resources and carrying capacity, sustained population growth at today's rates is unlikely.	The median age in 2000 was 37 (half the population was above 37 and half the population was below 37). By 2050, the median age is expected to rise to 46.
The population of juveniles will decline in the next 50 years. In 1950, juveniles aged 15 or less made up 38% of the population. In 2050, this figure is expected to decline to almost half that figure (20%).	The population of juveniles is declining. In 1950, juveniles aged 15 or less made up close to 30% of the population. In 2050, this figure is expected to decline to half that figure (15%).

For more information on population trends, visit the United Nations World Population Trends at www.undp.org or the CIA's World Factbook at www.odci.gov/cia/publications/factbook/index.html.

A good indicator of future trends in population growth is furnished by age structure diagrams. The age structure diagram in Figure 6.7 shows the age distribution of less-developed countries compared to more-developed countries for 2000 and projections for 2050. Notice that the bases of the diagrams (younger) become smaller and the areas near the top (older) become larger over time. When the base is large, there is a potential for an increase in population. While the more-developed nations of the world show pyramids that indicate a trend toward a slowing in the rate of increase in the population, the less-developed nations show a potential for increased population growth.

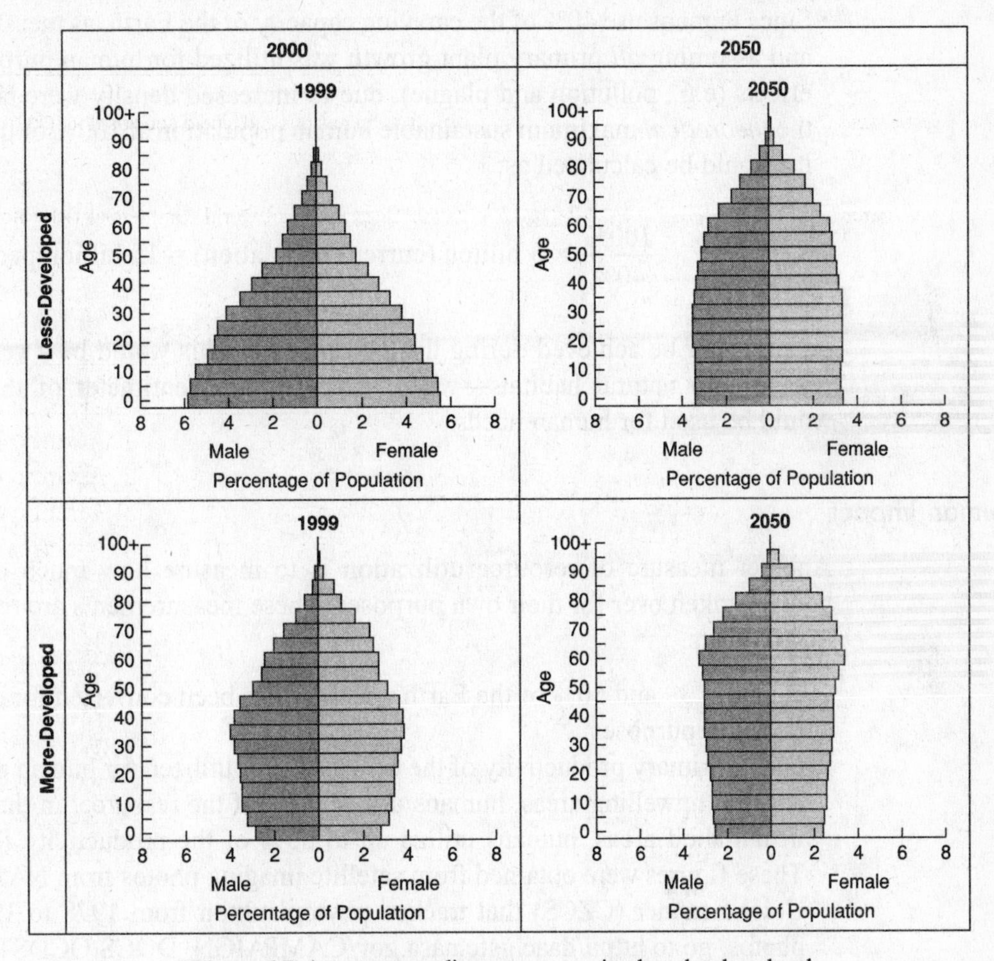

Figure 6.7. Age structure diagrams comparing less-developed and more-developed countries for the next 50 years. (Source: United Nations.)

Patterns of Resource Utilization

There are three ways to estimate the effect of humans on patterns of resource utilization:

1. Measure net primary productivity,
2. Estimate how much impact humans have had on the Earth,
3 Examine finite resources and from that, draw conclusions on increasing productivity.

Net Primary Productivity

- Net primary productivity (NPP) is the total amount of solar energy converted into biochemical energy through photosynthesis minus the energy needed by those plants for their own metabolic requirements. It can be a measure of the food resources available on Earth.
- NPP without human activity or influence has been estimated to be 150 billion tons of organic matter per year. Humans have caused a 12% decline in the NPP through deforestation. Humans utilize approximately 27% of the NPP for their own purposes (food, building material, energy, etc.) or by converting productive land to other uses. Together, this represents approximately 40% of the NPP on Earth committed to the use of humans. All other life forms are left with the 60% remaining.
- Since humans use 40% of the carrying capacity of the Earth, as measured through NPP, and assuming *all* primary plant growth was utilized for human purposes and no other effects (e.g., pollution and plague), due to increased density were taken into account, the *theoretical* maximum sustainable human population at 100% of the carrying capacity would be calculated as:

$$\frac{100\%}{40\%} \times 6 \text{ billion (current population)} = 15 \text{ billion people}$$

which *could* be achieved during the 21st century. This would be a very sad world—no wildlife, no natural habitats—where every square centimeter of the Earth available would be used for human needs.

Human Impact

Another measure of resource utilization is to measure how much of the Earth have humans taken over for their own purposes. These measurements are reflected in the following facts.

1. Between 39 and 50% of the Earth's surface has been converted for either agricultural or urban purposes.
2. Of the primary productivity of the oceans, 8% is utilized for human purposes (in nutrient-rich upwelling areas, humans utilize 25% of the resource; in the temperate continental shelf areas, humans utilize up to 35% of the productivity for human needs). These figures were obtained from satellite imaging photos from NASA's Coastal Zone Color Scanner (CZCS) that tracked phytoplankton from 1978 to 1986. To see these photos, go to http://daac.gsfc.nasa.gov/CAMPAIGN_DOCS/OCDST/OB_main.html.
3. CO_2 concentrations worldwide have increased 30% due to human activity.
4. More than 50% of all easily accessible fresh water sources are utilized by humans. This does not take into account glacial or groundwater sources.
5. More than 50% of all nitrogen fixation is caused by human activity. This includes using fertilizers extensively, raising nitrogen-fixing crops, and releasing nitrogen through combustion of fossil fuels (NO_x).
6. Humans have disrupted and introduced more than one quarter of all plant species into non-native continental areas. On some islands, this figure increases to more than one half.

7. Within the last 200 years, more than 20% of all bird species have become extinct due to human activity.
8. Almost one quarter of all marine fisheries are either depleted or exploited. One half are at the limit of becoming exploited.

Finite Resource Capacity

UTILIZATION OF NATURAL RESOURCES BY HUMANS
COMPARING 1990 TO PROJECTED 2010 VALUES.

Utilization of Natural Resources	1990	2010	Total Change (%)	Per Capita Change (%)
Population (millions)	5,290	7,030	33	—
Fish catch (million tons)	85	102	20	–10
Irrigated land (million hectares)	237	277	17	–12
Cropland (million hectares)	1,444	1,516	5	–21
Rangeland and pasture (million hectares)	3,402	3,540	4	–22
Forests (million hectares)	3,413	3,165	–7	–30

Source: Postel, S. "Carrying Capacity: Earth's Bottom Line." *State of the World*, 1994.

In summary, between 1990 and 2010,

- Human population should increase approximately one third.
- Fish catch from a limited resource (the ocean) should increase 20%, but the amount of fish catch per person will decline 10%.
- The amount of cropland, rangeland, and pasture will increase about 5%, but on a per person basis, the amount of these resources will decrease by about 22%.
- The world's forest will decline by about 7%, but with an increase in population, the amount of forestland per person will decline by almost 30%.
- Worldwide irrigated acreage should increase 17%, but the amount of irrigated acreage per person will decline 12%.
- Can you imagine what the figures would be if the world was at the carrying capacity for human population, 15 billion people?

FACTORS THAT AFFECT RESOURCE UTILIZATION

Factor	Characteristics
Population size	Large numbers of people lead to high rates of habitat loss and natural resource depletion. Africa has the largest growth rate (2.1%). The population is expected to double by 2015. Latin America is expected to double in 30 years, and Asia, in 36 years.
Population density	Most dense areas of human habitation occur in Asia and Europe. Density, more than population size, has a greater effect on the amount of pollution and use of energy. Japan, with a density of 331 people per square kilometer, imports three quarters of its grain and two thirds of its wood. It has exceeded its internal carrying capacity.
Carrying capacity	Carrying capacity is relative to the Earth as a whole, a continent, a country, and so on. Example: in 1900, 40% of the land of Ethiopia was covered by forest. Today, only 4% of Ethiopia is forested.
Technological development	More-developed countries consume more resources than less-developed countries. The United States represents about 5% of the world's population. However, the United States, a highly technologically developed country, consumes 25% of the entire world's resources and generates 25% of the world's waste. When comparing the utilization of resources between a less-technologically developed country such as India to the United States, a typical person in the United States uses: 50 times more steel 56 times more energy 170 times more synthetic rubber and newsprint 250 times more motor fuel 300 times more plastic
Famine	The population growth rate in Sub-Saharan African countries of Ivory Coast, Togo, Comoros, and Kenya are all above 3.5%, while the rate of food production is only increasing at about 1%.
Political unrest	Affects employment, food distribution, and standard of living that in turn affects utilization of resources.
Environmental degradation	Erosion, desertification, air pollution, water pollution, solid wastes, hazardous wastes, depletion of ozone layer, global warming, and the like. All these affect the amount of resources available and the cost of transforming resources into useful forms. Fewer resources, or higher costs to produce useful goods, affects how resources are distributed and utilized.
Extinction of animal and plant species	Close to 50% of the world's flora and fauna could be on a path to extinction within the next 100 years. All life forms are affected—fish, birds, insects, plants, and mammals. Out of the world's nearly 10,000 bird species, 1100, or 11%, are on the edge of extinction; and one in eight plants is at risk of becoming extinct.
Energy resources	On average, an American consumes five times the amount of grain as one Kenyan, and as much energy as 35 Indians or 500 Ethiopians.

FACTORS THAT AFFECT RESOURCE UTILIZATION (Continued)

Factor	Characteristics
Exploitation of natural resources as a function of gross domestic product	The richest 20% of the world's population contribute to resource depletion of energy, raw materials, and manufactured goods through over-consumption and waste, which is a function of pollution. The poorest 20% of the world's population deplete world resources by being forced to cut down forests, clear land, burn scarce wood, and so on.
Poverty	In 1960, the richest 20% of the world's population controlled 70% of the world income. Thirty years later, the same 20% of the population controlled 80% of world income. The poorest 20% of the world's population controls 1.5% of the world's income. The gross national product for Togo, Comoros, and Kenya averages around $500 per person as compared to about $23,000 per person in the United States.

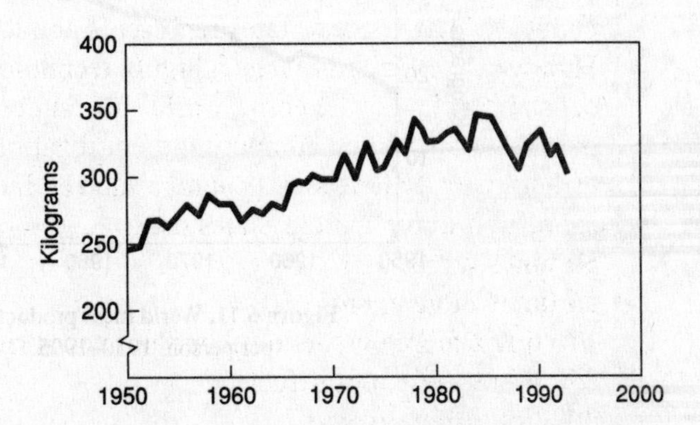

Figure 6.8. World grain production per person, 1950–1995.

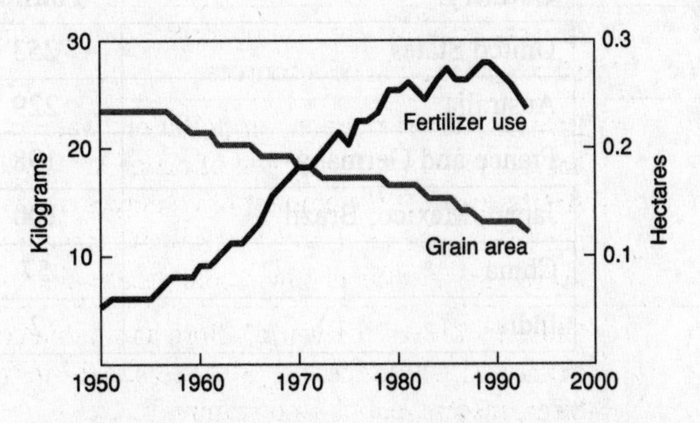

Figure 6.9. World grain harvest area and fertilizer use per person.

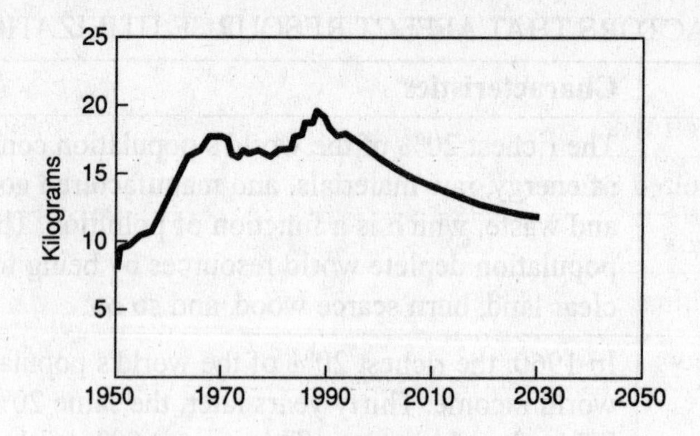

Figure 6.10. World fish catch per person, 1950–2030.

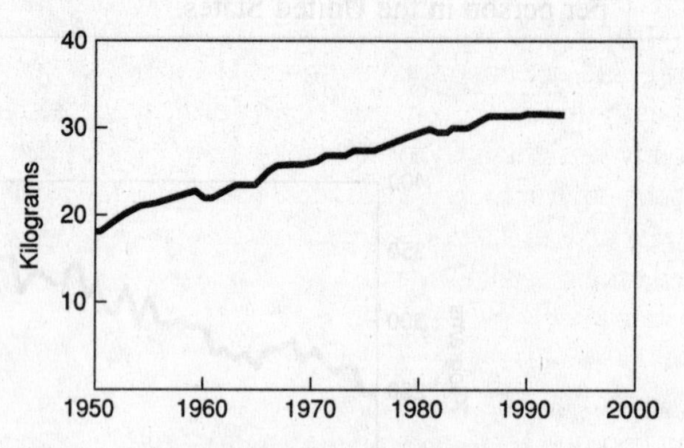

Figure 6.11. World meat production per person, 1950–1995.

ANNUAL PER CAPITA CONSUMPTION OF MEAT

Country	Pounds
United States	253
Australia	229
France and Germany	198
Japan, Mexico, Brazil	100
China	57
India	2

ANNUAL PER CAPITA GRAIN USE

Country	Pounds	Pounds Used As Animal Feed
United States	1800	880
Italy	900	550
China	660	81
India	440	97

Carrying Capacity

- Carrying capacity (K) refers to the number of individuals who can be supported in a given area within natural resource limits, and without degrading the natural social, cultural, and/or economic environment for present and future generations. The carrying capacity for any given area is not fixed. It can be altered by improved technology, but mostly it is changed for the worse by pressures that accompany a population increase. As the environment is degraded, carrying capacity shrinks, leaving the environment no longer able to support even the number of people who could formerly have lived in the area on a sustainable basis. No population can live beyond the environment's carrying capacity for very long.
- The average American's "ecological footprint" (the demands an individual with average amounts of resources, i.e., land, water, food, lumber, mining, waste assimilation and disposal, etc., puts on the environment) is about 12 acres.

Humans are K-Selected Organisms

- Humans mature slowly. Parental care is involved. Human population follows a logistic growth model (S or sinusoidal growth curve).
- Humans tend to live long lives, and there is usually a low juvenile mortality rate. Exceptions occur in extremely poor countries or countries where AIDS is prevalent (Sub-Sahara Africa).
- Humans compete well for resources. Advances in agriculture, disease and pest-resistant crops, higher-yield crops, food processing and distribution, advances in food preservation, widespread sharing and/or trading of technology (no one country can monopolize advances in technology), advances in health care, disease prevention, vaccinations, advances in family-planning services in many countries, and economic development and its ramifications make humans unique in their ability to share world resources. Most animals compete for resources, but humans tend to share or trade resources.
- Humans have few offspring at one time. Development is slower.
- Most endangered species are K-strategists. It is speculation and subject to conjecture as to whether humans are endangered and fit in this category, but there is evidence due to advances in life-destroying technologies; genocide; infanticide; terrorism; bioterrorism; wars over borders and territory; political and religious distortion and fanaticism; hate and racism due to religious, cultural, and life-style prejudices; intolerance; crime; rise in epidemic diseases such as AIDS; pollution; stress; and so on.
- Human population sizes stabilize near carrying capacity of their country resources. In fact, several countries are or will shortly experience negative growth rates.

Stages of Human Demographic Transitions

As seen in Chapter 5, increases in population follow characteristic states: exponential growth due to apparent unlimited resources, lag phases characterized by logistical growth patterns, death phases or crashes, and ultimate stabilization and population equilibrium resulting from reaching the carrying capacity of the environment. When these stages occur in human population, they are called demographic transitions. See Figure 6.12.

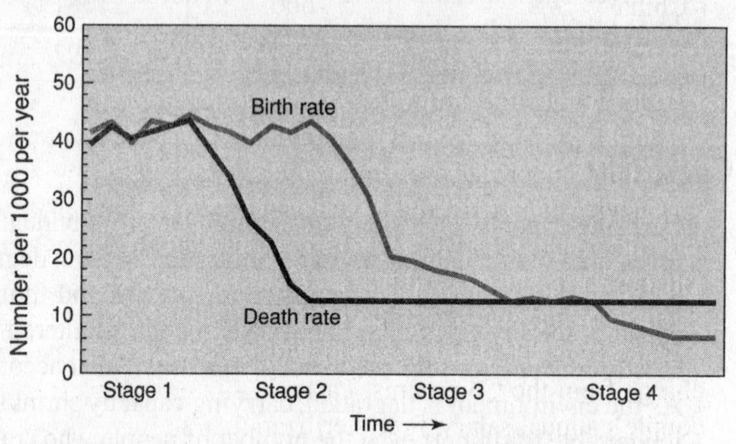

Figure 6.12. Demographic transitions occurring in human populations.

Stage	Description
Stage 1—Pre-industrial	Living conditions are severe, medical care is poor or nonexistent, and the food supply is limited due to poor agricultural techniques, preservation of food, pestilence, and distribution. Birth rates are high to replace individuals lost through high mortality rates. Net result is little population growth.
Stage 2—Transitional	This stage occurs after industrialization. For many less-developed countries of the world, this is the stage they are in now. Mortality rates drop as a result of advances in medical care, improved sanitation, cleaner water supplies, vaccinations, and higher levels of education. However, even though mortality has dropped, natality has not. The net result is a rapid increase in population. Examples include India and Pakistan. The risk is that economic growth and advancements may not allow these countries to advance to the next demographic stage (Industrial).
Stage 3—Industrial	In this stage, the country is industrialized. Most people live in urban areas and agriculture is highly advanced and run by large corporations. Food distribution is efficient, and the food supply is adequate and not a major focus of day-to-day life. Medical care, education, clean water, and improved sanitation are all in adequate supply. Retirement safety nets are in place reducing the need for extra children. Couples reduce the number of children that they want for economic reasons and leisure time is available where having more children would interfere with individual pursuits.
Stage 4—Postindustrial	This stage is reached when natality rates equal mortality rates and zero population growth is achieved, or in the case of some countries, natality rates fall below mortality rates and the population size decreases (Germany and Hungary and within a short time Japan, Russia, and many western European countries). Often these countries resort to liberal immigration policies to sustain entrenched systems.

Case Studies

China. Between 1972 and 2000, China has reduced its crude birth rate in half. This same period of time has seen the total fertility rate drop from 5.7 to 1.8. The family-planning program in China is one of the most efficient and strict programs in the world. Incentives such as extra food, larger pensions, better housing, free medical care, free school tuition, salary bonuses, and so on for parents who limit the number of children they have are responsible for the success of this program. Couples are encouraged to postpone marriage and to have only one child (if the first child is female, then the parents are allowed to have another child and still retain the financial incentives). After her first child is born, a woman is required to wear an intrauterine device, and removal of this device is considered a crime. Otherwise, one of the parents must be sterilized. Physicians also receive a bonus whenever they perform a sterilization. Couples are punished for refusing to terminate unapproved pregnancies, for giving birth when under the legal marriage age, and for having an approved second child too soon. The penalties include fines, loss of land grants, food, loans, farming supplies, benefits, jobs, and discharge from the Communist Party. In some provinces, the fines can be up to 50% of a couple's annual salary. Free sterilization, abortion, and birth control are provided by the government. An extensive mobile program also reaches into the most rural areas. China is a dictatorship and can impose the program.

India. In 1952, India began the first family-planning program. The population in India in 1952 was approximately 400 million people. Almost 50 years later, in 2000, India's population has more than doubled to 1 billion, which is 16% of the world population. Each day in India sees 50,000 new births. The per-capita income of India is near the bottom of all countries, with one third of the population earning less than 40 cents per day. In the 1970s, under Prime Minister Indira Gandhi and her son Sanjay, India instituted a forced sterilization program involving vasectomies. Reasons for the failure of family planning in India include poor planning, the low status of women, favoring male children, a democratic form of government that cannot force or coerce people, and extreme poverty.

Mexico. In some developing countries, large family size is necessary for farm labor and family income. Compounding the problem are cultural and religious issues. In these countries the death rates have decreased due to advances in social programs, but birth rates continue to remain high. In Figure 6.12, compare the age structure of Mexico in 1980 with the projected age structure diagram for 2050. For more age-structure diagrams visit http://www.census.gov/ipc/www/idbpyr.html.

Uganda. In 1988, Uganda had the highest rate of HIV infection in the world. By the late 1990s, 1.8 million Ugandans (out of a total population of 20.4 million) had died of AIDS, and 1.7 million children had lost their mothers or both parents to AIDS. Beginning in 1986, however, the government of Uganda created an AIDS commission to coordinate an extensive educational campaign to promote safer sexual behavior, prevent and treat sexually transmitted diseases, distribute condoms and teach people about their use, and develop local centers for HIV counseling and testing. The commission also encouraged religious organizations, unions, and businesses to carry out educational programs. By 1998, the adult rate of infection had fallen to 9.5%, a much lower rate than the other countries in Sub-Saharan Africa most affected by the AIDS epidemic.

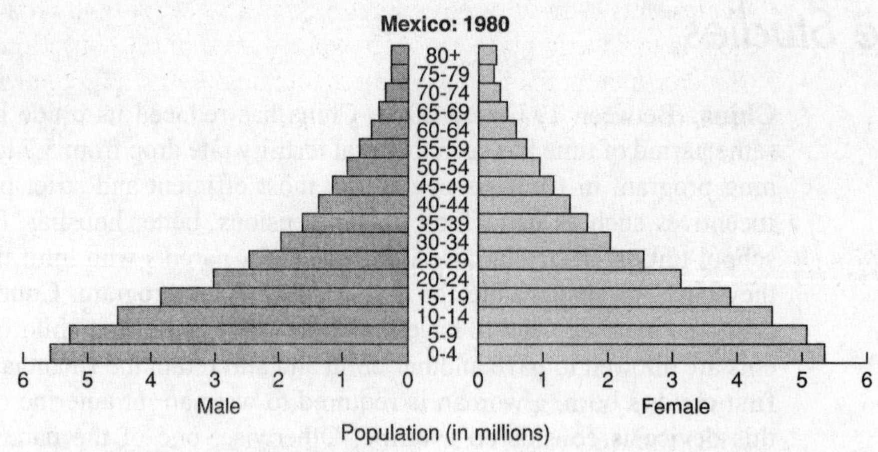

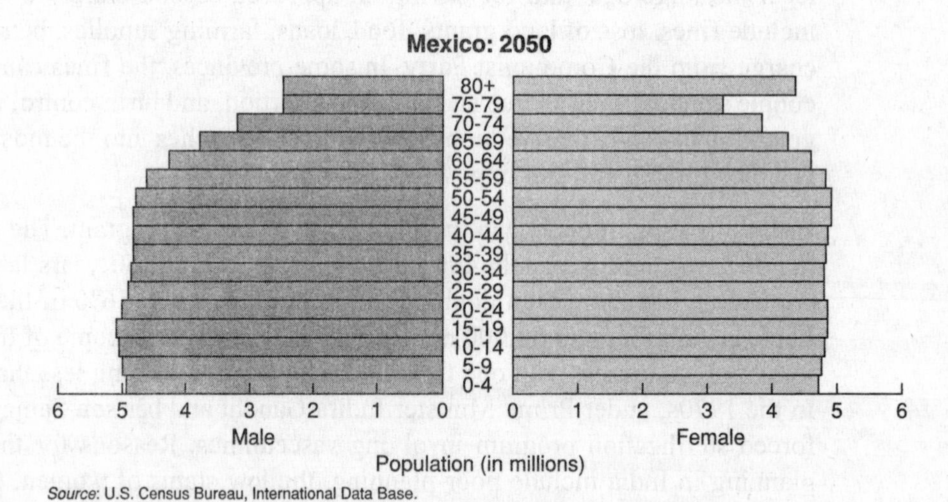

Figure 6.12. Age structure in Mexico, 1980 and 2050.

United Nations Conference on Population (Cairo, 1994). In 1994, 179 countries met in Cairo, Egypt, to develop an action plan to deal with population growth. The conference not only focused on ways of curbing population growth but also included social issues directly related to population growth such as poverty, and health care. The goal of the conference was to stabilize the human population at 7.8 billion by 2050. There were five basic components of the action plan:

1. Provide family-planning services
2. Promote free trade, private investment, and assistance to countries that need help. Recognize that raising the standard of living does not mean doing so at the cost of the environment. In fact, to achieve sustainable economic development, it must be in harmony with long-term environmental protection.
3. Address issues of gender equity, which can be achieved through health care, education, and employment opportunities.
4. Address issues of equal access to educational opportunity. Studies show that lack of or poor educational opportunity correlates with higher birth rates, less economic opportunity, and less access to medical care. The goal through education is to inform

adolescents of family planning, AIDS, and STDs (sexually transmitted diseases). The goal is to have in place by 2015, worldwide primary education.

5. Educate men in the following areas: conditions and requirements of healthy pregnancies, childcare, empowerment and rights of women, family planning, and STDs and AIDS.

Multiple-Choice Questions

1. Most populations in nature

   (A) show characteristics of a J-curve.
   (B) show characteristics of an S-curve.
   (C) show characteristics of dynamic equilibrium.
   (D) show wide fluctuations in population growth.
   (E) cannot be characterized.

2. A population showing a growth rate of 20, 40, 60, 80, . . . would be characteristic of

   (A) logarithmic growth
   (B) exponential growth
   (C) static growth
   (D) arithmetic growth
   (E) multiplying growth

3. If a population doubles in about 70 years, it is showing a _______% growth.

   (A) 1
   (B) 5
   (C) 35
   (D) 140
   (E) 200

Questions 4 and 5. On an island off the coast of Costa Rica, 500 black-throated trogons live. Population biologists determined that this bird population was isolated with no immigration or emigration of trogons. After one year, the scientists were able to count 60 births and 10 deaths.

4. The net growth rate for this population is

   (A) 0.5
   (B) 0.9
   (C) 1.0
   (D) 1.1
   (E) 1.5

5. The doubling time for this population of black-throated trogons would be

(A) 10 years.
(B) 13.3 years.
(C) 24.2 years.
(D) 63.6 years.
(E) 126 years.

6. Biotic potential refers to

(A) an estimate of the maximum capacity of living things to survive and reproduce under optimal environmental conditions.
(B) the proportion of the population or of each sex at each age category (prereproductive, reproductive, and postreproductive).
(C) the ratio of total live births to total population in a specified community or area over a specified period of time.
(D) a factor that influences population growth and that increases in magnitude with an increase in the size or density of the population.
(E) events and phenomena of nature that act to keep population sizes stable.

7. The number of children an average women would have assuming that she lives her full reproductive lifetime is known as the

(A) birth rate.
(B) crude birth rate.
(C) replacement-level fertility rate.
(D) zero population growth rate.
(E) total fertility rate.

8. The average American's "ecological footprint" (the demands an individual with average amounts of resources, i.e., land, water, food, lumber, mining, waste assimilation and disposal, etc., puts on the environment) is approximately

(A) 0.5 acres.
(B) 1 acre.
(C) 3 acres.
(D) 6 acres.
(E) 12 acres.

9. Which of the following statements is FALSE?

(A) The United States, while having only 5% of the world's population, consumes 25% of the world's resources.
(B) Up to 50% of all plants and animals could become extinct within 100 years.
(C) In 1990, 20% of the world's population controlled 80% of the world's wealth.
(D) The theoretical maximum number of people that the Earth could support is 15 billion.
(E) They are all true statements.

10. Pronatalists arguments would include all the following EXCEPT

(A) many children die early due to health and environmental conditions.
(B) children are expensive and time-intensive.
(C) children provide extra income for families.
(D) children provide security for parents when the parents reach old age.
(E) status of the family is often determined by the number of children.

11. The most successful method of controlling a country's population has been

(A) required sterilization.
(B) government quotas on children produced.
(C) birth control.
(D) financial incentives.
(E) all of the above.

For Question 12, examine the following age structure diagram:

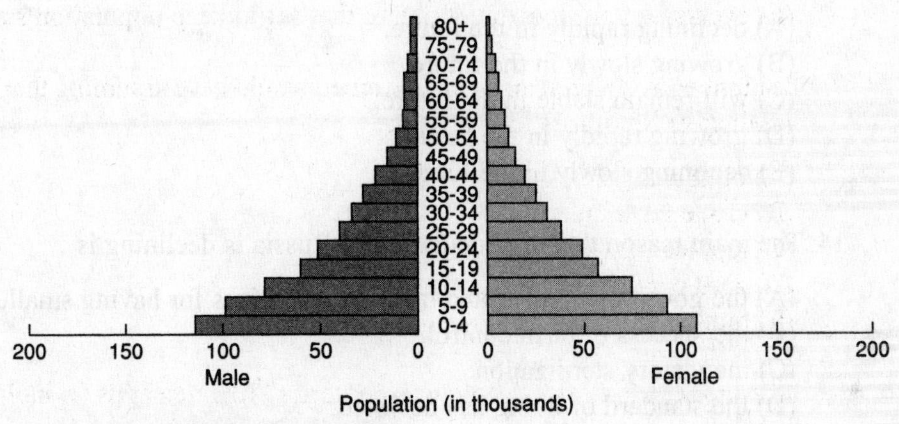

Source: U.S. Census Bureau, International Data Base.

12. This population would be typical of

(A) Russia.
(B) China.
(C) United States.
(D) Gaza Strip.
(E) Canada.

For Question 13, examine the following age structure diagram:

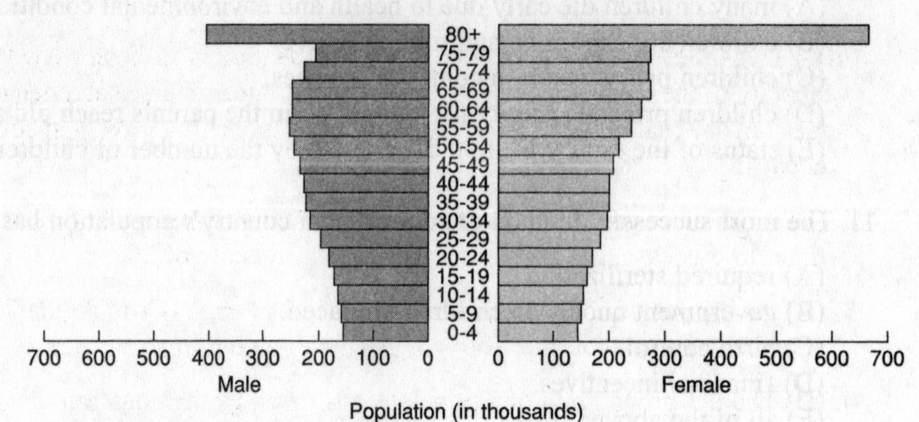

Source: U.S. Census Bureau, International Data Base.

13. This population will be

(A) declining rapidly in the future.
(B) growing slowly in the future.
(C) will remain stable in the future.
(D) growing rapidly in the future.
(E) declining slowly in the future.

14. The main reason that the population of Russia is declining is

(A) the government instituted taxation incentives for having smaller families.
(B) the success of birth control.
(C) mandatory sterilization.
(D) the standard of living has declined.
(E) massive emigration.

15. A density-independent factor would include all of the following EXCEPT

(A) drought.
(B) fires.
(C) predation.
(D) flooding.
(E) all the above are density-independent factors.

Answers to Multiple-Choice Questions

1. **C**
2. **D**
3. **A**
4. **D**
5. **D**
6. **A**
7. **E**
8. **E**
9. **E**
10. **B**
11. **C**
12. **D**
13. **A**
14. **D**
15. **C**

Explanations for Multiple-Choice Questions

1. **(C)** Dynamic equilibrium means a population that is in balance with the carrying capacity of the environment. Most populations in nature have attained a balance. This does not say that they have not gone through exponential growth at some time or had diebacks.

2. **(D)** Arithmetic growth is characterized by a constant amount of increase per unit of time. In this case, the constant amount is an increase of 20.

3. **(A)** At 1% per year, a population doubles in 70 years. If you divide 70 by the annual percentage growth rate, you will get the doubling time in years.

4. **(D)** The net growth rate of a population is given by the equation

$$R_0 = \frac{N_1}{N_0}$$

however, it is necessary to determine the final population size, N_1, before we can do the problem. Using the equation, $N_1 = N_0 + B - D + I - E$ we can substitute into the equation, $N_1 = 500 + 60 - 10 + 0 - 0 = 550$.

$$R_0 = \frac{N_1}{N_0} = \frac{550}{500} = 1.1$$

5. **(D)** Doubling time is given by the equation $D_t = 70 / R_0$.

   Therefore, $D_t = 70 / 1.1 = 63.6$ years

6. **(A)** The maximum reproductive rate of an organism is called the biotic potential. For the common housefly, *Musca domesita*, if allowed to breed and all its progeny survived and bred and produced flies, after one year there would be 5.6 trillion flies produced, or enough flies to cover the Earth many feet high. The reason that this does not occur is that there are many environmental factors (e.g., predation and juvenile mortality) that prevent this from happening.

7. **(E)** Replacement TFR is 2.1. Global TFR averages 2.9, developing countries have an average 3.2 TFR, and more-developed countries have an average 1.3 TFR. Within the United States (average 2.1), Hispanic TFR is ~2.9; Asian, TFR ~1.9; African American TFR, ~2.2, and white, TFR ~1.8.

8. **(E)**

9. **(E)**

10. **(B)** Pronatalists are people who urge people to have many children. Choice B is the only argument provided that does not promote having children.

11. **(C)** All of the choices listed have had some measure of success. China currently has in place choices B, C, D. India tried choice A, but it was a political disaster. Of all choices listed, birth control has the greatest success.

12. **(D)** This graph shows a wide base. Age structure diagrams with a wide base are populations that have a high proportion of young. Any country with many people below age 15 has a powerful built-in momentum to increase its population size, assuming that death rates do not unexpectedly increase. This age structure diagram is for the Gaza Strip in Israel in 2000.

13. **(A)** This graph is the reverse of the graph in Question 12. In this case, the top is the widest section of the structure. In this case, the majority of the population are elderly and are not reproductive. In this case, the death rate exceeds the birth rate and the population size decreases. This age structure diagram is for Hong Kong in 2050.

14. **(D)** Russia is declining in population at a rate of about 600,000 people per year. In 1992, when the Soviet Union broke up, many events occurred—hyperinflation, crime, corruption at all levels, and increased pollution and disease rates. As a result of these and other factors, the standard of living decreased dramatically (it was already low). This has resulted in an increase in the death rate at the same time birth rates are declining, resulting in a very large decrease in the population. The Soviet Union in 1950 was the fourth most populous nation in the world. Russia, only part of the former Soviet Union, is expected within 50 years to have a population similar to that of the Democratic Republic of the Congo today.

15. **(C)** Density-independent factors influence population growth and do not depend on the size or density of the population. Drought, fires, and floods do not depend on the size of the population. Predation rates are affected by population size; therefore, predation would be a density-dependent factor.

Free-Response Question

by William Aghassi, PE
Brooklyn Technical HS
Brooklyn, NY
BS (Mechanical Eng), Polytechnic Univ.
MS (Environmental Eng), New Jersey Institute of Technology

The 1990 United States census revealed that for the first time, a majority of Americans live in suburbs of major cities or in suburban-like communities.

a. List three factors why this trend may not be sustainable from a resource point of view.
b. Explain why each of the three reasons you have listed is not sustainable and the possible consequences of each.
c. What national public policy decision(s) made the United States into a suburban country as opposed to a Europe that is largely urbanized?
d. Many environmentalists think of the 21st century as the "Century of the City." State whether or not you agree or disagree with this statement and include your reasons.

Free-Response Answer

The 1990 United States census revealed that for the first time, a majority of Americans live in the suburbs of major cities or in suburban-like communities.

(a) List three factors why this trend may not be sustainable from a resource point of view.

One point for each factor listed. Total points available: 3

Three factors that may limit the trend to American suburbanization are:

1. Limit to the amount of open space available near cities.
2. Non-point source pollution (runoff) that could jeopardize the integrity of the region's watershed.
3. Traffic congestion could be a factor in limiting growth of the suburbs.
4. Diversity of plant and animal ecosystems could be jeopardized.
5. Degradation of air quality could be a limiting factor in the growth of suburbs.
6. Loss of agricultural lands.
7. Scarcity of available and cheap energy sources could be a limiting factor to the growth of suburbs.

(b) Explain why each of the three reasons you have listed is not sustainable and the possible consequences of each.

One point for each factor properly explained. Total points available: 3

1. The two features that distinguish land use in suburban towns is low-density housing and that over 60% of available land is devoted to the needs of the automobile. As suburbanization increases, these two factors tend to diminish the availability of open space quite rapidly.
2. Suburbanization near cities tends to encroach upon the watersheds of these regions. Suburban lawns, parking lots, and roads are responsible for a much greater nutrient and chemical loading than undeveloped land, and will greatly affect the water quality of the region's drinking water.
3. Suburban communities are low density and widely dispersed. Since suburban residents do not commute from or to a central location, they are usually ill-served by mass transit and car-pooling is not practical. Commuting times therefore lengthen to intolerable levels.
4. Suburban sprawl can destroy many unique and/or productive ecosystems such as wetlands and forests, thus limiting regional diversity.
5. Since suburbs are decentralized, residents must rely on the automobile for every errand, and all of their goods must be transported by truck. This often results in traffic congestion and air quality degradation that leads to respiratory distress in humans, damage to plants, and damage to buildings.
6. Land goes toward the highest value use. As suburbs encroach on agricultural areas, farmers may be forced to sell to developers due to rezoning and a resulting increase in real estate taxes, or to simply a higher rate of return than that of farming.
7. Low density single-family housing and exclusive reliance on the automobile to get around is terribly energy inefficient. If imported oil supplies become expensive or hard to obtain, this could limit suburban growth.

(c) What national public policy decision made the United States into a suburban country as opposed to a Europe that is largely urbanized?

Two points for proper listing of policy decision with explanation. Total points available: 2

After World War II, the decision to invest infrastructure dollars on roads, and to disinvest in mass transit made suburbanization possible.

The public policy decision to keep energy prices low through relatively low taxation compared with a Europe that has high-energy taxation and high-energy prices made suburban sprawl possible.

(d) Many environmentalists think of the 21st century as the "Century of the City." State whether or not you agree or disagree with this statement and include your reasons.

Two points for listing proper reason for agreement or disagreement. Total points available: 2

Agree: Cities are more energy efficient, and if well planned, are more convenient and offer a better sense of community than many suburbs. This will attract many Americans to move back to cities.

Disagree: Americans are reluctant to forgo the American dream of the single family home and the automobile to ever return to cities in large numbers.

UNIT III: RENEWABLE AND NONRENEWABLE RESOURCES: DISTRIBUTION, OWNERSHIP, USE, AND DEGRADATION

CHAPTER 7

Water

All rivers run into the sea, yet the sea is not full: Unto the place from which rivers come, thither they return again.

Ecclesiastes 1:7

Areas That You Will Be Tested On

A: Freshwater: Agricultural, Industrial, and Domestic
B: Oceans: Fisheries and Industrial

Key Terms

Note: Definitions for all key terms listed below can be found in the glossary starting on page 648. For additional definitions of relevant terms from this chapter, refer to *www.barronseduc.com/0764121618.html*.

aquifer	**intertidal zone**
biochemical oxygen demand	**renewable water supplies**
epilimnion	**reservoir**
erosion	**riparian**
estuary	**sustainable yield**
hypolimnion	**watershed**

Key Concepts

Freshwater: Agricultural, Industrial, and Domestic

- Water is essential for life
- Over 70% of the Earth's surface is covered by water:
  - 97% in oceans (not easily usable due to salt);
  - 2% in glaciers and ice caps (60% of all freshwater—not easily obtained);
  - 0.7% in groundwater;
  - 0.01% in lakes;
  - 0.005% as soil moisture;
  - 0.001% as atmospheric moisture;
  - 0.0001% in rivers and streams;
  - 0.00005% in living organisms.
- Human uses include drinking, sanitation, agriculture, transportation, industry, hydroelectric power.
- The total amount of water used by humans is 1.6×10^{15} gallons (6.1×10^{15} liters) per year
- The total amount of water used per person averages 2.6×10^{5} gallons (265,000 gallons)
- The amount of fresh water on Earth is finite.
- Since population is increasing, and increased population results in higher levels of pollution, and since freshwater sources are finite, the amount of freshwater per person is getting smaller every year.
- The primary source of freshwater is precipitation (mostly in the form of rain). About 10% of water vapor that comes from evaporation of oceans ends up as precipitation over land.
- Average rainfall of the Earth is 41 inches (104 cm) per year or about 0.1 inches per day. Annual *average* ranges of precipitation on Earth are between close to 0 inches per day up to 0.5 inch (almost half an inch) per day in some tropical rain forests.
- Areas on Earth with greatest amounts of rainfall occur:
  - Near the equator—constant solar heating and converging air masses resulting in frontal lifting
  - At midlatitudes, which fall between 30° and 60° north and south of the equator and which are influenced by Ferrel cells. Midlatitude cyclonic activity is significant and results in high precipitation in the southeastern United States.
  - On the windward side of mountain ranges near coastal regions where orographic uplift results in high amounts of precipitation.
- Areas that do not receive much precipitation include
  - Polar regions above 60° north and south latitudes: cold polar air cannot hold much water vapor.
  - Mid-continental areas—they are too far from oceans and clouds have released their moisture before they get to the deserts.
  - Subtropical deserts—air masses are subsiding.
  - Leeward side of mountains near coastal regions (rainshadow effect).

ALTERNATIVE SOURCES OF FRESHWATER

- **Desalination of ocean water** through distillation or reverse osmosis is very expensive (three to four times more expensive than any other method) and uses up a country's needed capital. Most desalination plants are located in the Middle East.
- **Massive water projects** to collect (dams), store (lakes behind dams), and transport (canals and pipelines) water. They also provide hydroelectric power and water for agricultural interests. Drawbacks include
  - Build-up of salts,
  - Build-up of silt,
  - Costs,
  - Environmental concerns and impacts,
  - Leakage and evaporation, and
  - Ownership of the water (water rights).

Examples include

1. Aswan Dam in Egypt. Loss of nutrients to fields needs to be replaced with expensive fertilizer. Decline of nutrients into Mediterranean has reduced fishing. Further complications include build-up of snail populations that harbor parasitic flatworms that cause schistosomiasis (bloodflukes) in humans.
2. Colorado River. Water rights are being disputed by California, Arizona, and Mexico, and silt build-up is a problem.
3. Hetch Hetchy in Yosemite National Park in the early 1900s. Interests of freshwater and electricity for San Franciso conflicted with John Muir and others who wanted to preserve nature. Controversy still exists.
4. James Bay project in Canada. Issues of flooding, melting of permafrost, leaching of mercury, destruction of wildlife habitat, and displacement of native people are involved.
5. Mono Lake in California. Its impact on wildlife is being challenged.
6. Sardar Sarovar Dam on the Narmada River in India. Millions of people being displaced.
7. Southeast Anatolia Project on the Tigris and Euphrates River in Turkey. Ownership of water is an issue.
8. Tehri Dam in Nepal. Fear of earthquakes is a problem.
9. Three Gorges Dam on the Chang Jiang (Yangtze) River. Building the dam caused millions of people to relocate.

- **Reuse and recycling of water** involves filtering and chemical treatment of gray and black water to be reused for agricultural, landscaping, and domestic use. California is the leading producer of recycled water.
- **"Seeding" of clouds** with dry ice, potassium iodide, or hydroscopic salts may bring water to needed areas, but it results in loss of water in other areas. Also, salts used for seeding are a source of pollution.
- **Towing icebergs** is expensive, and most would melt before reaching the destination.

AVERAGE DOMESTIC WATER USE IN THE UNITED STATES

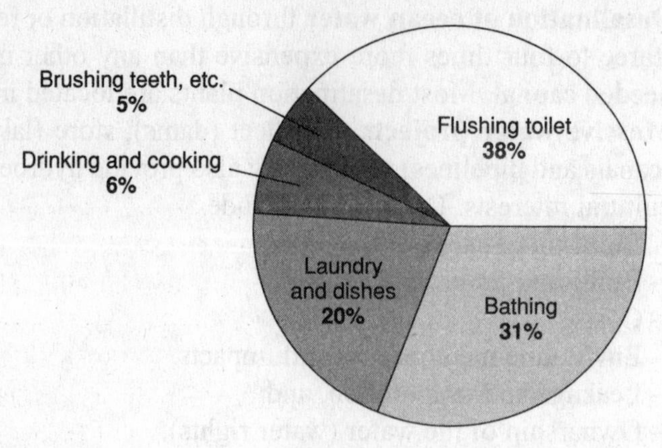

Figure 7.1.

- 4000 gallons of water produces one kilowatt-hour of hydroelectric power (enough to light a 100-watt bulb for 10 hours).
- To produce one gallon of milk, a dairy cow must drink four gallons of water. It takes eight gallons of water to grow a tomato.
- To produce a single day's supply of U.S. newsprint, 300 million gallons of water are needed.
- One penny buys 160 eight-ounce glasses of water in a typical U.S. community.
- A typical U.S. toilet uses between 3.5 and 7 gallons per flush; low-flow toilets use 1.6 gallons or even less.

CONSERVATION

- Use flow restrictors in showers. If you take a 5-minute shower, you can save 20 or more gallons.
  - Take shorter showers. Save 5 to 10 gallons for every minute you cut back.
  - Turn off water while showering, washing, and so on.
  - Use dishwashers only when full. Every load uses about 15 gallons.
- Use washing machines only for full loads. Washing machines use up to 60 gallons per load.
- Do not dump oil or other fluids down storm drains.
- Do not run water continuously when washing dishes by hand.
- Attach low-flow faucet aerators to faucets.
- Install low-flow toilets. Low-flow toilets typically use less than half the water (1.6 gallons) that standard toilets (5 gallons) use.
- Do not leave the water running while brushing your teeth and/or shaving. You can save up to 3 gallons.
- Check for leaking faucets and toilets, and repair them. A leaking tap, dripping once per second, wastes 6 gallons of water a day. A leaking toilet can waste up to 5,000 gallons per day. Put a few drops of food coloring in your toilet tank. If the coloring begins to appear in the toilet bowl without flushing, you have a wasteful leak that should be repaired at once.
- Use a nozzle on your hose that will stop the water flow when appropriate.
- Use a broom, not a hose, to clean driveways and sidewalks.
- Store drinking water in the refrigerator, rather than allowing the tap to run for a cold glass of water.
- Adjust sprinklers so that they do not water the pavement.
- Use plants and grasses that require less water.

- Water early in the morning or in the late afternoon or evening when evaporation is less.
- If you use water timers, adjust them for the season.
- Do not let the faucet run while you clean vegetables. Instead rinse them in a sink full of clean water.

CONSERVING HOT WATER

Water heaters are the second largest energy users in most homes—only space heating/cooling systems use more. Of total electricity used in an all-electric home, 25% is used to heat water for laundry, cleaning, and bathing. Here are some tips to save energy and money used for water heating:

- Lower the thermostat to between 120 and 130°F (49°–54°C). This simple action can save as much as $45 per year and reduce the risk of burns from tap water.
- Fix leaky faucets promptly. As you can see, a faucet that leaks 30 drops of water per minute uses 18-kilowatt hours per month:

  30 drops/minute = 84 gallons/month = 18 kWh/month
  60 drops/minute = 168 gallons/month = 37 kWh/month
  90 drops/minute = 253 gallons/month = 56 kWh/month
  120 drops/minute = 337 gallons/month = 74 kWh/month

- Wrap the water heater with insulation. If the water heater is located in an unheated area, wrapping it can save up to $1 per month.
- Take quick showers instead of baths. Using a flow restrictor also reduces the consumption of hot water.
- Turn off the electricity to the water heater at the main fuse box if you will be gone for three days or more.
- Use cold water with the disposal. Cold water solidifies grease so the disposal can get rid of it more effectively.
- Do all household cleaning with cold water if possible.
- When washing dishes in the sink, fill the sink and plug it instead of letting the water run constantly.
- Use cool water when washing clothes. Cold-water detergents can be used for much, if not all, of your laundry. If you must wash clothes in warm or hot water, rinse them in cold water.
- Use presoak cycle for heavily soiled loads to avoid two washings.
- Whenever possible, wash only full loads. If your washer has a water level selector, use the lowest practical level.

HOW MUCH FRESHWATER IS AVAILABLE?

- Oceans hold ~98% of all water.
- Freshwater constitutes ~2% of all water.
- Of the freshwater available, ~90% is trapped in ice and snow—85% of all ice is found in Antarctic glaciers; 10%, in Arctic ice; and 5%, on land.
- Of the ~0.3% of all water that is fresh and not trapped in snow, ~95% of it is found in groundwater, and ~3% is found in lakes, rivers and streams. Therefore, of all water on Earth, only about 0.01% is found in lakes, rivers, and streams.
- Most human settlement is determined by availability of freshwater being close by.
- Highest per capita supplies of fresh water are in countries with high rainfall and low populations (e.g., Iceland, 160 million gallons per person per year). Other water-rich countries include Surinam, Guyana, Papua New Guinea, Canada, Norway, and Brazil. These water-rich countries have very low water-withdrawal amounts. Lowest per capita

supplies are in areas with low rainfall and high populations. (e.g., Kuwait 3000 gallons per person per year). Other water-poor countries include Egypt, Israel, and most Middle Eastern countries. In these water-poor countries, water mining or withdrawal of water reserves exceeds 100% of their renewable supply.

- In the United States, renewable or replacement water averages 2.4 million gallons per person per year. The average amount withdrawn from water supplies in the United States is ~500,000 gallons per person per year.
- Use of freshwater worldwide is growing at twice the rate of population growth.

GROUNDWATER

Surface water percolates or infiltrates through the soil into aquifers—layers of porous rock, sand, and gravel where water is trapped above a nonporous layer of rock. The surface area in which water infiltrates into the aquifer is called the recharge zone. If pollution enters an aquifer, the aquifer for all practical purposes is no longer a source of water. Movement of water through aquifers is very slow. Artesian wells occur where this water breaks through to the surface. Of the freshwater available on Earth that is not trapped in glaciers, most of it (95%) is trapped underground. Aquifers in the United States hold 30 times more water than all U.S. lakes and rivers. Groundwater supplies almost 40% of freshwater in the United States. Ogallala aquifer lies between Texas and North Dakota. Removal of water from the Ogallala aquifer has occurred at a faster rate than recharge rate resulting in a drop in water table and subsidence (sinking of land) in some areas. Other areas of major subsidence include the San Joaquin Valley of California and Mexico City. Depletion of water in aquifers also leads to sinkholes and saltwater intrusion—a condition in which seawater replaces the freshwater in the aquifer making it unusable for human use.

HYDROLOGIC (WATER) CYCLE

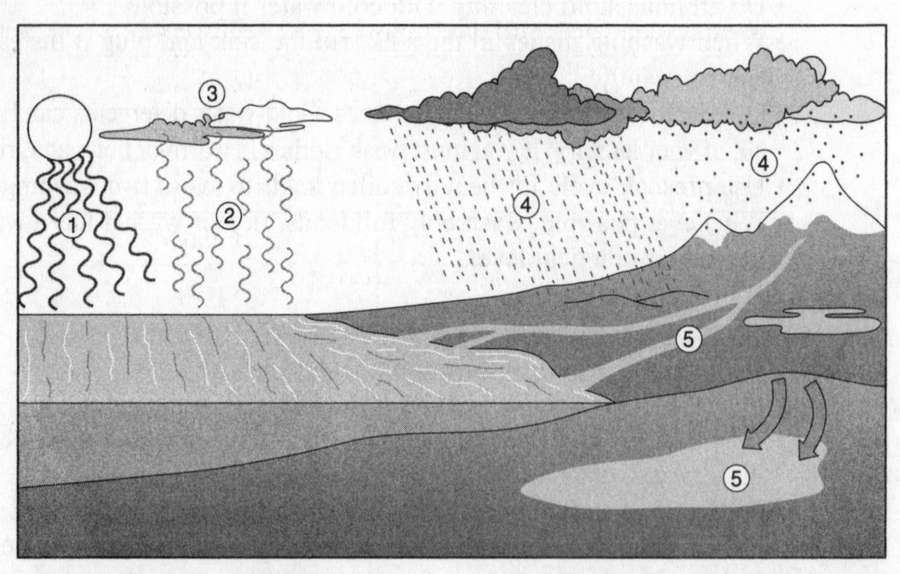

① The sun heats the ocean.

② Ocean water evaporates and rises into the air.

③ The water vapor cools and condenses to become droplets, which form clouds.

④ If enough water condenses, the drops become heavy enough to fall to the ground as rain and snow.

⑤ Some rain collects in groundwells. The rest flows through rivers back into the ocean.

Figure 7.2. Hydrologic water cycle.

WATER MOVEMENT THROUGH THE HYDROLOGIC CYCLE

Process	Description
Condensation	Changing from water vapor to liquid.
Deposition	Going directly from a gas to a solid as a result of cooling quickly.
Evaporation	Changing from liquid to water vapor below the boiling point.
Groundwater Flow	The underground flow of water due to gravity.
Infiltration	Water being absorbed by the soil.
Melting	Going from a solid to a liquid (ice to runoff).
Precipitation	Consists of rain, snow, hail, and so on.
Runoff	Excess water flowing on land as a result of gravity.
Sublimation	Going directly from the snow or ice to water vapor.
Transpiration	Water traveling through a plant from the roots and out the stomata of the leaves as a gas.

RENEWAL RATE OF H_2O FOR DIFFERENT SOURCES

Renewal rate (or residence rate) refers to the average amount of time an H_2O molecule remains in that source.

Source of H_2O	Average Renewal Rate
Groundwater (deep)*	~10,000 years
Groundwater (near surface)*	~200 years
Lakes*	~100 Years
Glaciers	~40 years
Water in the Soil	~70 days
Rivers	16 days
Atmosphere	8 days

*Realize the potential for contamination.

GENERAL PROPERTIES OF WATER

- Dissolves nutrients
- Distributes nutrients to cells
- Regulates body temperature
- Removes waste products
- Supports structures

PHYSICAL PROPERTIES OF WATER

1. Polar nature of the water molecule (Figure 7.3) makes it a good solvent (dissolver). It is able to transport molecules and inorganic compounds throughout living organisms and also through biogeochemical cycles on Earth.

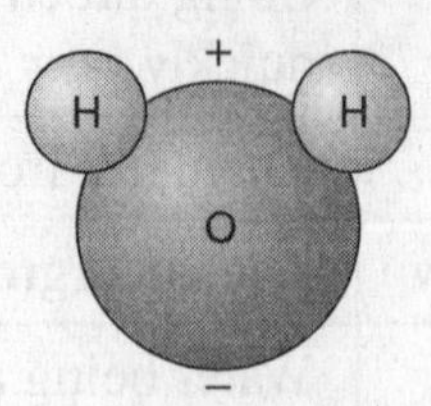

Figure 7.3. Water is a polar molecule—one side of the molecule is positively charged and the other side is negatively charged.

2. High surface tension (water forms drops) allows the formation of rain, that when reaching the Earth, ultimately dissolves nutrients in the soil and which are then transported to (and through) plants. See Figure 7.4.

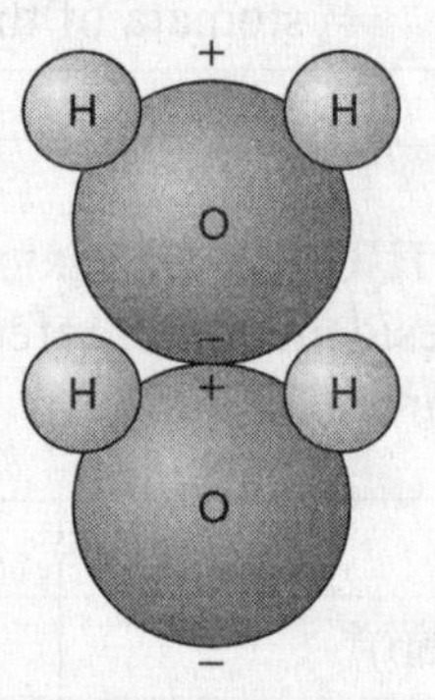

Figure 7.4. The polar nature of water causes the negative end of one water molecule to attract the positive end of another water molecule creating surface tension.

3. Water expands when it freezes, to about 9% of its volume. Ice floats. Lakes freeze on the surface, not the bottom. If water did not have this property, which is very unique, ice would build up on the bottom of oceans and lakes and eventually become a total solid block of ice. See Figure 7.5.

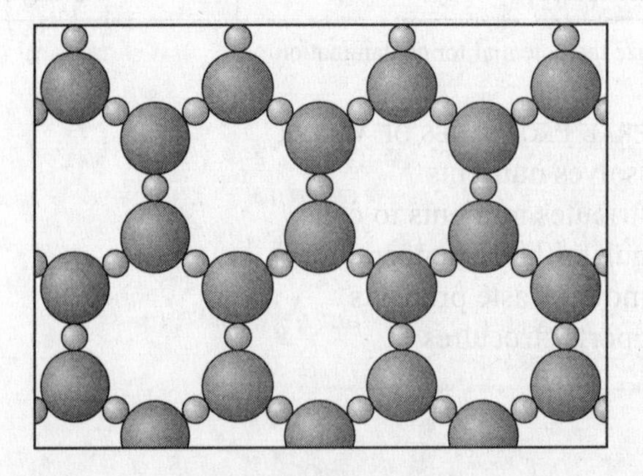

Figure 7.5. The ordered arrangement of water molecules in the solid state. Notice the hexagonal shape—the same shape that snowflakes take.

4. Water's high specific heat allows water to absorb large amounts of heat. Water also releases the latent heat very slowly. This property allows water to be a heat sink and moderates global climate. It also allows living organisms a mechanism to regulate internal heat and temperature.
5. Pure water has a pH of 7 (neutral).
6. Water conducts heat evenly. Temperatures at different depths of large bodies of water change temperature gradually.
7. Water remains a liquid over a wide temperature range, 32–212°F (0–100°C).
8. Water exists in all three phases on Earth—liquid, solid, and vapor. Evaporation of water is an endothermic process, and condensation is an exothermic process. These two processes help regulate and stabilize the temperature of the troposphere.

Uses of Water

- Agriculture uses about 70% of all freshwater. Use for agriculture depends upon national wealth, climate, and degree of industrialization. Canada uses ~10% of its water for agriculture (climate supplies most), whereas India uses about 90% of its freshwater for agriculture. Up to 70% of water intended for agricultural purposes may not reach crops in developing countries due to inefficiency—seepage, evaporation, and leakage. Drip-irrigation is used on less than 1% of crops worldwide.
- Industry uses about 25% of all freshwater, ranging from about 75% in Europe to less than 5% in developing countries. Water used for cooling of power plants is the biggest sector and creates thermal pollution. Water returns 60 times its economic value when used for industrial purposes rather than for agriculture.

WATER BUDGET

Water Leaving	Water Returning
496,000 km^3 (1.31×10^{17} gallons) of water evaporates from the Earth each year.	90% of evaporated water comes back to Earth as rain (90% of that over oceans—10% of it over land). 40,000 km^3 (1.1×10^{16} gallons) returns to ocean by runoff or underground flow.
Most water evaporates from oceans (86% comes from ocean water) and tropics than over land.	Total continental rainfall estimated to be about 110,000 km^3. Water that returns is (1) used by plants and animals, (2) flows into lakes and streams, and (3) percolates into the ground and is stored.
Latent heat is stored in ocean water and in water vapor as it evaporates and is redistributed around the Earth (heat of vaporization, ΔH_{vap}). The redistribution of heat helps to reduce temperature extremes on the Earth.	Latent heat is released as rain forms (condensation) and falls (precipitation).

Water Shortages

- Water shortages can be due to natural weather patterns, rivers changing course, flooding that contaminates existing supplies, competition for available water due to population size, overgrazing and the resulting erosion, pollution of existing supplies, and competing interests for available dollars that reduce money for water conservation programs.
- The rate of water consumption is growing faster than the population rate—almost twice as fast. With populations growing and the amount of available freshwater remaining fairly finite, the cost of freshwater will go up, impacting world economies. This in turn will affect the amount of agriculture that an area can support.
- People living in more-developed countries use about ten times more water for personal use (bathing, laundry, toilets, etc.) than people living in less-developed countries.
- About 10% of natural world runoff of freshwater (rivers, streams, etc.) is used by humans. Most of it is used to the extent that it is not directly reusable. In the United States, the largest use of water (~50%) is for agriculture.
- Water can be a limiting factor for people—it limits the amount of food that can be produced in a region. Without food being grown locally, food must be imported with associated cost factors. Economies are based on consuming products rather than producing them. Money that could be used to increase the standard of living is then spent on providing for basic survival.
- Countries in areas of low precipitation withdraw the highest percentage of water.

WAYS OF INCREASING SUPPLIES OF FRESHWATER

Method	**Advantages–Drawback(s)**
Construct dams and reservoirs.	Interferes with fish migration, destroys natural rivers and "white water." Also leakage, evaporation, and sediment buildup.
Use ground water.	Rate of use might exceed rate of recharge resulting in the ground sinking (subsidence), sinkholes, saltwater intrusion, earthquakes, and so on. Saudi Arabia obtains almost three quarters of its water this way.
Desalinate water.	Expensive; rate of production is low.
Use glaciers.	Expensive; loss due to melting in the sea if towed.
Reprocessing water	Generally the public is not supportive of drinking reprocessed toilet water. Reprocessed water could be used for irrigational purposes. However, this would require constructing separate pipelines or transporting water, which is expensive.
Making more freshwater available by using less wasteful means for agriculture. Because agriculture is the primary user of freshwater, use more efficient drip irrigation methods rather than open ditch irrigation.	Expensive. Large corporate farms can afford drip irrigation, but most single farmers in the world cannot. Sprinklers have high evaporative water losses.

WAYS OF INCREASING SUPPLIES OF FRESHWATER (Continued)

Method	Advantages–Drawback(s)
Plant crops that do not require as much water or water at night when evaporation is lower.	Market economies may make such crops not profitable due to supply-and-demand principles. No major drawbacks to watering at night.
Line irrigation channels and cover canals.	Initial start-up expense. Maintenance costs are involved. Who would pay for this? Increased costs would be passed on to consumers.
Reduce government subsidies.	Increased costs would be passed on to the consumer, driving inflation or making food less available to the poor.
Encourage recycling of products that use less water than producing the product the first time.	No drawbacks. Costs of collecting products would be outweighed by savings resulting from less water being used.
Because industry is the second largest user of freshwater, levy taxes or user fees if industries use more than their allotted share.	Prices would go up on products. Ability to compete with market products internationally would suffer when trying to compete with countries that do not levy user fees.
Reduce domestic use of water by using a tiered price scale. The more water a family uses, the more costs would go up disproportionately.	Larger families or families who provide care for other people in their home might have to pay more for water, reducing the effective family income and placing the burden of care on government institutions. Could be addressed through exemptions or allocated share of water per household member.
Reduce domestic use by switching to landscaping that does not require as much water (xeriscaping).	No drawbacks. Green lawns require large amounts of freshwater. Much of the water runs down the street as urban runoff.
Offer rebates for low-flush toilets, low-water consuming appliances, and the like.	No drawbacks. Rebates would be offered by water companies who over time, would recoup costs by having to buy less water from suppliers.
Educate the public on the costs of wasting water.	No drawbacks.
Meter all water used. Some communities still charge a flat fee for water, no matter how much is used.	Cities would reduce demand for freshwater if people knew they had to pay for every drop. Cities would recoup money spent on installing and buying water meters with having to pay less for water as people conserved more.
Engineer systems to collect more runoff.	Collecting water from urban areas and processing it to the point where it could be used for human purposes would be prohibitive due to heavy pollution and associated costs.
Seed clouds and create more rain in areas that need it.	Water distribution to other areas would be impacted. Who owns the clouds? Could be minimized through treaties and sharing arrangements.

Oceans: Fisheries and Industrial

- Although the ocean has often been looked to as an "unlimited" source of food for human populations, much of the ocean has low levels of productivity and therefore naturally low fish populations. Native fish populations are declining or are at historically low levels. This problem is evident in the growing list of threatened and endangered fish. Over 100 different types of fish are fighting for survival in the United States. Declining fish populations are indicators of broader problems in our environment. World ocean fishing has leveled off at between 80 million and 90 million tons annually. Low productivity usually results from a spatial separation of required plant nutrients, with light being restricted to surface waters and mineral nutrients being more abundant in deep waters. High-productivity hot spots occur when upwellings or other processes bring high levels of mineral nutrients to surface waters.
- Humans have a history of overfishing many commercially important ocean fisheries, depleting populations to levels where they are no longer economical to harvest.
- Of the total global catch of fish in 1989 (99,535 million tons), nearly 30% was used for purposes other than human consumption (e.g., fish meal and fish oil, which is then used as feed for livestock or cooking oil). In 1989, 130,000 tons of fish caught in the North Sea were used to produce fish meal that produced 28,000 tons of farmed salmon
- Almost one third of the total world catch—over 27 million tons—is made up of "non-target" fish, marine mammals, sea turtles, and seabirds that are accidentally ensnared in fishing nets. The vast majority of this "by-catch" is thrown overboard either dead or dying. Many of theses same practices, including bottom trawling and dredging, also damage coral reefs, rocky ridges and boulders, kelp forests, and other habitats that are used by marine life as breeding grounds, food sources, and protection from predators.
- Maximum sustained yield is the largest amount of marine organisms that can be continually harvested from a population, without causing the population to crash. For logistic growth, maximum sustained yield occurs when the population is maintained at half its carrying capacity. Fixed fish harvests at the maximum sustained yield can be unstable in the presence of environmental variability. Fixed effort or regulated escapement strategies reduce the likelihood of overharvesting the population but lead to variability in the economic return.
- The expense incurred in fishing exceeds the profit from the sale of fish. Fishing industries worldwide only survive through government subsidies.
- Fish and other aquatic species often disappear first when ecosystems are altered or the environment is polluted. Common threats to fish include land use effects (erosion, sedimentation, and altered stream flows), dams and obstructions, pollution, invasive species, and overfishing. Example: mangrove forest provides important habitats for over 2000 species of fish invertebrates and marine plants, many of which are unique to their particular area. The Philippines had 1469 km^2 (363,000 acres) of mangrove forest in 1980; by 1989 this was reduced to 380 km^2 (94,000 acres).
- Methods to restore habitats suitable for fish include planting native vegetation on stream banks, rehabilitating in-stream habitats, controlling erosion, controlling invasive species, restoring fish passage around man-made impediments, and monitoring and regulating recreational and commercial fishing to reduce impact on the habitat.
- Mariculture, is commonly termed "fish farming." It includes the commercial growing of marine animals and plants in water. The farming of aquatic organisms (aquaculture) including fish, mollusks and aquatic plants—requires some form of intervention in the rearing process, such as stocking, feeding, protection from predators, and fertilizing water.
- Mariculture is growing at about 6% annually. It provides more than 15% of total fish production worldwide and most of it (75%) comes from less-developed countries.

Currently, the main types of marine organisms being produced through mariculture include seaweeds, mussels, oysters, shrimps, prawns, salmon, and other species of fish. Clam production increased 60% between 1994 and 1995, with scallops increasing 500% the following year through mariculture.

- Fish (mostly salmon) make up about two thirds of total mariculture production. Seaweed (kelp) (*Porphyra* and *Eucheuma*) makes up about 17% of mariculture output. Kelp is used as food and as a colloid (binder) in food processing. Shellfish, including shrimp, mussels, oysters, and to a smaller extent, abalone make up about 16% of mariculture output.
- In comparing mariculture to wild-harvesting, 80% of mollusks that are consumed are raised through mariculture as compared to 20% wild-harvested. 75% of kelp that is used is produced through mariculture, 40% of shrimp, and only about 2% of all marine fish.
- Mariculture includes the following techniques: (1) capturing in the wild and rearing in controlled environments, (2) seeding natural habitats as with oysters and abalones, (3) ranching as in salmon hatcheries, (4) using enclosures in the sea in which animals live in confined pens, and (5) using land-based ponds and tanks.
- Mariculture offers several advantages to raising livestock in that cold-blooded marine organisms convert more feed to usable protein than do livestock. Example: for every million calories of feed required, a trout raised on a "farm" produces 30–40 grams of protein, chickens produces 15 g of protein, and cattle produce 2 g of protein. For every hectare of ocean, intense oyster farming can produce about 58,000 kg of protein while natural harvesting can only produce about 10 kg.
- For mariculture to be profitable, the species must be (1) marketable, (2) inexpensive to grow, (3) trophically efficient, (4) at marketable size within 1 to 2 years, and (5) disease resistant.
- Mariculture offers possibilities for sustainable protein-rich food production and for economic development of local communities. However, mariculture on an industrial scale may pose several threats to marine and coastal biological diversity due to (1) wide-scale destruction and degradation of natural habitats, (2) nutrients and antibiotics in mariculture wastes, (3) accidental releases of alien or modified organisms resulting from modern biotechnology, (4) transmission of diseases to wild stocks, and (5) displacement of local and indigenous communities.
- Mariculture creates dense monocultures that reduce biodiversity within habitats. It also requires large levels of nutrients in the water and thereby lowering the level of oxygen in the water through eutrophication. It also can have negative impact on wetlands as in the case of shrimp farming in Southeast Asia, which destroyed millions of hectares of mangrove swamps when they were cleared, destroying the natural habitats of plants, amphibians, reptiles, birds, fish, insects, and mammals.
- The oceans provide about 1% of all human food and about 10% of protein. South Korea obtains 70% of its protein requirements from the ocean; Japan, 45%; and the United States, about 3%.
- Pacific, Atlantic, and Mediterranean fisheries are generally in decline. Only in the Indian Ocean are the fish catches increasing. Species of fish in shallow water worldwide have already been exploited. Deep-water species are currently being exploited.
- Driftnet are made of invisible filament mesh. They are used in the open ocean and catch and hold fish by their gills. Drift nets catch marine mammals, birds, sharks, whales, dolphins, sea turtles, and nontarget fish. During the late 1980s, ten thousand dolphins and whales and millions of sharks were killed each year by drift nets. In 1993, the United Nations banned large-scale driftnetting in the open ocean with limited success. Smaller driftnets are used in coastal waters including those of the United States.
- Figure 7.6 shows the average annual catch of commercial fish distributed along lines of latitude.

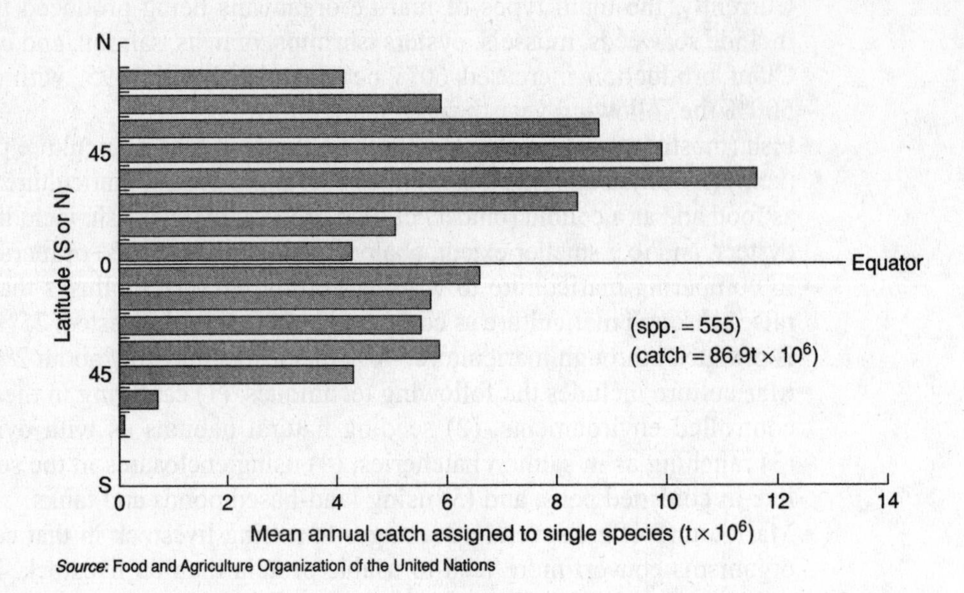

Source: Food and Agriculture Organization of the United Nations

Figure 7.6. Latitudinal distribution of world commercial fish catch.

COMMON COMMERCIAL CATCHES*

Area	1950	1990
Northeast (New England and Mid Atlantic)	1. Menhaden 2. Redfish 3. Atlantic herring 4. Haddock 5. Unclassified finfish for bait and animal food *90 species total*	1. Atlantic herring 2. Atlantic cod 3. Atlantic Surf Clam 4. American lobster 5. Ocean Quahog *132 species total*
Southeast (South Atlantic and Gulf)	1. Menhaden 2. Blue crab 3. Striped mullet 4. Marine shrimp (includes brown and white shrimp) 5. Eastern oyster *85 species total*	1. Menhaden 2. Brown shrimp 3. Blue crab 4. White shrimp 5. Striped mullet *146 species total*
Pacific† (not including Alaska or Hawaii)	1. Pacific sardine 2. Yellowfin tuna 3. Jack mackerel 4. Skipjack tuna 5. Albacore tuna *76 species total*	1. Chub mackerel 2. California market squid 3. Sea urchins 4. Ocean shrimp 5. Dover sole *146 species total*

Top five species landed, by weight, in three of the five major fisheries regions, for 1950 and 1990. Species may be more heavily targeted from one year to another because of changes in government regulations, fish populations, or market value.

†Kelp ranked second in 1950 and first in 1990 by weight landed in the Pacific.

Source: NOAA, National Marine Fisheries Service

COMMON OVERFISHED OR DEPLETED SPECIES

Alaska Pollock
Northeastern Pacific

Overfished

More pounds of Alaskan pollock were caught in 1993 than any other fish in the United States. Steller sea lions are declining in number. Biologists believe that food shortages due to pollock fishing may be one of the major causes of the sea lions' population decreases. In the Pacific, the pollock population has declined by one half since 1900. The Bering Sea Pollock Fishery is dominated by 20 factory trawlers and a highly efficient fleet of catcher boats, which continue to land nearly two billion pounds a year, even though pollock numbers have declined by 50% in 10 years.

Albacore Tuna
Pacific

Overfished

Albacore is overfished in the North and South Atlantic. In the Pacific and Indian Oceans, it is considered fully fished, and it is currently overfished in the Atlantic. The number of North Atlantic albacore catches has declined from 1964 to 1995. Peak catches occurred in 1964, when 64,400 metric tons were landed. The catch was 26,134 metric tons in 1995. Most of the recent catches were taken in surface fisheries, mainly by bait boats and trolled lines, although drift gillnets and pelagic trawls were also used. Before 1987, up to 50% of the catch was by longline gear. In 1993, the International Commission for the Conservation of Atlantic Tunas estimated the long-term potential yield for North Atlantic albacore to range from 37,340 to 68,930 metric tons. The recent average catch for the stock was 30,850 metric tons, of which the United States landed 376 metric tons.

Atlantic Cod
Northeastern Atlantic

Depleted

The numbers of Atlantic cod have fallen to historically low levels throughout the Atlantic. Overfishing has been the most important factor; however, it may have been aided by environmental changes. In 1992, the Canadian government closed fishing for Atlantic cod off of Newfoundland, and more than 50,000 people lost jobs. Since the moratorium was imposed, the northern cod stocks have remained at very low levels with little or no sign of improvement. Several problems have contributed to the decline in Atlantic cod: heavy fishing pressure, changes in water temperature and salinity, and decline in the cod's prey, a fish called capelin. The odds of a cod egg surviving to adulthood are roughly one in a million. To overcome this adversity, a single female cod may produce as many as 12 million eggs.

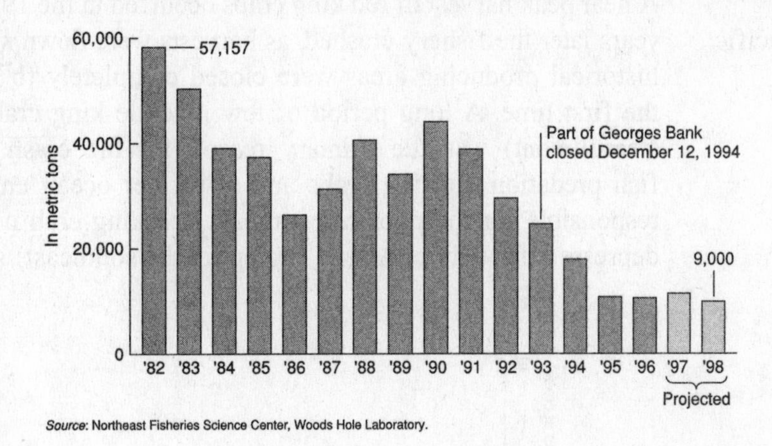

Figure 7.7. Atlantic cod catch, 1982–1996.

COMMON OVERFISHED OR DEPLETED SPECIES (Continued)

Atlantic Mackerel
Northeastern Atlantic
Overfished

Atlantic mackerel are iridescent blue-green above with a silvery white underbelly. Twenty to thirty black bars run across the top half of their body, giving them a distinctive appearance. The efficient spindle shape of their body and their strong tall fin give these fish their ability to move swiftly through the water. The catch reached a peak of roughly 430,000 metric tons in 1973. The catch for 1998 totaled 30,100 metric tons, of which 15,100 metric tons was taken by the United States (14,400 mt commercial; 700 mt recreational).

Atlantic Redfish (Ocean Perch)
Northeastern Atlantic
Overfished

Redfish (ocean perch) are viviparous, which means that the eggs are fertilized internally, and the eggs develop within the female's body. After hatching, up to 40,000 young are retained in the ovaries of the female until the yolk sac of each is absorbed. Then the young fish are born, alive and free swimming. Redfish commonly live to be about 40 years old. In 1959, 389,000 metric tons of redfish were caught by commercial fishermen. Today, strict quotas are imposed on the catch.

Blue-fin Tuna
Atlantic
Depleted

Atlantic blue-fin tuna are among the most powerful, as well as largest, fish inhabiting the ocean. Preying on mackerel, herring, and squid, giant blue-fin tuna may reach weights of more than 1400 pounds, grow as long as 15 feet, and live 30 years. Tuna swim at speeds approaching 55 miles per hour. Since 1970, there has been a 90% decrease in this species. Each female can release up to 30 million eggs. In 1973, the Japanese were paying 5 cents per pound. Today, prices are as high as $100 per pound. In 1964, 35,000 metric tons of blue-fin tuna were caught; today, one blue-fin tuna can be worth $20,000. There is no industry.

Haddock
Northwestern Atlantic
Depleted

Gulf of Maine haddock declined from about 5000 metric tons annually in the mid-1960s to less than 1000 metric tons in 1973. In 1998, 1000 metric tons of haddock were caught in the Gulf of Maine. From 1935 until 1960, haddock from Georges Bank averaged more than 150,000 metric tons per year. In 1998, 5200 metric tons were caught. Haddock prey primarily on small invertebrates, although adult haddock will occasionally consume fish. It is only within the last few years that a small increase in the haddock population has been noticed.

King Crab
Northeast Pacific Waters
Depleted

A near peak harvest of red king crabs occurred in the 1980–1981 season, but three years later the fishery crashed, as harvests were down sixty-fold, and the four top historical producing areas were closed completely to red king crab fishing for the first time. A long period of few juvenile king crabs surviving to adult size (recruitment) was the primary reason for the crash. Biologists theorize that fish predation on king crabs and a warmer ocean environment were probably responsible for the poor recruitment. Red king crab populations have remained depressed throughout Alaska (except in the southeast) since 1983.

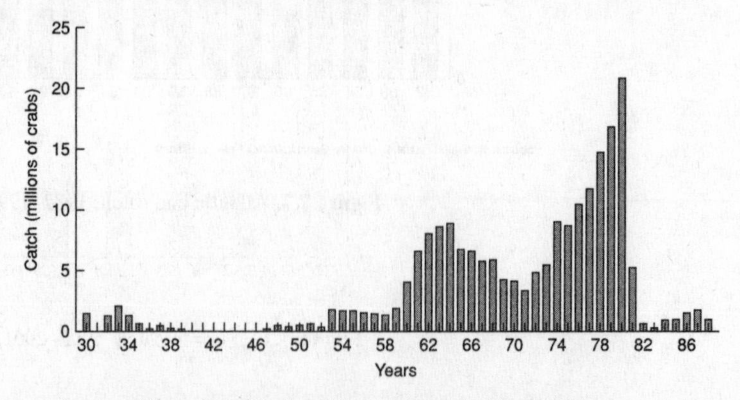

Figure 7.8. Alaskan king crab catch, 1930–1988.

COMMON OVERFISHED OR DEPLETED SPECIES (Continued)

Pacific Halibut
Northeastern Pacific
Overfished

In 1992, the entire U.S. commercial fishing season for Pacific halibut off Alaska had been reduced to two 24-hour openings per year. While the halibut fisheries of California, Oregon, and Washington are well managed in terms of the number of fish that can be caught, there is great concern about unacceptably high levels of bycatch when certain types of fishing gear are used. The state of California closed its 2000 halibut season early because of marine-mammal bycatch.

Salmon
Northeastern Pacific
Overfished

Wild Atlantic salmon (by definition being at sea for two winters) populations in the northwest Atlantic were at an all-time low of 80,000 in 1999. The aquaculture industry's proposed use of foreign (exotic species and nonlocal river strains) of farmed fish poses a major threat to wild salmon. When farmed fish escape from their sea cages, they invade the closest rivers, bringing with them the potential to transmit disease and parasites and to genetically pollute and destroy wild salmon populations. It is critical that appropriate regulations are in place to protect wild salmon against these invasions.

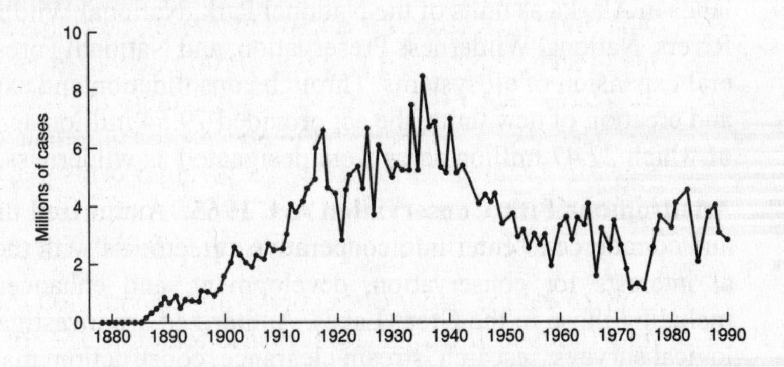

Figure 7.9. Wild salmon catch, 1880–1990.

Shrimp
East Central Pacific
Overfished

Nearly 70% of the world's fisheries are fully fished or overfished, and about 60 billion pounds of fish, sharks, and seabirds die each year as bycatch, caught accidentally as a result of wasteful fishing techniques. Shrimp fishing ranks the highest: for every pound of shrimp caught, between 4 and 10 pounds of marine life are discarded, dead or dying, back into the ocean. Shrimp farms are no better, spilling pesticides into surrounding waterways and destroying over a quarter of the world's mangrove forests. Currently, only a few types of shrimp are harvested somewhat more sustainably, and they are difficult to locate.

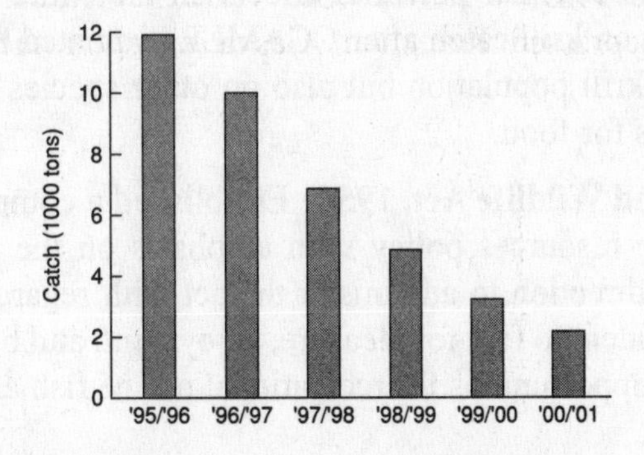

Figure 7.10. Shrimp catch, 1995–2001.

COMMON OVERFISHED OR DEPLETED SPECIES (Continued)

Silver Hake Northwestern Atlantic **Abundant**	Silver hake are strong, swift swimmers and voracious feeders. Their prey includes a variety of fish such as herring, mackerel, menhaden, and silversides. Silver hake normally measure around 14 inches in length. Following entrance of foreign fleets to the Northwestern Atlantic fishery in 1962, nominal catches increased rapidly to a peak of more than 350,000 metric tons in 1965, but declined to only 55,000 metric tons by 1970. With the 1977 Magnuson Fishery Conservation and Management Act (MFCMA) catches are currently averaging about 16,000 metric tons per year.

*For more information, visit http://www.mbayaq.org/.

LEGISLATION REGARDING AQUATIC ENVIRONMENTS

Alaska National Interest Lands Conservation Act, 1980. Designated certain public lands in Alaska as units of the National Park, National Wildlife Refuge, Wild and Scenic Rivers, National Wilderness Preservation, and National Forest Systems, resulting in general expansion of all systems. Through consolidation and expansion of existing refuges, and creation of new units, the act provided 79.54 million acres of refuge land in Alaska, of which 27.47 million acres were designated as wilderness.

Anadromous Fish Conservation Act, 1965. Authorized the secretaries of the interior and commerce to enter into cooperative agreements with the states and other nonfederal interests for conservation, development, and enhancement of anadromous fish, including those in the Great Lakes. Authorized are investigations, engineering and biological surveys, research, stream clearance, construction, maintenance and operations of hatcheries and devices and structures for improving movement, feeding and spawning conditions. Also included are provisions to make recommendations to the Environmental Protection Agency concerning measures for eliminating or reducing polluting substances detrimental to fish and wildlife in interstate or navigable waters, or their tributaries.

Atlantic Striped Bass Conservation Act, 1984. Recognized the commercial and recreational importance, as well as the interjurisdictional nature, of striped bass, and established a unique state-based, federally-backed management scheme.

Convention for the Conservation of Antarctic Marine Living Resources (CCAMLR), 1980. Treaty that required that regulations managing all southern ocean fisheries consider potential effects on the entire Antarctic ecosystem. In 1991, a limit was set on krill catch after CCAMLR evaluated the impact of the krill harvest not only on the krill population but also on other species that depend on these tiny shrimp-like animals for food.

Fish and Wildlife Act, 1956. Established a comprehensive national fish, shellfish, and wildlife resources policy with emphasis on the commercial fishing industry but also with a direction to administer the act with regard to the inherent right of every citizen and resident to fish for pleasure, enjoyment, and betterment and to maintain and increase public opportunities for recreational use of fish and wildlife resources.

Fish and Wildlife Coordination Act, 1980. Authorized the Secretary of the Interior and the Secretary of Commerce to assist in training of state fish and wildlife enforcement personnel to cooperate with other federal or state agencies for enforcement of fish and wildlife laws and to use appropriations to pay for rewards and undercover operations. Authorized financial and technical assistance to the states for the development, revision, and implementation of conservation plans and programs for nongame fish and wildlife.

Great Lakes Fish and Wildlife Restoration Act, 1998. Established goals for the U.S. Fish and Wildlife Service programs in the Great Lakes and required the service to undertake a number of activities specifically related to fishery resources.

Marine Mammal Protection Act, 1972. Established federal responsibility to conserve marine mammals, with management vested in the Department of Commerce for cetaceans and pinnipeds other than walrus. The Department of the Interior is responsible for all other marine mammals, including sea otter, walrus, polar bear, dugong, and manatee.

Non-indigenous Aquatic Nuisance Prevention and Control Act, 1990. Established a broad new federal program to prevent introduction of and to control the spread of introduced aquatic nuisance species and the brown tree snake.

Salmon and Steelhead Conservation and Enhancement Act, 1980. Established a salmon and steelhead enhancement program to be jointly administered by the Departments of Commerce and Interior and established a Washington State and Columbia River conservation area.

Case Studies

Aral Sea. In the 1970s the Amu Day'ya and Syr Dar'ya Rivers in Kazakhstan were diverted from emptying into the Aral Sea. Since that time, the Aral Sea has lost two-thirds of its volume.

Aswan High Dam, Egypt. Completed in the 1970s, the Aswan High Dam in Egypt was built to supply irrigation water. The water that is available is only half of what was expected due to evaporation and losses to due seepage in unlined canals, up to 15 billion cubic meters (4×10^{12} gallons) each year. Other problems that were encountered included the elimination of nutrients onto farmland that required the use of expensive fertilizer and the depletion of nutrients into the Mediterranean causing a decline in certain fish catches. Finally, the proliferation of snails caused an epidemic of shictosomiasis, in some areas up to 80% of the people are infected.

Bangladesh. In the 1960s, thousands of wells were dug in Bangladesh and West Bengal, India by foreign governments and humanitarian organizations in an effort to supply freshwater to the population. Shortly thereafter, arsenic poisoning began to appear in the population, and as many as 200 million people showed signs. Arsenic compounds from the soil apparently are leaching into the groundwater.

Chesapeake Bay, Maryland. Sediments in Chesapeake Bay have shown the presence of bacteria that are producing methylated tin.

Colorado River Basin. Diversion of water from the Colorado River has led to disputes between California, Arizona, and Mexico over water rights. Glen Canyon and Boulder Dam also trap large quantities of silt (10 million metric tons per year). Farm irrigation has resulted in high levels of sodium chloride in the alkaline soils to become incorporated in agricultural runoff that eventually ends up in the Colorado River. Millions of hectares of valuable farmland are now useless due to salt build-up in Colorado River water.

James Bay Project, Canada. Diversion of three major rivers into Hudson Bay to generate electrical power has resulted in massive flooding of lands that were used by the native Cree Nation. In addition, mercury has leeched out of rocks and into the water with nearby residents showing signs of mercury poisoning. Ten thousand caribou drowned during migration.

Ogallala Aquifer, United States. The Ogallala Aquifer underlies eight states from Texas to North Dakota. Before aquifer mining began, the Ogallala Aquifer held more water than all freshwater lakes, streams, and rivers on Earth. Due to aquifer mining, depletion of water in many locations is common, affecting entire communities.

Three Gorges Dam. At the time of its founding in 1949, the People's Republic of China had no large reservoirs and 40 small hydroelectric stations. By 1985, centrally planned projects to generate electricity through hydropower, increase irrigation coverage, and control flooding had resulted in the construction of more than 80,000 reservoirs and 70,000 hydroelectric stations. The world's largest hydroelectric dam is being built at the Three Gorges on the Yangtze River in China. The entire project, to be completed in 2009, has required the Chinese government to relocate an estimated 1.2 million people. Worldwide estimates are that 30 million to 60 million people have been forcibly moved from their homes to make way for major dam and reservoir projects. These "reservoir refugees" are frequently poor and politically powerless; many are from indigenous groups or ethnic minorities. The experience of more than 50 years of large dam building shows that the displaced are generally worse off after resettlement, and more often than not they are left economically, culturally, and emotionally devastated.

Multiple-Choice Questions

1. Water vapor returning to the liquid state is called

    (A) evaporation.
    (B) transpiration.
    (C) boiling.
    (D) condensation.
    (E) vaporization.

2. The temperature at which air becomes saturated and produces liquid is called

    (A) the saturation point.
    (B) the dew point.
    (C) the condensation point.
    (D) relative humidity.
    (E) absolute humidity.

Refer to the following diagram to answer Questions 3 and 4.

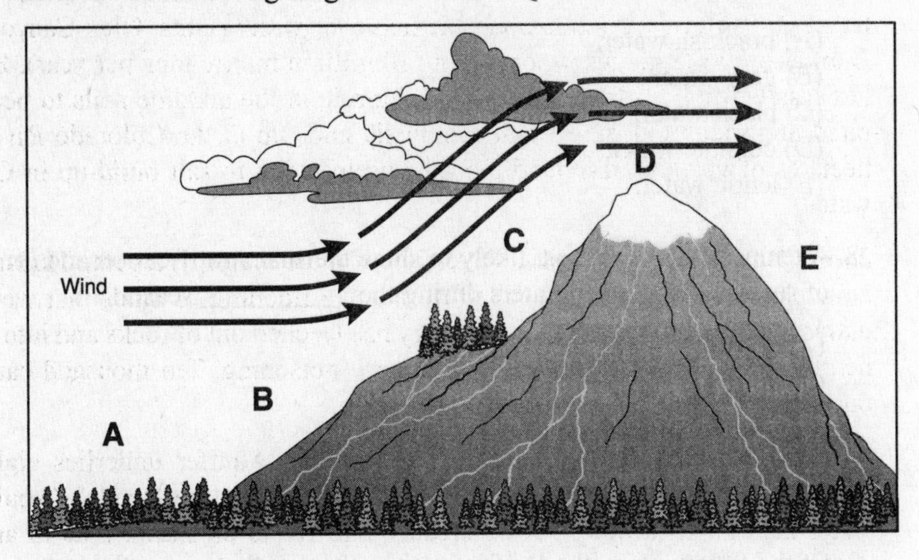

3. The area that would receive the most precipitation would be

(A) A.
(B) B.
(C) C.
(D) D.
(E) E.

4. The rain shadow effect would be located at point

(A) A.
(B) B.
(C) C.
(D) D.
(E) E.

5. Of the freshwater available on Earth that is not trapped in snow packs or glaciers, most of it (95%) is trapped in

(A) lakes.
(B) rivers.
(C) aquifers.
(D) dams.
(E) estuaries, marshes, and bogs.

6. The number one use of freshwater is for

(A) industry.
(B) domestic use.
(C) fishing.
(D) agriculture.
(E) landscape purposes.

7. A mixture of fresh- and saltwater is known as

   (A) brackish water.
   (B) gray water.
   (C) black water.
   (D) connate water.
   (E) lentic water.

8. A temperate lake is most likely to show thermal stratification and limited mixing of surface and deeper waters during the __________ season.

   (A) winter
   (B) spring
   (C) summer
   (D) fall
   (E) None of these answers; temperate lakes don't stratify

9. Of the following methods of irrigation, the one that currently conserves the most water is

   (A) flooding fields.
   (B) irrigation channels.
   (C) sprinklers.
   (D) drip irrigation.
   (E) misters.

10. The largest use for industrial water is for

    (A) cooling electrical power plants.
    (B) automobile manufacturing.
    (C) mining.
    (D) food and beverage industry.
    (E) aquaculture.

11. A country that would represent large per capita water use would be

    (A) China.
    (B) India.
    (C) Israel.
    (D) United States.
    (E) Iceland.

12. When compared to the rate of population growth, the worldwide demand rate for water

    (A) is about half.
    (B) is about the same.
    (C) is about double.
    (D) is about three times.
    (E) is about ten times.

13. The U.S. per capita use of water on a daily basis is closest to

    (A) 50 gallons.
    (B) 100 gallons.
    (C) 1500 gallons.
    (D) 5000 gallons.
    (E) 10,000 gallons.

14. Countries that are more likely to suffer from water stress would be located

    (A) in North America.
    (B) in South America.
    (C) in western Europe.
    (D) in the Middle East.
    (E) in Asia.

15. What fraction of the world population does not have access to adequate amounts of safe drinking water?

    (A) ½
    (B) ⅓
    (C) ¼
    (D) ⅙
    (E) ⅒

Answers to Multiple-Choice Questions

1. **D**	6. **D**	11. **D**
2. **B**	7. **A**	12. **C**
3. **C**	8. **C**	13. **C**
4. **E**	9. **D**	14. **D**
5. **C**	10. **A**	15. **D**

Explanations for Multiple-Choice Questions

1. **(D)** Evaporation is water changing from a liquid state to a gaseous state below the boiling point. Transpiration is water moving through a plant. Boiling occurs when the vapor pressure of the liquid equals the pressure over the liquid (boiling points are determined by the atmospheric pressure). Water boils at a lower temperature at higher altitudes, that is why it takes longer to cook food at the tops of mountains—not as much heat energy is getting into the food as it would at lower altitudes. Vaporization is going from the liquid state to the gaseous state.

2. **(B)** The saturation point is the maximum amount of water vapor that a particular volume of air at a given temperature can hold. The condensation point is the temperature and pressure at which water vapor turns into liquid water. Absolute humidity is the mass of water vapor in a given volume of air and is usually expressed in grams of water vapor per cubic meter of atmosphere. Relative humidity is the ratio of the actual amount of water vapor held in the atmosphere compared to the maximum amount that the air could hold. It is influenced by temperature and atmospheric pressure.

3. **(C)** As the air lifts (orographic lifting), it becomes cooler. Cooler air holds less water vapor. At the location marked C, the air is holding the maximum amount of water vapor, and given the fact that the temperature has decreased, it would receive the maximum amount of precipitation. At the top of the mountain (D), much of the water vapor has been depleted from the air.

4. **(E)** Point E is on the leeward side of the mountain. This side receives little precipitation because most of it has been deposited on the windward side, and the condition is called the rain shadow effect.

5. **(C)** Of all water on Earth, the oceans hold 98%. Freshwater makes up only 2% of all water on Earth. Of that 2%, 90% of it is trapped in ice and snow. Of what is left (0.3%), 95% of it is found in groundwater and 3% in lakes, rivers, and streams. Of the total amount of water on Earth, the water in lakes, rivers, and streams amounts to only about 0.01%.

6. **(D)** Agriculture uses about 70% of all freshwater. Use for agriculture depends upon national wealth, climate, and degree of industrialization. Industry uses about 25% of all freshwater, ranging from about 75% in Europe to less than 5% in developing countries. Water used for cooling of power plants is the biggest sector and creates thermal pollution.

7. **(A)** Gray water is sewage water that does not contain toilet wastes. Black water is sewage water that contains toilet wastes. Connate water is also known as fossil water. It is water that has been trapped within sediment or rock structure at the time the rock was formed. Lentic water is the standing water of lakes, marshes, ponds, and swamps.

8. **(C)** During the summer, the surface water warms up much faster than the deep water. The warmer surface water is less dense than the cooler, deep water, so it stays on the surface. The wind mixes the surface water a little, but only to a relatively shallow depth. The lake tends to become stratified, with a warmer upper layer, or epilimnion, and a cooler lower layer, or hypolimnion. The boundary between these layers is a fairly narrow layer called the thermocline, where the temperature drops sharply with increasing depth.

9. **(D)** Drip irrigation is one of the newer methods of irrigation.

   - Drip irrigation provides each plant with near-optimal soil moisture.
   - This method of irrigation can be automated as easily as a lawn-sprinkling system.
   - Prices for trickle irrigation emitters range from 10 cents to $2 for individual plants.

   Drip irrigation is popular because it can increase yields and decrease both water requirements and labor. The concept behind drip irrigation is to provide the plant with continuous, near-optimal soil moisture by conducting water directly to individual plants instead of providing water to the entire crop, as with flood or sprinkler irrigation. Drip irrigation saves water because only the plant's root zone receives moisture. Little water is lost to deep percolation if the proper amount is applied.

10. **(A)** Industry uses about 25% of all freshwater, ranging from about 75% in Europe to less than 5% in developing countries. Water used for cooling of power plants is the biggest sector, and creates thermal pollution. Water returns 60 times its economic value when used for industrial purposes rather than for agriculture.

11. **(D)** Highest per capita supplies of freshwater are in countries with high rainfall and low populations (e.g., Iceland, 160 million gallons per person per year). Other water-rich countries include Surinam, Guyana, Papua New Guinea, Canada, Norway, and Brazil. These water-rich countries have very low water-withdrawal amounts. Lowest per capita supplies are in areas with low rainfall and high populations (e.g., Kuwait, 3000 gallons per person per year). Other water-poor countries include Egypt, Israel, and most Middle Eastern countries. In these water-poor countries, water-mining or withdrawal of water reserves exceeds 100% of their renewable supply. The average amount withdrawn from water supplies in the United States is ~500,000 gallons per person per year.

12. **(C)** Since population is increasing, and increased population results in higher levels of pollution, and since fresh water sources are finite, the amount of freshwater per person is getting smaller every year. The total amount of water used by humans is 1.6×10^{15} gallons per year. The total amount of water used per person averages 2.6×10^5 gallons (265,000 gallons) per year.

13. **(C)** In the United States, renewable or replacement water averages 2.4 million gallons per person per year. The average amount withdrawn from water supplies in the United States is ~500,000 gallons per person per year, which is close to 1500 gallons per day.

14. **(D)** Areas that do not receive much precipitation include (1) polar regions above 60° north and south latitudes—cold polar air cannot hold much water vapor; (2) mid-continental areas—they are too far from oceans and clouds have released their moisture before they get to the deserts; (3) subtropical deserts—air masses are subsiding; and (4) the leeward side of mountains near coastal regions (rain shadow effect).

15. **(D)** It is estimated that over 1 billion people—about one sixth of the world's population—lack access to safe drinking water. A child dies every 8 seconds from contaminated water, with total deaths each year of over 5 million people.

Free-Response Question

Legal Notice

Initial Public Offering of Common Stock

"Shrimp Will Save The World"

Investors Needed

The Bubba Gump Shrimp Company, headquartered in New Orleans, Louisiana, is offering 5,000,000 shares of common stock at $3.00 per share to investors for the purpose of raising capital to construct shrimp farms in Southeast Asia. Raising shrimp in controlled environments, known as shrimp aquaculture or shrimp farms, provides needed food and employment for the local population. Excess shrimp that is raised is sold in open trading in world market adding to the profitability of the operation. Shrimp farming does not negatively impact the environment since shrimp are native to the area. Shrimp also supplies a valuable source of income and food supply (protein) to the local population. No guarantee of return on investment is implied. For further information on this investment opportunity, contact L.D. Breckenridge, Ltd., Shreveport, LA.

(a) Describe any negative or positive environmental impacts of shrimp aquaculture.

(b) Describe any parallels that have occurred in history in the development of aquaculture.

(c) Comment on the social implications of large-scale aquaculture and suggest possible modifications to aquaculture to make it more sustainable.

Free-Response Answer

by Gary Thorpe

Rubric

Question A: 4 points
Question B: 3 points
Question C: 3 points
Ten points possible for full credit.

RUBRIC	ESSAY
½ pt. Thesis statement	*Aquaculture, or the raising of marine organisms in confined areas, once promoted as a panacea for world hunger, has resulted in unexpected social and environmental consequences.* Aquaculture is not something new. In fact, it has existed for hundreds of years on a small-scale, local level.
½ pt. Comparison of historical techniques to current techniques.	*What once were small-scale family operations that supplied food for small communities of people or was sold in small, local markets and relied on the natural tidal action of water to flush out the ponds has changed most significantly due to the current large-scale operation.* Most aquaculture enterprises in less-developed countries today are funded by *large, multinational corporations with extensive distribution facilities and financial resources.*
½ pt. Recognition that mariculture is a corporate endeavor.	Shrimp farming, as an example of aquaculture and monoculture, utilizes vast coastal areas and is often *promoted and financially supported by third-world governments in lucrative contracts with multinational corporations.*
½ pt. Recognition that mariculture in third-world countries is sponsored by more-developed countries.	The shrimp that is raised, including the species known in the United States as "Tiger Prawns," is generally not raised to feed the people in the country in which the shrimp are raised. Instead, the shrimp are exported to countries in western Europe, the United States, and Japan, where market demand and available capital make the shrimp a readily available luxury commodity.
	In 1990, Asia produced close to half a million metric tons of commercially raised shrimp, which made up about 80% of the world output.
1 pt. Recognition that shrimp farming is damaging to the environment.	*The environmental cost to produce such large numbers of shrimp was almost one million hectares of land that is forever lost. Most of the wetlands that are used for shrimp farming purposes were once prime wetlands consisting of mangroves and other highly productive biomes.*
½ pt. Recognition that shrimp farming disrupts traditional community lifestyles.	Besides the serious environmental consequences are the *disruptions to the communities of people that make their living in these highly productive biomes* by fishing or operating small farms, particularly rice farming. Many of these small farmers and fishermen were forced from the land that

RUBRIC	ESSAY
2 pts. Parallel to historical account of displacement.	they had traditionally used and lived on for many years. This parallels in many ways the *exploitation of Native Americans,* who were forced against their will, to move to the least desirable of all lands from their once-rich prime land and hunting areas.*
1 pt. Description of physical setup of shrimp farm.	Shrimp farms consist of *very large ponds that are built near the ocean. Into these ponds (or tanks), which are filled with both seawater and groundwater, to create an artificial brackish environment, the farmers add pesticides, antibiotics, food, and other chemicals to increase production.* So much groundwater is extracted that land subsidence has been noted.
½ pt. Methods to increase production.	*Other methods to increase shrimp production include increasing the number of shrimp in the pond (increasing the density).* And yet, as in humans, increasing density has serious drawbacks.
½ pt. Response to increased density of shrimp.	As a response to the stress of increased density, *antibiotics and pesticides are added to the water.* Higher densities of shrimp also result in *higher waste build-up in the water with consequently lower oxygen levels.* As a result of these factors, the water must be changed periodically.
½ pt. Effect of polluted water on local area.	The farmers *release this large amount of polluted water either back into the sea (near the coastline where it can do the most damage) or allow it to escape onto prime agricultural lands to infiltrate back into the ground.* In areas near the coastline, the *shrimp farms have interfered with the daily routines of the fishing community.*
½ pt. Depletion of native population of shrimp.	Furthermore, active *capturing of small shrimp from native waters* to "seed" the ponds decreases the amount of shrimp in the natural environment, which has serious trophic level consequences.
½ pt. Soil salinity and its effect on groundwater.	The ground that the water is discharged onto, which in Asia is primarily rice fields, *increases the soil salinity* to such an extent that the productivity of rice drastically decreases. *The salt that is in this water also infiltrates into the groundwater supply and has a serious impact on the source of freshwater. Aquatic life living near where the shrimp water effluent is discharged also suffers from added pollutants, salinity differences, and oxygen demand in the water.*

*** I am Native American (mother's side) and trace my heritage to the Cherokee and Choctaw Nations. If you read between the lines of the essay I wrote, you may see how I feel about the issue of forcing people off the land that they have lived on for generations. My great-grandparents were forced to live on a squalid Indian Reservation. Do you also have issues in your life or in your family's history or ancestry that you too can relate to Environmental Science? Most of us do. The best essays always come from the passionate heart.**

—Gary Thorpe (Thorpe came from the Norwegian "Torbiørnsen" on my father's side)

RUBRIC	ESSAY
1 pt. Historical connection of land rights.	*Social disharmony* is rampant where the shrimp farms have been located. One sees the *parallel of this in examples that occurred in the Plains states during the late 1800s and early 1900s between the farmers who wanted barbwire and the ranchers who wanted open ranges.*
½ pt. Current problem in raising shrimp.	Recently, in some areas, the productivity of shrimp farms has decreased significantly. In Taiwan, almost 100,000 metric tons of shrimp were produced in 1987. One year later, less than 50,000 metric tons were produced primarily due to epidemic increases in pathogenic bacteria, viruses and protozoans. During the mid 1990s, a *virus invaded the shrimp farms* in India destroying the majority of the stock. Many less-developed countries have reexamined the issue of shrimp feeding the world.
2 pts. Alternatives to current practices, complete with specific example.	An *alternative to raising shrimp or fish near the coasts may be to raise it inland, far from coastal waters where wild fish feed and breed. Tilapia, a plant-eating fish, is easy to raise, and they produce protein for people without using wild fish as feed. Catfish and trout are raised inland in the United States. Carp have been pond-raised for centuries in China and Europe.*
	Will the little shrimp feed the world? The answer is clear. Not until the entire industry is reexamined in terms of its effect on the environment and the lives of the indigenous people it disrupts and displaces.
	Word Count: 887

CHAPTER 8

Food and Other Agricultural Products

One should eat to live, not live to eat.

Moliere (1622–1673)
The Miser, 1668
Act III, scene i

Areas That You Will Be Tested On

Note: The AP College Board has not specified what will be tested on in the area "Food and Other Agricultural Products."

Key Terms

Note: Definitions for all key terms listed below can be found in the glossary starting on page 648. For additional definitions of relevant terms from this chapter, refer to www.barronseduc.com/0764121618.html.

agribusiness	**genetic engineering (gene splicing)**
caloric intake	**Liebig's Law of the Minimum**
first green revolution	**second green revolution**

Key Concepts

Crops and Food Sources

- Only about 100 species of plants out of the 350,000 known are grown to meet human needs. Of these 100 species of plants, 15 of them supply over 90% of the world's needs.
- Eight terrestrial animal species supply over 90% of the world's need for meat. Producing meat and dairy products requires more energy—it takes about 16 pounds of grain to produce 1 pound of edible meat. Because producing meat requires more energy than producing crops, thereby being more expensive to produce, 20% of the world's richest countries consume 80% of the world's meat and dairy products. Benefits of consuming these products include that they are rich sources of proteins and essential amino acids, and they are concentrated in protein content. A disadvantage is the fact that they are generally high in saturated fats.
- Three major crops utilized by humans are wheat, rice, and corn (maize). Wheat and rice supply ~60% of human caloric intake.

- Other agricultural crop products listed in order of world production are potatoes, barley, sweet potato, cassava, grape, soybean, oats, sorghum, sugarcane, millet, banana, tomato, sugar beet, rye, orange, coconut, cottonseed, apple, yam, peanut, watermelon, cabbage, onion, bean, pea, sunflower seed, and mango.
- In northern Europe and northern Asia, the primary crops are barley, oats, potatoes, and rye.
- In Africa, Brazil, Melanesia, and South Pacific, primary crops are cassava and sweet potatoes. Cassava starch is also the source of tapioca.
- In the dry regions of Africa, primary crops are millet and sorghum.
- Eighty percent of all meat and milk are consumed by North America, Europe, and Japan, which constitute only 20% of the world's population. Meat and milk are rich sources of protein.
- About 90% of grain grown in the United States is used for animal feed. It takes about 16 kg of grain to produce 1 kg of animal protein. By eating the grain instead of eating the livestock that feed upon it, humans would get more than 20 times the calories and 8 times more protein. Inefficiency is due to energy used by animal and parts of the animal that humans do not eat. This is in agreement with the second law of thermodynamics.
- Fish and seafood overfishing and harvesting and habitat destruction are seriously impacting this source. One half of all protein in Japan comes from marine sources. According to the United Nations, 70% of marine sources of food are declining. Worldwide fishing costs more than the income produced—the difference is made up by government subsidies. One fourth of all organisms caught are not wanted and destroyed. This impacts many food chains. In 1998, 30 million marine mammals, sea birds and other animals not sought for, were killed by the fishing industry.
- The top five countries in order of producing the most amount (metric tons) of grains are China, United States, India, Canada, and Ukraine.

Malnutrition and Famines

- One quarter of the human population (1.5 billion) is malnourished.
- Areas of greatest malnutrition are Sub-Saharan Africa (~225 million), East and Southeast Asia (~275 million), South Asia (~250 million), and parts of Latin America. Improvements are expected in East, South, and Southeast Asia (see Figure 8.1).
- Malnutrition and famine stem from not getting enough calories per day in addition to not getting the necessary amounts of carbohydrates, protein, fats (lipids), minerals, and vitamins.
- People in countries that experience malnutrition generally have diets high in starch (carbohydrates), but they do not get enough of the other necessary dietary requirements.
- Conditions that cause famines include major droughts, increased population sizes, massive immigration into the country (as in people from Afghanistan flooding into Pakistan in 2001 due to war), floods, wars, chaos in the economy, political instability, mismanagement and oppression, land seizures, pestilence, breakdown in distribution networks, and panic-buying and hoarding.
- The problem is *not* that the world does not grow or raise enough food. The problem is that many people are too poor to buy food that is available. Enough wheat, rice and grain are produced on Earth each day to provide each human with 3500 calories per day. This does not count vegetables, beans, nuts, meats, and fish. Adding these sources would result in an average of 4.3 pounds (2 kg) of food per person per day *if* it was distributed equally. Remember, for men, 2500 is the minimum daily requirement of calories and for women it is 2000. So in reality, there is an abundance of food.

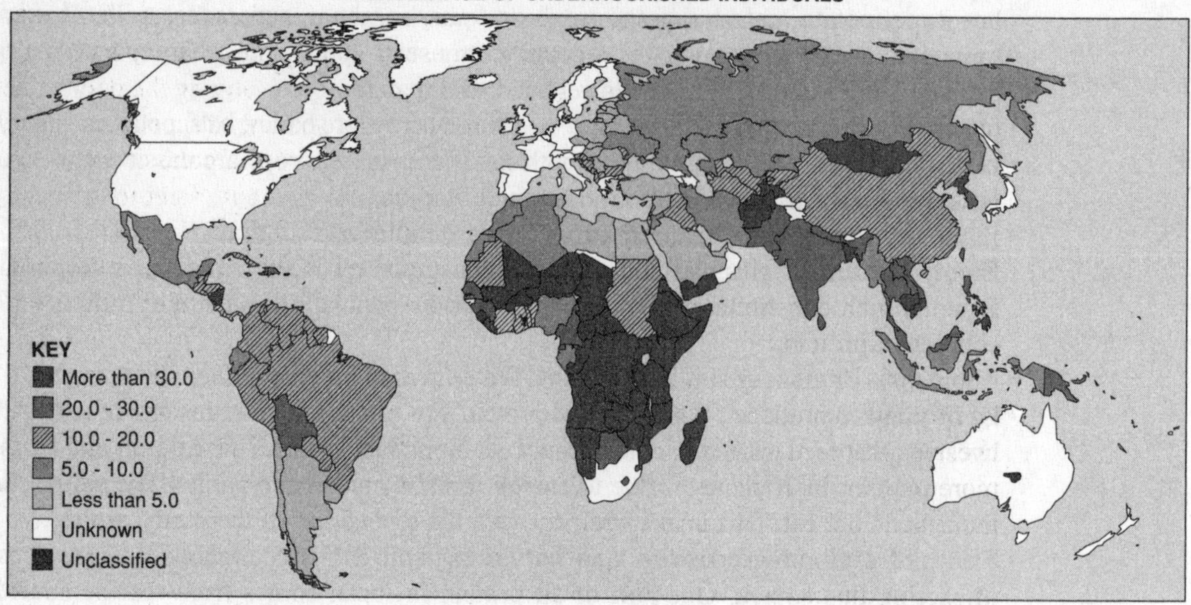

Figure 8.1. Worldwide malnutrition.

- Droughts, floods and other natural disasters contribute to famines, but are usually not the sole cause. They generally act to "tip the scale" in countries that are already suffering from food problems, which include land ownership, poverty, poor working wages, food distribution networks, and food preservation capacity.
- Population density may or may not be a contributor to famine. Population density must be linked to economic conditions in order for population density to contribute to famine. Example: Bangladesh has a very high population density (large amount of people in a small area). Famine is common in Bangladesh due to a low GNP. Famine is not an issue in Hong Kong where the GNP is much higher even though population density is comparable.
- Advances in food production due to the first and second green revolutions are not ending famine. In India, Mexico, and the Philippines, where large advances were made in crop production as a result of the agricultural practices of the first and second green revolutions, famine is still common. What is increasing along with famine is food exports from these countries (due to higher prices paid by more-developed countries) and degradation of the soil.
- Agribusiness or large-corporate ownership and management of farmland will not end famine. Agribusiness runs on profit. If market prices are unfavorable, large corporations can afford to leave farmland idle until the price increases. Research has shown that small, individual farmers typically achieve at least four to five times greater output per acre than do large corporations. However, if small farmers do not own the land, there is little incentive created to invest in agricultural practices that preserve the soil. Land reform in Japan, Zimbabwe, Taiwan, and Brazil where land has been redistributed into smaller holdings has raised output by up to 80%.
- Free-trade will not result in declines in famine; in fact, it may make it worse. Those involved in trading commodities are not the poor masses within the country. They are foreign investors whose motives are profit and positive dividends. Trading food from one country that needs it to another country that wants it as a luxury item does not solve famine. Soybean production in Brazil has dramatically increased. However,

rather than using the soybeans to add to the diets of Brazilians that need it, the soybeans are sold to Japanese and European corporations to feed their livestock. The livestock is then eaten by the Japanese and Europeans, leaving the Brazilians out of the picture. NAFTA and GATT treaties work to create competition among the poorest people of the world—"who is willing to work for the least amount of money and for the least benefits?"—the job going to the lowest bidder. Economic policies by the U.S. government, World Bank, and International Money Fund make it easier to send token relief packages of food to a fraction of a population needing it as a public-relations tool than to change policies that would allow people of that country to be self-sufficient. If they became self-sufficient, their lower overhead would allow them to "unfairly"(?) compete with more-developed countries. "Care Packages" only serve to reinforce the status quo. By reinforcing the status quo, it allows repressive regimes to continue.

- Keeping people in third-world countries in economic poverty may result temporarily in cheaper grapes, cheaper fruits, cheaper shoes, and cheaper TVs and computers for more-developed countries. However, in the long run, these practices eventually result in global corporations moving their manufacturing operations to these less-developed countries where wages are low and benefits nonexistent. This ultimately results in loss of jobs in the more-developed countries, which, in turn, creates a new class of people in the more-developed country called the "working poor." Working poor are forced to take part-time jobs at low wages and little benefits. Because they are "working," they do not qualify for assistance from the government.
- As a response to famine, already stressed people use up resources needed in the future. Example: eating seed, slaughtering livestock, and stripping the ground bare.
- Refugee camps cause dislocation, many die during the process of getting there, lack of sanitation and crowding causes disease, no employment exists in camps, social structure and family support breaks down, and homes and farm equipment are left behind and are usually looted so nothing remains when people do go back to their home.
- Ways to help prevent famines include:

  - Encourage more-developed countries to "adopt" a less-developed country and help it in all phases of food-production—providing seed, tools, technology, and irrigation capacity, and buying the products they produce at a fair-market value. Once that country is "on its feet," then the more-developed country picks another country to adopt. Improvement occurs one country at a time.
  - Teach farmers in less-developed countries how to grow crops more efficiently. China is an example of a country that produces crops very efficiently, although it is very labor-intensive. By being labor-intensive, it gives employment opportunities to people who may not have high educational levels—it puts people to work, rather than allowing them to sit in refugee camps eating the world's leftovers. Trying to teach highly refined and expensive techniques as found in the U.S. agribusiness to third-world countries that do not have the economic base to sustain such practices will not work. The model needs to be human-labor based.
  - Tie family-planning education to agricultural education and support. Fewer people means more food to go around. This reinforces the thoughts of Thomas Malthus.
  - Store any food surpluses produced by the less-developed countries in a food "credit" bank sponsored by the more-developed countries. These credits can then be redeemed during famines or food shortages. To build up "food credits" would be an incentive for less-developed countries to focus on increasing agricultural output. It is analogous to the fable of the ant that saved up its food all winter.

- Sign international treaties that guarantee the right of every human on Earth, the basic right to go to bed without being hungry, especially in a world that has an abundance of food. Coupled with that right would be required family-planning education and improvement in the status of women.

LOW-INCOME FOOD-DEFICIT COUNTRIES (LIFDCs)
AS LISTED BY THE UNITED NATIONS FOR 2001*

Area	Countries
Africa (42)	Angola, Benin, Burkina Faso, Burundi, Cameroon, Cape Verde, Central African Republic, Chad, Comoros, Congo (Republic), Congo (Democratic Republic), Côte d'Ivoire, Djibouti, Egypt, Equatorial Guinea, Eritrea, Ethiopia, Gambia, Ghana, Guinea, Guinea Bissau, Kenya, Lesotho, Liberia, Madagascar, Malawi, Mali, Mauritania, Morocco, Mozambique, Niger, Nigeria, Rwanda, Sao Tome & Principe, Senegal, Sierra Leone, Somalia, Sudan, Swaziland, Tanzania, Togo, and Zambia.
America (7)	Bolivia, Cuba, Ecuador, Guatemala, Haiti, Honduras, and Nicaragua.
Asia (24)	Afghanistan, Armenia, Azerbaijan, Bangladesh, Bhutan, Cambodia, China, Georgia, India, Indonesia, Korea Democratic People's Republic, Kyrgyzstan, Laos, Maldives, Mongolia, Nepal, Pakistan, Philippines, Sri Lanka, Syrian Arab Republic, Tajikistan, Turkmenistan, Uzbekistan, and Yemen.
Europe (3)	Albania, Bosnia and Herzegovina, and Macedonia.
Oceania (6)	Kiribati, Papua New Guinea, Samoa, Solomon Islands, Tuvalu, and Vanuatu.

*These countries also make up countries listed as less developed. *Source:* Special Program for Food Security Food and Agriculture Organization of the United Nations.

Genetically Engineered Foods

Biotech companies love to talk about feeding the world, but their products must pay off in a market that measures dollar demand, not human need. By far the greatest effort has gone into the potato that makes fast-food fries, not the yam grown by folks with no cash. The corn that feeds America's pigs and chickens, not the dry land millet that feeds Africa's children. The diseases of the rich, not the plagues of the poor. There is some public funding and corporate charity directed toward gene manipulations that might conceivably help feed the world, but the vast majority of minds and bucks are working on caffeine-free coffee beans, designer tomatoes, and seedless watermelons. They always will, if the market is the guide.

Donella Meadows, "Are Bioengineered Potatoes Organic?,"
Whole Earth, Summer 1999

- The first green revolution occurred between 1950 and 1970 and involved practices of planting monocultures and using high application of inorganic fertilizers, pesticides, and artificial irrigation systems.
- The second green revolution occurred during the 1970s and is continuing today. It involves growing genetically engineered crops. Practices of the first "green revolution" are still employed during the second green revolution.
- Genetic engineering involves moving genes from one species to another, or designing gene sequences with wanted characteristics.
- Advantages are that characteristics beneficial to humans could be incorporated into the food—pest resistance, drought resistance, mold resistance, saline resistant, higher protein yield, higher vitamin yield (golden rice).
- Disadvantages are that effects of altered gene sequences are not known.
- About 70% of all food processed food in the United States contains transgenic material.
- One third of all U.S. corn and soybeans are genetically engineered.
- Seventy-five percent of all cropland in the world is planted with genetically engineered crops.
- Canada and Argentina are also countries that produce a majority of genetically engineered food.
- Triticales are wheat-rye hybrids. Currently, triticale is grown primarily for animal feed as either a grain or forage crop. Breeders are attempting to improve grain quality to expand the market for human consumption. Agronomically, triticales are attractive due to a broad spectrum of disease resistance. Yield potentials can be significantly greater than wheat.

Food Production

- Sixty percent of global food stocks are in the hands of private companies, while 70% of world grain trade is carried out by just six companies.
- China and Indonesia have tripled food production in the last ten years due to upgraded irrigation, fertilizers, pesticides, high-yield crops, expanded cropland, and privatization of croplands in China.
- Food production in Sub-Saharan Africa is falling below population growth due to drought, war, poverty, government mismanagement, and high natality. Food production dropped in the former USSR due to privatization of croplands. Consequently, there was no money to pay for transportation, equipment, and labor, and customers could not buy the product because it was too expensive.
- About 11 million children die each year from starvation.
- Before the first green revolution, crop production was correlated with increases in acreage under cultivation. After the first green revolution, crop acreage increased about 25% (1950–1981). After 1981, the rate of increase reached a plateau. Reasons include the fact that it is easier and more economical to increase crop production through techniques of the first green revolution than to buy and clear new land.
- There was about a 25% increase in land converted to agriculture after the first green revolution but a 200% increase in grain yield (see Figure 8.3). This translated to a 2% increase per year, exceeding the rate of increase in world population (r = 1.3% in 2001). The reasons for such a dramatic increase in grain yield were
  - The use of green revolution hybrid crop varieties.
  - Changed agricultural practices.

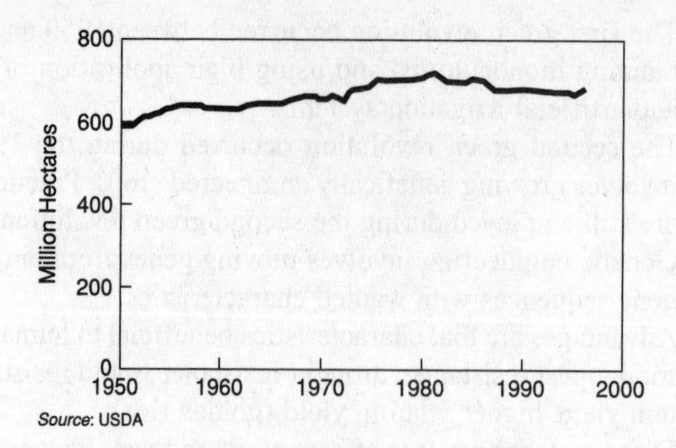

Figure 8.2. World grain acreage, 1950–1997.

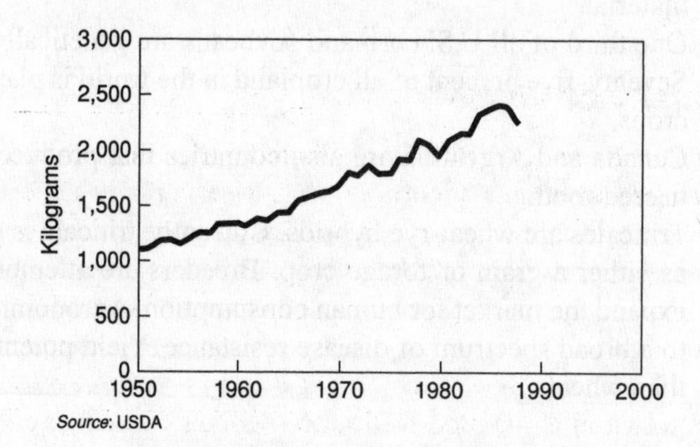

Figure 8.3. Grain yield per hectare, 1950–1990.

- When viewed as grain production per capita, world grain production increased about 40% between 1950 and 1984 (see Figure 8.4).

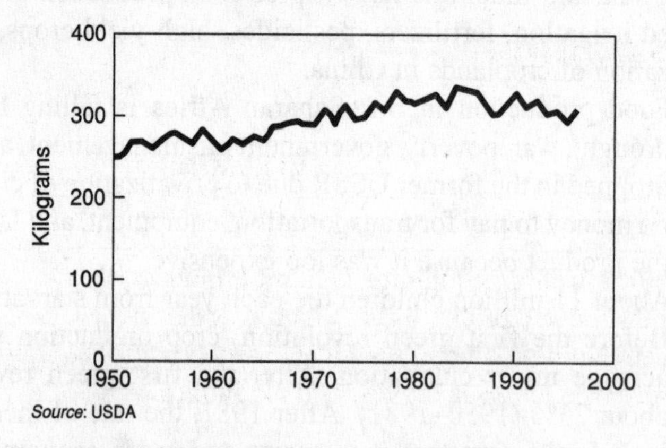

Figure 8.4. Grain production per person worldwide.

- When viewed as total grain produced worldwide, remembering that total acreage devoted to agriculture had reached a plateau in 1997, there was an average 2.6% increase between 1950 and 1990 (see Figure 8.5). This clearly is the result of new agricultural methods and hybrid crops.

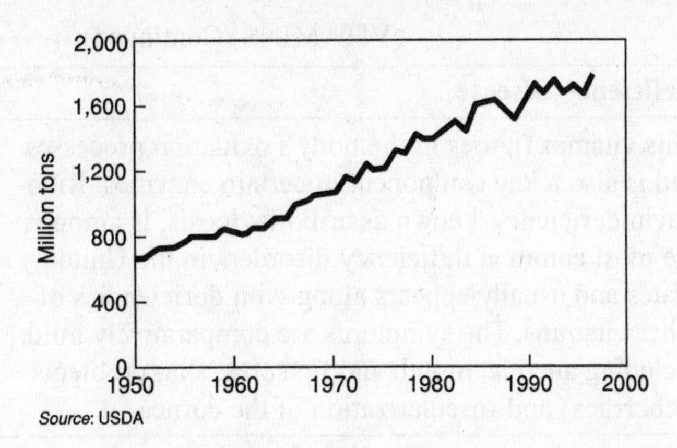

Figure 8.5. Worldwide grain production, 1950–1996

Genetic Diversity of Crops

- The green revolution has "singled out" and modified or created hybrids that produce the most output (yield) per acre. Motivation has been economic. This is in contrast to past agricultural practices in which farmers planted a variety of locally adapted strains. For example, to match growing conditions of water availability, soil type, climatic condition, and so on, farmers in India have planted more than 30,000 different varieties of rice. By 2005, however, this number will be reduced to ten different varieties. In 1990, 75% of all corn in the United States was limited to only six varieties. Of all wheat grown in the United States, 50% comes from nine different genotypes.
- Diversity in crop genotypes correlates with concepts of natural selection (see Chapter 5). A diverse mixture of genotypes ensures that environmental stresses such as pests, diseases, drought, climatic changes, etc. do not cause extinction of the species. See the case studies in this chapter on Texas Male Sterile Corn.

Vitamins and Minerals

VITAMINS

Vitamin	Deficiency Disease	Food Sources of Vitamin
A (retinol)	One of the early symptoms is night blindness, the inability to see in dim light. Vitamin A deficiency is also common among the elderly and urban poor, people who abuse laxatives, and alcoholics.	Eggs, whole milk, cream, cheese, liver, green and yellow vegetables (carrots, squash, sweet potatoes, spinach, kale, and broccoli).
B_1 (thiamine)	Beriberi usually begins with mild symptoms, including muscle cramps, irritability, loss of appetite, and prickling or burning sensations in the skin. As the disease advances it may affect either the heart or the nervous system. Death may result from heart failure or paralysis.	Ham, pork, milk, fortified cereals, peanuts, liver, and yeast.

VITAMINS (Continued)

Vitamin	Deficiency Disease	Food Sources of Vitamin
B_2 (riboflavin)	This vitamin figures in the body's oxidation processes and is also a key component in certain enzymes. Riboflavin deficiency, known as ariboflavinosis, is among the most common deficiency disorders in the United States and usually appears along with deficiencies of other vitamins. The symptoms are comparatively mild, including anemia, mouth and lip sores, skin problems (seborrhea) and vascularization of the cornea.	Liver and other organ meats, milk, green vegetables, fortified cereals, yeast.
Niacin	One of the B vitamins, niacin figures in the body's oxidation processes. The body can synthesize it from the amino acid tryptophan, so it is not a true vitamin. Niacin deficiency is called pellagra; once common in the southern United States, it is now often seen in chronic alcoholics and drug addicts. Severe cases include three basic symptoms—diarrhea, skin problems, and dementia or learning difficulties. Advanced pellagra can cause death.	Peanuts, lean meats, poultry, fish, bran, yeast, liver.
B_6	This vitamin is actually a group of substances that play a role in metabolism. Chronic alcoholism and interactions of certain medications, including oral contraceptives, can cause deficiency disorders of B_6. Mouth soreness, irritability, and weakness are among the early symptoms, while anemia, seizures, and other symptoms appear in advanced cases.	Whole-grain cereal, fish, legumes, liver and other organ meats, yeast.
B_{12}	This vitamin is a factor in the production of red blood cells. Vitamin B_{12} deficiency is called pernicious anemia and usually appears in the elderly because the small intestine is failing to absorb the vitamin. Numbness in the hands and feet, sore tongue, weight loss, weakness, and rapid heartbeat are among the symptoms of this disease. Left untreated, the deficiency can eventually cause death. Vitamin B_{12} is not found in vegetables, so strict vegetarians need to supplement their diet in some way.	Eggs, milk, liver.
C (ascorbic acid)	Also called ascorbic acid, vitamin C helps maintain healthy connective tissue, cartilage, and bone and is involved in metabolism. The elderly, who may neglect their diet, people who are too poor to feed themselves properly, and alcoholics are among those who may contract the vitamin C deficiency disease, called scurvy. The early symptoms of scurvy include swollen gums, loose teeth, and small black-and-blue spots on the skin. Untreated, scurvy leads to a decline in health and eventually death.	Fresh fruits and vegetables (oranges and other citrus fruits, brussel sprouts, cabbage, etc.).

VITAMINS (Continued)

Vitamin	Deficiency Disease	Food Sources of Vitamin
D	This vitamin is involved in calcium and phosphorus absorption by the body. When exposed to sunlight, the skin also manufactures vitamin D. Vitamin D deficiency in children is called rickets. The deficiency results in the body not having enough calcium to maintain healthy bone structure and leads to softening of bones and serious bone deformities.	Fortified milk, fish liver oil.
E	Vitamin E is believed to protect cell structures from damage by highly reactive groups of atoms in the body called free radicals. It may also help protect against cancer, cataracts, and heart disease. Vitamin E deficiency is usually due to problems with absorption by the intestines. It causes difficulty with walking, lack of reflexes, and paralysis of eye muscles.	Vegetable seed oil, egg yolk, cereals, beef liver.
K	Vitamin K helps promote clotting of blood. A deficiency may occur fairly quickly because the body stores so little vitamin K. The disorder may result from poor diet or failure of the intestines to absorb the vitamin, or when antibiotics administered for other medical reasons kill off bacteria that normally live in the intestines and that synthesize the vitamin. As the deficiency develops, the patient experiences bruising and bleeding that is difficult to stop.	Leafy green vegetables, liver.

MINERALS

Mineral	Characteristic
Calcium	Calcium is useful in preventing osteoporosis, the loss of bone mass that often occurs with aging. Although controversial, a number of drinks consumed by children are now fortified with calcium, especially orange juice. Calcium is one of the food supplements for which the U.S. Food and Drug Administration allows health claims in advertising.
Chromium	Chromium is involved in the metabolism of sugar and regulation of fat.
Cobalt	This mineral is needed because it is a part of vitamin B_{12}.
Copper	Although copper in large amounts is toxic (copper cooking pots are lined with tin or stainless steel), in small amounts it is needed to help form red blood cells, maintain communications in the nervous system, and create normal hair. Copper in appropriate amounts also lowers cholesterol levels. The genetic disease known as Menkes' syndrome results in low copper levels by interfering with normal transport of the metal across cell membranes. Injections of copper compounds at birth can prevent some of the worst symptoms of the disease.

MINERALS (Continued)

Mineral	Characteristic
Fluorine	Fluorine is essential in strengthening bones and preventing tooth decay. Water systems in which natural amounts of fluorides are low often have fluorides added to bring the amount up to four parts per million. Fluoridation has been in use in the United States since the 1940s, although there has always been some opposition to the practice. The only known effect besides strengthening teeth and bones is mottling or staining of teeth in about 10% of the exposed population.
Iodine	An essential ingredient in thyroid hormone, iodine helps regulate energy and promote growth. Chronic iodine deficiencies cause goiter swelling of the thyroid and hyperthyroidism. UNESCO estimates that 26 million children each year suffer brain damage due to iodine deficiency. Worldwide, 1.5 billion people are at risk.
Iron	The key mineral in hemoglobin is iron. Lack of iron results in too few effective red blood cells (anemia). Too much iron can occur in persons with a hereditary disorder that causes them to store iron in their organs, where it interferes with functioning.
Magnesium	Magnesium is believed to relax blood vessels and thus improve blood flow, but it may also have a role in regulating the heartbeat. Magnesium has also helped relieve symptoms of chronic fatigue syndrome; patients with the syndrome have slightly depressed levels of magnesium in their red blood cells.
Manganese	Manganese is believed to be necessary for bone formation and health of the nervous system.
Molybdenum	Molybdenum is needed for some enzymes to assume their correct shape.
Potassium and sodium	These metals form the two main electrolytes in blood and are also needed in large amounts for proper cell function and for nerve operations.
Selenium	Selenium is thought to promote the action of antioxidant vitamins. Some antioxidant vitamins are believed to reduce cell damage that can lead to cancer or heart disease. Populations from regions in which there is a low level of selenium in the soil have been shown to be more likely to have heart disease. Selenium is also thought to promote growth.
Zinc	Zinc is known to be a part of various enzyme molecules, including some important in sexual development and growth of sperm. Also, zinc appears to be involved in wound healing and the sense of taste. A severe zinc deficiency can lead to short stature and skin disorders.

Case Studies

BST Milk. Bovine somatotropin (BST) is one of the hormones involved in normal growth and development of the mammary gland and normal milk production in dairy cows. When cows are supplemented with increased amounts of BST, the mammary gland takes in more nutrients from the bloodstream and synthesizes more milk. This increase is accomplished by BST's effects on the synthesis and secretion of insulin-like growth factors (IGF-I). To support this synthesis, the cow voluntarily increases feed intake during the period of BST supplementation. Since 1985, the Food and Drug Administration has affirmed that milk from BST supplemented cows presents no health risk to consumers, although many consumer groups and scientists dispute the safety of BST treatment.

Genetically Altered Foods. In 1996, Pioneer Hi-Bred, a subsidiary of Dupont Corporation, inserted a gene from a brazil nut into the DNA of a soybean to create a soybean with a more complete protein. However, this soybean unexpectedly caused allergies in people who had never been allergic to soybeans before. Because this soybean looked identical to other soybeans, these individuals could have unknowingly eaten the soybeans, experienced a severe reaction and even died.

In 1989, the Showa Denko Company in Japan realized that they could speed up the production of a food supplement, tryptophan, by using genetic engineering. However, there was an unexpected toxin produced in the tryptophan, which killed at least 37 people and permanently disabled 1500 others when this product was marketed in North America before it was tested. Over $2 billion of litigation was instituted against Showa Denko.

In 2000, an illegal, likely allergenic variety of genetically engineered corn called StarLink had been detected in Kraft taco shells causing a massive recall of the product and costing the industry millions of dollars. In one case, an entire 55,000 ton shipload of U.S. corn destined for Japan was rejected after testing positive for StarLink. The yellow corn had been gene-spliced with a powerful Bt (*bacillus thuringiensis*) toxin. Developed by a subsidiary of the French-German biotech conglomerate Aventis, StarLink (also known as Cry9C) was approved only for animal feed because of fears that this variety of genetically engineered corn (50 to 100 times more potent than other Bt-spliced varieties) could set off food allergies in humans. As an indicator of the toxicity of Bt, critics have reported that the pollen from Bt corn has been shown to kill Monarch butterflies. Critics have also pointed out that Bt toxin is also killing lacewings, ladybugs, and beneficial soil microorganisms, thereby damaging the entire soil food web. Scientists also warn that bees and birds are likely being harmed by eating insects that have ingested the Bt toxin. In addition, organic farmers in the United States, two thirds of whom use a non-genetically engineered form of Bt spray as an emergency pest management tool, have pointed out that crop pests (beetles, boll worms, corn borers) will inevitably develop resistance to widely cultivated Bt-spliced crops, creating superpests that will overwhelm organic farmers and make organic agriculture more difficult, if not impossible.

Human health fears have arisen over antibiotic-resistant genes in genetically engineered cattle feed. Recently a government advisory board in Britain, revealed that antibiotic-resistant marker genes found in genetically engineered foods and animal feeds may be able to transfer antibiotic resistance to the bacteria in animals' guts, giving rise to dangerous pathogens in humans that can't be killed by traditional antibiotics. German scientists found that antibiotic-resistant genes from genetically engineered rapeseed plants were combining with bacteria in the stomachs and intestines of bees. As a result, Europe's leading food producers and supermarket chains have banned genetically engineered animal feeds in their meat and dairy production.

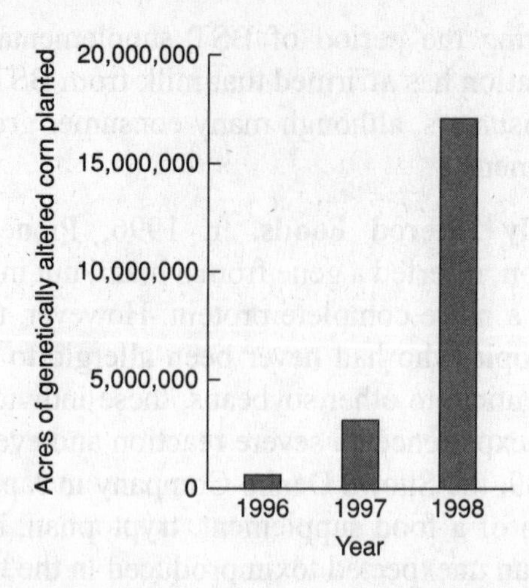

Figure 8.6. Increase in production of genetically modified corn.

Irish Potato Famine of 1845. In 1740–1741, there was a severe potato famine in Ireland. Between 1816 and 1842 there were 14 potato famines resulting in an unstable Irish economy. Land prices and rents for land fluctuated wildly. The country was primarily agricultural as opposed to England, which had begun to industrialize and diversify their economy. Civil unrest was common due to British anti-Catholic legislation.

In 1845, the fungus *Phytophthora infestons,* hit potato crops. Potatoes rotted in the field. The fungus reappeared in 1846. Between 1840 and 1911, the population decreased from 8,200,000 to 4,400,000—almost 50%. Deaths are estimated as high as 1.5 million, and emigration accounted for about 250,000 people leaving Ireland every year, many coming to the United States. Ireland was importing five times more grain than it exported.

Sir Robert Peel helped Ireland during this time by distributing "Indian Meal." He worked to build trust between the Irish and the English. He also helped pass the Catholic Emancipation Act reducing discrimination against Catholics, and helped abolish the Corn Laws, which prevented free trade with Ireland. Sir Edward Trevelyan also worked to aid Ireland through free-trade but thought the famine was rooted in a Malthusian concept that the famine would help relieve overpopulation and things would eventually stabilize.

The effects of the Irish potato famine were widespread. The famine accelerated a decline in population that had begun prior to the famine. The famine produced diversification in agricultural products (livestock). Many Irish felt that England could have done more and built up resentment that can be seen today between Northern Ireland and the Irish Republic.

Olestra. In 1968, Procter & Gamble, in their quest to create a way of increasing premature babies' intake of fat, synthesized a fat substitute called sucrose polyester. Procter & Gamble named its product Olestra. Chemically, olestra is a sucrose (table sugar) molecule to which are esterified as many as eight fatty-acid residues. Olestra molecules are so large and fatty that they cannot be metabolized by enzymes and bacteria in the gut and are neither absorbed nor digested. So, far from proving to be a means of increasing an infant's fat intake, olestra offered a means of replacing fat and producing fat-free or low-fat foods. Olestra, because it is fatty, can bind cholesterol, vitamins, and other fat-soluble molecules. In 1987, Procter & Gamble petitioned the Food and Drug Administration to approve Olestra as a general-purpose fat substitute, replacing some or all the fat in shortenings, fast foods, chips, and other products. In 1996, the FDA approved Olestra for use in savory snacks such as chips, crackers, and tortilla chips. Despite being approved as safe by the FDA, all snacks containing Olestra must carry a label that states: "This Product Contains Olestra. Olestra may cause abdominal cramping and loose stools. Olestra inhibits the absorption of some vitamins and other nutrients. Vitamins A, D, E, and K have been added." More than 18,000 people have submitted to the FDA reports of adverse reactions that they attribute to Olestra. That's more reports than for all other food additives in history combined.

Roundup Ready. Most genetically engineered crops are designed to withstand high amounts of weed killers or herbicides. Roundup Ready crops by Monsanto are genetically engineered plants containing the Roundup Ready gene, which allows Roundup Ultra herbicide (glyphosate) to be applied directly over the top of the crop for weed control. By applying heavy doses of specific herbicides, using Roundup Ready crops allows for conservation tillage and helps to prevent soil erosion. Risks of using crops designed to withstand large amounts of herbicides includes the risk of modified genes "escaping" or getting into wild species, with unpredictable consequences. Furthermore, much larger amounts of herbicides are needed. For example, in soybeans, traditional herbicide application is around 0.1 pound (0.05 kg) of active ingredient per acre per year. Using Roundup Ready soybean crops, herbicide application is 0.75 pounds (0.34 kg) of active ingredient per acre in a single spray, with several sprays occurring during the year. This will result in an additional 20 million more pounds (9 million kg) of herbicides being used this year. Roundup Ready genes have been spliced into soybeans, corn, canola, and cotton.

Texas Male Sterile Corn. Southern corn blight *was* a minor disease affecting corn. The pathogen known as *Bipolaris maydis* had been observed for several years, but it was never a serious problem and had no economic effect on yield. In 1957, a single cross hybrid corn, known as Texas Male Sterile (TMS), was introduced to the market. In 1970, nearly 80% of all hybrid field corn produced in the United States contained TMS cytoplasm. Then, a genetic change or a population shift occurred in the *B. maydis* population. A new race of *B. maydis* was found to be particularly virulent on corn with TMS cytoplasm. The new race was called Race T to differentiate it from Race O that caused a minor leaf spot disease. Race T was much more aggressive and could reproduce a new generation of infective spores within 50 hours. In many Southern states, entire fields were destroyed and 80–100% losses were common and estimated upwards of $1 billion.

U.S. Agriculture. U.S. agriculture is a $1 trillion industry, generates some 22 million jobs and $140 billion in annual exports, and comprises about 15% of the gross national product. The United States produces more corn and soybeans than any other country in the world, and feeds 30% of the global population. Americans spend only ten cents out of every dollar for food, a critical factor in the country's prosperity and security.

Multiple-Choice Questions

1. Chronically undernourished people are those who receive approximately ______ calories or less per day.

   (A) 500
   (B) 1000
   (C) 1500
   (D) 2000
   (E) 3000

2. Protein deficiencies can lead to what disease(s)?

   (A) Marasmus
   (B) Kwashiorkor
   (C) Pellagra
   (D) A & B
   (E) All of the above

3. Pellagra is a deficiency of

   (A) vitamin A.
   (B) vitamin B_6.
   (C) vitamin B_{12}.
   (D) thiamine.
   (E) niacin.

4. The second law of thermodynamics would tend to support

   (A) people eating more meat than grain.
   (B) people eating more grain than meat.
   (C) people eating about the same ratio of grain as meat.
   (D) people being undernourished.
   (E) people eating a balanced diet from all food groups.

5. The majority of nutrients and calories in the average human diet come from

   (A) potatoes, corn, rice.
   (B) wheat, rice, soybeans.
   (C) wheat, corn, rice.
   (D) wheat, corn, oats.
   (E) wheat, corn, soybeans.

6. Afghanistan, Bangladesh, India, and Pakistan are countries in

   (A) South Asia.
   (B) East Asia.
   (C) Southeast Asia.
   (D) Sub-Saharan Africa.
   (E) Northwest Asia.

7. Lack of _______ results in too few effective red blood cells, a condition called _______ .

   (A) calcium, osteoporosis
   (B) copper, Menke's syndrome
   (C) magnesium, chronic fatigue syndrome
   (D) iron, anemia
   (E) iodine, goiter

8. About how many children currently die in the world each year from starvation?

   (A) 2 million
   (B) 5 million
   (C) 11 million
   (D) 22 million
   (E) 37 million

9. According to the food pyramid, people should consume less _______ and more _______ .

   (A) meat, fats
   (B) fruit, breads and cereals
   (C) breads and cereals, meat
   (D) milk, bread
   (E) meat and fish, milk and cheese

10. The one area of the world that is not expected to increase food production soon is

    (A) South Asia.
    (B) East Asia.
    (C) Sub-Saharan Africa.
    (D) China.
    (E) India.

11. Most genetically engineered crops are designed to

    (A) reduce the amount of herbicide applied.
    (B) increase the ability to apply more herbicide.
    (C) reduce the amount of pesticide applied.
    (D) to increase the yield of the crop.
    (E) to resist plant disease.

12. What percentage of crops grown in the United States are transgenic, or genetically modified?

    (A) none
    (B) 15%
    (C) 35%
    (D) 75%
    (E) 100%

13. BST is used to

(A) replace fat in potato chips and other snacks.
(B) enhance the protein content in meat.
(C) reduce the fat content in meat.
(D) reduce the use of pesticides in corn, soybeans, and cotton.
(E) increase milk production in dairy cows.

14. How many chronically undernourished people are there in this world?

(A) 1 out of every 10 persons
(B) 1 out of every 8 persons
(C) 1 out of every 5 persons
(D) 1 out of every 3 persons
(E) 1 out of every 2 persons

15. Adding more fertilizer does not necessarily increase crop production is an example of

(A) law of supply and demand.
(B) Leibig's law of minimum.
(C) limiting factors.
(D) second law of thermodynamics.
(E) Gaia hypothesis.

Answers to Multiple-Choice Questions

1. **D**
2. **D**
3. **E**
4. **B**
5. **C**
6. **A**
7. **D**
8. **C**
9. **D**
10. **C**
11. **B**
12. **D**
13. **E**
14. **C**
15. **C**

Explanations for Multiple-Choice Questions

1. **(D)** People who live in wealthy nations receive an average of 3340 calories per day. Chronic malnutrition is defined as receiving fewer than 2060 calories per day.

2. **(D)** Kwashiorkor is severe protein malnutrition, especially in children after weaning, marked by lethargy, growth retardation, anemia, edema, potbelly, skin depigmentation, and hair loss or change in hair color. Pellagra is a disease caused by a deficiency of niacin and protein in the diet and is characterized by skin eruptions, digestive and nervous system disturbances, and eventual mental deterioration.

3. **(E)** See explanation for Question 2.

4. **(B)** The second law of thermodynamics states that with each successive energy transfer or transformation, less energy is available. Crop plants may convert 1–10% of sunlight to edible food. Grazing animals that feed on crop plants may convert only 1–10% of the plant to meat.

5. **(C)** Three major crops utilized by humans are wheat, rice, and corn (maize). Wheat and rice supply ~60% of human caloric intake. Other agricultural crop products listed in order of world production are potatoes, barley, sweet potato, cassava, grape, soybean, oats, sorghum, sugarcane, millet, banana, tomato, sugar beet, rye, orange, coconut, cottonseed, apple, yam, peanut, watermelon, cabbage, onion, bean, pea, sunflower seed, and mango.
6. **(A)** South Asia includes countries of Afghanistan, Bangladesh, Bhutan, India, the Maldives, Nepal, Pakistan, and Sri Lanka.
7. **(D)** Anemia is a pathological deficiency in the oxygen-carrying component of the blood, measured in unit volume concentrations of hemoglobin, red blood cell volume, or red blood cell number. Low iron intake leads to anemia. It is estimated that 3.6 billion people, more than half of the people in the world, suffer from iron deficiency. More than 80% of all pregnant women in India are anemic. Foods that are rich in iron include red meat, eggs, legumes, and green vegetables.
8. **(C)** Up to one-fifth of America's food goes to waste each year, with an estimated 130 pounds of food per person ending up in landfills. The annual value of this lost food is estimated at around $31 billion. Approximately 49 million people worldwide could be fed by those lost resources (11 million being children), more than twice the number of people in the world who die of starvation each year.
9. **(D)** The higher one goes up the pyramid, the less of that product one should consume per day. Fats, oils, and sweets should be consumed sparingly, they are at the top of the pyramid. Carbohydrates are at the bottom of the pyramid and one should have 6-11 servings of these items. Therefore, according to the choices provided, one should eat less milk and eat more bread.
10. **(C)** Areas of greatest malnutrition are Sub-Saharan Africa (~225 million people), East and Southeast Asia (~275 million people), South Asia (~250 million people), and parts of Latin America. Improvements are expected in East, South, and Southeast Asia.
11. **(B)** Most genetically engineered crops are designed to withstand high amounts of weed killers or herbicides. Roundup Ready crops by Monsanto are genetically engineered plants containing the Roundup Ready gene, which allows Roundup Ultra herbicide (glyphosate) to be applied directly over-the-top of the crop for weed control.
12. **(D)** Of all cropland in the world, 75% is planted with genetically engineered crops.
13. **(E)** Bovine somatotropin (BST) is one of the hormones involved in normal growth and development of the mammary gland and normal milk production in dairy cows. When cows are supplemented with increased amounts of BST, the mammary gland takes in more nutrients from the bloodstream and synthesizes more milk.
14. **(C)** One-fifth (20%) of the world's population today is malnourished. In 1970, one out of every three (33%) people were chronically undernourished.
15. **(C)** Crops are probably receiving as much fertilizer as they can effectively use. Applying more and more fertilizer is exceeding the amount they need or can metabolically utilize.

Free-Response Question

The following is an excerpt from a newspaper article describing a controversy.

Genetically Modified Crops—A Controversy
Recently, Smithville County inhabitants have been embroiled in a controversy regarding the use of genetically modified crops grown in the county. The *Smithville Express* interviewed county residents for opinions on genetically modified crops. Below is a sampling of their responses.

Sam Walters, farmer: "I need to feed my family. I can get more than double the crop I used to get if I use that new seed the salesman was talking about last week."

Jim Stevens, salesman: "Our new product, MiracleWheat, will result in less use of pesticides and fertilizer, and produce five times the amount of wheat that farmers used to get."

Kristin Thorpe, college student: "Genetically modified crops are not a panacea. They have many effects that are detrimental to the environment."

Maggie Rutherford, 88-year-old resident: "I wouldn't touch that stuff. God did not intend for us to be playin' around with genes and such. You'll end up poisoned if you eat that stuff."

Merle Robbins, high school environmental science teacher: "Genetically modified crops have produced great yields in agricultural production. And yet, there are effects of genetic diversity that need to be considered."

Jack Potter, contractor: "The world has too many people as it is. If we keep making more food, they'll just make more babies and we'll all end up starving."

(a) Select two of the people interviewed. Provide a concise argument, based on scientific principles, that supports or refutes each individual's viewpoint.
(b) Provide an example of a genetically modified organism (plant or animal). Describe both positive and negative consequences of using this altered species.

Free-Response Answer

Note to the teacher: This time I am not going to provide a canned answer to begin with. Have your students follow the directions and then, when they have finished, have them either voluntarily read their essays to the class or post them on the class website for constructive review by other students. If you have arrangements with other schools, share your class essays via the Internet.

Note to the student: Let's try something different in terms of writing *this* essay. This time, I am going to take the training wheels off and let *you* write the essay. I'll give you hints and suggestions along the way, but to really get the 5 that you deserve, *you* are the one that will ultimately write the essay in May. Let's begin.

First Step: Brainstorm

Write down as many key words or ideas as you can about genetically modified organisms. When you are through, come back to this page and compare your list with mine.

Remember, I got to peek back at the chapter for ideas and words. You're doing great if you got perhaps ten ideas down. If you get stuck after five, you are allowed to go back and peek at the chapter to find ideas.

MY LIST
first green revolution
second green revolution
who benefits from GMCs?
what is a GMC?
benefits—greater yield, pest resistant, drought resistant, mold resistant, higher protein content, vitamins—golden rice
how much food that we eat is from GMCs?
chronically malnourished stats
acreage devoted to agriculture
increase in land use vs. increase in crop production
increase in grain production per person
rate of population growth vs. rate of increase in crop production
increased energy requirements
more fertilizer
problems with fertilizer—excess nitrogen
law of diminishing return—limiting factors
s-curve
natural selection and crop diversity
soybean allergies
tryptophan
StarLink
Bt
Roundup Ready
Texas Male Sterile Corn (TMS)

Second Step: Organize Your Thoughts

Right now I have a bunch of words that all apply to GMCs, but if I wrote about them in the order I listed them, my essay would be disorganized to say the least. I also need to prioritize (or weed out) which of the concepts I want to discuss. If I discussed each one, I would definitely go past my 25-minute time limit. I also need to mind-map my response, that is, put it into some logical order that answers the questions I was given. So, right now, go back and look at *your* list. Organize major concepts around the people who were interviewed in the article. When you have finished, come back and compare your ideas with mine.

Sam Walters—economic and yield data
Jim Stevens—pesticide and fertilizer facts
Kristin Thorpe—unexpected results that occurred from GMCs
Maggie Rutherford—ethical issues
Merle Robbins—biodiversity, natural selection
Jack Potter—ethical and moral issues, carrying capacity

Third Step: Review the Directions

The directions said to pick two people who were interviewed and provide a concise argument, based on scientific principles that support or refute their viewpoints. I'm going to pick Sam Walters and Merle Robbins. I did not pick Maggie Rutherford or Jack Potter because I don't feel comfortable making ethical arguments, and I also don't have facts to back up such arguments. I think I would have ended up in a dead end very quickly. I did not pick Kristin Thorpe either because her statement really lent itself to part (b).

Sam Walters—economic and yield data

- Advances in food production due to the first and second green revolution are not ending famine. In India, Mexico, and the Philippines, where large advances were made in crop production as a result of agricultural practices of the first and second green revolution, famine is still common. What is increasing along with famine is food exports from these countries (due to higher prices paid by more-developed countries) and degradation of the soil.
- Due to depletion of soil nutrients, erosion, multiple cropping, and requirements of high-yield genetically modified crop species, fertilizer requirements have increased. From 1950 to 1984, the use of fertilizer increased almost 400%, while crop acreage *per person* fell by 50%. (Population is increasing at a faster rate than conversion of acreage to agriculture).

Merle Robbins—biodiversity, natural selection

- To match growing conditions of water availability, soil type, climatic condition, etc., farmers in India have planted more than 30,000 different varieties of rice. By 2005, however, this number will be reduced to 10 different varieties. In 1990, 75% of all corn in the United States was limited to only 6 varieties. 50% of all wheat grown in the United States comes from nine different genotypes.
- A diverse mixture of genotypes ensures that environmental stresses such as pests, diseases, drought, and climatic changes, do not cause extinction of the species.

Fourth Step: Begin Writing

Now that I have the points that I want to talk about matched to two people, it is time to start writing. Remember to begin with a thesis statement.

Thesis Statement

Since the agricultural revolution, farmers have battled insects, microorganisms, and weeds that have destroyed or competed with their crops—threatening their families with starvation. Indeed, many major events in history have resulted from devastating plant disease epidemics or insect infestations. The Irish potato famine of the mid-1800s, which was caused by a fungus, was responsible for the deaths of more than a million people and prompted a massive Irish emigration to the United States. Genetically modified crops hold the promise of increasing crop yields and ending world starvation.

(a) Sam Walters

Since the first genetically modified seeds were sold to farmers, the amount of acreage planted has grown exponentially. By the 1998 growing season, it was estimated that genetically altered crops planted in the U.S. accounted for approximately 70% of all food produced in the U.S. Corn alone, in its first season after winning regulatory approval, accounted for some 15 million acres. The percentage of genetically modified seed, some experts estimate, is now approaching 40 to 60 percent of all U.S. plantings. Recent studies have shown that:

- 1/3 of all U.S. corn and soybeans are genetically engineered.
- 75% of all cropland in the world is planted with genetically engineered crops.
- per capita grain production increased 40% between 1950 and 1984.

Merle Robbins
Farmers in India used to plant more than 30,000 different varieties of rice. Different varieties had adapted over time to acclimate to different climates, weather conditions, water, and nutrient availability. However, by the beginning of the 21st century, there were only 10 different varieties of rice grown on a commercial scale in India. In the United States, 75% of all corn grown comes from only 6 varieties and 50% of all wheat grown comes from only nine different genotypes. Having a wide diversity in the genotype ensures that environmental stresses such as pests, diseases, drought, climatic changes, etc., do not cause extinction of the species.

Fundamental to natural selection is variation in offspring. For natural selection, (1) the variations must be gene-expressed and capable of being inherited, and (2) variations that are advantageous to the individual in terms of survival allow more organisms possessing the trait(s) to survive and reproduce and pass on the characteristic(s) to future generations.

(b) Example of a GMC
Roundup Ready crops are genetically engineered crops that are designed to withstand large amounts of herbicides (specifically Roundup- manufactured by Monsanto). The gene conferring resistance to Roundup, a chemical known as glyposate, allows farmers to spray fields with a broad-spectrum herbicide that will spare only the selected crop. And this ability in turn lets farmers optimize their use of "no-till" agriculture, a tactic that can reduce soil erosion by an estimated 70 percent. Risks of using crops designed to withstand large amounts of herbicides includes the risk of modified genes "escaping" or getting into wild species, with unpredictable consequences. Furthermore, much larger amounts of herbicides are needed with Roundup Ready crops. Unforeseen consequences will probably occur when weeds that are naturally resistant to the effects of Roundup become predominant.

Fifth Step: Conclusion

I have looked at my watch and I have about 2 minutes to go, so it is time to start wrapping this up by writing a conclusion statement. The conclusion statement summarizes what I have been saying. This particular conclusion gives a slightly biased approach to the issue of genetically modified organisms.

Conclusion
The advent of plant biotechnology is the engine of a second green revolution, capable of providing farmers with hardier, higher-yielding, disease-resistant, and more nutritious crops that are needed to sustain a burgeoning world population. Modification of species is really nothing new; using tools such as selective breeding and hybridization, humans have been influencing the genetics of food crops for millennia. Indeed, present varieties of corn bear little resemblance to their historical progenitors. The contribution of

biotechnology is that the process can be sped up enormously and new traits incorporated from virtually any species. New varieties have been more extensively tested than any in history and their safety as food and in the environment is well proven.

Let's put the pieces together to see what the final essay looks like.

Since the agricultural revolution, farmers have battled insects, microorganisms, and weeds that have destroyed or competed with their crops—threatening their families with starvation. Indeed, many major events in history have resulted from devastating plant disease epidemics or insect infestations. The Irish potato famine of the mid-1800s, which was caused by a fungus, was responsible for the deaths of more than a million people and prompted a massive Irish emigration to the United States. Genetically modified crops hold the promise of increasing crop yields and ending world starvation.

Since the first genetically modified seeds were sold to farmers, the amount of acreage planted has grown exponentially. By the 1998 growing season, it was estimated that genetically altered crops planted in the U.S. accounted for approximately 70% of all food produced in the U.S. Corn alone, in its first season after winning regulatory approval, accounted for some 15 million acres. The percentage of genetically modified seed, some experts estimate, is now approaching 40 to 60 percent of all U.S. plantings. Recent studies have shown that:

- 1/3 of all U.S. corn and soybeans are genetically engineered.
- 75% of all cropland in the world is planted with genetically engineered crops.
- per capita grain production increased 40% between 1950 and 1984.

Farmers in India used to plant more than 30,000 different varieties of rice. Different varieties had adapted over time to acclimate to different climates, weather conditions, water, and nutrient availability. However, by the beginning of the 21st century, there were only 10 different varieties of rice grown on a commercial scale in India. In the United States, 75% of all corn grown comes from only 6 varieties and 50% of all wheat grown comes from only nine different genotypes. Having a wide diversity in the genotype ensures that environmental stresses such as pests, diseases, drought, climatic changes, etc., do not cause extinction of the species.

Fundamental to natural selection is variation in offspring. For natural selection, (1) the variations must be gene-expressed and capable of being inherited, and (2) variations that are advantageous to the individual in terms of survival allow more organisms possessing the trait(s) to survive and reproduce and pass on the characteristic(s) to future generations.

The advent of plant biotechnology is the engine of a second green revolution, capable of providing farmers with hardier, higher-yielding, disease-resistant, and more nutritious crops that are needed to sustain a burgeoning world population. Modification of species is really nothing new; using tools such as selective breeding and hybridization, humans have been influencing the genetics of food crops for millennia. Indeed, present varieties of corn bear little resemblance to their historical progenitors. The contribution of biotechnology is that the process can be sped up enormously and new traits incorporated from virtually any species. New varieties have been more extensively tested than any in history and their safety as food and in the environment is well proven.

Word Count: 482

CHAPTER 9

Land

Trees are poems that Earth writes upon the sky;
We fell them down and turn them into paper,
That we may record our emptiness.

Kahlil Gibran

Areas That You Will Be Tested On

A: Residential and Commercial
B: Agricultural and Forestry
C: Recreational and Wilderness
D: Erosion and Conservation
E: Natural Features, Buildings, and Structures
F: Wildlife
G: Mineral Resources and Mining

Key Terms

Note: Definitions for all key terms listed below can be found in the glossary starting on page 648. For additional definitions of relevant terms from this chapter, refer to *www.barronseduc.com/0764121618.html*.

acid mine drainage	**erosion**
alley cropping	**floodplain**
biomass	**forest succession**
clear-cutting	**intercropping**
competitive exclusion principle	**niche**
contour farming	**open-pit mining**
contour mining	**sustainable agriculture**
crop rotation	**traditional subsistence agriculture**
Dust Bowl	**urban renewal**

Key Concepts

Land Uses

- Since the beginning of human civilization, humans have been altering and changing the character of the Earth's land surface to suit the needs of humans. This change occurs either through changing the land from one type to another as in cutting down tropical rain forests and converting the land for monoculture agriculture or through modification of the land to better suit human needs—such as modifying grasslands to grow domestic grains. Biomes are changed or modified for agriculture, aquaculture, forestry, transportation and/or urban settlement (see Figure 9.1).

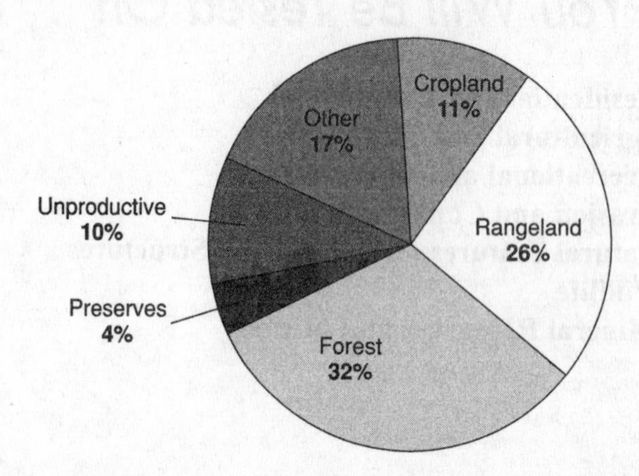

Figure 9.1. World land use. Unproductive land includes deserts, bare rock, ice, etc. Preserves include parks, wildlife refuges, and nature preserves. "Other" includes urban areas, wetlands, tundra, and taiga, among others.

- Since the Industrial Revolution and the mechanization of agriculture, all natural biomes have seen significant declines in land mass area and the conversion of said lands for agricultural purposes. In the last 200 years, the increase in the percent of land that has been converted for agriculture is close to 2000%, with humans either directly or indirectly influencing 40% of all terrestrial land mass.
- One tenth of the world's population is considered indigenous (native to a place or area, originating in and characterizing a particular region or country, and referred to as the Fourth World). They inhabit 25% of all land surface.

Residential and Commercial

Urbanization

- Urbanization refers to both movement of people from rural areas to cities and the change that accompanies it.
- People who live in rural areas are primarily dependent on agriculture. They may live in villages that are linked together by culture and customs.

- People who move to the urban areas may live in towns (smaller than cities and have decentralized areas of commerce) or cities (larger groups of people with centralized areas of commerce and residential living).

WORLD'S LARGEST CITIES
PROJECTED FOR 2015

City	Population (millions)
Tokyo	26.4
Bombay	26.1
Lagos	23.2
Dhaka	21.1
Sao Paulo	20.4
Karachi	19.2
Mexico City	19.2
New York	17.4
Jakarta	17.3
Calcutta	17.3

Source: United Nations—World Urbanization Prospects:
The 1999 Revision

- Urbanization began at about the same time as agriculture, 10,000 years ago. In 1900, 15% of the world's population lived in urban areas. Today, it is almost half, with 1% of all land surface devoted to urban areas.
- Organized agriculture allowed people to settle in one area instead of being nomadic. Food surpluses allowed people to engage in other activities for the benefit of the group.
- Urbanization was responsible for development of
  - Trade and commerce,
  - Politics and organized government,
  - Organized religion,
  - Technology,
  - Education,
  - Arts.
- Areas that are experiencing greatest growth in urbanization are countries in Asia and Africa. Asia alone has close to half the world's urban inhabitants, even if more than three fifths of its population is still in rural areas. Africa, which is generally considered overwhelmingly rural, now has a larger urban population than North America. Reasons include jobs, higher standard of living, health care, mechanization of agriculture, and education. In mostly developed countries, urbanization is leveling off. See Figure 9.2
- Reason for rapid rate of urbanization is primarily economic—the world's economy increased fivefold between 1950 and 1990. Nations with the most rapid increases in their levels of urbanization are generally those with the most rapid economic growth, while the nations with the highest per capita incomes are generally those with the highest proportion of their population in urban areas.

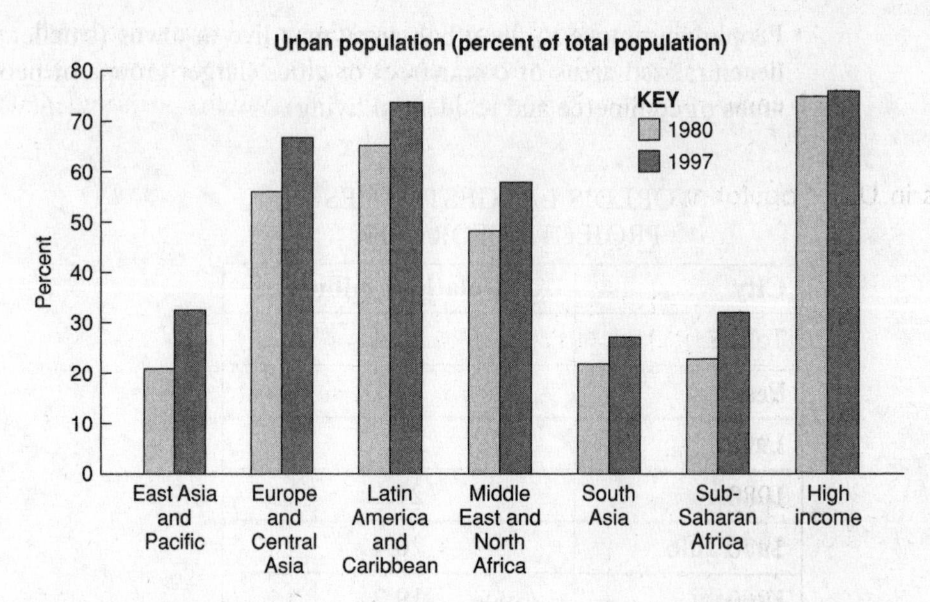

Figure 9.2. Urban population as a function of geography and income. Source: World Bank

Effects of Urbanization on the Environment

- Destruction of natural areas.
- Depletion of natural resources—requirements exceed capacity of the immediate area for supply.
- Pollution—because people and industries are concentrated in a limited area, so too are the effects of pollution (e.g., dust domes, smog, ozone).
- Creation of urban heat islands that impact local weather conditions.
- Air pollution and its effects including photochemical smog, greenhouse effect, and acid rain.
- Water runoff and flooding.
- Depletion of available clean water leading to land subsidence and salinization.
- Pollution of groundwater and surface water—in developing countries, 90% of untreated sewage is disposed into local surface water.

To create sustainable cities, the following concepts are required:

- Reduce urban sprawl by expanding upward. This preserves prime agricultural areas needed for the city and improves efficiency of city services. A reduction of 1 acre of prime agricultural land due to urban sprawl requires planting 3 acres of less desirable land to produce the same yield.
- Conserve natural habitats. Habitats provide many roles including purifying air and water and providing breeding grounds.
- Focus on energy and resource conservation. Reducing, reusing, and/or recycling resources diminish the impact of a more concentrated demand and its effect on the local environment.
- Develop methods of efficient mass transportation. This reduces congestion, fuel consumption, parking space, and air pollution.
- Provide ample green space. Besides the aesthetic benefits, green spaces such as parks provide habitat for wildlife, tend to moderate the effects of an urban heat island, stabilize soil, and improve air quality.

- Plan first, build later. When cities are being planned, modified, or upgraded, it is best to recognize environmental concerns first and then make plans to accommodate the concerns rather than to build and then react to the environmental consequences.

Shifts in U.S. Population

RURAL-URBAN SHIFTS IN THE UNITED STATES (1800–1990)

Year	% Urban	% Rural
1990	75.2	24.8
1980	73.7	26.3
1970	73.6	26.3
1960	69.6	30.1
1950	64.0	36.0
1940	63.1	36.9
1930	59.6	40.4
1920	56.5	43.5
1910	56.1	43.9
1900	51.2	48.9
1850	15.4	84.6
1800	6.1	93.9

Source: U.S. Census

- In 1850, 2% of the world population lived in urban areas (population of 10,000 +). In 2000, the percentage had increased to approximately 50%. In 2025, 63% of the world population will live in urban areas. The increase of urbanization in less-developed countries is estimated to be approximately 3.5% per year, with 38% of the population living in cities. This compares to less than 1% rate of increase in urbanization in more-developed countries.
- In the United States, the urban population was 5% in 1800. In 2000, the number had increased to 75%.
- 17% of the world's population that live in urban areas live in slums or shantytowns.
- In 1900, there were 19 cities worldwide that had populations greater than 1 million. In 2000, there were 400 cities and 19 cities worldwide had populations greater than 10 million.
- Growth in urbanization is the result of
  1. Births,
  2. Immigration,
  3. Movement from rural to urban areas where promises of a better lifestyle attract people,

4. Better paying jobs,
5. Access to education,
6. Access to better health care.

GROWTH IN URBANIZATION

Pro	Con
(1) Use less land—less impact on environment.	(1) Impact on land is more concentrated and more pronounced.
(2) Better educational delivery systems.	(2) Overcrowded schools.
(3) Mass transit systems decrease reliance on fossil fuels—commuting distances are shorter.	(3) Commuting times are longer because infrastructure not keeping up with growth.
(4) Better sanitation systems.	(4) Sanitation systems have greater volumes of wastes. Spills occur.
(5) Recycling systems are more efficient.	(5) Solid waste build-up is more pronounced. Total volume of wastes can overwhelm system. Landfill space becomes scarce and costly.
(6) Facilities such as museums, libraries, and concert halls that meet aesthetic and intellectual needs are possible.	(6) Infrastructure (roads, bridges, etc.) are crumbling and costly to retrofit.
(7) Large numbers of people generate high tax revenues.	(7) Large number of poor people put strain on social services and drains tax revenues. Wealthier people move away from city centers and decrease tax base.
(8) Large cities attract industry due to availability of raw materials, distribution networks, customers, and labor pool.	(8) Higher density creates more crime and less job opportunity—not enough jobs for all the people.

Green City Characteristics

A green city is an environmentally-sustainable city achieved by controlling pollution and waste, efficient use of energy, recycling programs, and using renewable energy sources. It has the following characteristics:

- Planned parks;
- Greenbelts;
- Planned trees and vegetation—reduces air pollution and noise and provides habitat for wildlife;
- Planned mass transportation alternatives—busses, light rail, bicycles, car-sharing networks, bike and ride systems, park and ride systems, etc.;
- Planned energy conservation measures in public buildings and facilities;
- Planned waste disposal methods—distributed recycling facilities and curbside collection;
- Solar energy rebates;

- Building codes that favor the use of environmentally-friendly building materials (no tropical wood), use of insulation, and high-efficient appliances;
- Areas of natural land set aside with vegetation for water runoff percolation and groundwater recharge, reduce urban heat dome, and reduce erosion;
- Numerous, small neighborhood shopping centers to discourage commuting and increase employment;
- Legislative limits to growth;
- Variety of housing available for all income levels.

Agricultural and Forestry

Forests

CLASSES OF FORESTS

Developed rural forest land	Forestland located close to rural areas that are not likely to be used for commercial timbering. Such forestland is usually associated with agricultural, recreational, or residential nonforest uses. Examples of such nonforest uses include cropland, home sites, camping areas, and farmsteads. The ecological character of the forest remains intact (i.e., the understory has not been removed or altered enough to preclude forest succession or replanting).
Other	Not capable of producing 20 cubic feet per acre per year of industrial wood products. Contains trees not currently utilized for industrial wood production or poor quality trees. Site conditions may be poor such as sterile soil, dry climate, poor drainage, high elevation, or rockiness.
Reserved timberland	Capable of producing in excess of 20 cubic feet per acre of industrial wood crops but which statutory restrictions prohibit the harvesting of trees.
Timberland	Capable of producing in excess of 20 cubic feet per acre per year of industrial wood products.
Urban	Forestland close to urban nonforest land making it unlikely to be used for commercial timbering. Uses include recreational and residential. The ecological character of the forest remains intact in that the understory has not been removed or altered enough to preclude forest succession or replanting. These urban forests surround golf courses, act as airport buffers, and are used as urban parks.

- There are a total of 192 million acres of National Forests and National Grasslands in the United States.
- Approximately 40% of all forests worldwide are "old growth"—forests that have been undisturbed and carry out natural life cycles.
- Each American uses the equivalent of a 100-foot tree a year.
- The average single-family home (2000 ft^2) can contain 15,824 board feet of timber.
- To grow a pound of wood, a tree uses 1.47 pounds of carbon dioxide and gives off 1.07 pounds of oxygen. An acre of trees might grow 4000 pounds of wood in a year, using 5880 pounds of carbon dioxide and giving off 4280 pounds of oxygen in the process.

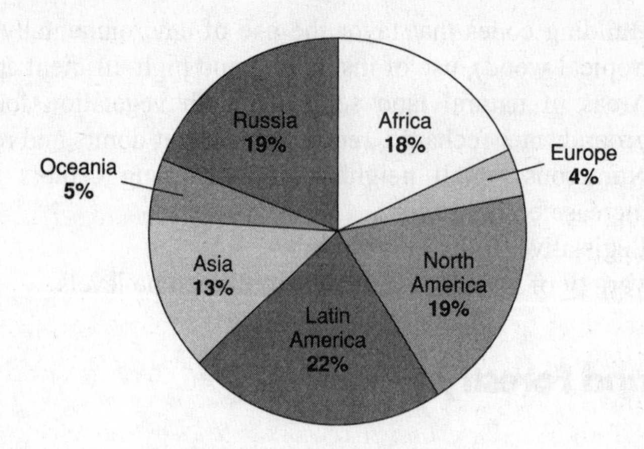

Figure 9.3. World forest distribution.

- Forestland accounts for 32% of the land in the United States, the largest of any land use category.
- Out of 747 million acres of U.S. forestland, 500 million acres (two-thirds) are non-federal.
- Forestry and wilderness issues have been distinguished by three related features:
  1. They attract significant middle-class popular support, particularly in the capital city electorates;
  2. They center on the preservation (or destruction) of aesthetic landscapes at considerable distance from these urban centers of greatest support; and
  3. They never directly challenge the dominant economic and material concerns of their supporters.
- Forests cover about one-third of all land surface. Four-fifths of these are *closed canopy* (tree crowns covering more than 20% of the ground); one-fifth of all forests are *open canopy* (tree crowns covering less than 20% of the ground area).
- Most forests (70%) are located in North America, Russian Federation, and South America (see Figure 9.3). In the United States, the largest area of timbering is in the Pacific Northwest, employing 150,000 people and representing a $7 billion a year business.
- Forests are used for fuel, construction materials, paper products, and wildlife habitat—3.7 billion metric tons of wood annually. Developed countries use 80% of wood products but produce less than half of the wood required. Japan is the largest importer of wood.
- Inexpensive ceramic stoves would save four times the amount of wood as cooking over an open fire. More than 50% of the world's population relies on firewood or charcoal for heating and/or cooking.
- It is estimated that at least 90% of the total bird, amphibian, and fish species of the United States and at least 80% of mammal and reptile species can be found on forested land.
- Forests have a role in regulating climate, controlling water runoff, producing oxygen, and providing food and shelter for wildlife, and they have scenic, historic, and cultural values.
- Forests are found in most humid and subhumid regions of the world, from the tropics to the tundra and are a renewable resource.

- 4% of forests are intensively harvested plantations. Most countries harvest far more than they replant.
- 14% of forests are actively used for a variety of goods and services.
- Approximately one-third of all forests have been converted for agriculture since it results in faster financial returns.
- Annual harvesting of forests for fuel and wood is currently about 5 billion cubic meters and is increasing by about 1.5% per year.
- Due to undervaluation, royalties, purchase prices, and/or stumpage fees have been set too low to recover sufficient capital for the proper management and reforestation. Logging roads in the United States are 10 times longer than the total interstate highway system.
- Of all U.S. Forest Service jobs, 75% are connected with recreation and 3% are connected with logging activities.
- In developed countries, many areas that were deforested have come back. Example: New York State has three times as much forest today as it did in 1850.
- In less-developed countries, due to high fuel prices and population growth, wood resources have declined rapidly.
- Accumulation of biomass in the forests causes

  1. Wildfires that destroy even the larger trees that withstand smaller fires;
  2. Invasion of insects, diseases, and weeds;
  3. Reduced native biological diversity by creating the unnatural forest conditions stated in 1 and 2 above.

TEMPERATE FORESTS

- In milder climates, deciduous trees (shed leaves during winter or dry seasons) are predominant. In colder climates, coniferous trees (mostly evergreen and that have needle-shaped or scale-like leaves to prevent water loss) are predominant.
- Most developed countries have completely harvested their original temperate forests. Second- or third-generation forests have replaced the original frontier forests. Eastern Russia, which contains 25% of the world timber reserves, is an area that is harvesting much of its temperate forests today to supply needed capital.
- Last areas of frontier forests in North America are in Washington, Oregon, northern California, and British Columbia, Canada. Much environmental activism is connected with preserving these trees because about 10% of original old-growth forest remains with about 80% of that being scheduled to be cut.
- One-quarter of all forest growth is ruined by fires, insects, and disease.
- Prescribed or natural fires are necessary for some species to reproduce. They also get rid of fuel that accumulates on the forest floor.
- Timber management is concerned with preserving frontier forests and environmentally sound methods to harvest trees.

TROPICAL FORESTS

- Approximately 6% of the Earth is covered with tropical forests. They contain more than two-thirds of the world's plant biomass and more than half of all species of plants and animals.
- Up to 36 million acres each year of tropical rain forest are destroyed; this represents approximately 0.8% each year. Brazil has the highest rate of tropical rain deforestation.
- Deforestation began in the 1700s due to demand of hardwoods (teak, mahogany, etc. for furniture and shipbuilding).

- Deforestation also made possible the growing of commercial crops—sugar in the West Indies, coffee and sugar in Brazil, and rubber in Malaysia.
- Demand for land to grow crops and settle people was matched by a rising demand for products from tropical forests. Clearing land for agriculture involves felling trees, and slashing and burning ground vegetation—especially in tropical deciduous forests. Fire releases nutrients trapped in vegetation to soil, but the nutrients are quickly leached out of the soil by rain and eventually pollute streams, rivers, and lakes or the nutrients are absorbed by the crops and removed from the land. When the nutrient levels fall below new crop requirements, the land is abandoned, and new areas are cleared in a process called "shifting agriculture."
- Tropical forests can be (1) rain forests, (2) deciduous forests with one or two dry seasons per year, or (3) dry, deciduous forests. Between 1980 and 1995, it is estimated that 200 million hectares of tropical rain forest was lost forever. Only 10% of it was replaced by reforestation.
- Five factors play a major role in reducing tropical rain forests:

  1. Expanding population and resettlement,
  2. Ranching and pasture development for cattle,
  3. Fuelwood and charcoal,
  4. Timber trade,
  5. Conversion of land to agriculture

- Harvesting of trees is accomplished by either selective or clear cutting. In selective cutting, intermediate-aged, mature, or diseased trees in a forest stand are either harvested as single trees or as small groups of trees. This harvesting technique encourages the growth of younger trees. In clear cutting, everything is cut down at once with the resulting loss of niches and the associated biodiversity.
- Amazon rain forest covers approximately 7 million square kilometers, 5 million of which are in Brazil. Brazil is intent on developing the commercial productivity of the rain forest by encouraging agriculture, cattle ranching, and logging and by generating fuel from biomass. A major highway going through the Amazon Basin was built during the 1960s.
- Tropical rain forests provide

  1. **Biodiversity.** A diverse gene pool can protect commercial plant strains against pests and can contain biological chemicals that may be used for drugs or other commercial products. The wild relatives of avocado, banana, cacao, cashew, cinnamon, coconut, coffee, grapefruit, lemon, paprika, oil palm, rubber, and vanilla are found in tropical forests. Approximately 170,000 plants species occur in the tropical rainforest many of which have never been classified or studied.
  2. **Climate.** Forests help to stabilize climate and create their own unique microclimates. Deforestation on a large scale releases carbon dioxide, methane, and nitrous oxide and accounts for approximately 25% of greenhouse gases released into the atmosphere (see Figure 9.4). This release of gases into the atmosphere enhances the greenhouse effect resulting in global warming.
  3. **Fisheries.** Forests protect fisheries in rivers, lakes, estuaries, and coastal waters by providing essential nutrients and sanctuaries for fish to spawn and develop.
  4. **Protection from erosion.**
  5. **Protection from flooding by absorbing excess water.**

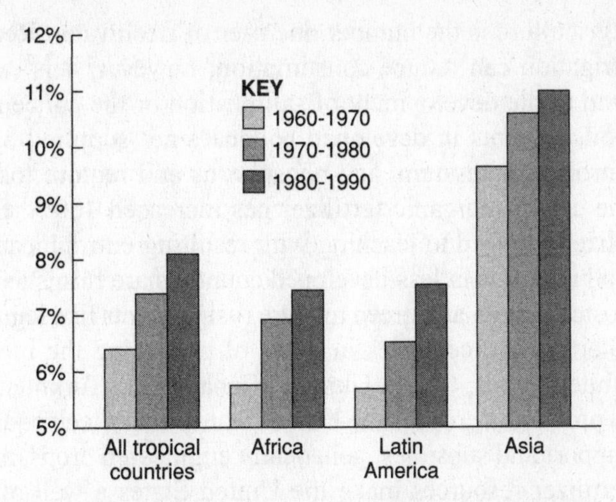

Figure 9.4. Rate of deforestation in tropical rain forests.

Agriculture

- The top five countries in order of having the most amount of land for agriculture include: Russian Federation, Kazakhstan, United States, India, and China. The Russian Federation, with a population estimated to be around 160 million, or 2.5% of the Earth's population, contains about 15% of the world's croplands (215 million hectares). Most of the crop is sold, or used as feed.
- The top five countries in order of having the most amount of land for agriculture per person are: Kazakhstan, Turkmenistan, Australia, Canada, and the Russian Federation.
- The top five countries in order of having the most land (%) irrigated are: China, India, Mexico, Uzbekistan, and the United States.
- The top five countries in order of using the most fertilizer per acre are: China, Ukraine, United States, India, and Mexico.
- Agriculture depends upon climate, nutrients, soil, and water.
- Two types of agriculture exist—industrial and subsistence.

1. **Industrial.** Mega-farms run by large corporations producing large quantities of a single product. Uses high-tech machinery, pesticides, herbicides, and fertilizers. Uses genetically-engineered crops and livestock. Common in developed countries and employs about 10% of the workforce.
2. **Subsistence.** Growing only enough to live on. Common in developing countries and employs about 65% of the population. Three types of subsistence agriculture are:
   - Shifting agriculture—see description of tropical forests earlier.
   - Intensive crop cultivation—small, privately owned farms; labor-intensive agriculture. Uses little machinery, inorganic fertilizers, herbicides, or pesticides. Common in parts of Africa, Central America, China, Japan, India, and Southeast Asia
   - Nomadic herding—agriculture involving the herding of livestock by a group of people who have no fixed home and move according to the seasons from place to place. Common in Afghanistan, Africa, Middle East, and northern China (Mongolia).

- Agriculture uses only 10% of all available land. Reuse of land can lead to erosion (wind and water facilitating the movement of top soil from one place to another) and desertification (conversion of marginal rangeland or cropland to a more desert like land type which can be caused by overgrazing, soil erosion, prolonged drought, or climate change) as well as use up available resources (nutrients, water, etc.) and make the land less productive.

- Agriculture is the number one user of freshwater. Recent developments including drip irrigation can reduce consumption; however, it is expensive to install. Irrigation can lead to the development of salinization or the concentration of salts in the soil.
- Soil nutrients in developed countries are supplied by inorganic fertilizers containing nitrogen, potassium, and phosphorus and require fossil fuels to produce. Since 1950, the use of inorganic fertilizer has increased 400%. Erosion or overapplication of fertilizer can lead to leaching with resulting eutrophication in aquatic systems.
- Soil nutrients in less-developed countries are furnished by manure, organic matter, composted matter, and green manure (using plants like legumes to supplement soil nutrients).
- The top five countries in order of producing the most amount of grain per acre are: United States, China, Ukraine, Canada, and Turkmenistan. The most efficient country in producing crops is the United States. High-tech equipment, agribusiness, government support and subsidies, genetically engineered crops, and well-established irrigation and fertilizer resources make the United States a well-oiled machine for producing food. Just slightly behind the United States in efficiency is China. Lacking high-tech equipment and advanced agribusiness, China makes up the difference by using intensive subsistence farming techniques. China, with almost 25% of the world's population, has only about 7% of the world's viable cropland. Although China is a large country, much of the land is not suitable for large-scale farming due to lack of water, mountainous terrain, etc. Their reliance on intensive subsistence farming requires much greater amounts of human labor, irrigation, and fertilizer than does the United States.
- Government subsidies (price supports) are payments made to farmers to guarantee a certain price for their products. Price supports cause farmers to generally produce more than their country needs. Surpluses are then sold on the open market, which tends to lower the price due to the higher supply. Government subsidies for agriculture products include the costs for fossil fuels, pesticides, and inorganic fertilizers, among others, which tend to perpetuate these environmentally less-friendly products and make it unattractive to switch to alternatives since switching would involve added expenses. For example, in more-developed countries, it takes more than a barrel of oil to produce a metric ton of grain. This is seven times the amount of oil it took more than fifty years ago. Agribusiness uses close to 10% of the world's oil output. In less-developed countries that do not have subsidies, these depressed prices make it uneconomical for farmers to make a profit growing their crops. In other words, it is cheaper to buy the artificially low-priced food than to grow it themselves. This makes the less-developed countries dependent on the more-developed countries for food. If situations occur where money in the less-developed country becomes scarce, then the natural result is famine. To break this cycle would require the end of government subsidies, which in turn would cause food prices to increase. If the rate of ending subsidies occurs too fast, it would cause the less-developed countries that had not caught up to suffer.
- The ability to produce agricultural resources is limited by fossil fuels and fertilizers.

FOSSIL FUELS

Listed in order of decreasing amount of energy required to produce 1 hectare of corn are

1. Nitrogen fertilizers,
2. Irrigation,
3. Gas and diesel fuels,
4. Machinery (including energy costs of manufacture),
5. Drying of harvested corn,
6. Seeds (includes all inputs required to produce the seeds),
7. Phosphorus fertilizers,
8. Herbicides.

The energy efficiency of production (i.e., the ratio of energy contained within the corn to the energy put into its production (energy out/energy in) was 3.5:1 in 1945 and 2.5:1 in 1983. This shows that *more* energy was being put into growing corn in 1983 than in 1945. Between 1910 and 1983, corn production increased 346% in the United States. In the same time period, energy required to grow the corn increased over 800%.

FERTILIZER

Due to depletion of soil nutrients, erosion, multiple cropping, and requirements of high-yield genetically modified crop species, fertilizer requirements have increased (see Figure 9.5). From 1950 to 1984, the use of fertilizer increased almost 400% while crop acreage *per person* fell by 50%. (Population is increasing at a faster rate than conversion of acreage to agriculture.)

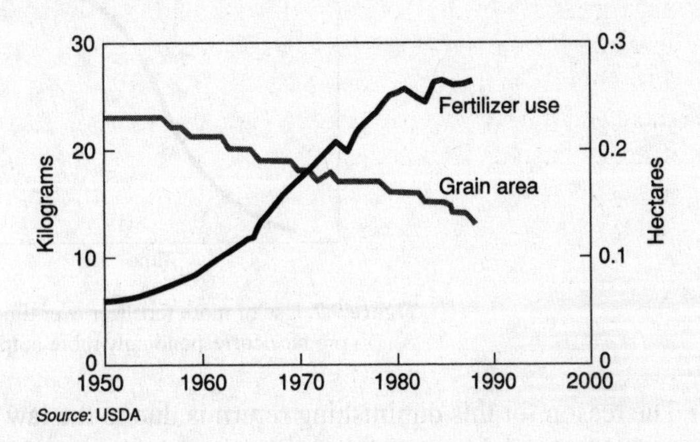

Figure 9.5. World use of fertilizer and grain acreage per capita, 1950–1988.

- When viewed as the total amount of fertilizer used in agriculture between 1950 and 1990, there was a 10-fold increase (15 million tons in 1950 as compared to 150 million tons in 1990—see Figure 9.6).

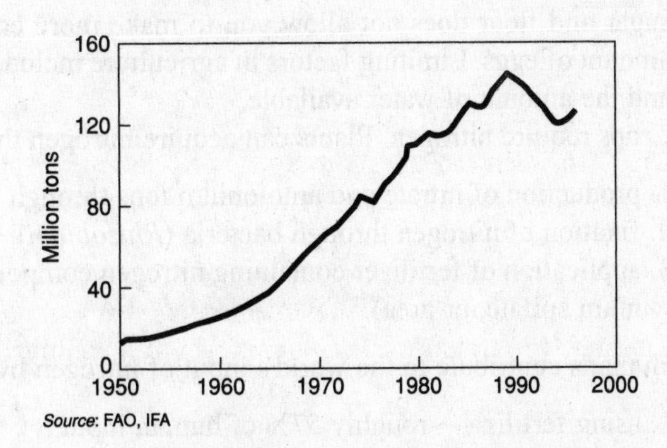

Figure 9.6. World use of fertilizer, 1950–1990.

- The rapid increase of the use of fertilizer as seen in Figure 9.6 has slowed in recent years due to

1. Weak prices for product as compared to increasing costs for fertilizer.
2. Third world debt. Many countries cannot afford massive amounts of fertilizer.

3. Most areas suitable for growing crops, as measured through such factors as soil type and suitability for agriculture, availability of water, and transportation requirements, have been utilized.
4. Environmental awareness.
5. Awareness of the law of diminishing return (more fertilizer does not produce corresponding proportional output). For example, in 1980, each additional ton of fertilizer applied to a corn crop produced a 20-ton increase in corn. Today, however, each additional ton of fertilizer produces only a 10-ton increase. If one were to graph this concept, it would be the familiar logistic or sigmoid curve. See Figure 9.7.

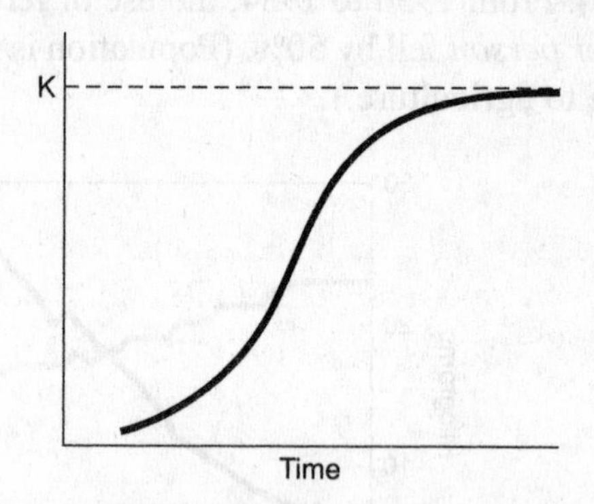

Figure 9.7. Use of more fertilizer over time does *not* produce correspondingly more output.

- The reason for this diminishing return is due to the law of limiting factor(s). Crops are probably receiving as much fertilizer as they can effectively use. Applying more and more fertilizer is exceeding the amount they need or can metabolically utilize. For example, you want to make cookies, and your recipe calls for 2 cups of flour, 2 cups of sugar, and 2 eggs to produce 2-dozen cookies. You go to collect your ingredients, and you find that you have 50 cups of flour in the house, 50 cups sugar, but only 2 eggs. How many dozen cookies can you make? Correct, still only 2-dozen. Having more sugar and flour does not allow you to make more cookies—you were limited by the amount of eggs. Limiting factors in agriculture include the amount of sunlight received and the amount of water available.
- Crops require nitrogen. Plants can acquire nitrogen through three methods:

  1. production of nitrate and ammonium ions through decay processes.
  2. fixation of nitrogen through bacteria (*Rhizobium*).
  3. application of fertilizer containing nitrogen compounds (ammonium nitrate, ammonium sulfate, or urea).

- Humans contribute to the world's input of nitrogen by:

  1. using fertilizer—roughly 57% of human input;
  2. growing nitrogen-fixing crops (legumes)—roughly 29% of human input;
  3. burning fossil fuel—14% of human input.

- This "extra nitrogen" ends up

  1. in rivers (25–33%), resulting in eutrophication of lakes due to an increase in NO_3^-.
  2. being given off as gas and returning to the Earth in precipitation (acid rain).
  3. as NO (nitric oxide)—results in reactions involved in increases in ozone levels (O_3) in the troposphere.

4. as N_2O (nitrous oxide)—a greenhouse gas and involved in the depletion of ozone in the stratosphere.

- Consequences of "extra nitrogen" include:

1. increased nitrate levels in lakes and coastal waters, resulting in eutrophication and die-offs.
2. depletion of stratospheric ozone (widening of the hole in the ozone layer).
3. increase in the amount of ozone in the troposphere.
4. climate change.
5. acid rain and its effect on wildlife.
6. leaching calcium and phosphate from the soil.

Rangeland and Prairie

- In 11 western states, 75% of all public land is grazed. Agencies responsible include the Forest Service and the Bureau of Land Management (BLM).
- Before 1995, grazing land policies by the BLM were determined by "rancher advisory boards" composed of grazing permit holders. After 1995, grazing land policies were determined by a diverse group representing different viewpoints and known as "resource advisory councils".
- 40% of all federal grazing permits are owned by 3% (about 2000) of all livestock operators.
- Federal-grazing permits issued by the BLM average about five cents per day per animal. Private lease rates (the true cost of doing business) average $10 per day per animal and can run up to $20.
- Grazing fee income only covers one-third the cost of managing the lands—the result is a government subsidy for cattle ranchers.
- Government subsidies to cattle ranching do not "trickle down." Only 0.06% of all jobs in areas with grazing are involved in cattle ranching, and only 0.04% of all jobs derive from the cattle ranching industry.
- 219 endangered species live on land administered by the BLM. Livestock grazing is the fifth rated threat to endangered plant species, fourth leading threat for all endangered wildlife, and the number one threat to all endangered species in arid regions of the United States. The greatest number of threatened plant species in North America occurs in rangelands (see Figure 9.8).

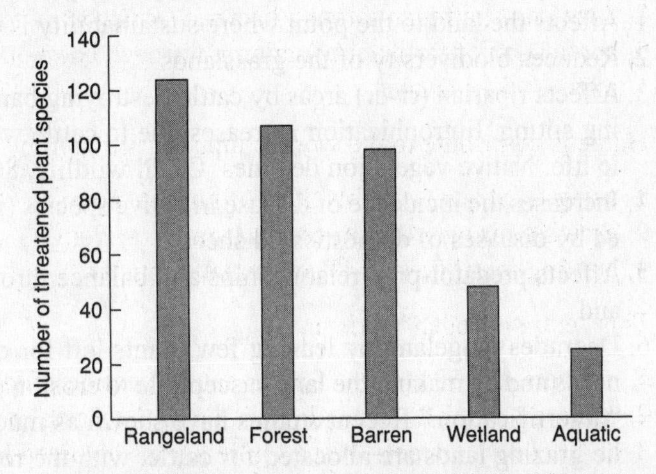

Figure 9.8. Endangered plant species as a function of biome.

- There are approximately 6 billion people on Earth; 3 billion grazing cattle, sheep, and goats; and 1.5 billion cows. There are more livestock than humans in the United States.
- More than one fourth of the Earth's land surface is used to support over 4.5 billion grazing animals that supply both meat and milk. This represents more land than twice that set aside for agriculture. The United States has 800 million acres of rangeland, 60% of which is privately owned. Of the 40% of U.S. rangeland that is owned by the U.S. government, 85% is rated in fair to very poor condition.
- Of grain produced in the United States, 70% is used as animal feed as compared to 37% worldwide.
- Open range and grasslands are being converted for agriculture and other human purposes at a rate three times that of tropical forest destruction. One-third of all rangeland in the world is threatened by overgrazing (see Figure 9.9).

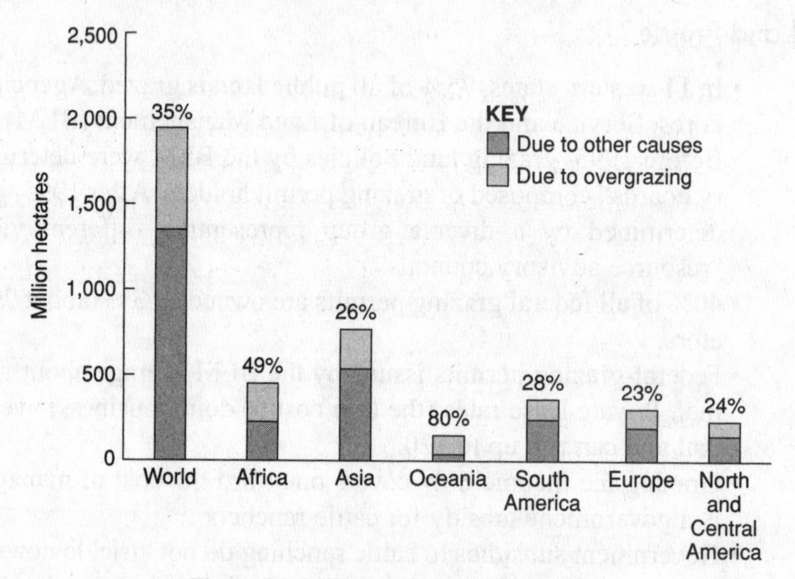

Figure 9.9. Soil desertification in rangelands.

- Nomadic herding can be environmentally sound if areas are not overused. Productivity can be ten times greater than farming in the same area. Factors that restrict nomadic herding include political problems, land rights, wars, changing climate, pressure to cultivate land, and growing populations of people inhabiting land.
- Overgrazing

1. Affects the land to the point where sustainability is threatened.
2. Reduces biodiversity of the grasslands.
3. Affects riparian (river) areas by cattle destroying banks and streambeds and increasing silting. Eutrophication increases due to cattle wastes, thereby destroying aquatic life. Native vegetation declines. Of all wildlife, 80% depend on riparian areas.
4. Increases the incidence of disease in native species (i.e., big horn sheep being affected by diseases of domesticated sheep);
5. Affects predator-prey relationships and balance through predator control programs; and
6. Degrades rangeland by leaving few plants left for other animals and their specific needs and by making the land susceptible to erosion producing a vicious cycle called "desertification." Recent studies have shown as much as 97% of resources on public grazing lands are allocated for cattle, with the remaining 3% allocated to native wildlife.

- Desertification follows the following steps:
  1. Overgrazing causes animals to eat up all available plant life.
  2. Rain washes away the trampled soil.
  3. Since nothing holds water anymore, wells and springs and other sources of water dry up.
  4. What vegetation is left dies from drought or is taken for firewood.
  5. Weeds that are unsuitable for grazing animals may grow.
  6. Ground is not suitable for seed germination.
  7. Wind and dry heat blow away top soil.
- Desertification is pronounced in parts of Africa, Mexico, Central America, South America, and the Middle East.

Sustainable Agricultural Methods:

1. Reduced erosion through soil conservation—see following discussion.
2. Management of water resources—reduces erosion.
3. Use of polyculture and crop rotation—increases biodiversity, reduces need for pesticides, protects from receiving low prices for a specific crop, and restores nutrients to soil.
4. Addition of organic matter to the soil and reduction of dependence on inorganic fertilizers—improves soil structure, enhances fertility, increases water storage capacity, and promotes the tilth (physical condition) of the soil making it easier for plants to emerge from the soil and extend roots downward.
5. Integrated pest management—uses a variety of techniques (listed above) to reduce reliance on pesticides.
6. Reduction of the use of fossil fuels.

Recreational and Wilderness

JURISDICTION OF WILDERNESS AND RECREATIONAL AREAS

Department	Agency	Areas Administered
Department of Interior	National Park Service	National parks (e.g., Death Valley National Park) National battlefields (e.g., Antietam National Battlefield) National monuments (e.g., Lava Beds National Monument) National seashores (e.g., Point Reyes National Seashore)
Department of Interior	U.S. Fish and Wildlife Service	National wildlife refuges (e.g., Salt Plains National Wildlife Refuge) National wildlife ranges (e.g., Sheldon National Wildlife Range)
Department of Interior	Bureau of Land Management	Areas of critical environmental concern (e.g., Moosehead Mountain) Conservation areas (e.g., Snake River Birds of Prey National Conservation Area) Recreation areas (e.g., Lassen Applegate Trail)
Department of Agriculture	U.S. Forest Service	National forests (e.g., Shasta-Trinity National Forest) National grasslands (e.g., Buffalo Gap National Grassland) Recreation areas (e.g., Redington Pass Recreation Area)

PARKS

- Of the world's land surface, 4% is protected as parks, preserves, and wildlife refuges.
- The United States has 376 national parks and recreation areas.
- Visitors to U.S. national parks are up 33% in the last 10 years with budgets down by 25%.
- Air pollution, crowds, pollution, stream pollution, concessions, and noise affect wildlife and the tranquility of the parks.
- In 2001, permission was granted to explore and drill for oil in the Arctic National Wildlife Refuge in Alaska.
- As of 2002, 6.4% of the Earth's surface (851 million hectares) is protected as 28,442 nature preserves. This is nowhere near enough to protect a major share of the world's endangered or threatened species. Most of these preserves are small—only 4% are larger than 100,000 hectares (250,000 acres), and only 0.5% are larger than 1 million hectares (250 million acres).
- Each year in the United States, 110 million people visit 150 zoos. Zoos used to just cage animals for the curious. Today, in developed countries, the emphasis is on research and species conservation. Most mammals in U.S. zoos are produced through captive breeding. Worldwide, zoos hold about 500,000 animals representing about 1000 species.
- Canada's Green Plan calls for 12% of Canadian land to be set-aside as national parks. Currently, Canada has 1400 national parks and protected areas known as ecological reserves.

WETLANDS

- Wetlands are areas that are covered by water and support plants that can grow in saturated soil.
- Countries with the most wetlands are Canada (14% of land area and includes Hudson Bay lowlands 320,000 km^2 or 80 million acres), Russia (West Siberian lowlands, 1,000,000 km^2 or 247 million acres), and Brazil (Amazon wetland, 800,000 km^2 or 198 million acres).
- Wetlands were once about 10% of the land area in the United States, but they are now less than 5% of the land area, mostly in Louisiana. Of wetland loss, 90% is due to conversion of land to agriculture; 10% is due to urbanization.
- One-third of all endangered species spend part of their life span in wetlands.
- Wetlands are home to many different types of species.
- High plant productivity supports large numbers of animal species.
- Wetlands act as natural water purification systems removing sediment, nutrients, and toxins from flowing water.
- Wetlands along lakes and oceans stabilize shorelines and reduce the damage caused by storm surges, reduce the effects of flooding, and reduce saltwater intrusion.
- Wetlands are important areas for recreation, hunting, fishing, bird watching and nature.

Erosion and Conservation

- Degradation of land occurs when its productivity is decreased. It can be caused by several factors, some of which include: (1) overgrazing by livestock; (2) soil erosion; (3) soil salinization; and (4) waterlogging. See Figure 9.10.
- Degradation of land is not a modern event. The following table summarizes land degradation through history.

TIMELINE OF LAND DEGRADATION AND LAND CONSERVATION

60,000 years ago	Evidence of fire used to clear forests in the Kalambo Falls site in Tanzania, Africa.
6000 B.C.	Deforestation leads to collapse of communities in southern Israel/Jordan.
2700 B.C.	Evidence of vast cedar forests in southern Iraq. By 2100, soil erosion and salinization have impacted agriculture to the point that the Sumerians migrated north to Babylonia and Assyria.
2700 B.C.	First recorded laws to protect remaining forests found in the city of Ur.
2600 B.C.	Commercial deforestation and timber harvesting in Phoenicia (Lebanon) and South India. Lumber was exported to Egypt and Sumeria.
1500 B.C.	City-states in Central America begin to collapse due to soil erosion.
500 B.C.	Greek cities begin to decline due to land degradation, specifically soil erosion. Silt fills in bays and rivers. Plato described it as "*All the richer and softer parts have fallen away and the mere skeleton of the land remains.*"
400 B.C.	Greek general Thucydides, in describing the Peloponnesian War, describes deforestation in northern Greece.
1560–1600	Industrialization in England results in heavy deforestation.
1690	Colonial Governor William Penn requires that settlers in Pennsylvania preserve one acre of trees for every five acres cut.
1748–1762	Essays by American colonist Jared Eliot (*Essays on Field Husbandry*) that described land degradation and soil erosion in colonial America. His essays promoted soil conservation through plowing organic matter back into the soil.
1784	Benjamin Franklin urges a switch from burning wood to burning coal to save forests.
1823	James Fennimore Cooper writing in *The Pioneers* writes that we should "govern the resources of nature by certain principles in order to conserve them."
1852	"Mother of the Forest," a giant sequoia which was 2,500 years old and located in what was to become Yellowstone National Park was cut down and displayed in carnivals. Horace Greeley, editor of the *New York Tribune,* called it "vandalism" and "villainous speculation." It sparked public outcry.
1872	Yellowstone National Park.
1872	Arbor day established. www.arborday.org/
1876	American Forestry Association lobbies to cut timber on federal preserves.
1891	Forest Protection Bill.
1906	Yosemite National Park
1913	Hetch Hetchy Dam approved in Yosemite National Park.
1916	National Park Service created.
1918	Leon Trotsky—"The proper goal of communism is the domination of nature by technology and the domination of technology by planning, so that raw materials of nature will yield to mankind all that it needs and more besides."

TIMELINE OF LAND DEGRADATION AND LAND CONSERVATION (Continued)

1934	Dust bowl storms begin in the Midwest United States.
1947	Everglades National Park established.
1964	National Wilderness Preservation System
1968	Wild and Scenic Rivers Act and National Trails System Act.
1969	Opening of Alaskan oil fields.
1989	Timbering in Alaska's Tsongass National Forest halted.
1991	Antarctic Treaty
2001	Bill Clinton protects 58 million acres of national forest from development and created eight million acres of land as new national monuments. The conservation record of Bill Clinton surpasses Teddy Roosevelt and Jimmy Carter, which up to that time had passed the greatest amount of pro-environment legislation.
2001	George W. Bush advocated oil exploration in the United States. Funding cuts were recommended for conservation programs.

- Each year, approximately 10 million hectares of cropland are lost to land degradation.
- According to the World Resource Institute (www.wri.org), the productivity of approximately 17% of all land surface on Earth has been decreased due to human activity.
- According to the United Nations Environmental Program, 60% of all agricultural land on Earth in nonhumid areas has suffered degradation. In the United States, the figure is 40%.

Soil Erosion

Soil erosion is the movement of soil components from one place to another. It is caused by flowing water, wind, and human activity such as cultivating inappropriate land, poor agricultural techniques, overgrazing, burning of vegetation, deforestation, and construction. There are three common types of soil erosion: (1) sheet erosion where soil moves off as a horizontal layer, (2) rill erosion where fast flowing water cuts channels in soil, and (3) gully erosion which is an extreme case of rill erosion where over time, channels increase in size and depth. Erosion causes damage to agriculture, waterways, and infrastructure.

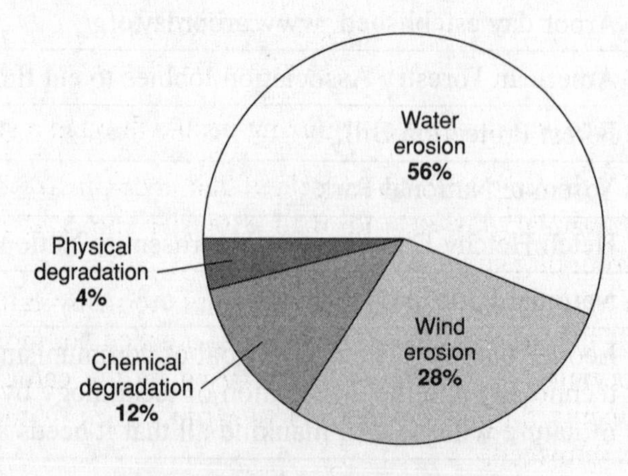

Figure 9.10. Soil degradation by type.

- One inch of topsoil can take 200 to 1000 years to form.
- Soil loss is one measurement of soil erosion. During the 1980s, estimates are that 240 billion tons of topsoil were lost due to erosion.
- Each year, soil loss averages are estimated to be
  - 5 to 10 tons per hectare in Africa, Australia, and Europe.
  - 10 to 20 tons per hectare in the Americas (extreme soil erosion in the United States can approach 250 tons per hectare per year. Due to extensive agriculture and overgrazing, more soil is lost each year in the United States than during the 1930s Dust Bowl. For example, in 1850, soils in the prairie states averaged 14 inches of topsoil. Today, the average is 7 inches. In the prairie states, the average topsoil erosion rate is 30 tons per hectare.
  - 30 tons or more per hectare in Asia.
- Up to 12 tons of soil per hectare can be replaced through natural soil formation processes.
- More than half of the fertilizer used on farmland in the United States is required to replace organic nutrients lost in topsoil through erosion. Soil in some states has lost up to 25% of their agricultural productivity due to erosion.
- Erosion (1) destroys the soil structure or profile, (2) decreases the water holding capacity, and (3) increases the soil's ability to become compacted. Because water cannot percolate through the soil, it runs off (runoff) the land, taking more soil with it (an example of a positive feedback loop). Because the soil cannot hold water due to compaction, and water does not percolate through the soil, crops grown in areas of soil erosion frequently suffer from drought. In areas of low precipitation, this becomes critical for crop productivity.
- Soil erosion occurs through
  - Overgrazing,
  - Monoculture,
  - Row cropping,
  - Improper tilling or plowing of the soil,
  - Removing crops instead of plowing the organic material back into the soil,
  - Development of new land.
- Soil that is washed away ends up being deposited as silt in rivers. In these areas, it interferes with, among others, wetland ecosystem cycles, reproduction, oxygen capacity of the water, and pH of the water.

Overgrazing

- Soil erosion through overgrazing occurs through a stepwise sequence:

1. Cattle are placed on rangeland that is dominated by perennial grasses.
2. Perennial grasses, with their extensive root systems, help stabilize the soil (keep it from blowing away).
3. Overgrazing occurs when there are too many cattle for too little area. Overgrazing of the grasses causes them to be crushed and nonproductive. The soil also suffers because of the compaction caused by the cattle. If the trampling of the grasses occurs during the reproductive life cycles of the grass, their seed production is diminished.

4. The cattle also crush other plants on the surface of the soil, such as lichens (fungus + algae) and algae. These primitive plants are involved in fixing nitrogen, retaining soil moisture, infiltrating water, and also preventing the soil from blowing away. They are also involved in natural processes of succession. The result is a less productive soil and a greater amount of water runoff, which increases soil erosion.
5. Conditions are created whereby "noxious weeds" are more suited to the environment. These weeds are usually annuals that produce large quantities of seed required for the next generation. These invasive weeds, which include wild oat, rush skeleton weed, cheat grass, and thistles:
   - Destroy wildlife habitat.
   - Displace many threatened and endangered species.
   - Reduce plant and animal diversity because of weed monocultures (single plant species that overrun all others in an area)
   - Disrupt waterfowl and neotropical migratory bird flight patterns and nesting habitats.
   - Cost millions of dollars in treatment and loss of productivity to private landowners.

For a list of noxious weeds go to http://www.aphis.usda.gov.

6. Soil begins to form a hard crust on the surface that causes less water to infiltrate into the soil. This results in significantly lower soil moisture content, making bad conditions even worse for new vegetation.
7. Drier conditions create potentials for fires of the vegetation that is left. At this point, even the annual noxious weeds are decreasing in number due to the low productivity of the soil.
8. Wind and rain now remove topsoil. Remember that it takes 200 to 1000 years to create one inch of topsoil. What was once productive grassland that could be sustainable with proper management has now become a nonproductive desert.

Benefits of Conservation Tilling

1. More sustainable farming due to dramatic reductions in soil erosion caused by water or wind.
2. More efficient conservation and utilization of water under semiarid conditions.
3. Improved energy efficiency as a result of reduced fuel requirements associated with fewer field operations.
4. Greater crop and farm profitability through reduced direct and indirect costs for chemicals, fuel, and labor.
5. Greater biodiversity than with standard cultivation practices where the surface has no crop residues.
6. Reduced build-up of soil sediments in reservoirs, drainage ditches, and the like caused by soil erosion.
7. Less pollution of drinking water sources caused by runoff of soil, fertilizers, and pesticides.
8. Lower risk of desertification arising from loss of productivity caused by soil erosion.
9. Reduced CO_2 emissions due to increased soil organic matter level; this offers the ultimate opportunity of converting agriculture from a net CO_2 source to a valuable carbon sink.

METHODS TO CONTROL SOIL EROSION

Method	Description
Alley Cropping (Agroforestry)	Planting trees and shrubs between rows of crops.
Conservation Tillage	Tillage planting system that leaves at least 30% of the field surface covered with crop residue after planting is completed and involves reduced or minimum tillage. Types of conservation tillage include: (1) no-till planting; (2) strip rotary tillage; (3) till planting; (4) annual ridges; (5) chiseling; and (6) disking.
Contour Cultivation	A special tillage practice carried out on the contour of the field. It can reduce the velocity of overland water flow.
Crop Rotation	Improves the overall efficiency of nitrogen uptake and utilization in the soil. If certain cover crops are planted in the winter, erosion and runoff is prevented when the ground thaws, and nutrients are trapped in the soil and released to the spring crops.
Diversion Structures	Channels that are constructed across slopes that cause water to flow to a desired outlet.
Drop Structures	Small dams used to stabilize steep waterways and other channels.
Grass Waterways	Storm runoff water is forced to flow down the center of an established grass strip. Grass waterways can carry very large quantities of storm water across a field without erosion.
Gully Reclamation	Planting gullies with soil-retaining plants, small dams, and building divergent channels.
Land Classification	Identification of areas that should not be planted.
Planting Windbreaks (Shelterbelts)	Retain soil from blowing away, provides habitats, and retains soil moisture.
Riparian Strips	Buffer strips of grass, shrubbery, plants, and other vegetation that grow on the banks of rivers and streams and areas with water conservation problems. The strips slow runoff and catch sediment. In shallow water flow, they can reduce sediment and the nutrients and herbicides attached to it by 30–50%.
Strip Cropping	Alternate strips of different crops are planted in the same field. There are three main types: contour strip cropping, field strip cropping, and buffer strip cropping.
Terraces	Enables water to be stored temporarily on slopes to allow sediment deposition and water infiltration. There are three types of terraces: bench terraces, contour terraces, and parallel terraces.

TYPES OF GRAZING

	Continuous	**Intensive**	**Rotational**
Vegetation	Shortgrass prairie. Northern mixed prairie. Annual grasslands.	Eastern Great Plains. Southern Pine Forest. Eastern improved pasturelands.	Tallgrass prairie. Bunchgrass
Climate	Semi-arid to arid. Annual precipitation < 12 inches.	Humid Annual precipitation >20 inches per year.	Temperate to humid. Annual precipitation 12–40 inches.
Terrain	Relatively level to flat.	Level to flat.	Hilly to rugged.
Pros	Not management- or labor-intensive. No expenditure required to stock correctly. Maintains range condition when properly stocked.	Improves livestock distribution, management, and herd health.	Improves range condition. Decreases erosion on riparian areas.
Cons	Can lead to overgrazing. Herd expansion only by leasing or purchasing more land. Damaging to streamside areas.	Requires more management. Requires more labor, fences, and water.	Requires more management than continuous grazing. May require more labor, fences, and water than continuous grazing. Can have lower animal gains than continuous grazing.

Agricultural Irrigation

- Worldwide, approximately 40% of all crop yield comes from 16% of all cropland that is irrigated.
- Three-quarters of all water used on Earth is used for agriculture. Use for agriculture depends upon national wealth, climate, and degree of industrialization. Canada uses ~10% of its water for agriculture (climate supplies most), whereas India uses about 90% of its freshwater for agriculture. Up to 70% of water intended for agricultural purposes may not reach crops in developing countries because of inefficiency—seepage, leakage, evaporation, and delivery to plants on a fixed schedule that doesn't match the plants' needs. Drip-irrigation is used on less than 1% of crops worldwide.
- Water rights often impose penalties the following year if farmers do not use their full allocation. This model is built on the "use it or lose it" philosophy and encourages waste.
- As more and more land is being converted to grow crops, less desirable land is being used. Much of this land is semiarid, requiring the use of irrigation. Water-intensive crops (cotton and corn) are often grown in these semiarid regions. The rate of increase

in irrigation is approximately 2% per year. The rate of growth in the human population is approximately 1.4%.

- Increases in irrigation as seen as a function of population has decreased (see Figure 9.11). Sustainable irrigation in the future will be limited as a result of increases in costs, depletion of current sources of irrigation water, competition for water by urban areas (cities can pay more for water than farmers), restoration of wetlands and fisheries, waterlogging, and salinization. Future water capacity will increase through increases in efficiency.

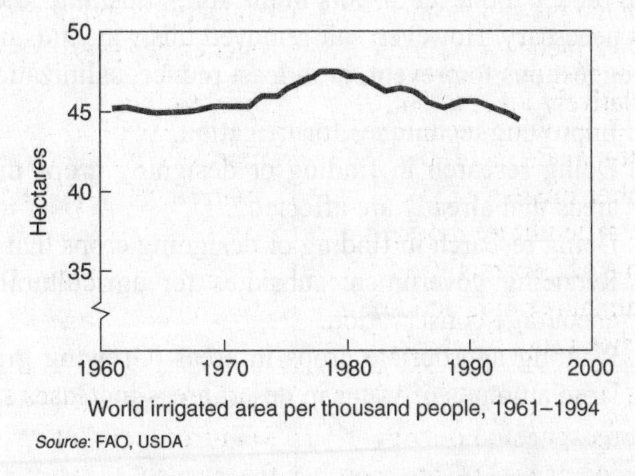

Figure 9.11. Irrigation capacity as a function of population growth.

- Examples of water sources that have become seriously depleted as a result of irrigation demands include the Colorado River, the Yellow River in China, and the San Joaquin River in California. The Aral Sea has decreased 75% in volume and the salinity has increased 300% due to demand for irrigation for cotton.
- Changes in water level, salinization, and silting are responsible for the listing of 122 endangered or threatened fish species in the western United States.
- Texas has lost approximately 15% of its irrigated land due to aquifer depletion. Of all water mining in the United States, 20% exceeds groundwater recharge.

Salinization and Waterlogging

- All water that is used for irrigation contains a low concentration of salts.
- Salinization is the accumulation, over time, of salts in the soil. The more land is irrigated, the greater the effect of salinization.
- As water evaporates, it leaves the salt behind. Over time, the salt builds up to levels that cause osmotic stress for plants (generally 3000 to 6000 ppm). The osmotic stress prevents plants from absorbing water.
- Areas that are most prone to salinization are semiarid regions that require a lot of irrigation. In the southwestern United States, build-up of calcium carbonate in the soil forms a cement-like deposit called caliche or calcrete.
- This impermeable layer of caliche retards the normal infiltration of water downward through the soil, and the soil soon becomes waterlogged. This condition prevents normal gas exchange with the atmosphere and is detrimental to healthy root structure.
- Waterlogged soil is estimated to affect 10% of all irrigated cropland.
- In the United States, salinization is estimated to reduce crop yields as much as one-third.

- In the San Joaquin Valley in California, 500,000 hectares of land will be nonproductive by 2080 due to salinization.
- In Egypt, the Nile River used to flush salts out of the soil each year when it flooded. After the Aswan High Dam was constructed in the 1960s, this flooding stopped, and soil salinization increased, lowering crop productivity.
- Other areas subject to high levels of soil salinization include the Colorado River Basin, the Middle East, and semiarid regions of Russia and China. It is estimated that up to 25% of all irrigated crops in the world are affected to some degree by soil salinization.
- To reduce the level of salts in the soil, "flushing" the soil with large amounts of water is necessary. However, salt removed this way ends up in estuaries and other wetlands.
- Suggestions to prevent, or at least reduce, salinization and waterlog include:

  1. Improving techniques for irrigation.
  2. Doing research in finding or designing crops that are saline-resistant to grow in areas that already are affected.
  3. Doing research in finding or designing crops that do not require as much irrigation.
  4. Reducing government subsidies for agricultural water. Raising the price would encourage conservation.
  5. Planting appropriate crops in areas. Growing green, leafy vegetables that require large amounts of water in desert areas increases salinization and waterlogging.

Natural Features, Buildings, and Structures

Natural Features

- Natural features include mountains, deserts, canyons, and seashores. The list of natural features is extensive and to cover information on *all* natural features is beyond the scope of this review book. Instead, one natural feature, mountains, will be looked at in detail.
- Mountains make up 25% of the Earth's land surface. Of the world's population, 10% live in mountainous terrains. Of the 185 countries in the world, 139 have mountains (most of the rest are small island nations). People depend on mountains for hydroelectric power, timber, mineral resources, water (80% of freshwater originates from mountains), and food.
- Half of all people who live in mountainous areas live in three major ranges: (1) Andes, (2) Hengduan-Himalaya-Hindu Kush system, and (3) the African mountain systems.
- Mountain habitats have many ecotones–biomes: glaciers, tundra, taiga, alpine grasslands, riparian areas, deciduous and coniferous forests, chaparral, deserts (rain-shadow side), and so on. Mountains are unique in that small changes in altitude result in large changes in the environment. A large variety of environments results in high biodiversity of plants and animals (e.g., over 15,000 different species of plants and animals reside in the Sierra Nevada mountain range in California).
- Many people who live in mountainous areas are indigenous and are geographically and culturally separated, resulting in their concerns and power being diluted by more-developed countries. They also tend to be among the world's poorest people (60% of people living in the Andes are classified as "extremely poor" and 98 million Chinese living in mountainous areas are classified as "absolute poor." Poverty in these examples results from cultural isolation and the lack of industry and jobs in these areas primarily

because of transportation issues. Isolation of these people results in discrimination (e.g., Tibetans; Kurds; Kashmiris, between Pakistan and India; "hillbillies," etc.). Discrimination results in conflict. For example, in 1995 there were 48 wars or armed conflicts that took place in 43 countries; 26 of the 48 wars occurred in mountainous regions.

- This cultural "separatism" of mountainous people results in less environmentally oriented governmental policies that are not proportionate to the number of people living in these areas. The results include
  1. Degradation of water systems, resulting in more landslides and less irrigational flow;
  2. Decreased crop yields compared to more productive and manageable croplands below. Reduced crop yields result in less food for growing populations that contribute to positive feedback cycles of malnutrition and disease;
  3. Usurpation of traditional landrights and exploitation of natural resources by multinational corporations including minerals, hydropower, and timber. The result is an eroded landscape, stripped of natural resources, which higher numbers of poorer people must utilize to survive;
  4. Isolation from the mainstream job training and job opportunities that one would find in metropolitan areas, contributing to the positive feedback cycle(s) of unemployment—poverty—lower educational opportunity—less access to health care—malnutrition—disease—higher mortality. For example, in 1993, coronary heart disease (CHD) rates were 19% higher for white men in Appalachia than white men outside of Appalachia. Similarly, CHD death rates were 21% higher among white women aged 35–64 years in Appalachia than among white women outside of Appalachia. In comparison, CHD death rates for blacks aged 35–64 years only differed slightly between Appalachia and the entire United States (*Source:* Center for Disease Control, *Morbidity and Mortality Weekly Report*, Washington, DC).

- Mountainous regions also fosters endemism, species of plants or animals that exist within narrow boundaries, specifically adapted to unique environmental conditions. Many tropical birds are endemic to specific regions in montane tropical forests. For example, 247 worldwide endemic nesting areas for birds have been identified; of this number, 131 occur in tropical mountains.
- In most less-developed countries, poorer people live on mountainsides. In more-developed countries this is reversed, and the more affluent in society live on mountainsides. Lacking the resources to purchase or rent housing, between one-third and two-thirds of urbanites in developing and less-developed countries become squatters on dangerously steep hillsides, flood-prone riverbanks, and other undesirable lands.
- Mass tourism and recreational loads are impacting mountainous areas almost as much as mining, hydropower, and timbering effects combined. For example, visits to mountainous U.S. National Parks has increased 12-fold since 1945. In 1995, more than 4.1 million people visited Yosemite, a new record and almost twice the number that visited in 1980. The resulting traffic problems are among the most severe in the entire national park system. During peak visitor times, the traffic jams in the Yosemite valley are the equal of those in major American cities. The Alps in Europe exceed 100 million visitor days each year).

Buildings and Structures

- "Green building" techniques focus on a whole-system perspective, including energy conservation, resource-efficient building techniques and materials, indoor air quality, water conservation, and designs that minimize waste while utilizing recycled materials. Designing green buildings also involves placing buildings, whenever possible, near public transportation hubs and other amenities, such as shopping areas, medical centers, and recreational facilities. Sites that are pedestrian-friendly encourage walking and bicycling and minimize the need for automobiles. Green building also preserves historical and cultural aspects of the community with designs blending into the natural feeling and aesthetics of a community.
- There are more than 76 million residential buildings and nearly 5 million commercial buildings in the United States. Together, these buildings use one-third of all the energy consumed in the United States, and two-thirds of all electricity. By the year 2010, another 38 million buildings are expected to be constructed.
- Buildings account for 49% of sulfur dioxide emissions, 25% of nitrous oxide emissions, 10% of particulate emissions, and 35% of the carbon dioxide emissions, the chief pollutant in global warming.
- A standard wood-framed home consumes over one acre of forest and the waste created during construction averages from 3 to 7 tons.
- If only 10% of homes in the United States used solar water-heating systems, it would avoid 8.4 million metric tons of carbon emissions each year.
- Green landscaping requires low water use. Xeriscaping reduces water needs. Natural landscaping is designed with plants that are appropriate for the site's microclimate and topography (e.g., using drought-tolerant plants in areas that are dry and windy, and using plants appropriate for wet areas to prevents water runoff). Proper landscaping can reduce heating energy consumption by 30%, air-conditioning energy consumption by 75%, and water consumption by 80%. Deciduous trees provide shade during the summer (when leaves are present) without interfering with solar gain during the winter (when leaves are absent).
- Use of recycled building materials reduces the negative environmental impact of constructing new buildings. For example, organic asphalt shingles contain recycled paper, and some shingles are made from remanufactured wood fiber. Cellulose insulation is manufactured from recycled newspaper. Lumber can be conserved by using stress-skin panels; engineered framing products (I-beams); glue-laminated products; and finger-jointed lumber.
- Energy efficient buildings are often "airtight." Without proper ventilation, these buildings can become "sick." Volatile organic compounds (VOCs) from paneling, carpets, and furniture can contribute to sick building syndrome. More than 1500 bacterial, fungal, and chemical air pollutants have been identified as contributing to sick building syndrome.

Wildlife

SPECIES EXTINCTION OVER TIME*

Time Period	% Species Lost	Species Affected
Present	Unknown	Birds—42% of extinctions are the result of hunting. Mammals—33% of extinctions are the result of hunting. 50,000 new species become extinct each year. In 25 years, 10% of all known species today will be extinct at current rates. By the end of the 21st century, it is estimated that 25% of all known plant species will be extinct.
65 million years ago (Cretaceous)	85%	Dinosaurs, plants (except ferns and seed-bearing plants), marine vertebrates, and invertebrates. Most mammals, birds, turtles, crocodiles, lizards, snakes, and amphibians were unaffected.
213 million years ago (Triassic)	44%	Marine vertebrates and invertebrates.
248 million years ago (Permian)	75–95%	Marine vertebrates and invertebrates.
380 million years ago (Devonian)	70%	Marine invertebrates.
450–440 million years ago (Ordovician)	50%	Marine invertebrates.

*Of all species that ever inhabited the Earth, over 99% of them are now extinct.

Conservation of wildlife has four possibilities:

***In Situ* Management** allows wildlife to exist in parks and preserves.

1. Wildlife exist in their own wild, natural habitat.
2. Wildlife exist free in protected and managed natural environments and preserves.

***Ex Situ* Management** allows conservation of wildlife in human-controlled settings.

3. Wildlife exist confined in zoos and preserves with the possibility of releasing the offspring back into their natural habitat. This is known as captive breeding. This technique is expensive and does not always take into account important ecological relationships that exist between species and the environment. Before releasing endangered wildlife back into the wild, it would be prudent to understand the factors that brought the species to the stage of being endangered and to determine if those same risks continue to exist.
4. Genetic samples maintained for future populations. Seeds kept in seed banks do not remain viable indefinitely and some plant species do not produce seeds. Being stored prevents the species from evolving in responses to changes in nature, which may mean that they are less successful when they are reintroduced to their native habitat. Accidents at seed banks can also threaten the stored material.

Categories of Threatened Wildlife.

About 6000 species fit in one of the three categories of threatened wildlife.

Category	Description
Vulnerable species	A species particularly at risk because of low or declining numbers or small range, but not a threatened species.
Threatened species	A species whose population is not yet low enough to be in immediate danger of extinction, but who certainly faces serious problems. If the problems affecting these species aren't resolved, it is probable that the species will become endangered. Examples of threatened species include the eastern indigo snake, freshwater sawfish, red kangaroo, Ethiopian wolf, and wandering albatross. To see the entire list of threatened species, visit http://www.redlist.org/
Endangered species	A species, plant or animal, that is in immediate danger of becoming extinct. Its numbers are usually low, and it needs protection in order to survive. The Siberian tiger, Galapagos sea turtle, marine iguana, the southern sea otter, the snow leopard, the green pitcher plant, and thousands of other plants and animals are endangered worldwide.
Extinct species	A species that is no longer living. Examples include passenger pigeon, blue pike, heath hen, Caribbean monk seal, Stellar's sea cow, Tasmanian tiger-wolf, moas, and the Stegosaurus.

NUMBER OF THREATENED AND ENDANGERED WILDLIFE SPECIES (1998).

Item	Mammals	Birds	Reptiles	Amphibians	Fishes	Snails	Clams	Crustaceans	Insects	Arachnids
Endangered Species Total	310	253	80	17	78	16	63	16	32	5
United States	59	75	14	9	67	15	61	16	28	5
Foreign	251	178	66	8	11	1	2	0	4	0
Threatened Species Total	23	21	34	8	41	7	8	3	9	0
United States	7	15	20	7	41	7	8	3	9	0
Foreign	16	6	14	1	0	0	0	0	0	0

Source: U.S. Fish and Wildlife Service, quarterly.

- In the United States, 735 species of plants and 496 species of animals are listed as threatened or endangered.
- Other life-forms threatened or at risk include 34% of all fish, 25% of all amphibians, 24% of all mammals, 20% of all reptiles, 14% of all plant species, 16% of all conifers, and 12% of all birds. These figures represent the species that have been identified. Most extinctions occur among the lifeforms that have not been identified—countless

insects, spiders, nematodes, mollusks, and bacteria of cleared rainforest, or the largely unrecorded biota of the ocean floor.

- Invertebrates make up 75% of all species that are endangered, and yet account for only 9% of species listed by provisions of the Endangered Species Act (1973) as "worthy" of protection. This is backbone racism!
- 266 of these listed species have recovery plans currently under development.
- There are more than 1000 animal species endangered worldwide.
- There are more than 3500 protected areas in existence worldwide. These areas include parks, wildlife refuges, and other reserves. They cover a total of nearly 2 million square miles (5 million km^2), or 3% of our total land area.
- Approximately 50% of all endangered species' habitat is on 80% of all private land in the United States. A Habitat Conservation Plan (HCP) is designed for private land owners, corporations, state or local governments, or other nonfederal landowners who wish to conduct activities on their land that might incidentally harm, or "take," a species listed as endangered or threatened so that they can utilize their land resources (as long as the specie(s) is not harmed). As of August, 2001, 360 HCPs had been approved, covering approximately 30 million acres and protecting 200 endangered or threatened species.
- Aquatic species, which are often overlooked, are facing serious trouble. One-third of the United States' fish species, two-thirds of its crayfish species, and almost three quarters of its mussel species are in trouble. Particular species in trouble include shark, billfish, shrimp, orange roughy, groupers, groundfishes, sea scallops, bluefin tuna, and red snapper.
- The Convention on International Trade in Endangered Species (CITES), 1975, is the only global treaty of 145 countries whose focus is the protection of plant and animal species from unregulated international trade. Problem stems in developing countries where wildlife is most threatened and economic incentives are most tempting. In 1997, CITES reported a world trade of 25,733 live primates, 235,000 parrots, 948,000 lizards, and 344,000 wild orchids, along with 1.6 million lizard skins, 1.5 million snake skins, and 850,000 crocodile skins, which represents just a very small fraction of the enormous trade in these commodities.
- Wildlife programs in the United States are financed through the sale of hunting and fishing licenses, and contributions to state wildlife programs making the management of preferred species the highest priority. Programs to support nongame wildlife use less than 10% of these sources of income.
- Certain species of wildlife in the United States that are preferred by hunters have, as a result of management, increased in population to the point that they are considered pests in some areas. They contribute to car accidents, carry Lyme disease, and forage agricultural crops and gardens. For example, in 1880, few deer were sighted in Iowa due to exploitation. In 2001, the population was over 200,000. Populations of raccoons, opossums, skunks, foxes, beaver, ducks, geese, and some squirrels are also at record levels.
- Management principles for saving wildlife include:

  – Protect enough land so that populations have the resources to survive,
  – Manage the land at a local level allowing for flexibility to changes,
  – Incorporate long-range objectives and goals,
  – Allow human use that is compatible with objectives and goals.

Characteristics of or Dangers for Endangered Species

1. Long-lived;
2. Large body size;
3. Low reproductive rates;
4. Large territory requirements but natural range may be limited;
5. Specialist species (exist only within narrow environmental conditions);
6. Small population size (populations may be isolated "islands");
7. Threat to humans (predator or livestock, eat agricultural products, etc.);
8. Habitat may be threatened (wetlands, tropical forests, coral reefs, etc.) may be diminishing in size, may become fragmented, or may be degrading in quality;
9. Have an economic value (fur, hides, tusks, horns, exotic meat, zoos, pets, cactus, ginseng, tropical fish, etc.);
10. Not popular due to competition, habitat destruction, or fear (bats, snakes, coyotes, prairie dogs, etc.);
11. High in the food chain;
12. Physiologically sensitive to effects of pollution (e.g., fish-eating birds and effects of DDT, waterfowl, swans, and cranes or scavengers who eat such birds and ingestion of lead shotgun pellets);
13. May be migratory—dependent on environmental conditions in many areas;
14. May inhabit areas close to commercial development, urbanization, or fishing-hunting areas;
15. May be subject and have no resistance to *introduced* disease organisms or pathogens (e.g., native Americans and smallpox, American chestnut and fungal blight from China, trout and *Myxobolus cerebralis*);
16. Genetic assimilation—crossbreeding with more rigorous, closely related species;
17. Radical climate or environmental change;
18. Introduction of nonindigenous species into the habitat.

The Endangered Species Act (ESA), 1973, identifies all endangered species and tries to save as much biodiversity as possible. Approximately 1500 species are listed by ESA with 500 waiting. It takes *years* for a species to make the list. Once a species makes the list, the Fish and Wildlife Service prepares a recovery plan. The Fish and Wildlife Service and other agencies spend approximately $150 million per year on managing and enforcing recovery plans. The Act's impact on society was *not* to be considered. For example, the Northern Spotted Owl requires old-growth forest habitats in the Pacific Northwest and will require $33 billion to save 2400 owls. Each owl would cost $14 million dollars to protect when the impact to the timber industry is taken into account. Another example will be the costs of saving the Columbia River salmon and steelhead that are endangered by hydropower power plants, reservoirs, and dams that interfere with migration. ESA addresses importing and exporting animals and possessing, selling, or harming endangered species. Fines for violations are as high as $100,000. The ESA expired in 1992. Industry trade associations formed a lobby, the NESARC (National Endangered Species Act Reform Coalition) http://www.nesarc.org/start.htm—to block and "water down" provisions for the renewal of the ESA in Congress. NESARCs provisions, which they refer to as "modernization," seeks to adopt the least costly, most cost-effective measure.

Justifications for Saving Endangered Wildlife

1. **Utilitarian.** Species may be useful for man someday for
    - Genetic diversity—As habitats change, and as the needs of man change, it is necessary to have genetic diversity in wildlife populations to meet the changes and be able to survive.
    - Medical and chemical research—Only 5,000 of the world's 250,000 plant species have been studied for possible medical use. Approximately 75% of the world's population relies on plants or plant extracts as sources of medicines. Wildlife supplies roughly half of the prescription and nonprescription drugs used in the world, and 25% of those used in the United States come from wild organisms, the rest being synthesized (e.g., Taxol).
    - New products—As many as 75,000 to 80,000 species of plants could be potentially utilized for human consumption. For example, palm, *Orbignya phalerata*, can be used to produce an oil substitute, and the Guayule shrub of the deserts of Texas can be used to produce latex for manufacturing rubber.
    - Needs of indigenous people.
    - Pollution control—Microbial, filter feeders, and trees such as *Leucaena*.
    - Ecotourism—84% of Canadians participate in bird watching, photography, and other forms of natural enjoyment. In Kenya, it is estimated that one lion may bring in more than $500,000 spent by tourists during its lifetime.
2. **Ecological.** Wildlife is necessary for the ecosystem and biosphere and the understanding that all species play a role or more than one role in a balanced ecosystem.
3. **Moral.** Species have a right to exist. It is not up to man to decide which species survive and should be allowed to continue. See Chapter 15.
4. **Aesthetic and Cultural.** Biological diversity adds to the beauty and understanding of nature.

EXAMPLES OF ENDANGERED SPECIES

Mammal	Location	Reason
African wild ass	Somalia, Sudan, Ethiopia	Habitat loss, poaching, interbreeding with domestic donkeys.
Bat (Indiana)	Eastern and mid-western U.S.	Loss of cave habitat due to commercialization of caves.
Cheetah	Africa to India	Loss of habitat, fur trade.
Deer (Key)	Southern Florida	Habitat loss to development, road kills.
Elephant (Asian)	South central & Southeast Asia	Habitat loss to agricultural development.
Gazelle, Arabian	Arabian Peninsula, including Israel	Poaching.
Gibbons	China, India, Southeast Asia	Loss of habitat.
Gorilla	Central and western Africa	Loss of habitat, collecting young, poaching.
Jaguarundi	Texas, Central America	Loss of habitat fur trade.

EXAMPLES OF ENDANGERED SPECIES (Continued)

Mammal	**Location**	**Reason**
Leopard	Africa, India, Southeast Asia	Loss of habitat, fur trade.
Monkey, Spider	Costa Rica, Nicaragua, Panama	Loss of habitat.
Mouse (Salt march harvest)	California	Drainage and filling of salt marshes.
Prairie dog	Western U.S.	Loss of habitat, poisoning.
Rhinoceros (White)	Central and eastern Africa	Poaching for horn.
Tiger	Temperate and tropical Asia	Loss of habitat, hunting.
Whale (Blue)	Oceans	Commercial hunting.
Whale (Finback)	Oceans	Commercial hunting.
Birds		
Akepa (Hawaiian)	Hawaii	Loss of habitat, disease-carrying insects.
Bobwhite (Masked)	Arizona, Mexico (Sonora)	Habitat loss, invasion of woody plants in habitat.
California Condor	California	Habitat loss, loss of carrion food source, low reproductive rates.
Whooping crane	Central U.S., Canada	Small population, low reproductive rates.
Duck, Laysan	Hawaii	Loss of habitat, small population.
Kite (Everglade)	Florida	Loss of habitat through drainage, loss of snails for food.
Parakeet (Orange-bellied)	Australia	Collecting for pet market.
Parrot, (Puerto Rico)	Puerto Rico	Loss of habitat, predators, nest destruction by rats.
Warbler (Kirtland's)	U.S., Canada, West Indies, Bahamas	Loss of habitat through fire control.
Reptiles-Fish-Other		
Caiman (Broad-snouted)	Brazil, Argentina, Paraguay, Uruguay	Killing for hides.
Crocodile (American)	Florida, Mexico, Central and South America	Killing for hides, loss of habitat.
Monitor (Komodo Island)	Indonesia	Reduction of food supply through human competition.

EXAMPLES OF ENDANGERED SPECIES (Continued)

	Location	Reason
Turtle (Leatherback sea)	Tropical and temperate seas and oceans	Killing, collecting eggs for food.
Salamander (Texas blind)	Texas	Loss of habitat—drainage.
Chub (Mohave)	California	Loss of habitat.
Pupfish (Devils Hole)	Nevada	Loss of habitat.
Trout (Arizona)	Arizona	Loss of habitat.
Mussel (Pearly)	Central and eastern United States	Loss of habitat—pollution.
Butterfly (Mission blue)	California	Loss of habitat.

Causes for Extinction

1. **Population risk** occurs during normal fluctuations of birth and death rates. It is most common for species that have only a single population in one localized habitat.
2. **Environmental risk and loss of habitat** is caused by changes in the environment or in the relationships among species (predator, prey, competitor, or symbiotic) that threatens the survival of a species. In the United States, tall-grass prairies have been reduced by 98%, virgin forests by 95%, and wetlands by 50%. About 50% of tropical rain forests were deforested by 1995. Human influences are either direct (cutting down forests, building roads and pipelines that fragment populations, draining wetlands for agriculture, etc.) or indirect (effects of air and water pollution, acid rain, thermal pollution of lakes and streams, soil erosion, siltation, etc.). Acid rain is believed to be responsible for a 40% decline in fish species in Canadian lakes. Nitrates from fertilizer have been shown to cause physical abnormalities and deaths among amphibians. DDT, a chlorinated pesticide, was responsible for the deaths of many birds due to its effect on eggshell formation. Coral reefs are productive marine ecosystems and have a species diversity only second to tropical rain forests. Of coral reefs, 25% have been destroyed by sediment pollution and climatic changes that affect ocean temperatures, with some estimates predicting that at current losses, most coral reefs will be destroyed by 2020.
3. **Natural catastrophe** is a major natural disaster that changes the environment to such an extent that a species can no longer survive there (e.g., fires, volcanoes, floods, and storms).
4. **Genetic risk** occurs when there are changes in the gene pool such as reduced genetic variation, mutation, and genetic drift. Especially susceptible are small populations that lack variation to withstand environmental change.
5. **Predation and destruction by humans** are two different things. Predation normally occurring in nature has checks and balances. A balanced ratio between prey and predator exists in equilibrium. Man, however, is able to intervene and destroy species either for sport or because he views the species as a pest—wolves in Europe; coyotes, passenger pigeons, and bison in North America; rhinoceros and tigers in Africa. Since

the Industrial Revolution, 83 species of mammals, 113 species of birds, 23 species of amphibians and reptiles, 23 species of fish, about 100 species of invertebrates, and over 350 species of plants have become extinct. A study in 1991 estimated that up to 50,000 unclassified species may have become extinct in the tropics due to human activity. This rate of extinction is up to 10,000 times greater than the natural rate of species extinction (2 to 10 species per year) prior to the appearance of human beings. The continued extinction of species on this planet by human activities is one of the greatest environmental problems facing nature.

Human Influence on Rate of Extinction

1. Hunting for food or sport—by the beginning of the 20th century, most states had some hunting and fishing restrictions and laws. The laws were anthropocentric rather than biocentric (save animals for mans future needs rather than save animals because they have a right to exist). Examples of species that have made a comeback due to regulations include: (1) white-tail deer—1890, half a million; today, 14 million estimated; (2) wild turkeys and wood ducks—fifty years ago they were almost extinct, today there are several million each.
2. Elimination of species considered to be pests.
3. Destruction of habitat.
4. Introduction of predators, parasites or competitors.
5. Pollution.

Introduction of Species to New Environments

Visit www.invasivespecies.gov or www.nybg.org/bsci/hcol/inva/ or www.invasive.org.

- Causes disruption to native species and food web. Established patterns of predator–prey relationships, parasitism, commensalisms, competition for limited resources, mutualism, disease-resistance, and other symbiotic relationships are disrupted.
- Disrupts competition patterns (e.g., kudzu, melaleuca, and Brazilian pepper). The Brazilian pepper *Schinus terebinthifolius*, sometimes called the "Florida Holly," is a large evergreen shrub that can grow as tall as 40 feet (12 meters). Brazilian peppers were introduced to Florida in the late 1800s. They grow rapidly in most habitats, are salt-tolerant, and have no natural predators.
- Disrupts food webs affecting many lifeforms.
- Introduces new diseases and pathogens.
- Disrupts and clogs rivers and lakes (e.g., hydrilla and water hyacinths).
- May threaten agricultural industry (e.g., Mediterranean fruit fly).
- May threaten livestock (e.g., fire ants).
- May damage property (e.g., termites).
- May disrupt biological cycles (e.g., pollination and surface fires required for conifers).

By Design Citrus was imported from China to Mediterranean climates worldwide. Corn, tomatoes, and potatoes were introduced to the Old World from the New World. Wheat and other grains from northern Africa and the Middle East were brought to North America. Estimates are that 50,000+ species of nonnative plants, animals, and microbes have been introduced into the United States on a large scale. It is also estimated that these introductions cause more than $150 billion worth of damage per year.

By Accident Human diseases (e.g., smallpox and chickenpox destroying many native Americans), plant diseases (e.g., Dutch elm disease), Norway rat, weeds (Russian thistle or tumbleweed), zebra mussel, sea lamprey, Asian walking catfish, and gypsy moth, among others, have been introduced accidentally.

- Other examples include:

Kudzu—covers everything in its habitat.

Leafy spurge—reduces carrying capacity of open range (it crowds vegetation and cattle can't eat it).

Purple loosestrife overtakes native vegetation in wetlands

Round goby eats eggs of any species in its territory; however, it eats zebra mussels.

Asian long-horned beetle larvae burrow into trees, destroying them.

- Natural background extinction rate is between 3 and 30 species per year. Human-caused extinctions are 50 to 1000 times the natural background rate. Worldwide, 11,000 plant and animal species are threatened with extinction. In the past 500 years, 816 species have become extinct. Some 103 of these are known to have occurred since 1800, which represents an extinction rate 50 times greater than the natural background rate. Estimates of losses expected over the next 25 years vary from 2 to 25%, but even the low end of this range is a thousand times the background rate of extinction.
- Every 1.8% loss in habitat area results in a 0.5% loss of species diversity. Loss of habitat affects nine out of ten threatened birds and plants and 83% of the threatened mammals.
- Most species need in the order of 10,000 individuals to sustain enough genetic variation to maintain enough diversity to adapt to environmental change. When a species is endangered, the numbers decrease to fewer than 1000 individuals of an animal species and fewer than 120 individuals of a plant species.

Harvesting, Hunting, and Poaching

Harvesting, hunting, and poaching are a $5 billion per year business. One-third of all animals, animal products and plants that enter the United States are brought in illegally (World Wildlife Fund). Uses include skins, furs, ivory, antlers, horns, meat, collecting, pets, and research. Examples include Bengal tigers (coats sell for more than $100,000), rhinoceros horn, guanacos, monkeys, and elephant tusks. In 1980, there were 1.3 million African elephants. In 1990, there were fewer than 650,000. Profits far outweigh penalties. Six chimpanzees die when captured or during transport for every one that survives for medical research. Many species captured for pets die in capture, transport, or are abandoned by owners who discover that they do not make good pets. Cacti and orchids are valuable black market commodities.

Types of hunting include

- subsistence—hunting for food,
- sport—killing animals for recreation,
- commercial—killing animals for profit.

EXAMPLES OF SUCCESS IN CONSERVATION OF WILDLIFE

Species	Population Size 1900	Population Size 1988
American Bison	1,000	75,000
Canada Goose	1 million	2.3 million
Pronghorn Antelope	13,000	1 million
Rocky Mountain Elk	41,000	1 million
Sea Otter	few	100,000
Trumpeter Swan	73	10,000
White-tailed Deer	500,000	15 million
Wild Turkey	30,000	3.8 million

What Can Be Done?

Endangerment, or worse extinction, is tied to such human activity as:

- Population growth—With finite resources for all life forms, there is competition.
- Exploitation of natural resources—Based on fundamental greed, natural resources are being depleted for luxuries,
- Indifference to the rights of other life forms.

Policies that could slow (not stop) the extinction rate might include:

1. International treaties addressing the issues;
2. Economic sanctions for violations of the treaty;
3. International tariffs on products produced from countries violating international agreements;
4. International enforcement agency (such as Interpol for crime);
5. International monitoring agency (such as UN Weapon Inspectors);
6. Increased budgets for enforcement of imported cargo and inspection of contraband;
7. Increased penalties for individuals or corporations involved in selling or trading contraband;
8. International fund for creating more sanctuaries and preserves;
9. Economic incentives for "wildlife land trusts" (http://www.wlt.org) that have provisions for protecting wildlife on privately owned land;
10. Increased budgets for wildlife rangers;
11. Increased research on species that might have human benefits—crops, medicines, and the like;
12. Provisions in K–12 curriculum for environmental awareness;
13. Clearinghouse for information on ecological worth of species and provide information on rumors of aphrodisiacs (e.g., rhinoceros horn powder cures impotency);
14. International limits on collecting species by zoos, botanical gardens, and so on, and to research the value of genetic diversity that is reduced by isolating wildlife from their natural habitats;
15. Creation of "seed banks" where sperm and egg from both animal and plant species are stored for possible future use;

16. Research on synthesizing and substituting products obtained from nature—for example, tax fur because there is absolutely no need for fur from animals when synthetic products are available (www.hsus.org/current/fur, www.banfur.com, http://worldanimal.net/fur-index.html);
17. Research cloning of wildlife to replace native populations decimated by man, creating "Jurassic Parks" for rhinos, condors, tigers, and so on, making it possible to bring back Dodo birds, dinosaurs, passenger pigeons, or your old dog Rover.

Mineral Resources and Mining

Mineral Resources

- Two billion tons of minerals are extracted and used each year in the United States (~10 tons for every American).
- Metallic mineral resources include iron, aluminum, and gold.
- Nonmetallic resources include salt, sulfur, and phosphate.
- Energy resources include uranium, coal, and oil.
- The United States imports more than 50% of its most needed minerals.
- As reserves become depleted, lower grades of ores are mined, which require more processing resulting in more pollution.
- The United States, Germany, and Russia represent 8% of the world's population, yet they use 75% of the most widely used metals. The United States alone uses 20% of the world's metals.
- Minerals extracted from public lands require no rents or royalties according to an 1872 mining law.
- The United States owns about 2.5% of the world's oil reserves, but uses about 30% of all oil extracted in the world each year; 65% of the oil is used for transportation (gasoline, jet fuel, diesel, etc.), and 26% of the oil is used by industry. Oil import dependency is increasing from 52% in 1996 to a projected 70% in 2010.
- Lead, steel, titanium, aluminum, and copper have the highest recycling rates in the United States. Recycling metals reduces the quantity of virgin minerals that must be mined and processed; reduces the environmental impacts associated with new mines; and reduces per capita energy and water (See Figure 9.12). For example, new steel from recycled scrap metal results in:
  - 90% reduction in costs compared to steel from virgin materials
  - 86% reduction in air pollution
  - 40% reduction in water use
  - 76% reduction in water pollution
  - 97% reduction in mining wastes
  - 105% reduction in consumer waste

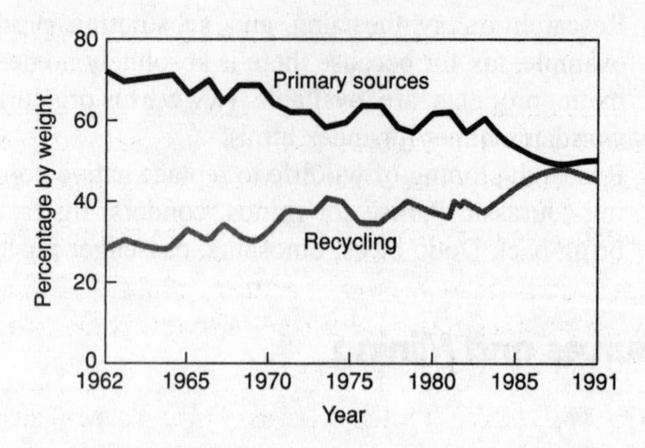

Figure 9.12. Half of all metals produced in the United States come from recycled sources and half come from primary sources (mining).

U.S. PRODUCTION OF VARIOUS MINERALS (IN METRIC TONS)

	Aluminum	Asbestos	Cement	Coal	Copper	Iron	Nitrogen-Ammonia	Zinc
1990	2,296	956	2,680,400	15,500,000	291,000	12,500,000	406,000‡	112,000
1998	3,713,000	136,000†	83,931,000	57,200,000	2,090,000	48,200,000	13,800,000	234,000

*For a complete list, visit http://pubs.usgs.gov/openfile/of01-006/
†1973
‡1943

DISCOVERING NONRENEWABLE MINERAL RESOURCES
1. Aerial photos to detect outcroppings characteristic of certain minerals;
2. Magnetometers to detect magnetic differences;
3. Gravimeters to detect density changes;
4. Drilling;
5. Sensors in wells;
6. Seismic surveys; and
7. Chemical analysis of soil, plants, and water.

TYPES OF SURFACE MINING
Approximately 50% of the coal produced in the United States is produced by one of the following surface mining operations:

1. Open-pit mining (coal, copper, iron, gypsum, potash, and phosphate rock);
2. Dredging (aggregates, stone, gravel, and sand);
3. Area strip mining; or
4. Contour strip mining.

ENVIRONMENTAL ASPECTS OF MINING
1. Scaring of land;
2. Collapse (subsidence) of ground above mines;
3. Erosion;
4. Acid mine drainage;
5. Release of chemical toxins into air; and
6. Wildlife being exposed to wastes.

IMPORTANT MINERALS AND THEIR USES

Mineral	Use
Barite	$BaSO_4$. Used in the process of oil, gas, and water exploration. Used in automobile paint primers, gloss agent for paint, leaded glass, and X-ray diagnostics.
Basalt	Igneous volcanic rock. Rich in iron and magnesium minerals.
Borate	Used in the manufacture of textiles, insulation, fiberglass, glass bakeware, laboratory glass, enamel glazes, fertilizers, fire retardants, soaps, detergents, magnetic alloys, electronics, and car air bags.
Calcite	$CaCO_3$. Used in Portland cement, lime, carpet backing, linoleum, soil stabilizers, and chewing gum.
Coal	Anthracite, bituminous, subbituminous, and lignite. Supplies over 50% of fuel to generate electricity in the United States. Used in cement, paper, automobile industry, textiles, plastics, paint, nylon, and aspirin.
Conglomerate	Sedimentary in origin, mixture of minerals including silica and lime. Used in the construction industry.
Copper	Malleable and ductile. Used in building and construction, pipes, wire, electronic equipment, industrial machinery and equipment, and the transportation industry.
Diatomite	Composed of diatoms. Used primarily as a filtering agent for water and beverages, polishing agents, cat litter, paints, fertilizers, and insecticides.
Feldspar	Used in glass making and pottery.
Fluorite	CaF_2. Used in the production of hydrofluoric acid (HF), basis for chemicals using fluorine, processing of aluminum and iron ore, added to toothpaste to reduce carries.
Gabbro	Used in making concrete and when polished known as black granite.
Galena	PbS. Lead ore is used in batteries, fuel tanks, solder, radiation shielding, protective coatings, ceramics, glass, foil, wire, chemicals, CRT glass, and plumbing.
Gneiss	Metamorphic rock composed of quartz, mica, feldspar, and hornblende. Used in the construction industry.
Gypsum	$CaSO_4 \cdot H_2O$. Used in wallboard, Plaster of Paris, paint, soil additive, glass, pottery, brown paper bags, and bakery items.
Gold	Properties include conductivity, long-wearing, and malleable. Used in jewelry, computer and electronic applications, dentistry, and coinage.
Granite	Igneous rock composed of quartz, feldspar, mica, and other minerals. Used in ornamental stone and construction.
Halite	NaCl. Used in the chemical industry, road deicing, industry, food and agriculture, water treatment, manufacture of chlorine and caustic soda, fabrics, and PVC.
Ilmenite	$FeO \cdot TiO_2$. Used in jet engines, airframes, space and missile parts, and white color for paint, paper, and frosting.
Iron	Used in making steel, automobiles, ships, machinery, tools, cookware, appliances, and construction supplies.

IMPORTANT MINERALS AND THEIR USES (Continued)

Mineral	Use
Limestone	$CaCO_3$ or $CaMg(CO_3)_2$. Used in cement, building blocks, refining of sugar, paint, brown paper bags, and glass.
Malachite	$Cu_2CO_3(OH)_2$. Used with tin to make bronze. Used with zinc to make brass.
Marble	Composed of calcite and/or dolomite that has undergone heat and pressure. Used for statuary, ornamental, and architectural purposes.
Mica	Able to exist in very thin sheets. Used in steam boilers, dielectrics in capacitors, electrical insulators, cements, paint, plastics, and joint compounds. Mica comes primarily from India.
Molybdenum	Used in steel alloys, machinery, electrical, transportation industry, chemicals, oil and gas refineries, automobiles, stainless steel, chemical processing, dies, paints, corrosion inhibitors, smoke and flame-retardants, and light bulb filaments.
Nickel	Used in stainless steel, alloys, electroplating, fabricated metal parts, petroleum industry, household appliances, chemical industry, aerospace industry, plumbing, and electrical equipment.
Oil Shale	Rock containing hydrocarbons capable of providing a fossil fuel.
Olivine	$(MgFe)_2SiO_4$. High melting temperature. Used in firebricks for furnaces.
Phosphate	PO_4^{3-}. Used in fertilizer, carbonated beverages, and in manufacture of phosphoric acid.
Potash	Used in fertilizer, chemicals, water softeners, drugs, textiles, dyes, glass, and soaps.
Pumice	Volcanic rock. Composed of SiO_2 and Al_2O_3. Used as an abrasive, polishing compound, and building blocks.
Quartzite	Used in growing cultured quartz crystals.
Realgar	AsS. Used as a wood preservative, semiconductors, and lead-acid batteries.
Sandstone	Sedimentary rock. Used in construction "brownstone."
Schist	Metamorphic rock used in construction.
Shale	Sedimentary rock composed of clay and silt. Used in bricks and cement.
Silver	Used in photography and X-ray film, computer equipment, dentistry, jewelry, coinage, and catalysts.
Sphalerite	ZnS. Produces zinc that is used in metal plating, galvanizing, alloys, brass, agricultural products, paint, rubber, medicinal ointments, sun blocks, and in making brass.
Spodumene	$LiAl(Si_2O_6)$. Source of lithium. Used in ceramics, glass, lubricants, greases, synthetic rubber, CRT screens, ceramics, and medication for mental illness.
Staurolite	$Fe_2Al_2O_7 \cdot (SiO_4)_4(OH)_9$. Used as an abrasive and blasting agent.
Sulfur	Used in food processing, textiles, sulfuric acid, paper, steel, photographs, synthetic fibers, explosives, plastics, matches, soaps and detergents, medicines, paints, and vulcanization of rubber.

IMPORTANT MINERALS AND THEIR USES (Continued)

Mineral	Use
Talc	Used in baby powder, plastics, paper, and specialized ceramics.
Titanium	See ilmenite.
Trona	$Na_2CO_3 \cdot NaHCO_3 \cdot 2H_2O$. Major source for soda ash. Used in glass manufacturing, chemicals, soaps and detergents, paper, baking soda, glass, ceramic tile, porcelain, and toothpaste.
Ulexite	Used for borax, nylon, glass, electronics, and automobile air bags.
Zircon	$ZrSiO_4$. Used in jewelry.

Mining Process

Before mining begins, economic decisions are made to determine whether the site will be profitable. Factors that enter the decision include current and projected price, amount of ore at site, concentration of commodity in the ore, type of mining required, cost to mine the commodity, cost of transporting ore to a processing facility, and cost of reclamation. After all factors are analyzed, and if the project is approved, the following steps are employed:

1. **Site development.** Samples are taken from an area to determine quality and quantity of minerals in a location. Roads are constructed and equipment is brought in.
2. **Extraction.** Generally three methods are employed:
    - Surface mining. Earth-moving equipment is employed. This is a safe form of mining and a cost-effective one; however, topsoil and vegetation is removed, air quality issues develop regarding dust, large amounts of waste material are involved, and pits are dug below water table so groundwater is pumped out. (Between 1990–1998, mining operations in the United States pumped out 1,150,000 acre feet of groundwater—enough water to supply a city of 4,600,000 people). The water that is pumped out affects fish and wetlands through erosion, sedimentation, dewatering of wetlands, and diverting and channelizing streams.
    - Underground mining. Large shafts are dug into the Earth. There is less surface destruction and waste rock produced than in surface mining, but it is unsafe. Acid mine drainage containing pyrites and sulfides are produced affecting air and water quality.
    - *In situ* leaching. Small holes are drilled into the site. Water-based chemical solvents are used to flush out desired minerals. Waste rock is minimal. Fluids injected into the Earth are toxic and enter the groundwater supply.
3. **Processing.** (also known as beneficiation). This step involves intensive chemical processing. For example, in cyanide heap leaching, gold ore is "heaped" onto a large pile up to 30 feet high (about 10 meters). Cyanide solution is sprayed on top of the pile. As the cyanide percolates downward, the gold leaches out of the ore and collects in pools at the bottom. The gold extracted may be only 0.01% of the total ore processed. Liquid wastes containing cyanide and other toxins are kept in holding ponds, which eventually leak and enter groundwater supplies.

4. **Site restoration.** Restoration involves returning the area mined back to as close to the original condition as possible. The Surface Mining Control and Reclamation Act established a nationwide program to protect society and the environment from the adverse effects of surface coal mining. Less-developed countries may be less environmentally conscious in this area due to other economic priorities. Grass seed may be planted in areas that were once forested and thereby said to be "restored." Radioactive waste material in the former Soviet block countries face problems of cost to transport material to remote areas, and disposal sites are not available.

WORLD MINERAL RESERVES

Material	World reserve (1,000 tons)	World lifetime years	U.S. reserves (1,000 tons)
Bauxite (Aluminum)	23,000,000	219	20,000
Chrome	1,400,000	109	0
Iron	150,000,000	178	16,100,000
Nickel	47,000	51	23
Platinum	56,000	190	250
Gold	44,000	20	4,770

Source: U.S. Geological Survey.

Case Study

SHADE GROWN COFFEE

- Most coffee is grown in or near Central and South America—Brazil, Colombia, and Costa Rica.
- Coffee (típica and borbón varieties) used to be grown exclusively on mixed plantations and family farms. Coffee trees were the understory with fruit trees, citrus trees, hardwood trees, and banana trees as overstory. Shade trees were required; otherwise, the leaves would burn. These areas provided wood, food, and work for local families. They also provided winter habitats for migrating bird species (e.g., Baltimore orioles).
- Much of the coffee that is now produced (particularly in Colombia, Brazil, and Costa Rica) is grown on large-scale plantations by multi-national corporations (agribusiness) where hybrid varieties (caturra, catuaí, mundo novo, and variedad colombiana) are grown as monocultures. These new varieties are shorter and denser and provide greater yield per acre.
- Large amounts of fertilizers and pesticides are now used and increased soil erosion and water runoff have been documented on coffee plantations.
- Biodiversity is nonexistent, and migrating birds cannot survive in these biological deserts. The Smithsonian Migratory Bird Center has documented sightings of up to 150 different bird species in a shaded coffee farm; however, in unshaded coffee farms, fewer than 20 species were counted. A large and diverse bird population is a source of insect and pest control, requiring less need for insecticides.

- Shade grown coffee farms produce greater biodiversity and are thus more "sustainable" than non-shade coffee farms. The variety of trees provides a nitrogen-rich mulch to the ground as their leaves fall and decompose. The presence of the mulch is also useful in soil moisture conservation, suppression of weed growth, and prevention of soil runoff during the rainy season. The result is less of a need for chemical fertilizers and herbicides.
- Shade trees also act as an economic buffer for the coffee farmer. The lowest layer of the canopy are citrus, avocado, and banana trees, which not only provide food but also allow the excess to be sold at local produce markets. The upper shade story is typically native hardwood trees or large nitrogen-fixing leguminous trees. These trees are thinned and pruned throughout the year, and their branches and timber serve as fuel and cooking wood, fencing material, and a viable source of timber for building and construction purposes.
- Coffees of southern Mexico (Chiapas), northern Nicaragua, El Salvador, Peru, Panama, Guatemala, Ethiopia, Sumatra, New Guinea, and Timor are virtually all shade grown.

Multiple-Choice Questions

1. Soil that is transported by the wind is called

   (A) alluvial.
   (B) feral.
   (C) gangue.
   (D) rill.
   (E) aeolian.

2. Most of the Earth's land area is

   (A) urban.
   (B) forest.
   (C) desert.
   (D) rangeland.
   (E) agricultural.

3. Cities experiencing the greatest urban growth are found in

   (A) Asia.
   (B) Africa.
   (C) Europe.
   (D) North America.
   (E) A and B.

4. Which of the following are NOT concepts designed to create a sustainable city?

   (A) Conserve natural habitats.
   (B) Focus on energy and resource conservation.
   (C) Design affordable and fuel-efficient automobiles.
   (D) Provide ample green space.
   (E) All are correct.

5. Approximately what percent of the world depends upon wood or charcoal for heating and/or cooking?

   (A) Less than 10%
   (B) Between 20 and 30%
   (C) Between 50 and 60%
   (D) More than 75%
   (E) Has not been determined

6. Of all the jobs in the U.S. Forest Service, 75% are connected with

   (A) lumber management.
   (B) fire management.
   (C) mining management.
   (D) recreation.
   (E) administrative functions.

7. In which of the following states or areas would you NOT find significant amounts of frontier forests?

   (A) New England
   (B) Appalachia
   (C) British Columbia
   (D) Washington
   (E) California

8. Most grain that is grown in the United States is used

   (A) for export to countries that need grain.
   (B) for cereals and baked good.
   (C) to feed cattle.
   (D) for trade with other countries.
   (E) to create fuel and liquor.

9. What is the number one source of soil erosion?

   (A) Physical degradation
   (B) Chemical degradation
   (C) Water erosion
   (D) Wind erosion
   (E) All forms equally contribute to soil erosion

10. A process in which small holes are drilled into the Earth and water-based chemical solvents are used to flush out desired minerals is known as

    (A) chemical leaching.
    (B) *ex situ* leaching.
    (C) beneficiation.
    (D) heap-leaching.
    (E) *in situ* leaching.

11. Planting trees and/or shrubs between rows of crops is known as

    (A) contour farming.
    (B) planting windbreaks.
    (C) strip cropping.
    (D) alley cropping.
    (E) tillage farming.

12. During the last 150 years, urban population has increased

    (A) 10%.
    (B) 20%.
    (C) 50%.
    (D) 100%.
    (E) 250%.

13. Population patterns in the United States are shifting

    (A) to the Pacific Northwest.
    (B) to north and west.
    (C) to the south and east.
    (D) to the west and east.
    (E) to the south and west.

14. The factor that is the greatest threat to the success of a species is

    (A) environmental pollution.
    (B) loss of habitat.
    (C) poaching.
    (D) hunting.
    (E) introduction of new predators into the natural habitat.

15. Which of the following is NOT a characteristic of an endangered species?

    (A) Large body size
    (B) High reproductive rate
    (C) Large territory requirement
    (D) Specialist species
    (E) Long-lived

Answers to Multiple-Choice Questions

1. **E**	6. **D**	11. **D**
2. **B**	7. **A**	12. **C**
3. **E**	8. **C**	13. **E**
4. **C**	9. **C**	14. **B**
5. **C**	10. **E**	15. **B**

Explanations for Multiple-Choice Questions

1. **(E)** Aeolian soils are sand-sized particles transported by wind action.

2. **(B)** 32% of the Earth's surface is covered by forests.

3. **(E)** The world's largest cities in the year 2015 are projected to be Tokyo, Japan (Asia); Bombay, India (Asia); Lagos, Nigeria (Africa); and Dhaka, Bangladesh (Asia).

4. **(C)** Designing affordable and fuel-efficient automobiles would have the effect of increasing the number of automobiles already within the city. Most cities are close to or have reached the limits of their infrastructure. Instead of policies creating more automobiles, sustainable cities should focus on strategies of reducing the number of cars—mass transit, mass transit subsidies, car-sharing networks, and flexible work schedules, and the like, would be some strategies that would reduce commuting time, not contribute to air pollution or fossil-fuel depletion, and would help to sustain a city.

5. **(C)** As late as the 1850's, wood supplied over 90% of the United States' energy requirements. Many countries in the developing world still use wood as their primary fuel. Half of the energy used in the continent of Africa is in the form of fuelwood utilized at a much faster rate than it is growing back, with the result that the total amount of forested land is shrinking rapidly. As forests disappear or recede, the ecosystems that are dependent upon the trees also recede or disappear. As ecosystems disappear, biodiversity is decreased. Another effect of deforestation is that it reduces the Earth's natural ability to regulate the amounts of carbon dioxide and oxygen in the air which affects global warming.

6. **(D)** Recreation.

7. **(A)** The Washington-based World Resources Institute, states "Only 20% of the world's original forests remain intact, and nowhere is the damage worse than in the continental United States, where 99% of its frontier forest has been destroyed or ruined." This effect is most pronounced in the New England area. This does not say however that this area is without trees. The trees that are in this region have replaced original old-growth forests.

8. **(C)** There are currently 1.28 billion cattle populating the Earth. They graze on nearly 24% of the landmass of the planet and consume enough grain to feed hundreds of millions of people. Their combined weight exceeds that of the human population on Earth. Cattle are also a major cause of global warming. They emit methane, a potent global warming gas, blocking heat from escaping the Earth's atmosphere. Cattle and other livestock consume over 70% of all the grain produced in the United States. Today about one-third of the world's total grain harvest is fed to cattle and other livestock while as many as a billion people suffer from chronic hunger and malnutrition. While millions of human beings go hungry for lack of adequate grain, millions more in the industrial world die from diseases caused by an excess of grain-fed animal flesh, and especially beef, in their diets.

9. **(C)** See Figure 9.10. Soil erosion due to water accounts for 56%.

10. **(E)** *In situ* leaching has several advantages over surface or subsurface mining.

    - Miners are not directly exposed to ore, dangers of working underground, or dust.

- It is less expensive to operate because large amounts of rock do not have to be broken up and removed. There are also shorter lead times to production, that is, it is quicker to produce an end product.
- There is little or no solid waste. Waste is confined to evaporation or confinement ponds. *If* constructed and processed properly, containment ponds can have less environmental effects than strip mining.
- *In situ* leaching mining is less costly to operate because it does not need the expensive infrastructure of open-cut and underground mining (i.e., shafts, tunnels, and crushers).
- There is much less ground disturbance. There are no open pits, shafts, tunnels, Earth moving equipment, or grinding and crushing facilities. *In situ* leaching operations take up less land; therefore, there is less visual impact.
- There is less rehabilitation required because there is less ground disturbance. Upon completion of mining, wells can be sealed and capped, process facilities removed, and the surface returned to its original contour and vegetation.
- Smaller, lower grade, and narrower ore bodies can be mined.

Disadvantages are the threat to the groundwater supply and the effect of toxic chemicals if they are not processed properly.

11. **(D)** Alley cropping is growing annual agricultural crops in alleys between rows of long-term perennial tree crops to provide an annual income while the tree crop matures.
    The purpose of alley cropping is to
    - Enhance or diversify farm products,
    - Reduce surface water runoff and erosion,
    - Improve the utilization of nutrients,
    - Reduce wind erosion,
    - Modify the microclimate for improved crop production, and
    - Improve wildlife habitat and enhance the aesthetics of the location.
12. **(C)** See the table entitled "Rural–Urban Shifts in the United States (1800–1990).
13. **(E)** Since the 1960s, the West ranked as the fastest-growing region of the United States. The U.S. Census Bureau shows the West held 7 of the 10 fastest growing cities in the 1990s and the 5 fastest-growing states. One of the reasons for this fast growth is globalization and more open international markets. Companies have tried to decentralize as much as possible in order to compete on costs, shifting jobs and people into Western cities. Nevada, Arizona, and Colorado were the fastest growing states. Las Vegas, Nevada, was the fastest-growing urban center. At the other extreme, 24 metro areas lost people, mainly from the industrial regions of Ohio, Pennsylvania, and New York.
14. **(B)** "Worldwide, the number one threat to plant and animal life is loss of habitat," E. O. Wilson, ecologist, author, and professor of biology at Harvard University.
15. **(B)** A high reproductive rate would be characteristic of a species that would generally not be in danger of extinction (houseflies are not in danger of extinction, but elephants and whales are). High reproductive rates also allow for a wider diversity in the gene pool since the number of individuals is greater and sexual reproduction (crossover and mutation) is a cornerstone of phenotypic variation—all of which allows for some recovery in the population from environmental influences.

Free-Response Question

by Sarah E. Utley
Environmental Content Specialist—Center for Digital Innovation
A.P. Environmental Science Division—University of California at Los Angeles

Deposits of coal were recently discovered in a small valley in the eastern foothills of the Rocky Mountains. These deposits were found both near the surface and deep underneath the ground.

(a) Describe a method that could be used to mine the coal.
(b) Using coal as an energy source can have serious environmental consequences. Discuss current technological methods employed to reduce the impact of using coal as an energy source.
(c) The residents of this small town have mixed opinions of the impact of the mine on their community. Some residents support the mine because of the increased employment opportunities and increased tax base that support projects such as schools and roads. Others are concerned about current or possible future environmental impacts on this region of the state. Discuss two of these possible environmental impacts.

Free-Response Answer

Section (a): (Choose either #1 or #2)

1. Coal that is located close to the surface is frequently extracted by strip-mining, a type of surface mining. In surface mining, the rock, soil, and vegetation layers that are on top of the coal are scraped away and removed. The removed rock and soil is called overburden. Specifically in strip mining, a long trench is dug, removing the overburden and the coal. Then a second trench is dug parallel to the first and the overburden from the second is used to fill the first trench. The hill of loose overburden that fills the previous trench is called a spoil bank. The process continues until the coal is depleted or the cost of future extraction is prohibitive.
2. Coal that is located deep underneath the surface of the Earth is removed through underground or subsurface mining techniques. A deep shaft must be dug or blasted into the area. An extensive network of tunnels (which are supported with wooden or metal beams) is then dug and these extend out from the main shaft. Sometimes part of the coal vein is left intact to support the shafts and tunnels. If the mine is large enough, electricity is wired, an elevator is installed to take workers to the active part of the mine, and tracks are laid on which the loaded bins of coal travel. The coal itself can be removed by hand (through the use of picks and shovels) or with special mining machines with large drills. The mined coal is then loaded onto the mining bins and removed to the surface.

Section (b):
Coal can contain contaminants such as rocks, ash, mineral dust, and sulfur. These contaminants decrease the efficiency of the coal and can increase the environmental impact of burning the fossil fuel. When coal with high sulfur content is burned, the sulfur can enter the atmosphere as sulfur dioxide that is a component of acid rain. There are several

techniques that can decrease the environmental impact of coal use. When coal is initially mined, it is usually washed. Washing involves separating the impurities from the coal based on density. The contaminants are generally denser than the coal and sink in the separation fluid. The process involves first grinding the coal into small pieces. Then the pieces are placed onto a separation table with a fluid. The coal is skimmed off the top and the contaminants, such as pyrite (a sulfur derivative), sink to the bottom.

In addition to the initial washing of coal, there are several remediation techniques that can decrease the amount of pollutants the result from the burning or usage of coal.

The Clean Air Act (revised 1990) requires that all new coal burning plants (built after 1978) must be equipped with air quality structures that remove the sulfur from the combustion gases before they are released through the smokestack. These structures are called scrubbers. In a scrubber, a mixture of water and crushed limestone is sprayed into the gases. The sulfur from the emissions and the limestone chemically combine to form a solid (calcium sulfate). This solid is then disposed in a solid waste disposal facility or can be used in the production of some building products.

In addition to scrubbers, there is additional technology that can decrease the environmental impact of coal burning. In a fluidized bed combustion plant, the coal particles are suspended within the boiler on jets of air. This tumbling of the burning coal allows limestone to be introduced into the plant prior to the smokestack. The limestone combines with the sulfur and rapidly chemically bonds with it to capture a greater amount of sulfur pollutants. In addition to removing the sulfur prior to entering the smokestack, a fluidized bed boiler burns at a cooler temperature than a conventional boiler. This cooler temperature prevents the smog-forming NO_x pollutants from volatizing and being released into the air.

Section (c): (Give two implications.)

1. Water pollution is frequently a concern with the mining of coal. Water is used throughout the mining process, from cooling machinery and smelters to washing coal. If this contaminated water is not contained and purified, it can seep into the local waterways or water table. In addition, precipitation that percolates through the area can leach soil contaminants from the mining process into the local water table or waterways. The water table of the areas located close to mines must be carefully monitored for quality.
2. Air pollution is also a concern in the mining industry. The mining machinery, which is frequently powered by diesel engines, contributes to air pollution in the mine vicinity. In addition, the burning of coal often releases air toxins such as sulfur dioxide, nitrogen oxides, and particulate matter into the atmosphere. New combustion units are now mandated by the Clean Air Act of 1990 to be fitted with scrubbers and other clean air apparatus. Many states are now requiring that even older equipment must meet new clean air requirements.
3. There is also the concern of habitat destruction and wildlife/native plant displacement. The process of surface mining, which requires the removal of many tons of soil, disturbs the natural ecosystem and displaces native animals. Subsurface mining is often less disruptive to soil surface, but has its own concerns such as subsidence and land instability. Though mining corporations are required by law to return the land to the original topography, since the soil structure has been altered, native plant revegetation techniques often fail. When the native plants fail to thrive, it is difficult to reintroduce wildlife.

4. Mining can also be harmful to human health. With subsurface mining, workers must be protected against unstable mine shafts, generally dangerous working conditions, and health concerns such as poor air quality. It can be difficult to safely ventilate the mines below the ground and care must be taken to protect the workers from respiratory diseases such as emphysema and lung cancer. With the large amount of mining equipment (large cranes, shovels, etc.) required in surface mining, human safety is always a concern.

Rubric

Ten points necessary for full credit.
Section 1: worth 2 points (Describe either surface or subsurface mining.)
Section 2: worth 4 points (Explain at least two remediation techniques.)
Section 3: worth 4 points (Two points awarded for each completely described environmental impact.)

POINTS ASSIGNED/EXPLAINED	EXAMPLE
Choose either #1 or #2. 1 point: description of mining type 1 point: specific description of coal removal	Section (a): (Choose either #1 or #2) 1. Coal that is located close to the surface is frequently extracted by *strip-mining, a type of surface mining*. In surface mining, the *rock, soil, and vegetation layers that are on top of the coal are scraped away and removed*. The removed rock and soil is called overburden. Specifically in strip mining, a *long trench is dug, removing the overburden and the coal*. Then a *second trench is dug parallel to the first and the overburden from the second is used to fill the first trench*. The hill of loose overburden that fills the previous trench is called a spoil bank. The *process continues until the coal is depleted* or the cost of future extraction is prohibitive.
OR 1 point: description of mining type 1 point: specific description of coal removal	2. Coal that is located *deep underneath* the surface of the Earth is removed through *underground or subsurface mining techniques*. A *deep shaft must be dug or blasted into the area*. An extensive *network of tunnels* (which are supported with wooden or metal beams) *is then dug* and these extend out from the main shaft. Sometimes part of the coal vein is left intact to support the shafts and tunnels. If the mine is large enough, electricity is wired, an elevator is installed to take workers to the active part of the mine, and tracks are laid on which the loaded bins of coal travel. The *coal itself can be removed by hand (through the use of picks and shovels) or with special mining machines with large drills*. The mined coal is then loaded onto the mining bins and removed to the surface.
Though an answer may contain more than two remediation techniques, at a minimum a full-credit answer should contain two techniques.	Section (b): Coal can contain contaminants such as rocks, ash, mineral dust, and sulfur. These contaminants decrease the efficiency of the coal and can increase the environmental impact of

POINTS ASSIGNED/EXPLAINED	EXAMPLE
	burning the fossil fuel. When coal with high sulfur content is burned, the sulfur can enter the atmosphere as sulfur dioxide that is a component of acid rain. *There are several*
½ point: introduction of techniques	*techniques that can decrease the environmental impact* of coal use. When coal is initially mined, it is usually washed.
1 point: purpose of washing	*Washing involves separating the impurities from the coal based on density*. The contaminants are generally denser than the coal and sink in the separation fluid. The process
1 point: description of washing (performed after excavation and prior to combustion)	involves first *grinding the coal into small pieces*. Then the pieces *are placed onto a separation table with a fluid. The coal is skimmed off the top* and the contaminants, such as pyrite (a sulfur derivative), sink to the bottom.
	In addition to the initial washing of coal, there are several remediation techniques that can decrease the amount of pollutants the result from the burning or usage of coal.
	The Clean Air Act (revised 1990) requires that all new coal burning plants (built after 1978) must be equipped with air quality structures that *remove the sulfur from the combus-*
1 point: purpose of scrubbers	*tion gases* before they are released through the smokestack. These structures are called *scrubbers*. *In a scrubber, a mixture of water and crushed limestone is sprayed into the gases*. The *sulfur from the emissions and the limestone*
1 point: description of scrubbers	*chemically combine to form a solid (calcium sulfate)*. This solid is then disposed in a solid waste disposal facility or can be used in the production of some building products.
	In addition to scrubbers, there is additional technology that can decrease the environmental impact of coal burning. In a
1 point: description of fluidized bed combustion	*fluidized bed combustion plant, the coal particles are suspended within the boiler on jets of air*. This *tumbling of the burning coal allows limestone to be introduced into the plant prior to the smokestack*. The *limestone combines with the sulfur and rapidly chemically bonds* with it to capture a greater amount of sulfur pollutants. In addition to *removing the sulfur* prior to entering the smokestack, a fluidized bed
1 point: purpose of fluidized bed combustion	boiler *burns at a cooler temperature than a conventional boiler*. This cooler temperature *prevents the smog-forming NO_x pollutants from volatizing* and being released into the air.
The question specifically asks for only two environmental implications.	Section (c): (Give two implications.)
	1. *Water pollution* is frequently a concern with the mining
½ point: name of concern	of coal. *Water is used throughout the mining process, from cooling machinery and smelters to washing coal*. If this
½ point: cause of impact	contaminated water is not contained and purified, it can

POINTS ASSIGNED/EXPLAINED	EXAMPLE
1 point: environmental result	*seep into the local waterways or water table*. In addition, precipitation that percolates through the area can *leach soil contaminants from the mining process into the local water table or waterways*. The water table of the areas located close to mines must be carefully monitored for quality.
½ point: name of impact ½ point: cause of impact	2. *Air pollution* is also a concern in the mining industry. The *mining machinery, which is frequently powered by diesel engines, contributes to air pollution* in the mine vicinity. In addition, the *burning of coal often releases air toxins such as sulfur dioxide, nitrogen oxides, and particulate matter into the atmosphere*. New combustion units are now mandated by the Clean Air Act of 1990 to be *fitted with scrubbers and other clean air apparatus*. Many states are now requiring that even older equipment must meet new clean air requirements.
1 point: remediation of impact ½ point: name of impact ½ point: cause of impact 1 point: result of impact	3. There is also the concern of *habitat destruction and wildlife/native plant displacement*. The process of surface mining, which requires *the removal of many tons of soil*, disturbs the natural ecosystem and *displaces native animals*. Subsurface mining is often less disruptive to soil surface, but has its own concerns such as *subsidence and land instability*. Though mining corporations are required by law to return the land to the original topography, since the soil structure has been altered, native *plant revegetation techniques often fail*. When the native plants fail to thrive, it is *difficult to reintroduce wildlife*.
cause of impact result of impact ½ point: name of concern 2 points: results of impact (½ point each)	4. Mining can also be *harmful to human health*. With subsurface mining, *workers must be protected against unstable mine shafts, generally dangerous working conditions, and health concerns such as poor air quality*. It can be difficult to safely ventilate the mines below the ground and care must be taken to protect the workers from respiratory diseases such as emphysema and lung cancer. With the *large amount of mining equipment* (large cranes, shovels, etc.) required in surface mining, human safety is always a concern.

UNIT IV: ENVIRONMENTAL QUALITY

CHAPTER 10

Air, Water, and Soil Pollution

"It isn't pollution that's harming the environment. It's the impurities in our air and water that are doing it."

U.S. Vice President Dan Quayle

Areas That You Will Be Tested On

A: Major pollutants
- **1. Types (SO_2, NO_x, pesticides, etc.)**
- **2. Thermal Pollution**
- **3. Measurement and Units of Measure (ppm, pH, micrograms)**
- **4. Point and Nonpoint Sources (Domestic, Industrial, Agricultural)**

B: Effects of Pollutants
- **1. Aquatic Systems**
- **2. Vegetation**
- **3. Natural Features, Buildings, and Structures**
- **4. Wildlife**

C: Pollution Reduction, Remediation, and Control

Key Terms

Note: Definitions for all key terms listed below can be found in the glossary starting on page 648. For additional definitions of relevant terms from this chapter, refer to *www.barronseduc.com/0764121618.html*.

acid deposition	**chlorofluorocarbons**
acid rain	**criteria air pollutants**
algae blooms	**integrated pest management**
biological oxygen demand (BOD)	**nitrogen fixation**
carbon dioxide	**NO_x**
carbon monoxide	**Organophosphates**
chlorinated hydrocarbon	**Polychlorinated biphenyls (PCB)**

Key Concepts

Air

- The term "smog" was first used in 1905 to describe emission of sulfur dioxide.
- Major man-made source of VOCs is automobiles. The threat from indoor pollution of VOCs (oil based paint and carpeting) is as great a threat as automobile smog.
- From 10,000 to 15,000 people in the United States each year are admitted to hospitals for ozone-related illnesses. Children are more susceptible because (1) their airways are narrower and (2) they spend more time outdoors. Every day, 12 million children are exposed to secondhand smoke in their homes. This leads to serious health consequences, ranging from ear infections and pneumonia to asthma.
- Radon is a cancer-causing, radioactive gas. Radon comes from the natural (radioactive) breakdown of uranium in soil, rock, and water and concentrates in homes and buildings. Radon can be found all over the United States. The Surgeon General has warned that radon is the second leading cause of lung cancer in the United States today. Only smoking causes more lung cancer deaths.
- Molds produce tiny spores to reproduce. Mold spores travel through indoor and outdoor air continually. When mold spores land on a damp spot indoors, they may begin growing and digesting whatever they are growing on in order to survive. Molds can grow on wood, paper, carpet, and foods. Potential health effects and symptoms associated with mold exposures include allergic reactions, asthma, and other respiratory complaints.
- The Clean Air Act was originally signed into law in 1963. Early version allowed individual states to set standards. Later versions switched responsibility of setting uniform standards to the federal government. In the 1970 version, criteria pollutants were identified. Standards were divided into (1) primary—to protect human health and (2) secondary—to protect materials, crops, climate, visibility, and personal comfort. The 1990 version addressed acid rain, urban smog, toxic air pollutants, ozone protection, marketing pollution rights, and VOCs. The 1997 version reduced ambient ozone from 0.12 to 0.08 ppm. Estimates were that measures would cost $15 billion per year but save 15,000 lives, reduce bronchitis cases by 60,000 per year, and reduce hospital admission for respiratory illnesses by 9000 per year. California has set standards higher than the Clean Air Act. As the table on page 327 shows, it was effective in reducing CO, VOCs, SO_2, and lead. The reduction in lead was due primarily to the phasing out of leaded gasoline.
- Acid rain and the dry deposition of acidic particles contribute to the corrosion of metals and the deterioration of paint and stone such as marble and limestone. These effects seriously reduce the value to society of buildings, bridges, cultural objects (e.g., statues, monuments, and tombstones), and cars. Dry deposition of acidic compounds can also dirty buildings and other structures, leading to increased maintenance costs. To reduce damage to automotive paint caused by acid rain and acidic dry deposition, some manufacturers use acid-resistant paints, at an average cost of $5 for each new vehicle (or a total of $61 million per year for all new cars and trucks sold in the United States).
- Air pollution returning to Earth (acid deposition) may return as solid, liquid, or gas (snow, rain, fog). Effects on plants and wildlife include
  - Damaged roots;
  - Destroyed or leached soil nutrients, making them unavailable for uptake by plants;

- Toxic aluminum and mercury that are leached from rocks and enter streams and lakes causing fish kills;
- Destruction of microorganisms necessary for fixation and decomposition (which break down organic matter);
- Destruction of waxy coating on leaves, increasing transpiration rates and making plant susceptible to disease;
- Weakened plants that are affected by strong winds, droughts, and heavy rainfall;
- Disrupted germination patterns;
- "Acid shock," which occurs during spring when water in lakes thaw and acid becomes soluble in water preventing reproduction of aquatic species;
- Weakened life forms because organisms exist in a "web." Waterfowl that depend on aquatic insects and fish have reduced food supplies, wetland mammals (e.g., beavers) and other animals begin to suffer as habitat erodes;
- Uptake of acidic water by trees causes trees to be weakened and affects organisms that live in them;
- Depleted fish stocks;
- Toxic fish which have increased levels of mercury due to acid precipitation;
- Acidic water that moves through metal pipes, corroding the pipes, increasing the lead and copper concentration, and making water even more unhealthy.

• Remediation of acidic precipitation includes

- Adding quicklime (calcium oxide) to small lakes raises pH (not effective on large lakes; process is expensive);
- Adding scrubbers and advanced filters to smokestacks;
- Installing catalytic converters on cars;
- Washing coal before burning it to reduce sulfur content;
- Utilizing low-sulfur oil and coal;
- Adopting accessible mass transit;
- Adopting emission trading policies;
- Pursuing international treaties on acidic effluents.

U.S. EMISSIONS OF COMMON AIR POLLUTANTS, 1970–1998

Pollutant	1970 (thousand short tons)	1998 (thousand short tons)	% Change
CO	129,444	89,454	–30.9%
NO_x	20,928	24,454	+16.8%
VOCs	30,982	17,917	–42.2%
SO_2	31,161	19,647	–36.9%
Pb	220,869	3,973	–98.2%
PM_{10}	13,042	34,741	+266.4%

Source: U.S. Environmental Protection Agency, National Air Pollution Trends.

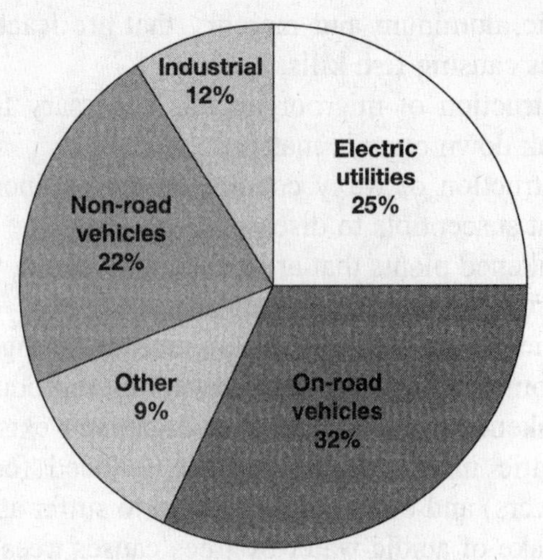

Figure 10.1. Nitrogen oxide emissions (thousands of short tons), 1998.
Source: EPA: National Air Pollution Emissions Trends, 1900–1998.

- As shown in Figure 10.1, transportation accounts for most of the NO_x emission (54%). Next is industrial fuel combustion (37%). "Other" represents petroleum and chemical processing.
- Up until 1970, air pollution legislation was at the local and state level. The first federal regulation was the Air Pollution Control Act of 1955. Federal money was allocated to help states deal with air pollution issues. In 1970, the Clean Air Act standardized enforcement and standards. The Air Quality Control Act of 1967 divided the country into Air Quality Control Regions and defined methods for setting emission standards.
- The goal of the Clean Air Act is to (1) protect public health from air pollution and (2) limit effects of air pollution on the environment. National Ambient (outdoor) Air Quality Standards (NAAQS) were established. Subject to regulation are six criteria pollutants: ozone (O_3), volatile organic compounds (VOCs), nitrogen dioxide (NO_2), carbon monoxide (CO), PM_{10}, and sulfur dioxide (SO_2).
- Hazardous Air Pollutants (HAPs) are less widespread than criteria air pollutants (e.g., toluene).
- The EPA estimates that mobile sources of air pollution (e.g., cars, buses, and trucks) may account for half of all cancers due to outdoor sources.
- Formaldehyde, acetaldehyde, diesel particulates, and 1,3-butadiene do not occur in gasoline but are produced through incomplete combustion of gasoline.
- Natural sources of air pollution include fires, volcanoes (hydrogen sulfide), fumaroles, sea spray, and decaying vegetation (sulfur), trees and bushes (VOCs), pollen, bacteria, spores, molds, dust, decaying cellulose in termites, and ruminants (methane).
- The United Nations estimates that 1.3 billion people live in areas of polluted air. Half of all people in the United States have some degree of alveolar (lung) deterioration (smoking is primary cause).
- Air pollution affects vegetation because it is (1) toxic and damages cell membranes and (2) disrupts normal growth and reproduction by interfering with metabolic regulators and hormones. Effects may be synergistic. Effects on plant life decrease gradually as the distance from source increases.
- Point sources are easier to identify for liability and easier to remediate than nonpoint sources (point sources are concentrated and generally localized).

AIR POLLUTION IN DIFFERENT COUNTRIES

Country	Description
Mexico	Mexico City's air quality is the worst in the world (3.3 million cars). Geography is a basin that traps pollutants. Averages 179 mg/m^3 of suspended particulates (World Health Organization uses 90 mg/m^3 recommended maximum). A 5-year National Environmental Program (1996–2000) invested $13.3 million to reduce air pollution. Federal tax incentive program for purchases of pollution control equipment. Many foreign industries have located in the border town of Juarez because *maquiladoras* or foreign factories are able to release pollutants due to economic incentives and lax environmental laws. Government proposals include (1) incentives for using cleaner fuels and smog control measures; (2) catalytic converters; (3) driving only on certain days of the week; (4) incentives for owners to upgrade to newer, more efficient cars; (5) companies to invest in cleaner vehicles; (6) government subsidies to outfit gasoline-powered delivery trucks with cleaner liquefied petroleum gas; (7) fuel efficient and cleaner taxis and buses; and (8) improving environmental regulation and monitoring.
Chile	Santiago is one of the most polluted cities in the world. It lies in the middle of a valley, surrounded by two mountain ranges—Andes and the Cordillera de la Costa. Small amounts of wind and rain during winter make the problem worse. Factors that make air pollution a problem include (1) geography; (2) weather; (3) growing economy; (4) rapid urban expansion; (5) dust from unpaved streets; (6) increasing rate of automobile use; and (7) emissions from factories. Chile is the world's largest producer of copper, the production of which contributes to air and water pollution (sulfur dioxide, arsenic, and suspended particulate matter). Efforts to control air pollution include: (1) a $1.2-million subsidy for natural gas buses; (2) paving the streets of Santiago's suburbs; and (3) an incentive program to relocate industries to other regions of the country.
China	The World Health Organization in 1998 reported that 7 of the 10 most polluted cities in the world are in China. Sulfur dioxide and soot caused by coal combustion are the two major air pollutants. Acid rain as a result of this pollution affects about one-third of the land area. Due to industry, 21 million tons of sulfur dioxide and 27 million tons of PM_{10} are emitted into the atmosphere. Industrial boilers and furnaces are the primary point sources for urban air pollution. To improve air pollution: (1) government vehicles, taxis, and buses are being converted to natural gas; (2) alternative and cleaner fuels are being made available; (3) heavy fines are being levied against polluters; and (4) alternatives to coal are being encouraged and subsidized. China, South Korea, and Japan have agreed to launch a five-year project to control transboundary air pollutants.
India	New Delhi is one of the top ten most polluted cities in the world, with respiratory diseases 12 times greater than other Indian cities. India's gross domestic product has increased 2.5 times in the last 20 years, while vehicular pollution has increased eight times, and industrial pollution has quadrupled. Vehicular emissions and untreated industrial smoke are the two major sources of air pollution. Uncoordinated government efforts in controlling air pollution has resulted in chaos and riots—banning public vehicles older than 15 years and mandating the conversion of taxis and buses to cleaner fuels has resulted in job losses, less public transportation, and commuter chaos and unemployment.

TECHNIQUES FOR IMPROVING AIR QUALITY

1. Emphasize pollution prevention and improvement with a reward system (tax incentives) rather than clean-up with a punishment system. Primary focus should be on automobile industries and industrial factories to prevent pollution rather than control it.
2. Set legislative standards for energy efficiency and reward those industries and individuals who meet goals with tax incentives. Cars in California must be inspected frequently to insure smog guidelines before registration is renewed.
3. Phase out the use of fossil fuels over a reasonable time-frame through incentives.
4. Enlist government support of research into increasing efficiency of alternative and renewable sources of energy such as solar, wind, and hydropower. Automobile manufacturers are conducting research into hydrogen-powered and electric vehicles.
5. Give government incentives for programs of reducing population growth, coupling worldwide assistance programs to countries who demonstrate effective and ethical population growth programs.
6. Incorporate into all trade policies incentive programs for addressing and meeting goals geared toward issues of pollution and land management.
7. Establish legislation that whatever penalties are assessed for not meeting air-quality goals be used strictly for tax incentives for those industries that do meet goals.
8. Distribute environmentally-friendly solar cookstoves to developing countries.
9. Establish worldwide programs of sharing antipollution technology with developing countries that do not have the resources for either research or adoption of newer technologies.
10. Phase out inefficient two-cycle gasoline engines (e.g., lawn mowers, jet skis, chain saws, and outboard motors).
11. Use filtering systems at industrial sites (e.g., baghouse filters, cyclone separators, and electrostatic precipitators).
12. Use catalytic converters in cars and other forms of transportation. They can remove up to 90% of nitrogen oxides, hydrocarbons, and carbon monoxide. A catalytic converter removes the pollutant gases from the exhaust by reducing or oxidizing them. Precious metals are used as catalysts in the process—platinum (37%) and rhodium (73%). Reactions (heterogeneous catalysis) take place in the converter. Both NO_x and CO are eliminated together by a redox reaction on a rhodium catalyst. NO_x oxidizes CO to CO_2, and is reduced to harmless nitrogen gas (N_2):

$$2NO\,(g) + 2CO\,(g) \xrightarrow{\text{rhodium}} N_2(g) + 2CO_2\,(g) \qquad \text{(reduction and oxidation)}$$

CO and C_xH_y are oxidized by air on a platinum catalyst. (C_7H_{16} = unburned hydrocarbon):

$$2CO\,(g) + O_2\,(g) \xrightarrow{\text{platinum}} 2CO_2(g) \qquad \text{(oxidation)}$$

$$C_7H_{16}\,(g) + 11O_2\,(g) \xrightarrow{\text{platinum}} 7CO_2\,(g) + 8H_2O\,(g) \qquad \text{(oxidation)}$$

The overall result is to convert the harmful CO, NO_x, and C_xH_y to relatively harmless N_2, CO_2, and H_2O. Though the exhaust gases are in contact with the catalyst for only 100 to 400 milliseconds, 96% of the hydrocarbons and CO are converted to CO_2 and H_2O. The emission of nitrogen oxides is reduced by 76%.

RESULTS OF EMISSION IMPROVEMENTS

Model Year	Emission/mile		
	CO	C_xH_y	NO_x
1970	34	4.1	4.0
1980	7.0	0.41	2.0
1983	3.4	0.41	1.0
1992	3.4	0.25	0.4

13. Modify building codes to control materials used in construction and to ensure proper ventilation in buildings.
14. Develop and encourage the use of efficient mass transit.
15. Include environmental costs in the pricing of energy resources and other activities that produce atmospheric pollution.

HUMAN IMPACT ON ATMOSPHERE

Cause	Effect
Burning fossil fuels	Adds CO_2 and O_3 to troposphere. Global warming. Altering climates. Produces acid rain. Increases acidity of lakes, streams, soil, harms wildlife, and damages structures.
Using nitrogen fertilizers and burning fossil fuels	Releases NO, NO_2, N_2O, and NH_3 into troposphere. Produces acid rain. Increases acidity of lakes, streams, soil, harms wildlife, and damages structures.
Refining petroleum and burning fossil fuels	Releases SO_2 into troposphere.
Manufacturing	Releases toxic heavy metals (Pb, Cd, and As) into troposphere.

INDOOR AIR POLLUTANTS

Type	Source	Effects on Humans
1,1,1-Trichloroethane	Aerosol sprays.	Dizziness, breathing irregularities.
Asbestos	Pipe insulation, ceiling materials, floor tiles, oven mitts, and hot pads.	Lung cancer and asbestosis.
Benzo-α-pyrene	Tobacco smoke, woodstoves.	Lung cancer.
Carbon monoxide	Faulty furnaces, stoves, fireplaces and vents. Cigarette smokers often have concentration of CO(g) in blood up to 10%.	Headache, heartbeat irregularity, death. Affinity of CO to hemoglobin is 250 times greater than oxygen itself—carboxyhemoglobin.
Chloroform	The amount of chloroform normally expected to be present in air ranges from 0.02 to 0.05 ppb (0.1 to 0.2 $\mu g/m^3$). Sources of chloroform in the atmosphere are pulp and paper mills and water and wastewater plants that use chlorine as a disinfectant. Chloroform appears to be a transported pollutant. Atmospheric half-life of 2 to 3 months.	Cancer.
Formaldehyde	Paneling, particleboard, furniture, foam insulation, carpeting, adhesives. The atmospheric half-life of formaldehyde is short, 4 to 10 hours.	Nausea, dizziness, irritation of throat, eyes, and lungs.
Methylene chloride	Paint strippers and paint thinner. Fairly long atmospheric half-life (3 to 4 months), indicating the fairly long persistence typical of transported pollutants.	Nerve disorder, diabetes.
Nitrogen oxides	Faulty furnaces, stoves, fireplaces, and vents.	Headaches, irritated lungs.
Para-dichlorobenzene	Air fresheners, mothballs.	Cancer.
Radon-222	Soil and rock near house foundation. Concrete blocks and cement.	Lung cancer.
Styrene	Carpets, plastics. Short atmospheric half-life (6 to 7 hours).	Kidney and liver damage.
Tetrachlorethylene	Dry-cleaning fluid on clothes. Atmospheric half-life is 70 to 100 days.	Nerve disorders, damage to liver and kidneys, cancer.
Tobacco smoke	Cigarettes and other smoking sources.	Lung cancer, heart disease.
Organic material	Dust mites, fungal and algal spores, dust, animal dander, hair, carpet fibers, fur.	Allergies, coughs, sneezing, eye irritation, sore throats, difficulty breathing.

COMMON AIR POLLUTANTS

Carbon Dioxide (CO_2)
Properties: Colorless, odorless gas. 0.0036% of atmosphere.
Effects: Impedes oxygen binding with hemoglobin. Slows reflexes, headaches, nausea, visual impairment, affects development of fetus, impairs breathing. Hazardous to people who have history of circulatory problems.
Sources: Incomplete combustion of fossil fuels and incinerators. 60% comes from auto exhaust; in cities, may be up to 95% of source.
Class: Carbon oxides. Examples: CO_2, CO. EPA standard is 9 ppm maximum concentration. 5.5 billion tons of carbon enter the atmosphere each year.

Lead (Pb)
Properties: Grayish metal.
Effects: Accumulates in tissue (blood, bone, etc.). Affects kidneys, liver, and nervous system, children are most susceptible. Mental retardation. Possible carcinogen. Affects enzymes and energy transfer reactions. May contribute to hypertension. 20% of all children living in the inner city have high lead levels in their blood.
Sources: Particulates, smelters, batteries, fuel in poorly developed countries
Class: Toxic metals. EPA national air standard is 1.5 micrograms per cubic meter ($\mu g/m^3$). Examples: Pb, Cd, Hg, Ni, Be, Tl, U, Cs, and Pu. 2 million metric tons of lead enter the atmosphere each year.

Nitrogen Dioxide (NO_2)
Properties: Reddish-brown gas that forms when fuel is burned at high temperatures. Strong oxidizing agent. Forms nitric acid in the air.
Effects: Produces acid rain (HNO_3) and causes eutrophication of coastal waters. Lung irritation and heart problems. Penetrates deeply into smaller airways and lung parenchyma, causes pulmonary edema at high concentrations. Damages structures. Decreased visibility due to yellowish color of NO_2. Suppresses plant growth. Decreased resistance to infection. May encourage the spread of cancer.
Sources: Combustion of fossil fuels. Power and industrial plants. Bacterial action in soil. Forest fires. Volcanic action. Lightning
Class: Nitrogen Oxides (NO_x) Examples: nitric oxide (NO), nitrogen dioxide (NO_2), nitrous oxide (N_2O). EPA air quality standard for NO_2 is 0.053 ppm.

Ozone (O_3)
Properties: Colorless. Unpleasant odor. Major component of photochemical smog.
Effects: Lung irritation (bronchial constriction, coughing, wheezing). Damages plants, rubber, fabrics, plants. Eye irritation. Decreased crop yields. 0.1 parts per million can reduce photosynthesis by 50%. Damages plastics and breaks down rubber. Harsh odor.
Sources: Created by sunlight acting on NO_x and VOC in the air. Cars. Industry. Gasoline vapors, chemical solvents, combustion products of fuels.
Class: Photochemical oxidants. Examples: ozone (O_3), peroxyacyl nitrates (PANs), hydrogen peroxide (H_2O_2), aldehydes (R-CHO).

Sulfur Dioxide (SO_2)
Properties: Colorless gas. Irritating odor.
Effects: Produces acid rain (H_2SO_4). Breathing difficulties. Precursor to sulfate formation—responsible for eutrophication of lakes and streams. SO_2 reduces the growth

of many plants. The tolerance of lichen and moss species to sulfur dioxide makes them an indicator species for measuring sulfur dioxide pollution.
Sources: Burning high-sulfur coal or oil. Smelting of metals. Paper manufacturing.
Class: Sulfur oxides. Examples: sulfur dioxide (SO_2), sulfur trioxide (SO_3). EPA's health-based national air quality standard for SO_2 is 0.03 ppm (measured on an annual arithmetic mean concentration) and 0.14 ppm (measured over 24 hours). SO_2 combines with water and ammonia in the soil to increase soil fertility and it acts as a fungicide.

Suspended particulate matter (PM_{10})
Properties: Particles of various sizes that remain suspended in air, (e.g., smoke, dust, haze). Particles with a diameter of 10 μm or less (0.0004 inches or one seventh the width of a human hair). EPA's health-based national air quality standard for PM_{10} is 50 μg/m^3 (measured as an annual mean) and 150 μg/m^3 (measured as a daily concentration).
Effects: Lung damage and irritation. May be mutagen, teratogen, and/or carcinogen. Elderly, children, and people with chronic lung disease are most susceptible. Particles smaller than 2.5 μm are the most dangerous. The World Bank has estimated that a given reduction in particulate matter smaller than 10 μm would produce health benefits 10 times greater than similar reductions in all other pollutants combined.
Sources: Burning coal. Burning diesel. Volcanoes. Factory exhaust. Unpaved roads. Plowing. Burning fields. Lint. Pollen. Spores.
Class: SPM. Examples: dust, soot, asbestos, lead, various nitrate and sulfate salts, PCBs, dioxins, pesticides.

Radon-222
Class: Radioactive substances. Examples: Radon-222, iodine-131, strontium-90, plutonium-239.

VOCs (Volatile Organic Compounds) 600 VOCs have been measured in air samples.
Properties: Organic compounds that have a high vapor pressure (evaporate easily). Usually aromatic-ring compound.
Effects: Eye irritation. Respiratory irritation. Carcinogenic. Decreased visibility due to blue-brown haze. Liver damage. Central nervous system damage. Kidney damage. Damages plants (formaldehyde and ethylene)
Sources: Evaporation of solvents. Evaporation of fuels. Incomplete combustion of fossil fuels. Naturally occurring compounds like terpenes from trees. Plants are largest source of VOCs! Aerosols. Paint thinners. Dry cleaning
Class: HAPs (Hazardous Air Pollutants). Examples: carbon tetrachloride (CCl_4), methyl chloride (CH_3Cl), chloroform ($CHCl_3$), benzene (C_6H_6), ethylene dibromide ($C_2H_2Br_2$), formaldehyde (CH_2O_2). Concentration indoors may be up to 1000 times higher than outdoors. 600 million tons of CFCs enter the atmosphere each year. Most air toxics are VOCs.

FACTORS THAT INFLUENCE FORMATION OF AIR POLLUTION AND ITS INTENSITY

- Local climate (inversions, air pressure, temperature, and humidity);
- Topography (hills and mountains);
- Population density;
- Amount of industry;
- Fuels used by population and industry for heating, manufacturing, transportation, and power;
- Rain;

- Snow;
- Wind;
- Presence of buildings (slows wind speed);
- Amount and type of mass transit;
- Economics.

FORMATION OF INDUSTRIAL SMOG

Procedure	Chemical Reaction
1. Carbon in coal or oil is burned in oxygen to produce carbon dioxide and carbon monoxide gas.	$C + O_2 \rightarrow CO_2$ $C + O_2 \rightarrow CO$
2. Unburned carbon ends up as soot.	C
3. Sulfur in oil and coal reacts with oxygen gas to produce sulfur dioxide.	$S + O_2 \rightarrow SO_2$
4. Sulfur dioxide reacts with oxygen gas to produce sulfur trioxide.	$SO_2 + O_2 \rightarrow SO_3$
5. Sulfur trioxide reacts with water vapor in air to form sulfuric acid.	$SO_3 + H_2O \rightarrow H_2SO_4$
6. Sulfuric acid reacts with atmospheric ammonia to form brown, solid ammonium sulfate.	$H_2SO_4 + NH_3 \rightarrow (NH_4)_2SO_4$

FORMATION OF PHOTOCHEMICAL SMOG

Time	Description
6–9 A.M.	As people drive to work, concentrations of nitrogen oxides and VOCs increase: $N_2 + O_2 \rightarrow 2NO$ $NO + VOC \rightarrow NO_2$ $NO_2 \xrightarrow{UV} NO + O$
9–11 A.M.	As traffic begins to decrease, nitrogen oxides and VOCs begin to react forming nitrogen dioxide: $2NO + O_2 \rightarrow 2NO_2$
11 A.M.–4 P.M.	As the sunlight becomes more intense, nitrogen dioxide is broken down and the concentration of ozone increases: $NO_2 \xrightarrow{UV} NO + O$ $O_2 + O \rightarrow O_3$ Nitrogen dioxide also reacts with water vapor to produce nitric acid and nitric oxide: $3NO_2 + H_2O \rightarrow 2HNO_3 + NO$
11 A.M.–4 P.M.	Nitrogen dioxide can also react with VOCs released by vehicles, refineries, gas stations, and the like to produce toxic chemicals such as PAN (peroxyacyl nitrates): $NO_2 + VOCs \rightarrow PANs$
4 P.M. to sunset	As the sun goes down, the production of ozone is halted. **Net Result: $NO + VOCs + O_2 + UV \rightarrow O_3 + PANs$**

FORMATION OF ACID DEPOSITION

Due to Sulfur Dioxide—70% of Source

1. Sulfur in the atmosphere comes from burning coal, smelting, organic decay, and ocean spray. 90%+ is from human sources.
2. Sulfur dioxide in atmosphere combines with water vapor to form hydrogen sulfite gas.

$$SO_2 + H_2O \rightarrow H_2SO_3$$

3. Hydrogen sulfite reacts with oxygen gas to form sulfuric acid (H_2SO_4).

$$H_2SO_3 + \frac{1}{2}O_2 \rightarrow H_2SO_4$$

Due to Nitrogen Oxides (NO_x)—30% of Source

1. Major sources of nitrogen oxides include combustion of oil, coal, natural gas, forest fires, bacterial action in soil, volcanic gases, and lightning-induced atmospheric reactions.
2. Nitrogen monoxide reacts with oxygen gas to produce nitrogen dioxide gas.

$$NO + \frac{1}{2}O_2 \rightarrow NO_2$$

3. Nitrogen dioxide reacts with water vapor in the atmosphere to produce hydrogen nitrite and hydrogen nitrate.

$$2NO_2 + H_2O \rightarrow HNO_2 + HNO_3$$

EFFECTS OF ACID RAIN

Effects of Acid Rain on Aquatic Ecosystems

1. May increase amount of plankton and moss.
2. Bottom of lakes and streams may accumulate undecayed organic material.
3. Streams, ponds, or lakes on acidic or neutral bedrock are sensitive to acidic deposition. Over time, heavy metals such as mercury, aluminum, and cadmium begin to leach from bedrock. Al^{3+} ions cause fish to produce excess mucous resulting in death. Streams, ponds, or lakes that are above bedrock that contain calcium and/or magnesium (limestone) are naturally buffered and may resist some effects of acid deposition.
4. Along with gases emitted by burning coal that contribute to acid rain, mercury may also be contained in effluent. Mercury is converted to methylmercury in aquatic ecosystems and is absorbed and accumulated into fatty tissues of fish. Humans eating fish suffer from nervous system and kidney ailments.
5. Acid shock (rapid melting of snow pack that contains acidic particles) results in acidic concentrations 5 to 10 times higher than acidic rainfall. Eggs and small fry are most sensitive to acid shock, pH < 5.0.
6. Acids also kill fish by altering body chemistry, reducing oxygen uptake, and disrupting muscle contractions.

Effects of Acid Rain on Soils and Plants

1. Leach essential plant nutrients from soil such as calcium, potassium, and magnesium causing a decline in plant growth.
2. Pb^{2+}, Cd^{2+} and Hg^{2+} may be leached from minerals and absorbed by plants that are harmful to the plant and harmful to consumers who eat the plant.
3. Al^{3+} ions leached from soil damages roots and can cause problems in uptake of essential nutrients (magnesium and potassium) and water by plants.

4. Mosses that thrive in acidic soil multiply and (a) retain water that kills feeder roots, (b) eliminate air from soil, and (c) kill mycorrhizal fungi that aid in the absorption of nutrients by roots.
5. Reduces health of trees and vegetation (including germination of seeds and young seedlings) so that they are susceptible to insects, diseases, drought, and severe weather.
6. High concentrations of nitric acid can increase the availability of nitrogen in the soil and reduce the availability of other nutrients necessary for plant growth. As a result, the plants become overfertilized by nitrogen (a condition known as nitrogen saturation).

Effects of Acid Rain on Humans

1. Toxic metals are released into environment and are ingested through water supplies, accumulation in fish and crops.
2. Effects on the fishing industry—commercial and sport.
3. Acid rain destroys statues, headstones, buildings, fountains, paint on cars—almost anything made of metal or stone.

PREVENTION AND CLEAN-UP OF THE EFFECTS OF ACID RAIN

1. Design more efficient engines that reduce NO_x emissions.
2. Increase efficiency of plants that burn coal to reduce SO_2, NO_x, and particulate emission.
3. Increase penalties on industries that do not meet air pollution guidelines.
4. Increase tax incentives to industries that do meet air pollution guidelines.
5. Increase funding in changing to less-polluting energy sources (e.g., tax incentives to buyers of hybrid vehicles).
6. Provide tax incentives to companies that use less-polluting sources of energy.
7. Add $CaCO_3$ (lime) to lakes that are suffering from acid deposition, although this is expensive and only temporary and has environmental impacts.

SPECIFIC AIR-POLLUTION TREATMENT TECHNOLOGY

Traditional ways to deal with air pollution were (1) to move the factory to a remote location or (2) to build a tall smokestack and have the wind blow the pollution somewhere else.

Biofiltration. Vapor-phase organic contaminants are pumped through a soil bed and filter to the soil surface where they are degraded by microorganisms in the soil.

High-energy destruction. The high-energy destruction process uses high-voltage electricity to destroy VOCs at room temperature.

Membrane separation. This organic vapor/air separation technology involves the preferential transport of organic vapors through a nonporous gas separation membrane (a diffusion process analogous to putting hot oil on a piece of waxed paper).

Oxidation. Organic contaminants are destroyed in a high-temperature combustor—1000°C (1832°F). Trace organics in contaminated air streams are destroyed at lower temperatures—450°C (842°F)—than conventional combustion by passing the mixture through a catalyst.

Vapor phase carbon adsorption. Off-gases are pumped through a series of canisters or columns containing activated carbon to which organic contaminants adsorb. Periodic replacement or regeneration of saturated carbon is required.

Electrostatic precipitators. Fly ash particles are attracted to electrostatically charged surfaces. Performance depends on particle size and chemistry, strength of the field, and gas velocity.

Sulfur removal. Using low sulfur fuel. Removing sulfur from fuel before burning. Mixing crushed limestone with coal before it is burned. Crushed limestone, lime slurry, or alkali can be injected into effluent gas to remove sulfur. Fluidized bed combustion suspends solid fuels on upward-blowing jets of air during the combustion process. The result is a turbulent mixing of gas and solids. The tumbling action, much like a bubbling fluid, provides more effective chemical reactions and heat transfer.

Nitrogen oxide control. Using staged burners where flow of air and fuel are controlled. Can reduce NO_x formation as much as 50%. Using catalytic converters. Raprenox (rapid removal of NO_x) injects nontoxic cyanuric acid into the exhaust and drives a reaction that breaks NO_x down into benign constituent elements by use of high temperature rather than a chemical catalyst.

Hydrocarbon control. Closed systems that prevent escape of fugitive gases through leaks prior to treatment. Use of afterburners. Use of PCV (positive crankcase ventilation systems) in cars—uses vacuum to draw vapors from the crankcase into the intake manifold. Vapor is then carried with the fuel/air mixture into the combustion chambers where it is burned.

Water

- Water costs are slightly more than $2 per 1000 gallons. Treatment accounts for about 15% of that cost.
- Because of high levels of toxic pollutants in the United States, 44% of lakes, 37% of rivers, and 32% of estuaries are unsafe for recreational activities.
- Pollutants include biological (pathogens); chemical, both water soluble and insoluble (heavy metals, nutrients, pesticides, wastes); and physical (sediments, radioactive materials, and heat).
- Pathogenic organisms are a major water pollutant. 2.5 billion people lack adequate sanitation, with about half of these lacking access to clean drinking water. Presence of *E. coli* (bacteria that live in intestines) is an indicator of waste products in drinking water. 1.5 million Americans become ill each year due to fecal contamination in drinking water.
- About 75% of water pollution in the United States comes from soil erosion, atmospheric deposition, and surface runoff (e.g., urban, feedlots, and agricultural).
- Less-developed countries discharge ~95% of all sewage directly into rivers, lakes, or the ocean. Reasons include lack of money, other priorities for money that is available, rapid urbanization, explosive population growth, and shift of industry by more-developed countries for cheaper labor.
- About 50% of the people in the United States depend on groundwater (95% in rural areas); 43% of U.S. agricultural water comes from groundwater. In the United States, 100,000 gasoline storage tanks are leaking chemicals into groundwater. 60% of the most hazardous liquid waste in the United States—34 billion liters per year of solvents, heavy metals, and radioactive materials—is injected directly into deep groundwater via thousands of injection wells. For example, in Denver, Colorado, 80 liters of several organic solvents contaminated 4.5 trillion liters of groundwater. Water that enters an

aquifer remains there for an average of 1400 years, compared to only 16 days for rivers. It is practically impossible to clean up groundwater pollution. Estimates from EPA are that 4.5 trillion liters of contaminated water seep into groundwater in the United States *each day.* Initial clean-up of contaminated groundwater at some 300,000 sites in the United States could cost up to $1 trillion over the next 30 years.

EFFECTS OF pH ON AQUATIC LIFE

pH	Effect
3.0–3.5	Toxic to most fish. Some plants and invertebrates can survive (waterbug, water boatmen, and white mosses).
3.5–4.0	Lethal to trout, whitefish, salmon, and smelt.
4.0–4.5	Harmful to trout, whitefish, salmon, smelt, bream, goldfish, and carp. Fish embryos die.
4.5–5.0	Harmful to fish eggs, fry, and carp. Lake is usually considered dead and unable to support a wide variety of life.
5.0–6.0	Ecology of the lake changes greatly. The number and variety of species begin to change. Salmon and minnow begin to become less diverse. Less diversity in algae, zooplankton, aquatic insects, and insect larvae. Rainbow trout do not occur, and mollusks become rare. Decline in trout, whitefish, salmon, and smelt fishing. High concentration of aluminum often present. Organic matter degrades slowly, and nutrients are trapped at the bottom and are not released back into the ecosystem. Green algae and diatoms disappear. The reduction in green plants allows light to penetrate further so acid lakes seem crystal clear and blue. Snails and phytoplankton disappear.
6.5–9.0	Harmless to most fish.
9.0–9.5	Harmful to trout, whitefish, salmon, perch, and smelt.
9.5–10.0	Slowly lethal to trout, whitefish, salmon, and smelt.
10.5–11.0	Lethal to trout, whitefish, salmon, smelt, carp, goldfish, and pike.
11.0–11.5	Lethal to all fish.

Drinking Water Treatment Methods

Adsorption. Contaminants stick to the surface of granular or powdered activated carbon.

Disinfection. Chlorine, chloramines, chlorine dioxide, or ozone.

Filtration. Removes clays, silts, natural organic matter, and precipitants from other treatment processes, using iron, manganese, and microorganisms. Filtration clarifies water and enhances the effectiveness of disinfection.

Flocculation-Sedimentation. Processes that combine small particles into larger particles, which settle out of the water as sediment. Alum, iron salts, or synthetic organic polymers (alone or in combination with metal salts) are generally used to promote coagulation.

Ion Exchange. Removes inorganic constituents if they cannot be removed adequately by filtration or sedimentation. Ion exchange can be used to treat hard water. It can also be used to remove arsenic, chromium, excess fluoride, nitrates, radium, and uranium.

Human Diseases Commonly Found in Polluted Water

Amoebic dysentery. Caused by the amoeba *Entamoeba histolytica.* Symptoms include diarrhea, often with blood and mucus, and abdominal pain.

Ancylostomiasis. Caused by the *Ancylostoma* worm. Symptoms include lung irritation, coughing, and severe anemia.

Cholera. Caused by the bacteria *Vibrio cholerae.* Symptoms include severe diarrhea, vomiting, abdominal cramps, and collapse.

Dysentery. Caused by the bacteria *Shigella dysenteriae.* Symptoms include diarrhea often accompanied by blood, abdominal pain, and cramps.

Enteritis. Caused by the bacteria *Clostridium perfringens.* Symptoms include loss of appetite, abdominal cramps, and diarrhea.

Infectious hepatitis. Caused by the Hepatitis A virus. Symptoms include inflammation of the liver, fever, nausea, vomiting, muscle aches, and loss of appetite.

Polio. Caused by the poliovirus. Symptoms include sore throat, fever, diarrhea, muscle aches, and eventually paralysis and atrophy of muscles.

Schistosomiasis. Caused by the *Schistosoma* fluke. Occurs primarily in the tropics with symptoms including diarrhea, weakness, blood in the urine, abdominal pain, and weakness.

Typhoid. Caused by the bacteria *Salmonella typhi.* Symptoms include fever, rash, and with advanced cases, hemorrhaging from the colon.

Common Water Pollutants

Bacteria. Fecal coliform bacteria is an indicator of human waste contamination in a water supply. See "Human Diseases Commonly Found in Polluted Water."

Heavy Metals. Heavy metals are naturally found in bedrock or sediment, or they can be introduced into water from industrial sources. The heavy metals are either consumed directly by humans through drinking water or ingested with organisms that have accumulated the metals into their tissues. The most famous case occurred in Japan when an industrial company dumped 27 tons of mercury-containing compounds into Minamata Bay in Japan from 1932 until 1968. The mercury was converted to methyl mercury and was ingested when people ate fish and shellfish from the bay. Symptoms included blurred vision, hearing loss, loss of muscular coordination, and congenital reproductive effects. 43 people died.

- Mercury (Hg)—leaching of soil due to acid rain, burning coal, industrial, household and mining wastes. Damage to nervous system, kidneys, vision, and congenital defects. Minimata disease.
- Lead (Pb)—paint, mining wastes, incinerator ash, water from lead pipes and solder, automobile exhaust. Damage to kidneys, nervous system (brain), and the ability to learn, depressed biosynthesis of protein, nerve and red blood cells, anemia, and irritability.
- Cadmium (Cd)—electroplating, mining, and plastic industries. Also found in sewage. Kidney disease.
- Arsenic (As)—found in herbicides and wood preservatives. A discharge of the mining industry. Damage to skin, eyes, GI tract, and liver. Cancer.
- Aluminum (Al)—leaching due to acid deposition. Anemia, loss of bone strength, and claims of a role in dementia and Alzheimer's disease.

Heat (thermal). From industrial and power plants. Reduces ability of water to hold dissolved oxygen. Causes death to organisms that cannot tolerate heat.

Nutrients. Primarily nitrogen and phosphorus. Sources include animal wastes, agricultural runoff, and sewage. When these nutrients enter a body of water they cause eutrophication—there is a large increase in photosynthetic and blue-green algae, which blocks light and decreases photosynthesis. As aquatic plants die, the biological oxygen demand (BOD) is increased causing fish and other organisms to die. If the source is from human sewage, the process is called cultural eutrophication.

Oil. Oil is introduced into aquatic (usually marine) environments through leaks from oil tankers or dumping oil down storm drains. Between 3 million and 6 million metric tons of oil are discharged into the world's oceans each year from land and sea operations. In 1989, the oil tanker *Exxon Valdez* spilled 11 million gallons of oil in Alaskan waters. The environmental damage was estimated to be over $15 billion. Up to 300,000 birds, 2500 sea otters, and untold number of other marine organisms were killed.

Organic material. Common sources are sewage, agricultural wastes and runoff, storm runoff, and so on. Organic matter is a source of nutrition for aerobic-decomposing bacteria in water. As bacteria decompose matter, they consume dissolved oxygen that results in a higher BOD. As the oxygen levels decrease, organisms that inhabit the body of water and depend on the oxygen die.

Radioactive wastes. Oceans become contaminated with nuclear waste by illegal dumping, atomic bomb tests, and accidents aboard nuclear powered vessels or at nuclear reactors near coastlines.

Sediment. The most significant water pollutant. Sources are runoff by human disturbances to natural habitats through forestry, clearing land, and agricultural and hydroelectric projects. Sediment chokes and fills lakes, reservoirs, harbors, and the like.

Water-Treatment Remediation Technologies

Adsorption/absorption. Solutes concentrate at the surface of a sorbent, thereby reducing their concentration in the bulk liquid phase.

Aeration. The area of contact between water and air is increased, either by natural methods or by mechanical devices.

Air stripping. VOCs are separated from extracted groundwater by increasing the surface area of the contaminated water exposed to air utilizing aeration methods.

Bioreactors. Contaminants in extracted groundwater are put into contact with microorganisms in attached or suspended-growth biological reactors for breakdown.

Cometabolic processes. Injection of a dilute solution of toluene or methane into the contaminated groundwater zone to support the co-metabolic breakdown of targeted organic contaminants.

Constructed wetlands. Uses natural geochemical and biological processes inherent in an artificial wetland ecosystem to accumulate and remove metals and other contaminants from contaminated water. See the discussion of the Living Machine in Case Studies in this chapter.

Deep well injection. Uses injection wells to place treated or untreated liquid waste into geologic formations that have no potential to allow migration of contaminants into potential potable water aquifers.

Directional wells. Wells are drilled horizontally, or at an angle, to reach contaminants not accessible by direct vertical drilling.

Enhanced bioremediation. The normal rate of bioremediation of organic contaminants by microbes is enhanced by increasing the concentration of electron acceptors and nutrients in groundwater, surface water, and leachate. Oxygen is the main electron acceptor for aerobic bioremediation. Nitrate serves as an alternative electron acceptor under low oxygen conditions.

Fluid/vapor extraction. High vacuum system is applied to remove liquid and gas simultaneously from low-permeability soil.

Granulated activated carbon (GAC)/liquid phase carbon adsorption. Groundwater is pumped through a series of columns containing activated carbon to which dissolved organic contaminants adsorb.

Hot water or steam flushing/stripping. Steam is forced into an aquifer through injection wells to vaporize volatile and semivolatile contaminants. Vaporized components rise to the unsaturated zone where they are removed by vacuum and then treated.

Hydrofracturing. Injection of pressurized water through wells causes cracks to develop in sediments. The cracks are filled with porous media that serve as substrates for bioremediation.

In well air stripping. Air is injected into a double-screened well, lifting the water in the well and forcing it out the upper screen. Simultaneously, additional water is drawn in the lower screen. Once in the well, some of the VOCs in the contaminated groundwater are transferred from the dissolved phase to the vapor phase by air bubbles. The contaminated air rises in the well to the water surface where vapors are drawn off and treated by a soil vapor extraction system.

Ion exchange. Removes ions from the aqueous phase by exchange with other ions.

Natural attenuation. Utilizes natural subsurface processes (dilution, volatilization, biodegradation, adsorption, and chemical reactions with subsurface materials) to reduce contaminant concentrations to acceptable levels.

Passive/reactive treatment walls. Uses barriers that allow the passage of water while causing the degradation or removal of contaminants.

Phytoremediation. Uses plants to remove, transfer, stabilize, and destroy organic/inorganic contamination in groundwater, surface water, and leachate.

Precipitation/coagulation/flocculation. Transforms dissolved contaminants into an insoluble solid, facilitating the contaminant's subsequent removal from the liquid phase by sedimentation or filtration. The process usually uses pH adjustment, addition of a chemical precipitant, and flocculation (clumping).

Separation. Concentrates contaminated wastewater through physical and chemical means.

Ultraviolet (UV) oxidation. Ultraviolet (UV) radiation, ozone, and/or hydrogen peroxide are used to destroy organic contaminants as water flows into a treatment tank.

Eutrophication

- Nitrates (NO_3^-) are water-soluble. Of the nitrates applied to fields as fertilizer, 82% (1) remain on field as accumulation; (2) leach into groundwater; (3) erode and end up eventually in surface runoff; or (4) volatilize into gas that enters the atmosphere and may end up contributing to acid rain or tropospheric ozone or acting as a greenhouse gas.
- Agriculture is the largest source of nonpoint water pollution in the United States.
- When nitrates become ingested (in drinking water), nitrate binds to hemoglobin and is converted to methemoglobin; the oxygen-carrying capacity of hemoglobin is reduced "nitrate poisoning." Nitrate poisoning is especially prevalent in amphibians and may be involved in the worldwide decline in biodiversity and numbers.
- Phosphates are also a component of fertilizer. They are not water-soluble. They adhere to soil particles. Since phosphates are not soluble, they accumulate in the soil and erode with soil into the oceans (see phosphate cycle, Chapter 3). Phosphates are accumulating in soils from fertilizer and animal wastes at a rate faster than they are removed from accumulating in crops or meat. Imbalance results in phosphate accumulating at a rate 75% higher than the pre-industrial period. Factors involved in soil erosion contribute to phosphate build up in streams, lakes, and wetlands.
- Nitrates and phosphates from human or animal sources is called cultural eutrophication. Up to 70% of nitrates and phosphates come from fertilizers and animal wastes. Sewage discharge (point source) also contributes to build-up of nitrates and phosphates. Nitrate build-up is more damaging in wetlands (nitrogen is the limiting factor), and phosphorus is more damaging in freshwater systems (phosphorus is the limiting factor).
- Nitrates and phosphates are algal nutrients. Increased concentrations of these nutrients increase the carrying capacity of lakes and streams. Explosion of algal population is called bloom. As a result, the following consequences occur:

  1. Increased algae decreases light penetration, killing off deeper plants and their supply of oxygen to water;

2. Oxygen concentration decreases in water due to action of decomposers with increased food supply;
3. Lower oxygen concentration causes fish to suffocate, and the decaying fish putrefy water for other organisms.
4. Decaying fish and some algal species produce toxins in water.

- Ponds, lakes, and reservoirs tend to stratify as a result of (1) light absorption, (2) limited vertical mixing, and (3) low replacement rates (time required to replace all the water with fresh inflowing water). The epilimnion (surface layer) is characterized by adequate light for plant growth, warmth due to solar insolation, and deficiency of plant nutrients, but it usually stays aerobic. The hypolimnion (lowest zone) is characteristically darker, cooler, and more nutrient-dense, but it becomes anaerobic due to consumption of oxygen by the "rain" of dead organic matter from the surface layer. Pollutants tend to accumulate in the hypolimnion and its sediments. Complete vertical mixing of the water column occurs only during the spring when the water is warming or in the autumn when the water is cooling.
- Methods to control eutrophication include (1) planting vegetation along streambeds to slow erosion and pick up some of the nutrients in their growth; (2) controlling application amount and timing of fertilizer; (3) controlling runoff from feedlots; and (4) researching the use of biological controls such as denitrification—specialized bacteria that convert nitrates to molecular nitrogen (N_2).

Soil

Pesticides

- Each year 2.5 million tons of 600 different types of pesticides are used for agricultural purposes.
- U.S. agriculture uses 77% of all pesticides. Between 1945 and 1990, pesticide usage increased 300% on American farms. Pesticides in 1990 were many times more toxic than those in 1945; yet, losses due to insects, plant pathogens, and weeds remained about the same (37%) as before the use of pesticides. It is estimated that 25 million farm workers are poisoned each year by pesticides resulting in over 220,000 deaths.
- North America uses about 30% of all pesticides; China and other developing nations use ~31%, whereas Europe uses ~27%, and Japan uses ~12%. In 1960, India used pesticides on 6 million hectares. By 1988, the acreage had increased to 80 million hectares using pesticides.
- Nonnative high-yield crops require more pesticides than native plants that have evolved natural adaptations for local pests and disease.
- Fruits and vegetables are labor-intensive. Workers are paid less in less-developed countries. Fruits and vegetables are a major import to the United States. Cosmetic appearance of fruit is important in the United States. Pesticides create "pretty" fruits and vegetables. Organically grown fruits and vegetables may not be as "pretty." There is less regulation in foreign countries on use of pesticides being used and residue levels. Pesticides affect laborers and also people who eat the fruits and vegetables. For example, 27% of grapes from Chile have fungicide residue compared to <1% for grapes grown in the United States. Additionally, 14,000+ truckloads of pesticide-tainted produce were stopped at the U.S.–Mexico border between 1985 and 1995. Most tainted

imported produce are apples, green beans, grapes, cucumbers, apricots, celery, strawberries, bell peppers, spinach, cherries, and peaches. Up to 13% of all imported produce into the United States may be tainted with pesticide levels above the legal limits (less than 1% of all produce is tested).

- Both the first and second green revolutions (see Chapter 8) required greater amounts of pesticides.
- The Environmental Protection Agency has ranked pesticide residue in foods as the third most serious cancer risk. More than 75 ingredients in pesticides are known carcinogens.
- UN Food and Agriculture Organization (FAO) reports nearly 500,000 tons of old and unused toxic pesticides have been abandoned on sites, mostly in the developing world (Africa, Middle East, Asia, Latin America, Eastern Europe, and the Soviet Union).
- Pests are naturally kept in check by predators, disease, genetic resistance of plants, adverse weather, and available food supply (host).
- Pesticides differ by: (1) their chemistry, (2) how long they remain effective in the environment, (3) how they accumulate in the food chain (bioaccumulation and biomagnification), (4) what type of organisms are affected, (5) how they work (affect nervous system, affect blood chemistry, etc.), (6) how fast they work, and (7) how they are applied.

PESTICIDE PROS AND CONS

Pros	**Cons**
Kill unwanted pests that carry disease (e.g., rats, mosquitoes).	Accumulate in the food chain.
Increase food supplies.	Pests develop resistance—500 species so far.
More food means food is less expensive.	Resistance creates pesticide treadmill.
Effective and fast-acting.	Estimates are $5–10 in damage done for each $1 spent on pesticide.
Newer pesticides are safer and more specific and genetically based.	Pesticides runoff and poison aquatic environments.
Reduces labor costs for farms.	Destroy bees—estimates are $200 million in losses due to reduced pollination.
Food looks better.	Threaten endangered species whose numbers are critical.
Agriculture is more profitable.	Affect birds by reducing eggshell thickness.
	Ineffectiveness in application—only 5% of pesticide reaches pest.
	Human and animal health risks—~20,000 human deaths worldwide per year.

Types of Pesticides

Biological. Living organisms that are used to control pests. Examples include bacteria, ladybugs, parasitic wasps, and certain viruses.

Carbamates. Urethanes that effect the nervous system of pests. Similar to organophosphates and include such chemicals as aldicarb (Temik), aminocarb (Zineb), carbaryl (Sevin), carbofuran (Baygon), and Mirex. Rate of application of aldicarb is 100 grams per hectare for same control as 2 kg DDT. Carbamates are more water-soluble than chlorinated hydrocarbons, resulting in potential for being dissolved in surface water and percolating into groundwater.

Chlorinated Hydrocarbons. Synthetic organic compounds that affect the nervous system of the pest. Examples include aldrin, chlordane, DDT, dieldrin, lindane, and paradichlorobenzene. Highly resistant to decomposition and can remain in ecosystems up to 15 years. During the 1950s and 1960s, several species of birds, including bald eagle, brown pelican, cormorant, osprey, and peregrine falcon, were severely affected by DDT by weakening the egg shells of these birds, thereby reducing the population. DDT rate of application was 2 kg per hectare.

Fumigants. Used to sterilize soil and prevent pest infestation of stored grain. Example: methyl bromide.

Inorganic. Broad-based. Uses poisons such as arsenic, copper, lead, and mercury. Highly toxic and accumulates in the environment.

Organic or Natural. Natural poisons derived from plants such as tobacco and chrysanthemum.

Organophosphates. Extremely toxic, but remain in the environment for only a short period of time. Examples include malathion, parathion, chlorpyrifos (Dursban, Trapper), acephate (Orthene, PT 280), propetamphos (Catalyst), and trichlorfon (Larva Lur).

Alternatives to Pesticides

- Using polyculture rather than monoculture;
- Intercropping—alternate rows of crops that do not have the same pests;
- Planting pest-repellent crops;
- Using mulch to control weeds;
- Using synthetic pyrethroids instead of toxic pesticides;
- Using natural insect predators (e.g., ladybugs, preying mantis, birds);
- Rotating crops often to disrupt insect cycles and numbers;
- Using of pheromones to attract insects into traps;
- Releasing of sterilized insects;
- Researching specific insect hormone interrupters;
- Developing genetically modified crops that are more insect-resistant.

Reasons Pests Have Increased Despite Increased Use of Pesticides

- Demand for blemish-free produce results in increased use of pesticides.
- Genetic resistance—natural selection for pesticide-resistance.
- Reduced crop rotation—does not disrupt insect life cycles.
- Increased mobility of pests due to travel—40% of insects, 40% of weeds, and 70% of plant pathogens in the United States are not native (e.g., Colorado potato beetle, and cheat grass).
- Conservation tillage—allows increases in pest population.
- Reduction in diversity of crops.

Pollutants

Evidence Required for Establishing Link to Pollutant

Concentration. Controlled experiments, statistical analysis, dose-response modeling, thresholds established. Do pollutants exist at levels to cause an observable effect? Residues found in tissue (fat soluble) or in soil consistent with history of exposure. Mathematical relationship(s) between concentration and effect(s). Is there a bioaccumulation effect?

Spatial consistency. Distance-effect relationship. Effects should normally decrease as distance from source increases. Other abiotic factors need to be analyzed—topography, winds, climate, rainfall patterns, etc.

Indicator species evidence. Quantitative data (quadrat analysis or sequential comparison index) on species whose status provides information on the overall condition of the ecosystem.

Mortality. Mortality rates can be directly related to a specific pollutant and are not the result of covariables.

Changes in species density and distribution. Pollution effects may be minimal but still cause change in population densities and distribution. Is there a mathematical relationship between history of species density and levels of pollution? Is there a change in the amount of plant groundcover over time correlated to pesticide use?

Changes in biomass. Data on changes in biomass over time. Can be measured through growth rates or photosynthetic activity (sugar analysis).

Temporal consistency. Issues relating to time between exposures to pollutant and time that observable evidence occurred. Pollen analysis (palynology), tree ring analysis (dendrochronology), soil or ice core samples can be used as historical background data.

Consistency between field and laboratory results. Results of pollution found in the field should match results obtained under controlled conditions. Cofactors should be ruled out. Difficult to match field conditions exactly. A particular nutrient level may be adequate under lab conditions but may be limited in field conditions, thereby causing different response mechanisms.

Mechanism. Physiological and biochemical pathways need to be demonstrated.

Historical consistency. Records of past studies are consistent with current observations.

Source of pollution. What is the source of the pollutant? Can it be analyzed from a point source or does it originate from nonpoint sources (ozone)?

Techniques for Studying Pollutants

Observational approach. Field studies. Establishment and identification of control areas or pollution gradients. Emphasis is to reduce variables between control area and area of study. Researcher attempts to draw conclusions from real-life conditions. Problems occur when external variables (sometimes unknown to the research) affect results and conclusions.

Experimental approach. Researcher manipulates variables. External co-factors are reduced as much as possible in an effort to establish causation. Weakness of approach is that co-factors and environmental interaction may have significant effects and that research projects are short-term whereas pollution effects are long-term.

Pollution Reduction, Remediation, and Control

- Advances and government support of removal technology.
- Government support of research.
- Financial incentives for pollution control and abatement.
- Tax on products that are not environmentally friendly.
- Education of the public on pollution and its consequences.
- Removal of subsidies and price supports for products and industries that pollute.
- True-cost pricing.
- Support of government treaties that address pollution.
- Government tax breaks for appliances and products that reduce pollution.
- Energy conservation—at home, at work, everywhere.
- Gasoline refueling instructions for efficient vapor recovery followed, being careful not to spill fuel and always tightening the gas cap securely.
- Car, boat and other engines kept tuned-up according to manufacturer's specifications.
- Proper tire inflation.
- Use of carpool, public transportation, bike, or walk whenever possible.
- Use of environmentally-safe paints and cleaning products whenever possible.

Case Studies

Bhopal, India (1984). In the early hours of December 3, 1984, gas leaked from a tank of methyl isocyanate at a plant in Bhopal, India, owned and operated by Union Carbide The Indian government reported that approximately 3800 persons died, 40 persons experienced permanent total disability, and 2680 persons experienced permanent partial disability.

Borneo and Interconnections. Malaria broke out in Borneo in 1950. The World Health Organization (WHO) sprayed large areas of Borneo with DDT to kill mosquitoes. DDT accumulated in cockroaches. Geckos (lizards) ate cockroaches. Geckos became sick. Cats ate geckos. Cats died. Rats increased. Sylvatic plague and typhus carried by rats increased. People became ill. DDT also killed parasitic wasp that infested caterpillars and kept caterpillar population in check. Caterpillars multiplied in thatch (straw). Thatch roofs fell in on people who were dying in their homes from plague and typhus. WHO dropped cats by parachutes onto island to restore order!

DDT. The first synthesized chlorinated organic pesticide (dichlorodiphenyltrichloroethane) was discovered in 1939 by Paul Müller (Nobel Prize 1944 for its discovery). It appeared to have low toxicity and was broad spectrum. It did not break down so it did not have to be reapplied often. It was water-insoluble (did not get washed away by rains) and was inexpensive. Crop production increased, mosquitoes decreased. In 1962, Rachel Carson published *Silent Spring,* which made connection between DDT and non-target organisms by (1) direct toxicity and (2) indirect toxicity (persistence in environment)—biomagnification (an increase in concentration up the food chain) and bioaccumulation (the tendency for a compound to accumulate in an organism's tissues). In Long Island Sound, concentration of DDT in the water was measured at 3 parts per trillion—at the top of the food chain, it had accumulated to 25 parts per million in fish-eating birds (a 10 million times increase). Furthermore, studies also showed reduction in calcium in eggshells, which caused eggs to crack, destroying chicks. It nearly wiped out bald eagles and peregrine falcons. DDT was showing up in Eskimos, seals, and human breast milk. It was pulled off U.S. markets in 1972. It is now being manufactured in Indonesia.

Donora, Pennsylvania (1948). Between October 26 and 31, 20 people were asphyxiated and died and over 7000 were hospitalized or became ill as the result of severe air pollution over Donora, Pennsylvania, a town of 14,000. The smog was a lethal mix of sulfur dioxide, carbon monoxide, and metal dust from steel mills in the vicinity, emissions from the local zinc works, pollutants from coke plants, and factories, and emissions from private, coal-burning homes that had become trapped in the narrow river valley by a thermal inversion. The investigation of this incident by state and federal health officials resulted in the first meaningful federal and state laws to control air pollution and marked the beginning of modern efforts to assess and deal with the health threats from air pollution.

***Exxon Valdez* 1989.** *Exxon Valdez* oil tanker hit rocks in Prince William Sound in Alaska, 1000 miles of shoreline were affected. At best, only 15% of oil from a spill can be recovered. Exxon initially spent $2.2 billion on cleanup, $1 billion in fines, and $5 billion in punitive damages. The total cost of spill is estimated to be $8.5 billion. Hot water high-pressure sprays used to clean rocks destroyed habitats. The accident may have been prevented if the ship had a double hull. By 2015, all new tankers must be have a double hull.

Living Machine. The Living Machine is a whole systems' approach to treating wastewater. It is a solar-powered, accelerated version of the water treatment facilities found in mature natural systems incorporating helpful microbes, plants, snails and fish into diverse, self-organizing and responsive communities. Living Machines are site-specific, biological solutions that re-route waste streams into resources. At the Ethel M chocolate factory in Las Vegas, Nevada, (makers of M & Ms and Mars), up to 32,000 gallons of highly concentrated wastewater are diverted each day from conventional waste treatment channels. The process begins when the water is routed from a grease trap into an aerobic digester where aerated bacteria are introduced. From there, wastewater is processed through four aerobic reactors that contain aerators and are planted with a variety of aquatic plants. The air and plants create an environment that hosts microbial communities that digest waste and minimize sludge production. Aided by gravity, a clarifier then separates microbial life and any remaining solids from the treated water. Plants like duck weed grow on the surface to shade out the sun and keep algae from growing. All solids that settle out are pumped to an on-site reed bed where they are composted and later added to the garden. After additional ultraviolet disinfection, it is used to irrigate their garden. For tour information, visit www.ethelm.com. *(They also give you delicious free candy! Another good APES tour if you are in Las Vegas, is Cranberry World at the Ocean Spray bottling facility, where you will learn a lot about cranberry bogs and New England wetlands.)*

London Fog. Most of the "fog" in London was pollution, caused by local factories and open coal fires used to heat houses. In the early 1950s, London had also just replaced its electric tramcars with diesel buses. In 1952, a four-day London fog, trapped by the surrounding hills and a stagnant mass of warm air above it, killed roughly 4000 Londoners and was the beginning of air pollution reform in the United Kingdom. Parliament enacted the Clean Air Act in 1956, effectively reducing the burning coal.

MTBE—Reformulated Gasoline. The Clean Air Act (1990) required 17 of the nation's urban centers with the worst air quality to add 2% oxygenates to gasoline. Oxygenates reduce VOCs and other toxic emissions from gasoline. In addition, gasoline must produce 20% less air toxic emission in 2000 as compared to 1990 values. This is achieved by reducing the volatility of gasoline and the benzene content. Compliance and administration is handled through the Federal Reformulated Gasoline Program. The most common reduction oxygenates are MTBE (methyl tertiary butyl ether) and ethanol (ethyl alcohol or EtOH). Since its addition to gasoline, issues of health concerns and groundwater contamination from leaking gasoline tanks have caused states to request the phasing out of MTBE. An alternative to MTBE is ethanol, which is produced from corn. Proponents of ethanol point out that ethanol (1) does not have major pollution issues; and (2) will help farmers and agriculture since it is produced from corn. Critics of ethanol bring up requirements of massive government subsidies required. Common Cause has estimated that federal subsidies for ethanol production are close to $7 billion. Research is showing that decreases in ozone by using oxygenates may not be as optimistic as first thought.

Multiple-Choice Questions

1. ___________ contributes to the formation of ___________ and thereby compounds the problem of ___________.

   (A) Ozone, carbon dioxide, acid rain
   (B) Carbon dioxide, carbon monoxide, ozone depletion
   (C) Sulfur dioxide, acid deposition, global warming
   (D) Nitrous oxide, ozone, industrial smog
   (E) Nitric oxide, ozone, photochemical smog

2. Photochemical smog does NOT require the presence of

   (A) nitrogen oxides.
   (B) ultraviolet radiation.
   (C) peroxyacyl nitrates.
   (D) volatile organic compounds.
   (E) ozone.

3. A natural component of the atmosphere, it comprises about 0.036% by volume of the atmosphere and is produced by the decay of vegetation, volcanic eruptions, exhalations of animals, the burning of fossil fuels, and deforestation. What is the gas?

   (A) Carbon monoxide
   (B) Carbon dioxide
   (C) Nitrous oxide
   (D) Nitrogen dioxide
   (E) Methane

4. Which of the following steps is NOT involved in the production of industrial smog?

   (A) $C + O_2 \rightarrow CO_2$
   (B) $C + O_2 \rightarrow CO$
   (C) $S + O_2 \rightarrow SO_2$
   (D) $NO_2 \rightarrow NO + O$
   (E) $SO_2 + O_2 \rightarrow SO_3$

5. Household water is most likely to be contaminated with radon in homes

   (A) that are served by public water systems that use a groundwater source.
   (B) that are served by public water systems that use a surface water source.
   (C) that have private household wells.
   (D) that use bottled water.
   (E) that are served by water agencies that use ozone to disinfect the water.

6. Which reaction is not involved in the formation of acid precipitation?

(A) $O_3 + C_xH_y \rightarrow$ PANs
(B) $SO_2 + H_2O \rightarrow H_2SO_3$
(C) $H_2SO_3 + ½O_2 \rightarrow H_2SO_4$
(D) $NO + ½O_2 \rightarrow NO_2$
(E) $2NO_2 + H_2O \rightarrow HNO_2 + HNO_3$

7. Normal rainfall has a pH of about

(A) 2.3
(B) 5.6
(C) 7.0
(D) 7.6
(E) 8.3

8. According to the Environmental Protection Agency, about _____ of all commercial building in the United States are classified as "sick."

(A) 5%
(B) 10%
(C) 17%
(D) 26%
(E) 32%

9. In developing countries, the most likely cause of respiratory illness would be

(A) photochemical smog.
(B) industrial smog.
(C) cigarette smoke.
(D) particulates.
(E) asbestos.

10. Humans least susceptible to the effects of air pollution are

(A) newborns.
(B) children from ages 2 through 10.
(C) teenagers.
(D) adult males.
(E) elderly.

11. Acid precipitation, leaching out the metal ____________, causes fish and other aquatic organisms to die from acid shock.

(A) Al
(B) Pb
(C) Hg
(D) Cd
(E) Fe

12. Which air pollutant best illustrates the effectiveness of legislation?

(A) NO_2
(B) SO_2
(C) CO_2
(D) O_3
(E) Pb

13. A type of pesticide that is highly toxic and that, in most cases, remains in the environment a relatively short period of time would be

(A) inorganic.
(B) organic.
(C) an organophosphate.
(D) a fumigant.
(E) a chlorinated hydrocarbon.

14. An effective method to decrease the amount of pesticide use would include all the following EXCEPT

(A) using monoculture techniques.
(B) rotating crops.
(C) using pheromones.
(D) using polyculture techniques.
(E) using insect-resistant crops.

15. The diagram below shows the range of organisms found within certain sections of a river in an industrial area. Which section of the river most likely has the LOWEST level of dissolved oxygen?

Effects of Sewage Discharge in a River
Organisms commonly found in discharge zone

Sewage pipe

Clean zone	Decomposition zone	Septic zone	Recovery zone	Clean zone
Trout	Carp	Worms	Carp	Trout
Perch	Catfish	Fungi	Catfish	Perch
Carp	Few perch	Bacteria	Blue-green algae	Carp
Catfish	Blue-green algae		Green algae	Catfish
Green algae	Green algae			Green algae

Direction of water flow →

(A) Clean zone
(B) Decomposition zone
(C) Septic zone
(D) Recovery zone
(E) None of the above

Answers to Multiple-Choice Questions

1. **E**	6. **A**	11. **A**
2. **C**	7. **B**	12. **E**
3. **B**	8. **C**	13. **C**
4. **D**	9. **D**	14. **A**
5. **C**	10. **D**	15. **C**

Explanations for Multiple-Choice Questions

1. **(E)** As the *sunlight* (photochemical) becomes more intense, nitrogen dioxide is broken down into nitric oxide (NO) and oxygen atoms and the concentration of ozone increases:

$$NO_2 \xrightarrow{uv} NO + O \qquad O_2 + O \rightarrow O_3$$

2. **(C)** Nitrogen dioxide can also react with volatile organic compounds released by vehicles, refineries, gas stations, and the like to produce toxic chemicals such as PAN (peroxyacyl nitrates). PANs are products, they are not required (reactants).

$$NO_2 + VOCs \rightarrow PANs$$

3. **(B)** Exhalations of animals was the key in this question. Carbon dioxide is a product of cellular respiration.

$$C_6H_{12}O_6 + 6O_2 \rightarrow 6CO_2 + 6H_2O$$

4. **(D)** This step is present in the formation of photochemical smog.

5. **(C)** Radon is a naturally occurring radioactive gas that causes cancer and may be found in drinking water and indoor air. Some people who are exposed to radon in drinking water may have increased risk of getting cancer, especially lung cancer, over the course of their lifetimes. Based on a National Academy of Science report, EPA estimates that radon in drinking water causes about 168 cancer deaths per year (89% from lung cancer caused by breathing radon released to the indoor air from water and 11% from stomach cancer caused by consuming water containing radon). The Safe Drinking Water Act, as amended in 1996, requires the U.S. Environmental Protection Agency (EPA) to develop a regulation to reduce radon in drinking water. Radon in soil under homes is the biggest source of radon in indoor air and presents a greater risk of lung cancer than radon in drinking water. Radon in indoor air is the second leading cause of lung cancer, responsible for 20,000 deaths a year in the United States. As required by the Safe Drinking Water Act, EPA has developed a proposed regulation to reduce radon in drinking water that has a multimedia mitigation option to reduce radon in indoor air.

6. **(A)** $O_3 + C_xH_y \rightarrow PANs$. See *Formation of Acid Deposition* on page 336.

7. **(B)** Rain is naturally acidic because carbon dioxide, found normally in the Earth's atmosphere, reacts with water to form carbonic acid. While "pure" rain's acidity is pH 5.6–5.7, actual pH readings vary from place to place depending upon the type and amount of other gases present in the air, such as sulfur dioxide and nitrogen oxides.

8. **(C)** The World Health Organization estimates that up to 30% of office buildings worldwide may have significant problems, with 10 to 30% of the occupants of the buildings experiencing health effects that are, or are perceived to be, related to poor indoor air quality. In the United States, with more stringent guidelines, the figure is about 17%. Remediation of sick building syndrome includes: making sure the rate of fresh airflow is adequate, eliminating tobacco smoke, maintaining proper humidities, removing or not using sources of pollution (carpets, furniture, and paneling that releases formaldehyde), not using janitorial supplies or pesticides that contain allergens, and maintaining ductwork and filters to reduce molds, spores, fungi, bacteria, and viruses.

9. **(D)** Developing countries around the world are experiencing increased level of particulate matter air pollution as a result of rapid increases in energy consumption and motor vehicle use, a product of rapid population and economic growth.

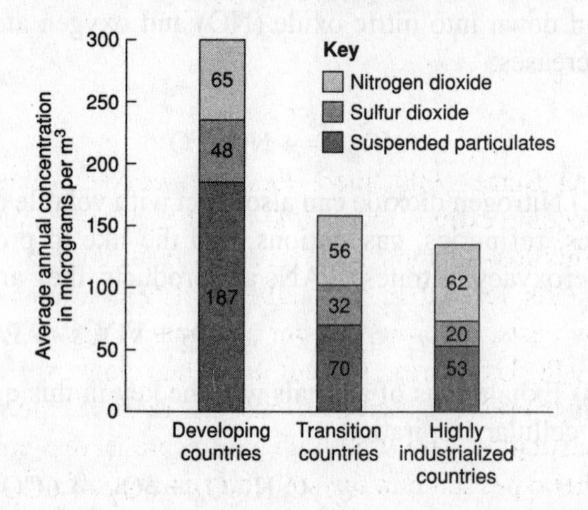

Particulate matter pollution in developing countries. *Source:* World Resources Institute.

10. **(D)** A recent study conducted by the American Lung Association shows that as many as 27.1 million children age 13 and under, and over 1.9 million children with asthma are potentially exposed to unhealthful levels of ozone based on a 0.08 ppm, eight-hour ozone level standard. Minority children are disproportionately represented in areas with high ozone levels. Approximately 61.3% of black children, 69.2% of Hispanic children and 67.7% of Asian-American children live in areas that exceed the 0.08 ppm ozone standard, while 50.8% of white children live in such areas. The elderly are also very susceptible to air pollution due to reduced lung function and other complicating medical conditions.

11. **(A)** There are two ways in which aluminum kills fish: (1) It is able to reduce the ion exchange through the gills and subsequently causes a salt depletion. Aluminum also precipitates in the gills and interferes with the transport of oxygen and other ions, so that the fish literally dies of suffocation. (2) The fish will exude mucus to combat the aluminum in their gills. This mucus builds up and clogs the gills so that oxygen and salt transport is inhibited. Research has shown that dead fish had low levels of Na^+ and Cl^- in their blood and were thus unable to regulate their body salts.

12. **(E)** Average levels of lead in the blood in the United States have fallen dramatically since the 1970s. In 1976–1980 the average amount in children was 15 μg/dL, whereas in 1991–1994, the average was 2.7 μg/dL. Legislation that banned leaded paint and leaded gasoline was responsible for the progress. However, some populations of

children continue to be disproportionately exposed to lead. In general, children who live in older housing are more likely to have elevated blood lead levels than the population of U.S. children as a whole. According to a national survey, from 1991–1994, 21.9% of black children ages 1 to 5 who were living in older housing had elevated blood lead levels (10 μg/dL or higher).

13. **(C)** Organophosphates are a group of closely related pesticides that affect functioning of the nervous system. Organophosphates are synthetic chemicals designed to be toxic, somewhat volatile, and only stable enough to remain in their toxic form for relatively short periods of time. Almost all organophosphates are esters of pentavalent phosphorous acids.

O or S
R—O
P—O—X
R—O

14. **(A)** Since 1900, more than 90% of vegetables and 80% of fruit varieties have become extinct. Over 6000 varieties of apples alone have become extinct since 1905. This loss of biodiversity results from the planting of monocultures that have high yields, less labor, and higher profits and that also may be more easily preyed upon by pests, requiring greater amounts of pesticides with diminishing effectiveness. In 1948, U.S. farmers used 50 million pounds of synthetic pesticides and lost 7% of their preharvest crops to insects. In 2002, more than 1 billion pounds of synthetic pesticides were used with a 13% preharvest crop loss—twenty times as much synthetic pesticide in the soil and running off into water supplies, while twice as many crops were destroyed by insects. Monoculture also results in soil depletion. Each year also results in 78 million more people without use of the 27 million tons of topsoil lost each year to runoff and degradation. The end result of decades of ever-increasing monocultural planting and pesticide use are (1) loss of farm acreage; (2) stronger, more adaptable pests; and (3) less adaptable crops due to lack of genetic variety.

15. **(C)** The level of dissolved oxygen in the septic zone is too low to support organisms that live by aerobic respiration. In this region, anaerobic organisms (especially bacteria) flourish and produce noxious waste products such as methane (CH_4) and hydrogen sulfide (H_2S).

Free-Response Question

by Dr. Ian Kelleher
Brooks School
North Andover, MA

(a) Study the following graph, which shows projected trends in annual CO_2 emissions, and then answer the following questions.

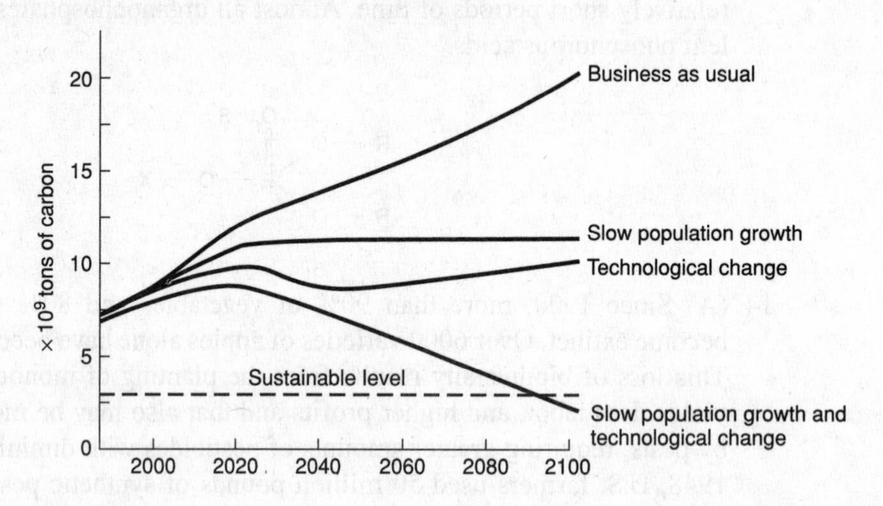

(i) In the "business as usual" model, what factors do you think might contribute to the increase in CO_2 emissions?

(ii) Given the shape of the graph, what do you think is meant in this case by "technological change"?

(iii) What is meant by a "sustainable level" of CO_2 emissions, and according to these predictions, what needs to happen for this level to be brought about?

(b) Use the example of acid deposition to illustrate the difference between remediation and alleviation of an environmental problem.

(c) Look at the graph of CFC production and account for the trends you observe.

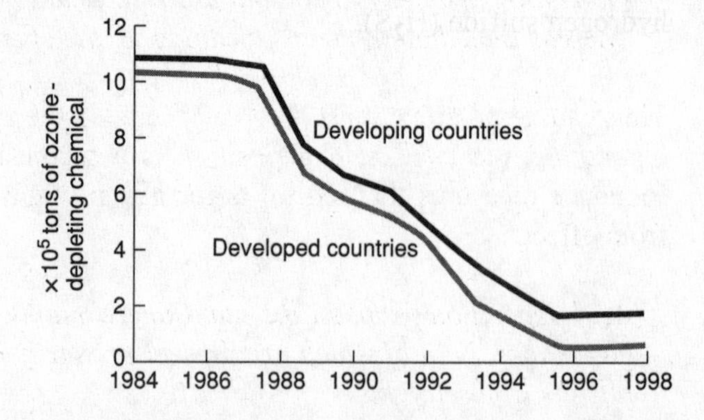

(d) In 1992, the World Bank designated indoor air pollution in developing countries as one of the four most critical global environmental problems. Use examples of major sources of indoor air pollutants to illustrate how the issue of indoor air quality in developing countries differs from that in developed countries.

Free-Response Answer

(a) (i) Cover all the relevant aspects of "business as usual," such as increasing size and development of populations.

Increased CO_2 emissions primarily come from increased burning of fossil fuels worldwide. The 'business as usual' model would include an increasing global population, which would mean an increase in the demand for energy. In the "business as usual" model, this need would primarily be met by increased use of fossil fuels, by far the most common source of energy in the world today. Increasing rates of development, particularly of developing countries, in this model would also lead to an increase in fossil fuel use.

(a) (ii) "Technological change" is a vague phrase that could mean many things. In this case, the curve thus labeled shows a dramatic reduction in CO_2 emissions, so the question is asking about technologies that reduce fossil fuel use.

"Technological change" reduces CO_2 emissions on the graph. Increased use of "alternative" sources of energy, such as nuclear power, hydroelectric, solar and tidal energy would all decrease the reliance on fossil fuels as the primary source of energy, and thus decrease fossil fuel emissions. Technologies that increased energy efficiency and saved power, such as more efficient car engines, would also decrease the use of fossil fuels and thus CO_2 emissions.

(a) (iii) A sustainable level of CO_2 emissions is one in which the amount absorbed by Earth's natural systems, such as oceans and plants, equals the amount released into the atmosphere. A combination of both technological change and slowed population growth is needed to bring CO_2 emissions to sustainable levels. As the graph shows, neither factor can produce sustainable levels on its own.

(b) Explaining the difference between these two similar terms would be a good way to start. These are basic terms that you should know. Starting like this should lessen the chance of mixing them up.

Alleviation of an environmental problem means stopping or lessening its cause, whereas remediation means cleaning up the effects of the problem.

Thus cause and effect must both be covered in the answer, and you must take care to distinguish the two. Acid deposition is a secondary pollutant, so explaining the sequence of events that led to its formation might be helpful to distinguish cause from effect.

Acid deposition forms in the atmosphere mainly from sulfur oxides and nitrogen oxides, both of which are present primarily as a consequence of the combustion of fossil fuels.

Anything that reduces the use of fossil fuels or limits the emission of sulfur and nitrogen oxides would be an example of alleviation. Anything that neutralizes the effects of acid in an environment would be an example of remediation. Be careful in giving just an appropriate level of detail.

Methods of alleviation may concentrate on decreasing the amount of sulfur and nitrogen oxide released into the atmosphere. Any method of reducing the rate of consumption of fossil fuels, such as increased use of nuclear power and "alternative" energy sources, or laws and education to help conserve energy, would decrease the amount of sulfur and nitrogen oxides released. "Clean fuel technologies," such as fluidized-bed combustion of coal, would result in less of these gases being produced on combustion. Scrubbers in smokestacks can remove much of what is produced. Perhaps the most important step in reducing emissions in the United States, however, was the passing of the Clean Air Act in 1970, and its subsequent amendments.

Methods of remediation may concentrate on neutralizing the acid deposited in an environment. Acids can be neutralized by adding a base. Since an abundant, low-cost, nontoxic material is often needed, limestone, $CaCO_3$, is commonly used. For example, limestone might be added to a lake to increase its pH. Powdered limestone could be spread over agricultural lands to increase the pH of the soil. Many nutrients are more soluble in acidic soils, and therefore might be washed away by rain so that the addition of fertilizer is also required. Deforestation caused by acid deposition may be addressed by treating the soil and then replanting.

(c) When asked to explain the trends shown in a graph, a good starting point is to describe what they are. Writing this down first should also help you compose your explanation.

Between 1984 and 1988 both developed and developing countries used ozone-depleting chemicals at a fairly consistent rate of approximately 1 million tons a year. After 1988, the amount used by both developed and developing countries decreased sharply, and at a fairly constant rate, for the next seven years. By 1996, developed countries used 50,000 tons a year and developing countries used 150,000 tons. Use remained at approximately these levels for the next two years.

This is perhaps more detail than necessary. The important point is that the amounts for both developed and developing countries fell dramatically at the same time, and basically at the same rate. This suggests that it comes as a result of the implementation of a new "law." Since we are dealing with a global situation, it is likely to be in the form of an international agreement.

The reduction is likely to be due to countries implementing technologies to comply with the Montreal Accord (1987), which set limits for the emission of chemicals that cause depletion of the ozone layer. The biggest reduction has come from using alternatives to CFCs, the principal ozone-depleting agent. Alternatives include HFCs and HCFCs.

(d) The wording of the World Bank designation suggests that the question is referring to the general masses of the population in developing countries rather than the small technologically-developed percentage. The answer is thus strongly focused on this difference in lifestyle.

Homes are perhaps the most important factor when considering indoor air quality since people tend to spend more time here than anywhere else. The majority of people in developing countries live in homes with much simpler technologies, so that many of the pollutants found in houses in developed countries are not present. The major source of indoor air pollutants, therefore, is the combustion of poor-quality fuels for heating, lighting, and cooking. Such dirty fuels include animal wastes, kerosene, and low-grade coal that may release large amounts of particulates, carbon monoxide, sulfur oxides, and other toxins on combustion. As fires are often burnt indoors in places with inadequate ventilation, the levels of these pollutants can be greatly concentrated.

Energy sources are much more technologically advanced in developed countries, so this is not such an important source of indoor air pollution. For example, poorly maintained furnaces may produce some carbon monoxide, but this problem is not nearly as widespread or generally as serious as the energy issue in developing countries. Instead, major sources of air pollution include lead (from lead paint), asbestos, and fumes from volatile organic compounds in paints, glues, plastic and furniture; these things will be less common in houses in developing countries. Houses in developed countries are often tightly sealed, more so than in developing countries. This leads to increased levels of concentration. For example, radon gas, which occurs naturally from radioactive decay in certain rocks, might accumulate to dangerous levels in a modern air-conditioned house, but not in a simpler hut with no glass in the windows.

CHAPTER 11

Solid Waste

The truth shall make you free, but first it shall make you angry.

Anonymous

Areas That You Will Be Tested On

A: Types, Sources, and Amounts
B: Current Disposal Methods and Limitations
C: Alternative Practices in Solid Waste Management

Key Terms

Note: Definitions for all key terms listed below can be found in the glossary starting on page 648. For additional definitions of relevant terms from this chapter, refer to *www.barronseduc.com/0764121618.html*.

backyard composting	**landfills**
biodegradable plastics	**low-waste society**
dioxins	**municipal solid waste**
heavy metals	**sanitary landfill**
incineration	**solid waste**
industrial waste	

Key Concepts

Types, Sources, and Amounts

- In 1998, approximately 220 million tons of *municipal* solid waste or garbage was generated in the United States. Each person generated an average of 4.5 pounds of solid waste per day: 24% of it is recycled or composted, but 76% is dumped in landfills or burned in incinerators (see Figure 11.1).
- The United States, with 5% of the world's population, generates 33% of the world's total waste, which amounts to 11 billion tons per year. This works out to approximately 10,000 pounds of waste per year for each American for *all* categories of waste.
- Six billion metric tons of solid *agricultural* wastes are generated each year in the United States, all of which can be recycled back into the soil.
- One third of *solid wastes* are mine-tailings and smelter residue.

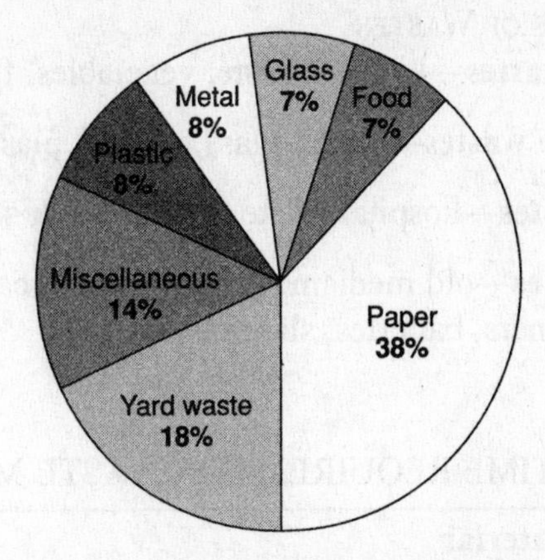

Figure 11.1. Amounts and types of municipal solid wastes (MSW) in the United States.

- There are 400 million metric tons of *industrial wastes* generated each year in the United States.
- There are 60 million metric tons of *hazardous materials* generated each year in the United States.
- There are 200 billion metric tons of metal, glass, and plastic food and beverage containers generated each year in the United States.
- Two hundred million liters of waste motor oil are dumped into sewers or onto the land each year in the United States; this is five times as much as that dumped by *Exxon Valdez.*
- One gallon of gasoline or oil can pollute one million gallons of freshwater.
- Each person in the United States averages ⅔ metric ton of waste per year—twice as much as Europe or Japan, five to ten times as much as in most developed countries.
- Since 1965, the amount of municipal solid waste each person creates has almost doubled from 2.7 to 4.5 pounds per day.
- Two-thirds of all aluminum cans are recycled.
- Minnesota has a 46% recycling rate, the highest rate in the United States.
- Two million trees a day are used for paper products.
- Each piece of litter picked up costs 32 cents.
- Of all heavy metal toxic wastes, 90% come from consumer electronics and batteries.

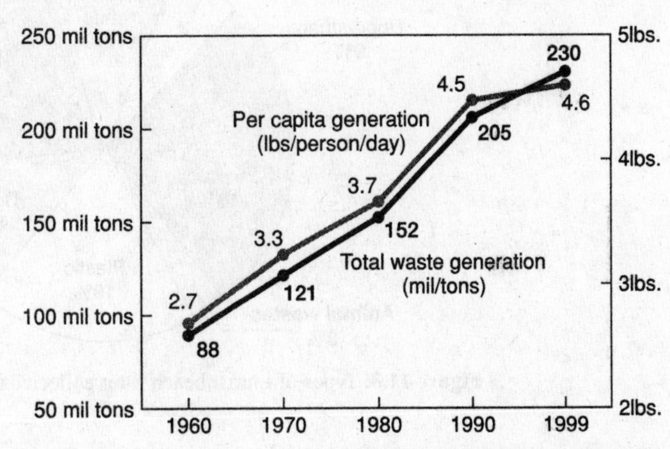

Figure 11.2. Trends in municipal solid waste generation, 1960–1999.

CATEGORIES OF WASTES

Organic wastes—kitchen waste, vegetables, flowers, leaves, fruits.

Recyclable wastes—paper, glass, metals, plastics.

Soiled wastes—hospital waste such as cloth soiled with blood and other body fluids.

Toxic wastes—old medicines, paints, chemicals, bulbs, spray cans, fertilizer and pesticide containers, batteries, shoe polish.

TIME REQUIRED FOR WASTE MATERIAL TO DECOMPOSE

Waste Material	Time
Organic waste such as vegetable and fruit peels and leftover foodstuff	1–2 weeks
Paper	10–30 days
Cotton cloth	2–5 months
Woolen items	1 year
Wood	10–15 years
Tin, aluminum, and other metal items such as cans	100–500 years
Plastic bags	one million years?
Glass bottles	Undetermined

Beach Litter

In 1995, a survey was conducted of beach litter collected on tourist beaches in Malta. The results of the survey are shown in Figure. 11.3.

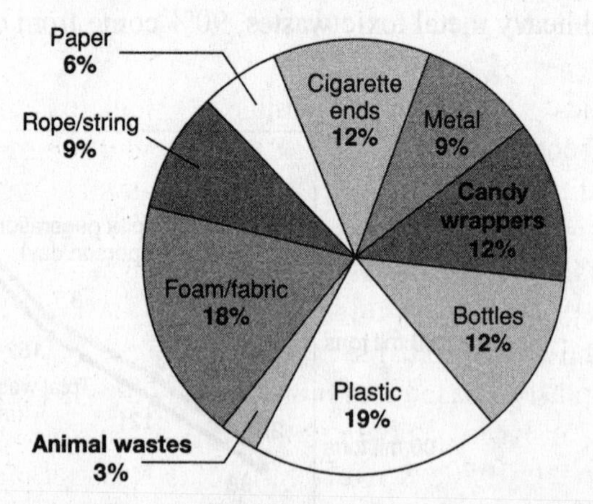

Figure 11.3. Types of tourist beach litter collected in a recent survey.

Biological Wastes

Biological wastes are generated during the diagnosis or treatment at hospital facilities, immunization of human beings or animals, research activities in these fields, or production or testing of pharmaceuticals. Biohazards may include wastes such as needles, soiled waste, disposables, anatomical wastes, cultures, discarded medicines, and chemical wastes that are in the form of disposable syringes, swabs, bandages, body fluids, human excrement, and the like. This waste is highly infectious and can be a serious threat to human health if not managed in a prescribed manner. It has been roughly estimated that for every 4 pounds of waste generated in a hospital, at least 1 pound is infected biological waste material.

Plastics

Sources of Waste Plastic

Household. Bags, bottles, containers, and trash bags.

Health Care. Disposable syringes, bottles, bags, blood and catheter bags, intravenous tubes, and surgical gloves.

Hotel and catering. Packaging items, water bottles, plastic plates, glasses, and spoons.

Travel. Bottles, plastic plates, glasses, spoons, and plastic bags.

- Plastics are light, strong, and most importantly from a waste perspective, inexpensive. The low cost has allowed plastic to invade every aspect of our daily lives. From an environmental perspective, most plastic decomposes at a very slow rate due to the fact that it does not decompose. Plastics are in all areas of human life—infrastructure, construction, agriculture, consumer goods, telecommunications, and packaging.
- Burning of plastics, especially PVC (polyvinyl chloride), releases dioxin and furan into the atmosphere. Dioxin, a highly-carcinogenic and toxic by-product of the manufacturing process of plastics, is one of the chemicals believed to be passed on through breast milk to nursing infants. Other research studies have shown dioxin to contribute to a decline in human sperm count and quality, genital abnormalities, and a rise in the incidence of breast cancer.
- Plastics (1) (bags) choke drains; (2) block the porosity of the soil and causes problems for groundwater recharge; (3) disturb the soil microbe activity; (4) once ingested, can kill animals; (5) (bags) can also contaminate foodstuffs due to leaching of toxic dyes and transfer of pathogens; and (6) get strewn on the ground, littered around in open drains, or in unmanaged garbage dumps and, because of slow decomposition, remain there piling up year after year.
- Partially biodegradable plastics have been developed and used; completely biodegradable plastics based on renewable starch rather than petrochemicals have only recently been developed and are in the early stages of commercialization.
- In the late 1970s, scientists from the National Marine Mammal Laboratory concluded that plastic entanglement was killing up to 40,000 seals a year. Annually, this amounted to a 4–6% drop in seal population. Over the last 30 years, a 50% decline in Northern Fur Seals has been reported.
- In 1975, the National Academy of Sciences estimated that 14 billion pounds of garbage (much of it plastic) was being dumped into the ocean every year. This was about 1.5

million pounds per hour. More than 85% of this trash was estimated to come from the world's merchant shipping fleet in the form of cargo-associated wastes. According to the academy, the United States was the source of approximately one-third of this ocean pollution. In 1987, The Marine Plastic Pollution Research and Control Act (MARPOL) was passed restricting the dumping of plastics into the ocean.

Figure 11.4. Sign commonly posted on California beaches.

- Plastic six-pack holders, plastic sandwich bags, Styrofoam particles, and plastic pellets are often mistaken by marine mammals, sea turtles, and other organisms as authentic food. The plastic blocks their intestines allowing them to miss out on vital nutrients, thus they starve to death. Seabirds are also affected by plastic in that they mistake small pieces of plastic for fish eggs, small crabs, and other prey, and even feed the pellets to their young. Only 0.05% of plastic pieces from surface waters are pellets, but they comprise 70% of the plastic eaten by seabirds. Small plastic particles have been found in the stomachs of 63 of the world's approximately 250 species of seabirds.
- *Photodegradable* plastics are made to become weak and brittle when exposed to sunlight for prolonged periods. Many states have passed laws requiring six-pack holders be biodegradable (look for a small diamond between the rings).
- *Biodegradable* plastics are made with cornstarch, so bacteria and other organisms eat away at the plastic, breaking it up into smaller pieces.

Important: Both photodegradable and biodegradable plastic only make the plastic smaller; they *do not* eliminate it. By being in smaller pieces, it *increases* the negative effect of plastic in the food chain because it affects *more* organisms lower in trophic levels.

Current Disposal Methods and Limitations

METHODS OF DISPOSING OF SOLID WASTES

Burning, incineration, or energy recovery

Pros:
- Heat can be used to supplement energy requirements.
- Reduces impact on landfills and other sources of storage.
- Mass burning is inexpensive.
- What is left over is 10–20% of original volume.
- United States incinerates about 15% of its wastes.
- France, Japan, Sweden, and Switzerland incinerate more than 40% of their municipal wastes and use the heat to generate electricity.

Cons:
- Air pollution including lead, mercury, nitrogen oxides, cadmium, sulfur oxides, hydrogen chloride, sulfuric acids, fluorides, and dioxins.
- Refuse-derived fuel (wastes such as batteries and plastics are sorted before it is burned) is expensive.
- Mass burn—everything is burned. No way of knowing toxic consequences.
- Ash is more concentrated with toxic material (dioxins, furans, lead, and cadmium). Initial costs of incinerators are high.
- Tipping fees are high.

Composting

Pros:
- Creates nutrient-rich soil additive.
- Aids in water retention.
- Slows down soil erosion.
- No major toxic issues to deal with.

Cons:
- Public reaction to odor, vermin, and insects.
- NIMBY—increased traffic and noise

Demanufacturing

Pros:
- Recovers materials that would have been discarded.
- Beneficial to inner cities as an industry because material is available and jobs are needed.

Cons:
- Toxic material may be present (CFCs such as Freon).
- Electronics contain heavy metals.

Detoxifying

Pros:
- Reduces impact on the environment.

Cons:
- Expensive.

Exporting

Pros:
- Gets rid of problem immediately.
- Source of income for poor countries.

Cons:
- Garbage imperialism or environmental racism.
- Long-term effects not known.

Land disposal—sanitary landfills

Pros:
- Waste is covered each day with dirt to help prevent insects and rodents.
- Plastic liners, drainage systems, and other methods help control leaching of material into groundwater.
- Geological studies and environmental impact studies are performed prior to building.
- Collection of methane and use of fuel cells to supplement energy demands.
- Use of anaerobic methane generators reduces dependence on other energy sources.

Cons:
- Rising land prices. Current costs are approximately $1 million per hectare.
- Transportation costs to the landfill.
- High cost of running and monitoring landfill.
- Legal liability.
- Suitable areas are limited.
- NIMBY (Not in My Backyard).
- Degradable plastics don't decompose completely—in breaking down, they release toxic chemicals. Don't degrade very much in sanitary landfills.

Land disposal—open dumping

Pros:
- Inexpensive.
- Provides a source of income to the poor by providing recyclable products to redeem.

Cons:
- Trash blows away in wind.
- Vermin and disease.
- Leaching of toxic materials into soil.
- Aesthetics.

Ocean dumping

Pros:
- Inexpensive.
- Illegal in the United States.

Cons:
- Debris floats to unintended areas.
- Marine organisms and food webs are impacted.

Recycling

Pros:
- Takes a waste and turns it into an inexpensive resource.
- Reduces impact on landfills.
- Reduces need for raw materials and the costs associated with it.
- Reduces energy requirements to produce product. Example: recycling aluminum cuts energy use 95%; producing steel from scrap reduces energy requirements 75%.
- Reduces dependence on foreign oil.
- Reduces air and water pollution.
- "Bottle bills" provide economic incentive to recycle.

Cons:
- Poor regulation—asphalt or concrete filler—what happens as material wears away? "Fertilizer supplement"—no regulation as to what the "supplement" is.
- Fluctuations in market price for commodities.
- "Throw-away packaging" more popular.
- Current policies and regulations favor extraction of raw materials. Energy, water, and raw materials sold below real cost to stimulate new jobs and economy.

Reuse

Pros:
- Most efficient method of reclaiming materials.
- Industry models already in place—auto salvage yards, building materials, etc.
- Refillable glass bottle can be reused 15 times. Saves money that would have been spent making new ones.
- Reusable diapers (diaper service) do not impact landfills.

Cons:
- Cost of collecting material on a large scale may be too high.
- Cost of washing and decontaminating containers may be prohibitive.
- Only when items are expensive and labor is cheap is it economical to reuse.

- The EPA has ranked the most environmentally sound strategies for MSW:

  1. Source reduction (including reuse).
  2. Recycling and composting.
  3. Disposal in combustion facilities and landfills.

- Currently, in the United States, 28% of MSW is recovered and recycled or composted, 15% is burned at combustion facilities, and the remaining 57% is disposed of in landfills.

SPECIFIC WASTE TREATMENT STRATEGIES

Biopiles. Contaminated soils are mixed with organic matter and placed in above-ground enclosures. It is a composting process in which compost is aerated.

Bioventing. Oxygen is injected into soils to stimulate biodegradation.

Chemical extraction. Waste-contaminated soil and solvents are mixed, dissolving the contaminants. The extracted solution is then treated to separate the waste material from the solvent.

Chemical reduction/oxidation. The chemical process of reduction/oxidation (redox) chemically converts hazardous contaminants to nonhazardous or less toxic compounds that are more stable, less mobile, and/or inert. The oxidizing agents most commonly used are ozone, hydrogen peroxide, hypochlorites, chlorine, and chlorine dioxide.

Composting. Contaminated soil is removed from a site and mixed with organic amendments such as wood chips, hay, manure, and vegetative wastes. This mixture promotes thermophilic (heat-loving) microbial activity and speeds up the breakdown of organic compounds.

Dehalogenation. Chemicals are added to soils contaminated with halogenated organics (containing Cl, F, Br, or I) to either replace the halogen molecules or decompose and partially volatilize them.

Electrokinetic separation. Metals and polar organics are removed from low permeability soil, mud, sludge, and marine dredgings. This *in situ* soil processing technology using electricity is primarily a separation and removal technique for extracting contaminants from soils.

Enhanced bioremediation. Water-based solutions containing microbes are circulated through contaminated soils to enhance *in situ* biological degradation of organic contaminants. Nutrients, oxygen, or other amendments may be added to enhance the process.

Fungal biodegradation. The degradation of a wide variety of organic pollutants by using lignin-degrading or wood-rotting fungus *in situ* or in bioreactors.

Hot gas decontamination. This process involves heating the contaminated material to a very high temperature. Gases are often fed into an afterburner to further destroy contaminants.

Incineration. Extremely high temperatures (1600–2200°F) are employed to destroy organic contaminants.

Landfarming. Contaminated soil is dug up, placed in lined pits, and then aerated to increase microbial decomposition.

Landfill cover enhancements. Methods of reducing the amount of water percolating through the soil. Two specific methods are (1) water harvesting, which manages the amount of runoff flowing from a site, and (2) vegetative cover, which reduces the amount of water entering the soil.

Land treatment. Contaminated surface soil is treated by tilling the soil to increase aeration, and if necessary, by adding chemicals. Tilling the soil increases the microbial activity, which enhances breakdown.

Natural attenuation. Natural processes such as adsorption, biodegradation, chemical reactions, dilution, dispersion, and volatilization are employed to reduce contaminant concentrations to acceptable levels.

Pyrolysis. Chemical decomposition is induced in organic materials by heat in the absence of oxygen. Organic materials are vaporized and a solid residue remains. It is known as "coke," which contains carbon and ash.

Separation. Separation techniques are used to detach contaminated solids through physical and chemical means from their medium (i.e., soil, sand, and/or binding material that contains them).

Slurry phase biological treatment. An aqueous slurry is created by combining soil, sediment, or sludge with water and other additives. The slurry is mixed to keep solids suspended and microorganisms in contact with the soil contaminants. Upon completion of the process, water is removed from the slurry, and the treated soil is disposed of.

Soil flushing. Water is injected into the groundwater to raise the water table in the contaminated soil zone. The contaminated water is then extracted and treated.

Soil vapor extraction. A network of above-ground piping connected to a vacuum is employed to encourage vaporization of organics from a landfill. The process also includes a system for handling the collected gas. This technology also is known as soil venting, volatilization, enhanced volatilization, or soil vacuum extraction.

Soil washing. Contaminants are separated from bulk soil in a water-based system based on the size and properties of the particles. The wash water may be augmented with leaching agents, surfactants, buffers, or chelating agents to help remove the contaminants.

Solidification/stabilization. Contaminants are solidified, or chemical reactions are employed to reduce the ability of contaminants to migrate.

Solar detoxification. Contaminants are destroyed by photochemical and thermal reactions using the ultraviolet energy in sunlight.

Thermal desorption. Wastes are heated to vaporize water and other organic contaminants. A carrier gas or vacuum system transports the organic vapors to a gas treatment system.

Composting

- Composting involves converting vegetable matter to compost (a mixture of decaying organic matter, as from leaves and manure) and is used to improve soil structure and provide nutrients. The result of this decomposition process (averages around 45 days) is compost, a crumbly, earthy-smelling, soil-like material rich in carbon and nitrogen.
- Compost piles should be 30 parts carbon sources (leaves, straw, and woody material) to one part nitrogen sources (food scraps and grass).
- Various invertebrates inhabit compost piles and aid in decomposition. See Figure 11.5 below.
- Temperatures between 90 and 140°F (32–60°C) are common in compost piles due to biological processes occurring within the pile.
- More than two-thirds of the municipal solid waste produced in the United States (including paper) is compostable material.
- There are more than 3800-yard trimmings composting facilities in the United States..
- Compost allows the soil to retain more plant nutrients over a longer period.

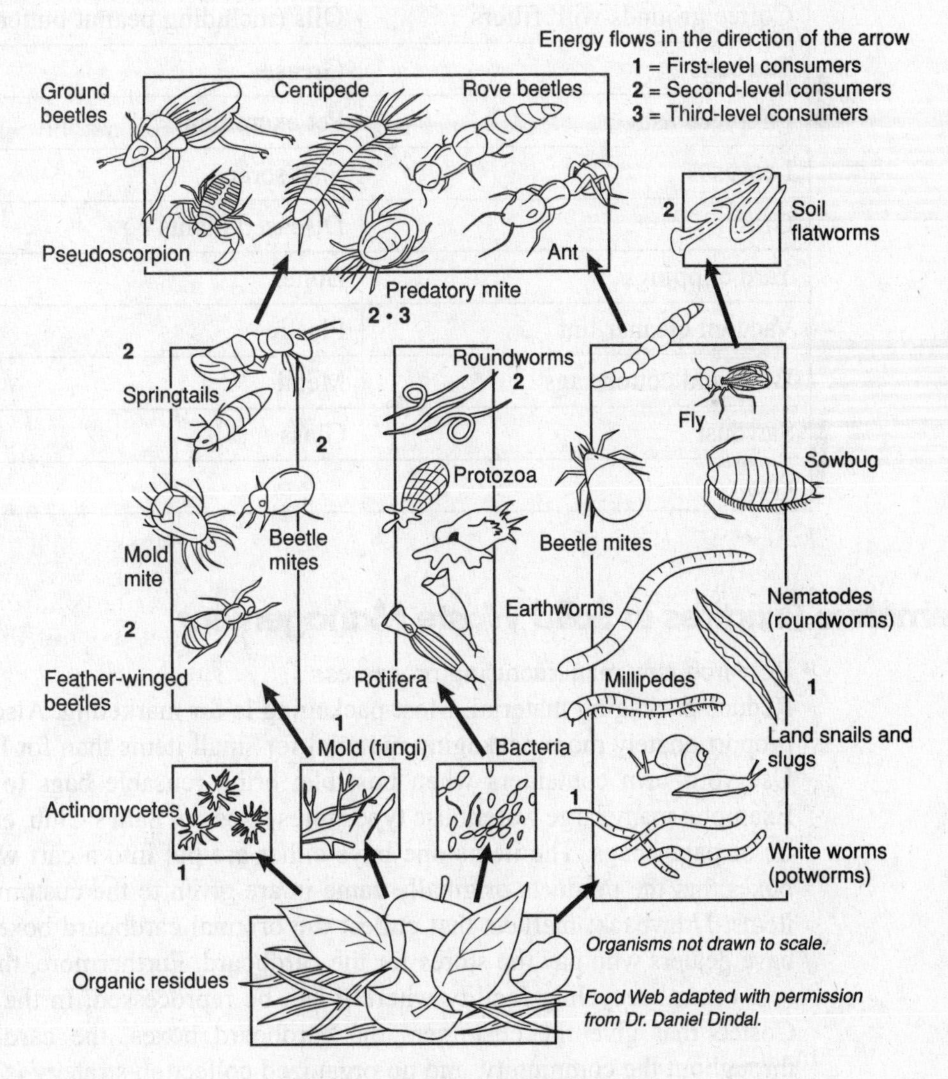

Figure 11.5. Food chain of a compost pile.

- Red wigglers (i.e., the worms used in vermicomposting) eat their weight in organic matter each day.
- Compost supplies part of the 16 essential elements needed by the plants.
- Compost helps reduce the adverse effects of excessive alkalinity, excessive acidity, or the excessive use of chemical fertilizer.
- Compost makes soil easier to cultivate.
- Compost helps keep the soil cool in summer and warm in winter.
- Compost aids in preventing soil erosion by keeping the soil covered.
- Compost helps in controlling the growth of weeds in the garden.

ITEMS FOR/NOT FOR COMPOSTING

Items for Composting	**Items Not for Composting**
Fruit and vegetable scraps	Meats
Egg shells	Dairy foods
Coffee grounds with filters	Oils (including peanut butter and mayonnaise)
Tea bags	Grease
Fireplace ash	Pet excrement
Leaves	Fish scraps
Grass	Diseased plants
Yard clippings	Bones
Vacuum cleaner lint	Plastic
Wool and cotton rags	Metal
Sawdust	Glass
Paper	

Alternative Practices in Solid Waste Management

- Research new manufacturing processes.
- Reduce packaging material. Most packaging is for marketing. Also, it generally takes proportionately more packaging material for small items than for larger items.
- Use your own containers when possible; bring reusable bags to the grocery store. Example: many large warehouse type stores (Costco, Sam's Club, etc.) do not use plastic or paper bags. The items one buys either are put into a cart without a bag or the boxes that the products originally came in are given to the customers to hold smaller items. Drawback: markets that collect the original cardboard boxes and bundle them have dealers who pay the stores for the cardboard. Furthermore, the cardboard is sent to a central recycling facility where it can be reprocessed. In the case of stores like Costco that give the customers the cardboard boxes, the cardboard is dispersed throughout the community, and no organized collection strategy is employed.
- Recycling paper can only be done about eight times at the most due to degradation of the fibers. Recycled papers contain only a fraction of reused fibers.

- Buy and support either recyclable products or products in refillable containers. New designs in aluminum cans and plastic packaging have reduced the weight (which is a function of the material used) of these containers by about one third. Glass containers can generally be refilled up to 15 times before the glass is recycled. Approximately one-third of all glass containers in the United States have been recycled. Making glass the first time requires twice the amount of energy as making it from recycled material. Fiberglass can be made from recycled glass.
- Purchase plastic products that are degradable—either photodegradable or biodegradable. For example, in 1990, McDonald's Corporation instituted biodegradable paper-based wrapping material and containers. They had been made out of nondegradable foam.
- Plastics come in different physical properties (melting temperatures, strengths, elasticity, etc.). Consequently, plastics need to be sorted to be recycled. One cannot just melt assorted plastics together. Other physical properties of plastics cause them to have a limited or fixed number of cycles in which they can be recycled. Furthermore, temperatures needed to kill bacteria and viruses are generally not reached when plastics are melted, limiting them to nonconsumable uses.
- Separate trash for recycling.
- Support community recycling programs.
- Wash and reuse packaging for own use.
- Compost yard wastes.
- Become aware of the leaders in your community. Who are your representatives? What programs are currently in place for solid waste management? What future programs are being considered? Are solid waste issues being addressed in trade agreements? Does your state export wastes to other states or countries—"Out of sight-out of mind?"
- Observe the three Rs: Reduce–Reuse–Recycle. Currently, only about one-fourth of the solid municipal wastes are being recycled. Research shows that this amount could be doubled.
- Secure landfills for hazardous wastes.
- Use permanent retrievable storage for hazardous wastes.
- Practice bioremediation.
- Establish a "trash tax"—a municipal tax that is assessed to the homeowner or business for either (1) having more trash than a set maximum allowable amount or (2) not separating the trash into categories for collection. In cities that have adopted a trash tax (Victoria, British Columbia, Canada), household wastes decreased almost 20%.

Steps You Can Do To Reduce Solid Wastes

Reduce

1. Reduce the amount of unnecessary packaging.
2. Adopt practices that reduce waste toxicity.
3. Reduce the amount of unnecessary mail by paying bills over the Internet.
4. Reduce the amount of paper you use by getting your news from TV, radio, or the Internet.

Reuse

5. Consider reusable products.
6. Maintain and repair durable products.
7. Reuse bags, containers, and other items.

8. Borrow, rent, or share items used infrequently.
9. Sell or donate goods instead of throwing them out.

RECYCLE

10. Choose recyclable products and containers and recycle them. Recycled aluminum uses 97% less water, produces 95% less air pollution, and requires 95% less energy to produce than producing aluminum from bauxite ore. In 1994, the worldwide recycling rate for aluminum was 33%, and for the United States it was 40%; 74% of all (108 billion) aluminum cans were recycled (80 billion).
11. Select products made from recycled materials. In 1994, the United States recycled 40% of its waste paper. This compares to a 97% recycling rate for Denmark. Paper requires more water in the manufacturing process than any other product. If the United States just recycled the Sunday newspaper, it would save 500,000 trees every week. Each ton of paper requires 2 tons of trees and 89 tons of other materials. The paper industry is the fifth largest user of energy in the world. Recycled paper reduces water pollution up to 35%. Recycled paper reduces air pollution by up to 95% as compared to pollution produced in making paper from raw materials. Recycled paper saves space in landfills. Recycling paper creates as much as five times as many jobs as what is required to harvest trees.

    Plastics are reusable and recyclable. However, they produce hazardous wastes during their manufacture. Plastics represent 8% by weight and 20% by volume of all MSWs. Plastics also represent approximately 60% of all trash on beaches. Plastics may take up to 500 years or longer to decompose.
12. Compost yard trimmings and some food scraps.

RESPOND

13. Educate others on source reduction and recycling practices. Make your preferences known to manufacturers, merchants, and community leaders.
14. Be creative. find new ways to reduce waste quantity and toxicity.
15. Support legislation that
    - Includes the environmental and health costs of raw materials in the true market price of consumer items;
    - Provides more tax break incentives and subsidies for reuse and postconsumer recycling industries and products, rather than for resource-extracting industries;
    - Develops large, steady markets for recycled materials;
    - Assesses households by the amount of waste they generate and give credits for reducing waste through recycling;
    - Demands more scrutiny and honest labeling of eco-friendly materials and services;
    - Encourages government and public agencies to use postconsumer recycled materials to increase demand and reduce the price of these materials to the public.

Recycling

- Typical materials that are recycled include batteries, recycled at a rate of 94%; paper and paperboard, at 42%; and yard trimmings, at 45%. These materials and others (see Figure 11.6) may be recycled through curbside programs, drop-off centers, buy-back programs, and deposit systems.
- Recycling, including composting, diverted 64 million tons of material away from landfills and incinerators in the United States in 1999, up from 34 million tons in 1990 (see Figure 11.7).

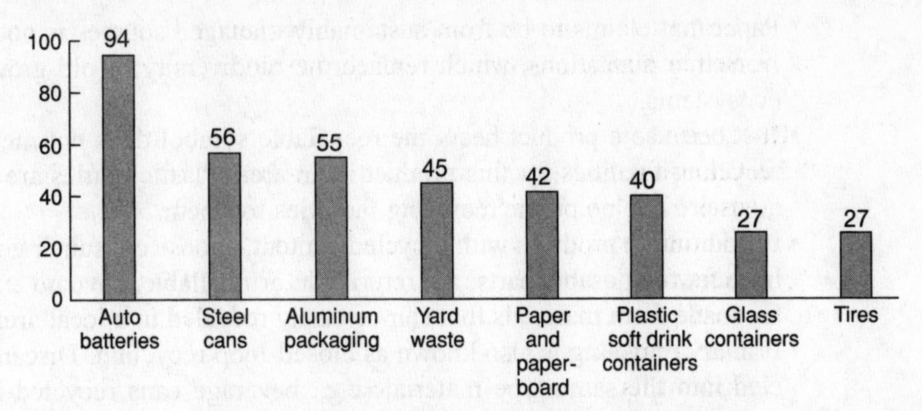

Figure 11.6. Recycling rates of selected materials, 1999.

- Recycling prevents the emission of many greenhouse gases and water pollutants, saves energy, supplies valuable raw materials to industry, creates jobs, stimulates the development of greener technologies, conserves resources for the future, and reduces the need for new landfills and combustors.
- Recycling creates new products such as aluminum cans, newspapers, cereal boxes, paper towels, egg cartons, carpeting, motor oil, car bumpers, nails, trash bags, glass containers, books, and laundry detergent bottles. Steps in the recycling process include collecting the recyclable components of municipal solid waste, separating materials by type, processing them into reusable forms, and purchasing and using the goods made with reprocessed materials.
- Using recovered material generates less solid waste (see Figure 11.7). Recycling helps to reduce the pollution caused by the extraction and processing of virgin materials. Also, when products are made using recovered rather than virgin materials, less energy is used during manufacturing and fewer pollutants are emitted.

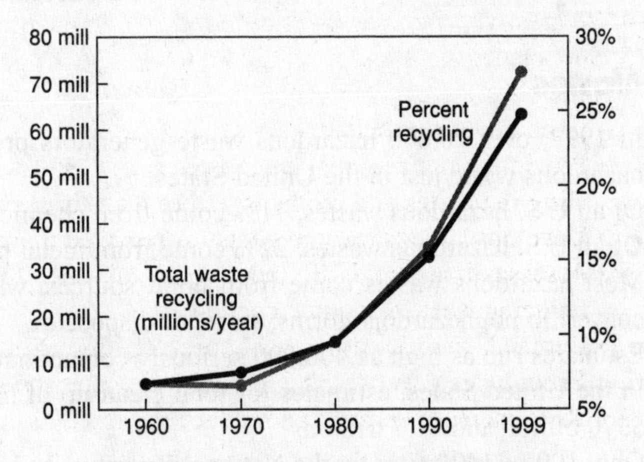

Figure 11.7. Waste recycling rates, 1960–1999.

- Recycling reduces greenhouse gas emissions that affect global climate. In 1996, recycling of solid waste in the United States prevented the release of 33 million tons of carbon into the air—roughly the amount emitted annually by 25 million cars.
- A recycling symbol on a product may not mean that it is made from recycled materials. It may only indicate that the material used is recyclable.

- Paper that claims to be from sustainably managed sources is not recycled but comes from tree plantations, which replace the biodiversity of old growth forests and their ecosystems.
- Just because a product bears the recyclable symbol does not mean that there will be recycling facilities for this product in an area. Plastic bottles are recyclable but some areas provide no public recycling facilities for them.
- In addition to products with recycled content, choose consumer goods that are durable, have few disposable parts, are returnable or refillable, have no excess packaging, and are made from materials that can be easily recycled in a local area.
- Primary recycling is also known as closed-loop recycling. Discarded wastes are recycled into the same type material (e.g., beverage cans recycled into beverage cans). Reduces consumption of original materials 20 to 90%. Benefits of closed-loop recycling include:

  1. Reduced use of virgin resources;
  2. Reduced throughput of material and energy resources; and
  3. Reduced pollution and environmental degradation.

- Secondary recycling is also known as open-loop recycling. Waste material is converted into different products (e.g., plastic bottles are recycled into plastic lawn furniture). Reduces consumption of original materials by <25%.

WASTE RECYCLING ADVANTAGES

1. Leads to less utilization of raw materials.
2. Reduces environmental impacts arising from waste treatment and disposal.
3. Makes the surroundings cleaner and healthier.
4. Saves on landfill space.
5. Saves money.
6. Reduces the amount of energy required to manufacture new products.

Hazardous Wastes

- In 1999, over 20,000 hazardous waste generators produced over 40 million tons of hazardous waste just in the United States.
- Of all U.S. hazardous wastes, 71% come from chemical and petroleum industries.
- Of all U.S. hazardous wastes, 22% come from metal processing and mining.
- Most hazardous wastes come from point sources, which makes it easier to recycle, convert to nonhazardous forms, store, or dispose of.
- Estimates run as high as 400,000 seriously contaminated sites in the United States.
- In the United States, estimates for total clean-up of hazardous waste sites is between $370 billion and $1.7 trillion.
- Only 100 of 1400 sites on the National Priority List have been handled.
- Ten most Superfund-hazardous materials are lead, trichloroethylene (TCE), toluene, benzene, PCBs, chloroform, phenol, arsenic, cadmium, and chromium.
- Areas most likely to be a hazardous waste site are smelters, mills, petroleum refineries, chemical manufacturing plants, mining facilities, and gasoline stations.
- Methods to dispose of hazardous wastes include

  1. Discharge into streams and "dilute it." (Remember that the material is still there; the same amount enters the biosphere whether it is diluted or not.)

2. Locate the material to deep wells, salt caverns, or specially designed landfills. (There are issues of leakage and groundwater contamination.)
3. Process it, detoxify it, recycle it, and so on. (Expensive. Who pays for it?)
4. Store the materials in pits (groundwater contamination, seepage through groundwater can deliver the hazardous material away from the site; volatile compounds can catch fire or vapors enter the atmosphere). 70% of containment ponds have no liners to prevent material from seeping into the ground, 90% have been assessed as a threat in contaminating ground water. Waste lagoons cause migrating birds problems when they land in them.
5. Incinerate it. (Ash may contain toxins.)
6. Use state-of-the-art landfills. Toxic wastes are sealed in drums, and the drums are placed in hazardous waste landfills. Problems include the leaking of the drums.

- Environmental Protection Agency (EPA) or a state hazardous waste agency enforces the hazardous waste laws in the United States. EPA encourages states to assume primary responsibility for implementing the hazardous waste program through state adoption, authorization, and implementation of the regulations.
- Many types of businesses, both large and small, generate hazardous waste. Examples: dry cleaners, auto repair shops, hospitals, exterminators, photo processing centers, chemical manufacturers, electroplating companies, and petroleum refineries.
- Americans generate 1.6 million tons of hazardous household wastes per year. Proper household waste management:

1. Conserves resources and energy that would be expended in the production of more products;
2. Saves money and reduces the need for generating hazardous substances; and
3. Prevents pollution that could endanger human health and the environment.

- The average home can accumulate as much as 100 pounds of hazardous household wastes.
- During the 1980s, many communities started special collection days or permanent collection sites for handling hazardous household wastes. In 1997, there were more than 3000 hazardous household waste permanent programs and collection events throughout the United States.
- Only 6% of all hazardous wastes produced in the United States are *legally* defined as "hazardous waste." Materials *not* included in the U.S. regulatory definition of "hazardous waste" are:

1. Radioactive waste,
2. Hazardous and toxic materials discarded from households,
3. Mining wastes,
4. Oil and gas-drilling wastes,
5. Liquid wastes containing hydrocarbons,
6. Cement kiln dust, and
7. Wastes from businesses that produce less than 220 lb (100 kg) of hazardous waste per month.

DEPARTMENT OF TRANSPORTATION HAZARDOUS MATERIAL PLACARDS

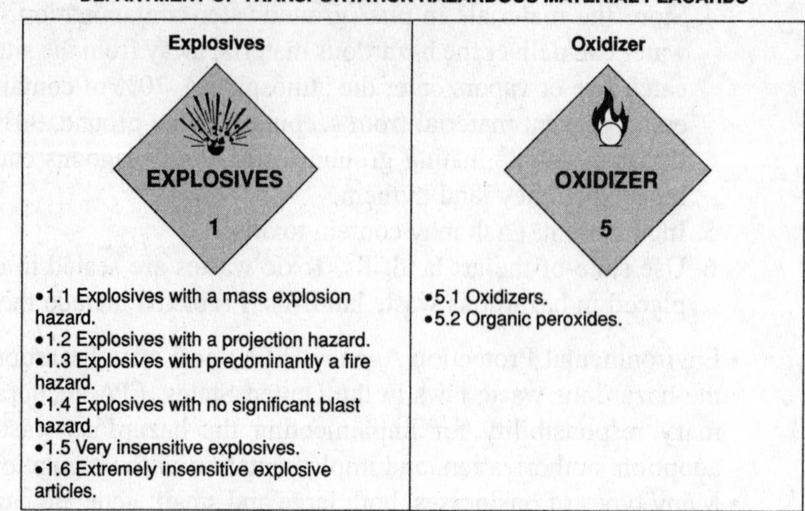

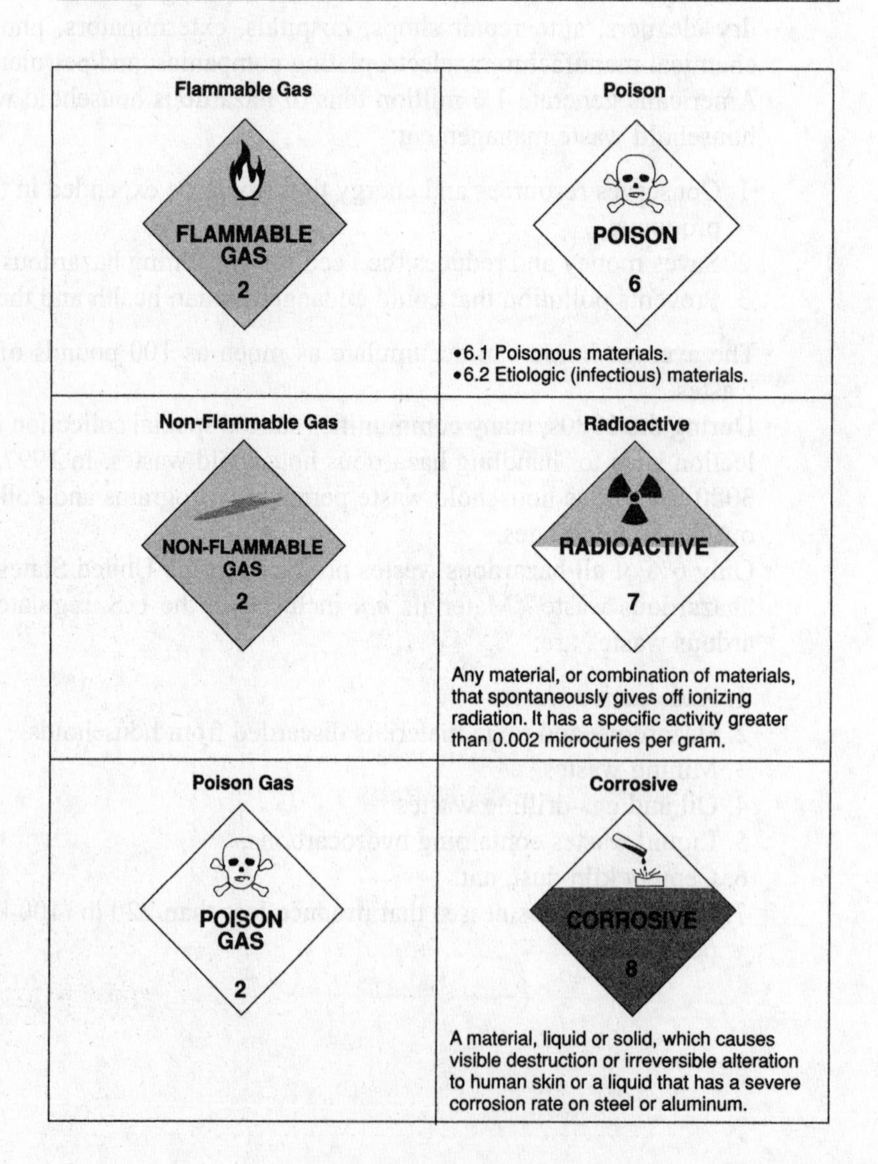

Flammable Liquid

- 3.1 Flashpoint below -18°C (0°F).
- 3.2 Flashpoint -18°C and above, but less than 23°C (73°F).
- 3.3 Flashpoint 23°C and up to 61°C (141°F).

Miscellaneous

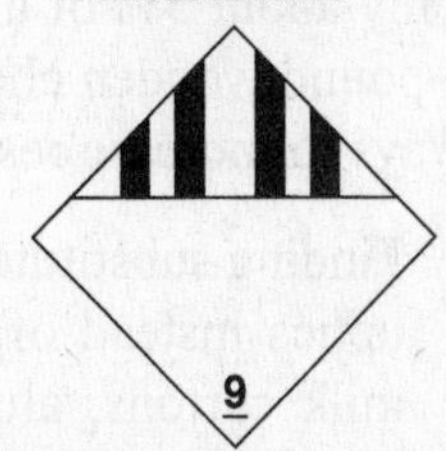

A material which presents a hazard during transport, but which is not included in any other hazard class (such as a hazardous substance or hazardous waste).

Flammable Solid

Dangerous

Applies to placarding only and is required when 1001 lbs or more of material is transported and it is composed of 2 or more hazard categories.

Spontaneously Combustible

Identification Number

The identification number, when required, may be displayed separately from the warning placard.

1090

Dangerous When Wet

Fire Hazard

Orientation (Upright)

Liquid hazardous material in non-bulk packaging must be packaged with the closures upward and be legibly marked with package orientation markings pointing in the correct upright direction.

Resource Productivity

- Only about 5% of an original resource ends up in a finished product. For example, a 5-pound wooden chest, requires 100 pounds of trees.
- Ways to increase resource productivity include

  1. Finding substitute material (e.g., put printed material onto CDs or Internet, use fiber optics instead of copper wire). Advances have been made in plastic bags, plastic milk cartons, aluminum drink cans, disposable diapers, and plastic frozen food bags.
  2. Committing to new technology and research (e.g., use of stronger and lighter steel, use of composites and cermets to replace wood and metals). "Space-age" stronger plastics have (a) reduced the weight of cars which increases fuel efficiency, and (b) reduced the weight of appliances, which reduces demand for mining, smelting, and processing—all of which require great amounts of energy and which produce air and other forms of pollution.

- In the last 10 years, increased material technology, use of substitute material in manufacturing and packaging, and recycling have reduced rate of growth of these materials almost 50%.

Case Studies

Bet Trang, Cambodia—1999. A Taiwan plastics factory shipped 3000 tons of toxic waste including mercury to Bet Trang, Cambodia. Bribes were involved. Villagers used wastes for personal use. Villagers became sick and some died.

Disposable diapers. Single-use disposable diapers consist of a waterproof polyethylene outer layer, an inner layer made from wood pulp and synthetic polyacrylate *(*a super-absorbent crystal), and a water-repellant liner. Most brands also have fragrances and perfumes.

- Flushing disposable diapers down the toilet causes 95% of clogged sewer lines in the United States.
- Disposable diapers create 43,000 tons of extra sludge per year and adds 84 million pounds of raw fecal matter to the environment each year.
- Disposable diapers dumped in landfill sites create breeding grounds for viruses and bacteria. As many as 100 different viruses can survive in soiled diapers for up to two weeks, including live polio virus excreted by recently-vaccinated babies. They threaten the health of sanitary workers, water supplies, and wildlife.
- Each baby that uses disposable diapers consumes 4.5 trees and puts two tons of solid waste into the environment (based on 2 years in diapers).
- Disposable diapers are the single largest nonrecyclable part of household garbage.
- One billion trees worldwide are used per year, just to manufacture single-use diapers.
- The potential for recycling disposable diapers is low, and they contain chemicals whose long-term effects on users and the environment remain unknown.
- Bleaching the wood pulp with chlorine gas to give disposables a desirable bright, white look produces toxic chemicals such as dioxin and furans.
- It takes less than 22 pounds of cotton to supply all the reusable cotton diapers required by one baby during the same 2 years spent wearing diapers. This compares with 400–900 pounds of fluff pulp and 300 pounds of plastic *(including packaging)* each year to supply one baby with disposable diapers.

- The super absorbent gel, sodium polyacrylate, has been linked to toxic shock syndrome, allergic reactions, and is potentially lethal to pets.
- 1,265,000 metric tons of wood pulp from trees and 45 metric tons of plastics were used in 1994 solely for the making of disposable diapers.
- It has been estimated that it will take 500 years for a disposable diaper to decompose in a landfill.
- U.S. landfills receive 18 billion disposable diapers a year, enough to stretch to the moon and back nine times.
- Disposable diapers use twice as much water to manufacture as cotton diapers.
- Disposable diapers use three times as much energy as cotton diapers.
- Disposable diapers generate 60 times more solid waste than cotton diapers.
- Disposable diapers use 20 times as much raw materials as cotton diapers.
- One ton of garbage is created for each baby who uses disposable diapers.
- One cup of crude oil is used for the plastic in one disposable diaper.
- Studies have shown (*Archives of Disease in Childhood*) that plastic disposable diapers trap heat, raising scrotal temperatures in boys as much as 2°F, and may have some role to play in decreased male fertility.

Khian Sea—1986. City of Philadelphia needed to get rid of 14,000 tons of toxic incinerator ash. It had always dumped it in a landfill in New Jersey. New Jersey refused to accept it. It was loaded onto a ship, the *Khian Sea,* which roamed the world for two years looking for someplace to dump the load. The waste "mysteriously" disappeared at sea.

Love Canal, Niagara Falls, New York—1970s. More than 20,000 tons of toxic wastes by Hooker Chemical were buried here during the late 1940s. The toxic wastes were covered with soil. Hooker Chemical sold the land to the Niagara Board of Education for $1. In the 1970s, the area became a housing development. Serious health problems arose including miscarriages, birth defects, and cancers. More than 80 different hazardous chemicals were released into ground water, coming up from the soil, seeping into basements, and so on. It led to the passage of the Resource Conservation and Recovery Act, Comprehensive Environmental Response, Compensation, and Liability Act, and the creation of the Hazardous Substances Superfund. Almost 1000 families were relocated at a cost close to $300 million.

Surat, India. The rapid urbanization and rise in population in Surat, India, led to the growth of slums, an increase in garbage, and overflowing drains. In 1994, Surat was struck by an outbreak of a virulent disease somewhat like the plague. The disease caused panic within the entire country of India, and while the citizens blamed the municipality, the civic authorities in turn blamed the citizens for their lack of civic sense.

Within 18 months the city made a complete reversal from a dirty, garbage-strewn city to one of the cleanest cities in India. This transformation was due to the Surat Municipal Corporation and the efforts of the community. Community participation played a key role in the rapid implementation of decisions taken by the corporation.

Institutional changes were the first thing to happen. The city was divided into six zones to decentralize the responsibilities for all civic functions. A commissioner was appointed for each zone. The officials responsible for solid waste management were made accountable for their work, and field visits were made mandatory for them each day. Grievance cards were issued to people so that complaints could be registered. Each complaint was attended to within 24 hours, and the card was returned to the citizen. In addition to the administrative changes, the changed laws had an important role to play in improving the conditions by also making the citizens aware of and responsible for certain preventive actions. Penalizing littering was adopted. Fines were doubled for every subsequent offense. The city roads were swept twice a day. Private contractors were, and continue to be, actively involved in the transport, collection, and disposal of solid waste. Surat has thus become a model city and the working of its municipality an example for other city governments to follow and implement.

Multiple-Choice Questions

1. Which one of the following statements is TRUE?

    (A) The United States generates approximately 230 million tons of municipal solid waste a year, about 4.6 pounds per person per day.
    (B) Food waste is the third largest component of generated waste (after yard waste and corrugated boxes) and the second largest component of discarded waste, after yard waste.
    (C) It takes 3 to 12 months to produce compost, depending on the process and type of food wastes being composted.
    (D) The average American college student produces 640 pounds of solid waste each year, including 500 disposable cups and 320 pounds of paper.
    (E) All statements are true.

2. The major source of solid waste in the United States comes from what source?

    (A) Homes
    (B) Factories
    (C) Agriculture
    (D) Petroleum refining
    (E) Mining wastes

3. Using the phrase "strict, joint and several," who is liable for clean-up or containment of a toxic site?

    (A) The federal government
    (B) The state government
    (C) The corporation that created the toxic site
    (D) All parties, past and present, who have or had anything to do with the site
    (E) Environmental Protection Agency

4. What act established federal authority for emergency response and clean-up of hazardous substances that have been spilled, improperly disposed of, or released into the environment?

    (A) Resource Recovery Act
    (B) Resource Conservation and Recovery Act
    (C) Solid Waste Disposal Act
    (D) Superfund
    (E) Hazardous Materials Transportation Act

5. What state is "leading the way" in recycling efforts?

   (A) California
   (B) Arizona
   (C) Minnesota
   (D) Rhode Island
   (E) Oregon

6. What is the largest type of domestic solid wastes in the United States?

   (A) Yard wastes
   (B) Paper
   (C) Plastic
   (D) Glass
   (E) Metal

7. Which of the following is the most readily recyclable?

   (A) Paper
   (B) Plastic
   (C) Metal
   (D) Glass
   (E) all are equally and readily recyclable

8. Which of the following statements is TRUE?

   (A) Recycling is more expensive than trash collection and disposal.
   (B) Landfills and incinerators are more cost-effective and environmentally sound than recycling options.
   (C) The marketplace works best in solving solid waste management problems; no public sector intervention is needed.
   (D) Landfills are significant job generators for rural communities.
   (E) None of the above are true.

9. In the 1970s, houses were built over a toxic chemical waste disposal site. This case study is known as

   (A) Love Canal.
   (B) Bet Trang.
   (C) Bhopal.
   (D) Brownfield.
   (E) Chernobyl.

10. A piece of industrial or commercial property that is abandoned or underused and often environmentally contaminated, and could be considered as a potential site for redevelopment is known as a

    (A) redevelopment site.
    (B) business opportunity zone.
    (C) strategic development site.
    (D) Superfund site.
    (E) brownfield.

11. Estimates run as high as ______ seriously contaminated sites in the United States.
    (A) 10,000
    (B) 50,000
    (C) 100,000
    (D) 400,000
    (E) 1,000,000

12. Of the 1400 sites on the Superfund National Priority List,
    (A) only 50 have been handled.
    (B) about 100 have been handled.
    (C) about half have been handled.
    (D) all have been handled.
    (E) none have been handled.

13. Which of the following methods of handling solid wastes is against the law in the United States?
    (A) Incineration
    (B) Dumping it in open land fills
    (C) Burying it underground
    (D) Exporting the material to foreign countries
    (E) Dumping the material in the open ocean

14. About 70% of U.S. hazardous wastes come from
    (A) agricultural pesticides.
    (B) smelting, mining, and metal manufacturing.
    (C) nuclear power plants.
    (D) chemical and petroleum industries.
    (E) households.

15. Contaminated surface soil is treated by tilling the soil to increase aeration and, if necessary, by adding chemicals. Tilling the soil increases the microbial activity that enhances breakdown. This method of treating solid waste is known as
    (A) landfarming.
    (B) landfill cover enhancement.
    (C) land treatment.
    (D) natural attenuation.
    (E) soil flushing.

Answers to Multiple-Choice Questions

1. **E**	6. **B**	11. **D**
2. **E**	7. **A**	12. **B**
3. **D**	8. **E**	13. **E**
4. **D**	9. **A**	14. **D**
5. **C**	10. **E**	15. **C**

Explanations for Multiple-Choice Questions

1. **(E)** Food waste includes leftover portions of meals and trimmings from food preparation activities in kitchens, restaurants, and fast food chains. Food waste is the third largest component of generated waste (after yard waste and corrugated boxes) and second largest component of discarded waste, after yard waste. About 4% of food wastes are composted. Composting is the controlled decomposition of organic matter by microorganisms into a humus-like product. Techniques such as windows, static piles, and in-vessel systems generate energy and heat and destroy pathogens. Water and carbon dioxide dissipate into the atmosphere during this process. Windows are the most popular technique for composting food waste. It takes 3 to 12 months to produce compost, depending on the process and type of food wastes being composted. Food waste compost is not a fertilizer. It is a useful soil conditioner that improves texture, air circulation, and drainage. Compost can moderate soil temperature, enhance nutrient and water-holding capacity, decrease erosion, inhibit weed growth, and suppress some plant pathogens. High-quality compost can find a marker as a soil amendment and as mulch for landscaping, farming, horticulture, and home gardens. Compost also can be used as landfill cover or in land reclamation projects. Paper products make up over 40% of municipal solid waste, by far the largest contributor. Every year, nearly 900,000,000 trees are cut down to provide raw materials for American paper and pulp mills.

2. **(E)** Mining wastes, along with oil and gas production constitute about 75% of all solid wastes generated in the United States. Agriculture is second, contributing about 13%. Industry is third at around 10%, and MSW is about 2%.

3. **(D)** Any one or all of the potentially liable parties can be held responsible for clean-up costs regardless of whether they are at fault. One party can thus be held responsible for another party's liability. For example, company X is responsible for 95% of the waste at a site, but the company is insolvent. Company Z now owns the site and generated 5% of the waste. By law, "strict, joint, and several," company Z can be held liable for 100% of the clean-up costs. Industry and potentially liable parties have long sought to address such discrepancies.

4. **(D)** The Superfund program began in 1980 to locate, investigate, and clean up the most polluted sites nationwide. The EPA administers the Superfund program in cooperation with individual states and tribal governments. The office that oversees management of the program is the Office of Emergency and Remedial Response (OERR).

5. **(C)** The Minnesota Pollution Control Agency (MPCA) was established in 1967. Its purpose is to protect Minnesota's environment through monitoring environmental quality and enforcing environmental regulations. Visit http://www.pca.state.mn.us/netscape.shtml.

6. **(B)** Paper makes up 38% of MSW.

7. **(A)** All items listed are recyclable. However, metal, glass, and plastic generally have to be sorted before they can be reused. This step is labor-intensive. Paper, however, can, for the most part, be broken down into fibers and reused without sorting for cardboard or other similar paper products that do not require high quality.

8. **(E)** In choice (A), when designed right, recycling programs are cost-competitive with trash collection and disposal. In choice (B), recycling programs, when designed properly, are cost-competitive with landfills and incinerators, and provide net pollution prevention ben-

efits. Recycling materials not only avoids the pollution that would be generated through landfilling and incinerating, but also reduces the environmental burden of virgin materials extraction and manufacturing processes. In choice (C), the solid waste system has always operated under public sector rules. Currently these rules encourage unchecked product consumption and disposal. In choice (D), recycling creates many more jobs for rural and urban communities than landfill and incineration disposal options.

9. **(A)** Love Canal is a neighborhood in Niagara Falls, New York. The nickname "Love Canal" came from William Love, who in 1896 began digging a canal connecting Lake Ontario and Lake Erie (bypassing Niagara Falls) to serve as a waterpower conduit. It was never completed, but the Hooker Chemical Company, located west of the canal, turned the uncompleted canal into a dumping ground for the chemical by-products of its manufacturing process. Once the canal was filled with waste, the land was covered over and sold to the Niagara Falls city school board for $1.00 and a school and subdivision of homes was built right on top of the waste. The chemicals were detected leaking out of the site in 1977, and many health problems were also reported. Residents were evacuated after a lengthy fight with the New York State government. Today, it remains a ghost town.
10. **(E)** Brownfields are abandoned, idled, or underused industrial or commercial facilities where expansion or redevelopment is complicated by real or perceived environmental contamination.
11. **(D)** The National Research Council estimates that in the United States, the costs of cleaning up the known 300,000 to 400,000 heavily contaminated sites where groundwater is polluted will be as high as $1 trillion over the next 30 years alone. One-third of the wells tested in California's San Joaquin Valley in 1988 contained the pesticide DBCP (dibromochloropropane) at levels 10 times higher than the maximum allowed for drinking water—more than a decade after the chemical was banned. The U.S. Environmental Protection Agency estimates that about 100,000 underground storage tanks for gasoline are leaking chemicals into soil and groundwater. In Santa Monica, California, wells supplying half the city's water have been closed because of dangerously high levels of the gasoline additive MTBE (methyl tertiary-butyl ether). Since 1943, billions of gallons of radioactive wastes have been dumped into soils and aquifers in Washington state by the Department of Energy's Hanford Nuclear Reservation. Some of this waste has a half-life of 250,000 years.
12. **(B)** Of the more than 1,400 sites on Superfund's National Priority List, 160 are on lands controlled by federal agencies—including one under EPA authority. Only two of the federally-controlled sites, or 1.25%, have been cleaned up—versus nearly 5% on private sites, for a total close to 100.
13. **(E)** The Ocean Dumping Act (ODA) regulates the dumping of materials into the United States territorial ocean waters.
14. **(D)** In 1999, over 20,000 hazardous waste generators produced over 40 million tons of hazardous waste regulated by the Resource Conservation and Recovery Act (RCRA), which was enacted by Congress in 1976. RCRA's primary goals are to protect human health and the environment from the potential hazards of waste disposal, to conserve energy and natural resources, to reduce the amount of waste generated, and to ensure that wastes are managed in an environmentally sound manner. RCRA regulates the management of solid waste (e.g., garbage), hazardous waste, and underground storage tanks holding petroleum products or certain chemicals.
15. **(C)** Refer to the chart entitled "Specific Waste Treatment Strategies."

Free-Response Question

City Council Debates Sanitary Landfill Tonight

The Metropolis City Council will meet tonight to discuss views on a proposed landfill to be built on the vacant property known as the "Wetlands." A group known as "Vigilantes for the Earth" plan on protesting tonight's meeting. Spokesman for the environmental group, Josh Wagner, told *The Metropolitan*, "We will do whatever it takes to stop this project—there is no way this city is going to dump trash on the Wetlands." Mayor Jane Robbins responded that tonight's debate will hear from many people representing many different viewpoints. However, the mayor said that the city must deal with ever-increasing amounts of trash each year and that the city's current facility will soon reach its maximum capacity and will be closed.

Tonight's debate will be held at Metropolis High School's auditorium at 7:30 P.M. Members from the audience are requested to fill out question cards prior to the start of the meeting.

You are on the City Council. You have been handed the following question cards. Respond to each in clear, coherent, and factual statements.

(a) **REPLY CARD**

What is a sanitary landfill?

(b) **REPLY CARD**

Are there alternatives to building a sanitary landfill on the Wetlands? Could you also give me three alternative methods of dealing with our city's waste problems and the pros and cons of each alternative?

(c) **REPLY CARD**

I'm from Vigilantes for the Earth. Could you give me your ethical viewpoints, backed up by established ethical theory, of your thoughts on building a landfill on the Wetlands?

(d) **REPLY CARD**

Are there things that we can do today to reduce the amount of solid wastes that Metropolis is generating?

Free-Response Answer

Part (a): 1 point
Part (b): 4 points
Part (c): 3 points
Part (d): 2 points
Ten points necessary for full credit.

RUBRIC	ESSAY
½ pt. Lined pit. ½ pt. Covered each day—shows difference between a sanitary landfill and an open dump. ½ pt. Problem of contaminants from landfill leaching into groundwater.	(a) Sanitary landfills are generally located in urban areas where a large amount of waste is generated and has to be dumped in a common place. Unlike an open dump, it is a pit that is dug in the ground. The landfill is lined with heavy plastic liners. The garbage is dumped and the pit is covered, thus preventing the breeding of flies and rats. At the end of each day, a layer of soil is scattered on top of it and is compressed. After the landfill is full, the area is covered with a thick layer of mud and the site can thereafter be developed as a parking lot or a park. Here in Metropolis, your City Council has decided to turn the sanitary landfill into a park, for all people to enjoy. However, landfills have some problems. All types of waste are dumped in landfills and sometimes when water seeps through them, it gets contaminated and in turn pollutes the groundwater. This contamination of groundwater and soil through landfills is known as leaching.
1 pt. Incineration as an alternative. 1 pt. Facts regarding the efficiency of incineration. ½ pt. Concept of cogeneration. 1 pt. Problem of fly ash. 1 pt. Pollution control devices to combat problems of incineration. 1 pt. Mentioning items that should not be burned.	(b) There are several alternatives to sanitary landfills. The city could look into the possibility of incineration that would reduce the impact of needing a new landfill. This method would reduce the volume of our waste material by 90% and decrease the weight by up to 75%. The heat generated by burning the wastes could be captured to power steam turbines that could cogenerate electricity. Fly ash is a common concern of communities that live near incinerators. Fly ash contains high levels of heavy metals, dioxins, and furans. However, if the city went with the option of incineration rather than a sanitary landfill, it would be necessary in my opinion to adopt pollution control devices on the effluent gases. Such control devices would be electrostatic precipitators and filters to collect ash and scrubbers to reduce the impact of noxious gases. We would also need to screen the trash before it is burned to make sure that materials are not being burned that would have an adverse affect on our air quality. Such items would include batteries, hazardous wastes, and plastics.
1 pt. Another option to sanitary landfill. 1 pt. Criticizing option.	Another option that we, the city, could adopt would be to export our MSW to another country or state. This method is criticized as "garbage imperialism" or environmental

RUBRIC	ESSAY
1 pt. Benefit of option	racism; however, it does provide needed income and jobs to less-developed countries or poorer states. The costs of exporting our MSW would be a significant drawback and require further study.
1 pt. Argument of the interest of the city vs. the interest of the wetlands. 1 pt. Importance of Wetlands. 1 pt. Identification of an ethical viewpoint—ecocentrist in this case. 1 pt. Expansion of ecocentrism in mentioning a "school."	(c) Let me begin by saying that I have been called a Promethean environmentalist. That is, I believe that the great city of Metropolis can solve any environmental problem we face by working together. The problem we face with the issue of the landfill is the ethical dilemma between the rights of the Wetlands to exist versus the rights of this city to utilize the land for a demonstrable need. It is obvious that we need to address the issue of our waste problem. However, we also need to keep in mind that the Wetlands provide a unique biological habitat, unequaled in productivity. And once lost, our wetlands are gone forever. I oppose the sanitary landfill. My ethical compass is that of an ecocentrist, in particular more in line with the *"Environmental-Wisdom School,"* which expounds that resources are limited, that technology and economic growth can both benefit and damage the environment, and, most importantly, that Nature does not exist for man; that we should not attempt to adapt the Earth for our needs but instead, learn to adapt to the environment.
1 pt. Mentioning famous environmentalists that subscribe to an ethical framework.	I believe that the protection of the Wetlands deserves moral consideration on its own, not associated with this city's interest. My thoughts are similar to those of John Muir, Rachel Carson, and Henry David Thoreau. There are several alternatives that I will address in my next response.
1 pt. Adoption of an integrated plan encompassing many solutions. 1 pt. Concept of recycling.	(d) I believe that a better approach to solving this city's MSW problem is to adopt an IWM—an integrated waste management program. IWMs are comprehensive management plans of handling solid wastes that incorporate their collection, processing, and disposal combined with a plan to reduce, reuse, and recycle. This plan would incorporate the following actions:
1 pt. For each plausible idea mentioned (3 max)	(1) Provide tax breaks for industry and citizens who compost material. More than two-thirds of all MSWs that cities generate are compostable material. The compost would also be available to franchised brokers who could buy it from the people who produce it and market it. (2) Adopt minimum packaging laws. For example, citizens would be required to bring their own reusable bags to the store. Extra charges would be assessed to citizens who required new bags. (3) Establish community recycling programs with economic incentives to the consumer.

RUBRIC	ESSAY
	(4) Establish an advisory committee to establish guidelines for the city and other government facilities to reduce as much paper as possible by either using recycled paper or using Internet resources. (5) Provide bidding preference to vendors wishing to do business with the city who incorporate environmental friendly policies and actions into their bids. (6) Adopt a hierarchy for tax levies on generated waste (trash tax). Those that produce the most waste, pay the most in fees.
This is the conclusion statement of the essay. It wraps it up and does not leave the reader "hanging."	There is no magic bullet, no one answer to the solid waste problems that this city faces. Sacrificing irreplaceable wetland resources is out of the question. If we agree to that, then developing a comprehensive, detailed plan that addresses this issue from many perspectives is in line with the purpose of this city—living and solving problems together for the benefit of all.
	Word Count: 917

Now, it's your turn to practice *this* question. We have four questions to address. Let's begin by brainstorming each question separately. Remember to break the problem down. In each box, write as many Key Terms or Key Concepts as you can that apply to the question asked.

Brainstorm: (a) What is a landfill:

Brainstorm: (b) Alternative methods for handling solid wastes:

Brainstorm: (c) Ethical viewpoint:

Brainstorm: (d) Ways to reduce MSW:

Next, look at each box. Above each term that you wrote, place a number that indicates the order in which you wish to talk about that topic. For example, if you wrote "covered pit" in the brainstorm box for "what is a sanitary landfill" and you want to talk about that concept first, place a 1 above the words "covered pit." Now, look back at your words in that box. What do you want to talk about next? Place a 2 above that term, and so on.

When you have finished this task, you can take a relaxing breath. You've got everything you want to say down on paper and it is ordered. The tough part is over. Now, on your own, go ahead and construct your paragraphs. Remember to write concise thesis statements, include as much data or as many facts as you can, and conclude the paragraph(s) with a conclusion statement.

CHAPTER 12

Human Health

The right dose differentiates a poison and a remedy.

Paracelsus (1493–1541)

Areas That You Will Be Tested On

A: Agents: Chemical and Biological
B: Effects: Acute and Chronic, Dose-Response Relationships
C: Relative Risks: Evaluation and Response

Key Terms

Note: Definitions for all key terms listed below can be found in the glossary starting on page 648. For additional definitions of relevant terms from this chapter, refer to *www.barronseduc.com/0764121618.html.*

acceptable risk	**infectious disease**
acute toxicity	**LD_{50}**
biomagnification	**risk assessment**
carcinogen	**risk management**
dose-reponse curve	**toxicity**

Key Concepts

Chemical and Biological Agents

Biological Agents

The Centers for Disease Control (CDC) has classified viruses and other disease-producing agents into four biosafety levels. Biosafety level 1 applies to agents that do not ordinarily cause human disease. Biosafety level 2 is appropriate for agents that can cause human disease, but whose potential for transmission is limited and immunization and antibiotics are available. Biosafety level 3 applies to agents that may be transmitted by the respiratory route that can cause serious or potentially lethal infection. Biosafety level 4 is used for the diagnosis of exotic agents that pose a high risk of life-threatening disease, which may be transmitted by the aerosol route and for which there is no vaccine or therapy.

Level 1. *Escherichia coli, Klebsiella oxytoca, Lactobacillus, Saccharomyces.*

Level 2. Measles virus, *Salmonellae*, *Toxoplasma spp.*, Hepatitis B and C, Pathogenic *E. coli, Streptococcus* sp.

Level 3. *Mycobacterium tuberculosis*, St. Louis encephalitis virus, *Coxiella burnetii,* HIV.

Level 4. Ebola virus, Lassa fever, Monkey B virus, Marburg virus.

RISK ASSESSMENT OF BIOLOGICAL AGENTS INVOLVES:

1. ***Pathogenicity*** includes disease incidence and severity (i.e., mild morbidity versus high mortality and/or acute versus chronic disease). *E. coli* would be considered to be a low risk, whereas ebola would be a serious pathogen due to its high mortality rate and no known cure.
2. ***Route of transmission*** (e.g., airborne, ingestion, skin contact, etc.). Aerosol routes of transmission have the highest risk (see Figure 12.1).
3. ***Agent stability*** is the ability to survive over time in an environment exposed to such factors such as desiccation, exposure to sunlight or ultraviolet light, or exposure to chemical disinfectants.
4. ***Infectious dose*** refers to how much of the agent is required to produce an infection.
5. ***Concentration*** is the number of infectious organisms per unit volume. Risk factors increase as the working volume of high-titered microorganisms increases because additional handling of the materials is often required.
6. ***Origin*** refers to geographic location (e.g., domestic or foreign); host (e.g., infected or uninfected human or animal); or nature of source (potential zoonotic or associated with a disease outbreak).
7. ***Availability of data from animal studies*** provides additional information. In the absence of human data, it may provide insight into the pathogenicity, infectivity, and route of transmission.
8. ***Availability of an effective prophylaxis or therapeutic intervention*** is desirable. Risk assessment runs from extreme risks where no therapeutic intervention is available to low risk where established immunization is effective.

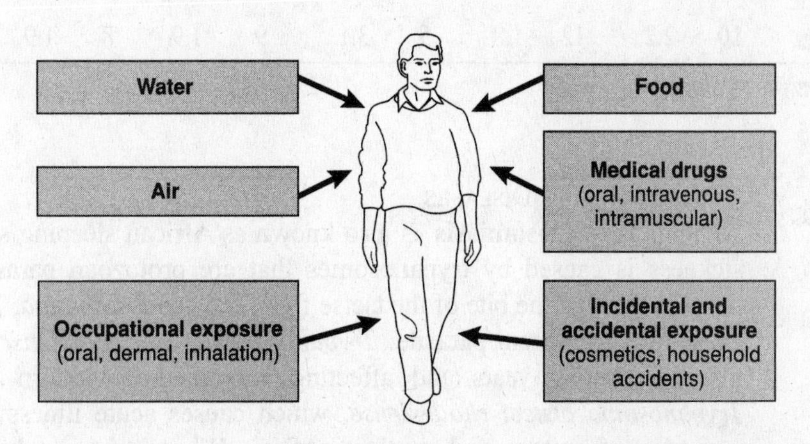

Figure 12.1. Exposure routes to toxic and hazardous environmental factors.

LEADING CAUSES OF DEATH WORLDWIDE AND RANK PER REGION

	All Countries		**Africa**		**The Americas**		**Eastern Mediterranean**		**Europe**		**Southeast Asia**		**Western Pacific**	
Deaths	Rank	% of Total	Rank	% of Total	Rank	% of Total	Rank	% of Total	Rank	% of Total	Rank	% of Total	Rank	% of Total
Ischematic heart disease	1	13.7%	9	2.9%	1	17.9%	1	13.6%	1	25.5%	1	13.8%	3	11.1%
Cerebrovascular disease (stroke)	2	9.5	7	4.7	2	10.3	5	5.3	2	13.7	4	6.5	1	14.3
Acute lower respiratory infections	3	6.4	3	8.2	3	4.2	2	9.1	4	3.6	2	9.3	4	4.0
HIV/AIDS	4	4.2	1	19.0	13	1.8	27	0.4	42	0.2	8	2.2	42	0.2
Chronic obstructive pulmonary disease	5	4.2	14	1.1	6	2.8	10	1.7	5	2.7	11	1.6	2	12.0
Diarrheal diseases	6	4.1	4	7.6	10	2.0	3	7.4	22	0.7	3	6.6	17	1.2
Perinatal conditions	7	4.0	5	5.5	7	2.6	4	7.3	13	1.2	5	6.0	10	2.2
Tuberculosis	8	2.8	11	2.2	19	1.0	7	3.7	23	0.6	6	5.1	9	2.9
Cancer of trachea-bronchus/lung	9	2.3	38	0.3	4	3.2	20	1.0	3	4.2	15	1.2	6	3.6
Road traffic accidents	10	2.2	12	1.8	5	3.1	9	1.9	8	1.9	7	2.5	12	2.0

Source: The World Health Report, 1999.

COMMON HUMAN DISEASES

African Trypanosomiasis is also known as African sleeping sickness. African sleeping sickness is caused by trypanosomes that are protozoan parasites. It is transmitted to humans through the bite of the tsetse fly of the genus *Glossina*. There are two forms, each caused by a different parasite: *Trypanosoma brucei gambiense*, which causes a chronic infection lasting years and affecting countries of western and central Africa; and *Trypanosoma brucei rhodesiense*, which causes acute illness lasting several weeks in countries of eastern and southern Africa. When a person becomes infected, the trypanosome multiplies in the blood and lymph glands, crossing the blood-brain barrier to invade the central nervous system where it provokes major neurological disorders. Infection by trypanosomes causes neurological alterations that are often irreversible even after successful treatment. Psychomotor and neurological retardation even among cured children is frequent. Without treatment, the disease is invariably fatal.

Sleeping sickness is a daily threat to more than 60 million men, women, and children in 36 countries of Sub-Saharan Africa, 22 of which are among the least-developed countries in the world. Sleeping sickness has a major impact on the development of rural areas by decreasing the labor force and hampering production and work capacity. It remains a major obstacle to the development of entire regions. In countries such as Angola, Democratic Republic of Congo, or Sudan, the operational capacity to respond to the epidemic situation is largely surpassed and in certain endemic areas the observed prevalence is huge.

Anthrax is an acute infectious disease caused by the spore-forming bacterium *Bacillus anthracis*, which is commonly found in cattle, sheep, and other herbivores. The spores have recently been used as a bioterrorism weapon, being distributed through the mail. Cross-contamination results when other items come in contact with the spores. Common forms include cutaneous (skin) and inhalation. Symptoms for inhalation anthrax, the most serious and usually fatal form, include initial fatigue, fever, difficulty breathing, profuse sweating, cyanosis, and shock. The cutaneous form begins with swelling and boils on the skin.

Botulism is a rare but serious illness caused by toxin produced by *Clostridium botulinum* bacterium. The bacterium may enter the body through wounds, or they may live in improperly canned or preserved food. *Clostridium botulinum* is found in soil and untreated water throughout the world. It produces spores that survive in improperly preserved or canned food, where they produce toxin. When eaten, even minute amounts of this toxin can lead to severe poisoning. Foods most commonly contaminated are home-canned vegetables, cured pork and ham, and smoked or raw fish. Botulism may also occur if the organism enters open wounds and produces toxin there. Approximately 20 cases of food-borne botulism in adults and 250 cases in infants are reported per year in the United States. Respiratory failure caused by weakness in the muscles that control breathing can cause death in up to 10% of food-borne illness and 2% of infant disease.

Cholera is generally a disease spread by poor sanitation, resulting in contaminated water supplies. The few U.S. cases between 1973 and 1991 were associated with the consumption of raw shellfish or of shellfish either improperly cooked or recontaminated after proper cooking. Environmental studies have demonstrated that strains of this organism may be found in the temperate estuarine and marine coastal areas surrounding the United States. Cholera is endemic in India and countries in South America, Southeast Asia, and Africa. Symptoms include abdominal cramps, nausea, vomiting, dehydration, and shock; after severe fluid and electrolyte loss, death may occur. Illness is caused by the ingestion of viable bacteria, which attach to the small intestine and produce cholera toxin. The production of cholera toxin by the attached bacteria results in the watery diarrhea associated with this illness.

Cryptosporidiosis is an infection of the large and small intestines caused by the organism (parasite) *cryptosporidium*. Transmission occurs through ingestion of fecally-contaminated materials typically from animals to humans, humans to humans, and through contaminated water. Risk factors are travel to areas with poor sanitation, and having a suppressed immune system. In the last few years, several large outbreaks have occurred in major cities when municipal water supplies became contaminated. These outbreaks affected thousands of people.

Dengue fever is a mild viral illness transmitted by mosquitoes and characterized by fever, rash, and muscle and joint pains. Dengue fever is caused by several related viruses (four different arboviruses) and is transmitted by the bite of mosquitoes, most commonly *Aedes aegypti*, found in tropic and subtropic regions. Early symptoms of Dengue hemorrhagic fever are similar to those of Dengue fever, but after several days the patient becomes irritable, restless, and sweaty followed by a shock-like state. Bleeding may appear as pinpoint spots of blood on the skin and larger patches of blood under the skin. Bleeding may occur from minor injuries. Death generally occurs from shock. Using personal protection (clothing, mosquito repellent, netting) and traveling during periods of minimal mosquito activity can be helpful. Mosquito abatement programs may reduce the risk of infection, but vaccination is the only sure method of prevention.

Ebola hemorrhagic fever (ebola fever) is a virus-caused disease that is limited in area to parts of Africa. The exact mode of transmission is not understood. The incubation period appears to be up to 1 week, at which time the patient develops fatigue, malaise, headache, backache, vomiting, and diarrhea. Within a week, a raised rash appears over the entire body; the rash often contains blood. Hemorrhaging generally occurs from the gastrointestinal tract, causing the patient to bleed from both the mouth and rectum. Mortality is high, reaching 90%.

Encephalitis is an infection of the brain, most often caused by a virus. The specific viruses involved may vary. Exposure to viruses can occur through insect bites, food or drink, or skin contact. In rural areas, arboviruses that are carried by mosquitoes or ticks, or that are accidentally ingested, are the most common cause. In urban areas, enteroviruses are most common. After the virus has entered the bloodstream, it can localize in the brain causing inflammation of the brain cells and surrounding membranes. White blood cells invade the brain tissue as they try to fight off the infection. The brain tissue swells (cerebral edema) and can cause destruction of nerve cells, bleeding within the brain (intracerebral hemorrhage), and brain damage. Public health measures to control mosquitoes (a mosquito bite can transmit some viruses) can reduce the incidence of some types of encephalitis.

Giardiasis is an infection of the small intestine caused by a protozoa, *Giardia lamblia*. Giardia outbreaks can occur in communities in both developed and developing countries where water supplies become contaminated with raw sewage. It can be contracted by drinking water from lakes or streams where water-dwelling animals such as beavers and muskrats, or where domestic animals such as sheep, have caused contamination. It is also spread by direct person-to-person contact, which has caused outbreaks in day-care centers. Campers and hikers are at risk if they drink untreated water from streams and lakes. Water purification methods such as boiling, filtration, and iodine treatment should be used when surface water is used. Hikers or others using surface water should consider all sources as potentially contaminated.

Hanta virus a distant cousin of the Ebola virus, is a disease characterized by flulike symptoms followed by respiratory failure. Hanta virus is carried by rodents, particularly deer mice, and is present in their urine and feces. An effective treatment for Hanta virus is not yet available. Even with intensive therapy, over 50% of the diagnosed cases have been fatal.

Hepatitis is a general term meaning inflammation of the liver and can be caused by a variety of different viruses such as hepatitis A, B, C, D and E. The development of jaundice is a characteristic feature of liver disease. Hepatitis B is a highly contagious virus that attacks the liver. In the mildest case, infected people never know they have it, and it may be gone in six months. But some people become carriers for the rest of their lives. Others go on to have chronic liver disease. Carriers may develop cirrhosis, a disease that scars the liver, or liver cancer. Chances of liver cancer are 300 times higher for hepatitis B carriers. The Centers for Disease Control and Prevention (CDC) estimates that each year about 240,000 Americans get hepatitis B. One out of 20 people in this country will get hepatitis B at some time during their life. Hepatitis B is spread through unsafe sex or from contact with infected blood or body fluids. Hepatitis C (HCV) is the most common chronic blood-borne infection in the United States. The incidence of hepatitis C infection is 3.9 million to 5 million people in the United States or approximately 1 in 70 to 100 people. Persons who may be at risk for hepatitis C are those who received a blood transfusion prior to July 1992, injected street drugs or shared a needle with someone who has hepatitis C, have been on long term kidney dialysis, as a healthcare worker had frequent contact with blood on the job, or had sex with a person who has hepatitis C or shared personal items, such as toothbrushes and razors, which may have blood on them.

Influenza is spread from person to person by inhaling infected droplets from the air. There are three types of influenza virus. Type A is usually responsible for the large outbreaks and is a constantly changing virus. New strains of Type A virus develop regularly and result in a new epidemic every few years. In an average year, influenza is associated with more than 20,000 deaths in the United States. The incidence rate of influenza in the United States is 7 out of 1000 people. In 1918–1919 "Spanish flu" caused the highest known influenza-related mortality: approximately 500,000 deaths occurred in the United States, 20 million worldwide. In 1957–1958 "Asian flu" was responsible for 70,000 deaths in the United States; in 1968–1969 "Hong-Kong flu" killed 34,000 people in the United States.

Lassa fever is an acute viral illness of one to four weeks duration caused by lassa virus, a member of the arenavirus family of viruses. Consequences range widely in severity, from asymptomatic infection without illness to extremely severe illness that may have a fatal outcome. It is transmitted to humans from wild rodents (the multimammate rat, *Mastomys natelensis*). Lassa infection in rodents persists, and the virus is shed throughout the life of the animal. Disease transmission is primarily through direct or indirect contact with excreta of infected rodents deposited on surfaces such as floors or beds, or in food or water. Person-to-person and laboratory infections occur, especially in the hospital environment, by direct contact with blood (including inoculation with contaminated needles), pharyngeal (throat) secretions or urine of a patient, or by sexual contact. Person-to-person spread may occur during the acute phase of fever when the virus is present in the throat. Because Lassa fever may have a long (up to 21 day) incubation period, it is possible that travelers from endemic areas may be incubating the disease.

Legionnaire's disease is an acute respiratory infection caused by the bacterium *Legionella pneumophilia*, which causes a serious pneumonia. The bacteria have been found in water delivery systems and can survive in the warm, moist, air conditioning systems of large buildings including hospitals. The infection is transmitted through the respiratory route. Most infection occurs in middle-aged or older people. Risk factors include cigarette smoking; underlying diseases such as renal failure, cancer, diabetes or chronic obstructive pulmonary disease; and suppressed immune systems.

Leprosy an infectious disease known since Biblical times, and is characterized by disfiguring skin lesions, neurological damage, and progressive debilitation. Leprosy is caused by the organism *Mycobacterium leprae*. It is a difficult disease to transmit and has a long incubation period. Leprosy is common in many countries in the world, and in temperate, tropical, and subtropical climates. Approximately 270 cases per year are diagnosed in the United States. Most cases are limited to the South, California, Hawaii, and U.S. island possessions. Recently, the emergence of drug-resistant *Mycobacteria leprae* has caused increased concern about this disease. Prevention consists of avoiding close physical contact with untreated people.

Lyme disease is an acute inflammatory disease characterized by skin changes, joint inflammation, and flu-like symptoms, caused by the bacterium *Borrelia burgdorferi* transmitted by the bite of a deer tick. It has now been reported in most parts of the United States. Most cases occur in the Northeast, upper Midwest, and along the Pacific coast. Mice and deer are the most commonly infected animals that serve as host to the tick. Most infections occur in the summer. The disease is difficult to diagnose because the symptoms mimic other diseases. A characteristic red rash usually occurs at the site of the bite; however, the bite may go unnoticed. A few months after the bite, muscle paralysis, joint inflammation, neurological symptoms, and sometimes heart symptoms may occur.

Malaria is caused by a parasite that is transmitted from one human to another by the bite of infected *Anopheles* mosquitoes. In humans, the parasites (called sporozoites) migrate to the liver where they mature and release another form, the merozoites. These enter the bloodstream and infect the red blood cells. The disease is a major health problem in much of the tropics and subtropics. The CDC estimates that there are 300 million to 500 million cases of malaria each year, and more than one million people die. It presents the greatest disease hazard for travelers to warm climates. In some areas of the world, mosquitoes that carry malaria have developed resistance to insecticides, while the parasites have developed resistance to antibiotics. This has led to difficulty in controlling both the rate of infection and spread of this disease. *Falciparum* malaria, one of four different types, affects a greater proportion of the red blood cells than the other types and is much more serious. It can be fatal within a few hours of the first symptoms.

Plague an infectious disease of animals and humans caused by the bacterium *Yersinia pestis*. It is carried by fleas that generally are associated with rodents. It is known as pneumonic plague if it infects the lungs and bubonic plague if it infects the lymph system. During the Middle Ages, 25 million people died from the plague (one fourth of the population of Europe). It is also known as "black death." The last urban plague epidemic occurred in Los Angeles, California, in 1924–1925. The World Health Organization reports 1000 to 3000 cases of plague every year worldwide and about 20 cases in the United States. Plague exists in Africa, Asia, South America, and in the southwestern United States.

Rift Valley fever (RVF) is a zoonosis (a disease which primarily affects animals, but occasionally causes disease in humans). It may cause severe disease in both animals and humans leading to high morbidity and mortality. The death of RVF-infected livestock often leads to substantial economic losses. Many different species of mosquitoes are vectors for the RVF virus. There is, therefore, a potential for epizootics (epidemics amongst animals) and associated human epidemics following the introduction of the virus into a new area where these vectors are present.

Salmonella is a common form of "food poisoning." *Salmonella enterocolitis* can range from mild to severe diarrhea illness. The infection is acquired through ingestion of contaminated food or water. Any food can become contaminated during preparation if conditions and equipment for food preparation are unsanitary. The incubation period is 8 to 48 hours after exposure, and the acute illness lasts for 1 to 2 weeks. The bacteria are shed in the feces for months in some treated patients. A carrier state exists in some people who shed the bacteria for 1 year or more following the initial infection. The incidence is 1 out of 1000 people. Two thirds of patients are less than 20 years of age. Infants and children under the age of 9 years are most commonly affected. Proper food handling and storage are preventive measures. Good hand washing is important especially when handling eggs and poultry.

Smallpox is caused by variola virus. Smallpox was responsible for the deaths of Queen Mary II of England, Emperor Joseph I of Austria, King Luis I of Spain, Tsar Peter II of Russia, Queen Ulrika Elenora of Sweden, and King Louis XV of France. The effects of smallpox and its extreme contagious nature are illustrated by the fact that in 1509 the Spanish came to the island of Hispañola (Haiti and Santa Domingo) to raise sugarcane. The Spanish brought smallpox with them. Nine years later, 2.5 million original native people of the island who had been enslaved by the Spanish had died of smallpox, wiping out the labor force. As a result, the labor was replaced with slaves from Africa, setting the precedent for black slavery in the Caribbean and the New World. In the early 1950s—150 years after the introduction of vaccination—an estimated 50 million cases of smallpox still occurred in the world each year. Due to massive determination to end smallpox through worldwide vaccination, the last known case occurred in Somalia in 1977. Initial symptoms include high fever, fatigue, and head and back aches. A characteristic rash, most prominent on the face, arms, and legs, follows in 2 to 3 days. Because of its high case-fatality rates (~30%) and transmissibility, smallpox represents one of the most serious bioterrorist threats to the civilian population. As few as 50 cases would likely invoke large-scale, perhaps national emergency control measures. The Bush administration, in 2001, responded to the bioterrorist threat by ordering smallpox vaccine for every American.

Tuberculosis (TB) is a chronic, contagious bacterial infection caused by *Mycobacterium tuberculosis*, which has spread to other organs of the body by the blood or lymph system. The infection can develop after inhaling droplets sprayed into the air as from a cough or sneeze by someone infected with *Mycobacterium* tuberculosis. The disease is characterized by the development of tumors in the infected tissues. The usual sites of the disease are the lungs, but other organs may be involved. The risk of contracting TB increases with the frequency of contact with people who have the disease, in crowded or unsanitary living conditions, and with poor nutrition. Hispanics, Native Americans, and blacks are at higher risk for developing the disease. An increased incidence of TB has been seen recently in the United States primarily in people with AIDS and HIV infection and increasing numbers of homeless people. Incomplete treatment of TB infections (not taking medications for the prescribed length of time) contributes to the proliferation of drug-resistant strains of TB.

One-third of the world's population is estimated to be infected with *Mycobacterium tuberculosis.* The disease kills three million people annually—of these, 98% are in the developing world. It is estimated that eight million new cases occur each year. The number of cases of TB rose by 25% in Russia and in 18 of the 26 Eastern European countries from 1994 to 1996. Nearly 22% of all TB cases in Latvia are multidrug resistant (MDR), and over 25% of all cases in Russia and Estonia are resistant to at least one drug.

Typhoid fever is a bacterial infection characterized by diarrhea, and a rash; it is most commonly caused by *Salmonella typhi.* Typhi are spread by contaminated food, drink, or water. Following ingestion, the bacteria spread from the intestine to the intestinal lymph nodes, liver, and spleen via the blood where they multiply. Early symptoms are very general and include fever, fatigue, and abdominal pain. As the disease progresses, the fever becomes higher (greater than 103°F), and diarrhea becomes prominent. Weakness, profound fatigue, delirium, and an acutely ill appearance develops. Adequate water treatment, waste disposal, and protection of food supply from contamination are important public health measures. The disease is not common in the United States, but it is common in Third World Countries.

Yellow fever is an insect-borne tropical disease characterized by fever, jaundice (yellow skin and eyes), and hemorrhage. Yellow fever is caused by an arbovirus (*Flaviviridae*), a small RNA virus that is transmitted by the bite of mosquitoes (*Aedes aegypti*, *Aedes africanus*, and *Haemagogus*). Areas of high incidence are Central America, the northern half of South America, and Central Africa. Mild sub-clinical infection occurs and may be more common in children. The acute form of the disease causes headache, muscle aches, fever, loss of appetite, and vomiting. The infected person may also develop a red tongue, flushed face, and reddening of the eyes. By the fifth day, jaundice (the yellow color of the skin for which the disease is named), stomach pain, and bleeding (which may appear as bloody vomiting) develops. Delirium and seizures followed by coma are common. Death occurs at the end of the first week of symptoms and as many as half of the infected people may die. Mosquito control has decreased the risk in countries where the disease was once prevalent. The risks remain, however, because the mosquito has increasing resistance to insecticides.

SEXUALLY TRANSMITTED DISEASES (STDS)

AIDS A deficiency of the body's immune system to fight a variety of infections and cancers. AIDS is the final (and fatal) stage that results from the infection of the Human Immunodeficiency Virus (HIV). HIV is the most dangerous of all the STDs because there is still no cure or vaccine.

Candidiasis Also called yeast infection, thrush, or moniliasis. Caused by a yeastlike fungus *Candida albicans.* Candidiasis is a very common cause of vaginitis.

Chlamydia The most common sexually transmitted disease caused by the bacteria *Chlamydia trachomatis*. Affects between 3 million and 5 million people a year. It can lead to sterility. The infection can spread up the fallopian tubes and leaves scar tissue so that pregnancy is no longer possible.

Genital herpes Caused by the herpes simplex virus II (HSV-II). It is estimated that 500,000 to 1 million new cases occur each year in the United States alone. Once infected, the virus remains in the body for life.

Genital warts The most common cause of STD in the world is caused by the human papillomavirus (HPV). Estimates are that as many as 26 million Americans are infected, with rates of infection increasing.

Gonorrhea An infectious STD caused by the bacteria *Neisseria gonorrhoeae* that affects primarily the mucous membrane of the genital tract, the urinary tract, the rectum, and sometimes the eyes. Discharges from the involved mucous membranes are the source of the infection.

Hepatitis B Caused by the hepatitis B virus (HBV) virus. Transmitted from one person to another through blood and other bodily fluids. About 80% of all liver cancer is believed to be caused by the HBV. Only STD that can be effectively prevented by a vaccine.

Pelvic inflammatory disease (PID) A serious infection (usually chlamydia or gonorrhea) of a woman's reproductive organs. PID is a leading cause of infertility in the United States. PID can cause scarring in the fallopian tubes, which can lead to tubal pregnancy or infertility.

Pubic lice—"crabs" *Pediculosis pubis* or "crabs" are lice that infest the pubic hair and survive by feeding on human blood.

Scabies An itchy rash from a skin infestation caused by a tiny female mite, *Sarcoptes scabiei,* that burrows into the skin to lay eggs.

Syphilis Caused by a bacteria called *Treponema pallidum*. The disease progresses through several stages: primary, secondary, latent, and sometimes tertiary.

Trichomoniasis—"trick" Caused by the protozoan *Trichomonas vaginalis.* Causes an inflammation of the mucous membranes of the vagina and of the urethra in both males and females.

Chemical Agents

Arsenic is combined with oxygen, chlorine, and sulfur in the environment, to form inorganic arsenic compounds. Arsenic in animals and plants combines with carbon and hydrogen to form organic arsenic compounds. Inorganic arsenic compounds are mainly used to preserve wood. Organic arsenic compounds are used as pesticides, primarily on cotton plants. Exposure to higher than average levels of arsenic occurs mostly in the workplace, near hazardous waste sites, or in areas with high natural levels. At high levels, inorganic arsenic can cause death. Exposure to lower levels for a long time can cause a discoloration of the skin and the appearance of small corns or warts. Arsenic has been found at 1014 of the 1598 National Priority List sites identified by the Environmental Protection Agency.

Asbestos causes asbestosis, a respiratory disease. Inhaling asbestos fibers can cause scar tissue (fibrosis) to form inside the lung. Scarred lung tissue does not expand and contract (elasticity) normally. The severity of the respiratory disease depends upon the duration of exposure and the amount inhaled. Asbestos fibers were commonly used in construction before 1975. Asbestos exposure occurs from asbestos mining and milling industries, construction, fireproofing, and other industries. In families of asbestos workers, exposure can also occur from particles brought home in the worker's clothing. More than 9 million workers are at risk of developing this disease. Cigarette smoking increases the risk of developing the disease. The incidence is 4 out of 10,000 people.

Benzene is a clear, colorless liquid with a sweet odor. It burns readily. Benzene is obtained from crude petroleum. Small amounts may be found in products such as paints, glues, pesticides, and gasoline. Breathing benzene vapor in small amounts may cause headache, euphoria (a "high"), a light-headed feeling, dizziness, drowsiness, or nausea. With more serious exposure, benzene may cause sleepiness, stumbling, irregular heartbeats, passing out, or even death. Benzene vapors are mildly irritating to the skin, eyes, and lungs. If liquid benzene contacts the skin or eyes, it may cause burning pain. Liquid benzene splashed in the eyes can damage the cornea of the eyes.

Cadmium can be found in all soils and rocks, including coal and mineral fertilizers. Most cadmium used in the United States is extracted during the production of other metals like zinc, lead, and copper. Cadmium does not corrode easily and has many uses, including batteries, pigments, metal coatings, welding, and plastics. Exposure to cadmium happens mostly in the workplace where cadmium products are made. The general population is exposed from breathing cigarette smoke or eating cadmium-contaminated foods. Cadmium damages the lungs, can cause kidney disease, and may irritate the digestive tract. This substance has been found in at least 776 of the 1467 National Priorities List sites identified by the Environmental Protection Agency. It causes itai-itai disease.

Chlorine is used extensively in manufacturing, especially the plastics, solvents, and paper industries. Chlorine compounds are persistent, can accumulate in body fat, and are toxic.

Dioxins (mainly 2,3,7,8- tetrachlorodibenzo-*p*-dioxin) may be formed during the chlorine bleaching process at pulp and paper mills. They are also formed during chlorination by waste and drinking water treatment plants. They can occur as contaminants in the manufacture of certain organic chemicals. Dioxins are released into the air in emissions from municipal solid waste and industrial incinerators. Exposure to chlorinated dioxins occurs mainly from eating food that contains the chemicals. It causes effects on the skin and may cause cancer in people. This chemical has been found in at least 91 of 1467 National Priorities List sites identified by the Environmental Protection Agency.

Formaldehyde exposures generally occur by inhalation or skin/eye contact. In cases of acute exposure, formaldehyde will most likely be detected by smell; however, persons who are sensitized to formaldehyde may experience headaches and minor eye and airway irritation at levels below the odor threshold.

Lead has many different uses. It is used in the production of batteries, ammunition, metal products (solder and pipes), and devices to shield X-rays. Because of health concerns, lead from gasoline, paints and ceramic products, caulking, and pipe solder has been dramatically reduced in recent years. Exposure to lead can happen from breathing workplace air or dust, eating contaminated foods, or drinking contaminated water. Children can be exposed from eating lead-based paint chips or playing in contaminated soil. Lead can damage the nervous system, kidneys, and reproductive system. Lead has been found in at least 1026 of 1467 National Priorities List sites identified by the Environmental Protection Agency.

Metallic mercury is used to produce chlorine gas and caustic soda, and is also used in thermometers, dental fillings, and batteries. Mercury salts are sometimes used in skin lightening creams and as antiseptic creams and ointments. Mercury also combines with carbon to make organic mercury compounds. The most common one, methylmercury, is produced mainly by microscopic organisms in the water and soil. More mercury in the environment can increase the amounts of methylmercury that these small organisms make. Exposure to mercury occurs from breathing contaminated air, ingesting contaminated water and food, and having dental and medical treatments. Mercury, at high levels, may damage the brain, kidneys, and developing fetus. This chemical has been found in at least 714 of 1467 National Priorities List sites identified by the Environmental Protection Agency.

Particulates in the air cause health problems. Sources include farming, mining, smelting, industries, burning fossil fuels, and volcanoes. In the United States, particulates contribute to approximately 60,000 deaths per year. People who live in urban areas have 15 to 25% higher mortality rates than those who live in rural areas in terms of particulate risk.

Effects—Acute and Chronic, Dose-Response Relationships

Dose-Response Curves

- Dose-reponse models tell whether the level of response increases or decreases with dose and how rapidly it changes as a function of the dose.
- The *x*-axis plots concentration of a drug or hormone. The *y*-axis plots response (e.g., enzyme activity, accumulation of an intracellular second messenger, membrane potential, secretion of a hormone, heart rate or contraction of a muscle).
- There are three types of agonists:

1. **full agonist**—a drug that appears able to produce the full tissue response;
2. **partial agonist**—a drug that provokes a response, but the maximum response is less than the maximum response to a full agonist; and
3. **antagonist**—a drug that does not provoke a response itself, but blocks agonist-mediated responses.

- The standard dose-response curve follows an S-curve. In Figure 12.2, note that at the beginning of the curve, there is a dose below which no response occurs or can be measured and that the plateau shows that once a maximum response is reached, any further increases in the dose will not result in any increased effect.

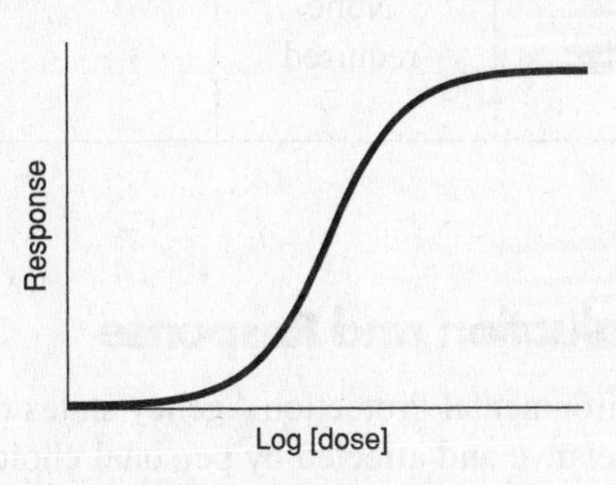

Figure 12.2. Standard dose-response curve follows an S-curve.

- Allergic responses do *not* follow a dose-response curve; they are not toxic responses. Allergic responses are the result of a chemical stimulating the body to release natural chemicals that are in turn directly responsible for the effects seen.
- Concentrations of chemicals in the environment are most commonly expressed as ppm (parts per million) or ppb (parts per billion).

MEASUREMENTS FOR EXPRESSING LEVELS OF CONTAMINANTS IN FOOD AND WATER

Dose	Metric Equivalent	Approx. Amount in Water
parts per million (ppm)	milligrams per kilogram mg/kg	1 teaspoon per 1000 gallons
parts per billion (ppb)	micrograms per kilogram (μg/kg)	1 teaspoon per 1,000,000 gallons

- LD_{50} (lethal dose for 50% of the organisms tested) is usually expressed in milligrams of chemical per kilogram of body weight (mg/kg). Chemical with a low LD_{50} (~5 mg/kg) are highly toxic. Chemical with large LD_{50} (~1000 to 5000 mg/kg) are generally nontoxic, however, effects may be noticed at any LD_{50} level.

TOXICITY RATING SCALE AND LABELING REQUIREMENTS FOR PESTICIDES.

Category	Required on Label	LD_{50} Oral mg/kg (ppm)	LD_{50} Dermal mg/kg (ppm)	Probable Oral Lethal Dose
I Highly toxic	DANGER—POISON (skull and crossbones)	Less than 50	Less than 200	A few drops to a teaspoon
II Moderately toxic	WARNING	51 to 500	200 to 2000	Over 1 teaspoon to 1 ounce
III Slightly toxic	CAUTION	Over 500	Over 2000	Over 1 ounce
IV Practically nontoxic	None required			

Relative Risks—Evaluation and Response

- The Environmental Protection Agency states that an acceptable risk is 1 in 1 million.
- Risk is relative and affected by personal choice. For example, the risk of getting cancer from carcinogens in drinking water is 1 or 2 in 1 billion, but the risk of dying in an auto accident is 1 in 5000. And yet, people drive under the influence, don't wear seat belts, and drive at unsafe speed but demand that drinking water be 100% pure.
- Risk assessment is

Influenced by self or group interest;

Affected by the information the person(s) have;

Affected by personal experience(s);

Influenced by a "it won't happen to me" syndrome; and

Influenced by the media.

Four Steps of Risk Assessment

Hazard identification. Estimate damage from varied or chronic doses. Attempts to determine cause and effect relationship between a substance, agent, or event, and an injury to human health (e.g., cancer, birth defects, neurological effects) or the environment (e.g., fish kills, forest die-off, or reduced reproduction rates).

Dose-response assessment. Determine what damage will occur and which bodily systems will be damaged as the dose of a chemical increases. Involves extrapolation techniques.

Exposure assessment. Determine how much of a chemical is absorbed from all sources (food, water, air, or skin). Identify and describe exposure in potentially exposed populations. Involves identification of other causative factors and alternative pathways of exposure.

Risk characterization. Use data from hazard identification, dose-response assessment, and exposure assessment to determine a hazard risk. Involves modeling and probability theory.

- Risk assessment may not be totally accurate due to:

  Lack of understanding of natural phenomena;

  Improper collection of data;

  Improper experimental design;

  Improper analysis of data and conclusion(s).

Risks of Life

Risk of Dying

In an accident on the job . . . 40 in 1 million
From an earthquake or volcano . . . 1 in 11 million
From a dog bite . . . 1 in 20 million
From a snakebite . . . 1 in 36 million
Once infected with flesh-eating bacteria . . . 1 in 4
Once infected with Ebola . . . 2 in 3 (risk of getting Ebola . . . 1 in 14 million)
A person aged 14–25 dying from a car accident . . . 1 in 3500
A person aged 14–25 dying from murder . . . 1 in 4500
A person aged 14–25 dying from suicide . . . 1 in 7700
While giving birth in Ireland . . . 1 in 50,000
While giving birth in the United States . . . 1 in 12,500
While giving birth in Mali . . . 1 in 50

Risk of Crime

Murdered in the next 12 months . . . 1 in 11,000
Robbed in the next 12 months . . . 1 in 400
Burglarized in the next 12 months . . . 1 in 50
Someone stealing your car . . . 1 in 145
Someone stealing your bicycle . . . 1 in 280
Someone stealing your wallet or purse . . . 1 in 1560

Risk of Substance Abuse

That the average teenager will drink hard liquor daily . . . 1 in 60
That the average teenager will drink hard liquor once or twice a month . . . 1 in 6
That a teenager will smoke marijuana daily . . . 1 in 30
That a teenager will smoke marijuana once or twice a month . . . 1 in 17
That a teenager will use cocaine daily . . . 1 in 100
That a teenager will use cocaine once or twice a month . . . 1 in 140

Of U.S. Children Who Die

46% die in traffic accidents
15% die in fires
3% die from choking
3% die from guns
1% die from poisoning
13% die from other causes (disease, etc.)

- DALY (Disability-Adjusted Life Year) represents the number of years lost from premature death and/or disability. The following chart illustrates DALY values for various diseases in poor households including the possible or feasible reduction percentage. As measured through DALY measurements, 90% or more of losses in life quality occur in less-developed countries and poor households and are results of infections, sanitation, clean water, childhood inoculations, and so on.

DISABILITY-ADJUSTED LIFE YEAR

Disease	**Source**	**DALY**	**% Reduction**
Respiratory infections	Air pollution	119	15
Diarrhea	Sanitation	99	40
Tuberculosis	Malnutrition, crowding	46	10
Chronic respiratory disease and cancer	Indoor air pollution	45	10
Parasites	Sanitation	26	30

Source: World Bank, *World Development Report,* 1998.

A COMPARISON OF COUNTRIES AND SELECTED CAUSES OF DEATHS

Country	**All Causes**	**Infectious and Parasitic Diseases**	**Traffic Accidents**	**Suicide**	**Homicide**	**Heart Disease**
United States	577.7	14.9	15.3	10.5	7.6	106.5
Russian Federation	1084.4	19.5	17.8	34.1	22.2	259.9
Japan	405.6	8.4	8.3	14.4	0.6	30.0

*Age-standardized death rate per 100,000 of the standard population for various countries.
Source: World Health Organization.

Case Studies

Agent Orange. Agent Orange was used as a defoliant during the Vietnam War. 250,000 veterans from the Vietnam War may have been affected.

AIDS. As of 2002, 40 million people are infected worldwide. Areas that are experiencing the greatest increase are Eastern Europe and former Soviet bloc countries. It costs up to $10,000 a year for treatment, making it unavailable for most people of the world.

Ebola. Ebola-Zaire, Ebola-Ivory Coast, and Ebola-Sudan are members of a new family of negative-stranded RNA viruses known as *Filoviridae.* It was first recognized in 1976 near the Ebola River in the Democratic Republic of the Congo. Most research indicates that the Ebola strains are zoonotic (i.e., normally found in animal hosts). Humans do not carry the ebola virus. The natural reservoir of the virus has yet to be determined. Current thinking is that the initial patient in an ebola outbreak becomes infected through contact with an infected animal. After this first "index case" becomes infected, contact with any blood or bodily secretion will invariably spread the disease to new hosts. Rapid spread of the virus occurs through amplification—a rapid increase in cases due to cross-contamination of bodily fluids by caregivers and members of a dense community. Ebola-Reston, the strain of ebola that occurs only in monkeys, may be contagious through aerosol particles (air-borne). Within one week of becoming infected, most patients (66%) will die. The latest ebola outbreak (Ebola-Sudan strain) occurred in Uganda in 2000–2001 in which 425 people became infected and 225 died (53% mortality rate).

Lead poisoning. Lead poisoning affects approximately 900,000 children ages 1 to 5 in the United States, which represents almost 5% of the population. Levels of lead in the blood greater than 10 μg/dl are dangerous. Most of the children affected come from low-income families (a risk factor 8 times greater than higher-income families). African-American children are 5 times more likely to be affected than white children. Approximately 22% of all African-American children living in older housing suffer from lead poisoning, and in some areas, the figures are even higher. The effects of lead poisoning are greatest among the young because lead is absorbed into tissues at this age. Lead poisoning causes damage to the nervous system resulting in mental retardation, reduced attention span, hyperactivity, impaired growth, learning disabilities, and behavioral problems. Exposure is through long-term, chronic, low-level exposure. There is no effective cure for children who have been chronically exposed. Most of the lead comes from lead dust that originated from lead-based paints that were once used to paint walls. Lead dust that flakes off and contaminates the soil around the exterior of the house is another major pathway for poisoning. In 1978, the Federal government banned the use of lead in household paint. Prior to 1940, lead-based paints often contained 10% lead, with concentrations as high as 50%. The reason that African-American and poorer children suffer disproportionately higher rates of lead poisoning is an economic issue—a higher proportion live in older homes, homes that were painted with lead-based paint.

Minamata. A vinylchloride manufacturing plant on a Japanese island allowed mercury to be discharged into the bay. Mercury was converted to methylmercury and was an example of biomagnification. Human symptoms varied but included nervous system damage, blurred vision, and loss of muscular coordination. It was determined that a threshold amount of methylmercury was required for symptoms to appear. Some of the effects cleared up, but others did not.

Times Beach, Missouri. Oil sprayed on roadways contained dioxin. The town was evacuated and destroyed by the U.S. government because dioxin disrupts the endocrine, immune, and reproductive systems.

Timeline

460–377 B.C.	Hippocrates ("Father of Medicine") writes about causes of disease. Begins the scientific study of medicine by maintaining that diseases have natural causes.
280 B.C.	Herophilus studies the nervous system and distinguishes between sensory nerves and motor nerves.
250 B.C.	Erasistratus studies the brain and distinguishes between the cerebrum and cerebellum.
200 B.C.	Galen (Greek physician) observes cause-and-effect relationship between lung disease and mining. Studies the connection between paralysis and severance of the spinal cord.
100 A.D.	Romans decree that due to dangers of mining, only slaves will be allowed in the mines.
100–400	Decline of Roman Empire. Theories exist that lead was used to "sweeten" wine.
1347–1350s	Bubonic or "black" plague of Europe.
1473	First book on occupational diseases.
1546	Girolamo Fracastoro postulates on nature of contagious disease.
1628	William Harvey explains the vein-artery system and structure of the heart in *De Motu Cordis et Sanguinis*.
1662	First book on vital statistics published. Beginning of the science of epidemiology.
1701	Giacomo Pylarini gives the first smallpox inoculations.
1712	Bernardo Ramazzini, the father of occupational medicine, publishes a book describing the hazards of 52 occupations.
1741	First hospital for children opened in London.
1750	Typhus epidemics in London.
1754	Relationship between citrus fruits and scurvy established in England.
1789	French Revolution—concept of health care for all people was a major cornerstone of the Revolution.
1796	Discovery of smallpox vaccination.
1821	Virchow develops "cell theory."
1847	Cholera outbreak in London.
1848	American Medical Association formed.
1860	Louis Pasteur's "germ theory."
1867	Joseph Lister introduces antiseptic surgery, reducing surgical mortality rates.
1878	American quarantine laws passed.
1881	Louis Pasteur develops an anthrax vaccine.

1882	Louis Pasteur develops a rabies vaccine.
1882–1883	Robert Koch discovers that tuberculosis and cholera are caused by bacteria.
1890	U.S. Public Health Service begins inspection of immigrants.
1891	W. T. Sedgwick developed the first sewage treatment techniques.
1895	First diphtheria antitoxin issued. Daily testing of water supply begins.
1899	Mosquito transmission of yellow fever demonstrated.
1904	Lead in paint determined toxic.
1905	Mandatory smallpox vaccinations in the United States.
1906	Pure Food and Drug Act passed.
1906	Wasserman introduces serum diagnosis of syphilis.
1910	Mandatory reporting of venereal disease.
1917	Immunization against diphtheria in public schools and institutions.
1918	Influenza pandemic.
1921	Edward Mellanby discovers vitamin D and shows that its absence causes rickets.
1922	Insulin introduced as treatment for diabetes.
1928	Alexander Fleming discovers penicillin.
1930	National Institute of Health started.
1932	First sulfa drug, prontosil, discovered.
1952	Jonas Salk develops the first polio vaccine.
1960	First oral contraceptives.
1963	Measles vaccine.
1973	Abortion legalized, Roe vs. Wade.
1976	First case of Ebola recognized.
1981	First AIDS case described.
1984	HIV virus discovered.

Note: No attempt was made to make this a complete or current timeline. To do so would have made this timeline larger than this book. Recent medical advances are growing exponentially. This timeline only attempts to outline *some* major advances or historical events in medicine and public health.

Multiple-Choice Questions

1. Which of the following statements regarding malnutrition is TRUE?

   (A) Hunger is caused when food-producing resources in much of the world are stretched to the limit so that there's simply not enough food to go around.
   (B) A child who eats enough to satisfy immediate hunger may still be severely malnourished.
   (C) The surest sign of starvation is emaciated, thin bodies.
   (D) Hunger affects the young and old, men and women, boys and girls equally.
   (E) All statements are true.

2. Effects produced from a long-term, low-level exposure are called

   (A) acute.
   (B) chronic.
   (C) pathological.
   (D) symptomatic.
   (E) synergistic.

3. The greatest disease hazard for travelers to warm climates is

   (A) cholera.
   (B) encephalitis.
   (C) hepatitis C.
   (D) AIDS.
   (E) malaria.

4. A disease spread by poor sanitation, resulting in contaminated water supplies would be

   (A) botulism.
   (B) cholera.
   (C) shistosimiasis.
   (D) typhoid.
   (E) yellow fever.

5. The disease that causes the most number of deaths worldwide is

   (A) malaria.
   (B) AIDS.
   (C) cancer.
   (D) heart disease.
   (E) alcoholism-related.

6. Chemicals that can cause changes in DNA are known as

   (A) teratogens.
   (B) mutagens.
   (C) antigens.
   (D) carcinogens.
   (E) gametogens.

7. A measure of the premature deaths and losses due to illness and disability in a population as a means of assessing a disease is known as

(A) DALY.
(B) LD_{50}.
(C) LD_{100}.
(D) managed risk.
(E) disability index.

8. Which disease(s) listed here is/are caused by a flagellated protozoan in which water supplies become contaminated by raw sewage?

(A) Giardiasis
(B) Cholera
(C) Cryptosporidosis
(D) Trypanosomiasis
(E) Giardiasis and cholera

9. The following dose-response curve shows that

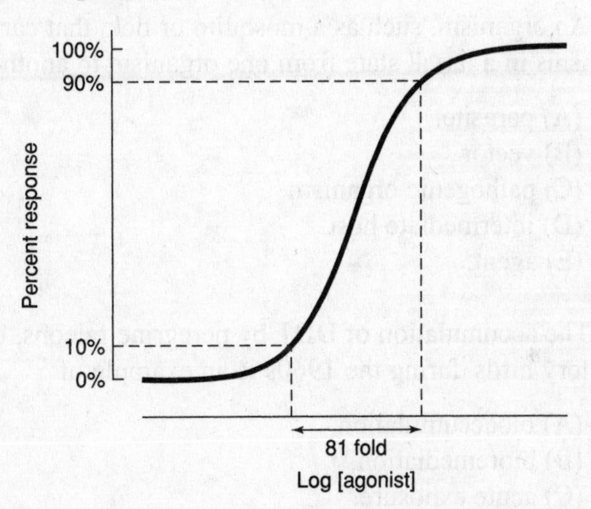

(A) larger amounts of agonist produce a corresponding increase in response.
(B) 80% more agnonist is required to achieve 80% more response.
(C) 81 times more agonist is needed to achieve a 90% response than a 10% response.
(D) an 80% higher response is achieved with a ten-fold increase in agonist.
(E) none of the above are true.

10. A dose that is represented as LD_{50}

(A) shows a response in 50% of the population.
(B) kills half of the study group.
(C) is a dose that has an acceptable risk level of 50%.
(D) is a dose that has threshold response of 50%.
(E) is a dose that is administered to 50% of the population.

11. Problem(s) associated with risk assessment include
    (A) people making risk assessments vary in their conclusions of long-term versus short-term risks and benefits.
    (B) some technologies benefit some groups and harm others.
    (C) there is consideration of cumulative impacts of various risks rather than consideration of each impact separately.
    (D) there may be conflict of interest in those carrying out the risk assessment and review of the results.
    (E) all these answers.

12. Currently, the single-most important threat to human health is
    (A) toxic chemicals.
    (B) accidents.
    (C) pathogenic organisms.
    (D) pollution.
    (E) nontransmissible disease such as cancer and cardiovascular disease.

13. An organism, such as a mosquito or tick, that carries disease-causing microorganisms in a larval state from one organism to another is called a(n)
    (A) parasite.
    (B) vector.
    (C) pathogenic organism.
    (D) intermediate host.
    (E) agent.

14. The accumulation of DDT by peregrine falcons, brown pelicans, and other predatory birds during the 1960s is an example of
    (A) bioaccumulation.
    (B) bioremediation.
    (C) acute exposure.
    (D) biomagnification.
    (E) a case-controlled study.

15. 50 parts per million (ppm) is equivalent to
    (A) 5×10^{-2} ppb.
    (B) 5×10^{-3} ppb.
    (C) 5000 ppb.
    (D) 50,000 ppb.
    (E) 500,000 ppb.

Answers to Multiple-Choice Questions

1. **B**	6. **B**	11. **E**
2. **B**	7. **A**	12. **E**
3. **E**	8. **E**	13. **B**
4. **B**	9. **C**	14. **D**
5. **D**	10. **B**	15. **D**

Explanation for Multiple-Choice Questions

1. **(B)** The world produces enough grain alone to provide every human being on the planet with 3500 calories a day, more than the minimum amount of food required. This does not include all other sources of food—meat, dairy, fruits, vegetables, and so on. With all forms of food, if equally distributed, each person on Earth would receive 4.3 pounds of food per person per day. It's not a simple matter of whether a child can satisfy his or her appetite. It's about getting the right quantities and combinations of nutrients. Three quarters of the children who die worldwide of causes related to malnutrition have no outward signs of malnutrition. Starving people's bodies often swell so they look surprisingly healthy—the ballooning effect (edema) is a build-up of water, not tissue. Today, nearly 12 million children under 5 die of malnutrition each year. The vast majority of the 24,000 people that die each day because of hunger or diseases related to hunger are often children under 5, the elderly, and women. Women will often give the food to the men and children first, and feed themselves only after that.

2. **(B)** Chronic diseases are of long duration denoting a disease of slow progress. Examples of chronic diseases would be congestive heart failure, Parkinson's disease, and cerebral palsy. Acute disease is a sudden disease of short duration. An example of an acute disease would be acute renal failure.

3. **(E)** Malaria is a public health problem today in more than 90 countries, inhabited by a total of some 2.4 billion people—40% of the world's population. Worldwide prevalence of the disease is estimated to be in the order of 300 million to 500 million cases each year. More than 90% of all malaria cases are in Sub-Saharan Africa. Mortality due to malaria is estimated to be over 1 million deaths each year. The vast majority of deaths occur among young children in Africa, especially in remote rural areas with poor access to health services.

4. **(B)** Cholera is an acute, diarrheal illness caused by infection of the intestine with the bacterium *Vibrio cholerae*. In the western hemisphere, 1,099,882 cases and 10,453 deaths were reported between January 1991 and July 1995. In the first six months of 2001, there were 106,159 cases of cholera in South Africa.

5. **(D)** Heart disease (ischematic) was responsible for 13.7% of all deaths in the world in 1999. Review table entitled "Leading Causes of Death Worldwide and Rank Per Region."

6. **(B)** Mutations in a DNA sequence can be: (1) transitions (purine to purine or pyrimidine to pyrimidine); transversions (purine to pyrimidine or pyrimidine to purine); or (3) frameshift mutations—the insertion or deletion of one or more (not in multiples of three) nucleotides in the coding region of a gene.

7. **(A)** The Disability-Adjusted Life Year (DALY) is the only quantitative indicator of burden of disease that reflects the total amount of healthy life lost, to all causes, whether from premature mortality or from some degree of disability during a period of time.

8. **(E)** Giardia has become recognized as one of the most common causes of water-borne disease (drinking and recreational) in humans in the United States. Giardia may be found in soil, food, water, or surfaces that have been contaminated with the feces from infected humans or animals. Cholera, produced from flagellated bacteria of the family *Vibrio*, are one of the most common organisms in surface waters of the world. In its extreme manifestation, cholera is one of the most rapidly fatal illnesses known. A healthy person may

become hypotensive within an hour of the onset of symptoms and may die within 2 to 3 hours if no treatment is provided. Cryptosporidium is similar in description to giardia and cholera except that it is not produced from a flagellated protozoan; instead, it is produced from sporulated oocysts—bacterial spores that have thick walls and that are able to withstand varying temperatures, humidity, and other unfavorable conditions.

9. **(C)** The science of toxicology is based on the principle that there is a relationship between a toxic reaction (the response) and the amount of poison received (the dose or agonist). An important assumption in this relationship is that there is almost always a dose below which no response occurs or can be measured. A second assumption is that once a maximum response is reached, any further increases in the dose will not result in any increased effect.

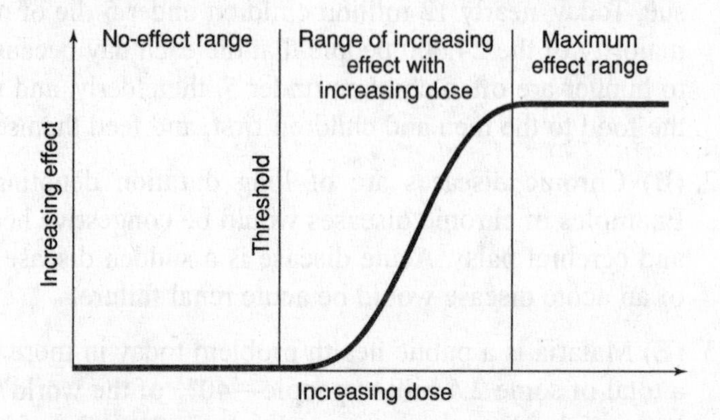

10. **(B)** The LD_{50} (Lethal Dose, 50%) value is typically expressed in milligrams of material per kilogram of subject-body-weight, and indicates the quantity of material that, if administered to a population of subjects, will cause 50% of the subjects to perish.

11. **(E)** Risk assessment is the process evaluating the likelihood of an adverse health effect, with some statistical confidence for various levels of exposure. Risk assessment does not determine what level of risk is allowable or acceptable—that is a part of risk management.

12. **(E)** Examine table entitled "Leading Causes of Death Worldwide and Rank Per Region."

13. **(B)** A host is a living organism that affords subsistence or lodgment to an infectious agent. An intermediate host is a host in which the infectious agent is in a larval or asexual state. A definitive host is a host in which the infectious agent attains maturity or passes its sexual stage.

14. **(D)** Rachel Carson, in her book *Silent Spring*, first brought the problem of biomagnification to the public's attention. She described the devastating effect that human produced toxins have on wildlife. Biomagnification is the increase in the concentration of toxic substances as one moves up the food chain. With each increment up in the food chain the concentration level of toxins may increase tenfold (10×). As a result, those animals at the top of the food chain get the highest concentration of toxins and experience the worst effects.

15. **(D)** This question allows us to go back and review the factor-label method. Recall that it takes 1,000 million to equal 1 billion ($10^3 \times 10^6 = 10^9$).

$$\frac{50 \text{ parts}}{\cancel{\text{million}}} \times \frac{1000\ \cancel{\text{million}}}{1 \text{ billion}} = \frac{50{,}000 \text{ parts}}{\text{billion}}$$

UNIT V: GLOBAL CHANGES AND THEIR CONSEQUENCES

CHAPTER 13

Cause and Effect Relationships

We give you thanks, most gracious God, for the beauty of the Earth and sky and sea;
for the richness of mountains, plains, and rivers;
for the songs of birds and the loveliness of flowers.
We praise you for these good gifts and pray that we may safeguard them for our posterity.
Grant that we may continue to grow in our grateful enjoyment of your abundant creation,
to the honor and glory of your name, now and forever.
Amen.

Book of Common Prayer

Areas That You Will Be Tested On

A: First-Order Effects

1. Atmosphere: CO_2, CH_4, Stratospheric Ozone
2. Oceans: Surface Temperatures, Currents
3. Biota: Habitat Destruction, Introduced Exotics, Overharvesting

B: Higher-Order Interactions

1. Atmosphere: Global Warming, Increasing Ultraviolet Radiation
2. Oceans: Increasing Sea Level, Long-Term Climate Change, Impact on El Niño
3. Biota: Loss of Biodiversity

Key Terms

Note: Definitions for all key terms listed below can be found in the glossary starting on page 648. For additional definitions of relevant terms from this chapter, refer to *www.barronseduc.com/0764121618.html.*

aerosols
biodiversity
Clean Air Act (Title VI)
Global Warming Potential
Intergovernmental Panel on Climate Change (IPCC)
Kyoto Protocol
National Oceanic and Atmospheric Administration (NOAA)
ozone depletion potential
polar stratospheric clouds

Key Concepts

The Atmosphere

CO_2

Visit http://cdiac.esd.ornl.gov/trends/trends.htm for detailed information on CO_2

- Joseph Black (Scotland) discovered CO_2 in 1754.
- The relationship between global temperatures and carbon dioxide concentration is credited to Svante Arrhenius in 1896.
- CO_2 was initially released into the atmosphere from the Earth's molten surface during cooling. Current sources of CO_2 include burning fossil fuels (~5.5 billion metric tons of carbon per year), decay of vegetation, volcanic eruptions, cellular respiration, and deforestation (~1.5 billion metric tons of carbon per year, mostly from Africa and Asia, which deplete carbon sinks). CO_2 may be offset by an increase in forests (aggrading) occurring in Europe and North America. Timber harvesting is balanced by regrowth in forests so it does not significantly contribute to depletion of carbon sink and increase in flux (rate) to atmospheric carbon dioxide.
- CO_2 concentrations have approximately doubled in the last 150,000 years (see Figure 13.1). Volume of CO_2 in the atmosphere has increased ~25% in the last 300 years, primarily through burning of fossil fuels and deforestation. Current level is 0.0360% by volume (360 parts per million) and increasing at a rate of about 0.4% per year (1.5 ppm per year). Human activity is responsible for approximately 5500 million tons of CO_2 added to the atmosphere annually. Residence time of a carbon dioxide molecule in the atmosphere is approximately 100 years. The rapid increase in CO_2 coincided with the Industrial Revolution. In 1992, the United States produced 5.4 tons of carbon per person per year; Canada produced 4.2 tons; Russia, 4.0 tons; and China, 0.6 tons of carbon per person per year.

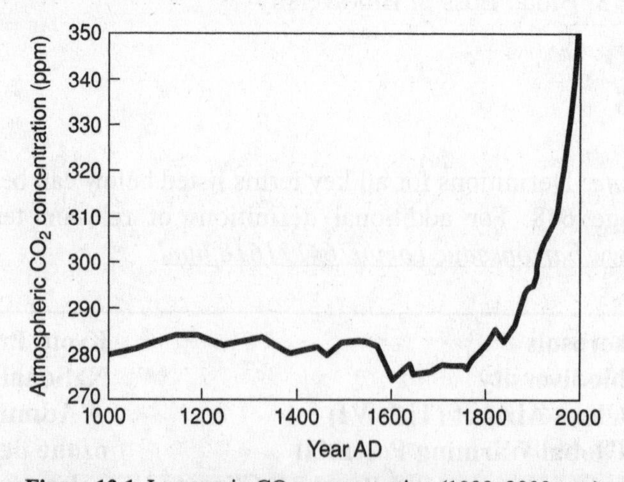

Figure 13.1. Increase in CO_2 concentration (1000–2000 A.D.).

- Sources of CO_2 output include burning of fossil fuels for industry, transportation, space heating, cooking, and generation of electricity by using fossil fuels. 65% of increase in CO_2 is due to activities associated with fossil fuels; the balance is due to deforestation and conversion of natural ecosystems for agricultural purposes. Natural wetlands, forests, and grasslands are able to absorb up to 100 times more carbon dioxide per acre than the same acre converted to agriculture. For the immediate future, the amount of CO_2 contributed by less-developed countries should increase as percentage from more-developed countries should decline. In 1950, the United States contributed over 50% of the world's CO_2, whereas, in 1980, the amount had decreased to around 25%. Likewise, in 1950, the less-developed countries contributed less than 10% of the world's CO_2, and in 1980, the figure increased to over 20%.

SPECIFIC SOURCES FOR CO_2 PRODUCTION IN THE UNITED STATES

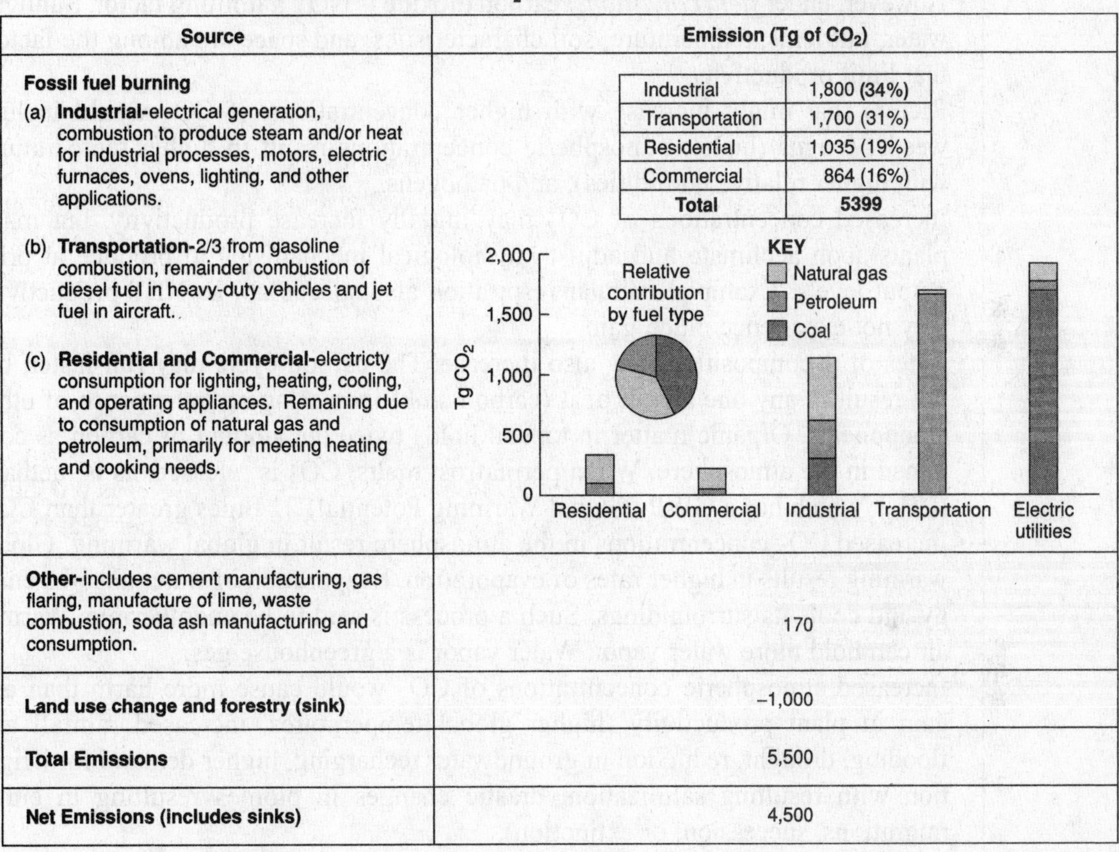

Source	Emission (Tg of CO_2)
Fossil fuel burning (a) **Industrial**-electrical generation, combustion to produce steam and/or heat for industrial processes, motors, electric furnaces, ovens, lighting, and other applications. (b) **Transportation**-2/3 from gasoline combustion, remainder combustion of diesel fuel in heavy-duty vehicles and jet fuel in aircraft.. (c) **Residential and Commercial**-electricty consumption for lighting, heating, cooling, and operating appliances. Remaining due to consumption of natural gas and petroleum, primarily for meeting heating and cooking needs.	Industrial 1,800 (34%) Transportation 1,700 (31%) Residential 1,035 (19%) Commercial 864 (16%) **Total 5399**
Other-includes cement manufacturing, gas flaring, manufacture of lime, waste combustion, soda ash manufacturing and consumption.	170
Land use change and forestry (sink)	–1,000
Total Emissions	5,500
Net Emissions (includes sinks)	4,500

- Cellular respiration produces carbon dioxide: $C_6H_{12}O_6 + 6O_2 \rightarrow 6CO_2 + 6H_2O$ + Energy.
- Photosynthesis requires carbon dioxide: $6CO_2 + 6H_2O$ + energy $\rightarrow C_6H_{12}O_6 + 6O_2$.
- For most of Earth's modern history, there was a balance between consumption of CO_2 through photosynthesis and production through cellular respiration. An exception was the Carboniferous Period, which occurred from about 360 to 286 million years ago during the late Paleozoic Era when most of the sources of our present fossil fuels were produced. Carboniferous conditions (warm and moist) allowed profusion of plants to store atmospheric carbon, which we release back into the environment today by burning fossil fuels. The biological carbon cycle (photosynthesis-respiration) involves 20 times the amount of carbon as released by the burning of fossil fuels each year.
- Approximately 100 million years ago, during the Cretaceous Period, estimates are that there was ten times the amount of CO_2 in the atmosphere as there is today and that global temperatures were significantly higher than at present. Theories include massive plate tectonic movement that released CO_2 through volcanic activity.

Free-Response Outline

Practice brainstorming *possible solutions to the question:*

If photosynthesis requires carbon dioxide, wouldn't *more* carbon dioxide result in higher crop yields and be beneficial?

- Under *laboratory conditions,* with carbon dioxide held as the limiting factor, it is true, plant productivity increases with higher concentrations of carbon dioxide (there are limits).
- However, under *field conditions,* carbon dioxide is NOT a limiting factor. Sunlight, water, nutrients, temperature, soil characteristics, and space are among the factors that limit productivity
- Factors that might increase with higher concentrations of CO_2 would include weeds, fungus (higher atmospheric concentrations result in higher temperatures with higher relative humidities), and pathogens.
- Increased concentrations of CO_2 may initially increase productivity, but many plants soon acclimate and adjust physiological mechanisms to produce at prior output levels. Example: cellular respiration also speeds up, so NET productivity may not experience much gain.
- Rates of decomposition may also increase. The carbon cycle may run faster, but not result in any one aspect of it (carbon sink) increasing at the expense of other components. Organic matter in topsoil holds twice the amount of carbon as contained in the atmosphere. When permafrost melts, CO_2 is released, as is methane (CH_4), which has a GWP (Global Warming Potential) 21 times greater than CO_2.
- Increased CO_2 concentrations in the atmosphere result in global warming. Global warming results in higher rates of evaporation. Evaporation of water absorbs energy and cools its surroundings. Such a process is said to be endothermic. Warmer air can hold more water vapor. Water vapor is a greenhouse gas.
- Increased atmospheric concentrations of CO_2 would cause more harm than any gain in plant productivity (higher global temperatures, increased rainfall and flooding, drought, reduction in groundwater recharging, higher demand for irrigation with resulting salinization, drastic changes in biomes resulting in either migrations, succession, or extinction).
- Increased atmospheric CO_2 concentration raises global temperatures (CO_2 is a greenhouse gas). Any effect of increased photosynthetic productivity due to increased amounts of CO_2 would be offset by the stress plants reacting to higher temperatures (especially in areas that do not receive additional precipitation). Higher heat stress would place additional demands on finite freshwater sources.
- Higher atmospheric concentrations of CO_2 result in higher global temperatures. Higher global temperatures also mean higher ocean temperatures. CO_2 would diffuse out of the oceans more readily at higher temperatures, feeding and compounding a positive feedback cycle (oceans hold 50 times more carbon than the atmosphere).

- Major carbon sink is in sedimentary rock. However, this cycle, known as the geochemical carbon cycle, is a very slow cycle in which carbon is released through volcanic activity. Other major sinks are forests, soil, ocean water (forms carbonic acid $CO_2 + H_2O \Leftrightarrow H_2CO_3$), and fossil fuels including oil, methane, peat, and coal.
- CO_2 is a greenhouse gas responsible for approximately 55% of the effect of global warming.
- CO_2 absorbs infrared radiation that is emitted from the Earth's surface by solar heating (see Figure 13.2). Reradiation of infrared energy back to Earth's climatic system keeps global temperatures at an average of 60°F (16°C). If not for this effect, global temperatures would average 0°F (–18°C), and life as we know it would not exist. During glacial periods, CO_2 levels were about 30% lower. Carbon dioxide diffuses in ocean water. As temperature increases, less CO_2 is able to be dissolved. A cycle exists, known as the Milankovitch cycle, in which it is theorized that less solar radiation caused ocean temperatures to decrease, which allowed more CO_2 to be dissolved, resulting in less CO_2 in the atmosphere, which resulted in less of the greenhouse effect, which caused global temperatures to decline, which resulted in ice ages.

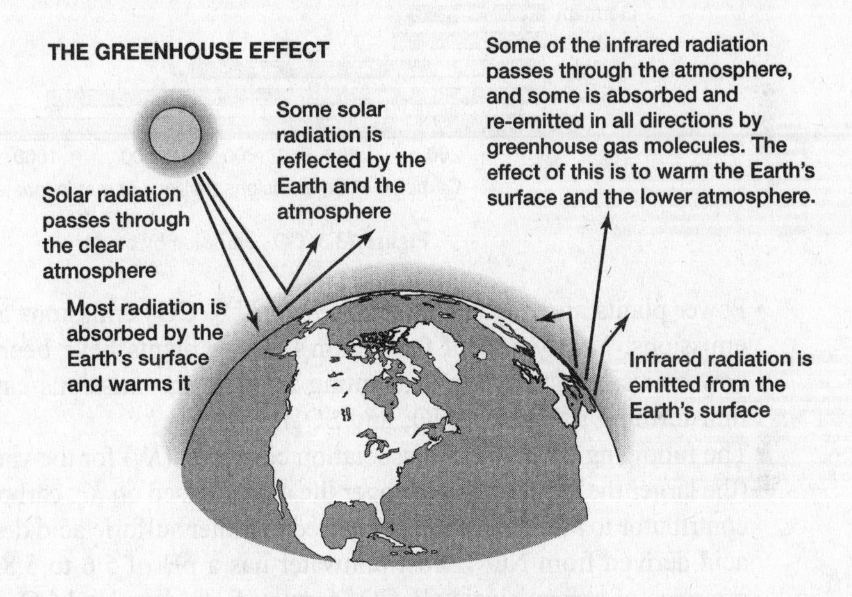

Figure 13.2. The greenhouse effect.

- Computer models show that doubling CO_2 concentration correlates to an increase in average global temperatures from 1 to 3°C. At current rates of CO_2 output, estimates indicate that CO_2 may increase 80% over present levels by 2050, up to 600 ppm, resulting in an increase in global temperatures between 1.5 and 4.5°C.
- The 1997 Kyoto Protocol would have required the United States to reduce greenhouse emissions by 7%, against 1990 levels, over a five-year period (see Figure 13.3). Under the protocol, the United States would have faced penalties if it did not meet its emissions cuts. The United States saw this as an unattainable target since CO_2, and greenhouses gases overall, continues to increase and are projected by the Energy Information Administration to increase for the next 20 years. The Bush administration felt that the protocol held developed nations responsible for meeting the cuts but did not apply the same standards to developing nations. President Bush said he withdrew support for the Kyoto Protocol because the cost of meeting the emissions targets would have been too high and the timeframe too short for implementation. He also said he

did not see a definite correlation between greenhouse gases, which include CO_2, and global warming. In a letter to Senator Chuck Hagel, Bush said, "I do not believe . . . the government should impose on power plants mandatory emission reductions for carbon dioxide, which is not a 'pollutant' under the Clean Air Act."

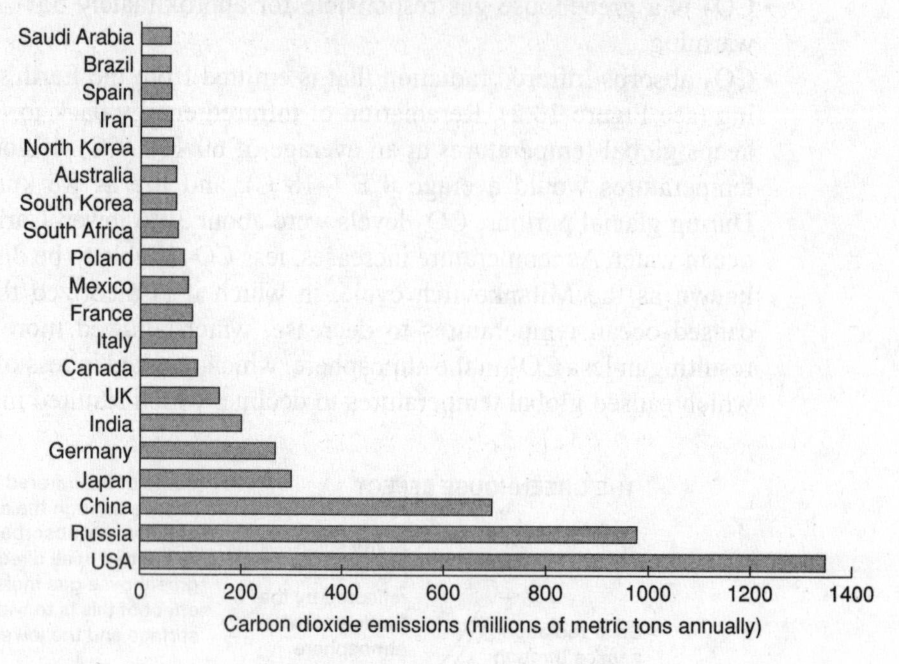

Figure 13.3. CO_2 emission by country.

- Power plants are responsible for 36% of U.S. CO_2 emissions and 32% of all mercury emissions. Nearly 600 of the nation's power plants have been "grandfathered" from new emissions standards, exempting them from emissions caps for smog- and acid-rain-forming pollutants (NO_x and SO_x).
- The following table shows dissociation constants (K_a) for the various atmospheric gases (the larger the number, the stronger the acid). Based on K_a, carbon dioxide is not a major contributor to acid rain when compared to either sulfuric acid derived from SO_2 or nitric acid derived from NO_2. Most rainwater has a pH of 5.6 to 5.8, simply because of the presence of carbonic acid (H_2CO_3 formed from dissolved CO_2 gas and H_2O).

Gas	Acid Formation	Acid Name	Dissociation Constants (K_a) @ 25°C
CO_2	$CO_2 + H_2O = H_2CO_3$	Carbonic	4.3×10^{-7}
SO_2	$2SO_2 + O_2 = 2SO_3$ $SO_3 + H_2O = H_2SO_4$	Sulfuric	1
NO_2	$NO_2 + H_2O = HNO_2 + HNO_3$	Nitrous Nitric	4.5×10^{-4} 1

- Ocean water holds up to 60 times more CO_2 than the atmosphere and is a major sink. Algae consume CO_2 and are either consumed by higher trophic organisms or die and sink to the ocean floor, eventually forming carbonate deposits.
- The equatorial Pacific is the largest oceanic source of carbon dioxide to the atmosphere. Interannual variability of CO_2 fluxes (rate of movement) to the atmosphere in

the central and eastern equatorial Pacific are primarily affected by changes in the rate of upwelling of CO_2-enriched waters from the equatorial undercurrent and advection of CO_2-depleted waters from the west at the surface along the equator. Both processes are affected strongly by El Niño. Net result or influx: oceans are absorbing more CO_2 than they are releasing, about 2 billion tons per year of carbon. This results in the oceans absorbing about 38% of the CO_2 released by human (anthropogenic) activity.

- Scientific evidence for CO_2 concentrations in older atmospheres comes from analysis of air bubbles trapped in glacial ice. Estimates are that up to 50% of past climatic changes may be accounted for by changes in greenhouse gases.

CH_4

- Methane concentration in the atmosphere is about 0.00017% by volume or 1.7 ppm. Methane has about a 12-year residence time in the atmosphere. Since 1800, there has been about a 150% increase in methane concentration in the atmosphere as compared to about a 30% increase in carbon dioxide. In 1900, the CH_4 concentration was 900 ppb; in 2030 it is projected to be 2200 ppb (see Figure 13.4).

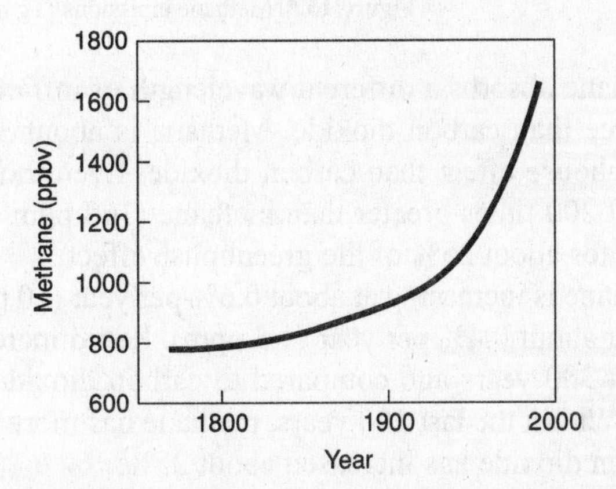

Figure 13.4. Methane concentration in the atmosphere (1750–2000 A.D.).

- During the Archean (3.8 billion to 2.5 billion years ago), methane was probably the most abundant gas (1000 times greater concentration than today) and was produced by methanogenic bacteria—organisms that create methane from organic material or hydrogen and carbon dioxide. Oxygen that was produced by cyanobacteria was immediately reduced. Sunlight acting on the methane produced polyacetylenes or smog—the early atmosphere was smoggy!
- Sources of methane include decay of organic matter including landfills; wetlands (low oxygen swamps, ~120 Tg · yr^{-1}); termites; natural gas, coal and oil extraction (100 Tg · yr^{-1}); burning of biomass; rice cultivation (flooding) (50 Tg · yr^{-1}); cattle and other animals (enteric fermentation, ~80 Tg · yr^{-1}), and bacterial (methanotrophic) action that occurs when wetlands thaw. Human (anthropogenic) activity produces about 340 million metric tons (340 Tg) of methane per year, whereas nonhuman activity releases about 160 Tg per year. Wetland source of methane production is being reduced because of wetland destruction, but all other sources are increasing (see Figure 13.5).

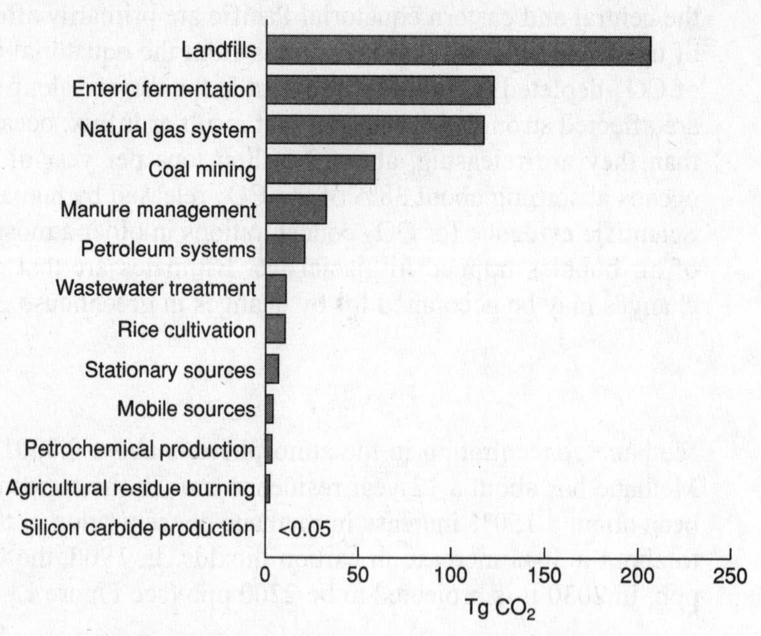

Figure 13.5. Methane emissions (Tg of CO_2 equivalents).

- Methane absorbs a different wavelength of infrared radiation coming from the Earth's surface than carbon dioxide. Methane is about 30 times more effective in causing a greenhouse effect than carbon dioxide. Even though carbon dioxide concentration is about 200 times greater than methane (360 ppm versus 1.7 ppm), methane is responsible for about 25% of the greenhouse effect.
- Methane is increasing at about 0.6% per year (10 ppb), while carbon dioxide is increasing at about 0.4% per year (1.5 ppm). Rapid increase in methane concentration began about 300 years ago compared to carbon dioxide rapidly increasing about 150 years ago. Within the last 100 years, methane has more than doubled in concentration, while carbon dioxide has increased about 25%.
- The main loss of CH_4 in the atmosphere is in its reaction with the OH radical and its loss to the stratosphere (about 12%). Other losses are due to uptake by dry soils (about 10%).

Stratospheric Ozone

- Christian Friedrich Schönbein first described ozone in 1840 as a product of electrolysis of water, and in 1845 as a "component of atmospheric air, which may play a part in slow oxidation."
- The stratosphere contains most (97%) of the ozone in the atmosphere and lies between 6 and 31 miles (10–50 km) above the Earth's surface (see Figure 13.6). Ozone is found in only about one molecule out of every 100,000 air molecules. Most of the ozone in the stratosphere occurs about 15 miles (24 km) above the Earth's surface. Above 22 miles, the densities of gases are so low that oxygen atoms rarely collide with oxygen atoms. Below 15 miles in height, not enough UVC radiation penetrates to dissociate molecular oxygen. Ozone is thinnest above the tropics and thickest above the poles. The increase in temperature with height occurs because of absorption of ultraviolet (UV) radiation from the sun by this ozone. Temperatures in the stratosphere are highest over the summer pole and lowest over the winter pole.

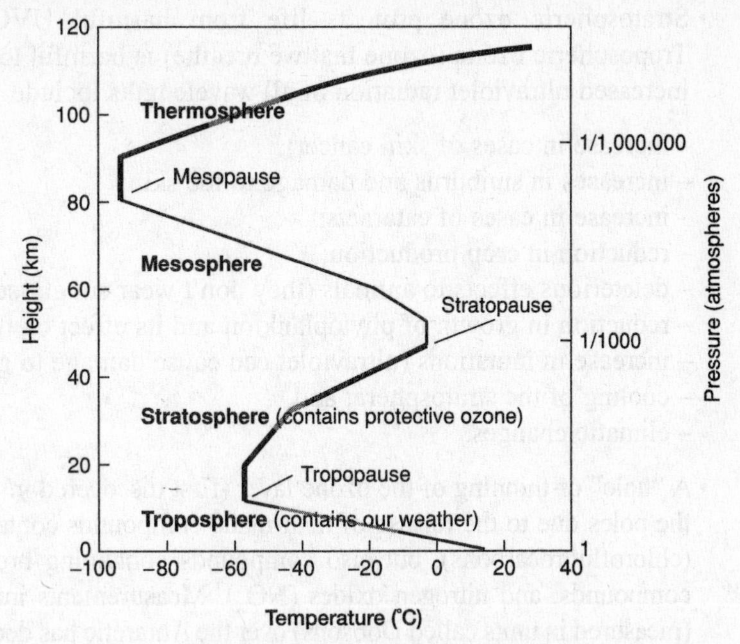

Figure 13.6. Layers of the atmosphere.

- Ozone is made in the stratosphere by the reaction of atomic oxygen combining with oxygen molecules in the presence of sunlight. When ozone absorbs ultraviolet, it degrades to oxygen and atomic oxygen through the following step-reactions known as Chapman reactions:

$$O_2 + hv \rightarrow O + O \text{ (1)}$$

$$O + O_2 \rightarrow O_3 \text{ (2)}$$

(wavelength < ~ 240 nm)

Although the UVC radiation splits the ozone molecule, ozone can reform through the following reactions resulting in no net loss of ozone and is the reaction that protects living organisms on Earth from harmful ultraviolet radiation:

$$O_3 + hv \rightarrow O_2 + O \quad \text{(3)}$$

$$O + O_2 \rightarrow O_3 \quad \text{(2)}$$

Ozone is also destroyed by the following reaction:

$$O + O_3 \rightarrow O_2 + O_2 \quad \text{(4)}$$

Reaction (2) becomes slower with increasing altitude, while reaction (3) becomes faster. The concentration of ozone is a balance between these competing reactions. In the upper atmosphere, atomic oxygen dominates where UVC levels are high. Moving down through the stratosphere, the air gets denser, UVC absorption increases, and ozone levels peak at approximately 12 miles (19 km). Closer to the ground, UVC levels decrease, and ozone levels decrease.

- Stratospheric ozone protects life from harmful UVC radiation from the sun. Tropospheric ozone (ozone that we breathe) is harmful to life. Harmful effects due to increased ultraviolet radiation of all wavelengths include

  – increase in cases of skin cancer;
  – increases in sunburns and damage to the skin;
  – increase in cases of cataracts;
  – reduction in crop production;
  – deleterious effects to animals (they don't wear sunglasses);
  – reduction in growth of phytoplankton and its effect on the food web;
  – increase in mutations (ultraviolet can cause damage to genes);
  – cooling of the stratosphere; and
  – climatic changes.

- A "hole" or thinning of the ozone layer (first discovered in 1985) seasonally occurs over the poles due to the release of man-made compounds containing chlorine such as CFCs (chlorofluorocarbons), but also compounds containing bromine, other related halogen compounds, and nitrogen oxides (NO_x). Measurements indicate that the loss of ozone (measured in units called Dobsons) over the Antarctic has decreased as much as 60% since the late 1970s, with an average net loss of about 3% per year worldwide (see Figure 13.7). CFCs were first manufactured by General Motors Corporation in the late 1920s. Primary use for CFCs is as a refrigerant (Freon). They are also used as aerosol propellants, as electronic part cleaners, and in the manufacture of foam products. The largest single source of CFCs to the atmosphere is leaking car air conditioners. The average residence time of CFC is 200 years, and by 1988, over 300,000 metric tons had been produced.

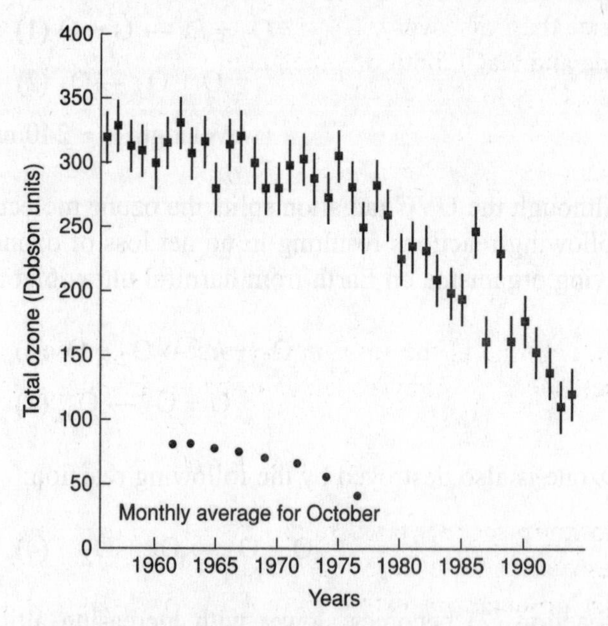

Figure 13.7. Ozone levels over the Antarctic, 1955–1995.

- The 1987 Montreal Protocol, and subsequent meetings—London Agreement 1988 and Copenhagen Treaty 1992—sped up time tables and called for complete elimination of CFCs of more-developed countries by 1996, with complete elimination in less-developed countries by 2010. Europe and North America used two-thirds of all CFCs in 1991. Result and compliance of world treaties means that chlorine and bromine levels will begin to decrease in the stratosphere by 2000, with increase of stratospheric ozone to the 1979 level.

- Ultraviolet radiation causes the decomposition of CFCs in the stratosphere, producing atomic chlorine (Cl).

$$CCl_2F_2 \xrightarrow{\text{photon of UV radiation}} \cdot CClF_2 + \cdot Cl$$

The atomic chlorine reacts with ozone in the stratosphere to produce the free radical, chlorine monoxide, which then reacts with ozone to produce more atomic chlorine and molecular oxygen.

$$\cdot Cl + O_3 \longrightarrow \cdot ClO + O_2$$
$$\cdot ClO + O_3 \longrightarrow \cdot Cl + 2\,O_2$$

The chlorine acts as a catalyst to destroy vast amounts of ozone during these propagation steps. Generally, one chlorine atom destroys about 100,000 ozone molecules. The reaction will not stop until two radicals collide and form a nonradical species. Because there is a low concentration of free radicals and the reaction is in the gas phase, termination is rare, consequently, the chlorine radicals have a long atmospheric residence. In fact, destruction of ozone that is occurring now is a result of CFCs released since the late 1920s because the residence time of CFCs in the stratosphere is 100 years. The problem will get much worse before it gets better since it takes up to 8 years for a CFC molecule to reach the stratosphere and CFC use was growing at about 5% per year since their discovery. The only other sources of Cl in the atmosphere are methyl chloride and NaCl, both of which come from ocean water, and HCl from volcanoes, all of which are in low atmospheric concentrations, and get washed out of the atmosphere in the troposphere.

- Chlorine, in the form of chlorine monoxide, is removed from the atmosphere by reacting with nitrogen dioxide forming chlorine nitrate:

$$ClO + NO_2 \rightarrow ClONO_2$$

Or, chlorine in the form of the radical, atomic chlorine, is removed by reacting with methane to form hydrogen chloride (hydrochloric acid) and a methyl radical, CH_3.

$$CH_4 + \cdot Cl \rightarrow HCl + \cdot CH_3$$

However, these two reactions that would tie up chlorine radicals and slow down ozone destruction is severely inhibited by the presence of a polar vortex (swirling cold air) that produces an extremely cold (–85°C or –121°F), nitrogen-rich cloud above the Antarctic polar regions during the winter and early spring, known as polar stratospheric clouds (PSC). The PSCs prevent gaseous reactions and provide a surface for reactions to occur involving free chlorine. PSCs tend to slow down the removal of chlorine from the stratosphere and reduce the effectiveness of the last two equations, resulting in a significant removal of ozone during this time of year. Feedback loops that exacerbate the problem include (1) decreased stratospheric ozone results in less ultraviolet being absorbed in the stratosphere which results in colder temperatures—

prolonging PSCs; (2) increased CFCs add to the greenhouse effect and result in increased tropospheric temperatures and decreased stratospheric temperatures; and (3) sulfuric acid molecules (which come from SO_2 released during the burning of fossil fuels and that contribute to acid rain) create surfaces where reactions occur that tend to slow down the removal of chlorine from the stratosphere and reduce removal of free chlorine radicals. In other words, a pollutant (SO_2) that is responsible for acid rain and tropospheric air pollution, increases the rate of ozone decomposition but at the same time decreases the rate of global warming.

- Ozone depletion is not as severe over the Arctic because (1) the Antarctic is colder, thus prolonging PSCs, and (2) Arctic stratosphere warms faster in spring. Ozone loss increases as latitude increases, with winter and spring being the most severe time of the year.
- Bromine, found in much less quantities than chlorine, is about 50 times more effective than chlorine in its effect on stratospheric ozone depletion. It is responsible for about 20% of the "ozone hole." Bromine is found in halons, which are used in fire extinguishers, methyl bromide, which is used in fumigation and agriculture and is naturally released from marine phytoplankton and biomass burning.
- Alternatives to CFCs are (1) HCFC, which replaces a chlorine with a hydrogen but is still capable of destroying ozone albeit less effectively because it breaks down more readily in the troposphere; (2) helium, ammonia, propane or butane as a coolant (helium-cooled refrigerators use 50% electricity); (3) environmentally friendly products (terpenes from citrus rinds); and (4) alternatives to halons used in fire extinguishers.
- On an individual level, remediation takes the form of (1) using pump sprays instead of aerosol spray cans when possible; (2) complying with disposal requirements of the Clean Air Act for old refrigerators and car air conditioners; (3) reading labels and, when choices are available, "ozone friendly" products; and (4) supporting legislation that reduces ozone-destroying products.

Global Warming

- The Earth is 5 to 9°F warmer today than it was 10,000 years ago during the last ice age. Global mean surface temperatures have increased 0.5 to 1.0°F since the late 19th century. The 20th century's 10 warmest years all occurred in the last 15 years of the century. Scientists expect that the average global surface temperature could rise 1 to 5°F (0.5 to 3°C) in the next fifty years, and 2 to 10°F (1.5 to 6°C) in the next century, with significant regional variation. Before the Industrial Revolution, increases in global mean temperatures were significantly less (1 to 2°C per 1000 years) than current or projected rates. (see Figure 13.8.)
- Data show that the increases in global temperatures are more pronounced at night than during the day. Possible reasons include (1) burning fossil fuels releases particulate matter and sulfates into the atmosphere that tend to reflect incoming solar radiation during the day; and (2) warmer temperatures increase rates of water evaporation, which form more clouds that tend to trap heat close to the ground at night.
- "Proxy data" from objects that are sensitive to climatic phenomena can be used to provide estimates of past climate conditions, such as temperature, precipitation, or wind speed. Proxy data includes tree ring widths, ice cores, pollen deposits, glacier lengths, coral samples, CO_2 trapped in layered lake sediments, and deep-sea sediments. Furthermore, ratios of oxygen-16 to oxygen-18 and other isotopes of hydrogen are dependent on temperatures prevalent at the time the ice was formed. Ice core samples

(Vostoc ice core sample from Antarctica, which was collected over 1 mile down through the ice) give information on climatic and atmospheric conditions as far back as 160,000 years ago. Data show that temperatures dropped as CO_2 and methane levels dropped and that temperatures were about 5°C warmer during interglacial periods with CO_2 concentrations being about 25% higher.

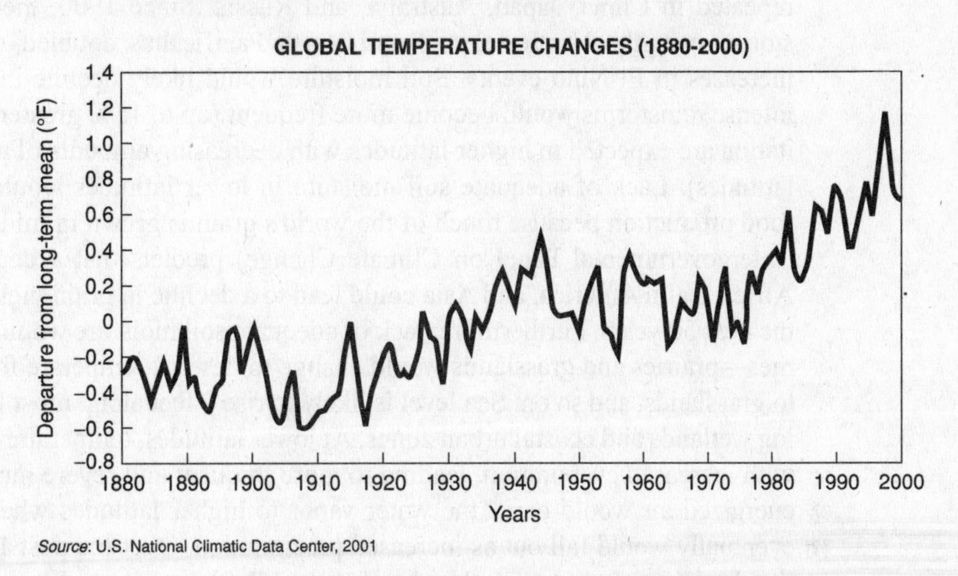

Figure 13.8. Global temperature change, 1880–2000.

- Water vapor (0 to $4\%_{vol}$), carbon dioxide ($0.036\%_{vol}$), ozone ($0.000004\%_{vol}$), methane ($0.00017\%_{vol}$), nitrous oxide ($0.00003\%_{vol}$), and chlorofluorocarbons make up greenhouse gases. Solar radiation penetrates the atmosphere and warms up the Earth's surface. The earth's surface radiates this energy in the form of infrared (heat) energy, which either (1) escapes into space or (2) is absorbed by molecules of greenhouse gases that reradiate it back to the lower atmosphere, causing the air temperature to increase. Without greenhouse effect, the Earth's surface temperature would average 0°F.
- Global warming potential (GWP) has been developed to compare the ability of each greenhouse gas to trap heat in the atmosphere relative to another gas. Carbon dioxide (CO_2) was chosen as the reference gas. GWPs are not provided for the criteria pollutants CO, NO_x, VOCs, and SO_2 because there is no agreed-upon method to estimate the contribution of these gases to the greenhouse effect.

VARIOUS GREENHOUSE GASES

Greenhouse Gas	**GWP**
Carbon dioxide (CO_2)	1
Methane (CH_4)	21
Nitrous oxide (N_2O)	310
HFC-23 (a replacement for CFC)	11,700
CF_4	6,500
SF_6	23,900

- Higher temperatures would cause increased evaporation from the Earth's surface that leads to declines in the amount of precipitation at lower latitudes and increases in the amount of precipitation at higher latitudes. Since 1910, there has been an overall increase of 10% in precipitation in the continental United States, mostly during winter, and the intensity of storms has likewise increased (see Figure 13.9). This pattern is repeated in China, Japan, Australia, and Russia. Since 1900, the number of intense storms over the North Atlantic and North Pacific has doubled, with corresponding increases in El Niño events. Soil moisture would likely decline in many regions, and intense rainstorms would become more frequent (up to 15% greater amounts of precipitation are expected in higher latitudes with decreasing amounts of precipitation in midlatitudes). Lack of adequate soil moisture in lower latitudes would drastically reduce food production because much of the world's grain is grown in midlatitudes. The IPCC (Intergovernmental Panel on Climate Change) predicts that a decrease in rainfall in Africa, Latin America, and Asia could lead to a decline in grain yield of 10 to 15% over the next 50 years. Furthermore, lack of adequate soil moisture would cause shifts in biomes—prairies and grasslands would change to deserts, temperate forests would change to grasslands, and so on. Sea level is likely to rise 2 feet along most U.S. coasts, destroying wetlands and coastal urban zones. At lower latitudes, temperatures would be warmer with increased evaporation, leading to more frequent and severe hurricanes. The highly energized air would carry the water vapor to higher latitudes where the excess water eventually would fall out as increased precipitation. Over the past 100 years, precipitation has increased globally by about 1%, while the amount of precipitation in tropical areas has declined. At the same time, there has been nearly a 5% net increase in precipitation across the United States. Projections are that Arctic temperatures could rise 18–27°F (10–15°C), which would lead to melting of glaciers and permafrost (Europe's Alpine glaciers have lost half their volume since 1850. The U.S. government predicts that there will be no more glaciers left in Montana's Glacier National Park by 2030). As permafrost melted, conifers would begin to grow in the tundra as temperatures became milder. In lower latitudes, coniferous forests would convert to prairies due to higher temperatures and less water availability. Entire species would die out as the climatic change would be faster than the adaptation or evolutionary rate. Lake levels would drop in lower latitudes affecting shipping, commerce, and hydroelectric potential. Water levels in the Great Lakes could decrease by as much as 8 feet.
- Plants and animals that cannot adapt to new conditions will become extinct. Plants and animals around the world are shifting their ranges in an effort to escape changing climatic patterns. The first documented extinction due to global warming was the Costa Rican Golden Toad. North Pacific salmon populations crashed after ocean temperatures in the region soared 9°F (6°C) above normal. Warming seas led to hundreds of thousands of seabird deaths off the coast of California. Coral reefs around the world have been severely damaged by unusually warm ocean temperatures and at the current rate of degradation, the entire Great Barrier Reef could be nonexistent within 75 years. Break up of pack ice means that polar bears in some areas are finding it more difficult to survive.
- Global warming may already be causing the spread of infectious diseases and increasing heat-wave deaths. As temperatures rise, disease-carrying mosquitoes and rodents move into new areas, infecting people in their wake. Scientists at the Harvard Medical School have linked recent outbreaks of dengue ("breakbone") fever, malaria, hanta virus, viral encephalitis, and other diseases to climate change. Since 1990, outbreaks of malaria have occurred in California, Florida, Georgia, Michigan, New Jersey, and

New York. Global warming will likely put as much as 65% of the world's population at risk of malaria—an increase of 20%. West Nile virus, encephalitis, and dengue fever, all of which are spread by mosquitoes, is also increasing in the United States.

- Global warming will worsen heat waves, increasing urban death rates.

HEAT WAVE DEATHS DUE TO GLOBAL WARMING

City	1997 Climate Deaths	2020 Climate Average Deaths*	2050 Climate Average Deaths*
Chicago, IL	191	401 (210%)	497 (260%)
Cincinnati, OH	14	52 (371%)	67 (481%)
Kansas City, MO	49	115 (234%)	127 (260%)
Newark, NJ	26	122 (469%)	146 (562%)
Tampa, FL	28	64 (229%)	81 (288%)
St. Louis, MO	79	160 (200%)	185 (235%)

*Based on computer simulation and modeling.
Source: United Kingdom Meteorological Model, Global Fluid Dynamics Laboratory Model, and Max Planck Institute.

- The cost of property damage is rising at around 10% a year, and between 1970 and 2000, economic losses caused by natural disasters doubled.

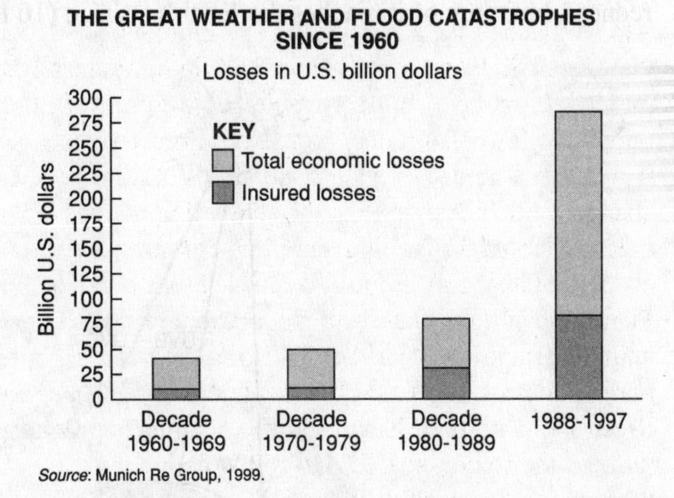

Figure 13.9. Weather and flood catastrophes, 1960–1997. *Source:* Munich Research Group.

- Remediation for global warming includes (1) consuming energy more efficiently; (2) supplying energy cleanly and more efficiently ("cogeneration" that generates electricity and makes use of the waste heat by pumping it to factories and homes); (3) shifting to less-polluting fuels (from coal to natural gas, especially in the power sector); and (4) increasing the share of renewable energy supply for fuels and electricity.
- Advantages of remediation include: (1) avoidance of unnecessary loss of human life and increasingly costly damage from extreme weather events such as floods, droughts, and hurricanes; (2) savings on energy bills; (3) improved comfort and quality of life; (4) simultaneous reductions in other air pollutants, including sulfur dioxide and

nitrogen oxides; (5) opportunities for cleaner industries to create new jobs; (6) more rapid technological innovation and increasing industrial competitiveness; and (7) reduced risk of large-scale species extinctions being triggered by global warming.

- Curbing emission of greenhouse gases is the most effective remediation policy for stemming global warming. Other methods to reduce greenhouse gas emissions include reducing energy use, increasing efficiency (vehicles, appliances, etc.), utilizing alternative energy sources (solar, wind, alternative fuels, etc.), reducing CFCs, slowing rates of deforestation and encouraging reforestation, decreasing tropospheric ozone, reducing dependence on inorganic nitrogen-based fertilizers, and utilizing conservation tillage techniques. Natural sinks operate at much lower rates of absorption than current levels of emissions. To stabilize the situation (not increase or decrease, but to remain the same), the following emissions would need to be instituted:

  – a decrease in methane (CH_4) emission by 8%,
  – a decrease in nitrous oxide (N_2O) by 50%, and
  – a decrease in carbon dioxide (CO_2) between 50 and 80%.

Increasing Ultraviolet Radiation

- The sun emits a wide variety of electromagnetic radiation, including infrared, visible, ultraviolet A (UVA; 320 to 400 nm, longer wavelength radiation, close to the blue in the visible spectrum, and that usually causes skin tanning and browning), ultraviolet B (UVB; 290 to 320 nm, shorter wavelength radiation that causes blistering sunburn and is associated with skin cancer), and ultraviolet C (UVC; 10 to 290 nm, found only in the stratosphere). UVA radiation is 1000-fold less effective than UVB in producing skin redness, but more of it reaches the Earth's surface (10 to 100 times more than UVB).

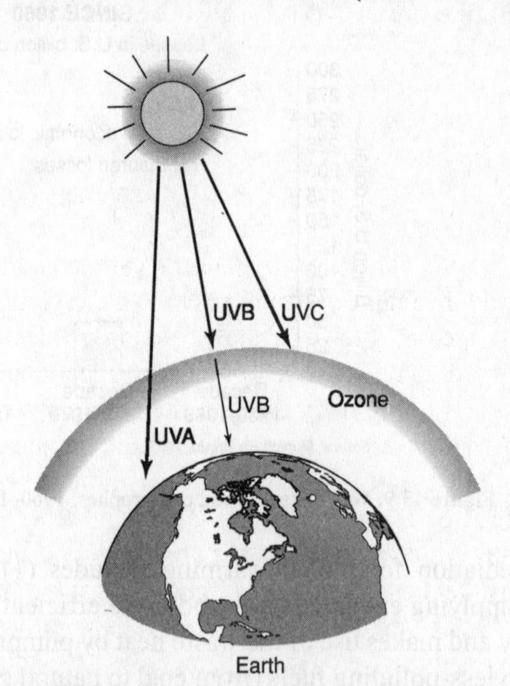

Figure 13.10. Ultraviolet radiation reaching the Earth.

- Average Americans have a 1 in 7 chance of getting some form of cancer during their lifetimes. One case of severe sunburn can increase the chance of skin cancer, the most common form of cancer, by 50%. A 10% decrease in ozone in the stratosphere results in about a 25% increase in melanoma (skin cancer). There has been a 20 to 40% rise in skin cancer in the human population since the 1970s. The incidence of skin cancer rose 90% in the United States from 1974 to 1990. Basal and squamous cell carcinomas affect 600,000 Americans each year primarily because of increased exposure to the sun but also due to a growing population of elderly people and better health screening.

- Susceptibility to damage by ultraviolet radiation may be influenced by genetic and acquired disorders, genetic traits, age-related factors, the use of some medications (oral antibiotics, antihypertensives, immunosuppressive agents, and nonsteroidal anti-inflammatory drugs), race, ethnicity, eye and hair color, and the tendency toward formation of freckles.

CATEGORIES OF CANCER

Carcinomas. 85% of cancers—surface tissue surrounding all organs including skin cancer.

Sarcomas. 2% of cancers—connective tissues such as cartilage, tendons, muscle, and bone.

Leukemias. 4% of cancers—bone marrow and lymphatic system.

Lymphomas. 5% of cancers—grow specifically in the lymph nodes. Later they spill into the bloodstream and spread rapidly.

- Basal skin cancer is the most common type of skin cancer. It arises in the cells at the base of the epidermis (the outer layer of the skin). There are a number of basal cell cancer varieties: (1) resembling a black mole; (2) appearing as an eczema rash or as flat-yellow lesions (morphea); and (3) appearing as a small raised bump that has a smooth, pearly appearance; looks like a scar, and is firm to the touch. Basal skin cancers occur on areas of the skin that have been in the sun. Basal cell cancers may spread to tissues around the cancer, but it usually does not spread to other parts of the body.
- Squamous skin cancer occurs on areas of the skin that have been in the sun, often on the top of the nose, forehead, lower lip, and hands. They may also appear on areas of the skin that have been burned, exposed to chemicals, or had X-ray therapy. Often this cancer appears as a firm red bump. Sometimes the tumor may feel scaly or bleed or develop a crust. Squamous cell tumors may spread to the lymph nodes in the area.
- Melanoma skin cancer comprises malignant cells that are found in the cells and color the skin (melanocytes). Melanoma can spread (metastasize) quickly to other parts of the body through the lymph system or the blood. Warning signs include change in the size, shape, or color of a mole, oozing or bleeding from a mole, or a mole that feels itchy, hard, lumpy, swollen or tender to the touch. If untreated, it can spread quickly to other parts of the body through the lymph system or through the blood. Four kinds of treatments are used to treat melanoma: surgery, chemotherapy, radiation therapy, and biological therapy (using the body's immune system to fight cancer).
- The lens of the eye absorbs UVB. Evidence links increased UVB to cataracts. A cataract is a cloudy or opaque area in the normally transparent lens of the eye. As the opacity thickens, it prevents light rays from passing through the lens and focusing on the retina. Symptoms include blurred vision, sensitivity to light and glare, increased nearsightedness, or distorted images. Livestock in areas near the Antarctic (Chile) show increased rates of blindness.
- Evidence exists between diminished immune systems and increases in UVB. Low doses of UVB can reduce antigen-presenting capacity of Langerhans cells, block normal effector pathways, and invoke inappropriate cellular responses by activating T-suppressor networks that cause keratinocytes to release damaging compounds. UVB also

causes unique dermal damage such as alterations in architecture, matrix composition, vascular structure and function, and cellular activities.

- Plants do not utilize UVB in photosynthesis. The UVB has a negative impact on photosynthesis and damages nucleic acids. Soybeans, legumes, and loblolly pines are particularly sensitive to UV radiation. Soybeans are the third largest food crop in the United States. Loblolly pines are the source of more than two-thirds of the wood pulp formed for paper manufacturing. Conifers are more tolerant.
- UVB penetrates to about 30 feet down in water. Phytoplankton (oxygen producers) and fish larvae and their role in the food web are negatively impacted by UVB. Evidence shows about a 10% decrease in phytoplankton productivity near the Antarctic.
- Significant decreases in amphibian populations may be linked to increases in UVB exposure. Amphibian eggs exposed to UVB experience mortality (up to 100%) or increased incidences of deformities (due to UVBs influence on nucleic acids).
- The UV Index describes the next day's likely levels of exposure to UV rays. The index predicts UV levels on a 0 to 10+ scale:

Index Number	Exposure Level
0–2	Minimal
3–4	Low
5–6	Moderate
7–9	High
10–15	Very High

The UV index is computed using forecasted ozone levels, a computer model that relates ozone levels to UV incidence on the ground, forecasted cloud amounts, and the elevation of the forecast cities. For every 1000 feet in elevation, there is a 4% increase in UV exposure. For every 1% decrease in stratospheric ozone levels, there is a 2% increase in UVB radiation reaching the Earth's surface that further results in up to a 3% increase in non-melanoma skin cancers. A dose-response relationship (Chapter 12) exists between the amount of UVB exposure and the incidence of nonmelanoma skin cancers. Other factors that influence UV exposure are heat, wind, humidity, pollutants, cloud cover, snow, season, reflectivity off water and other surfaces, and time of day.

- Solar flare patterns (11-year cycles) may increase ultraviolet radiation 400%. When there is little solar flare activity, there is a decrease in ozone concentration that allows more UVB to reach the Earth's surface.
- Reducing exposure to ultraviolet radiation and reducing risk includes

Using sunscreen with SPF (sun protection factor) 15+ and UVA and UVB protection and reapply every 2 hours;

Wearing protective clothing (hats, sleeves, sunglasses with UV protective lenses);

Seeking shade;

Reducing exposure time in the sun (10 A.M. to 4 P.M.);

Keeping young children out of the sun (young children are more sensitive)—most people receive 80% of their lifetime exposure to the sun by 18 years of age;

Avoiding exposure to sunlamps or tanning parlors (1 million Americans go to tanning facilities every day).

The Oceans

Surface Temperatures

- The ocean has warmed significantly during the past 40 years. The largest warming has occurred in the upper 300 meters (985 feet) of the ocean—an average of 0.56°F. The water in the upper 3000 meters (2 miles) of the world ocean warmed an average of 0.11°F (roughly equivalent to the energy consumed by 100 trillion 100-watt light bulbs burning for a year). The Pacific and Atlantic Oceans have been warming since the 1950s; the Indian Ocean has warmed since the 1960s. The warming patterns of the Pacific and Indian Oceans are similar, which suggests that the same phenomenon is causing the changes to occur in both oceans. Since the 1950s, the California Current that runs southward along the western boundary of the United States has risen in temperature about 2.7°F (1.5°C). Since that time, zooplankton in those waters has decreased over 80% with further negative ramifications up the food chain (declines in rock fish populations that are not commercially harvested). Possible reasons include oil spills, pollution, increased temperatures, and the decrease in upwelling (warmer surface temperatures lower the thermocline). The same phenomena occurred in 1998 during an El Niño in the Caribbean—surface temperatures increased 1.8°F (1°C), and zooplankton populations crashed.
- The North Atlantic is the warmest and most saline of the world's oceans, having a mean potential temperature of 5.08°C and mean salinity of 35.09 parts per thousand, compared to the global average of 3.51°C and salinity of 34.72. Most of the warmer and more saline waters of the world are concentrated in the upper kilometer (0.6 mile) of the subtropical and tropical circulation regimes in what is called the main thermocline (a region of rapid decrease in temperature with depth, in the North Atlantic typically the upper kilometer—see Figure 13.11). About 77% of world-ocean volume is colder than 4°C, with salinities in the relatively narrow range 34.1 to 35.1°C. At the sea surface, only about 26% of the surface area is colder than 4°C, and it is within this area that the large volume of cold water acquires its characteristics before sinking.

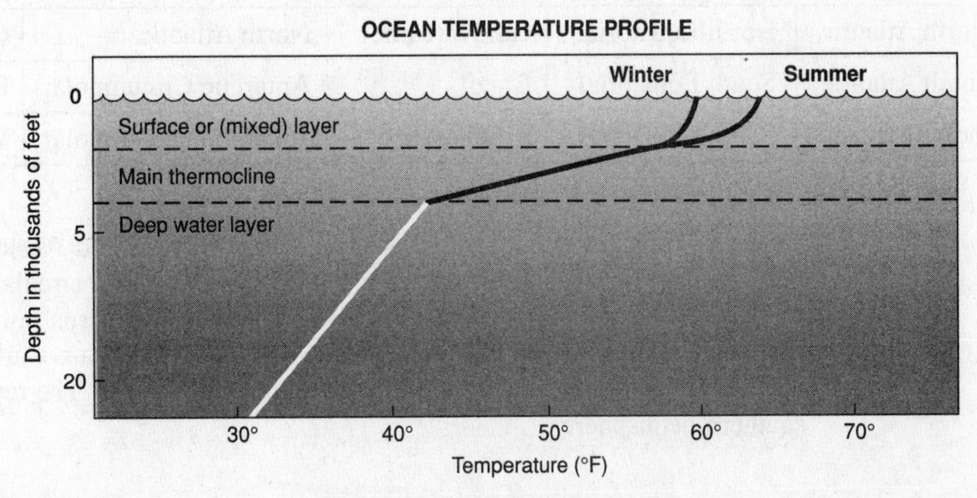

Figure 13.11. Ocean temperature profile.

- The northern hemisphere is dominated by land and the southern hemisphere, by oceans. Temperature differences between summer and winter are more extreme in the northern hemisphere (the land warms and cools more quickly than the ocean).
- Compared to air, water has an extremely high heat capacity, and so it takes much more sunlight to warm up. Warm seawater is lighter than cool seawater, so the warmed water stays on top and is reluctant to pass its heat downward. As a result, the sea warms slowly but cools more quickly. During summers, a thermocline develops between the warm surface water and the cooler bottom water. As the sea warms further, this sharp boundary moves deeper.

Currents

- Surface ocean currents are driven by wind patterns that result from the flow of high thermal energy sources generated at the tropics (higher pressure) to low energy sources in polar areas (lower pressure). They distribute the heat generated near the tropics. Prevailing winds are named for the direction from which they originated, and the moving water is named for the direction it travels (a west wind is responsible for an eastern current). Deepwater, density-driven currents are controlled primarily by differences in temperature and salt content (thermohaline circulation)—dense water sinks, and less dense water rises. Only about 10% of the ocean volume is involved in wind-driven surface currents. The other 90% circulates due to density differences in water masses (primarily caused by differing temperatures and salinities). Currents are restricted by continents and create circular patterns known as gyres. North–south currents within a gyre flow near the landmass, east–west currents flow at the top and bottom of the gyre. Western boundary currents are warm, fast, and have sharp boundaries. Eastern boundary currents are cold, slower, and have diffuse boundaries.
- Because of the Coriolis effect, water will be deflecting toward the right of the direction it is traveling in the northern hemisphere (left in the southern hemisphere), resulting in water that will be moving toward the center of the ocean and forming a "hill" in the center of the gyres. These hills are offset to the west due to the rotation of the Earth.

Gyre	West-moving	Pole-moving	East-moving	Equator-moving
North Pacific	North Equatorial	Kuroshio	North Pacific	California
South Pacific	South Equatorial	East Australian	Antarctic Circumpolar	Peru
North Atlantic	North Equatorial	Gulf Stream	North Atlantic	Canary
South Atlantic	South Equatorial	Brazil	Antarctic Circumpolar	Benguela
Indian Ocean	South Equatorial	Agulhas	Antarctic Circumpolar	West Australian

- Easterlies near the equator push surface water westward, and Westerlies in the higher latitudes move surface water eastward, resulting in clockwise circulation patterns in the northern hemisphere and counterclockwise circulation patterns in the southern hemisphere. In the northern hemisphere, north-flowing currents are warm (from the equator) and south-flowing currents are colder (from the Arctic). The reverse is true in the southern hemisphere.

CURRENT CHARACTERISTICS

North and South Equatorial Currents. Both flow westward between 2 and 4 miles per day and generally run between 325 feet and 650 feet below the surface. The returning flow is called the Equatorial Counter Current and flows toward the east. During El Niño, the Equatorial Current is a stronger current.

Western Boundary Currents. Flow from the equator to the poles. They are narrow currents, extend very deep from the surface (more than half a mile down), and travel much faster than the equatorial currents, up to 65 miles per day. They are named for the area in which they are located (see preceding chart). The most famous current, the Gulf Stream, is a western boundary current that transports 150 million cubic meters of water per second (by comparison, all rivers flowing into the Atlantic move 0.6 million cubic meters per second, the Amazon about 0.2 million of that, and the Mississippi about 0.02 million).

Eastern Boundary Currents. Cold water currents that flow from the poles to the equator. They are much broader than Western Boundary Currents, shallower, and travel much more slowly (up to 4 miles per day). They also have specific names depending on where they are found (see preceding table).

North Pacific Current, North Atlantic Current, South Pacific Current, South Indian Current, and South Atlantic Current. Move water from Western Boundary Currents to starting points of Eastern Boundary Currents. South Pacific, South Indian, and South Atlantic currents are associated with the Antarctic Circumpolar Current (West Wind Drift) that flows around Antarctica due to no land restrictions. Return of water to the southern currents is limited.

- Subsurface currents travel more slowly than surface currents and are affected by water densities that are a function of temperature and salinity. In the North Atlantic, cold, highly saline seawater sinks to the ocean floor (downwelling). Moving south, along the eastern coasts of North and South America, the water travels to Antarctica. Here this bottom water then travels eastward and then splits into two currents, both of which move north. One subsurface current moves north along the east coast of Africa and the other moves north off the east coasts of Australia and Asia. Traveling northward, the deep moving waters begin to surface (upwelling) and eventually return to the North Atlantic. The entire trip takes 1000 years. Sites of upwelling bring cold water and nutrients from depths of 165 to 1000 or more feet below the surface, up to the surface and provide rich feeding grounds for fish and marine mammals. They occur from the divergence associated with currents flowing parallel to a coast that get deflected by the Coriolis force and are characteristic in areas along coasts where air is moving toward the equator.
- Antarctic Bottom Water is cold (–0.5°C or 31.1°F) and salty (34.65 ppt). It forms at the edge of the Antarctic continent and slowly flows under all other water masses into the deep basins as it moves toward the equator, staying close to the bottom. Antarctic Bottom Water travels far from its origin, penetrating into the North Atlantic and North Pacific basins. A cold, salty deepwater mass also forms in the Arctic, but the Arctic basin keeps most of the water contained.
- Mediterranean Outflow Water is a deepwater (4920 to 13,120 feet) mass that results from high salinity (not cooling). The high evaporation rate in the Mediterranean increases salinity. As the water leaves the Mediterranean basin it spreads into the Atlantic. Mediterranean Outflow Water is saltier (38 ppt) than the North Atlantic Deep Water, but much warmer, so it floats above it.

- North Atlantic Deep Water forms in the region around Iceland. It actually is modified from another water mass (North Atlantic Intermediate Water) that has come near the surface and has been cooled by the contact with the air. The cooling increases the density of the water mass, and it sinks and moves slowly. North Atlantic Deep Water is the driving force of the Gulf Stream and runs around the southern end of Greenland and then follows the coast of Canada down to the coast of the United States where it turns slightly eastward, out from the coast and then continues southeast, past the eastern tip of South America. The North Atlantic Deep Water has shut down in the past (thousands of years ago) and causes a decrease in the strength of the Gulf Stream and the North Atlantic Drift, in turn cooling the climate of northwestern Europe. There is concern that global warming might cause this to happen again.
- When surface waters move away from each other, vertical currents upwell. When surface waters converge upon each other, vertical currents downwell. Both of these situations occur along coastlines when prevailing winds push the water away from the coast (causing upwelling) or toward the coast (causing downwelling).

Sea Level

- Causes for changes in sea level include

  – Isostatic adjustments,
  – Tectonic effects,
  – Sedimentation,
  – Groundwater and oil extraction,
  – Changes in glaciers and ice sheets,
  – Ocean currents and tides,
  – Hydrologic cycle changes, and
  – Expansion and contraction (steric effects).
- Global sea level rose 6–8 inches (15–25 cm) in the 20th century. Approximately 1–2 inches (2–5 cm) of the rise resulted from the melting of mountain glaciers. Since 1995, more than 5400 square miles, an area equal to Connecticut and Rhode Island combined, have broken off of the Antarctic ice shelves and melted. Another 1–2 inches resulted from the expansion of ocean water that resulted from warmer ocean temperatures. Best scientific estimates indicate that sea levels will rise 7 inches (18 cm) by 2030 and 23 inches (58 cm) by 2090 (see Figure 13.12). Flooding will probably occur in lowlands, and many cities along the coast may be under seawater. Along most of the U.S. coast, sea level has been rising 10–12 inches per century.
- During the end of the last ice age (~18,000 years ago), when global temperatures had risen about 5°C, the sea level rose approximately 430 feet. Much of the rise was due to the fact that when ice on land melts, it enters the sea with a direct correlation to increase in sea level, and much of the ice at that time was on land. However, when ice that is floating in water melts, most of the seawater has already been displaced by the ice, resulting in less of a factor in sea-level rise. In other words, when the Arctic sea ice melts in summer, sea level doesn't rise. The sea ice is already floating on water and is in equilibrium with it. However, sea ice plays an important role in controlling climate, and the amount of open water and ice affects how much energy is absorbed and reflected back to space. With more open water, more energy would be absorbed by the water and less reflected back to space. Consequently, temperatures would be expected to increase even more.

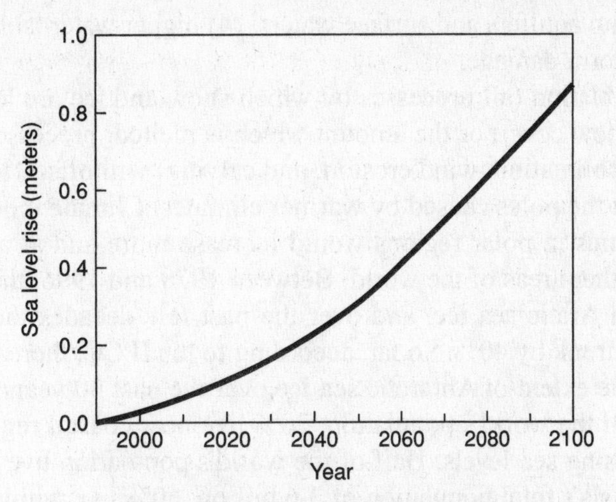

Figure 13.12. Projected rise in sea level.

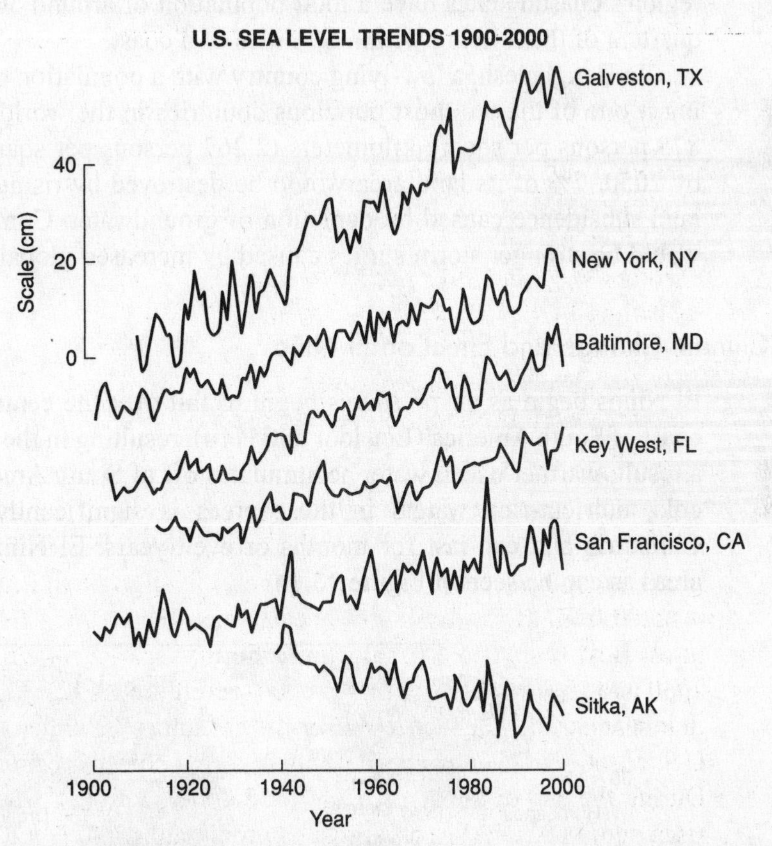

Figure 13.13. Sea levels in the United States, 1900–2000. In Galveston, the removal of groundwater led the land above the water table to sink. In areas that were covered by glaciers during the last ice age (Sitka, Alaska) by contrast, the land is rising because of the removal of the weight of the ice, which had previously compressed the land downward. As a result, the sea is dropping relative to these coasts. *Source:* EPA.

- Wetlands would be most affected. A one-foot increase in sea level would result in up to 40% of U.S. wetlands being destroyed. Other environmental impacts include (1) inundation of low-lying areas, especially in Louisiana and Florida (the rapid rate in Louisiana resulted from the settling of newly created land formed by the sediments that washed down the Mississippi River); (2) erosion of beaches and bluffs; (3) salt intrusion into aquifers and surface waters; (4) higher water tables; and (5) increased flooding and storm damage.

- Ablation (all processes by which snow and ice are lost from a glacier, floating ice, or snow cover; or the amount which is melted; processes include melting, evaporation or sublimation, wind erosion, and calving) will offset, to some extent, increased snowfall at the poles caused by warmer climates. Climate models have suggested that temperatures in polar regions would increase more and at a faster pace in the Arctic than in other areas of the world. Between 1978 and 1996, there has been about a 6% decrease in Arctic sea ice, and over the past few decades the depth of the Arctic ice cap has shrunk by 40%. So far, according to the IPCC, there has been no discernible change in the extent of Antarctic sea ice over the past 30 years.
- Of the world's population, 20% live near coastal regions that will become affected by rising sea levels. Half of the world's population live within 120 miles of the coast. Of Asia's total population of 3.6 billion, 60% live within 250 miles of a coast. The population of Latin America and the Caribbean are even more clustered on the coasts. The region's coastal states have a total population of around 502 million with a full three quarters of them living within 60 miles of a coast.

  In Bangladesh, a low-lying country with a population of 129 million people, making it one of the ten most populous countries in the world with an overall density of 875 persons per square kilometers (2,267 persons per sqare mile), it is expected that by 2050, 7% of its land area would be destroyed by rising sea levels combined with land subsidence caused by depletion of groundwater. Compounding this loss of land would be stronger storm surges caused by increased global temperatures.

Long-Term Climate Change and Effect on El Niño

- El Niños begin as air pressures begin to fall over the central Pacific and the western coast of South America (Ecuador and Peru), resulting in the trade winds decreasing. As a result, warmer ocean water accumulates off of South America. Normal upwelling of cold, nutrient-rich waters in these areas is significantly reduced starting around Christmas and can last for months or even years. El Niños climatically affect wide areas as can be seen in Figure 13.14

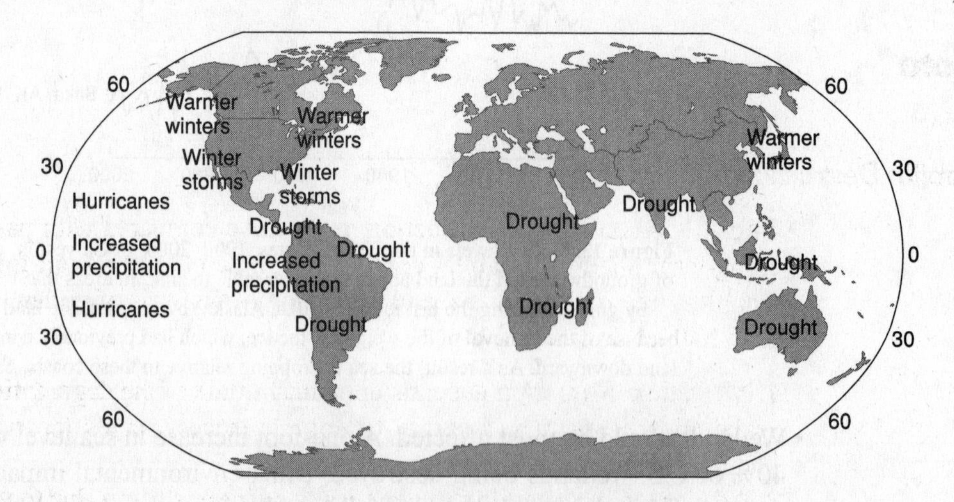

Figure 13.14. Climatological effects of El Niño.

- El Niños normally occur every 10 to 20 years and generally increase global mean temperature between 0.1 and 0.2°C. However, since 1980, they have occurred much more frequently, have been much more dramatic, and have lasted much longer than normal. El Niños developed in 1982–1983, 1986–1987, 1991–1992, 1993, 1994, and 1997–1998. The El Niño that developed in 1982–1983 produced ocean temperatures off the coast of South America that were up to 11°F (6°C) warmer than normal. This resulted in fish catches that were less than half of normal amounts. Effects of El Niños are widespread, and Australia, India, and southern Africa suffer from extreme droughts during El Niño years. Major flooding and crop damage caused by too much rainfall affects California. Temperatures in the winter are warmer than normal in the North Central States, and cooler than normal in the Southeast and the Southwest.
- La Niñas develop when air pressures increase over the central Pacific resulting in the trade winds increasing. This causes large quantities of cold water to build up in the central and eastern Pacific. A La Niña that developed in 1998 persisted through the winter of 2000. La Niñas result in larger than normal hurricanes in the Atlantic. One of the costliest years for major hurricanes in the last 100 years was 1998 because Hurricane Mitch grew to become the Atlantic basin's fourth strongest hurricane ever with sustained winds of 180 mph and an estimated 10,000 dead, thousands more missing, and billions of dollars in damage. It was the strongest storm in the western Caribbean since Hurricane Gilbert in 1988. As with El Niños, La Niñas also have widespread climatic effects such as abnormally heavy monsoons in India and southeast Asia, and winter temperatures that are warmer than normal in the Southeast and cooler than normal in the Northwest United States.
- Volcanoes affect climate by introducing large quantities of sulfur dioxide (SO_2) into the atmosphere that later is converted to sulfate (SO_4^{2-}) particles in the stratosphere. The sulfate particles reflect solar radiation (primarily shorter wavelengths) rather than absorb longer wavelengths emitted from the Earth from heating, and serve as condensation nuclei for high clouds. In 1991–1992, the effect of stratospheric sulfate particles decreased the average global temperature by about 0.5°C. The particles settle out of the atmosphere usually within 2 years.
- Atmospheric temperature increases are greatest between 40°N and 70°N latitude, with the rate of warming in the Antarctic being twice the average rate of global warming.

Biota

Habitat Destruction

- Information on habitat destruction needs to be compared with past habitat conditions. Long-term history of habitat conditions can be obtained from fossil records, ice core samples, tree rings analysis, and palynology—analysis of pollen. Organisms can cope with habitat destruction by

  1. Migration. Migration depends upon magnitude of the degradation, rate of the degradation, the organism's ability to migrate, access routes (corridors), and proximity and availability of suitable new habitats.
  2. Adaptation or the ability to genetically adapt or survive due to variations in the gene pool. Adaptation depends upon magnitude of degradation, rate of degradation, birth rate, length of generation, population size (genetic variability), and gene flow between populations as a function of genetic diversity.

3. Acclimatization or the ability to adjust to environmental changes slowly on an individual or population level. Acclimatization depends upon magnitude of degradation, rate of degradation, and physiological plasticity, and/or behavioral plasticity.

- Plants are (initially) more susceptible to environmental degradation than animals:
  - Plants do not "migrate."
  - Dispersal rates of seeds are usually slow events. Spruce trees are able to increase their range about 1 mile every 100 years (a very fast rate), due to the fact that they have very light seeds that are easily wind-dispersed. Seed dispersal rates are slower than what is required in moving geographically to maintain stable environmental conditions as weather patterns change.
  - Plants cannot seek nutrients or water.
  - Seedlings must survive and grow in degraded conditions.
  - Stressed plants become prone to disease and infestation (animals can migrate).

- Coral reefs are among the most diverse, biologically complex, and ancient ecosystems on Earth (250 million years). Coral reefs, which cover less than one quarter of 1% of the marine environment, are home to more than a quarter of all known marine fish species. They also buffer waves and protect shorelines from erosion; they help transfer nutrients from the land to the open ocean; they provide feeding, breeding, and nursery areas for many commercially important species of fish and shellfish; and they are a source of potential medicines. Nearly 70% of the world's reefs are at risk from human impacts such as destructive fishing impacts, pollution, increased ocean temperatures, and coastal development, and many have already been degraded beyond recovery. "Bleaching" occurs when corals undergo stress from pollution, sedimentation, or thermal stress. Bleaching results when corals expel zooxanthallae (a dinoflagellate) that exist in a symbiotic relationship with the coral (zooxanthallae are photosynthetic and in return they receive shelter from the coral).
- Low-frequency active sonar used by the Navy, commercial shipping (supertankers), and industrial activities like oil exploration all emit noise. Sound is transmitted in water much more effectively than in air. A 3500 times greater power level (35.5 dB) is necessary in air versus water to produce an equivalent sound level. Scientists are concerned that noise pollution could interfere with biological functions, such as mating, feeding, navigating, nursing, and communicating by marine species whose survival depends on their ability to hear and be heard.
- Marine debris threatens hundreds of species of seabirds, marine mammals, turtles, and fish through entanglement, smothering, and interference with digestive systems. Most of marine debris is plastic, which is buoyant and indestructible. Close to 80% of debris is washed, blown, or dumped from shore, while 20% comes from boats, ships, fishing vessels, and ocean platforms. Lost synthetic fishing gear alone may total 150,000 tons a year, and as many as 1000 boxcar-sized shipping containers fall off ships annually.
- Pollution eventually finds its way to the oceans: 44% of all ocean pollution originally stems from land-based discharges, such as runoff from pesticides and fertilizer (14 million tons in 1950 to 130 million tons today), and sewer discharges. 33% comes from acid rain, 12% from marine sources such as oil spills, 10% from ocean dumping, and 1% from oil exploration and production, such as offshore oil drilling.
- More than 50% of the world's coastlines are degraded from moderate to severe development pressures using four indicators: cities and population density, major ports, road density, and pipeline density.

- More than 50% of coastal wetlands, including mangrove swamps and salt marshes, have been lost (25 million hectares of mangrove swamps alone). For example, in the Philippines, the mangrove swamps have decreased in size by 90%—from one million hectares in 1960 to around 100,000 in 1998. Mangrove swamps provide a rich, biodiverse habitat for over 2,000 species of fish, shellfish, invertebrates, and plants.
- Of the world's beaches, 70% are eroding at greater than natural rates due to coastal construction, dredging, mining for sand, and harvesting of coral reefs for building material.

Introduced Exotics

- Invasive species are animals and plants that are transported to an area where they do not naturally live. A major marine source is marine ballast. Every minute ships discharge 40,000 gallons of foreign ballast water, containing foreign plant and animal species, into U.S. harbors. Often these plants and animals grow at an uncontrolled rate because they have no natural predators. These plants and animals can become pests, disrupt the balance of life, and crowd out native plants and animals. Over 240 invasive species are found in San Francisco Bay alone. Common examples on land include kudzu and gypsy moths, and common examples in water are zebra mussels and water hyacinth.
- Zebra mussels, *Dreissena polymorpha*, originally came from eastern Europe and western Asia and were first discovered in U.S. waters in 1988. They arrived in the ballast of ocean-going vessels that emptied their tanks in Great Lakes ports and have now spread to all of the Great Lakes and major river systems in the Midwest, moving through waterways by attaching to boats and commercial barges. They spread to the Mississippi River in 1991, and today the exploding zebra mussel population has carpeted some parts of the Mississippi River bed with 10,000 to 20,000 mussels per square yard.
- Kudzu was introduced to the United States from Japan in 1876 at the Centennial Exposition in Philadelphia, Pennsylvania. During the Great Depression of the 1930s, the Soil Conservation Service promoted kudzu for erosion control. The vines grow as much as a foot per day during summer months, climbing trees, power poles, and anything else they contact. Under ideal conditions kudzu vines can grow 60 feet each year. While they help prevent erosion, the vines can also destroy valuable forests by preventing trees from getting sunlight.
- The gypsy moth, *Lymantria dispar*, is one of North America's most devastating forest pests. The species originally evolved in Europe and Asia and has existed there for thousands of years. In 1869, the gypsy moth was accidentally introduced near Boston. The gypsy moth is known to feed on the foliage of hundreds of species of plants in North America, but its most common hosts are oaks and aspen. Gypsy moth hosts are located through most of the United States.
- Water hyacinth, *Eichhornia crassipes,* was introduced into Florida in the 1880s. Its growth rate is among the highest of any plant known: hyacinth populations can double in as little as 12 days. Besides blocking boat traffic and preventing swimming and fishing, water hyacinth infestations also prevent sunlight and oxygen from getting into the water. Decaying plant matter also reduces oxygen in the water. Thus, water hyacinth infestations reduce fisheries, shade out submersed plants, crowd out immersed plants, and reduce biological diversity.

- The mongoose was purposely brought to Hawaii to kill off rats in the sugar cane fields. However, the mongoose sleeps at night and hunts in the day, and the rat sleeps in the day and hunts at night, so they never met. To survive, the mongoose and the rats prey upon ground nesting birds such as the Hawaii's state bird, the Nene goose.

Overharvesting

- Worldwide, more than three million fishing boats remove between 70 million and 90 million tons of fish and shellfish from the oceans each year. The United Nations Food and Agriculture Organization estimated that in 1999, up to 80% of worldwide marine fish stocks require urgent intervention to halt population declines and to rebuild species depleted by overfishing. Of the world's marine fish stocks, 70% are either fully exploited, overfished, or depleted. The American Fisheries Society recently identified 82 species at risk of extinction in North American waters. Among this list of severely depleted species are some of the world's most prized food and game fish—several species of shark, skates, sturgeons, groupers, Atlantic salmon, Atlantic halibut, and Pacific rockfish.
- Twenty million metric tons of fish and shellfish are destroyed each year as by-catch (not economical) and represents one-fourth of all catch taken from the sea. These organisms are not "trash" and their destruction results in loss in the world's biodiversity. The International Whaling Commission estimates that between 65,000 and 80,000 whales, dolphins, seals, and other marine mammals perish as by-catch each year. Shrimp trawlers catch, and throw back dead, an estimated 5.2 pounds of marine life for every pound of shrimp landed. In all, shrimp fishing causes 35% of the world's by-catch. Today, in the South Atlantic and Gulf of Mexico, shrimping operations reportedly discard as much as 2.5 billion pounds of fish annually, while drowning hundreds of endangered and threatened sea turtles.
- Longline fishing consists of fishing lines up to 40 miles long and containing hundreds of hooks spread out along the line. The Food and Agriculture Organization reports that 40,000 sea turtles are killed annually in the use of global longline fishing. Drift gill-nets are designed to catch fish by their gills. They also kill more than 30 different species of marine mammals.
- 11 of the 15 major fishing regions in the world have reported serious declines in catch. Since 1970, there has been a 25% decrease in catches of cod, tuna, and haddock. From 1960 to 1990, the number of commercially valuable fish species in the Black Sea decreased from 26 to 5. In terms of catch, the tonnage fell from 1 million metric tons in 1982 to less than 100,000 metric tons 10 years later. Reasons for the decline include pollution and introduction of alien species from the Atlantic. Eutrophication is so bad in the Black Sea that only 10% of the surface waters have enough oxygen to support life above the microorganism level. Marine organisms provide up to 90% of protein sources for Southeast Asia and the South Pacific, and yet, water within 9 miles of land in Southeast Asia is classified as overfished. Practices such as trawling, poisons, dynamite, and fine mesh nets contribute to this condition.
- In 1997, about 30 million metric tons of fish, shellfish, and seaweed was farmed, 70% by China alone. Of this farmed resource, 85% is exported and does not meet the needs of its own people.
- Deepwater fishing trawls may crush, grind, catch, and discard up to 20 pounds of corals, tube worms, sponges, anemones, and other deep sea bottom-dwelling creatures for every pound of commercial fish caught. Trawl damage can harm the entire ocean

ecosystem because these bottom creatures form the basis of the marine food chain and provide structures in which fish and other sea life hide.

- Pollution is a major contributor to declining fish catches. Up to 90% of all fishing is done along coasts—the area of greatest pollution.

Loss of Biodiversity

- Biodiversity includes (1) ecological diversity (the variety of ecosystems and ecological communities); and (2) genetic diversity (the range of genetic differences found within and between species).
- Diversity of life forms increased dramatically during the Cambrian and Ordovician (543 million to 443 million years ago). The number of species remained fairly constant until the end of the Permian (248 million years ago) when a mass extinction occurred that resulted in over 90% of life becoming extinct. There have been five massive extinctions of life since life began (see Figure 13.15). Species average between 5 million to 10 million years before becoming extinct.

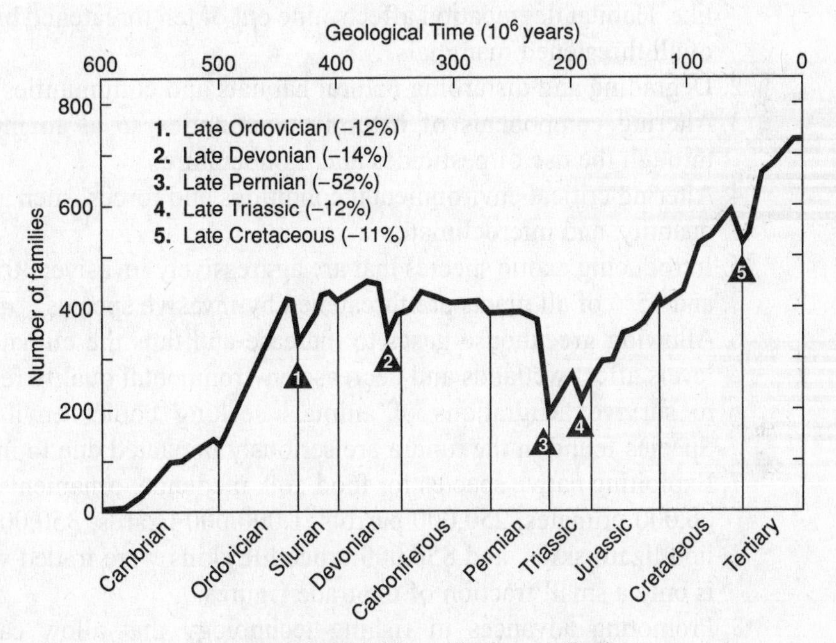

Figure 13.15. Major extinctions on Earth.

- Since 1500, 816 species have become extinct, 103 of them since 1800, a rate 50 times greater than the natural background rate. In the next 25 years, extinction rates are expected to be as high as 25%. Endangered mammals rose from 484 in 1996 to 520 in 2000, with primates increasing from 13 to 19. Endangered birds increased from 403 species to 503 species. Freshwater fish more than doubled, going from 10 species to 24 species in only 4 years. One-fourth of all mammals and reptiles, one-fifth of all amphibians, one-eighth of all birds, and one-sixth of all conifers are in some manner endangered or threatened of extinction. And these figures represent only the number of species identified. Most species extinctions are unknown since the species had yet to be identified. Millions of spiders, insects, nematodes, mollusks, bacteria, and organisms that live in deep marine areas have not been identified or studied. When these are included, scientists estimate an average of 137 species of life forms are driven into extinction every day; or 50,000 each year (Edward O. Wilson. 1992. *The Diversity of Life,* Harvard University Press, Cambridge, MA).

- The number of different species on Earth is at its maximum during the current Tertiary Period. Estimates are that up to 13 million different species of life exist on Earth today; 60% of the total are insects. Plants and vertebrates make up only 3%.
- Diversity is highest in tropical regions and diminishes toward the poles. Every second, 2.4 acres (1 hectare or the equivalent to two U.S. football fields) of tropical rain forest are lost. In 1 year, this is about the size of Poland.
- Biodiversity increases when (1) diversity of the physical habitat increases; (2) small environmental disturbances increase; (3) variations in available nutrients, precipitation, and temperature decreases; (4) succession is in the middle stages; (5) environmental stress decreases; (6) large environmental disturbances decrease; (7) there are no competing exotic (nonindigenous) species; and (8) there is decreasing geographic isolation.
- Biodiversity has been decreased by
  1. Removing natural habitats and communities by land clearing for agriculture, pastures, mining, logging, roads, human settlement and expansion, industry, and the like. Habitat degradation affects nine out of ten threatened birds and plants, and 85% of all threatened mammals.
  2. Degrading and disturbing natural habitats and communities by pollution.
  3. Altering components of natural communities so as to interfere with food webs through the use of pesticides and monoculture.
  4. Altering critical environmental conditions and forces, such as fire, water quality and quantity, and microclimate.
  5. Introducing exotic species that are aggressively invasive; 30% of all threatened birds and 15% of all plants are threatened by invasive species.
  6. Allowing greenhouse gases to increase and thus the climate to change. Rising sea levels affect wetlands and decrease environmental quality required for many species to survive. Migrations of animals seeking cooler environments are restricted. Species found in the tundra are seriously impacted due to increasing temperatures.
  7. Exploiting native species for food, folk medicine, ornaments, and pets. For example, 26,000 primates, 250,000 parrots, 1,000,000 lizards, 350,000 wild orchids, 1.5 million lizard skins, and 850,000 crocodile skins were traded worldwide in 1997. This is only a small fraction of the trade figures.
  8. Promoting advances in fishing technology that allow capture rates faster than species can reproduce.
- Less than 7% of the Earth's surface is classified as "protected," consisting of about 30,000 preserves covering about 900 million hectares. Most of these preserves are small with few larger than 100,000 hectares (386 square miles). Less than 1% is larger than 1 million hectares (3900 square miles).

Case Study

Biosphere 2—1991. Biosphere 2 is a $200 million, 3.2-acre self-contained facility built near Tucson, Arizona, to study possible self-sustaining stations in space and to increase understanding of environmental processes on Earth. It contains a lake, desert, stream, freshwater and saltwater wetlands, tropical rain forest, and a small coral reef, and is stocked with over 4000 species of plants, animals, and microorganisms. Energy is provided by solar and external natural gas-powered generators. It was designed to be self-sufficient for 2 years; however, the project failed because the following problems occurred:

- Oxygen disappeared,
- Nitrous oxide rose to toxic levels,
- Carbon dioxide rose to dangerous levels and caused weedy vines to choke out food crops,
- Nutrients leached from the soil and polluted the water system,
- Tropical birds died due to cold,
- Arizona ant species entered the facility and killed off soft-bodied insects,
- Due to imbalance in other insect populations, cockroaches and katydids proliferated,
- 19 of 25 small animal species died, and
- Plant-pollinating insects died and doomed many plant species.

Multiple-Choice Questions

1. Which of the following represents the greatest contribution of methane emission to the atmosphere, ultimately resulting in global warming?

   (A) Enteric fermentation or flatulence from animals
   (B) Coal mining
   (C) Landfills
   (D) Rice cultivation
   (E) Burning of biomass

2. Ozone depletion reactions that occur in the stratosphere are facilitated by

   (A) NO_2^-.
   (B) NH_3.
   (C) NH_4^+.
   (D) N_2O.
   (E) NO_3^-.

3. In addition to absorbing harmful solar rays, how do ozone molecules help to stabilize the upper atmosphere?

   (A) The release heat to the surroundings.
   (B) They create a buoyant lid on the atmosphere.
   (C) They create a warm layer of atmosphere that keeps the lower atmosphere from mixing with space.
   (D) All of the above.
   (E) None of the above.

4. What are two conditions that may indicate an El Niño?

    (A) Sea surface warms, trade winds strengthen.
    (B) Sea surface warms, trade winds weaken.
    (C) Sea surface cools, trade winds strengthen.
    (D) Sea surface cools, trade winds weaken.
    (E) None of the above.

5. The bromine released from methyl bromide is about 40 times more damaging to the ozone layer than chlorine on a molecule-to-molecule basis. Under the Montreal Protocol, methyl bromide in developed countries will be phased out in 2005. What is methyl bromide used for?

    (A) Sterilizing soil in fields and greenhouses
    (B) Killing pests on fruit, vegetables, and grain before export
    (C) Fumigating soils
    (D) Killing termites in buildings
    (E) All the above

6. Rising sea levels due to global warming would be responsible for all the following EXCEPT

    (A) destruction of coastal wetlands.
    (B) beach erosion.
    (C) increased damage due to storms and floods.
    (D) increased salinity of estuaries and aquifers.
    (E) all would be the result of rising sea level.

7. The Intergovernmental Panel on Climate Change (IPCC) is a United Nations-sponsored group of 2500 scientists around the world and serves as a resource and authority on dealing with the global warming issue. According to this organization, what is the expected increase in the world population's risk of infection due to global warming?

    (A) 10%
    (B) 20%
    (C) 30%
    (D) 40%
    (E) 50%

8. What does La Niña bring to the southeastern United States?

    (A) Warm winters
    (B) Extremely cold winters
    (C) Hot summers
    (D) Cooler than normal summers
    (E) None of the above

9. One of the reasons the vortex winds in the Antarctic are important in the formation of the "ozone hole" is

   (A) they prevent warm, ozone-rich air from mixing with cold, ozone-depleted air.
   (B) they quickly mix ozone-depleted air with ozone-rich air.
   (C) they harbor vast quantities of NO_2.
   (D) they bring in moist, warm air, which accelerates the ozone loss process.
   (E) none of the above.

10. Which biome is experiencing the greatest loss in biodiversity?

    (A) Wetlands
    (B) Grasslands
    (C) Tropical rain forests
    (D) Temperate deciduous forests
    (E) Deserts

11. Although the Montreal Protocol curtailed production of ozone-depleting substances, the peak concentration of chemicals in the stratosphere is only now being reached. A recovery will only occur if all nations implement the controls of the Montreal Protocol. When do scientists expect a recovery in the ozone to occur?

    (A) 2010
    (B) 2030
    (C) 2050
    (D) 2090
    (E) It will never recover.

12. Which of the following ocean currents flows unimpeded (without obstruction or barriers) around the Earth?

    (A) Gulf Stream
    (B) California Current
    (C) Antarctic Circumpolar Current
    (D) Aghulas Current
    (E) They all flow unimpeded.

13. A project aimed at producing paleoclimatic maps showing sea-surface temperatures in different parts of the globe, at various times was

    (A) Global Sun-Temperature Project.
    (B) PALEOMAP.
    (C) NOAA.
    (D) NIMBUS.
    (E) CLIMAP.

14. Which of the following is NOT a symptom of melanoma?

    (A) Asymmetry in a mole or pigmented spot.
    (B) Irregular border in a mole or pigmented spot.
    (C) Variation in color in a mole or pigmented spot.
    (D) Diameter of a mole or pigmented spot greater than 6 mm (pencil eraser).
    (E) They are all symptoms.

15. Which international convention is in force to restrict commercial trade in endangered species?

(A) CITES
(B) Agenda 21
(C) Convention on Biological Diversity
(D) Endangered Species Act
(E) Environmental Protection Act

Answers to Multiple-Choice Questions

1. **C**	6. **E**	11. **D**
2. **D**	7. **B**	12. **C**
3. **D**	8. **A**	13. **E**
4. **B**	9. **A**	14. **E**
5. **E**	10. **C**	15. **A**

Explanations for Multiple-Choice Questions

1. **(C)** Methane, produced during decomposition in landfills, produces the most methane that is released to the atmosphere. Since 1800, there has been about a 150% increase in methane concentration in the atmosphere as compared to about a 30% increase in carbon dioxide. Methane is about 30 times more effective in causing a greenhouse effect than carbon dioxide. When methane is completely burned, it produces carbon dioxide and water vapor (both contribute to global warming). Methane hydrates (methane locked in ice) are a recently discovered source of methane which form at low temperature and high pressure and are found in two types of geologic settings: (1) on land in permafrost regions where cold temperatures persist in shallow sediments and (2) beneath the ocean floor at water depths greater than about 500 meters where high pressures dominate. The hydrate deposits themselves may be several hundred meters thick. Methane bound in hydrates amounts to approximately 3000 times the volume of methane in the atmosphere. Some believe there is enough methane in the form of hydrates to supply energy for hundreds, maybe thousands, of years. Conventional natural gas reserves in the United States are estimated to be 1400 trillion cubic feet while estimates of methane hydrates in waters belonging to the United States are estimated to be 200,000 trillion cubic feet. Worldwide, estimates of the natural gas potential of methane hydrates approach 400 million trillion cubic feet compared to the 5000 trillion cubic feet that make up the world's currently known gas reserves. Natural gas is expected to take on a greater role in power generation, largely because of increasing pressure for clean fuels and the relatively low capital costs of building new natural gas-fired power equipment. The United States will consume increasing volumes of natural gas well into the 21st century. U.S. gas consumption is expected to increase from almost 23 trillion cubic feet in 1996 to more than 32 trillion cubic feet in 2020—a projected increase of 40%. Also, gas demand is expected to grow because of its expanded use as a transportation fuel and potentially, in the longer-term, as a source of alternative liquid fuels (gas-to-liquids conversion) and hydrogen for fuel cells.

2. **(D)** Nitrous oxide is responsible for about 7% of the anthropogenic greenhouse gas climate effect. It has a long lifetime in the atmosphere (about 120 years) and is mainly destroyed by solar ultraviolet radiation in the stratosphere resulting in the formation of chemically reactive nitrogen compounds, which react with ozone. It is responsible for about 70% of the natural removal of stratospheric ozone. N_2O has an ozone depletion potential comparable to that of many of the hydrofluorocarbons (HCFCs). The main anthropogenic sources of N_2O are agricultural fertilization and industrial processes, for example nitric acid production. Natural sources of N_2O, arising from biological processes in soils and oceans, are about twice as large as anthropogenic sources. N_2O has increased about 15% since preindustrial times.

3. **(D)** Solar energy that is absorbed by ozone molecules and turned partly into heat creates a warm region in the stratosphere. This creates a stable air mass that resists sinking and mixing with the lower atmosphere, effectively forming a barrier. In the ozone layer, temperature increases with height, creating a stable and buoyant air mass that keeps an effective lid on the lower atmosphere, preventing it from mixing with outer space.

4. **(B)** The warm waters of El Niño are a part of the event called the El Niño Southern Oscillation (ENSO), which encompasses the central Pacific region. Normally a warm pool of ocean water builds up in the western equatorial Pacific Ocean, kept at bay from the trade winds, blowing east to west. The easterly trade winds are driven by a surface pressure pattern of lower pressure in the western Pacific and higher pressure in the eastern Pacific. When this pressure pattern weakens, the trade winds decrease or even reverse. This causes the warmer water to propagate eastward. The warmer water causes the thermocline to sink, which causes the upwelling to occur at a deeper level in the ocean.

5. **(E)** All of the above.

6. **(E)** Sea level has fluctuated by an order of 100 meters over the last 18,000 years. Sea level is rising worldwide and is caused by both natural and human factors. Most research indicates that sea level is currently rising approximately 2 mm/yr. More than 75% of the human population lives within 37 miles of a coast. A 1-centimeter rise in sea level erodes beaches about 1 meter horizontally. It is predicted that within 100 years there will be a net loss of 17 to 43% of coastal wetlands due to rising sea levels. A one-meter rise in sea level would enable a 15-year storm to flood areas that today are only flooded by 100-year storms. Rising sea level would allow saltwater to penetrate further inland and upstream.

7. **(B)** The world's leading authority on global warming, the IPCC, has concluded that unchecked global warming will cause a significant increase in human mortality due to extreme weather and infectious disease. They concluded that global warming will likely put as much as 65% of the world's population at risk of infection—an increase of 20%.

8. **(A)** La Niña can bring warm winters to the Southeast and cooler than normal winter temperatures in the Northwest. It is the cold counterpart of El Niño, which is a mass of warm water along the equator in the tropical Pacific. La Niña's strong easterly winds bring cold ocean water to the surface in the eastern Pacific and fuels increased rainfall in the western Pacific. The jet stream is also affected, often coming in over Alaska and into the Great Lakes rather than through the Pacific Northwest.

9. **(A)** Ozone depletion follows an annual cycle that corresponds to the amount of sunlight that reaches the Antarctic. The cycle begins every year around June when the vortex winds develop in the Antarctic. Cold temperatures produced by these winds create polar stratospheric clouds that capture floating chlorofluorocarbons (CFCs). For the next two months, a reaction occurs on the cloud surface that frees the chlorine in the CFCs but keeps the chlorine contained within the vortex area. In September, sunlight returns to the Antarctic and triggers a chemical reaction, causing chlorine to convert ozone to normal oxygen. Measured ozone levels usually are lowest in October. November brings a breakdown in the vortex that allows the ozone-rich air to combine with the thinning ozone. Wind currents carry this mixture over the southern hemisphere and carry the "hole" over other areas of the Earth.
10. **(C)** There is a direct relationship between the size of an area and the number of species that it contains. A square yard of temperate forest habitat may have ten species of plants while an acre will often number in the hundreds. In general, the larger the area, the more species are encountered. Of equal importance is the size of the area occupied by each species. Species that are restricted to small geographic areas are much more likely to become extinct than are those with widespread distributions. Also, the smaller the population, the higher the probability of extinction. Tropical species commonly have smaller populations and much more restricted distributions than species in other biomes. Thus, destroying an acre of tropical forest will likely have a much higher extinction impact than the loss of an acre of temperate forest. If 1% of the world's tropical rain forests are destroyed each year (a conservative estimate based on current rates of deforestation), then 0.2 to 0.3% of all species would become extinct per year. Over 100 years, this would be a loss of at least 20% of all species, if extinction rates remain constant. Based on an estimated total of 10 million species, the current loss is calculated to be 20,000 to 30,000 species per year.
11. **(D)** Concentrations of ozone-depleting chlorofluorocarbons (CFCs) have leveled off in the stratosphere and have actually declined in the lower atmosphere. However, the largest Antarctic ozone hole ever recorded occurred in 2000, and the effects of global climate change may exacerbate the problem. It can take a CFC molecule about 2 years after being released at the ground to make it to the stratosphere where the ozone is. And it can take decades for it to be converted by sunlight into a form that is harmful to ozone. Furthermore, while higher carbon dioxide concentrations are thought to cause a warming of the atmosphere's lowest layer (the troposphere), this same carbon dioxide actually causes the stratosphere to cool down, accelerating ozone destruction.
12. **(C)** The Antarctic Circumpolar Current is the most powerful current system on Earth. It exerts a strong influence on climate. It circles the world in the Southern Hemisphere between 40° and 60° South latitude and connects the three great ocean basins—Atlantic, Indian and Pacific, allowing water, heat, salt and other properties to flow from one to the other. Unlike the Northern Hemisphere, there are no landmasses to break up this great continuous stretch of seawater.
13. **(E)** CLIMAP was a project developed in the 1970s by the National Oceanic and Atmospheric Administration (NOAA) Paleoclimatology Program and the World Data Center, using computer models to develop a detailed climatological map of the world 18,000 years ago.

14. **(E)** The choices listed form the ABCD's of recognizing melanoma.

    Asymmetry—one half unlike the other half.

    Border irregular—scalloped or poorly circumscribed border.

    Color varied from one area to another—shades of tan and brown; black; sometimes white, red or blue.

    Diameter larger than 6 mm as a rule (diameter of a pencil eraser).

15. **(A)** CITES (Convention on International Trade in Endangered Species of Wild Fauna and Flora) is the first international convention on conservation that can prosecute offenders. It accords varying degrees of protection to more than 30,000 species of animals and plants, whether they are traded as live specimens, fur coats, or dried herbs. CITES is an agreement by member governments, currently numbering 150, to ban commercial trade on an agreed list of endangered species, and to regulate and monitor trade in others. International wildlife trade is estimated to be worth billions of dollars and to include hundreds of millions of plant and animal specimens. The trade is diverse, ranging from live animals and plants to a vast array of wildlife products derived from them, including food products, exotic leather goods, wooden musical instruments, timber, tourist curios, and medicines. Levels of exploitation of some animal and plant species are high and the trade in them, together with other factors, such as habitat loss, is capable of heavily depleting their populations and even bringing some species close to extinction.

Free-Response Question

by Annaliese Beery, B.A.
California Institute of Technology
Williams College

A.P. Environmental Science Teacher, Marin Academy
San Rafael, CA

Life on Earth has been punctuated by several mass extinctions. Humans are playing a role in another mass extinction, potentially the largest ever. As we attempt to create a sustainable future, efforts are being taken to slow the loss of endangered species.

(a) How can scientists assess the current population size of a species? Explain how "tag and recapture" methods could be used to estimate the number of monarch butterflies in an area too large to sample exhaustively.
(b) Give one example each of a direct and an indirect threat to biodiversity.
(c) Explain a piece of legislation designed to preserve the biodiversity of our planet.
(d) After a population of organisms has been reduced to a small size, it has greater risk of going extinct, even if the population returns to its original size (a phenomenon sometimes referred to as the bottleneck effect). Why would the same sized population be more likely to go extinct after a population "bottleneck"?

Free-Response Answer

(a) One way to estimate the size of a large population is to "tag and recapture" individuals. This process involves catching several individuals from the population—the more the better—and marking them in some way you can identify later. It is important for this mark not to interfere with their functioning or likelihood of survival. In the case of monarch butterflies, a small dot of nail polish applied to the top of the thorax would suffice. After marking the butterflies, you would release them into the population and wait long enough for them to become randomly dispersed. Next you would capture several butterflies and count how many of them have your mark. You would repeat this procedure, releasing, waiting, and capturing the same number of butterflies until you had several pieces of data to average. Based on the average percentage of marked butterflies you captured, you can estimate the total population size. If 5% of the butterflies you capture are marked, then the number you originally marked represents about 5% of the total population size.

(b) A direct threat to biodiversity is something that affects organisms by interfering with them; for example, cars might run over them. An indirect threat to biodiversity impacts them through a chain of events. If development fragments the habitat of a large mammal, for instance, then the mobility of the animals will be decreased, limiting their access to resources and mating partners. If habitat fragmentation or loss is severe enough, extinction of the population can result.

(c) The Endangered Species Act is a piece of legislation that makes it illegal to injure, kill, or collect any listed "threatened" or "endangered" species within the United States. This legislation also prohibits the import (for any purpose other than preservation or research) of any endangered species or any product made of endangered species. This clause is intended to keep organisms in their native habitats.

(d) When the size of a population is severely reduced, many of the genetic variations present in the original population will be lost. For example, a world population of 20 humans could not represent all of present human diversity. Even if the "bottlenecked" population reproduces enough to return to the original population size, the genetic diversity will still be limited to that of the parents, since new genetic variation will arise extremely slowly. Thus, the population will be more vulnerable to selection pressures such as disease because the likelihood of the population having resistant individuals is much lower. This would make the population more likely for the population to go extinct.

Rubric (Max score for this essay is 10)

Part (a): 6 points possible; max score for this section is 4 points

1 pt. Capture an original sample of several butterflies and mark them.

1 pt. The mark must not interfere with their functioning or survival.

1 pt. You should wait for the marked organisms to mix in with the population before recapturing.

1 pt. You should recapture several individuals and count how many are marked.

1 pt. You should repeat the recapture multiple times and average the data.

1 pt. The average percentage of marked Monarchs you capture represents the percentage of the total population you originally marked. From this you can calculate the total population size.

Part (b): 2 points possible; max score for this section is 2 points

1 pt. Direct threat (hunting, accidentally injuring, etc.).

1 pt. Indirect threat (habitat loss, contamination of habitat, etc.).

Part (c): 2 points possible; max score for this section is 2 points

1 pt. Name of act (the Endangered Species Act and the Lacey Act are two possibilities).

1 pt. Explanation (the Endangered Species Act prohibits injury to listed species or import of these species; the Lacey Act prohibits carrying wild animals across state borders without a federal permit).

Part (d): 3 points possible; max score for this section is 3 points

1 pt. The "bottleneck" reduces variation in the population.

1 pt. Even if the population size is achieved, the *variations* will still be limited.

1 pt. Decreased variation implies a less significant likelihood of surviving a selection pressure.

UNIT VI: ENVIRONMENT AND SOCIETY: TRADE-OFFS AND DECISION MAKING

CHAPTER 14

Economic Forces

"This president is going to lead us out of this recovery."
"Bank failures are caused by depositors who don't deposit enough money to cover losses due to mismanagement."
"Our party has been accused of fooling the public by calling tax increases 'revenue enhancement.' Not so. No one was fooled."

U.S. Vice President, Dan Quayle

Areas That You Will Be Tested On

A: Cost-Benefit Analysis
B: Marginal Costs
C: Ownership and Externalized Costs

Key Terms

Note: Definitions for all key terms listed below can be found in the glossary starting on page 648. For additional definitions of relevant terms from this chapter, refer to www.barronseduc.com/0764121618.html.

abatement costs
balance of trade
capital
capitalistic market economy
cost-benefit analysis
demand
deregulation
direct costs
durable goods
economic resources
effluent fees
emission taxes
environmental indicators
external costs
gross domestic product
gross national product
internal costs
law of demand
law of supply
law of supply and demand
less-developed country
market economy
more-developed country
pollution costs
subsidy

Key Concepts

Macroeconomics

Comparative Economics

Comparative economics compares various factors of a country's economy with other countries to gain insight into the wealth and the distribution of wealth. Furthermore, comparative economics is fundamental in the study of environmental science to understand a country's economic priorities. For example, if North Korea believes that it is acceptable to spend 27.5% of its Gross National Product on defense, then there may be little available resources left to deal with issues of environmental degradation.

GROSS DOMESTIC PRODUCT (1998) PER CAPITA FOR SELECTED COUNTRIES

Country	GNP ($/person)	Rank	Average Annual Growth
Switzerland	$39,980	3	1.5
Japan	$32,350	7	–2.9
United States	$29,240	10	1.5
Mexico	$3,840	75	3.0
Russian Federation	$2,260	97	–6.4
Belarus	$2,180	99	10.8
China	$750	145	6.4
Indonesia	$640	149	–18.0
Ethiopia	$100	206	–4.2

Source: The World Bank, Washington, DC, *World Development Indicators.*

COMPARATIVE PRICE LEVELS

Country	U.S. Dollars
Japan	$164
Switzerland	$124
Iceland	$125
United States	$100
Mexico	$70
Czech Republic	$40

Source: Organization for Economic Cooperation and Development, Paris, France, *Main Economic Indicators.*

MILITARY EXPENDITURES AND ARMED FORCES PERSONNEL (1997) OF SELECTED COUNTRIES

Country	% of GNP	$ per Capita	Armed Military Personnel per 1000 Population
Korea (North)	27.5	281	51.6
Angola	20.5	147	9.0
Saudi Arabia	14.5	1,050	9.0
Israel	9.7	1,690	33.4
Russia	5.8	283	8.8
Iraq	4.9	59	19.0
United States	3.3	1,030	5.7
Iran	3.0	74	9.1
China	2.2	61	2.1
Argentina	1.2	103	1.8
Mexico	1.1	44	2.6
Japan	1.0	325	2.0

Source: U.S. Department of State, Bureau of Verification and Compliance, *World Military Expenditures and Arms Transfers.*

ANNUAL PERCENT CHANGE IN CONSUMER PRICES (1998 TO 1999)

Country	Change
Argentina	–1.2%
Japan	–0.3%
Mexico	16.6%
Russian Federation*	85.7%
Turkey	64.9%
United States	2.2%

Source: International Monetary Fund, Washington, DC, International Financial Statistics.

*Inflation is a major problem in the Russian Federation.

RELATIVE HOURLY COMPENSATION COSTS FOR PRODUCTION WORKERS BY COUNTRY (1998)

Country	Compensation
Germany	147
Switzerland	123
United States	100
Mexico	10
Sri Lanka	3

Source: U.S. Bureau of Labor Statistics, News Release 00-254.

Note for Essays: Often the APES essays require *you* to provide possible alternatives to environmental issues. For example, "Provide three methods that could be utilized to reduce acid deposition." The following platforms from the Green Party, a grassroots political action organization, on economic reform will get you through these type of questions should they appear for economic issues.

Green Party on Employment

The goal of full employment is compatible with environmentalism. Full employment increases the GDP and GNP. It increases the standard of living. It increases and makes available the societal resources to be able to focus on environmental issues. The converse, high unemployment, saps governmental resources and delays attention and focus on the environment (i.e., when it comes down to spending limited resources on food for hungry people or spending resources to clean up a toxic dump site, the resources are spent on food). Only in more-developed countries can government budgets be prioritized, with environmental issues on the table.

All too frequently, in the American mixed economy, investors have one priority—to make money. How money is made is generally not a major issue. Often, the investors and managers live in communities far away from where the goods and services are produced. They do not have a stake in the community. More often than not, these manufacturing sites are being relocated to foreign countries (global economy) where the sense of investment in the well-being of a community is even more distant. In fact, the decision to move manufacturing to a foreign country, which results in (a) higher unemployment in the originating country, (b) lower tax base for the local area, (c) moving the issues of toxic wastes to countries that have little or no regulation, and (d) paying the foreign workers only a fraction of the wages paid to workers in the home country with no health or retirement benefits, is seen as a positive move by investors since profits increase. The Green Party (www.gpus.org), a grassroots political organization based on environmentalism, has outlined the following platform on creating full employment that is compatible with environmental issues:

- "Revamp the tax, fee, and regulatory burdens on small business. We advocate fiscal policies that encourage the development of smaller scale, appropriate-level technology and the establishment of locally-owned businesses that are ecologically sound. We should also reduce the costs a business incurs from hiring people.
- "Set up local non-profit development corporations. Large businesses whose ownership is outside the community, or whose profits are transferred outside the community, should be required to invest in these local corporations.

- "Establish 'enterprise zones' to create jobs in poor neighborhoods. These should, at a minimum, be developed through profit-sharing arrangements or, ideally, as employee-owned businesses.
- "Establish local Economic Conversion Commissions in areas where the decline of the defense industry has caused unemployment.
- "Change our foreign trade policies to discourage the exportation of jobs to countries that have weak labor and environmental laws—all of which result in workers and resources being easily exploited.
- "The Federal Government should reduce military spending by 75%. A portion of these funds should be distributed to state and local governments to pay for infrastructure and public works projects."

Green Party on Restructuring the Economy

The Green Party believes that the current mixed economy of the United States is a controlling and repressive force—it works on a sole goal, that of profit; and only responds to sanctions imposed by the government (i.e., because pollution control involves initial investment, it is done as a response, not an initiative). And yet companies who have implemented massive pollution control technology have actually seen an increase in their profit margin. The Green Party advocates the following policies for restructuring the American economy in line with sound environmental practices.

- Encourage ecologically-sound and employee-owned or profit-sharing businesses of appropriate scale. Such businesses would primarily serve a regional market to keep money circulating within the community, rather than sending it to distant corporate headquarters. To further such enterprises, we advocate "incubator programs" and other forms of assistance, as well as "buy local, buy green" campaigns. Establishing work place democracy must also be promoted.
- Encourage neighborhood nonprofit development corporations that work to establish community-based economics, rather than the usual redevelopment fixation on huge projects in the downtown areas of cities.
- Support the creation of cooperative and collective businesses. One way to encourage this would be to free nonprofit and locally owned businesses from the overly complex tax and regulatory structures designed for profit-making corporations.
- Encourage the development of an informal economy, including volunteerism. The unemployed and underemployed could participate in the local economy through a "credit barter" system. In this system, the medium of exchange would be individual credit–debit accounts. People would extend credits and collect debits from one another for the purchase of each other's goods and services. These credits and debits could then be transferred among community members, circulating much like a currency. Because local members would redeem the currency with their goods and services, this medium of exchange would have a local value. Changes in banking regulations would be required to support such a program.
- Further develop the American "intentional community" movement—a residential community composed of people who have come together for a common purpose and live with some degree of economic sharing. In the United States, over 200,000 people now live in some type of intentional community.

Green Party on True-Cost Pricing (Including External Costs)

Internal costs are costs paid by those that use a resource. It usually involves the direct cost of accessing the resource and converting it to a useable product or service. It does not include external costs—harmful effects of a product not included in the market price and paid for by those who do not necessarily use the resource (see section entitled "Externalized Costs" later). The mixed economy of the United States, which is fueled by profit, does not generally allow for true-cost pricing—adding the environmental and social costs of pollution, deforestation, and toxic wastes created during production, use, or disposal. Internal costs (costs of Earth resources, manufacturing, labor, etc.) are paid for by the manufacturer and are reflected in the market price. The true costs, which are usually delayed, are paid for by the society. These costs result in higher taxes, adverse health effects, and a reduced quality of life. Externalized costs are subsidized by the society for products used by consumers. Artificially low market prices encourage over-consumption or products that are environmentally unsound and underconsumption of products that are environmentally sound. The Green Party advocates the following policies for true-cost pricing:

- "Implement product labeling to inform consumers of the total cost of the product's ingredients and manufacturing process.
- "Provide education to explain that true-cost pricing incorporates the true life-cycle cost of a product and, therefore, is not a tax—it is a method of internalizing appropriate costs that will result in a net decrease in consumer prices. As true-cost pricing is implemented and less damaging practices are adopted, taxation for environmental clean-up and the other externalized costs will be dramatically reduced.
- "Establish an information clearinghouse, consultant's network, and other communication channels to facilitate the exchange of information. Many ecologically benign techniques already exist on a small scale or in limited locations. Sharing this expertise is crucial.
- "Integrate the concept of true-cost pricing into domestic industrial policies and regulations, and likewise, promote it in international trade agreements.
- "Recognize that some items exist for which true-cost pricing and/or the market may not be appropriate. These include, but are not limited to, health care, education and National Parks. True-cost pricing is one option available to address the failure of the market to include the economic, social and environmental costs and benefits created by the production, use, and disposal of goods and services.
- "Recognize that true-cost pricing may have short-term impact on people of lesser financial means. We must explore and implement measures to mitigate these effects."

Green Party on Economic Measurement

The Gross Domestic Product is the primary value used to measure economic growth. It is measured by combining

- Personal consumption,
- Government expenditures,
- Private investment,
- Inventory growth, and
- Trade balance.

It is calculated using a "chain-weighted" method. The system recognizes that business has been globalized, deregulation is increasing, and business activity and relative prices for goods change quickly and dramatically. The chain system also recognizes that output for computers, telecommunications equipment, and health services are growing much faster than other parts of the economy. The "chain" method forces the government to recalibrate the relative prices of these goods—and their relative importance to the economy—every year. The GDP does show positive gains when goods and services increase but does not reflect externalized costs. For example, the GDP increased due to the *Exxon Valdez* oil spill (see Free-Response Question at the end of this chapter) due to the flow of money during the clean-up operation. The Green Party advocates the following policies for alternative indexes for economic growth:

- "Devise an economic monitoring system that measures productive enterprises' total costs to the environment and society. Some of these costs cannot be expressed in monetary terms, but various accounting techniques are being developed to represent such costs. We support these efforts and will encourage their implementation to augment or replace the GDP.
- "Account for not only environmental costs but also social costs such as substandard wages and working conditions.
- "Classify activities such as volunteerism, domestic work, and child rearing as contributions to the economy.
- "Require businesses and government agencies to determine what social and environmental effects their activities are having and to make that information public."

Green Party on Government Spending

The State of California spends 5.2% of its state budget on prisons and correctional programs—it is the fastest growing segment of the budget (increasing at a 15% rate per year and double the rate of any other state budget category) and represents almost five times the amount budgeted for environmental protection (see Case Studies section in this chapter). Social welfare programs account for 24%—almost 20 times the amount for environmental protection. The Green Party advocates the following policies for reforms on government spending.

- "Conduct cost-effectiveness studies of the major departments in state and local government.
- "Make strategic social investments to avoid much greater future costs. For example, investing in quality education and social programs will avoid future spending on prisons and welfare. Providing effective family planning services will avoid later costs associated with neglected children.
- "Allow citizen organizations to have input into the formation of county, state and, ideally, federal budgets. Preliminary budget drafts should be made widely available to allow discussion and feedback.
- "Review the salary structures of public employees. Currently the average state employee salary is $40,000. We must examine whether this average represent a fair distribution of income between the lower and higher paid employees. We must also determine if there is an efficient proportion between the number of management positions and functional lower-level positions.
- "Grant tax expenditures (exemptions, deductions, etc.) only to achieve socially desirable purposes, such as environmental protection or job creation. Expenditures should have periodic review to determine if they are meeting their goals.

- "Stop the enormous expansion of the prison industry, which will result in this being the largest item in the state budget in the near future. Not only is this an ineffective crime deterrent, but it is also more expensive than putting money into preventative programs such as education, training, and job creation."

Microeconomics

Cost-Benefit Analysis

- Cost-benefit analysis is a technique for comparing the costs and benefits of a project over a time period. The technique has its origins in economic feasibility studies of public infrastructure projects such as dams.
- Cost-benefit analysis began as a result of the Federal Navigation Act of 1936, which required the U.S. Corps of Engineers to implement improvements to the nation's waterways when the total benefits of the project exceeded the costs of the project.
- Cost-benefit analysis requires quantifying all aspects of a project into a common, comparative unit (dollars).
- Private companies use profitability to weigh projects. Cost-benefit analysis cannot use profitability since the projects being analyzed are governmental projects that do not run on a capitalistic model (i.e., dams are not built and then sold for a profit). Factors such as public good and benefits to the environment are difficult to place a dollar value on. Cost-benefit analysis attempts to find the most economical method of meeting a public need.
- Cost-benefit analysis is an objective management tool to make decisions regarding public actions such as public safety, public health, and environmental projects and to quantitatively compare competing projects.
- Throughout much of the history of the United States, the private sector was not held responsible or accountable for negative consequences or aftermaths of public projects. Environmental or societal consequences were either ignored or paid for by taxpayers at either a local or national level. Since the 1960s and the use of cost-benefit analysis, all factors of a project are carefully identified, and negative consequences are quantitatively factored into decision making.
- Cost-benefit analysis applies to three economic situations:

1. To judge whether public services provided by the private sector are adequate (e.g., President George W. Bush's platform that private charities can do a better job in meeting social welfare programs than government-run programs).
2. To judge and assess inefficiencies ("market failures") in the private sector and their impact on the health, safety, and environmental needs of the country.
3. To meet societal needs in a cost-effective manner in areas that only government can address (defense, preservation of scenic areas, environmental protection, etc.)

- Cost-benefit analysis can be used to: (1) evaluate policy alternatives; (2) shape regulatory strategies; and (3) evaluate specific regulations.
- Cost-benefit analysis requires (1) gathering all information and data about a public issue, including history and background; (2) defining the possible solutions to solving the public issue; (3) brainstorming the possible environmental and societal consequences of the alternatives; (4) quantifying the benefits and costs; and (5) making decision and balancing concerns.

- The alternative to a project must be explicitly specified and considered in the evaluation of the project. For example, a city must decide on either expanding its bus service or building a light-rail system. The number of rides that would have been taken on an expansion of the bus system should be deducted from the rides provided by the light-rail system and likewise the additional costs of such an expanded bus system would be deducted from the costs of a light-rail system.
- The long-range impact of a project must be considered. For example, the U.S. Corps of Army Engineers is planning to expand irrigation into a semi-arid region of California to increase corn production. At the same time, due to market conditions, the U.S. Department of Agriculture limits the corn production in the Midwest by issuing quotas and incentives for Midwest farmers not to grow corn. The result of greater corn production in California might be offset by a reduction in the corn production for Nebraska. Thus the impact of the irrigation project might be zero rather than being the amount of corn produced by the project. Such benefits or costs are usually a function of the lobbying group representing a certain interest. In this example, lobbyists for California would claim a benefit since more corn has been produced. Lobbyists for Midwestern farmers, however, would claim no benefit and perhaps even a cost due to lower prices generated by higher output.
- Careful consideration needs to be taken in how benefits or costs are counted. For example, the State of Missouri builds a new state highway. Studies show that travel time has been decreased and fewer fatalities are occurring due to safer design. At the same time, increases in commuters bring more shoppers to a city. The city attracts more businesses and property taxes increase. When counting and assessing benefits, one would count only the increase in property tax *or* the decrease in commuter time—but not both.

FRAMEWORKS OF COST-BENEFIT ANALYSIS

Health or environmental protection standards. Reducing risk to the public whatever the cost. Examples: setting national goals and standards that affect all citizens. Asbestos abatement in schools or the Clean Air Act.

Technology. To achieve results that are predictable and certain. Example: focusing only on the individuals in society who are directly affected such as mine safety laws and regulations, which only affect miners.

Risk-benefit or cost-risk. Balancing health or environmental protection with the costs of providing the protection. Example: allowing a certain level of acceptable risk and/or negative consequences to reduce costs. The standards for front-end collision on cars are set to allow for acceptable risk. The standards could be set higher and technology is available, but then the cars would not be economical or competitive in price with cars produced in other countries. Another example is automobile speed laws.

Cost-effectiveness. Implementing a specific environmental, health, or safety objective at the least cost. Emphasis is on achieving the objective. Flexible regulatory guidelines are adapted to find the lowest cost to solve a problem. Each source of stationary pollution would be allowed to adapt mechanisms that would be the most cost-effective given their situation.

Cost-benefit. Determine an action and levels of action that achieve the greatest net economic benefit. Exploring options and determining incremental levels of remediation that provide the most benefit for the least cost. National health insurance might have different levels of coverage (senior citizen coverage as compared to working adult) with different premiums and different levels of prescription drug coverage to achieve the greatest benefits for the least cost.

Marginal Costs

- **Marginal cost is the cost of making one unit of a product. The average cost of producing a unit is determined by dividing the cost of production by the quantity of items produced. If production increases, the cost of an article will decrease.**

Example: A company produces 50 lamps. It costs $400 to produce these lamps. If the company produces 51 lamps, it will cost the company $404. Calculate the marginal cost to produce an extra lamp.

Solution:

Production costs can go down when more units are produced, because the fixed costs stay the same.

The cost of one lamp if 50 lamps are produced: $\frac{\$400}{50 \text{ lamps}} = \$8$

The *average* cost of one lamp if 51 lamps are produced: $\frac{\$404}{51 \text{ lamps}} = \$7.92$

The average cost per lamp decreased because the fixed costs stayed the same. However, if you have to employ more workers to increase production it will increase production costs.

The marginal cost to make one extra lamp is $404 – $400 = **$4.00**

- **Standard costs include costs for raw materials and labor for producing one item.**

Items Produced	Fixed Costs	Variable Costs	Total Costs
0	$1000	0	$1000
1000	$1000	$500	$1500
2000	$1000	$1000	$2000

In this case, fixed costs are the same, regardless of how many items are produced. For example, the rent of the building is a constant expense; it is not determined by how many items are produced. Variable costs increase as production increases. Examples of variable costs might include the cost of raw materials to produce the item. The more items produced, the more materials are needed.

When 1,000 units are made, the unit cost is $\frac{\$1500}{1000 \text{ units}} = \$1.50$

When 2,000 units are made, the unit cost is $\frac{\$2000}{2000 \text{ units}} = \$1.00$

Therefore, when more units are produced the cost of producing one unit decreases.

Marginal Costs of Producing Electricity

Marginal costs for producing electricity are defined as the operations and maintenance costs of the most expensive generating plant needed to supply the immediate demand for electricity (the marginal cost of generation). Average costs are defined as the total costs of production divided by sales to the consumers (total costs divided by total sales). Competition drives prices to marginal costs if there are many producers and consumers. In the case of electricity, this means that competitive prices would be based on the costs of producing one more kilowatt-hour of electricity. This is different from the cost-of-service, which uses average costs as the basis of prices. The application of marginal costs as the basis of price assumes that no supplier or consumer exercises market power. Market power exists when a supplier or consumer influences prices by virtue of its size or control over important aspects of the market, such as in this example, access to transmission lines. If suppliers exercise market power, prices could be higher than marginal costs. If a consumer segment exercises market power, then that segment could have a price advantage over other customers. Regulated prices for generating electricity are based on average costs, and competitive prices for generating electricity are based on marginal costs. During periods of high demand, when demand approaches the limits of generating capacity, prices may rise above the marginal cost of generation.

Breakeven Analysis

The breakeven point occurs when the price the item sells for equals the cost of production. At this point, there is no profit or loss. Anything above this value is considered profit, and anything below is a loss.

Example: Examine the following spreadsheet for the Widget Company. Each Widget sells for $20. Determine the breakeven point.

# of Widgets	Fixed Costs	Variable Costs	Total Costs	Sales @ $20 per Widget	Profit/Loss (Sales – Costs)
0	10,000	0	10,000	0	(10,000)
1000	10,000	10,000	20,000	20,000	0
2000	10,000	20,000	30,000	40,000	10,000
2500	10,000	25,000	35,000	50,000	15,000
3000	10,000	30,000	40,000	60,000	20,000

Solution: Breakeven is the point of zero profit. If the Widget Company sells 1000 widgets, it will break even. Anything above 1000 widgets sold is profit.

Cost of Production

Cost of production is determined by:

1. Direct costs—the cost of raw material used for a product.
2. Indirect costs—costs not directly linked to the product or the production process. Examples: rent of the factory, administration, cost of the business, and the wages and salaries of the administration staff.

3. Standard costs—raw material cost and standard labor costs of producing one item.
4. Fixed costs—costs that do not change according to the level of output.
5. Variable costs—costs that change according to level of output.

See Figure 14.1.

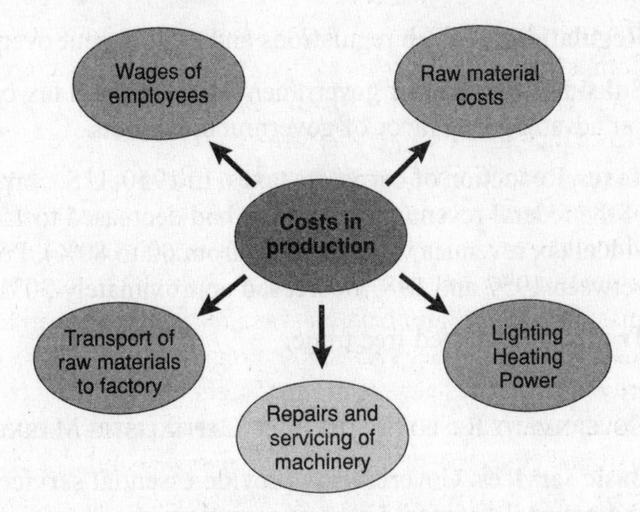

Figure 14.1. Costs of production.

Production Decisions

Production decisions can be either long term, which determine the nature of the business, or short term which influence the daily running of the business.

Short Term	Long Term
Programs—What inventories are necessary to maximize production and minimize cost?	**Product**—Is there a need for the product?
	Place—Where will the product be produced?
People—What is the least number of people required to produce the product? What level of training is required?	**Processes**—What range of products will be produced? How will the product be produced?

Methods of Production

- Job Production—a single product is made according to customer specifications.
- Batch Production—similar items are made in blocks or batches.
- Flow Production/Mass Production—production passes from one stage straight on to the next.
- Just-in-Time Production—rush orders.

Capitalist Market Economy Principles

Competition. To gain a monopoly and gain control of market price. Five international firms today control more than 50% of consumer durable goods, automobiles, aerospace, airlines, electronics, and steel and 40% of computer technology, oil, and media.

Goal. To produce the highest profit possible.

Liability. Pass on costs associated with product safety and environmental degradation to the public through higher prices or to the responsibility of the government.

Obligation. Corporations have no legal obligation to any one particular country. Issues of environmental protection, worker safety, and product safety should be left to a free-market economy and is the business of the corporation.

Product information. Restrict information on product safety.

Regulation. To keep regulations and government oversight to a minimum.

Subsidies. In favor of government subsidies and tax breaks that give companies a market advantage. In favor of government bailouts.

Taxes. Reduction of corporate taxes. In 1950, U.S. corporate taxes made up close to 40% of the federal revenue. By 1995, it had decreased to 15%. During the same period, individual tax revenues went up 20% (from 60 to 80%). Property taxes paid by corporations between 1957 and 1987 decreased approximately 30%.

Trade. Unrestricted free trade.

GOVERNMENT RESPONSES TO PURE CAPITALISTIC MARKET PRACTICES

Basic services. Governments provide essential services such as health care, education, and national defense. Large corporations are increasingly taking control of health care services.

Compensation. Governments provide disaster relief.

Economic stability. Adjusting interest rates. Interest rates are the costs of borrowing money for companies to expand.

Monopolies. Antitrust regulations. Made difficult because many corporations are transnational and no one country has complete jurisdiction. Of all world trade, 75% is controlled by 500 transnational corporations.

Natural resources. Government passes laws to protect natural resources such as air, water, soil, and the ozone layer.

Pollution. Government passes laws to reduce or prevent pollution and degradation of the environment.

Protection. Government passes laws on environmental protection, health and safety, fraud, consumer product safety, and other laws to protect individuals.

Public land. Government passes laws to protect and manage public lands.

Welfare. Government provides resources to people who are unable to work due to health, age, or other factors.

BASIC CORPORATE DECISION MAKING

- Administrative costs
- Advertising
- Manufacturing
- Profit
- Raw material costs
- Research and development
- Salaries
- Waste disposal

ECONOMIC RESOURCES

- Natural Resources
- Human resources
- Financial resources
- Manufactured resources

COMPARISON OF MAJOR TYPES OF ECONOMIC SYSTEMS

Capitalistic market. A real-world economic system in which the means of production and distribution are privately or corporately owned and development is proportionate to the accumulation and reinvestment of profits gained in a free market. The goal of a capitalistic market is to eliminate competition and gain a monopoly. Techniques for maximizing profits include: lobbying against global free trade and governmental price control, lobbying for subsidies and tax breaks, seeking regulations against competitors; and withholding information about harmful health and environmental conditions. Generally found in developed and moderately developed countries. Examples include North and South America, Europe, and Australia. Approximately half of the world's population lives in a capitalist market economy.

Command economic system. A real-world economic system in which the means of production and distribution are privately or corporately owned and development is proportionate to the accumulation and reinvestment of profits gained in a free market. Generally found in developed and moderately developed countries. Examples include North and South America, Europe, and Australia. Approximately half of the world's population lives in a capitalist market economy. Environmental pollution and degradation is seen as the result of economic activity with the solution being that the government should control such activity and dictate measures to correct it. Countries that are based on this system include China and North Korea.

Communal resource management system. Resources that are managed and sustained by a small group of people living together. Members of a community have inhabited the area for a long time and expect their children to do so also; the community is well-established and has definite boundaries; resources are relatively scarce, which requires members to cooperate to sustain the resource; and rules for utilizing the resource fairly by all members are well-defined.

Ecological economics. Incorporates ecological principles and moral principles into economic theory. Areas of issue in ecological economics include who owns resources, how they can be shared fairly among all people, how resources can be preserved and protected for the future, and the rights of all species to exist. Other issues focus on finite capacity, recycling, resource consumption, and carrying capacity.

Economic localization system. Regional or local economic systems that take into account the needs and concerns of a country that unites local communities and preserves national customs and traditions. Example: Norway decided not to join the European Union and to keep its own economic system.

Free-market economy. All economic decisions are made in markets where buyers freely interact with sellers without governmental interference. Price is based on supply and demand. No regulations, taxes, subsidies, or barriers. Prices include harmful environmental effects.

Frontier or pioneer economy. Occurs when supplies of natural resources seem unlimited and supply far exceeds demand. Example: frontier America when land, forests, and wild animals seemed inexhaustible. Characterized by waste and massive damage to the environment. Prices are characteristically low for the resource.

Global Market. An economic arrangement that consists of countries around the world investing money and resources into other countries, forming one large economic system. This means the economies of all countries are linked, so they all rely heavily on the success of the other to prosper. Developments in technology have made it easier for countries to do business with each other. For more information on the global market, visit www.theglobalist.com.

Market-based. Allocates resources primarily through the interaction of individuals (households) and private firms.

Mixed economic system. An economic system that lies between a pure command and a pure market economy. Almost all of the world's economic systems are mixed, being controlled more by private individuals and institutions than by government. Pure competition mostly determines buying and selling, with interactions of demand, supply, and price. Government intervention involves subsidies, taxes, and the prevention of unfair development of monopolies.

Natural resource economy. Nature is seen as a deposit of raw materials. The Earth is seen as a sink for wastes.

Neoclassical economics. Applying objective and value-free principles of modern science and mathematical predictive analysis to economic theory. Subscribe to principles of supply and demand in determining price and resource allocation. Economic growth is required to maintain full employment. Natural resources are just the means for production—substitutes can be found if necessary.

Political economy. A focus on economic theory incorporating a moral philosophy that deals with issues of benefits to members of the society, class equality and structure, distribution of assets within the community, and so on.

Survival economy. An economy whereby basic needs are obtained directly from nature. Approximately half of the world's population exists in a survival economy and includes China, India, Indonesia, and Sub-Saharan Africa.

Management Types
- Ecoindustrial networks
- Life cycle management
- Process design management
- Totally life quality management

Resource Scarcity (refer to Figure 14.2)
- Prices go up according to law of supply and demand.
- Competition drives discovery of new sources.
- Competition drives research and development of looking for alternatives.
- When prices are high due to low supply, conservation becomes important.
- Research provides more productive and cost-effective ways to utilize resources.
- Trading of the commodity becomes more important.
- Investment and potential profits may increase due to higher prices.
- Recycling becomes an economical alternative and profitable.

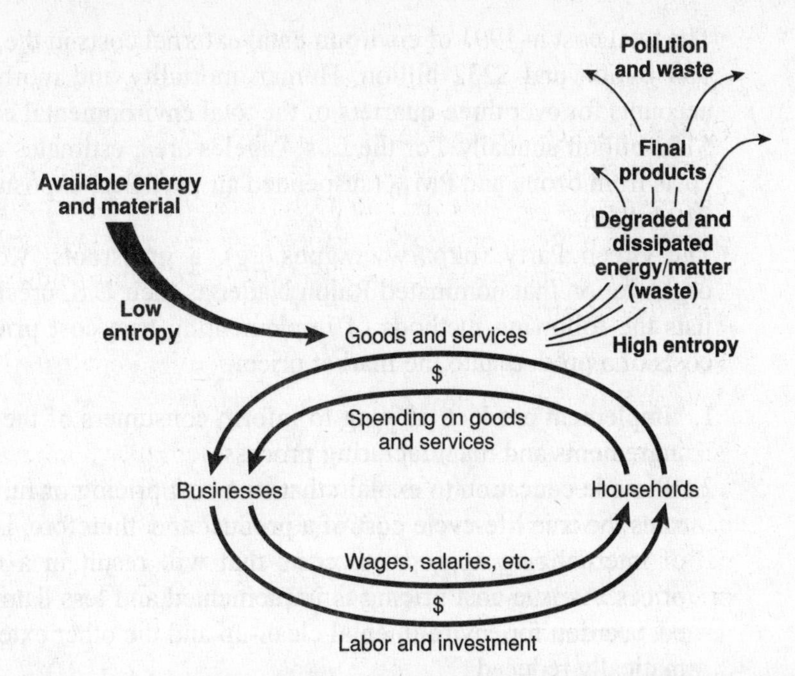

Figure 14.2. The linear throughput of low-entropy energy and matter (upper part of diagram) sustains the economy and drives the circular flows of exchange value (lower part of diagram).

Externalized Costs

- External costs are harmful effects of a product not included in the market price of the item and paid for by those who do not necessarily use the resource. Current accounting and pricing systems used in the United States and most other countries do not include hidden environmental and social costs such as air and water pollution, deforestation, and toxic waste into the price of the product. These hidden costs are created during the production, use, or disposal of the products. The revenue and profits from these products are internalized by the producer; however, the external costs are paid by the society in diminished resources, environmental pollution, and, in general, a lower quality of life. Externalizing costs allow consumer prices to be artificially low, and competitive in a global market, but it eventually is seen and paid for in adverse health effects, higher health costs, increased solid waste costs, increased energy costs, increased costs for national clean-up of Superfund sites, etc.
- Examples of external costs include:

1. Paper companies that dumped their effluent wastes into rivers and streams lowering their costs of clean-up but passing the cost of clean-up to the public.
2. Tobacco companies concealing the risks of their product and passing the costs of health care onto the public.
3. The cost of burning fossil fuels does not cover the external costs of required health care, crop damage, and global warming. Conservative estimates put the externalized costs of California's current natural gas power plants at $1.9 billion per year.
4. The cost of timber, which does not account for the external costs of degraded scenery, lost recreation sites, water pollution, and the loss of biodiversity.
5. The price of gasoline, which does not cover the increased costs of health care and insurance rates due to asthma and other respiratory ailments. It also does not cover the costs of days of illness; and the cost of the damage to the environment and fisheries due to oil spills.

- The total cost in 1991 of environmental external costs in the United States ran between $54 billion and $232 billion. Human mortality and morbidity due to air pollution accounts for over three-quarters of the total environmental cost and could be as high as $182 billion annually. For the Los Angeles area, estimates of the annual health-based costs from ozone and PM_{10} (suspended air particles) exposure alone may be as high as $10 billion.
- The Green Party (http://www.gpus.org), a grassroots, worldwide environmentalist organization that nominated Ralph Nader as their U.S. presidential candidate in 2000, lists the following methods of implementing true cost pricing, that is, including all costs of a product into the market price:

1. "Implement product labeling to inform consumers of the total cost of the product's ingredients and manufacturing process.
2. "Provide education to explain that true cost pricing or full-cost accounting incorporates the true life-cycle cost of a product and, therefore, is not a tax—it is a method of internalizing appropriate costs that will result in a net decrease in consumer prices. As true-cost pricing is implemented and less damaging practices are adopted, taxation for environmental clean-up and the other externalized costs will be dramatically reduced.
3. "Establish an information clearinghouse, consultant's network, and other communication channels to facilitate the exchange of information. Many ecologically benign techniques already exist on a small scale or in limited locations. Sharing this expertise is crucial.
4. "Integrate the concept of true-cost pricing into domestic industrial policies and regulations, and likewise, promote it in international trade agreements.
5. "Recognize that some items exist for which true-cost pricing and/or the market may not be appropriate. These include but are not limited to health care, education and National Parks. True-cost pricing is one option available to address the failure of the market to include the economic, social, and environmental costs and benefits created by the production, use and disposal of goods and services.
6. "Recognize that true-cost pricing may have short-term impact on people of lesser financial means. We must explore and implement measures to mitigate these effects."

Tragedy of the Commons—An Economics Lesson

Garrett Hardin wrote the *Tragedy of the Commons* in 1968. It has proven to be a very popular and useful work for understanding how, as humans, we have come to be at the brink of numerous environmental catastrophes. People face a dangerous situation created not by malicious outside forces but by the apparently appropriate and innocent behaviors of many individuals acting alone.

The *Tragedy of the Commons* is a parable and can be traced back to Aristotle who noted that "what is common to the greatest number has the least care bestowed upon it." The story is of a *common* grassy area used by a community of local herdsmen for their grazing cattle. Each herdsman owns his own cattle, but no one person owns the common. Each herdsman can use the common as he so chooses—it is open to all. Hardin now describes the situation:

It is to be expected that each herdsman will try to keep as many cattle as possible on the commons. Such an arrangement may work reasonably satisfactorily for centuries because tribal wars, poaching, and disease keep the numbers of both man and beast well below the carrying capacity of the land. Finally, however, comes the day of reckoning, that is, the day when the long-desired goal of social stability becomes a reality. At this point, the inherent logic of the commons remorselessly generates tragedy.

The tragedy is that each herdsman tries to make the most amount of money that he can. There is no problem with that. The herdsman then asks himself, "What harm will one more cow grazing on the common cause—the amount is insignificant." The benefit to the herdsman is that he makes more money by having one more cow eating from a common area and he reaps the rewards. The tragedy is the damage done to the commons by overgrazing and the price of decreased vegetation and soil erosion that is shared by all of the herdsmen.

The rational herdsman concludes that the only sensible course for him to pursue is to add another animal to his herd. And another; and another. . . . But this is the conclusion reached by each and every rational herdsman sharing the commons. Therein is the tragedy. Each man is locked into a system that compels him to increase his herd without limit—in a world that is limited.

This story parallels what is happening worldwide in regard to resource depletion and pollution. The seas, air, water, animals, minerals—they are the commons. They are there for humans to use. Those who exploit them become rich. The price of depleting the resources and of polluting the world is paid for by all of us. For example, if a boat takes just one more haul of fish, the captain and crew become richer. And yet, the depletion of the stock, and its reduced reproductive capacity, is something we all share (in economic terms and in higher prices). If there were only one captain, the fish would recover. If the captain only took one more haul, the fish would recover. The trouble is, there are millions of captains, each taking as much as they can get.

LIMITS TO THE TRAGEDY OF THE COMMONS THEORY

- Economic decisions are generally short term, based on reactions in the market. Environmental decisions are long term.
- Land that is privately owned is subject to market pressure. If privately owned timberland is increasing in value at an annual rate of 3%, but interest rates on loans to purchase the land could be 7% or higher, this could lead to the land being sold or the timber being harvested for short-term profit.
- Some "commons" are easier to control than others. Land, lakes, rangeland, deserts, and forests are easier to control because they are geographically defined. Other commons, such as air, and open oceans do not belong to any one group and are thus harder to achieve international cooperation in their use (e.g., Kyoto Protocol).
- Incorporating discount rates into the valuation of resources (incorporating the future value) would be an incentive for investors to bear a short-term cost for a long-term gain. However, future valuation is difficult and subject to error.
- Breaking a "commons" into smaller privately owned parcels fragment the policies governing the entire commons. Different standards and practices used in one parcel affect all of the other parcels.

Case Studies

Silicon Chips—A Study in Externalized Costs. The production of *each* 6-in. silicon wafer uses the following resources:

- 2275 gallons of deionized water,
- 3200 cubic feet of bulk gases,
- 22 cubic feet of hazardous gases,
- 20 pounds of chemicals, and
- 285-kilowatt hours of electrical power.

Silicon Valley in Santa Clara County, California, has more Superfund sites (29) than any other county in the United States (80% of which were caused by the high-tech electronics industry). As of 1996, 20 of the 29 Superfund sites in Santa Clara County were directly caused by the process of producing silicon wafers and other high-tech electronic components. Another five Superfund sites were caused by related industries that support chip manufacturing (equipment manufacturers, chemical suppliers, and waste disposal).

In addition, vast amounts of groundwater reserves are contaminated when chips are made. In the city of Santa Clara, for example, the high-tech electronics industry alone used almost 24% of the city's water in 1994/1995. Of the top wastewater discharging companies in Santa Clara County in 1994, 65% were electronics companies.

Over the years, California passed tougher legislation strengthening environmental laws resulting in major high-tech manufacturing firms looking elsewhere to relocate their plants. Austin, Texas; Phoenix, Arizona; and Albuquerque, New Mexico became major relocation centers for the chip manufacturers. For example, in September 1995, Sumitomo Sitix, a huge Japanese silicon wafer manufacturer, chose north Phoenix for a $400 million plant. They were lured to Phoenix by a promised property tax cut of 80%, duty free import/export privileges, $7 million in immediate infrastructure improvements, $5.5 million for off site sewer and water systems, and $1.5 million for street improvements. This same Sumitomo plant is projected to use 2.4 million gallons of water per day (750,000,000 gallons per year). This is in Phoenix, Arizona, where water supply is already a critical issue (i.e., water rights between California and Arizona over Colorado River Water). Sumitomo will also be discharging contaminated water to the public treatment plant, requiring additional investments by the city of Phoenix.

In Albuquerque, New Mexico, residential customers pay $1.75 for 1000 gallons of water. Intel Corporation, another major chip manufacturer, pays 41 cents for the same amount of water. Water demand by Intel is expected to reach 6 million gallons of water per day. Again, the issue of locating a high-water-demand industry in the arid southwestern United States arises. Other incentives to move to Albuquerque have included a $250 million tax incentive and an $8 billion dollar industrial revenue bond.

It becomes clear that the cost of a computer chip, especially in an economically-competitive market such as the electronics industry, does not include the external costs of damage to the environment or the depletion of natural resources. To account for all of these costs would be an example of full-cost or true-cost pricing.

The California State Budget—A Matter of Priorities. The total revenue expected by the State of California in 2001–2002 is projected to be approximately $80 billion. California Governor Gray Davis', 2001–2002 state budget has proposed 1.4% for environmental issues. See Figure 14.3.

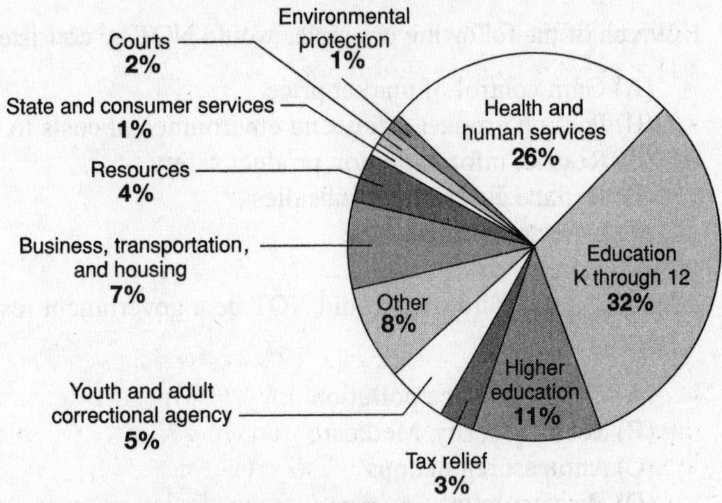

Figure 14.3. Total Expenditures (including selected bond funds).
Source: Department of Finance, State of California.

A breakdown of the 1.4% budget allocation for environmental protection shows:

Ecosystem restoration	\$101.5 million
Environmental water account	\$30.2 million
Water use efficiency	\$9.8 million
Water transfers	\$1.3 million
Watershed management	\$20.0 million
Drinking water quality	\$26.1 million
Levees	\$23.3 million
Storage	\$18.9 million
Conveyance	\$19.2 million
Science	\$30.3 million
Program management	\$13.9 million

Multiple-Choice Questions

1. Which of the following principles would NOT be consistent with a capitalist market?

   (A) Gain control of market price
   (B) Pass on product safety and environmental costs to the public
   (C) Restrict information on product safety
   (D) Reduce government subsidies
   (E) Unrestricted free trade

2. Which of the following would NOT be a government response to a capitalist market practice?

   (A) Laws to reduce pollution
   (B) Social security, Medicare
   (C) Antitrust regulations
   (D) Disaster relief
   (E) All are responses

3. What would gross national product per capita show?

   (A) What kinds of goods and services the country produces
   (B) If all people share equally in the wealth of a country
   (C) Unreported cash payments for goods and services, bartering, or black market trading
   (D) The general standard of living enjoyed by the average citizen
   (E) All of the above

4. When college students leave town for the summer, the demand for books and supplies at the local stationery store declines. This results in

   (A) a decrease in equilibrium price and an increase in quantity.
   (B) an increase in equilibrium price and quantity.
   (C) a decrease in equilibrium price and quantity.
   (D) an increase in equilibrium price, and a decrease in quantity.
   (E) none of the above.

5. Preserving the value of a resource for the future is a(n)

   (A) aesthetic value.
   (B) cultural value.
   (C) existence value.
   (D) use value.
   (E) option value.

6. It costs a copper smelter $200 to reduce its emissions by 1 ton, and $250 for an additional ton. It costs an electric utility $100 to reduce its emissions by 1 ton, and $150 for an additional ton. What is the cheapest way of reducing total emissions by 2 tons?

   (A) Legislate that both firms must reduce emissions by 1 ton.
   (B) Charge both firms $251 for every ton they emit.
   (C) Allow both firms to buy a permit to pollute that costs $151.
   (D) File an injunction to halt production until the firms reduce emissions by 2 tons.
   (E) None of the above

7. In economic analysis, the optimal level of pollution

   (A) is always zero.
   (B) is where the marginal benefits from further reduction equals the marginal cost of further reduction.
   (C) occurs where demand crosses the private cost supply curve.
   (D) should be determined by the private market without any government intervention.
   (E) is none of the above.

8. The supply curve is upward sloping because

   (A) as the price increases, so do costs.
   (B) as the price increases, consumers demand less.
   (C) as the price increases, suppliers can earn higher levels of profit or justify higher marginal costs to produce more.
   (D) all the above are correct.
   (E) none of the above is correct.

9. Most governments of the world have a

   (A) capitalist market system.
   (B) command economic system.
   (C) mixed economic system.
   (D) free-market economy.
   (E) survival economy.

10. The gross national product is a measure of

   (A) the economic growth of a country.
   (B) a measure of how much wealth a country has.
   (C) a measure of the quality of life in a country.
   (D) how many people are gainfully employed and producing goods and services.
   (E) the sum of the goods and services a country produces in one month.

11. In a developing country, approximately what percentage of the population lives in poverty?
    (A) less than 10%
    (B) 10%
    (C) 25%
    (D) 35%
    (E) more than 50%

12. Which of the following countries is NOT a member of the G7?
    (A) Japan
    (B) Russian Federation
    (C) Italy
    (D) Canada
    (E) France

13. A system utilized by the United Nations to track and compare development in social areas such as life expectancy, education, and standard of living is known as the
    (A) Human Development Index (HDI).
    (B) gross national product (GNP).
    (C) per capita GNP.
    (D) gross domestic product (GDP).
    (E) Index of Sustainable Economic Welfare (ISEW).

14. Externalized costs of nuclear power include all the following EXCEPT
    (A) disposing of nuclear wastes.
    (B) government subsidies.
    (C) costs associated with Three Mile Island.
    (D) Price-Anderson Indemnity Act.
    (E) all are external costs.

15. Which of the following is NOT part of cost-benefit analysis?
    (A) Judging whether public services provided by the private sector are adequate
    (B) Juding and assessing inefficiencies in the private sector and their impact on health, safety, and environmental need
    (C) Determining external costs to society
    (D) Meeting societal needs in a cost-effective manner
    (E) All are part of cost-benefit analysis

Answers to Multiple-Choice Questions

1. **D**	6. **C**	11. **C**
2. **E**	7. **B**	12. **B**
3. **D**	8. **C**	13. **A**
4. **C**	9. **C**	14. **E**
5. **E**	10. **A**	15. **C**

Explanations for Multiple-Choice Questions

1. **(D)** Subsidies are monetary assistance granted by a government to a person or group in support of an enterprise regarded as being in the public interest. A capitalistic form of government favors business and passes legislation to foster it. To reduce subsidies, would be a type of economics known as free market.

2. **(E)** Refer to page 463, *Capitalist Market Economy Principles.*

3. **(D)** Gross national product (GNP) per capita is the dollar value of a country's final output of goods and services in a year, divided by its population. It reflects the average income of a country's citizens. Countries with a GNP per capita in 1998 of $9,361 or more are described as high income, between $761 and $9,360 as middle income, and $760 or less as low income. Low- and middle-income countries produce about 20% of the world's goods and services, but have more than 80% of the world's population. This trend results in people in low- and middle-income countries having a smaller share of the world's goods and services than people in high-income countries.

   Of the world's 6 billion people, more than 1.2 billion live on less than $1 a day. Two billion more people are only marginally better off. About 60% of the people living on less than $1 a day live in South Asia and Sub-Saharan Africa. In high-income countries, farmers—men and women—make up less than 6% of the workforce, while in low- and middle-income countries combined, they represent nearly 60% of all workers. And developing countries account for almost $1 out of every $4 that industrial countries earn from their exports.

4. **(C)** Equilibrium is a situation in which (1) there is no inherent tendency to change, (2) quantity demanded equals quantity supplied, and (3) the market just clears. At the market equilibrium (where supply intersects with demand), every consumer who wishes to purchase the product at the market price is able to do so, and the supplier is not left with any unwanted inventory.

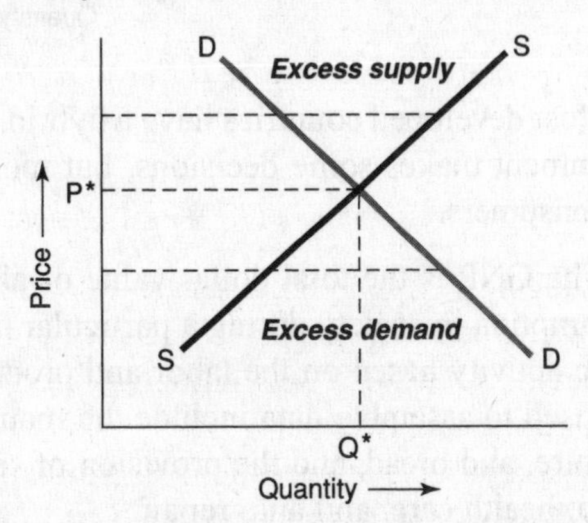

5. **(E)** An option value is the value that people place on having the option to enjoy something in the future, although they may not currently use it.

6. **(C)** The law of diminishing return refers to the diminishing amount of extra output that we get when we successively add extra units of varying input (labor) to a fixed amount of some other input (land). Nature's critical limit function, like that of carrying capacity, is defined as the point where the expansion of an economy is equal to the contraction of an ecology, (i.e., zero-sum outcome), resulting from the the law of diminishing return; the origin of which is based on a Malthusian population growth-to-land ratio model.

7. **(B)** The marginal benefit of pollution control declines as the level of environmental quality goes up, and at the same time, the marginal cost of pollution control tends to increase. The optimal level of pollution from an economic standpoint is where the marginal benefit equals the marginal cost. As long as the marginal benefit exceeds the marginal cost, there is economic incentive for cleaning up. Past that point, from an economic viewpoint, people should tolerate pollution. In other words, the cost of cleaning up every single molecule of toxic compounds on the Earth would far exceed the economic benefit that might be received. A certain level of acceptable pollution must be allowed, balanced by the environmental benefits versus the costs of clean-up.
8. **(C)** The law of supply holds that other things being equal, as the price of a good rises, its quantity supplied will rise, and vice versa. Supply is derived from a producer's desire to maximize profits. When the price of a product rises, the supplier has an incentive to increase production because he can justify higher costs to produce the product, increasing the potential to earn larger profits. Profit is the difference between revenues and costs. If the producer can raise the price and sell the same number of goods while holding costs constant, then profits increase.

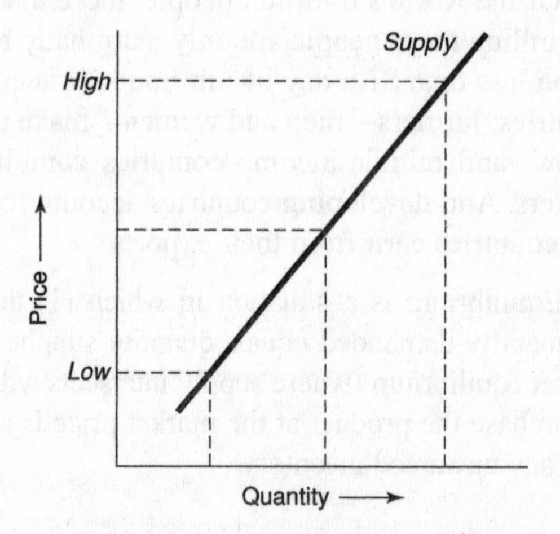

9. **(C)** Most developed countries have a hybrid, or mixed economic system, wherein the government makes some decisions, but most are made in the market by producers and consumers.
10. **(A)** The GNP is the total dollar value of all final goods and services produced for consumption in society during a particular time period. Its rise or fall measures economic activity based on the labor and production output within a country. The figures used to assemble data include the manufacture of tangible goods such as cars, furniture, and bread, and the provision of services used in daily living such as education, health care, and auto repair.
11. **(C)** The percentage of the world's population living in poverty has declined sharply over the last several decades. Still, as total global population has climbed, the absolute number of poor has remained unchanged at nearly 1.2 billion. The percentage of poor living in developing countries is approximately 25%.
12. **(B)** The Russian Federation is part of the G8.

13. **(A)** The Human Development Index (HDI) is based on health, education, and the standard of living. A summary of the 1998 HDI follows.

	Life Expectancy	Adult Literacy Rate	GDP per capita	HDI
All developing countries	62.2	70.44	3068	0.5864
Least-developed countries	51.16	49.2	1008	0.3439
Industrial countries	74.17	98.63	16337	0.9114
World	63.62	77.58	5990	0.7715

Source: United Nations Human Development Program.

14. **(E)** External costs are the costs that are borne by people other than the producer of a good.

15. **(C)** Refer to page 459, *Frameworks of Cost-Benefit Analysis.*

Free-Response Question

by Annaliese Beery, B.A.
California Institute of Technology
Williams College

A.P. Environmental Science Teacher, Marin Academy
San Rafael, California

In 1989, the oil tanker *Exxon Valdez* struck an offshore reef in Alaska and spilled more than 11 million gallons of crude oil into Prince William Sound. It was the largest spill in U.S. history, and although Exxon spent $2.2 billion partially dispersing and removing oil, many seabirds and mammals died as a result of this catastrophe.

Ironically, environmental accidents such as the *Exxon Valdez* oil spill actually stimulate our nation's economy by causing an increase in gross domestic product and gross national product. Using this information, answer the following questions.

(a) Explain what gross domestic product (GDP) or gross national product (GNP) is a measure of, and give a specific example of a possible cause for an increase in the GDP or GNP following the *Exxon Valdez* oil spill.

(b) Give at least one critique of GDP and GNP as progress indicators and mention an alternative progress indicator. What does that alternative attempt to measure?

(c) Name and explain one internal cost and one external cost that might be associated with the oil industry. Include in your answer an explanation of what it means to be an internal or external cost.

(d) How would internalizing external costs (sometimes called full-cost analysis or true-cost analysis) affect the pricing and economic competitiveness of petroleum products?

Free-Response Answer

(a) Gross domestic product (GDP) is a measure of the market value of goods and services transacted within a country during a year, so it increases any time products or services are paid for. This means that the additional expenses Exxon incurred by controlling the spill (up to $2.2 billion) increased the GDP. For example, money was spent to hire boats with skimmers in an attempt to capture and contain floating oil.

(b) GDP and GNP have been critiqued as progress indicators because economic growth is only indirectly related to "quality of life." The GDP and GNP is often stimulated by events that directly lower quality of life, such as the *Exxon Valdez* oil spill, the Oklahoma city bombing, and war. Alternative progress indicators attempt to capture quality of life or standard of living, although these are often harder to define and measure than flow of dollars. One such indicator is the net national product (NNP), which accounts for depletion and destruction of natural resources along with changes in GNP. Other progress indicators also account for inequalities in income distribution, loss of farmland, and air quality.

(c) An internal cost is a cost that is paid for by the organization producing a product. This cost is typically passed along to the consumer when they purchase the product. One internal cost of the petroleum industry could be the cost of extracting petroleum from the ground using oil drills.

An external cost is a "cost," usually to the world at large, that is not paid for by the producer and is not part of the price of a product. In other words, it is a harmful effect that the producer is not financially responsible for remediating. An example of an external cost of the petroleum industry would be the water contamination that occurs when oil tankers leak oil that is not cleaned up.

(d) Internalizing external costs would mean that organizations producing a product would pay for any harmful effects (the external costs) of their products. This would allow the market price of a product to reflect the full cost of a product to produce and clean up, and it would increase the price of any product that has external costs. A product with high external costs would become substantially more expensive and less economically competitive when this happened. In the energy industry, sources of power such as oil and gas have relatively high external costs and could become more expensive than lower-impact sources of energy such as wind power.

Part (a): 2 points
Part (b): 2 points
Part (c): 4 points
Part (d): 2 points
Ten points necessary for full credit.

RUBRIC	ESSAY
1 pt. Explanation of GDP or GNP. 1 pt. Example of how cleaning up the spill or replacing lost equipment/oil causes goods or services to be transacted.	(a) *Gross domestic product (GDP) is a measure of the market value of goods and services transacted within a country during a year*, so it increases any time products or services are paid for. This means that the *additional expenses Exxon incurred by controlling the spill (up to $2.2 billion) increased the GDP. For example, money was spent to hire boats with skimmers* in an attempt to capture and contain floating oil.
1 pt. GDP and GNP are economic measures, not direct quality of life measures. ½ pt. Example of alternative indicator. Some specific examples include the NNP, ISEW, GPI, and NEW. A general alternative is to take into account any standard of living data (literacy rate, percent of population below poverty line, etc.). ½ pt. Explanation of what example indicator measures.	(b) GDP and GNP have been critiqued as progress indicators because economic growth is only *indirectly related to "quality of life."* The GDP and GNP are often stimulated by events that directly lower quality of life, such as the *Exxon Valdez* oil spill, the Oklahoma city bombing, and war. Alternative progress indicators attempt to capture quality of life or standard of living, although these are often harder to define and measure than flow of dollars. One such indicator is the *net national product (NNP), which accounts for depletion and destruction of natural resources along with changes in GNP*. Other progress indicators also account for inequalities in income distribution, loss of farmland, and air quality.
1 pt. Naming of an internal cost. Costs may be expenses related to clean-up, ship or oil replacement, cost of labor, and the like. 1 pt. Explanation that internal costs are costs paid for by the organization producing the product.	(c) *An internal cost is a cost that is paid for by the organization producing a product.* This cost is typically passed along to the consumer when they purchase the product. One internal cost of the petroleum industry could be the *cost of extracting petroleum from the ground using oil drills.*
1 pt. Naming of an external cost Such as water pollution left behind after a spill, air or water pollution from operating tankers, and increase of CO_2 in the atmosphere when the petroleum products are consumed. 1 pt. Explanation that external costs are harmful effects of the product that are not paid for by the organization.	*An external cost is a "cost,"* usually to the world at large, *that is not paid for by the producer and is not part of the price of a product*. In other words, it is a harmful effect that the producer is not financially responsible for remediating. An example of an external cost of the petroleum industry would be the *water contamination that occurs when oil tankers leak oil that is not cleaned up.*
1 pt. Price of product would increase as external costs are paid for by the producer and passed along to the consumer. 1 pt. Petroleum products would become relatively less economically competitive.	(d) Internalizing external costs would mean that organizations producing a product would pay for any harmful effects (the external costs) of their products. This would allow the market price of a product to reflect the full cost of a product to produce and clean up, and *it would increase the price of any product that has external costs.* A product with high external costs would become substantially more expensive and less economically competitive when this happened. In the energy industry, sources of power such as oil and gas have relatively high external costs and could become *more expensive than lower-impact sources of energy such as wind power.*

Word Count: 419

CHAPTER 15

Environmental Ethics

The more clearly we can focus our attention on
the wonders and realities of the universe about us,
the less taste we shall have for destruction.

Rachel Carson, 1954

Areas That You Will Be Tested On

Note: The AP College Board has not specified what will be tested on in the area "Ethics."

Key Terms

Note: Definitions for all key terms listed below can be found in the glossary starting on page 648. For additional definitions of relevant terms from this chapter, refer to *www.barronseduc.com/0764121618.html*.

deep ecology	**positivists**
ecofeminism	**poststructuralists or postmodernists**
environmental racism	**relativists**
malthusian	**stewardship**
morals	**universalists**
naturalism	**utilitarianism**
neo-luddites	**values**
nihilism	

Key Concepts

Theories of Moral Responsibility to the Environment

Anthropocentric (human-centered)

- Role of humans—to be a master
- Intrinsic value—humans
- Instrumental value—nature

Environmental responsibility and duties are derived from human interests.

- *Free-market School*—manage the Earth through a free-market global economy and competition, minimal government interference; Milton Freidman.

- *No problem School*—All environmental and resource problems can be solved through better management and technology.
- *Responsible planetary management School*—A mixture of free-market, better technology and government regulation; enlightened self-interest.
- *Spaceship Earth School*—Earth is a single, complex machine that can be managed.
- *Stewardship School*—Humans have ethical responsibility to be good managers.
- Possible concerns: lack of knowledge or wisdom, focus on the short term, based on increased losses of natural resources.

Biocentric (life-centered)
John Muir and Aldo Leopold

- Role of humans—to be one of many
- Intrinsic value—species
- Instrumental value—abiotic nature

All forms of life have an intrinsic right to exist. Controversy arises in assigning hierarchical rights—animals vs. plants, rights of pests to humans, etc. Controversy over rights of individual organism vs. rights of its species.

Ecocentric (Earth-centered)
Aldo Leopold, Saint Francis of Assisi, Ralph Waldo Emerson, Henry David Thoreau, John Muir, Rachel Carson

- Role of humans—destroyers
- Intrinsic value—processes
- Instrumental value—individuals

Environment deserves moral consideration on its own, not associated with human interests.

- *Environmental-wisdom school*—Resources are limited, technology and economic growth can benefit and damage the environment, and nature does not exist for man, do not attempt to adapt the Earth for man's needs—instead, learn to adapt to the environment.
- *Deep ecology school*—Recognizes both intrinsic and instrumental value of species, ecosystems, and biosphere. Also believe in reducing human population size, appreciating inherent life quality (not material standards), and humans have no right to reduce or interfere with environmental interdependence, richness, or diversity.

CONCEPTS OF NATURE

Beautiful. To understand nature is to discover the meaning of existence. The purpose of nature is for man to enjoy its beauty and create meaning from that experience.

Capricious. Nature is unpredictable and has no sense of caring. Humans must struggle against nature to survive.

Chaotic. People are needed to add order to nature. It is the duty of man to harness, tame, and manage nature.

Commodity. Nature exists as a storehouse of resources for man to use.

Dangerous. Nature is wild and dangerous. Man can test himself against nature and emerge victorious. Nature can be a threat to survival.

Machine. Nature works on predictable laws and principles. To understand nature, as one would understand the principles of sound or light, means that man could control nature.

Necessary for survival. Both man and nature are tied to each other for survival. Rise and fall of past civilizations was based on use and misuse of nature. George Perkins Marsh, Henry David Thoreau.

Ordered. Nature is perfectly balanced and in perfect harmony. Humans upset this balance. Leave nature alone, and it returns to balance or a state of equilibrium. Aristotle, Cicero.

Scenic wonder. Nature is to provide pleasure, curiosity, and entertainment for man. Man can derive or obtain peace and understanding by watching nature. Henry David Thoreau, Ralph Waldo Emerson, John Wesley Powell, John Muir.

Separate. Early man and cultures felt that they were one with nature. As agriculture developed, distance between nature and man increased. Nature is an obstacle to overcome.

ENVIRONMENTAL REVOLUTION COMPONENTS

1. Increasing energy and efficiency.
2. Pollution prevention techniques.
3. Economics—financially reward positive environmental practices.
4. Demographics—focuses on bringing human birth rate into balance with world resources.
5. Sufficiency—bringing into balance the basic needs of people worldwide.

Ethics seeks to define what is fundamentally right and wrong, regardless of cultural differences.

1. **Development ethic**—based on individualism or egocentrism. Earth exists for human. Progress, work ethic, bigger-better-faster is good. Upward mobility. Progress at all costs.
2. **Preservation ethic**—nature has intrinsic value. Views stem from religious beliefs, aesthetics, recreational, or scientific.
3. **Conservation ethic**—stresses balance between development and preservation.

JUSTIFICATION FOR ENVIRONMENTAL CONSERVATION

Aesthetic. Biological diversity is beauty and adds to the quality of life.

Ecological. We do not understand the full ecological roles of all organisms. Organisms are necessary to maintain the functions of ecosystems.

Moral. Species have the moral right to exist. Our role is to be a steward for organisms and to conserve biological diversity. Has roots in religion and culture.

Utilitarian. We do not know the full potential of all species of plants and animals in the world. Species are becoming extinct each day. New medicines and drugs could come from plants in the rain forest (e.g., Taxol, rosy periwinkle, Mexican yams, serpent-wood, willow bark). Plants are the source of 25% of prescription medicines. Genetic variation is important. As new disease strains develop, genetic variation is essential—natural selection. New crops are potential, and products are derived from the crops. A wide, diverse ecosystem has built-in mechanisms to deal with pollution (e.g., filter-feeders, soil fungi removes carbon monoxide). Diverse and healthy habitats provide tourism.

LEVELS OF ETHICAL CONCERNS:
(a) Self–Family–Friends–Community and Neighbors–State–Nation–All Countries
(b) Individual Animals–All Animals of a Particular Species–All Animals
(c) Plants of a Particular Species–All Plants
(d) Other (Individuals–Populations–Communities–Ecosystems–Biomes–Planet)

MAJOR CULTURAL GROUPINGS IN THE UNITED STATES
- **Moderns** represent ~50% of U.S. population. Materialistic, cynical view of idealism and caring, accept planetary management model. Not open to new ideas. Nature is to be conquered.
- **Traditionals** represent ~25% of U.S. population. Religious conservatism, nationalistic, wants to return to traditional ways. Pro-environment, anti-big-business.
- **Cultural Creatives** represent ~25% of U.S. population. Reject materialism and intolerance. Commitments to education, equality, personal growth. Optimistic and open to different viewpoints.

NATURALIST PHILOSOPHERS

Ralph Waldo Emerson. *Nature* (1836) and *Journals* (1840). Critic of rampant economic development.

Henry David Thoreau. *Walden* (1854). "Truth in nature and wilderness over the deceits of urban civilization."

John Muir. Leave cities occasionally to enjoy wilderness. Wilderness was threatened by man. Organized Sierra Club. Government control needed to protect wilderness areas.

Aldo Leopold. *A Sand County Almanac.* Founded field of game management. The amount of space and type of forage of a wildlife habitat determines the number of animals that can be supported in an area. Regulated hunting can maintain proper balance of wildlife.

Rachel Carson. *Silent Spring* (1962). Focused attention on pesticides. Led to changes in pesticide use in the United States.

REPRESENTATIVE ORGANIZATIONS

Coalition for Environmentally Responsible Economies (CERES). A coalition of religious, labor, and environmental groups which focuses on environmental policies of industries. www.ceres.org.

Commission for Environmental Cooperation (CEC). Promotes agreement between Canada, United States, and Mexico to resolve environmental disputes. www.cec.org.

Defenders of Wildlife. Dedicated to species protection and species preservation. www.defenders.org.

Ducks Unlimited. Promotes knowledge of waterfowl and wetlands conservation. www.ducksunlimited.org.

Earth Island Institute. Concerned with environmental quality and wildlife protection. www.earthisland.org.

Environmental Defense Fund. Concerned with legal, economic, and scientific issues. Lead efforts to ban DDT. www.environmentaldefense.org.

Friends of the Earth Concerned with conservation. Dedicated to the preservation, restoration, and wise use of natural resources. www.foe.org.

Green Peace. Highly aggressive environmental organization. Concerned with human domination of nature and industrialism. www.greenpeace.org or www.greenpeaceusa.org.

National Audubon Society. Founded in 1905. Encourages natural resource and wildlife conservation. www.audubon.org.

National Wildlife Federation. Promotes sustainable management of wildlife and natural resources. www.nwf.org.

The Nature Conservancy. Promotes conservation of plant and animal species through protection of natural habitats. Owns extensive amounts of land and keeps them as sanctuaries. http://nature.org.

Sierra Club. Active in public education, lobbying, and defending environmental issues through the courts. www.sierraclub.org.

World Conservation Union. Promotes conservation of wildlife and natural resources. www.iucn.org.

Worldwatch Institute. Concerned with population trends, energy use, natural resources, pollution, and species conservation. www.worldwatch.org.

World Wildlife Fund. Dedicated to wildlife protection. www.worldwildlife.org.

RESOURCES ON ENVIRONMENTAL ETHICS

Amazing Environmental Organization Web Directory
www.webdirectory.com/

A Land Ethic
www.wilderness.org/ethic/

Alliance for Sustainable Jobs and the Environment
www.asje.org/

Assisi Nature Council
www.assisinc.ch/

Association of Forest Service Employees for Environmental Ethics
www.afseee.org/

Cascade Policy Institute
www.cascadepolicy.org/env.asp

Cato Institute
www.cato.org

Center for Environmental and Regulatory Reform (CERR)
www.pacificresearch.org

Center for Ethics and Toxics
www.cetos.org/

Corporate Watch
www.corporatewatch.org/

Ethics Institute
www.dartmouth.edu/artsci/ethics-inst/

Environmental Ethics Journal
/www.cep.unt.edu/enethics.html

Ethics and Value Studies, National Science Foundation
www.cep.unt.edu/EVS.html

Environmental Ethics
www.phil.gu.se/Environment.html

International Society for Environmental Ethics
www.cep.unt.edu/ISEE.html

International Society for Environmental Ethics Newsletter
www.phil.unt.edu/ISEE/

National Library for the Environment
www.cnie.org/nle/

New Environmentalism
www.newenvironmentalism.org/

Sand County Foundation
www.sandcountyfoundation.net/

Second Nature
www.secondnature.org

VALUES

Aesthetic value. The value of appreciating things for their beauty.

Cultural value. Values of held by a society.

Existence value. Valuing things that we know to exist but may never use.

Option value. Preserving the value of a resource for the future.

Scientific-educational value. Values of knowledge.

Use value. The price paid to use or consume a resource.

Case Studies

Sierra Club v. Disney Corporation. In 1965, the Sierra Club launched a campaign to protect Mineral King Valley in the Sierra Nevada Mountains in California. The opponent was Walt Disney Productions, which intended to turn this secluded valley into an elaborate ski resort. After a number of unsuccessful attempts to halt the project through the political system, the Sierra Club Legal Defense Fund (now known as Earth Justice (www.earthjustice.org), filed a lawsuit to block development. The case went all the way to the Supreme Court. The result was the preservation of Mineral King and the confirmation of citizens' rights to seek review of environmental disputes in courts of law. P.S.—Walt Disney was a member of the Sierra Club!

Tourism in Goa. Goa is located on the west coast of India. It has recently seen a tremendous increase in tourism. However:

- Much of Goa's natural ecological assets have already been squandered on tourists.
- Sand dunes have been leveled or built upon. Thousands of fruit-bearing trees were cut to make way for resorts.

- Millions of liters of water are used to maintain hotel lawns and golf courses, even when the local river, the Assnoda ran dry. Farmers in the region lost their crops when irrigation water was diverted for the tourist industry.
- There are more than 50 swimming pools between Baga and Siquerim (a 2-km stretch of beach), all within 300 meters of the sea.
- Electricity is sporadic. Locals have to put up with frequent cuts or low voltage while hotels are provided with transformers.

The brunt of the tourism burden is being borne by ordinary people in whose name development is promoted. Should the government improve basic services and facilities for locals first instead of pandering to the foreign tourist?

Oil in Nigeria and Environmental Racism. Nigeria is ranked tenth among countries with major gas deposits—estimated at 120 trillion cubic feet. When large, multinational corporations drill for oil, flammable petroleum gas is a by-product. Rather than capturing the gas, liquefying it, and selling it, or reinjecting it back into the wells, it is much more economical to burn the gas off, a process known as flaring. Flaring has occurred in Nigeria for over 40 years. Nigeria flares more than 76% of its total gas output, which is the highest anywhere in the world. The World Bank estimates that gas flaring in the Niger delta releases some 35 million metric tons of carbon dioxide annually into the air. Global estimates indicate that the flaring of petroleum associated gas in Nigeria alone accounts for 28% of total gas flared in the world. Libya is a distant second, flaring only 21% of its total gas production, and Mexico only flares 5%. Some of the diseases that have begun to manifest themselves among villagers exposed to gas flaring include cancer, cardiac diseases, tuberculosis, brain disorders, stroke, anemia, and high blood pressure. The locals are also exposed to perpetual heat and constant daylight conditions with the physiological and psychological disorder that goes with them. The valuable wetlands of the Niger delta that once were comprised of mangrove forests, freshwater swamps, and intertidal mudflats have been destroyed. Most of the creeks in the region are today polluted; endangering fishing that is the main livelihood of the locals. Farmers in the area have also been affected because animals and plants have withered as much as their farmlands that have been laid to waste by acid rain. The ethical problem: the world needs energy.

Animal Rights. Every year, thousands of wild animals are destroyed, injured, or orphaned as a result of human activities. For example, traps break their legs and their wings are fractured after colliding with a car. They sustain head trauma from hitting windows and are poisoned after eating insects and plants treated with pesticides and herbicides. During spring and summer, newborns and infants are orphaned when their parents suffer any of these injuries. They are shot for sport and entertainment. They are executed for their fur. They are killed for their tusks. Elephant feet are made into wastebaskets. They are tortured in the name of science and progress. Their young offspring are eaten (veal and lamb). They are killed for entertainment (bullfighting). Humans have for the most part ended slavery. Humans, in some areas, have attempted to end discrimination. Are humans alone worthy of the moral status that is the basis of rights? Do animals have any rights at all? I recommend reading *Defending Animal Rights* by Tom Regan. Also, visit http://www.peta-online.org.

Multiple-Choice Questions

Select the philosophy that best exemplifies the statement(s) in Questions 1 through 5.

(A) Malthusian
(B) Neo-Luddite
(C) Nihilism
(D) Utilitarianism
(E) Moral extensionism

1. The belief that all members of society deserve equal treatment.

2. Limits on natural resources will result in catastrophic consequences in the world economy.

3. The world makes no sense at all. Everything is arbitrary. The only principle or truth is the instinct to survive. Morals do not exist.

4. Something is right if it produces the greatest good for the greatest number of people. Goodness is equated with happiness and happiness with pleasure. The good life is the one that gives the greatest pleasure.

5. Believe that all large-scale human endeavors eventually fail, and that science and technology cause more problems than they solve. Believe in going back to a low-tech pastoral or hunter-gatherer society.

6. Which of the following statements is consistent with an ecofeminist viewpoint?

   (A) Male-dominated societies are responsible for environmental imbalance.
   (B) The environment should be nurtured, not conquered.
   (C) The role of humans is to be caretakers.
   (D) Intrinsic values are represented by relationships whereas the instrumental values are represented by roles.
   (E) All of the above

7. "All environmental and resource problems can be solved through better management and technology" would be represented by the

   (A) free-market School.
   (B) no problem School.
   (C) responsible planetary management School.
   (D) spaceship Earth School.
   (E) stewardship School.

8. Which major cultural group would support the planetary management model, are characteristically materialistic, and are generally not open to new ideas?

   (A) Cultural creatives
   (B) Moderns
   (C) Traditionals
   (D) Orthodox
   (E) Conservatives

For Questions 9 through 13, select the person who matches the description provided.

   (A) Aldo Leopold
   (B) Henry David Thoreau
   (C) Ralph Waldo Emerson
   (D) John Muir
   (E) Rachel Carson

9. Focused attention on pesticides.

10. Organized Sierra Club.

11. Founded concepts of game management. *A Sand County Almanac.* Hunting can maintain proper balance of wildlife.

12. Truth and understanding are found in nature. *Walden.*

13. Critic of rampant economic development.

14. The fact that we do not know the full potential of species and that species are needed for biodiversity to meet potential changes in pathogens would be what type of justification?

   (A) Aesthetic
   (B) Ecological
   (C) Moral
   (D) Utilitarian
   (E) Rationalization

15. "The environment deserves moral consideration on its own, not associated with human interests" would be characteristic of what type of moral responsibility to the environment?

   (A) Anthropocentric
   (B) Biocentric
   (C) Ecocentric
   (D) Hedonistic
   (E) Relativism

Answers to Multiple-Choice Questions

1. **E**	6. **E**	11. **A**
2. **A**	7. **B**	12. **B**
3. **C**	8. **B**	13. **C**
4. **D**	9. **E**	14. **D**
5. **B**	10. **D**	15. **C**

Explanations for Multiple-Choice Questions

1. **(E)** Moral extensionism is a movement advocating the extension of human-centered concepts of morality and values onto "nature," representing a rupture with anthropocentic worldviews.
2. **(A)** Malthusians believe that a population constantly tends to outrun subsistence but that it is held in check by "vice"—abortion, infanticide, prostitution—and by misery in the form of war, plague, famine, and unnecessary disease. If all persons were provided with sufficient sustenance, and these checks were removed, the relief would be only temporary; for the increase of marriages and births would soon produce a population far in excess of the food supply.
3. **(C)** Nihilism is the belief that all values are baseless and that nothing can be known or communicated. It is often associated with extreme pessimism and a radical skepticism that condemns existence. A true nihilist would believe in nothing, have no loyalties and no purpose other than, perhaps, an impulse to destroy. While few philosophers would claim to be nihilists, nihilism is most often associated with Friedrich Nietzsche who argued that its corrosive effects would eventually destroy all moral, religious, and metaphysical convictions and precipitate the greatest crisis in human history.
4. **(D)** Utilitarianism is a modern form of the hedonistic ethical theory that teaches that the end of human conduct is happiness, and that consequently the discriminating norm that distinguishes conduct into right and wrong is pleasure and pain. In the words of one of its most distinguished advocates, John Stuart Mill,

   *The creed which accepts as the foundation of morals, utility or the greatest happiness principle, holds that actions are right in proportion as they tend to promote happiness, wrong as they tend to produce the reverse of happiness. By happiness is intended pleasure and the absence of pain; by unhappiness, pain and the privation of pleasure (*Utilitarianism, *ii, 1863)*
5. **(B)** Ned Ludd was a man who fought against the change of his time. He saw the industrial revolution and mechanization of the 19th Century as a threat to the way of life of many people, and took action to prevent the catastrophe. The Neo-Luddites constitute a growing movement in America, encompassing a broad spectrum of critics and criticism, from back-to-nature environmentalists who live in adobe huts, to those who use the World Wide Web to sound their warnings. Their targets include computerization, biotechnology, and excesses of the global economy.

6. **(E)** Ecofeminism connects ideas about social justice with those of environmental degradation by using domination as a common theme; domination of women by men, of people of color by white people, of Third World people by First World people, and domination of nonhumans by human beings.

7. **(B)** Refer to *Theories of Moral Responsibility to the Environment* on page 481.

8. **(B)** Moderns make up 50% of the U.S. population. They are materialistic, have a cynical view of idealism and caring, and accept the planetary management model. They are not open to new ideas and believe that nature is to be conquered.

9. **(E)** Carson's work, *Silent Spring,* exposed the dark side of science. It showed that DDT and other chemicals that were being using to enhance agricultural productivity were poisoning our lakes, rivers, oceans, and ourselves. Thanks to her, progress can no longer be measured solely in tons of wheat produced and millions of insects killed. Thanks to her, the destruction of nature can no longer be called progress. She is my environmental hero.

10. **(D)** In 1892, John Muir, wrote to the editor of *Century Magazine,* "Let us do something to make the mountains glad." He accomplished that by founding the Sierra Club. Throughout his life, Muir was concerned with the protection of nature both for the spiritual advancement of humans and, as he said so often, "for Nature itself." Though the arguments in favor of ecological thinking are often couched in scientific terms, the basic impetus remains as Muir stated it: "When we try to pick out anything by itself, we find it hitched to everything in the universe."

11. **(A)** Leopold's influence is based mostly on a series of articles he wrote for magazines such as *American Forests, Journal of Forestry,* and *Journal of Wildlife Management*. These, published after his death as parts of *A Sand County Almanac,* is a volume of nature sketches and philosophical essays recognized as one of the enduring expressions of an ecological attitude toward people and the land. The notion of a land ethic was rooted in Leopold's perception of the environment, and that perception was deepened and clarified throughout his life.

12. **(B)** Henry David Thoreau moved to Walden Pond, "to live deliberately." His time at Walden, slightly over two years, demonstrated the natural harmony that was possible when a thinking man went to live simply, reading books, writing in his diary, and walking in the woods. During his life, Thoreau was little known outside his small social and intellectual circle. Yet his reputation as a prophet for ecological thought and the value of wilderness, born at Walden, now grows with each passing year. He articulated the idea that humans are part of nature and that we function best, as individuals and societies, when we are conscious of that fact.

13. **(C)** The major premise of transcendental eco-wisdom is that connection with nature is essential for a person's intellectual, aesthetic, and moral health and growth. This connectedness is the basis of the self-reliance that determines how a person lives with integrity in nature and society. In *Nature,* Emerson takes an unabashedly anthropocentric view, seeing nature as a great and holy teacher of the self-reliant man who will look beyond its uses as mere commodity and see it as infused with spirit, with a wonder known to few adults. Nature's purpose was clear to Emerson: "All the parts incessantly work into each other's hands for the profit of man" and "the endless circulations of divine charity nourish man."

14. **(D)** Utilitarianism, according to Jeremy Bentham, states that an action is right if its results are superior to those of any other action. The basic idea is to generate the greatest possible amount of happiness among the greatest number. Utilitarianism is a powerful force in support of many environmentalist philosophies. Rather than believe in the absolute "rights" of animals and nature, many environmentalists instead contend that their program maximizes utility.

15. **(C)** Ecocentrism, or Earth-centered philosophy, believes that the Environment deserves moral consideration on its own, not associated with human interests. There are two schools of thought in ecocentrism. In the *environmental wisdom School* resources are limited, technology and economic growth can benefit and damage the environment, and nature does not exist for man. It would not have us attempt to adapt the Earth for man's needs but to learn to adapt to the environment. The *deep ecology School* recognizes both intrinsic and instrumental value of species, ecosystems, and the biosphere. Proponents also believe in reducing human population size and appreciating inherent life quality (not material standards). Humans have no right to reduce or interfere with environmental interdependence, richness, or diversity. Ecocentrists believe that the direction in which humanity is moving cannot be sustained. They believe that our technologies, our food production, our global economy, our hectic consumer life style, and our huge population are all based on rapid consumption and destruction of living systems and minerals that have taken the Earth a billion years to produce. In particular, they believe that humans have in one hundred years destroyed half of the oil that was created by natural systems over hundreds of millions of years. They believe that current human culture reduces or destroys other species of life at an unprecedented rate, without understanding how we and other life forms may depend on those species and the ecosystems they function in.

Free-Response Question

The green revolution has increased worldwide life expectancy an estimated 10 to 20 years depending on the country. It has allowed people who would have died of famine to survive and produce children. This causes some people to have ethical and/or moral concerns because some believe man should not interfere with nature, while others feel that we already have too many people on Earth and should not be trying to add more.

(a) Describe the first and second green revolution.
(b) Discuss the green revolution from a perspective of carrying capacity and depletion of natural resources.
(c) Compare two ethical perspectives to discuss this issue. Include major proponents of each perspective.

Free-Response Answer

Another technique I teach in writing answers to APES essays is to "box it." By this I mean break the question down into pieces. When some students read a long question with many parts as this one, they may know the material, but they get tense and lose their train of thought trying to sort out how *all* of it should fit together. Don't worry about making the perfect essay in 25 minutes where everything fits together. Rarely does it. Instead, take each question, brainstorm it for a minute, organize the concept map, and then write.

When I began writing this book it seemed like an impossible task. Where does one begin in preparing students to "ace" the AP exam in such a broad topic as environmental science? My AP Chemistry book was easy. Gas laws—explain them, give a few examples, move on to the next topic. But how can I cover the atmosphere, human population, cycling of matter, energy, pollution, economics, the Earth, legislation, water, land, food, health, and so on? There are a million concepts! How can I make them all somehow connect? Well, I have learned that the best way to do any job is to break the job into small pieces. Don't get overwhelmed by the totality of the problem. If you are working on a multipart question, just focus for a few minutes on the first question. When you have mastered it, go onto the next. Writing one essay in which you interconnect all questions is very difficult to say the least. Try answering each question separately without necessarily focusing on all questions at once. Let's try it here.

The first question is about the green revolution. BRAINSTORM: fertilizer, irrigation, hybrids, gmo (genetically modified organisms), monoculture. Now WRITE.

(a) The first green revolution occurred between 1950 and 1970 and involved practices of planting monocultures, high application of inorganic fertilizers, pesticides, artificial irrigation systems, and high-yield hybrid crops. The second green revolution began during the 1970s and involves growing and designing genetically engineered crops. After the first green revolution, there was a 200% increase in grain yield.

The next question is carrying capacity and natural resources. BRAINSTORM: definition, degradation of environment, pollution, waterlogging, nitrogen fixation, sustainability, raised K, Hardin. Now WRITE.

(b) Carrying capacity refers to the number of individuals who can be supported in a given area within natural resource limits, and without degrading the natural social, cultural and/or economic environment for present and future generations. The carrying capacity for any given area is not fixed. It can be altered by improved technology, but mostly it is changed for the worse by pressures that accompany a population increase. As the environment and natural resources are degraded through pollution, salinization, waterlogging, excess nitrogen from fertilizer, etc., the carrying capacity shrinks, leaving the environment no longer able to support even the number of people who could formerly have lived in the area on a sustainable basis. No population can live beyond the environment's carrying capacity for very long. Garrett Hardin, in *Tragedy of the Commons,* discusses this issue as "life boat ethics" or bioregionalism. He argues that a region should only have as many people as the environment can support without artificial food subsidies. By growing food in richer countries and shipping it to poorer countries, it only hastens the richer country in reaching its own carrying capacity as it depletes and degrades its own resources. The green revolution has therefore artificially raised the carrying capacity of the world, with significant postponed effects.

There is one more to go. The question is about ethical perspectives, 2 + people. BRAINSTORM: ecocentrism, Emerson, Thoreau, Muir, Carson, moral consideration, natural resources, green revolution, environmental wisdom, deep ecology, anthropocentric, define, No Problem, Spaceship Earth. *Check time*—Now WRITE:

(c) Ecocentrism, an Earth-centered perspective, whose proponents include Ralph Waldo Emerson, Henry David Thoreau, John Muir, and Rachel Carson believe that the environment deserves moral consideration on its own, not associated with the needs or interests of humans. Ecocentrists believe that natural resources are limited. They also believe that technology and growth, as evidenced through the green revolution, especially when not fully understood or poorly managed, can damage the Earth; that rather than trying to increase the number of humans on the Earth; that we should be learning to adapt to the inherent natural capacity of the Earth. Within ecocentrisim, there are two schools: (1) the Environmental Wisdom School, which believes that natural resources are limited and (2) the Deep Ecology School, which believes in reducing human population size, believing that it is quality, not quantity of humans that is important. The Deep Ecology School adheres to what is considered to be an emerging ecological sensibility, where the attitudes of humans to "nature" are based on a new understanding of humanity as one part, equal with other parts, of "nature" in an interconnected and mutually supportive system and consists of the following points:

1. All life has value in itself, independent of its usefulness to humans.
2. Richness and diversity contribute to life's well being and have value in themselves.
3. Humans have no right to reduce this richness and diversity except to satisfy vital needs in a responsible way.
4. The impact of humans in the world is excessive and rapidly getting worse.
5. Human lifestyles and population are key elements of this impact.
6. The diversity of life, including cultures, can flourish only with reduced human impact.
7. Basic ideological, political, economic, and technological structures must therefore change.
8. Those who accept the forgeoing points have an obligation to participate in implementing the necessary changes and to do so peacefully and democratically.

Anthropocentric, or human-centered ethical beliefs, believe that humans can overcome any problem. Anthropocentrism says that the world exists for humanity. Believers in this philosophy would say that humans could rightfully try to benefit as much as possible from the environment. A tradition of disregard for the environment, and the fact that plants and animals do not vote, has led to a primarily human-centered perspective in some decision makers who follow an anthropocentric philosophy.

The No Problem School (all environmental and resource problems can be solved through better management and technology) and the Spaceship Earth School (Earth is a single, complex machine that can be managed), within the framework of Anthropocentrism, come closest to the feelings that the green revolution benefits mankind. The Anthropocentrics believe that the role of the human is to be a master over his environment, such as increasing crop productivity through the techniques utilized through the Green Revolution.

Word Count: 734

Assignment

It is now your turn to write an essay on environmental ethics (bioethics).

South African game preserves are a big business in Africa. More than 560 South African national parks occupy more than 7 million acres. They have been responsible for preserving many endangered species. They are a sanctuary for wildlife. They are so successful that many populations have to be culled or thinned out to preserve the balance between population size and resources. They are thinned out by charging hunters up to $50,000 for a three-week safari. And yet, the land that the animals live, or die on, once belonged to villagers, whose land the government seized and refused to compensate the villagers for. These displaced villagers now live in shacks. They receive little education, and poor health care.

(a) Compare and contrast the rights of endangered species to survive with the rights of the villagers to survive.
(b) Compare two ethical perspectives in discussing this issue. Include in each comparison at least one historical person that would accept your argument(s).
(c) Propose some possible solutions that might offer a compromise between the needs and rights of the animals with the needs and rights of the villagers.

Practice the technique of BOXING IT. Remember, break the question into pieces. BRAINSTORM each piece. Then WRITE. When you finish, share your answers in class. Break into small groups of three or four people. Each person should provide enough copies of his or her essay for everyone in the group. Take about 10 minutes going over each person's answer. What points are being made? Were all the questions answered? How was the organization? Was it smooth and flowing or did it skip from one point and poorly transition into another? Have the group devise a rubric on how you would grade the essay. You'll learn a lot if the group comes up with what is important. Share the rubrics with the class. Have the class synthesize all rubrics into one. How would your divide up the answer to award 10 points?

CHAPTER 16

Environmental History, Laws, and Regulations: Regional, National, and International

Remember that the most beautiful things in the world are the most useless; peacocks and lilies, for instance.

John Ruskin (1819–1900)

Areas That You Will Be Tested On

Note: The AP College Board has not specified what will be tested on in the area "Environmental History, Laws, and Regulations." Environmental history can be found in timelines throughout this book.

Key Terms

Note: Definitions for all key terms listed below can be found in the glossary starting on page 648. For additional definitions of relevant terms from this chapter, refer to *www.barronseduc.com/0764121618.html*.

administrative courts
administrative law
anti-environmental groups
arbitration
case law
environmental impact report
environmental justice principle
Environmental Protection Agency
General Agreement on Tariffs and Trade (GATT)
legislative riders
lobbying
mediation
North American Free Trade Agreement (NAFTA)
policy life cycle
precautionary principle
statute of limitations

Key Concepts

Key People

Abbey, Edward. 20th Century writer. *Desert Solitaire* focused on the desert and how it was being destroyed as society developed. *Monkey Wrench* inspired vandalism and attacks against logging industry.

Adams, Ansel. 20th-century photographer and conservationist who introduced many people to the American landscape and its diversity. Famous for his pictures of Yosemite National Park. Adams also used photography to display negative images, such as over-grazing in the southwest, and used his prints to further the cause of conservation. His pictures helped establish Kings Canyon National Park. He was on the Board of the Sierra Club for some time and played a large role in many of its decisions.

Audubon, John. One of the most acknowledged artists and naturalists in 19th-century America. He is most famous for his collection of paintings of birds, *The Birds of America.*

Bennet, Hugh. American soil conservationist. One of the first scientists to recognize the process of soil erosion.

Berry, Wendell. Proponent of traditional rural life. *Home Economics and The Unsettling of America—Culture and Agriculture.* Criticized industrial farming for its effect on the environment and culture.

Borlaug, Norman. American environmental activist. Known as the father of the "green," or agricultural, revolution—high genetic-yield crops, better means of disease resistance, use of fertilizers and pesticides, mechanized threshing, and new crop varieties that could be grown with varied amounts of daylight. Awarded Nobel Peace Price in 1970.

Bush, George W. 43rd president of the United States.

- Wants more drilling and exploration for oil, at the same time, giving incentives for SUVs.
- Wetland preservation restricted.
- Make less stringent regulations regarding arsenic levels in water and CO_2 levels in the atmosphere.
- More lead emission-reporting requirements.
- Kyoto Protocol is ignored. Does not address issues of clean water.
- Wants to work with landowners to achieve environmental goals.
- Wants federal guidelines but local government to be watchdog and regulator. No command and control philosophy. More of a hands off approach—let the locals do it.
- The old "mandate/regulate/litigate" to be replaced with decentralized efforts.
- Wants Amtrak to be competitive in mass transit and more cost-effective.
- The Federal government will provide economic incentives for land stewardship and conservation.
- Believes Superfund is too costly and not effective. Let the local government deal with local problems.
- Remove federal roadblocks to state's moving to clean energy.

Brownfields and Parks
- The federal government will provide $450 million annually for wildlife and open spaces.
- The federal government will provide close to $5 billion to repair "crumbling" national parks.
- 6-point plan for brownfield clean-up.
- Repair national parks.

Global warming
- Does not think CO_2 emissions are a serious threat. "More research is needed."
- The Kyoto Protocol is too much of an economic burden on the United States. It is unfair.

• Drilling for oil in the Arctic National Wilderness Preserve will leave only "footprints."
• Work with industry to clean up the air.

Carson, Rachel. *Silent Spring* (1962)—effect of DDT on environment. Beginning of the environmental movement in the United States.

Commoner, Barry. Environmental scientist whose work focused on ozone depletion. His writings, *Closing Circle* and others, have focused on nuclear testing, profit over environmental concern, preventing environmental pollution in the future rather than focusing on clean-up, and renewable energy sources (particularly solar), and Third World debt forgiveness and redistribution of wealth. Ran for U.S. president in 1980 as a Citizen's Party candidate.

Ehrlich, Paul. American ecologist and strong population control advocate. *The Population Bomb* (1968) is an aggressive attack on the causes of overpopulation and on the problem of world hunger and contends that countries should balance their birth and death rates in some instances by discouraging human reproduction and making sterilization mandatory In 1968, founded Zero Population Growth, Inc. which strives to gain support for balanced population levels.

Hamilton, Alice. Founder of occupational medicine and first woman professor at Harvard Medical School. In the typhoid fever epidemic in Chicago in 1902, made a connection between improper sewage disposal and the role of flies in transmitting the disease. She noted that the health problems of many of the immigrant poor were due to unsafe conditions and noxious chemicals, especially lead dust, to which they were being exposed in the course of their employment. At the time there were no laws regulating safety at work and employers routinely fired sick workers and replaced them with new ones looking for jobs.

Hardin, Garrett. American ecologist, best known for his controversial beliefs about population control. *Tragedy of the Commons* (1986) makes an analogy between overgrazing the common pastures in medieval England and the general attitude of modern individuals toward the environment. In both cases, people rely on others to make sacrifices and eventually they destroy the environment because people are inclined to act for themselves, rather than for the good of society as a whole. Employs many key environmental concepts, such as overgrazing, air and water pollution, and population growth.

Gibbs, Lois Marie. American environmentalist who was a resident of Love Canal and wrote *Love Canal: My Story.* Uses forceful and effective tactics to ensure the protection of communities dealing with toxic waste problems. Founded the Citizen's Clearinghouse for Hazardous Wastes, which gives aid to communities dealing with toxic waste problems. Because of her efforts, the United States has not opened a hazardous waste site in over 20 years.

Leopold, Aldo. Early 20th-century American ecologist and environmental writer best known for his book, *A Sand County Almanac,* which has been called the bible of the environmental movement. It outlines "biological community" and "land ethic"—an extension of many human ethical codes incorporating ideas of community and interdependence—and broadens these to include the land and its living elements. Helped form the Wilderness Society and an early pioneer of the value of wilderness and forest preservation.

Marsh, George Perkins. 19th-century American diplomat, author, lawyer, scholar, and congressman. Many have credited him with founding the science of ecology. *Man and Nature: Physical Geography as Modified by Human Action* stated that every human action disturbs some aspect of nature, thus producing an unfit home for its inhabitants, therefore, humans should restore the harmonies of nature.

Muir, John. Influential 19th-century American naturalist. Instrumental in the wilderness preservation movement and in the move to create national parks. Proposed the idea that glaciers can cause changes in landform. Was influential in the establishment of Yosemite National Park in 1890. Also convinced the U.S. Government to get involved in forest conservation, which resulted in the creation of 21 million acres of forest preserves. Influenced such figures as Theodore Roosevelt in their environmental decision making, including the decision to found the national wildlife refuge system.

Nader, Ralph. American consumer advocate. *Unsafe at Any Speed* targeted U.S. automakers, condemning their practices and values and played a part in the passage of the 1966 Traffic and Motor Vehicle Act, which granted the U.S. Government increased power over the safety standards of American cars. Promoted passage of the Wholesome Meat Act of 1967. Helped establish the Environmental Protection Agency in 1970. In 1974, he was influential in the passage of the Freedom of Information Act.

Pinchot, Gifford. Believed that conservation meant "the greatest good to the greatest number for the longest time" and that people could live off the environment around them, and yet renew that same environment for generations to come. Helped establish the U.S. Forest Service, the national forest system, and forest reserves and saw these lands for public use, not as locked reserves.

Reagan, Ronald. 40th president of the United States. Cut EPA research and enforcement budgets by 50%. Appointed Anne Gorsuch (Burford) as head of the EPA—who was against environmental protection. Enforcement actions by the EPA dropped by more than 70%.

Roosevelt, Theodore. 26th president of the United States. Used his position to pave the way for environmentalists of the future. Known for setting aside land for national forests, establishing wildlife refuges, developing the farmlands of the American West, and advocating protection of natural resources. Throughout his life and work, Roosevelt remained focused on future generations and on the condition of the Earth that they would inherit.

Simon, Julian. Professor of Business Administration at the University of Maryland and a senior fellow at the Cato Institute. Best known for his work on population, natural resources, and immigration. His book *The Ultimate Resource,* recently reissued as *The Ultimate Resource 2,* challenged the neo-Malthusian notion that an increase in population has negative economic consequences, that population is a drain on natural resources, and that we stand at risk of running out of resources through overconsumption.

Sinclair, Upton. In 1906, described in his novel *The Jungle,* the unwholesome working environment in the Chicago meatpacking industry and the unsanitary conditions under which food was produced. Public awareness dramatically increased and led to the passage of the Pure Food and Drug Act.

Thoreau, Henry David. 19th-century American writer, philosopher, humanist, ecologist, naturalist, and conservationist. Most influential works were *Walden, Civil Disobedience,* and his *Journal,* which totaled 14 volumes. In these works, Thoreau sets out the philosophy that he saw as necessary for the good of the planet. He pioneered the concept of human ecology—of the relationship between humans and nature. He saw unity and community as important aspects of nature, and he saw all disturbances in these links as caused by human beings. One of Thoreau's greatest themes was simplicity, believing that modern materialism would lead to the destruction of the environment needed for humans and other living things.

Key American Organizational Platforms on the Environment

Democratic Party

Encourage open space and rail travel—"We want to transform sprawl into growth. We support tax credits to build more livable communities. We should acquire lands for forests and recreation sites and set aside wildlife preserves. We support the building of high-speed rail systems. High-speed rail reduces highway and airport congestion, improves air quality, stimulates the economy, and broadens the scope of personal choice for traveling between our communities. We support grants to Amtrak and the states for improving rail routes." (*Source:* Democratic National Platform, August 15, 2000)

We do not have to choose between economy and environment—"Al Gore is committed to restoring the Everglades; protecting the coasts and the Arctic National Wildlife Refuge from oil and gas drilling; and preserving our forests. Democrats believe that communities, environmental interests, and government should work together to protect resources while ensuring the vitality of local economies. Once Americans were led to believe they had to make a choice between the economy and the environment. They now know this is a false choice." (*Source:* Democratic National Platform, August 15, 2000)

Invest in technology and transportation friendly to Earth—"We must give incentives to invest in fuel-efficient cars, trucks, and sport utility vehicles; and energy-efficient homes. We need to clean up power plants. We should invest in roads, bridges, light rail systems, cleaner buses, the aviation system, our national passenger railroad (Amtrak), and trains that would give Americans choices—freeing them from traffic, smog-choked cities, and being held hostage to foreign oil. With the right investments, environmentally-friendly technologies can create new jobs." (*Source:* Democratic National Platform, August 15, 2000)

Green Party

Tax industrial pollution—"We call on new approaches to taxation, such as environmental taxes, as a partial substitute for income taxes. Taxing industrial pollution is an idea long overdue. Environmental taxes of this type, and "true-cost pricing," will aid in transforming major industries from being non-sustainable in their use of natural resources to being sustainable in character." (*Source:* Green Party Platform, at 2000 National Convention, June 25, 2000)

Reduce greenhouse gas emissions now; more later—"Greens believe the following are possible if we are to make a start on protecting our global climate. It is imperative that we strive for no less:

- An early target must still be set to prevent emissions rising so far that future reductions become even more difficult.
- Allowing sinks and trading within the protocol will create such loopholes that no real reductions will occur.

- It is said that U.S. industries emit over 20% of greenhouse gases globally. As a nation, we must implement public and private initiatives at every level to support the "GLOBAL CLIMATE TREATY" signed at the Earth Summit in 1992, committing industrial nations within a time framework to reducing emissions to 1990 levels.
- GREENHOUSE GASES and the threat of global warming must be addressed by the international community in concert, through international treaties and conventions, with the industrial nations at the forefront of this vital effort." (*Source:* Green Party Platform, at 2000 National Convention, June 25, 2000)

Libertarian Party

Government is the worst polluter—"Who's the greatest polluter of all? The oil companies? The chemical companies? The nuclear power plants? If you guessed 'none of the above,' you'd be correct. Our government, at the federal, state, and local levels, is the single greatest polluter in the land. In addition, our government doesn't even clean up its own garbage!

"Government, both federal and local, is the greatest single polluter in the U.S. This polluter literally gets away with murder because of sovereign immunity. Libertarians would make government as responsible for its actions as everyone else is expected to be. Libertarians would protect the environment by first abolishing sovereign immunity.

"The environment would benefit immensely from the elimination of sovereign immunity coupled with the privatization of 'land and beast.'" (*Source:* Libertarian Solutions; Mary Ruwart on LP Web site, November 7, 2000)

The parties responsible for pollution should be held liable—"Pollution of other people's property is a violation of individual rights. Strict liability, not arbitrary government standards, should regulate pollution. We demand the abolition of the Environmental Protection Agency. Rather than making taxpayers pay for toxic waste clean-ups, the responsible managers and employees should be held strictly liable for material damage done by their property. Claiming that one has abandoned a piece of property does not absolve one of the responsibility." (*Source:* National Platform of the Libertarian Party, July 2, 2000)

Natural Law Party

Help farmers stay in business while promoting health—"Support legislation that will ensure social, economic, and environmental sustainability of agriculture while balancing the following goals:

1. Ensuring high-quality, healthy food for consumers;
2. Promoting health in the population as a whole;
3. Protecting natural resources and the environment;
4. Cushioning farmers from the instability unique to agriculture;
5. Enabling farmers to better pursue financial profitability."

(*Source:* Natural Law Party's "50-point Action Plan," November 7, 2000)

Create new jobs in energy conservation—"Support the development of environmentally clean energy sources. Clean up polluted air, rivers, wetlands, and oceans by eliminating industrial pollution at its source and using waste products productively. Lead the effort to prevent the destruction of the Earth's forests, the decimation of the diversity of species, and damage from ozone depletion and the greenhouse effect." (*Source:* Natural Law Party's "50-point Action Plan," November 7, 2000)

Reducing global warming will help the economy—"America should lead the effort to prevent the destruction of the earth's forests, the decimation of diversity, and damage from ozone depletion and the greenhouse effect. The programs supported by the Natural Law Party will stimulate the economy. Cleaning the air will reduce medical costs. Protecting forests and reducing exhaust emissions will avert the disastrous expenses of adapting to climatic changes due to global warming." (*Source:* Natural Law Party Platform 2000, November 7, 2000)

Republican Party
Encourage market-based solutions to environmental problems
- Economic prosperity and environmental protection must advance together.
- Environmental regulations should be based on science.
- The government's role should be to provide market-based incentives to develop the technologies to meet environmental standards.
- We should ensure that environmental policy meets the needs of localities.
- Environmental policy should focus on achieving results.

(*Source:* Republican Platform adopted at GOP National Convention, August 12, 2000)

Provide tax incentives for energy production—America needs a national energy strategy:
- Increase domestic supplies of coal, oil, and natural gas.
- Provide tax incentives for production.
- Promote environmentally-responsible exploration and development of oil and gas reserves on federally-owned land, including the Coastal Plain of Alaska's Arctic National Wildlife Refuge.
- Offer a degree of price certainty.
- Advance clean coal technology.
- Expand the tax credit for renewable energy sources.

(*Source:* Republican Platform adopted at GOP National Convention, August 12, 2000)

Problem Addressed in Major Environmental Legislation
1. Increasing size and rapid growth of the human population;
2. Increasing use of natural resources by humans and developing legislation that encourages resource conservation;
3. Loss, fragmentation, and degradation of natural ecosystems (wildlife habitat);
4. Human-caused premature extinction of plant and animal species and loss of species diversity;
5. Poverty and poverty-related human health problems;
6. Pollution of air, water, and soil and setting standards;
7. Screening of new substances for safety threats;
8. Complete analysis and evaluation of activities that affect the environment; and
9. Protection of habitats and wildlife from environmental damage.

Principles of Environmental Policy Decision
Ecological design. Incorporate good ecological principles into decisions and laws.

Environmental justice. No single group takes on a disproportionate share of environmental risks.

Humility. Humans working together have a duty and capacity to coexist with and manage nature properly.

Integrative. Solutions require more than a simple answer; they require input from all sources and analysis of effects.

Precautionary. Caution is necessary when making decisions about issues when all the facts are not known.

Prevention. Being careful not to make decisions that will make the situation worse.

Reversibility. Being careful not to make decisions that will be difficult or impossible to reverse later.

Parts of Government

Article III of the U.S. Constitution establishes the executive, legislative, and judicial branches of the federal government.

Executive Branch

The executive branch is the branch of the U.S. government that is responsible for carrying out the laws.

EXECUTIVE BRANCH AREAS OF ENVIRONMENTAL RESPONSIBILITY

Council on Environmental Quality
Environmental policy, environmental impact statements.
www.whitehouse.gov/ceq

Department of Agriculture
Soil conservation, forestry.
www.usda.gov

Department of Commerce
Oceanic and atmospheric monitoring.
www.commerce.gov

Department of Energy
Energy policy, petroleum allocation.
www.energy.gov

Department of Health and Human Services.
Human health issues.
www.os.dhhs.gov

Department of the Interior
Endangered species, energy, minerals, and national parks, public lands.
www.interior.gov

Department of Justice
Environmental law and litigation.
www.justice.gov

Department of Labor
Occupational health.
www.dol.gov

Department of State
International treaties dealing with the environment.
www.state.gov

Department of Transportation
Mass transit, roads, oil spills, and airplane noise.
www.dot.gov

Environmental Protection Agency
Air and water pollution, noise, pesticides, solid waste management, radiation, hazardous wastes.
www.epa.gov

Nuclear Regulatory Commission
Licensing nuclear power plants.
www.nrc.gov

Office of Management and Budget
Agency coordination and budget.
www.whitehouse.gov/omb

White House Office
Overall coordination of all departments.
www.whitehouse.gov

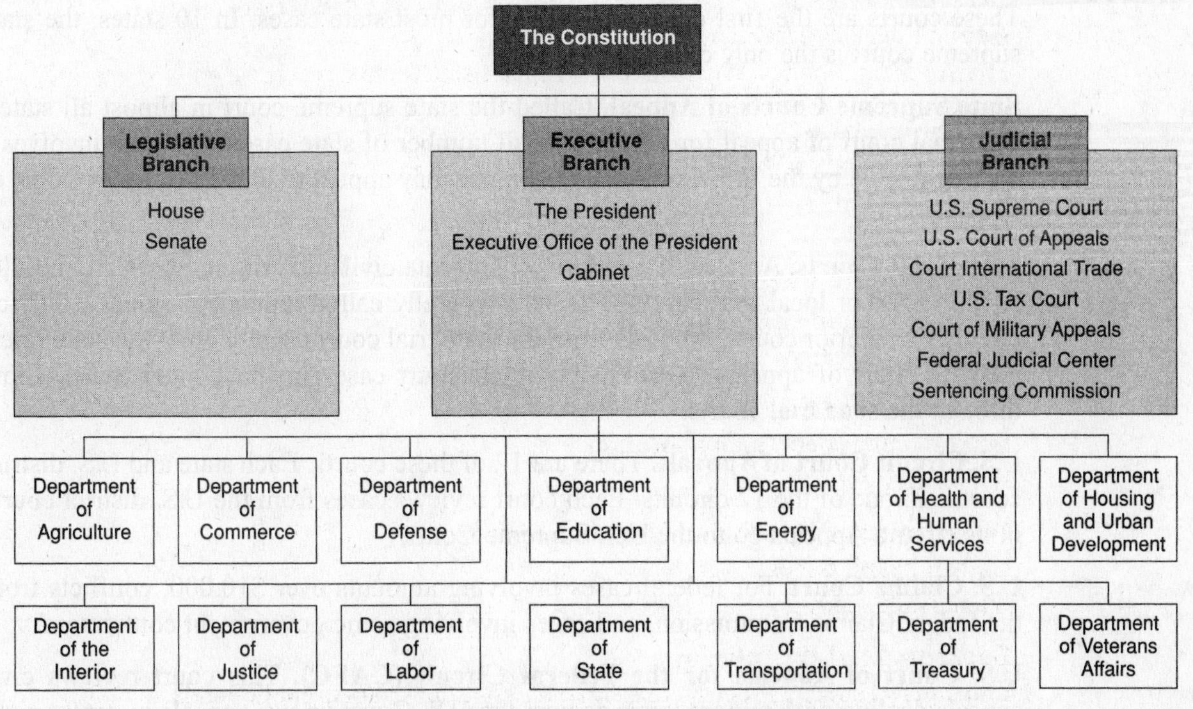

Figure 16.1. Organizational Chart—Executive Branch.

Under President George W. Bush, cabinet-level rank also has been accorded to the Administrator, Environmental Protection Agency; the director, Office of Management and Budget; the director, National Drug Control Policy; Office of Homeland Security; and the U.S. Trade Representative.

Judicial Branch

Types of Law

Administrative law. The body of rules and regulations and orders and decisions created by administrative agencies of government.

Case law. Law based on judicial decision and precedent rather than on statutes.

Civil law. The body of laws dealing with the rights of private citizens.

Common law. The system of laws originated and developed in England and based on court decisions, on the doctrines implicit in those decisions, and on customs and usages rather than on codified written laws.

Criminal law. Law that deals with crimes and their punishments.

Environmental law. Law that addresses environmental issues such as energy, transportation, environmental markets, forest, land, pollution, and water issues.

Statutory law. The body of laws created by legislative statutes.

Tort law. Law dealing with damage, injury, or a wrongful act done willfully, negligently, or in circumstances involving strict liability, but not involving breach of contract, for which a civil suit can be brought.

Court Systems

State Intermediate Courts of Appeal. 40 states have intermediate courts of appeal. These courts are the first court of appeals for most state cases. In 10 states, the state supreme court is the only court of appeals.

State Supreme Courts of Appeal. Called the state supreme court in almost all states. The final court of appeal for all but a small number of state cases. If a case involves a right protected by the U.S. Constitution, a party may appeal to the U.S. Circuit Court of Appeals.

State Trial Courts. Almost all cases involving state civil and criminal laws are initially filed in state or local trial courts. They are typically called municipal, county, district, circuit, or superior courts. Appeals from the state trial court usually go to the state intermediate court of appeals. About 95% of all court cases in the United States come through the state trial courts.

U.S. Circuit Court of Appeals. There are 12 of these courts. Each state and U.S. district court is in one of the 12 circuits. Each court reviews cases from the U.S. district courts in its circuit. Appeals go to the U.S. Supreme Court.

U.S. Claims Court. For federal cases involving amounts over $10,000, conflicts from the Indian Claims Commission, and cases involving some government contractors.

U.S. Court of Appeals for the Federal Circuit (CAFC). This court reviews civil appeals dealing with minor claims against the U.S. Government; appeals in patent-right cases, and cases involving international trade disputes.

U.S. Court of International Trade. Specializes in cases that involve international trade. Appeals go to the U.S. CAFC.

U.S. District Courts. There are 94 federal district courts, which handle criminal and civil cases involving:

- Federal statutes
- The U.S. constitution
- Civil cases between citizens from different states and the amount of money at stake is more than $75,000 (This is the most common type of case in the U.S. District Court.)

Most appeals from here go to the U.S. Circuit Court of Appeals; some go to the U.S. CAFC.

U.S. Supreme. The U.S. Supreme Court is free to accept or reject the cases it will hear. It must, however, hear certain rare mandatory appeals and cases within its original jurisdiction as specified by the Constitution.

For more information regarding U.S. courts, visit www.uscourts.gov.

Steps of a Civil Lawsuit

1. Complaint is filed by plaintiff.
2. Defendant replies to complaint (answer, counterclaim, motion to dismiss).
3. May file for motion for judgment on the pleadings or motions for preliminary relief.
4. Discovery (interrogatories, request for production of documents, depositions).
5. May file motion for summary judgment.
6. Pretrial conference.
7. Trial—first plaintiff, then defendant.
8. Jury is instructed at conclusion of trial.
9. Jury deliberates.
10. Verdict is issued.
11. Possible motion for a new trial or motion for a judgment notwithstanding the verdict.
12. Judge hands down judgment.
13. Possible appeal.

Environmental lawsuits are difficult because:

1. The individual plaintiff must be directly harmed by an action.
2. Lawsuits are expensive, especially when suing large corporations or the government.
3. Corporations can deduct legal expenses as tax write-offs, individuals cannot.
4. Legal burden of showing proof that the defendant is solely responsible.
5. Process is lengthy.

Legislative Branch

Federal Legislative Steps

Introduction of the Bill. Any member of Congress can introduce a bill. Cosponsors can add their names to the bill after it has been introduced on the floor. H.R. means that the bill originated in the House of Representatives; S means that it originated in the Senate.

Referral to Committee. The bill is then referred to a committee or subcommittee that has jurisdiction over the issue.

Committee Hearing. The committee chairperson decides whether the bill will be heard by the committee.

Mark Up. Mark up occurs when members of the committee officially meet to offer amendments to make changes to the bill as introduced.

Committee Report. A report of the bill is written describing the intent of legislation, the legislative history such as hearings in the committee, the impact on existing laws and programs, and the position of the majority of members of the committee. The members of the minority may file dissenting views as a group or individually.

Floor Debates. Speaker of the House and the majority leader of the Senate determine if and when a bill comes before the full body for debate and amendment.

Vote. The bill is presented to the chamber for a vote. A majority vote is required for an amendment and for final passage.

Referral to Other Chamber. If the bill is passed, it is referred to the other chamber where it usually follows the same route through committee and floor action. This chamber may approve the bill as received, reject it, ignore it, or amend it before passing it.

Conference. If there are minor changes made by the house the bill has been referred to, the bill returns to the originating chamber for a concurring vote. If there are major changes, a conference committee is appointed to reconcile the differences. A conference report is written. Both chambers must approve the conference report. If either chamber rejects the conference report, the bill is defeated.

Presidential Action. If the president approves the bill or the conference report, the bill becomes a law. If during session, the president does nothing, the bill becomes a law. The president can veto the bill. To override the veto requires a two-thirds majority vote of Congress. A pocket veto occurs if the president takes no action on the bill when Congress is not in session.

House Environmental Committees and Subcommittees

Committee on Agriculture

Subcommittees:

- Nutrition
- Forestry
- Livestock and Horticulture
- Specialty Crops and Foreign Agriculture Programs
- Conservation, Credit, Rural Development and Research
- General Farm Commodities and Risk Management

Committee on Energy and Commerce

Subcommittees:

- Commerce, Trade, and Consumer Protection
- Energy and Air Quality
- Environment and Hazardous Materials
- Oversight and Investigations
- Telecommunications and the Internet

Committee on Resources

Subcommittees:

- Energy
- Fisheries
- Forests
- Parks
- Water

Committee on Transportation and Infrastructure
Subcommittees:
- Aviation
- Coastguard and Maritime Transportation
- Economic Development
- Public Building
- Emergency Management
- Highways and Transit
- Railroads
- Water Resources and Environment

Senate Environmental Committees and Subcommittees
Committee on Agriculture, Nutrition, and Forestry
Subcommittees:
- Production and Price Competitiveness
- Marketing, Inspection, and Product Promotion
- Forestry, Conservation, and Rural Revitalization
- Research, Nutrition, and General Legislation

Committee on Banking, Housing, and Urban Affairs
Subcommittees:
- Housing and Transportation
- Economic Policy
- International Trade and Finance

Committee on Commerce, Science, and Transportation
Subcommittees:
- Oceans, Atmospheres, and Fisheries
- Surface Transportation and Merchant Marine
- Science, Technology, and Space

Committee on Energy and Natural Resources
Subcommittees:
- Energy
- Public Lands and Forests
- National Parks
- Water and Power

Committee on Environment and Public Works
Subcommittees:
- Clean Air, Wetlands, and Climate Change
- Transportation, Infrastructure, and Nuclear Safety
- Fisheries, Wildlife, and Water
- Superfund, Toxics, Risk, and Waste Management

Committee on Health, Education, Labor and Pensions
Subcommittee:
- Public Health

Key Legislation

Air Quality

Asbestos Hazard and Emergency Response Act (AHERA), 1986. Requires the inspection of U.S. schools for asbestos-containing material and either their management or removal.

Clean Air Act, 1955. Established national primary and secondary air quality standards. Required states to develop implementation plans.

Kyoto Protocol, 1997. Agreement among 150 nations requiring greenhouse gas emission reduction. Developing nations' reductions are voluntary.

Pollution Prevention Act, 1990. Requires facilities to reduce pollution at its source. Reduction can be either in terms of volume and/or toxicity.

Energy

Energy Policy and Conservation Act, 1975. Authorizes the president to draw from the strategic petroleum reserve as well as established a permanent home-heating oil reserve in the Northeast. Clarifies when the president can draw from these reserves, expands the Department of Energy Weatherization Assistance Program, and requires the secretaries of the Interior and Energy to undertake a national inventory of onshore oil and natural gas reserves.

Energy Policy Act, 1992. Provides for improved energy efficiency. It includes provisions to allow for greater competition in energy sales.

National Energy Security Act, 1978 and Energy Tax Act, 1978. The United States first experienced the economic impact of international oil disruptions in 1973 when an Arab Oil Embargo caused long lines at gas stations, lost productivity, declines in the stock market, and economic recession. The first congressional response to the petroleum crisis was the National Energy Security Act and the Energy Tax Act of 1978. The acts created favorable tax legislation and research and development commitments to expand the use of fuel ethanol in the United States to stimulate domestic production of ethanol.

National Appliance Energy Conservation Act, 1987. Set minimum efficiency standards for numerous categories of appliances, including residential comfort equipment.

Price-Anderson Act, 1957. Limits the liability of the nuclear industry in the event of a nuclear accident in the United States. It covers incidents that occur through operation of nuclear plants as well as transportation and storage of nuclear fuel and radioactive wastes.

U.S. Public Utility Regulatory Act (PURPA), 1978. Requires utilities to interconnect with small-scale independent power producers and pay fair market price for the electricity produced.

General

Environmental Education Act, 1990. Established a program within the Environmental Protection Agency to increase public understanding of the environment. The program awards grants for developing environmental curricula and training teachers, supports internships and fellowships to encourage the pursuit of environmental professions, selects individuals for environmental awards, and sponsors workshops and conferences.

International Environmental Protection Act, 1983. Authorized the president to assist countries in protecting and maintaining wildlife habitat and provides an active role in conservation by the Agency for International Development (AID). It further provided that AID shall use the World Conservation Strategy as an overall guide for actions to conserve biological diversity. Funds are explicitly denied for actions that significantly degrade national parks or similar protected areas or introduce exotic plants or animals into such areas.

National Environmental Policy Act (NEPA), 1969. Authorized the Council on Environmental Quality as the oversight board for general environmental conditions; directs federal agencies to take environmental consequences into account in decision making; and requires Environmental Impact Statement (EIS) be prepared for every major federal project having environmental impact.

Administrative Procedure Act (APA), 1946. Requires regulatory agencies to follow standardized procedures when issuing regulations or standards and granting permits to facilities.

Pollution Prevention Act, 1990. Focuses industry, government, and public attention on reducing the amount of pollution through cost-effective changes in production, operation, and raw materials use.

Freedom of Information Act (FOIA), 1966. Provides specifically that "any person" can make requests for government information. Citizens who make requests are not required to identify themselves or explain why they want the information they have requested. The position of Congress in passing FOIA was that the workings of government are "for and by the people" and that the benefits of government information should be made available to everyone.

Occupational Safety and Health Act, 1970. Ensures worker and workplace safety. Goal is to make sure employers provide their workers a place of employment free from recognized hazards to safety and health, such as exposure to toxic chemicals, excessive noise levels, mechanical dangers, heat or cold stress, or unsanitary conditions.

Land Use and Conservation

Alaskan National Interests Lands Conservation Act, 1980. Protects 100 million acres of parks, wilderness, and wildlife refuges in Alaska.

Antarctic Conservation Act, 1978. Protects native mammals, birds, and plants and their ecosystems. The law applies to all U.S. citizens and to all expeditions to Antarctica that originate from the United States. The act makes it unlawful, unless authorized by permit, to take native mammals or birds, engage in harmful interference, enter specially designated areas, introduce species to Antarctica, introduce substances designated as pollutants, discharge designated pollutants, or import certain Antarctic items into the United States.

California Desert Protection Act, 1994. Designates two new national parks in California—Death Valley and Joshua Tree—and one national preserve—the Mojave.

Coastal Barrier Resources Act, 1982. Eliminated federal development incentives on undeveloped coastal barriers, thereby preventing the loss of human life and property from storms, minimizing federal expenditures, and protecting habitat for fish and wildlife. Coastal barriers are landscape features that protect the mainland, lagoons, wetlands, and salt marshes from the full force of wind, wave, and tidal energy. The major types of coastal barriers include fringing mangroves, tombolos, barrier islands, barrier spits, and bay barriers. Composed of sand and other loose sediments, these elongated, narrow landforms are dynamic ecosystems and are vulnerable to hurricane damage and shoreline recession. Coastal barriers also provide important habitat for a variety of wildlife and are an important recreational resource.

Coastal Development Act, 1990. Opened Florida coastal areas to development.

Coastal Zone Management Act, 1972. Established an extensive federal grant program within the Department of Commerce to encourage coastal states to develop and implement coastal zone management programs. Activities that affect coastal zones must be consistent with approved state programs. The act also establishes a national estuarine reserve system.

Emergency Wetlands Resources Act, 1986. Authorized the purchase of wetlands, established a National Wetlands Priority Conservation Plan, required the states to include wetlands in their comprehensive outdoor recreation plans, and transferred to the Migratory Bird Conservation Fund amounts equal to the import duties on arms and ammunition. It also continued the National Wetlands Inventory and established entrance fees at National Wildlife Refuges.

Endangered American Wilderness Act, 1978. Prohibits commercial activities, motorized access, and infrastructure developments in congressionally-designated areas.

Federal Land Policy and Management Act (FLPMA), 1976. Along with the Taylor Grazing Act, outlines policy concerning the use and preservation of public lands in the United States. Grants federal government jurisdiction on consequences of mining on public lands. Grants Bureau of Land Management (Department of Interior) responsibility to manage all public lands not within national forest or national parks—multiple use policy.

Food Security Act (Farm Bill), 1985. Attempts to reverse the declining economic environment on the American farm and to conserve and restore wetlands, prairies, and other habitats. The "sodbuster" and "swampbuster" provisions help conserve the natural values of wetlands and reduce soil erosion and commodity surpluses. The legislation also provides for the establishment of conservation reserve, conservation set aside, and conservation easement programs on existing farmlands.

Forest and Rangeland Renewable Resources Planning Act (FRRRPA), 1976. Also known as National Forest Management Act. Requires the Secretary of Agriculture to develop a management program for national forest lands based on multiple-use and sustained yield principles. Also addresses timber harvesting rates, methods, and locations.

Forest Reserve Act, 1891. Gave the president authority to establish forest reservations from public domain lands.

Homestead Act, 1862. Allowed anyone to file for a quarter section of free land (160 acres). The land was yours at the end of five years if you had built a house on it, dug a well, plowed 10 acres, fenced a specified amount, and actually lived there. Additionally, one could claim a quarter section of land by timber culture (commonly called a tree claim). This required that you plant and successfully cultivate 10 acres of timber.

Land and Water Conservation Act, 1965. Provides for Congress to appropriate for planning, acquiring land and water areas for outdoor recreational use, and constructing outdoor recreation facilities. Allows collection of user fees at recreation areas. Designed to assist state and federal agencies in meeting present and future outdoor recreational demands, primarily in planning; acquisition of land, water, or interests in land or waters; or development.

Multiple Use and Sustained Yield Act, 1960, 1968. Directs U.S. Secretary of Agriculture to manage national forests for recreation, wildlife habitat, and timber production through principles of multiple use and sustained yield.

National Coastal Zone Management Act, 1972. Responsible for advancing national coastal management objectives and maintaining and strengthening state and territorial coastal management capabilities. It supports states through financial assistance ($58 million in 2000), mediation, technical services and information, and participation in priority state, regional, and local forums.

National Forest Management Act, 1976. Authorized the creation and use of a special fund "in situations involving salvage of insect-infested, dead, damaged, or down timber, and to remove associated trees for stand improvement." See Forest and Rangeland Renewable Resources Planning Act (FRRRPA).

National Park Service Act, 1916. Created the National Park Service within the Interior Department and made it responsible for national parks and monuments. In managing these areas, the Park Service was directed "to conserve the scenery and the natural and historic objects and the wildlife therein and to provide for the enjoyment of the same in such manner and by such means as will leave them unimpaired for the enjoyment of future generations."

National Trails System Act, 1968. Established to provide for recreation, public access, enjoyment, and appreciation of the "open-air, outdoor areas and historic resources of the Nation."

Outer Continental Shelf Lands Act, 1953. Authorized the Secretary of the Interior to promulgate regulations to lease the outer continental shelf to prevent waste and conserve natural resources and to issue leases through competitive bidding. The 1978 amendments provide for cancellation of leases or permits if continued activity is likely to cause serious harm to life, including aquatic life. These amendments also stipulate that economic, social, and environmental values of renewable and nonrenewable resources are to be considered in managing the outer continental shelf.

Public Rangelands Improvement Act, 1978. "Established and reaffirmed" a commitment to "manage, maintain and improve" rangelands so that they "become as productive as feasible for all rangeland values in accordance with management objectives and the land use planning process." Congress rejected grazing cutbacks as "undesirable because of their insensitivity to economic factors and potential to totally disrupt or shut down ranching operations" and because "studies" indicated "that range conditions show far greater, and more rapid, improvement with regulated livestock grazing than without."

Renewable Resources Planning Act (RPA), 1974. Mandates periodic assessments of forests and range lands in the United States. Directs that the assessments be conducted by the U.S. Forest Service, and consider a broad range of renewable resources, including outdoor recreation, fish, wildlife, water, range, timber, and minerals.

Soil and Water Conservation Act, 1977. Provides for a continuing appraisal of U.S. soil, water, and related resources, including fish and wildlife habitats, and a soil and water conservation program to assist landowners and land users in furthering soil and water conservation.

Soil Conservation Act, 1935. Established the Soil Conservation Service, which deals with soil erosion problems, carries out numerous soil surveys, and does research on soil salinity. It also provides computer databases for scientific research on such topics as pesticides.

Surface Mining Control and Reclamation Act (SMCRA), 1977. Established a program for regulating surface coal mining and reclamation activities. It establishes mandatory uniform standards for these activities on state and federal lands, including a requirement that adverse impacts on fish, wildlife, and related environmental values be minimized. The act creates an Abandoned Mine Reclamation Fund for use in reclaiming and restoring land and water resources adversely affected by coal mining practices.

Taylor Grazing Act, 1934. Governs management and preservation of federal public land, excluding national forests and national parks. Established grazing districts and permits.

Wild and Scenic Rivers Act, 1968. Established a National Wild and Scenic Rivers System for the protection of rivers with important scenic, recreational, fish and wildlife, and other values. Rivers are classified as wild, scenic, or recreational. The act designates specific rivers for inclusion in the system and prescribes the methods and standards by which additional rivers may be added.

Wilderness Act, 1964. Allowed U.S. government to set aside sections within national forest system, national parks, and national wildlife refuges as wilderness areas to be administered by National Park Service, Forest Service, Fish and Wildlife Service, and Bureau of Land Management.

Noise Control

Noise Control Act, 1965. Promotes a national environment free from noise that jeopardizes health and welfare. Establishes research, noise standards, and information dissemination.

Quiet Communities Act, 1978. Provides for coordination of federal research and activities in noise control. It also authorized FAA funds for development of noise abatement plans around airports. Eligible projects included constructing barriers and acoustical shielding, soundproofing buildings, and acquiring land and air easements to achieve compatibility with noise standards.

Pesticides

Federal Insecticide, Fungicide, and Rodenticide Control Act (FIFRA), 1947. Regulates manufacture and use of pesticides. Pesticides must be registered and approved. Labels require directions for use and disposal.

Food, Drug, and Cosmetics Act, 1906. Concerns sanitary conditions and safety of food, including food additives, and the efficacy and safety of drugs and cosmetics.

Food Quality Protection Act (FQPA), 1996. Emphasizes the protection of infants and children in reference to pesticide residue in food.

Resources and Solid Waste Management

Marine Plastic Pollution Research and Control Act, 1987. Discharge of plastics, including synthetic ropes, fishing nets, plastic bags, and biodegradable plastics into the water is prohibited. Food waste or paper, rags, glass, metal, bottles, crockery, and similar refuse cannot be discharged in the navigable waters or in waters offshore inside 12 nautical miles from the nearest land. Finally, food waste, paper, rags, glass, and similar refuse cannot be discharged in the navigable waters or in waters offshore inside 3 nautical miles from the nearest land.

Mining Act, 1872. Allows individuals and corporations the right to stake claims on prospected mineral deposits on public land. If the deposit is found to be economically recoverable, the holder of a claim can patent the deposit and thus have title of the minerals and land.

Resource Conservation and Recovery Act (RCRA), 1976. Comprehensive management of nonhazardous and hazardous solid waste. Minimal standards for all waste disposal facilities and for hazardous wastes. Regulates treatment, storage, and transport.

Resource Recovery Act, 1970. Authorized a 2-year study to determine a national system for the disposal of radioactive and other hazardous wastes. In addition, the act called for the secretary to promote and conduct research into financing waste disposal programs, ways to reduce the amount of waste nationally, methods of waste disposal, proposals for recovering materials and energy from waste, and the health effects of exposure to waste. The EPA gained jurisdiction over these functions upon its creation later that year. The bill further authorized grants for demonstration recycling efforts and innovative waste management facilities.

Solid Waste Disposal Act, 1965. First federal law that required environmentally sound methods for disposal of household, municipal, commercial, and industrial waste.

Waste Reduction Act, 1990. Requires the Environmental Protection Agency to establish an Office of Pollution Prevention, develop and coordinate a pollution prevention strategy, and develop source reduction models. In addition to authorizing data collection on pollution prevention, the act requires owners and operators of manufacturing facilities to report annually on source reduction and recycling activities.

Medical Waste Tracking Act, 1988. Legislation in response to medical wastes washing up on several beaches on the East Coast of the United States. Required EPA to create a 2-year Medical Waste Demonstration Program. For the purposes of this 2-year program, the act defined medical waste and those wastes to be regulated; established tracking system; required management standards for segregation, packaging, labeling and marking, and storage of the waste.

Toxic Substances

Comprehensive Environmental Response, Compensation, and Liability (Superfund) Act (CERCLA), 1980. Established federal authority for emergency response and clean-up of hazardous substances that have been spilled, improperly disposed, or released into the environment.

Federal Environmental Pesticides Control Act, 1972. Required registration of all pesticides in U.S. commerce.

Hazardous Materials Transportation Act (Hazmat), 1975. Also known as the federal hazmat law. Governs the transportation of hazardous materials and wastes. Regulations issued by U.S. Department of Transportation. Covers containers, labeling, and marking standards.

Low-Level Radioactive Policy Act, 1980. Made states responsible for disposing of their own low-level radioactive waste and set forth the federal policy that waste disposal is best handled on a regional basis.

Nuclear Waste Policy Act, 1982. Established a schedule to identify a site for, and construct, an underground repository for spent fuel from nuclear power reactors and high-level radioactive waste from federal defense programs. Currently Yucca Mountain, Nevada, is most feasible.

Toxic Substances Control Act (TOSCA), 1976. Gives the Environmental Protection Agency (EPA) the ability to track the 75,000 industrial chemicals currently produced or imported into the United States. EPA repeatedly screens these chemicals and can require reporting or testing of those that may pose an environmental or human health hazard. EPA can ban the manufacture and import of those chemicals that pose an unreasonable risk.

Federal Hazardous Substances Act (FHSA), 1960. Requires that certain hazardous household products (hazardous substances) bear cautionary labeling to alert consumers to the potential hazards that those products present and to inform them of the measures they need to protect themselves from those hazards. Any product that is toxic, corrosive, flammable or combustible, an irritant, a strong sensitizer, or generates pressure through decomposition, heat, or other means requires labeling.

Water Quality

Clean Water Act, 1972. Set national water quality goals and created pollutant discharge permits. The Santa Barbara, California, oil spill of 1969, which was the first major televised environmental disaster, helped to pass this act.

Coastal Zone Management Act, 1972. Provided funds for state planning and management of coastal areas.

Federal Water Pollution Control Acts, 1948. Authorized the surgeon general of the Public Health Service, in cooperation with other federal, state, and local entities, to prepare comprehensive programs for eliminating or reducing the pollution of interstate waters and tributaries and improving the sanitary condition of surface and underground waters. Authorizes additional water quality programs, standards, and procedures to govern allowable discharges and funding for construction grants or general programs.

Great Lakes Critical Programs Act, 1990. Authorized the Fish and Wildlife Service to establish and implement a fishery resources restoration, development, and conservation program including hatchery production at a specified minimum level and to conduct a wildlife species and habitat assessment survey in the lake's basin, including threatened and endangered species, migratory nongame species, migratory bird populations, and related habitat needs. The Fish and Wildlife Service is also authorized to conduct general activities to control sea lampreys and other nonindigenous aquatic animal nuisances, improve the health of fishery resources, and update surveys of fishery resources.

Marine Protection, Research, and Sanctuaries Act, 1972. Regulated dumping of waste into oceans and coastal waters.

Ocean Dumping Ban Act, 1988. Makes it unlawful for any person to dump or transport for the purpose of dumping sewage, sludge, or industrial waste into ocean waters.

Oil Spill Prevention and Liability Act, 1990. Streamlined and strengthened the Environmental Protection Agency's ability to prevent and respond to catastrophic oil spills. A trust fund financed by a tax on oil is available to clean up spills when the responsible party is incapable or unwilling to do so. The EPA requires oil storage facilities and vessels to submit to the federal government, plans detailing how they will respond to large discharges.

Safe Drinking Water Act, 1974. Established to protect the quality of drinking water in the United States. This law focuses on all waters actually or potentially designed for drinking use, whether from above ground or underground sources. Establishes safe standards of purity and required all owners or operators of public water systems to comply with primary (health-related) standards.

Water Resources Development Act, 1986. This act provides money to establish and maintain dam safety programs. The act authorizes funding for a research program to develop improved techniques for dam inspections and publish updates for the National Inventory of Dams.

Water Resources Planning Act, 1965. The act provides for a plan to formulate and evaluate water and related land resources projects. Establishes a Water Resources Council, establishes River Basin Commissions, and maintains a continuing assessment of the adequacy of water supplies in each region of the United States. Establishes principles and standards for federal participants in the preparation of river basin plans and in evaluating federal water projects. The council reviews these plans with respect to agricultural, urban, energy, industrial, recreational, and fish and wildlife needs. Establishes a grant program to assist states in developing related comprehensive water and land use plans.

Water Quality Act, 1965. Established the first clear water purity standards, a reform Congress hoped would prevent pollution before it became a problem. Created the Water Pollution Control Administration under the Department of Health, Education, and Welfare (HEW). States retained initial responsibility for water purity in their interstate bodies of water, but if HEW determined that a state had not taken proper precautions, the federal government would assume jurisdiction and set new standards.

National Estuary Program (NEP), 1987. To identify nationally significant estuaries that are threatened by pollution, land development, or overuse, and to award grants that support the development of comprehensive management plans to restore and protect them.

Source Water Assessment Program (SWAP), 1996. Requires states to identify the areas that are sources of public drinking water, assess water systems' susceptibility to contamination, and inform the public of the results.

Source Water Protection Program (SWPP), 1996. After completing a SWAP, states are encouraged to participate in a complete Source Water Protection Program (SWPP). A SWPP is a proactive, community-based approach aimed at preventing pollution of groundwater, lakes, rivers, and streams that serve as sources of drinking water.

Surface Water Treatment Rule (SWTR), 1996. To improve control of microbial pathogens, including specifically the protozoan *Cryptosporidium,* in drinking water; and address risk trade-offs with disinfection by-products.

Wildlife Conservation

Alaska National Interest Lands Conservation Act, 1980. Designated certain public lands in Alaska as units of the National Park, National Wildlife Refuge, Wild and Scenic Rivers, National Wilderness Preservation, and National Forest Systems, resulting in general expansion of all systems. Through consolidation and expansion of existing refuges and creation of new units, the act provided 79.54 million acres of refuge land in Alaska, of which 27.47 million acres were designated as wilderness.

Anadromous Fish Conservation Act, 1965. Authorizes the Secretaries of the Interior and Commerce to enter into cooperative agreements with the states and other nonfederal interests for conservation, development, and enhancement of anadromous fish, including those in the Great Lakes. Authorized are investigations, engineering and biological surveys, research, stream clearance, construction, maintenance and operations of hatcheries, and devices and structures for improving movement, feeding and spawning conditions. Also included are provisions to make recommendations to EPA concerning measures for eliminating or reducing polluting substances detrimental to fish and wildlife in interstate.

Atlantic Striped Bass Conservation Act, 1984. Recognized the commercial and recreational importance, as well as the interjurisdictional nature, of striped bass, and established a unique state-based, federally-backed management scheme.

Convention for the Conservation of Antarctic Marine Living Resources (CCAMLR), 1980. A treaty that requires that regulations managing all southern ocean fisheries consider potential effects on the entire Antarctic ecosystem. In 1991, a limit was set on krill catch after CCAMLR evaluated the impact of the krill harvest not only on the krill population but also on other species that depend on these tiny shrimp-like animals for food.

Endangered Species Act, 1973. Provides a program for the conservation of threatened and endangered plants and animals and the habitats in which they are found. The U.S. Fish and Wildlife Service of the Department of the Interior maintains the list of 632 endangered species (326 are plants) and 190 threatened species (78 are plants).

Fish and Wildlife Act, 1956. Established a comprehensive national fish, shellfish, and wildlife resources policy with emphasis on the commercial fishing industry but also with a direction to administer the act with regard to the inherent right of every citizen and resident to fish for pleasure, enjoyment, and betterment and to maintain and increase public opportunities for recreational use of fish and wildlife resources.

Fish and Wildlife Improvement Act, 1978. Authorized the Secretaries of the Interior and Commerce to assist in training state fish and wildlife enforcement personnel to cooperate with other federal or state agencies for enforcement of fish and wildlife laws and to use appropriations to pay for rewards and undercover operations. Authorized financial and technical assistance to the states for the development, revision, and implementation of conservation plans and programs for nongame fish and wildlife.

Fish and Wildlife Conservation Act, 1980. Commonly known as the Nongame Act, it encourages states to develop conservation plans for nongame fish and wildlife of ecological, educational, aesthetic, cultural, recreational, economic or scientific value.

Fishery Conservation and Management Act (Magnuson-Stevens Fishery Conservation and Management Act), 1976. Established a U.S. exclusive economic zone that ranges between 3 and 200 miles offshore, and created eight regional fishery councils to manage the living marine resources within that area. The act was passed principally to address heavy foreign fishing, promote the development of a domestic fleet, and link the fishing community more directly to the management process. U.S. Coast Guard crews may board domestic and foreign vessels in U.S. coastal waters to enforce fishery laws. Violators can be fined, and their gear and catch seized. The Coast Guard checks on sizes of fish and lobsters caught, gear and net-mesh size, use of turtle excluder devices on shrimp nets, and other activities.

Fur Seal Act, 1966. Generally prohibits the taking of fur seals. An exception is provided for Indians, Aleuts, and Eskimos who dwell on the coasts of the North Pacific Ocean.

Great Lakes Fish and Wildlife Restoration Act, 1998. Established goals for the U.S. Fish and Wildlife Service programs in the Great Lakes and requires the service to undertake a number of activities specifically related to fishery resources.

Lacey Act, 1990. Authorized the Secretary of the Interior to adopt measures to aid in restoring game and other birds in parts of the United States where they have become scarce or extinct and to regulate the introduction of birds and animals in areas where they had not existed.

Marine Mammal Protection Act, 1972. Established federal responsibility to conserve marine mammals, with management vested in the Department of Commerce for cetaceans and pinnipeds other than walrus. The Department of the Interior is responsible for all other marine mammals, including sea otter, walrus, polar bear, dugong, and manatee.

Marine Protection, Research, and Sanctuaries Act, 1972. Part I: Marine Sanctuaries: Authorized the Secretary of Commerce, with significant public input, to designate and manage national marine sanctuaries based on specific standards. It provides for supervision by the secretary over any permitted private or federal action that is likely to destroy or injure a sanctuary resource and requires periodic evaluation of implementation of management plans and goals for each sanctuary.

Part II: Ocean Dumping: Prohibits transporting any material from the United States for the purpose of dumping it into ocean waters, or dumping any material into ocean waters, except as authorized by permit. The act sets controls on materials and sites for dumping and requires fees and compliance with agreements for alternative waste management and disposal. The act also provides for the establishment of monitoring and research programs on the effects of dumping into ocean waters, including the possible long-term effects of pollution, overfishing, and human-induced changes in ocean ecosystems.

Migratory Bird Conservation Act, 1929. Established a Migratory Bird Conservation Commission to approve areas of land or water recommended by the Secretary of the Interior for acquisition as reservations for migratory birds.

Migratory Bird Hunting Stamp Act, 1934. Commonly known as the Duck Stamp Act, this act requires use of a migratory bird stamp for hunting and raises funds for the conservation of migratory waterfowl.

Migratory Bird Treaty Act, 1918. Implements various treaties and conventions between the United States and Canada, Japan, Mexico, and the former Soviet Union for the protection of migratory birds. Under the act, taking, killing, or possessing migratory birds is unlawful.

National Wildlife Refuge System Act, 1966. Provides for the administration and management of the National Wildlife Refuge System, including wildlife refuges, areas for the protection and conservation of fish and wildlife threatened with extinction, wildlife ranges, game ranges, wildlife management areas, and waterfowl production areas.

Nonindigenous Aquatic Nuisance Prevention and Control Act, 1990. Established a broad new federal program to prevent introduction of and to control the spread of introduced aquatic nuisance species and the brown tree snake.

Pittman-Robertson Act (Federal Aid in Wildlife Restoration Act), 1937. Provides federal aid to the states for the management and restoration of wildlife. The aid, funded through an excise tax on sporting arms and ammunition, may be used to support a variety of wildlife projects, including acquisition and improvement of wildlife habitat.

Salmon and Steelhead Conservation and Enhancement Act, 1980. Established a salmon and steelhead enhancement program to be jointly administered by the Departments of Commerce and Interior and established a Washington State and Columbia River conservation area.

Species Conservation Act, 1966. In the opening provision of the Endangered Species Preservation Act of 1966, Congress declared protection of endangered species its official policy. The bill then directed the secretary of the interior to compile a list of endangered species and allocated $15 million to acquire lands and waters for preserving and restoring the species identified. In addition, it combined federal lands already maintained for this purpose into the National Wildlife Refuge System. It also addressed the growing problem of international extinction. With this legislation, Congress banned the importation of any fish or wildlife on a new list of species in danger of worldwide extinction.

Key Environmental Agencies of the U.S. Government

Agency for Toxic Substances and Disease Registry—charged under the Superfund Act to assess the presence and nature of health hazards at specific Superfund sites and to help prevent or reduce further exposure and the illnesses that result from such exposures. www.atsdr.cdc.gov/atsdrhome.html

Bureau of Land Management (BLM)—to sustain the health, diversity, and productivity of the public lands for the use and enjoyment of present and future generations. The BLM manages over 264 million acres of land—about one-eighth of the land in the United States—and more than 560 million acres of subsurface mineral resources. Most of these lands are located in the West, including Alaska, and are dominated by extensive grasslands, forests, high mountains, Arctic tundra, and deserts. The BLM is responsible for the management and use of a wide variety of resources on these lands, including energy and minerals, timber, forage, wild horse and burro populations, fish and wildlife habitat, recreation sites, wilderness areas, and archaeological and historical sites. www.blm.gov/nhp/index.htm

Centers for Disease Control (CDC)—the lead federal agency for protecting the health and safety of people—at home and abroad, providing credible information to enhance health decisions, and promoting health through strong partnerships. CDC serves as the national focus for developing and applying disease prevention and control, environmental health, and health promotion and education activities designed to improve the health of the people of the United States. The CDC is located in Atlanta, Georgia. www.cdc.gov

Consumer Product Safety Commission—to protect the public against unreasonable risks of injuries and deaths associated with consumer products. www.cpsc.gov

Council on Environmental Quality (CEQ)—to coordinate federal environmental efforts and to work closely with agencies and other White House offices in the development of environmental policies and initiatives. The council's chair, who is appointed by the president with the advice and consent of the Senate, serves as the principal environmental policy adviser to the president. In addition, CEQ reports annually to the President on the state of the environment; oversees federal agency implementation of the environmental impact assessment process; and acts as a referee when agencies disagree over the adequacy of such assessments. www.whitehouse.gov/ceq

Department of Energy (DOE)—to foster a secure and reliable energy system that is environmentally and economically sustainable, to be a responsible steward of the Nation's nuclear weapons, to clean up our nuclear facilities, and to support continued United States leadership in science and technology. www.energy.gov

Environmental Council of the States (ECOS)—national association of state and territorial environmental commissioners that works closely with the EPA to emphasize the rights of the states in environmental issues. www.sso.org/ecos

Fish and Wildlife Service—to conserve, protect, and enhance the nation's fish and wildlife and their habitats for the continuing benefit of people. www.fws.gov

Food and Drug Administration (FDA)—responsible for approving food and drugs for widespread use. www.fda.gov

General Accounting Office (GAO)—congressional investigative agency that examines the use of public funds, evaluates federal programs and activities, and provides analyses, options, recommendations, and other assistance to help the Congress make effective oversight, policy, and funding decisions. www.gao.gov

Global Change Research Program—to increase the skill of predictions of seasonal-to-interannual climate fluctuations and long-term climate change. www.globalchange.gov

National Institute for Environmental Health Sciences (NIEHS)—multidisciplinary biomedical research programs, prevention and intervention efforts, and communication strategies that encompass training, education, technology transfer, and community outreach relating to human health and disease. www.niehs.nih.gov

National Institute for Occupational Safety and Health (NIOSH)—responsible for conducting research on the full scope of occupational disease and injury ranging from lung disease in miners to carpal tunnel syndrome in computer users. NIOSH also investigates potentially hazardous working conditions when requested by employers or employees; makes recommendations and disseminates information on preventing workplace disease, injury, and disability; and provides training to occupational safety and health professionals. www.cdc.gov/niosh/homepage.html

National Institute of Standards and Technology (NIST)—established by Congress to assist industry in the development of technology, to improve product quality, to modernize manufacturing processes, to ensure product reliability, and to facilitate rapid commercialization of products based on new scientific discoveries. www.nist.gov

National Oceanic and Atmospheric Administration (NOAA)—conducts research and gathers data about the global oceans, atmosphere, space, and sun. www.noaa.gov

National Response Center (NRC)—federal point of contact for reporting the release of oil and hazardous chemical spills. www.nrc.uscg.mil/index.htm

National Response Team (NRT)—consists of 16 federal agencies with interests and expertise in various aspects of emergency response to pollution incidents. www.nrt.org

National Science Foundation (NSF)—independent U.S. government agency responsible for promoting science and engineering through programs that invest money in research and education projects in science and engineering. www.nsf.gov

National Strike Force (NSF)—established in 1973 as a direct result of the Federal Water Pollution Control Act of 1972. The NSF's role and responsibilities in supporting the National Response System have expanded under subsequent major environmental legislation, including the Clean Water Act of 1977 and the Oil Pollution Act of 1990. The NSF's mission is to provide highly-trained, experienced personnel and specialized equipment to Coast Guard and other federal agencies to facilitate preparedness and response to oil and hazardous substance pollution incidents to protect public health and the environment. www.uscg.mil/hq/nsfcc/nsfweb

National Technical Information Service (NTIS)—the official resource for government-sponsored U.S. and worldwide scientific, technical, engineering, and business-related information. www.ntis.gov

National Toxicology Program (NTP)—established in 1978 by the Department of Health and Human Services to coordinate toxicological testing programs; strengthen the science base in toxicology; develop and validate improved testing methods; and provide information about potentially toxic chemicals to health, regulatory, and research agencies, the scientific and medical communities, and the public. http://ntp-server.niehs.nih.gov

Nuclear Regulatory Commission (NRC)—to regulate commercial nuclear power reactors; nonpower research, test, and training reactors; fuel cycle facilities; medical, academic, and industrial uses of nuclear materials; and the transport, storage, and disposal of nuclear materials and waste. www.nrc.gov

Occupational Safety and Health Administration (OSHA)—to save lives, prevent injuries and protect the health of American workers. To accomplish this, federal and state governments must work in partnership with the more than 100 million working men and women and their 6.5 million employers who are covered by the Occupational Safety and Health Act of 1970. www.osha.gov

Office of Solid Waste and Emergency Response (OSWER)—provides policy, guidance, and direction for safely managing waste, preparing for and preventing chemical and oil spills, and cleaning up contaminated property. OSWER provides technical assistance to all levels of government to establish programs that safeguard air, water, and land from the uncontrolled spread of waste. www.epa.gov/swerrims

Natural Resources Conservation Service (NRCS) (formerly the Soil Conservation Service)—USDA federal agency that works in partnership with citizens to conserve and sustain natural resources. www.nrcs.usda.gov

United States Geological Survey (USGS)—provides reliable scientific information to describe and understand the Earth; minimize loss of life and property from natural disasters; manage water, biological, energy, and mineral resources; and enhance and protect quality of life. www.usgs.gov

How Do Americans Feel About the Environment?

How important is the environment to you?
92.8% of Republican voters felt the environment was important.
Zogby International

Will environmental problems get better or worse?
51% of those asked said that pollution and other environmental problems will get worse, up from 44% in 1996; 13% said that pollution and environmental problems would not get worse.
Washington Post

How do you feel about clean air legislation?
86% strongly favor stricter clean air health standards.
American Lung Association

Does the government care enough about environmental issues?
57% believe that the government does not care enough about the environment.
Gallup

Would you pay more for cleaner air?
7 of 10 respondents said that they would pay 5 cents more per gallon for cleaner gasoline.
American Lung Association

Which political party is more concerned about the environment?
Of adults polled, 54% stated that Democrats are more active in helping the environment, compared to 25% for Republicans.
New York Times/CBS News survey

Who should set air quality standards?
75% of voters trust the Environmental Protection Agency to set health-based air quality guidelines. Only half trusted Congress to do the same.
American Lung Association

Source: 1999 surveys.

Steps to Affect or Influence Environmental Legislation

WHILE WORKING WITHIN THE SYSTEM

1. Become informed. Read newspapers, read books, research the issue(s) using the Internet, visit independent political analysis websites, and read independent political literature (e.g., League of Women Voters).
2. Attend meetings or "coffee hours" that legislators conduct within your community. Ask questions at these meetings. Don't be afraid to ask difficult questions.
3. Visit websites such as www.opensecrets.org and www.sierraclub.org. Confront your representatives with questions you have after you have looked at the facts.
4. Follow up on how politicians vote. Be careful of the glib "one-liners" that have been rehearsed and staged for the news media. What a politician says and how they vote (knowing that few people will know how they voted, but many will know what they say) are two different things.

5. Follow the money. Investigate who is paying your representative, where are the sources of money coming from, and so on. Prepare a brief report, citing references, and submit the information to your local newspaper and media. Supply a copy to the representative. Remember, unless a candidate has a lot of money to run, they usually don't win. Therefore, they seek money to win office. This begins the spiral of "owing favors to the devil."
6. Find out in an election what the issues are. Identify the candidate that most closely feels about the environment as you do. Ask them for their experience in dealing with this issue? Are they a member of any environmental organization? Once you find out background on the candidate, and it is compatible with your feelings, contact the candidate's office and volunteer—if nothing else, contribute $10 or more.
7. Send thank-you notes to a candidate when they keep their promise on voting on a particular environmental issue. If they do not keep their promise, send another type of letter. Send a copy to "Letters to the Editor" section of your local newspaper.
8. Consider joining an organization that has organized political action committees. Remember, PACs are not a bad thing. Without PACs from the Sierra Club and other environmental organizations to keep a watch on Washington, D.C., and the other Big Money PACs such as the oil industry, mining industry, and health industry, there would be no balance.
9. Bring the issue up in class. Ask your teacher if your class can take on the issue as a project, assigning different jobs to different teams in the class: media contacts, letters to the community, press conferences, rallies, lunch meetings for students in the school interested in helping out, and so on. Example: Spain and Mexico promote bullfighting. A cruel and unnecessary "sport" that desensitizes people to the rights of animals to be treated with respect. Contact the embassies, hold demonstrations, contact media to cover the demonstration, send letters to students in those countries explaining how your class feels, contact trade groups, contact animal rights groups for speakers to come in, contact your local representative from Congress to address the issue, boycott products from those countries until they outlaw this primitive and cruel sport (one person not buying products from Spain or Mexico will have no effect). You need to make headlines. Other countries that need "education" include Norway and Japan for hunting whales and Canada for clubbing to death baby harp seals. But whatever you do, be respectful and informed. Attack the issue, not the country. Attack the issue, not an individual. You will lose all credibility if you scream, insult, and don't know the facts. In other words, you will look stupid and do more harm than if you had done nothing.
10. Contact environmental action groups that you can identify with or support. Working through them will save a lot of work, and the work will be unified.
11. Be informed of your opposition—who are they, what group do they front for, what tactics have they used in the past, what tactics could they use in the future. Often, they are headed by former top-level cabinet executives who are hired for their knowledge on how to defeat or get around legislation. They work to:

   - confuse the issues (e.g., National Endangered Species Act Reform Coalition sounds like a group that supports endangered species doesn't it? Wrong. They are organized to weaken or defeat the Endangered Species Act. They are funded by large land-owning corporations).

- confuse the public and turn the public against the environment. They are skilled and ruthless. Their tactics are sometimes illegal and violent. "Facts don't matter. In politics, perception is reality" (R. Arnold, a leader in the Wise-Use movement).
- file lawsuits to bankrupt opposition groups or delay outcomes through the courts.

Case Study

Campaign Financing. The following statistics show the amount of money that U.S. senators facing reelection in 2002 received on average from various sources to run their campaign. All figures are based on FEC reports filed by members of Congress through July 31, 2001. They cover financial activity that took place between January 1, 2001 and June 30, 2001. PACs are political action committees or special interest groups. To see sources of influence for *all* politicians in Congress and the specific sources of money, visit http://www.opensecrets.org, a *fascinating* site and a good site for research! My conclusion after visiting the site—top PAC money does *not* go to senators from populous states, it goes instead to senators from states with few people. Why? Because each senator has the same voting power as other senators. Furthermore, raising campaign funds is more difficult in less-populated states so it is easier to have influence with politicians if a contribution is not diluted by a large populous representing many different viewpoints. Smaller states also have narrower interests.

SENATORS FACING REELECTION IN 2002

Average raised	$1,067,458
Average spent	$233,403
Average from PACs	$310,593
Average from individuals	$705,867
Average cash on hand	$1,441,382

SOURCES OF SELECTED PAC MONEY TO POLITICAL PARTIES

Rank	Industry	Total	Democrat %	GOP %
1	Lawyers/law firms	$8,383,065	68%	32%
4	Real estate	$4,148,221	52%	48%
16	Oil and gas	$1,875,320	31%	69%
21	Crop production	$1,670,985	41%	59%
34	Agricultural services	$978,079	39%	61%
44	Building materials	$708,688	27%	73%
47	Chemicals	$660,991	26%	74%
48	Forest products	$638,519	23%	76%
49	Tobacco	$613,744	24%	75%

The U.S. senator in the 2000 election year that raised the most Political Action Committee money (political influence) was Senator Max Baucus, Democrat of Montana. Further breakdown, such as what specific types of agribusiness PACs (dairy, cattle, hog, corn, etc.), can be found at www.opensecrets.org. A breakdown of his sources of PAC money follows.

A BREAKDOWN OF PAC MONEY FOR SENATOR MAX BAUCUS, D-MONTANA

Agribusiness	$46,500
Communications/electronics	$36,376
Construction	$23,000
Defense	$5,000
Energy/natural resource	$53,458
Finance/insurance/real estate	$223,618
Health	$123,332
Lawyers and lobbyists	$43,026
Transportation	$70,185
Misc. business	$88,000
Labor	$84,500
Ideology/single-issue	$67,600
Other	$2,500
Unknown	$3,000

Multiple-Choice Questions

1. The first congressional response to the oil shortage crisis was the

   (A) Energy Policy Act of 1992.
   (B) National Appliance Energy Conservation Act.
   (C) Environmental Education Act.
   (D) National Energy Security Act.
   (E) Administrative Procedure Act (APA).

2. Which Act regulates the manufacture and use of pesticides?

   (A) Food Quality Protection Act
   (B) Federal Insecticide, Fungicide, and Rodenticide Control Act
   (C) Federal Environmental Pesticides Control Act
   (D) Toxic Substances Control Act (TOSCA)
   (E) Federal Hazardous Substances Act (FHSA)

3. The type of court where civil cases between citizens from different states and the amount of money at stake is more than $75,000 would be

(A) U.S. District Court.
(B) U.S. Claims Court.
(C) U.S. Supreme Court.
(D) U.S. Court of Appeals for the Federal Circuit.
(E) U.S. Circuit Court of Appeal.

4. Written questions to a party to a lawsuit asked by the opposing party as part of the pre-trial discovery process are known as

(A) torts.
(B) tombolos.
(C) riders.
(D) depositions.
(E) interrogatories.

5. A resource one would use to discover the progress of proposed legislation would be the

(A) *Federal Register.*
(B) *Lobbyist Daily.*
(C) *Code of Federal Regulations.*
(D) *Congressional Quarterly.*
(E) *Environmental Update.*

6. Which of the following statements is NOT a principle of the "precautionary principle"?

(A) People have a duty to take anticipatory action to prevent harm.
(B) The burden of proof of harmlessness of a new technology, process, activity, or chemical lies with the general public, not with the proponents.
(C) Before using a new technology, process, or chemical, or starting a new activity, people have an obligation to examine "a full range of alternatives" including the alternative of doing nothing.
(D) Decisions must be open, informed, and democratic and must include affected parties.
(E) All the statements are components of the "precautionary principle."

7. Which of the following could not be a lobbying group?

(A) Charity
(B) Think tank
(C) Public relation firm
(D) Law office
(E) All could potentially be a lobbying group

8. The agency that has jurisdiction over oceanic and atmospheric monitoring is the

(A) Department of Commerce.
(B) Department of the Interior.
(C) Department of State.
(D) Department of Agriculture.
(E) Council on Environmental Quality.

9. The type of law that is based on judicial decisions and legal precedents is

   (A) administrative law.
   (B) case law.
   (C) civil law.
   (D) common law.
   (E) statutory law.

10. A voluntary action plan that addressed environmental problems, promoted sustainable development, and provided a framework for the United Nations in developing future environmental policies was (the)

   (A) Rio Declaration on Environment and Development.
   (B) Agenda 21.
   (C) GATT.
   (D) NAFTA.
   (E) Montreal Protocol.

For Questions 11 through 15, select the person who is best described by the philosophy or life.

(A) Ansel Adams
(B) Rachel Carson
(C) Garrett Hardin
(D) Aldo Leopold
(E) Gifford Pinchot

11. 20th-century photographer and conservationist. Introduced many people to the American landscape and its diversity and is famous for his pictures of Yosemite National Park.

12. Author of *A Sand County Almanac,* which has been called "the bible of the environmental movement" and which outlines "biological community" and "land ethic"—an extension of many human ethical codes incorporating ideas of community and interdependence and broadening these to include the land and its living elements.

13. Author of *Tragedy of the Commons* (1986), which draws an analogy between overgrazing the common pastures in medieval England and the general attitude of modern individuals towards the environment.

14. Believed that conservation meant "the greatest good to the greatest number for the longest time" and that people could live off the environment around them, and yet renew that same environment for generations to come. Also helped establish the U.S. Forest Service.

15. The book *Silent Spring,* and the discussion of the effects of DDT, was the beginning of the environmental movement in the United States.

Answers to Multiple-Choice Questions

1. **D**	6. **B**	11. **A**
2. **B**	7. **E**	12. **D**
3. **A**	8. **A**	13. **C**
4. **E**	9. **B**	14. **E**
5. **D**	10. **B**	15. **B**

Explanations for Multiple-Choice Questions

1. **(D)** The United States first experienced the economic impact of international oil disruptions in 1973 when an Arab Oil Embargo caused long lines at gas stations, lost productivity, declines in the stock market, and economic recession. The first congressional response to the petroleum crisis was the National Energy Security Act and the Energy Tax Act of 1978. The acts created favorable tax legislation and research and development commitments to expand the use of fuel ethanol in the United States to stimulate domestic production of ethanol. A new Energy Security Act (2001) is currently before Congress and is sponsored by Alaskan Senator Frank Murkowski. This new act has the following negative environmental provisions as analyzed by the Sierra Club: (1) opens the Arctic National Wildlife Refuge to drilling; (2) restricts the U.S. Bureau of Land Management's and Forest Service's authority to set conditions on oil and gas drilling projects that are needed to protect fish and wildlife, water quality, and other environmental values; (3) increases air pollution by weakening standards for power plants; (4) increases global warming pollution by encouraging dependence on fuels that produce the most carbon dioxide (oil and coal); (5) increases subsidies to coal, oil, gas, and nuclear industries; (6) drives up demand for gasoline by encouraging the production of gas-guzzling SUVs; (7) repeals major electricity regulation laws without ensuring adequate safeguards; and (8) provides inadequate gas pipeline safety. The following is a graph of campaign income for Senator Murkowski (courtesy of www.opensecrets.org).

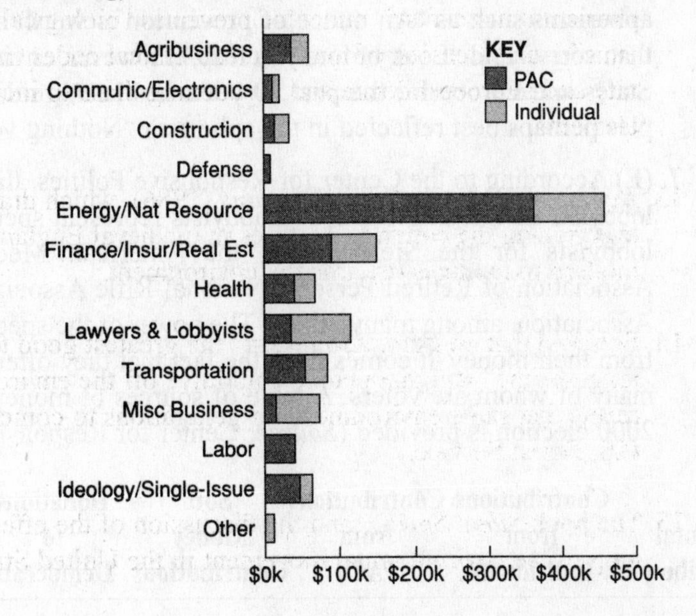

2. **(B)** The primary focus of the Federal Insecticide, Fungicide, and Rodenticide Control Act (FIFRA) is to provide federal control of pesticide distribution, sale, and use. The EPA was given authority under FIFRA not only to study the consequences of pesticide usage but also to require users (farmers, utility companies, and others) to register when purchasing pesticides. Through later amendments to the law, users also must take exams for certification as applicators of pesticides. All pesticides used in the United States must be registered and licensed by EPA. Registration assures that pesticides will be properly labeled and, if in accordance with specifications, will not cause unreasonable harm to the environment.
3. **(A)** The U.S. district courts are the trial courts of the federal court system. Within limits set by Congress and the Constitution, the district courts have jurisdiction to hear nearly all categories of federal cases, including both civil and criminal matters. There are 94 federal judicial districts, including at least one district in each state, the District of Columbia, and Puerto Rico. The 94 U.S. judicial districts are organized into 12 regional circuits, each of which has a U.S. court of appeals. A court of appeals hears appeals from the district courts located within its circuit, as well as appeals from decisions of federal administrative agencies.
4. **(E)** Interrogatories are written questions designed to discover key facts about a case. Witnesses can only be questioned in person at a deposition. Interrogatories are part of the pretrial discovery stage of a lawsuit, and must be answered under penalty of perjury. Lawyers can write their own sets of questions, or can use form interrogatories, designed to cover typical issues in common lawsuits.
5. **(D)** *Congressional Quarterly*, despite its name, is published each day and on-line (www.cq.com) and provides daily updates to *proposed* legislation. The *Federal Register* is the official daily publication for Rules, Proposed Rules, and Notices of Federal agencies and organizations, as well as executive orders and other presidential documents.
6. **(B)** The statement should have read "The burden of proof of harmlessness of a new technology, process, activity, or chemical *lies with the proponents, not with the general public.*" The essence of the precautionary principle is captured in common-sense aphorisms such as "An ounce of prevention is worth a pound of cure," "Better safe than sorry," and "Look before you leap." However, environmental policy in the United States and Europe for the past 70 years has been guided by entirely different principles perhaps best reflected in the aphorism "Nothing ventured, nothing gained."
7. **(E)** According to the Center for Responsive Politics, there are over 20,000 registered lobbyists in Washington, D.C. Lobbyists represent special interest groups. There are lobbyists for the Sierra Club, the American Medical Association, American Association of Retired Persons, National Rifle Association, and National Education Association, among many others. The power of the special interests doesn't come just from their money. It comes from the fact that they often have thousands of members, many of whom are voters. A table of sources of money to federal politicians for the 2000 election is provided (*Source:* Center for Responsive Politics).

Election	Total Contributions	Contributions from Individuals	Contributions from PACs	Soft Money Contributions	Donations to Democrats	Donations to Republicans	% to Dems	% to Repubs
2000	$16,238,589	$13,293,271	$693,275	$2,252,043	$8,066,492	$8,145,185	50%	50%

8. **(A)** The U.S. Department of Commerce is responsible for the National Oceanic and Atmospheric Administration (NOAA), and can be found at www.noaa.gov. NOAA's mission is to describe and predict changes in the Earth's environment and to conserve and wisely manage the nation's coastal and marine resources. NOAA's strategy consists of seven interrelated strategic goals for environmental assessment, prediction, and stewardship. They include:

   Environmental assessment and prediction
   1. advance short-term warnings and forecast services
   2. implement seasonal to interannual climate forecasts
   3. assess and predict decadal to centennial change
   4. promote safe navigation

   Environmental stewardship
   5. build sustainable fisheries
   6. recover protected species
   7. sustain healthy coastal ecosystems

9. **(B)** Case law is the law of reported judicial opinions. *Stare decisis* (or precedent) is the basic concept of case law (common law) in which courts look to statute, regulation, and/or prior court decisions to render opinions.

10. **(B)** Agenda 21 is a blueprint for sustainable development into the 21st century. Its basis was agreed upon during the Earth Summit at Rio in 1992 and was signed by 179 heads of state and government. Agenda 21 is a guide for individuals, businesses, and governments in making choices for development that help society and the environment. To read the full text of Agenda 21, visit http://www.un.org/esa/sustdev/agenda21text.htm.

11. **(A)** See the table entitled "Key People" starting on page 495.

12. **(D)** See the table entitled "Key People" starting on page 495.

13. **(C)** See the table entitled "Key People" starting on page 495.

14. **(E)** See the table entitled "Key People" starting on page 495.

15. **(B)** See the table entitled "Key People" starting on page 495.

Free-Response Question

You have turned 18. You are excited to be able to vote. You get a Voter's Guide Pamphlet in the mail that has the following proposition.

Official Title and Summary prepared by the Attorney General

Proposition 7

AIR QUALITY IMPROVEMENT. TAX CREDITS. INITIATIVE STATUTE.

Authorizes State Air Resources Board and delegated air pollution control districts to award $218 million in state tax credits annually until January 2011, to encourage air-emissions reduction through acquisition, conversion, and retrofitting of:

- vehicles, buses, and heavy-duty trucks;
- health products;
- construction vehicles and equipment;
- lawn and garden equipment;
- ambient air pollution destruction technology;
- off-road, nonrecreational vehicles;
- port equipment;
- agricultural waste and rice straw conversion facilities; and through research and development.

Requires study of air quality market-based incentive program.

(a) Describe the issues and facts relative to this proposition in terms of air quality.
(b) Describe what is meant by "state tax credits."
(c) Choose a current political party. Describe that party's history and current philosophy of dealing with the issue of air pollution.

Free-Response Answer

No answer is provided for this question. This question is to be answered during your AP Environmental Science class and turned into the instructor for analysis. You may research the question prior to arrival at class, but do not have any notes or materials when you write the answer. Write the answer under simulated test-taking conditions.

PART III: Practice Exams

Practice Exam 1

Section I—Multiple-Choice Questions

Time: 90 minutes
100 questions
60% of total grade
No calculators allowed.
This section consists of 100 multiple-choice questions. Mark your answers carefully on the answer sheet.

General Instructions

Do not open this booklet until you are told to do so by the proctor.

Be sure to write your answers for Section I on the separate answer sheet. Use the test booklet for your scratch work or notes, but remember that no credit will be given for work, notes, or answers written only in the test booklet. Once you have selected an answer, blacken thoroughly the corresponding circle on the answer sheet. To change an answer, erase your previous mark completely, and then record your new answer. Mark only one answer for each question.

EXAMPLE	SAMPLE ANSWER
The Pacific is	Ⓐ Ⓑ ● Ⓓ Ⓔ

(A) a river.
(B) a lake.
(C) an ocean.
(D) a sea.
(E) a gulf.

To discourage haphazard guessing on this section of the exam, a quarter point is subtracted for every wrong answer, but no points are subtracted if you leave the answer blank. Even so, if you can eliminate one or more of the choices for a question, it may be to your advantage to guess.

Because it is not expected that all test takers will complete this section, do not spend too much time on difficult questions. Answer first the questions you can answer readily, and then, if you have time, return to the difficult questions. Don't get stuck on one question. Work quickly but accurately. Use your time effectively.

Answer Sheet: Practice Exam 1

1. A B C D E
2. A B C D E
3. A B C D E
4. A B C D E
5. A B C D E
6. A B C D E
7. A B C D E
8. A B C D E
9. A B C D E
10. A B C D E
11. A B C D E
12. A B C D E
13. A B C D E
14. A B C D E
15. A B C D E
16. A B C D E
17. A B C D E
18. A B C D E
19. A B C D E
20. A B C D E
21. A B C D E
22. A B C D E
23. A B C D E
24. A B C D E
25. A B C D E
26. A B C D E
27. A B C D E
28. A B C D E
29. A B C D E
30. A B C D E
31. A B C D E
32. A B C D E
33. A B C D E
34. A B C D E
35. A B C D E
36. A B C D E
37. A B C D E
38. A B C D E
39. A B C D E
40. A B C D E
41. A B C D E
42. A B C D E
43. A B C D E
44. A B C D E
45. A B C D E
46. A B C D E
47. A B C D E
48. A B C D E
49. A B C D E
50. A B C D E
51. A B C D E
52. A B C D E
53. A B C D E
54. A B C D E
55. A B C D E
56. A B C D E
57. A B C D E
58. A B C D E
59. A B C D E
60. A B C D E
61. A B C D E
62. A B C D E
63. A B C D E
64. A B C D E
65. A B C D E
66. A B C D E
67. A B C D E
68. A B C D E
69. A B C D E
70. A B C D E
71. A B C D E
72. A B C D E
73. A B C D E
74. A B C D E
75. A B C D E
76. A B C D E
77. A B C D E
78. A B C D E
79. A B C D E
80. A B C D E
81. A B C D E
82. A B C D E
83. A B C D E
84. A B C D E
85. A B C D E
86. A B C D E
87. A B C D E
88. A B C D E
89. A B C D E
90. A B C D E
91. A B C D E
92. A B C D E
93. A B C D E
94. A B C D E
95. A B C D E
96. A B C D E
97. A B C D E
98. A B C D E
99. A B C D E
100. A B C D E

Directions: Each group of lettered answer choices refers to the numbered statements of questions that immediately follow. For each question or statement, select the one lettered choice that is the best answer and fill in the corresponding circle on the answer sheet.

1. Which of the following would be most likely to increase competition among the members of a squirrel population in a given area?

(A) An epidemic of rabies within the squirrel population
(B) An increase in the number of hawk predators
(C) An increase in the reproduction of squirrels
(D) An increase in temperature
(E) An increase in the food supply

2. The least expensive and most feasible short-term energy replacement for nuclear power would be

(A) solar power.
(B) wind power.
(C) hydrogen fuel cells.
(D) biomass.
(E) nuclear fusion.

3. Which one of the following statements is FALSE?

(A) The greenhouse effect is a natural process that makes life on Earth possible with 98% of total global greenhouse gas emissions being from natural sources (mostly water vapor) and 2% from man-made sources.
(B) The United States is the number one contributor to global warming.
(C) The Kyoto Protocol would have allowed the United States to increase its greenhouse gas emissions—primarily carbon dioxide (CO_2), methane (CH_4), and nitrous oxide (N_2O)—by only 2% per year based on 1990 levels.
(D) Effects of global warming on weather patterns may lead to adverse human health impacts.
(E) Global warming may increase the incidence of many infectious diseases.

4. Which of the following would be an external cost?

(A) The cost of steel in making a refrigerator
(B) The cost of running a refrigerator for one month
(C) The cost of labor in producing refrigerators
(D) The taxes paid by consumers in purchasing refrigerators
(E) The costs associated with health care when the refrigerator leaks refrigerant into the atmosphere

5. The tickbird–rhinoceros association is an example of

(A) facultative commensalism.
(B) obligatory commensalism.
(C) obligatory mutualism.
(D) facultative mutualism.
(E) obligatory parasitism.

6. Which of the following is NOT an example of environmental mitigation?

(A) Promoting sound land use planning, based on known hazards
(B) Relocating or elevating structures out of the floodplain
(C) Constructing living snow fences
(D) Organizing a beach clean-up
(E) Developing, adopting, and enforcing effective building codes and standards

7. The upthrusting of a mountain barrier

(A) is an instance of diastrophism.
(B) is an example of gradation.
(C) stimulates desert formation of the ocean.
(D) has no effect on the local climate.
(E) has never occurred.

8. Laws dealing with damage, injury, or a wrongful act done willfully, negligently, or in circumstances involving strict liability, but not involving breach of contract, for which a civil suit can be brought are known as

(A) penal codes.
(B) statutes.
(C) torts.
(D) common law.
(E) case law.

9. Which theory explains evolution as a result of long, stable periods of time interrupted by geologically brief periods of major change?

(A) Punctuated equilibrium
(B) Gradualism
(C) Natural selection
(D) Reproductive isolation
(E) Use and disuse

10. What has been the approximate rise in sea level over the past century?

(A) 2–4 in. (5–10 cm)
(B) 6–8 in. (15–20 cm)
(C) 10–12 in. (25–30 cm)
(D) 14–16 in. (35–40 cm)
(E) more than 16 in.

11. Branching evolution, producing many branching lines of descent from a single ancestral source, is called

(A) divergent evolution.
(B) convergent evolution.
(C) straight-line evolution.
(D) macroevolution.
(E) adaptive evolution.

12. Place the following economic activities in order—starting with those activities closest to natural resources and ending with those farthest away.

I. Use raw materials to produce or manufacture something new and more valuable.
II. Professions that process, administer, and disseminate information. (e.g., computer engineers, professors, and lawyers).
III. Agriculture, fishing, hunting, herding, forestry, and mining.
IV. Include all activities that amount to doing service for others. (e.g., doctors, teachers, and secretaries).

(A) I, II, III, IV
(B) I, III, IV, II
(C) III, I, II, IV
(D) III, I, IV, II
(E) II, IV, I, III

13. The first heterotrophs that evolved on the primitive Earth must have been able to obtain energy from

(A) enzymes present in the environment.
(B) radiant energy of the sun.
(C) inorganic compounds in the atmosphere.
(D) organic compounds present in the environment.
(E) molecules of ATP in the environment.

14. The film *Erin Brockovich* focused on

(A) pesticide accumulation.
(B) toxic landfill wastes.
(C) nuclear wastes.
(D) clean drinking water.
(E) air pollution.

15. Which of the following statements regarding coral reefs is FALSE?

(A) Modern reefs can be as much as 2.5 million years old.
(B) Coral reefs capture about half of all the calcium flowing into the ocean every year, fixing it into calcium carbonate rock at very high rates.
(C) Coral reefs store large amounts of organic carbon and are very effective "sinks" for carbon dioxide from the atmosphere.
(D) Coral reefs are among the most biologically diverse ecosystems on the planet.
(E) Coral reefs are among the most endangered ecosystems on Earth.

16. Which of the following strategies to control pollution would incur the greatest governmental cost?

(A) green taxes
(B) government subsidies for reducing pollution
(C) regulation
(D) charging a "user fee"
(E) tradable pollution rights

17. A gene pool is

(A) the number of recessive genes available in an ecosystem.
(B) the sum total of all genes of an individual.
(C) the genes that are available to a community.
(D) the sum total of all the genes of all the individuals in the population.
(E) the sum total of all the mutated genes of the various species that share an ecosystem.

18. The *Exxon Valdez* spilled 10.8 million gallons of crude oil into Prince William Sound in Alaska. What happened to most of the oil?

(A) It was cleaned up by Exxon.
(B) It eventually evaporated into the air.
(C) It sank into the ground.
(D) It biodegraded and photolyzed.
(E) It dispersed into the water column.

19. Which of the following contributes LEAST to speciation?

(A) Sexual reproduction
(B) Asexual reproduction
(C) Selection
(D) Variation
(E) Isolation

20. Which Act's primary goal is to protect human health and the environment from the potential hazards of waste disposal and calls for conservation of energy and natural resources, reduction in waste generated, and environmentally-sound waste management practices?

(A) RCRA
(B) FIFRA
(C) CERCLA
(D) OSHA
(E) FEMA

21. The United States produces about ________ of the world's solid waste.

(A) 25%
(B) 33%
(C) 50%
(D) 66%
(E) 75%

22. In 1994, President Clinton signed an executive order that directed federal agencies to identify and address "disproportionately high and adverse human health and environmental effects of programs, policies, and activities on minority populations and low-income populations." This executive order focused on the principle of

(A) civil rights.
(B) equal rights.
(C) environmental justice.
(D) human rights.
(E) equal protection under the law.

23. All the statements about trophic levels and food webs are incorrect EXCEPT

(A) they describe the general routes of energy flow in ecosystems.
(B) they measure the amounts of energy in ecosystems.
(C) they signify the kind of materials cycled in ecosystems.
(D) they measure the amounts of materials that pass through ecosystems.
(E) they serve as storehouses for the chemical energy initially trapped by plants.

24. When present in large numbers, the plant most likely to cause a lake to become filled gradually is the

(A) lichen.
(B) elodea.
(C) maple tree.
(D) duckweed.
(E) mushroom.

25. A group of 100 daphnia was placed into each of three culture jars of three different sizes. The following graph shows the average number of offspring produced per female each day in each jar.

Key:
(A) Daphnia in 1000 mL of pond water
(B) Daphnia in 500 mL of pond water
(C) Daphnia in 250 mL of pond water

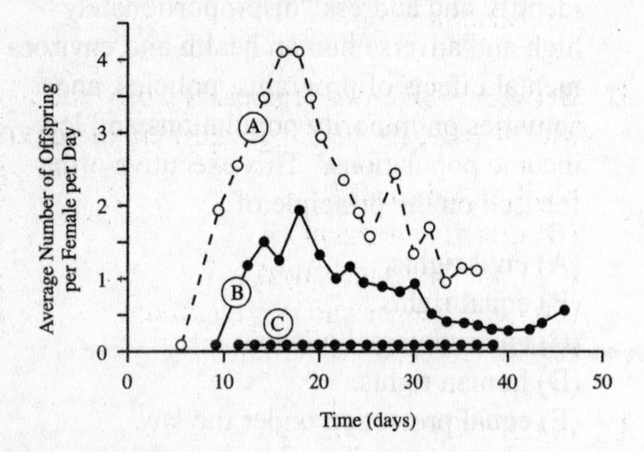

The information suggests that the population growth rate

(A) is density-dependent.
(B) is density-independent.
(C) depends on a predator–prey relationship.
(D) increases exponentially after 20 days.
(E) levels off when the carrying capacity is reached.

Questions 26–28

Base your answers to Questions 26 through 28 on the following diagram.

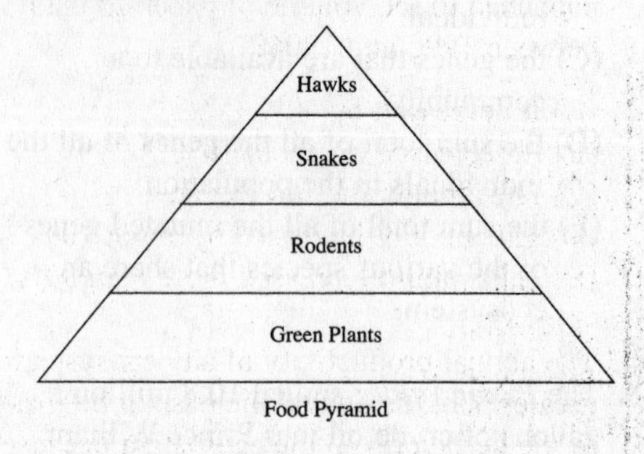

Food Pyramid

26. The greatest amount of energy present in this pyramid is found at the level of the

(A) hawks.
(B) snakes.
(C) rodents.
(D) green plants.
(E) decomposers.

27. The pyramid implies that, in order to live and grow, 1000 pounds of snakes would require

(A) less than 1000 pounds of green plants.
(B) 1000 pounds of rodents.
(C) more than 1000 pounds of rodents.
(D) no rodents.
(E) less than 1000 pounds of hawks.

28. Which of the following represents the food chain shown in the pyramid?

(A) Green plant, rodent, snake, hawk
(B) Rodent, green plant, snake, and hawk
(C) Snake, green plant, rodent, hawk
(D) Hawk, green plant, rodent, snake
(E) Green plant, snake, hawk, rodent

29. In 1990, Americans sifted through 3.8 million tons of advertising mail, which amounted to 2% of U.S. municipal solid waste. With the advent of e-mail in the 1990s, what happened to the volume of paper ad mail between 1990 and 1998?

(A) It decreased by 50%.
(B) It decreased by about 10%.
(C) The amount did not change.
(D) The amount increased by about 10%.
(E) The amount increased more than 30%.

30. The annual productivity of any ecosystem is greater than the annual increase in biomass of the herbivores in the ecosystem because

(A) plants convert energy input into biomass more efficiently than animals.
(B) there are always more animals than plants in any ecosystem.
(C) plants have a greater longevity than animals.
(D) during each energy transformation, some energy is lost.
(E) animals convert energy input into biomass more efficiently than plants.

Questions 31–33

Following is a diagram showing the relationships that exist in an arid ecosystem. Base your answers to Questions 31 through 33 on the diagram and your knowledge of environmental science.

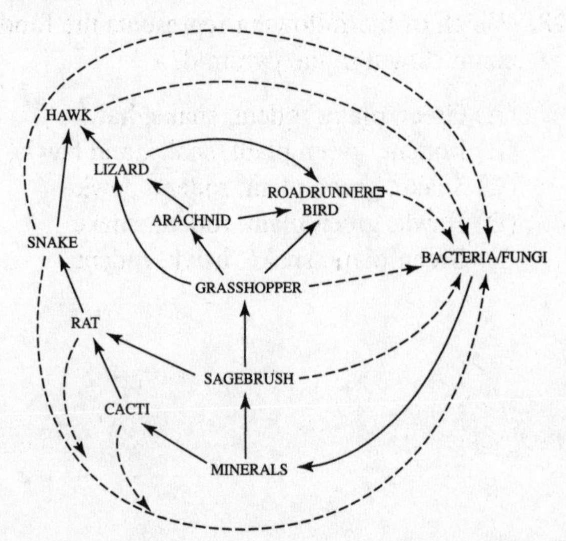

31. Which organisms are CORRECTLY ordered to demonstrate a secondary nutritional relationship?

(A) Snake → rat → arachnid → grasshopper
(B) Cacti → rat → sagebrush → grasshopper
(C) Grasshopper → arachnid → grasshopper → bacteria/fungi
(D) Grasshopper → roadrunner → sagebrush → rat
(E) Snake → hawk → roadrunner → bacteria/fungi

32. Between which two organisms would there MOST LIKELY be the greatest competition?

(A) Rat and snake
(B) Lizard and arachnid
(C) Cacti and sagebrush
(D) Grasshopper and bacteria/fungi
(E) Arachnid and roadrunner

33. In the diagram, which statement CORRECTLY describes the role of bacteria/fungi?

(A) The bacteria/fungi convert radiant energy into chemical energy.
(B) The bacteria/fungi directly provide a source of nutrition for animals.
(C) The bacteria/fungi are saprophytic agents restoring inorganic material to the environment.
(D) The bacteria/fungi convert atmospheric nitrogen into minerals and are found in the nodules of cacti.
(E) The bacteria/fungi consume live plants and animals.

34. When resources are scarce,

(A) prices go down.
(B) recycling is not profitable.
(C) investment and potential profits decrease.
(D) there is an impetus for conservation.
(E) there is little competition to discover new or substitute products.

35. Genetic drift, which can lead to the loss of certain genes by chance, is particularly significant

(A) in determining the fitness of a species.
(B) in the evolution of small populations.
(C) in determining the gene pool of a large population.
(D) in accounting for the total gene pool of a species.
(E) in determining the survival of mutations

36. Choose the correct order for the passage of a federal environmental bill.

I. Floor debates and vote
II. Referred to the other chamber
III. Presidential action
IV. Bill is introduced—Bill is referred to a committee, and a hearing is conducted.
V. Committee reports

(A) IV–III–II–I–V
(B) IV–V–I–II–III
(C) I–II–III–IV–V
(D) IV–I–II–III–V
(E) IV–III–I–V–II

37. Sympatric populations are those

(A) closely resembling species that live in the same place.
(B) closely resembling species that live in different climes.
(C) species that have traveled different lines of evolution.
(D) species that have evolved from two ancestral lines.
(E) closely resembling species that have migrated away from each other.

38. What is a condition that does NOT occur during an El Niño-Southern Oscillation (ENSO)?

(A) Trade winds blowing on the Pacific Ocean at the equator normally push surface waters west into a pool of warm water.
(B) Rain clouds form closer to South America.
(C) Increased rainfall occurs across the southern tier of the United States and in Peru.
(D) Plankton populations decrease off the South American coast.
(E) The Midwest and the Northeast have winters with more snowfall.

39. Which of the following statements would NOT be consistent with the deep ecology movement?

(A) The well-being and flourishing of human and nonhuman life on Earth have intrinsic value. These values are independent of the usefulness of the nonhuman world for human purposes.
(B) Richness and diversity of life forms contribute to the realization of these values and are also values in themselves.
(C) Our success depends on how well we can understand and manage the Earth's life-support systems.
(D) The flourishing of human life and cultures is compatible with a substantial decrease in human population. The flourishing of nonhuman life requires such a decrease.
(E) Present human interference with the nonhuman world is excessive, and the situation is rapidly worsening.

40. A type of fishing that involves dragging a chain across the seafloor and pulling a huge net behind it in which a single pass can remove up to a quarter of seafloor life and with repeated passes can remove nearly all seafloor life, including not only sessile animals and plants, but also many species of fish and marine invertebrates, is known as

(A) trawling.
(B) dredging.
(C) scraping.
(D) scoop netting.
(E) drag lining.

41. Which item listed here COULD be placed in a compost pile?

(A) Chemically treated wood products
(B) Pet wastes
(C) Manure
(D) Pernicious weeds
(E) Bones

42. Two thirds of Iceland's energy sources come from clean, renewable hydroelectric and geothermal sources. Research in Iceland is currently underway in developing hydrogen fuel cells. Iceland is an example of a country that is practicing

(A) sustainability.
(B) remediation.
(C) conservation.
(D) preservation.
(E) mitigation.

43. The diagram represents evolutionary pathways for two hypothetical organisms (X and Y).

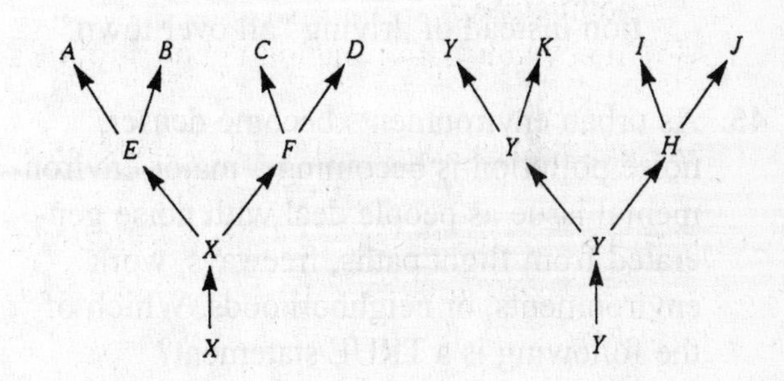

Which species was probably the most successful in the environment over time?

(A) X
(B) Y
(C) K
(D) J
(E) C

44. A group promoting greenbelts would believe in all ideas EXCEPT

(A) encouraging smart growth instead of sprawl.
(B) designing compact, pedestrian-friendly, transit-oriented neighborhoods.
(C) including affordable housing in all city plans for expansion.
(D) directing public and private investment into existing communities instead of into wasteful sprawl development.
(E) encouraging large retail chains such as Wal-Mart and Costco to locate within their city so that driving distances could be reduced by allowing customers to find everything they need in one location instead of driving "all over town."

45. As urban environments become denser, noise pollution is becoming a major environmental issue as people deal with noise generated from flight paths, freeways, work environments, or neighborhoods. Which of the following is a TRUE statement?

(A) Loud sound is not dangerous, as long as you don't feel any pain in your ears.
(B) Hearing loss after sound exposure is temporary.
(C) Hearing loss is mostly caused by aging.
(D) Loud sound damages only your hearing.
(E) If you have a hearing loss already, you still have to protect your hearing.

46. Depending on weather conditions, smoking three packs of cigarettes a day would be the equivalent of living in which city?

(A) Los Angeles
(B) Mexico City
(C) Athens
(D) Rome
(E) London

47. Due to CITES, elephant populations in Africa

(A) have grown to record numbers.
(B) have stabilized.
(C) are near the point of collapse.
(D) are being moved to safe wildlife sanctuaries.
(E) no longer exist.

Questions 48–50

Use this information for Questions 48 through 50. Two varieties of the same species of voles (meadow mice), albino and red-backed, were used in an experiment. Both varieties were subjected to the predation of a hawk, under controlled laboratory conditions. During the experiment, the floor of the test room was covered on alternate days with white ground cover that matched the albino voles and red-brown cover that matched the red-backed voles. The results of 50 trials are shown in the following tabulation.

NUMBER OF VOLES CAPTURED

Variety	White Cover	Red-Brown Cover	Total
Albino	35	57	92
Red-backed	60	40	100
Total	95	97	192

48. Which result would have been most likely if only red-brown floor covering had been used?

(A) Ninety-seven red-backed voles would have survived.
(B) Ninety-five albino voles would have survived.
(C) The survival rate of the albino voles would have decreased markedly.
(D) A greater number of red-backed voles would not have survived.
(E) There would have been no change in the results.

49. The results of this experiment lend credence to the concept of

(A) the female species being more deadly than the male.
(B) use and disuse of organs and tissues.
(C) overproduction as a means of species perpetuation.
(D) adaptation permitting the survival of the species.
(E) spontaneous generation of lethal mutation.

50. The purpose of the experiment was to

(A) measure the visual acuity of hawks.
(B) decrease the vole population.
(C) compare the agility of the several varieties of voles.
(D) determine the basic intelligence of hawks and voles.
(E) compare the effects of protective coloration.

Questions 51 and 52

For Questions 51 and 52, choose the term that best fits each symbiotic description.

(A) Commensalism
(B) Mutualism
(C) Parasitism
(D) Saprophytism
(E) Competition

51. Termites eat wood but are unable to digest it without the aid of certain flagellated protozoans that live within the bodies of termites.

52. Certain young clams attach themselves to the gills of a fish. In a short time, each clam becomes surrounded by a capsule formed by cells of the fish. The clam feeds and grows by absorbing nutrients from the fish's body.

53. Natural capitalism is associated with

(A) Paul Hawken.
(B) Garrett Hardin.
(C) Barry Commoner.
(D) Rachel Carson.
(E) Aldo Leopold.

54. Which federal agency does NOT manage designated wilderness areas in the United States?

(A) National Park Service
(B) Forest Service
(C) Bureau of Land Management
(D) Fish and Wildlife Service
(E) All share management.

Questions 55 and 56

Answer Questions 55 and 56 on the basis of the following diagram. Each circle represents the range for a pair of owls. X represents the area where reproduction and nesting take place.

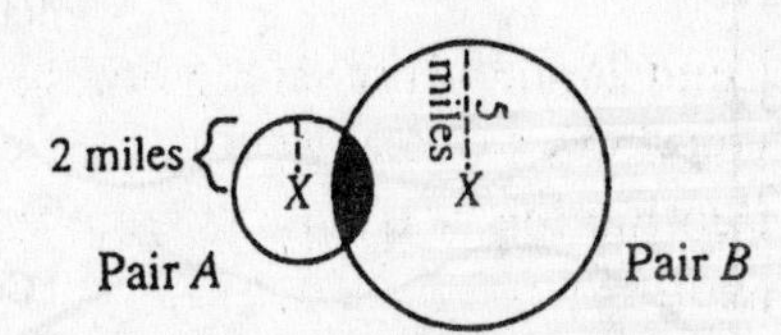

55. The shaded area represents the area where

(A) most of the predators of the owls are located.
(B) most of the owls' food is located.
(C) competition is likely to occur between the pairs.
(D) mating occurs.
(E) there is no competition between the two pairs.

56. What is the most likely reason why the range of pair B is larger than the range of pair A?

(A) Pair B is younger.
(B) Pair B can fly faster than pair A.
(C) Pair B has more prey near its nest than pair A.
(D) Pair B has less food available near its nest than pair A.
(E) Pair A does not need as much living space as pair B.

Questions 57 and 58

In 1940, ranchers introduced cattle into an area. The following graph shows the effect of cattle ranching on the populations of two organisms present in the area before the introduction of the cattle. Base your answers to Questions 57 and 58 on the graph and your knowledge of environmental science.

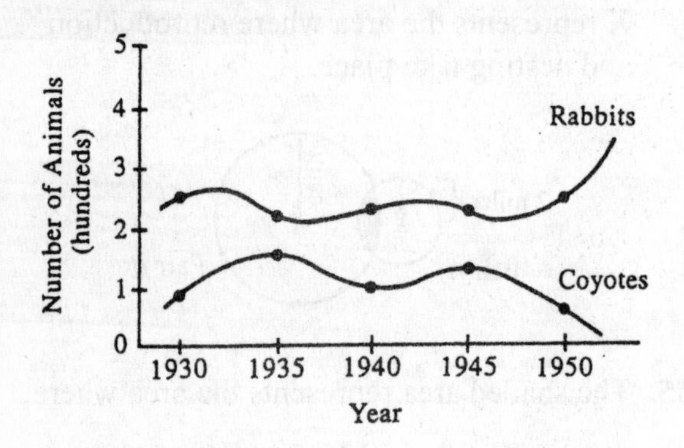

57. The most probable reason for the increase in the rabbit population after 1950 was

(A) more food became available to the rabbits.
(B) the coyote population declined drastically.
(C) the cattle created a more favorable environment for the rabbits.
(D) the coyotes and cattle competed for the same food.
(E) the coyote population increased.

58. If the interrelationship of rabbits and coyotes was once in balance, what is the most probable explanation for the decline of the coyotes?

(A) Mutations
(B) Starvation
(C) Disease
(D) Increase in reproductive rate
(E) Removal by human beings

Questions 59 and 60

Questions 59 and 60 refer to the following graphs of temperature and rainfall for six major ecosystems. A year's temperature from January through December (the line) and rainfall pattern (the shaded area) of each ecosystem are shown.

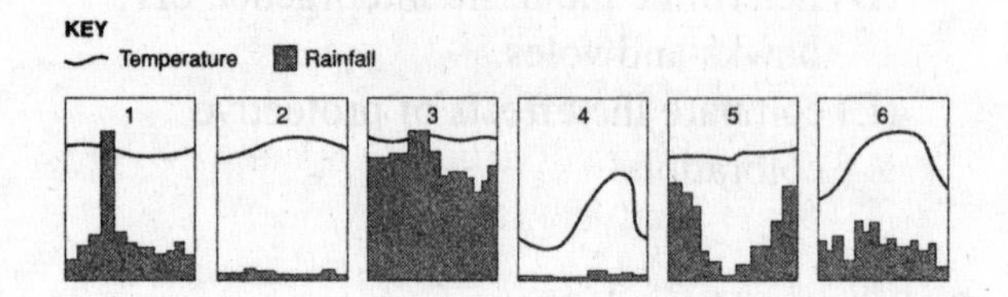

59. Which of the climatograms represents a savanna?

(A) 1
(B) 2
(C) 3
(D) 5
(E) 6

60. Which represents a tundra?

(A) 1
(B) 2
(C) 3
(D) 4
(E) 5

61. Areas of terrestrial and coastal ecosystems that are internationally recognized within the framework of UNESCO's Man and the Biosphere (MAB) program are known as

(A) wilderness preserves.
(B) wildlife sanctuaries.
(C) habitat sanctuaries.
(D) biosphere reserves.
(E) preservation zones.

62. Rank the following biomes in order of most productive to least productive, as measured by biomass produced per acre.

(1) Desert
(2) Tropical Rain Forests
(3) Tundra
(4) Grassland

(A) 1, 2, 3, 4
(B) 4, 3, 2, 1
(C) 3, 1, 4, 2
(D) 2, 3, 4, 1
(E) 2, 4, 3, 1

63. You work for a company that has violated federal environmental statutes regarding dumping of hazardous wastes. Your company's case would initially be heard in

(A) U.S. District Court.
(B) U.S. Claims Court.
(C) U.S. Circuit Courts of Appeal.
(D) State Trial Courts.
(E) U.S. Supreme Court.

64. Refer to data taken from a stream. The Roman numerals indicate collection sites.

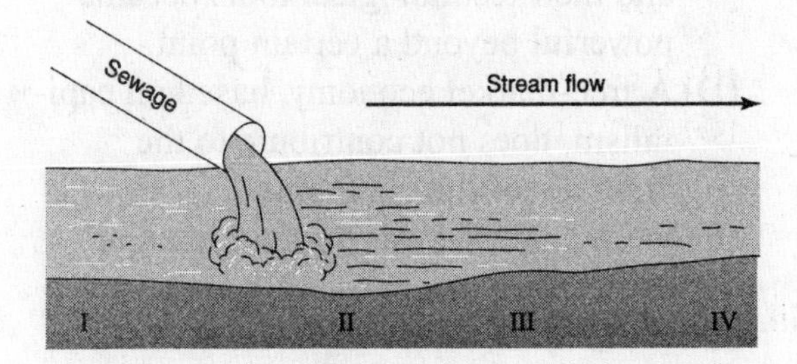

Where would the lowest DO (dissolved oxygen) content be expected?

(A) I
(B) II
(C) III
(D) IV
(E) Dissolved oxygen would not be affected by sewage effluent; therefore, DO would be equal at all points.

65. The theory that great disasters serve to maintain a population and its food supply balance was initially proposed by

(A) Darwin.
(B) Wallace.
(C) Hardy and Weinberg.
(D) Malthus.
(E) Lyell.

Questions 66 and 67

The following graph shows survival rates for five animal populations. When survival curves are calculated, the following assumptions are made:

(a) All individuals of a given population are the same age.
(b) No new individuals enter the population.
(c) No individuals leave the population.

These curves show the relationship of the number of individuals in a population to units of physiological life span. Base your answers to Questions 66 and 67 on the graph and your knowledge of environmental science.

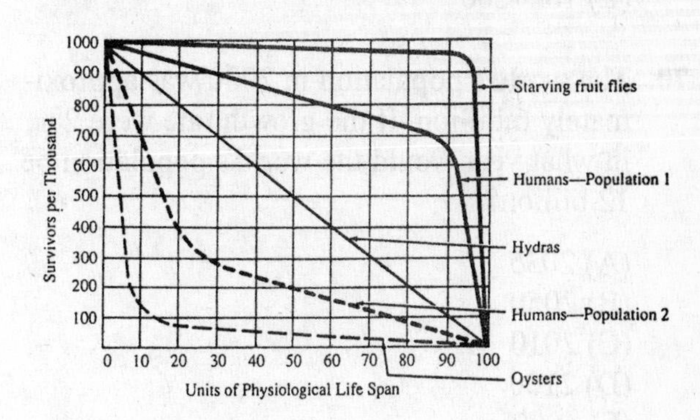

66. According to the data, it can be assumed that

(A) fruit flies live longer than humans.
(B) oysters outlive fruit flies.
(C) the population of hydras is steadily declining.
(D) the life span of human populations is related to that of oysters.
(E) there is a high mortality rate among young oysters.

67. The survival curves indicate that

(A) starving fruit flies live out their full life span.
(B) human populations are more vulnerable than hydras.
(C) human population 2 has a greater rate of survival than human population 1.
(D) the hydra has a longer life span than the oyster.
(E) fruit flies die directly after pupation.

68. The average requirement for drinking water per person per day is approximately

(A) 1 pint.
(B) 1 quart.
(C) ½ gallon.
(D) 1 gallon.
(E) 3 gallons.

69. The average American uses how many gallons of water per day for domestic use?

(A) 1–10
(B) 10–20
(C) 10–80
(D) 50–100
(E) 100–200

70. The world's population in 2000 was approximately 6 billion. If the growth rate were 2%, in what year would the world's population be 12 billion?

(A) 2035
(B) 2050
(C) 2010
(D) 2100
(E) 4000

71. You are on a trip visiting Rwanda in Africa. Which of the following would you notice?

(A) A larger number of people between 30 and 45 than any other age category
(B) A population that is equally distributed in age categories
(C) Greater number of people aged 65 and more compared to children under 15
(D) Fewer people aged 65 compared to children under 15
(E) A population wiped out by AIDS

72. Which of the following is NOT a component of an Environmental Impact Report?

(A) Alternatives to the proposed action
(B) A statement of positive and negative environmental impacts of the proposed activities
(C) A statement of financial liability
(D) Purpose and need for the project
(E) The relationship between short-term resources and long-term productivity

73. Which of the following statements is NOT consistent with *Tragedy of the Commons* by Garret Hardin?

(A) We will always add one too many cows to the village commons, destroying it.
(B) The destruction of the commons will not be stopped by shame, moral admonitions, or cultural mores anywhere near so effectively as it will be by the will of the people expressed as a protective mandate; in other words, by government.
(C) The "tragedy of the commons" is a modern phenomenon. Humans were not capable of doing too much damage until the population exceeded certain numbers and their technological tools became powerful beyond a certain point.
(D) A free-market economy, based on capitalism, does not contribute to the "tragedy of the commons."
(E) We will always opt for an immediate benefit at the expense of less tangible values such as the availability of a resource to future generations.

74. The most common method of disposing of municipal solid wastes in the United States is

(A) incineration.
(B) ocean dumping.
(C) sanitary landfills.
(D) recycling.
(E) exporting.

75. What would be the best environmental solution for beach erosion?

(A) Bring in sand from other sources to replace sand that had been washed away
(B) Build sea walls
(C) Build breakwaters
(D) Restrict development near beaches
(E) Restrict access to the beach

76. Which one of the following statements is FALSE?

(A) Every week more than 500,000 trees are used to produce the two-thirds of newspapers that are never recycled.
(B) The equivalent of ten city blocks of rainforest is destroyed every minute; an area the size of Pennsylvania is lost every year.
(C) Of the Earth's dry land surface, 7% is rainforest, home to more than 50% of the world's plants and animals.
(D) About one-fifth of the United States—442 million acres—is forested.
(E) Private owners account for 59% of the nation's 490 million acres of commercial forestland; the government owns 27%, and the forest industry owns 14%.

77. The densest populations of most organisms that live in the ocean are found near the surface. The most probable explanation is that

(A) the surface is less polluted.
(B) the bottom contains radioactive material.
(C) salt water has more minerals than freshwater.
(D) the light intensity that reaches the ocean decreases with increasing depth.
(E) the largest primary consumers are found near the surface.

78. Which of the following would be classified as a "point source" for pollution?

(A) Sulfur oxides released from an electrical generating plant
(B) Oil, grease, and toxic chemicals from urban runoff and energy production
(C) Sediment from improperly managed construction sites, crop and forestlands, and eroding stream banks
(D) Salt from irrigation practices and acid drainage from abandoned mines
(E) Bacteria and nutrients from livestock, pet wastes, and faulty septic systems

79. Which of the following would NOT be an example of accelerated erosion?

(A) Rill erosion
(B) Geological erosion
(C) Sheet erosion
(D) Gully erosion
(E) Wind erosion

80. John Muir was responsible for developing the

(A) Green Peace.
(B) Nature Conservancy.
(C) Audubon Society.
(D) Sierra Club.
(E) Environmental Defense Fund.

81. In response to the Great Depression, President Franklin D. Roosevelt created many programs designed to put America back to work. Which agency, listed here, was established in early 1933 with a twofold mission: to reduce unemployment, especially among young men, and to preserve the nation's natural resources.

(A) WPA
(B) EPA
(C) CCC
(D) ACLU
(E) TVA

82. Which American president dealt with an energy shortage by establishing a national energy policy by decontrolling domestic petroleum prices to stimulate production and expanding the national park system by including protection of 103 million acres of Alaskan lands?

(A) Jimmy Carter
(B) Ronald Reagan
(C) Bill Clinton
(D) George Bush
(E) George W. Bush

83. What was the name of the movement, in the late 1970s, that tried to gain land back from the federal government?

(A) State's Rights Rebellion
(B) Public Land Rebellion
(C) Homestead Movement
(D) Land Rush Rebellion
(E) Sagebrush Rebellion

84. In what area of the continental United States are the most endangered and threatened species of wildlife located?

(A) Northeast
(B) Southwest
(C) Central
(D) Northwest
(E) Southeast

85. Which legislation was signed into law almost 25 years ago by President Nixon that sets out the conditions under which a species can be listed as endangered or threatened, calls for the development of a plan for the recovery of the species' population, and allows for designating and protecting habitat critical to the species' survival?

(A) National Wildlife Refuge System Act
(B) Species Conservation Act
(C) National Forest Management Act
(D) Endangered Species Act
(E) National Environmental Policy Act

86. One of the greatest successes of the Endangered Species Act has been the

(A) passenger pigeon.
(B) whooping crane.
(C) bald eagle.
(D) condor.
(E) auk.

87. If global temperature warmed for any reason, atmospheric water vapor would increase due to evaporation, which would increase the greenhouse effect and thereby further raise the temperature causing the ice caps to melt back making the planet less reflective so that it would warm further. This is an example of

(A) positive coupling.
(B) negative coupling.
(C) synergistic feedback.
(D) positive feedback loop.
(E) negative feedback loop.

88. As of 2002, approximately how many people have become infected with the AIDS virus since it was discovered in the early 1980s?

(A) 5 million
(B) 25 million
(C) 60 million
(D) 120 million
(E) 350 million

89. What is the most frequent cause of beach pollution?

(A) Polluted runoff and storm water
(B) Sewage spills from treatment plants
(C) Oil spills
(D) Ships dumping their holding tanks into coastal waters
(E) People leaving their trash on the beach

90. Atmospheric CO_2 levels

(A) are about 10% higher than they were at the time of the Industrial Revolution.
(B) are about 25% higher than they were at the time of the Industrial Revolution.
(C) are about the same as they were at the time of the Industrial Revolution.
(D) are slightly lower than they were at the time of the Industrial Revolution.
(E) are significantly lower than they were at the time of the Industrial Revolution.

91. 550 parts per million (ppm) would be equivalent to

(A) 5.5 ppb (parts per billion).
(B) 55 ppb.
(C) 5,500 ppb.
(D) 55,000 ppb.
(E) 550,000 ppb.

92. The major sink for phosphorus is

(A) marine sediments.
(B) atmospheric gases.
(C) seawater.
(D) plants.
(E) animals.

93. Volcanoes that are built almost entirely of fluid lava flows that pour out in all directions from a central summit vent, or group of vents, building a broad, gently sloping dome-shaped cone and that are built up slowly by the accretion of thousands of highly fluid basalt lava flows that spread widely over great distances, and then cool as thin, gently dipping sheets are known as

(A) lava domes.
(B) composite volcanoes.
(C) cinder cones.
(D) shield volcanoes.
(E) stratovolcanoes.

94. Which of the following would NOT be a likely source for seismic activity?

(A) Along midoceanic ridges
(B) Faults associated with volcanic activity
(C) Boundaries between oceanic and continental plates
(D) Interior of continental plates
(E) Boundaries between continental plates

95. The first reptiles appeared on Earth

(A) during the Cambrian (550 million to 500 million years ago).
(B) during the Devonian (410 million to 360 million years ago).
(C) during the Carboniferous (360 million to 280 million years ago).
(D) during the Permian (280 million to 250 million years ago).
(E) during the Jurassic (210 million to 140 million years ago).

96. Examine the following weather map.

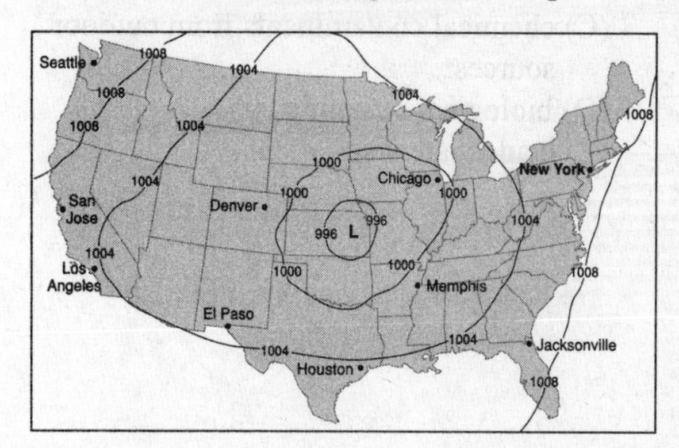

Which of the following would be TRUE?

(A) It is likely to be fair weather in the Midwest.
(B) It is likely to be raining in the Midwest.
(C) It is likely to be raining in the Northeast.
(D) Rain would be expected in the western United States.
(E) Not enough information is provided.

97. An AP Environmental Science class was investigating soil texture. Each student placed 2 teaspoons of soil in their hands and mixed a small amount of water into the soil until it was the consistency of moist putty. The students were able to make a small ball of mud that stayed the shape of a ball after they squeezed it. However, they were unable to make the ball of soil form a ribbon using their thumb and index finger. Most likely the type of soil that they were testing was

(A) gravel.
(B) silty clay.
(C) sand.
(D) clay.
(E) loamy sand.

98. Causes of sick building syndrome include all EXCEPT

(A) radon and asbestos.
(B) chemical contaminants from indoor sources.
(C) chemical contaminants from outdoor sources.
(D) biological contaminants.
(E) inadequate ventilation.

99. Which of the following statements is FALSE?

(A) Evidence exists of a dose–response relationship between nonmelanoma skin cancer and cumulative exposure to UV-B radiation.
(B) Individuals, usually those living in areas with limited sunlight and long dark winters, may suffer severe photoallergies to the UV-B in sunlight.
(C) Increased absorption of UV-B triggers a thickening of the superficial skin layers and an increase in skin pigmentation.
(D) There is a relationship between skin cancer prevalence and increases in ultraviolet radiation due to the depletion of tropospheric ozone.
(E) Acute exposure to UV-B causes sunburn and chronic exposure results in loss of elasticity and increased skin aging.

100. Which of the following would NOT be consistent with frontier ethics?

(A) The Earth is an unlimited supply of resources for exclusive human use.
(B) Humans are a part of nature, not apart from it.
(C) Success comes through domination and control of nature.
(D) Anthropocentrism
(E) Greater value assumption

Section II—Free-Response Questions

Time: 90 minutes
Four questions
40% of total grade
No calculators allowed.

Directions: Answer all four questions, which are weighted equally; the suggested time is about 22 minutes for answering each question. Write all your answers on the pages following the questions in the pink booklet. Where calculations are required, clearly show how you arrived at your answer. Where explanation or discussion is required, support your answers with relevant information and/or specific examples.

1. The environmental impact of washing a load of clothes in an electric washing machine is different than washing the same clothes by hand. Use the following information to answer the questions that follow. Show your calculations.

Assume the following:

1. All the clothes can be washed in one load in the washing machine.
2. The water entering the water heater is 60°F.
3. The water leaving the water heater is 130°F.
4. The electric washing machine uses 20 gallons water. It uses 110 volts of electricity at an average of 1500 watts for 30 minutes.
5. Washing the clothes by hand requires 35 gallons of hot water.

Other information:

1 gallon of water = 8 pounds

1 BTU = amount of energy required to raise the temperature of 1 pound of water by 1°F

1 kilowatt-hour = 3400 BTUs

(a) Calculate the total amount of energy (in BTUs) to wash the clothes using the washing machine.

(b) Calculate the total amount of energy (in BTUs) to wash the clothes by hand.

(c) Discuss the environmental impact of washing clothes.

2. An AP Environmental Science class did an investigation on competition. Part I of the investigation focused on intraspecific competition to assess the effect of growth among radish plants at different population densities. Part II of the investigation focused on the relative competitiveness of two species of plants (radish and wheat) when they were planted together. The results follow.

PART I: INTRASPECIFIC COMPETITION AMONG RADISH PLANTS

Seeds Per Pot	Total Biomass Per Pot (g)
1	5.0
10	70.0
20	75.0

PART II: INTERSPECIFIC COMPETITION BETWEEN RADISH AND WHEAT PLANTS

Seeds Per Pot	Total Biomass Per Pot (g)
1 radish	5.0
1 wheat	3.0
10 radish	50.0
10 wheat	25.0
20 radish	75.0
20 wheat	40.0

(a) Discuss the results of Part I of the investigation.

(i) At what population density was the biomass per plant highest?
(ii) What resources may have been limited?
(iii) Discuss the results obtained from the class for Part I in terms of biological laws or principles.

(b) Discuss the results of Part II of the investigation.

(i) Which plant was most affected by the competition between the two species?
(ii) What are possible reasons why the plant chosen in (i) may have been more successful.
(iii) Discuss the results obtained from the class for Part II in terms of biological laws or principles.

3. Examine the following age structure diagrams of Sweden and Kenya and answer the questions.

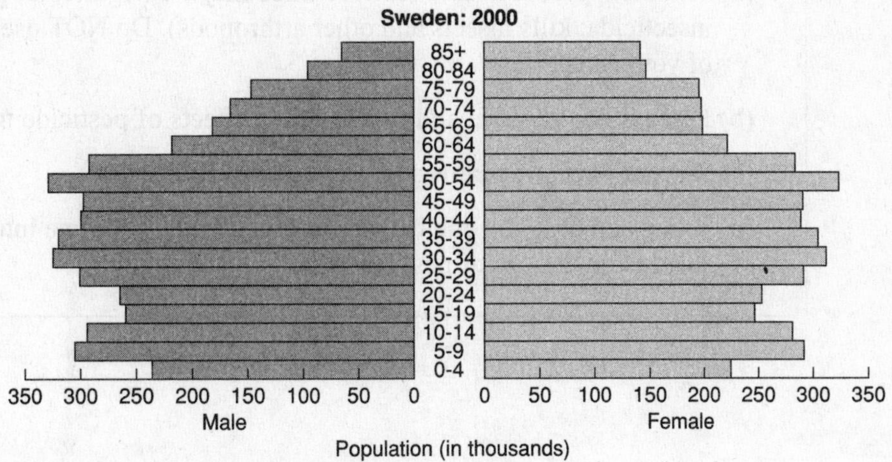

Source: U.S. Census Bureau, International Data Base.

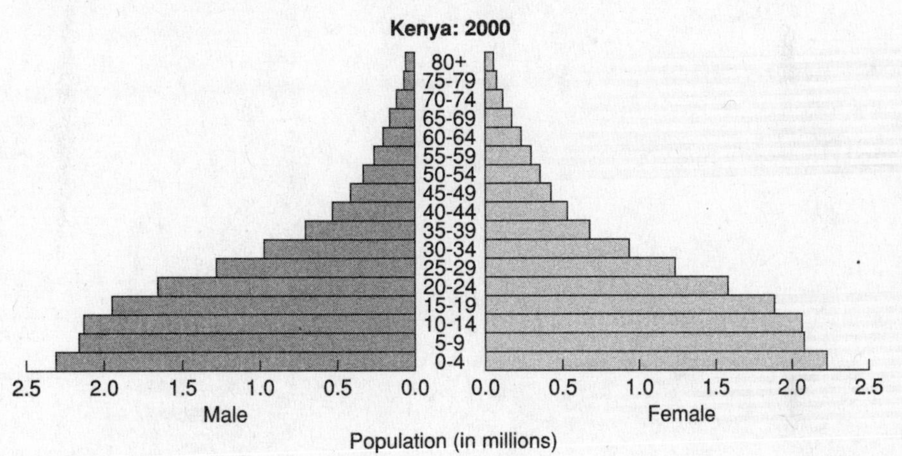

Source: U.S. Census Bureau, International Data Base.

(a) Compare and contrast the two age structure diagrams in terms of two population dynamics—birth rate and death rate.

(b) What factors affect birth rates and death rates?

(c) Discuss methods that have been employed in another country to curb population growth.

4. The use of pesticides has become the most widely used form of pest management and is a controversial topic in environmental science.

(a) What is a pesticide and describe three major categories of pesticides (Example: insecticide: kills insects and other arthropods). Do NOT use "insecticide" as one of your categories.

(b) Discuss two positive and two negative effects of pesticide use.

(c) Discuss three alternatives to the use of pesticides.

(d) Name and describe one United States federal law or one international treaty that focuses on the use of pesticides.

Answers and Explanations for Practice Exam 1

Answer Key for Multiple-Choice Questions

1. **C**	26. **D**	51. **B**	76. **D**
2. **D**	27. **C**	52. **C**	77. **D**
3. **C**	28. **A**	53. **A**	78. **A**
4. **E**	29. **E**	54. **E**	79. **B**
5. **D**	30. **D**	55. **C**	80. **D**
6. **D**	31. **A**	56. **D**	81. **C**
7. **A**	32. **B**	57. **B**	82. **A**
8. **C**	33. **C**	58. **E**	83. **E**
9. **A**	34. **D**	59. **A**	84. **B**
10. **B**	35. **B**	60. **D**	85. **D**
11. **A**	36. **B**	61. **D**	86. **C**
12. **D**	37. **A**	62. **E**	87. **D**
13. **D**	38. **E**	63. **A**	88. **C**
14. **D**	39. **C**	64. **C**	89. **A**
15. **C**	40. **A**	65. **D**	90. **B**
16. **C**	41. **C**	66. **E**	91. **E**
17. **D**	42. **A**	67. **A**	92. **A**
18. **D**	43. **B**	68. **C**	93. **D**
19. **B**	44. **E**	69. **C**	94. **D**
20. **A**	45. **E**	70. **A**	95. **C**
21. **B**	46. **B**	71. **D**	96. **B**
22. **C**	47. **B**	72. **C**	97. **E**
23. **A**	48. **C**	73. **D**	98. **A**
24. **D**	49. **D**	74. **C**	99. **D**
25. **A**	50. **E**	75. **D**	100. **B**

Explanations for Multiple-Choice Questions

1. **(C)** An increase in the population of squirrels would increase the competition for food and space.

2. **(D)** Biomass currently provides about 4% of the energy produced in the United States, and it could supply up to 20%. Due to the available land and agricultural infrastructure already in place, biomass could, sustainably, replace all the power that nuclear plants generate. Furthermore, biomass is used to produce ethanol, which could reduce oil imports upwards of 50%. The U.S. Department of Agriculture estimates that 17,000 jobs are created per every million gallons of ethanol produced, and the Electric Power Research Institute has estimated that producing 5 quadrillion BTUs of electricity on 50 million acres of land would increase overall farm income by $12 billion annually (the United States consumes about 90 quadrillion BTUs annually). Gasification is the newest method to generate electricity from biomass. Instead of simply burning the fuel, gasification captures about 65 to 70% of the energy in solid fuel by converting it first into combustible gases. This gas is then burned as natural gas is, to create electricity, fuel a vehicle, be used in industrial applications, or be converted to synthetic fuels.

3. **(C)** For (C) to be true, it would have read, "The Kyoto Protocol would have required the United States to reduce its greenhouse gas emissions—primarily carbon dioxide (CO_2), methane (CH_4), and nitrous oxide (N_2O)—to 7 percent below 1990 levels by the year 2012." Global warming is a pollution problem, created by the burning of oil, coal, and, to a lesser extent, natural gas. As concentrations of greenhouse gases have continued to rise, global average temperatures have risen by approximately 1°F. Unless greenhouse gas emissions are curbed, the Earth's mean temperature is projected to rise by 2 to 6°F within 100 years—a rate faster than any observed during the last 10,000 years. With only 4% of the world's population, the United States accounts for 22% of global emissions.

4. **(E)** External costs are the costs that are borne by people other than the producer of a product.

5. **(D)** The tickbird and the rhinoceros can live without each other. However, when in close association, they are mutually helpful. Armed with a pointed beak and sharp, curved claws, the tickbird (oxpecker) is well equipped for its life of scavenging parasites from the backs of large, thick-skinned animals such as the rhinoceros. It rids its host of unwelcome pests and provides early warning of danger by calling loudly when alarmed.

6. **(D)** Mitigation involves taking steps to lessen risk by lowering the probability of a risk event's occurrence or reducing its effect should it occur. Organizing a beach clean-up is remediation—reacting after the beach had been polluted.

7. **(A)** Diastrophism can be divided into two processes: (1) orogenesis or mountain building in which rock is forced to buckle under horizontal pressure (folding) and can happen to a degree significant enough to create entire mountain ranges and (2) epeirogenesis or continent building in which pressures may cause an entire continent to warp or bow.

8. **(C)** Torts are civil wrongs recognized by law as grounds for a lawsuit. These wrongs result in an injury or harm constituting the basis for a claim by the injured party. The primary aim of tort law is to provide relief for the damages incurred and deter others from committing the same harms. The injured person may sue for an injunction to prevent the continuation of the tortious conduct or for monetary damages. Torts fall into three general categories: intentional torts; negligent torts; and strict liability torts (e.g., liability for making and selling defective products).

9. **(A)** Punctuated equilibrium is a theory of evolution that proposes that species are stable and remain unchanged for long periods of time. Instead of a slow, continuous movement, evolution tends to be characterized by long periods of virtual standstill ("equilibrium"), "punctuated" by episodes of very fast development of new forms. Speciation occurs when

the stability is interrupted by brief periods of major changes.

10. (B) According to the United States Environmental Protection Agency. 1–2 in. (2–5 cm) of this increase results from melting mountain glaciers, and 1–3 in. (2–7 cm) results from ocean water expansion due to warmer ocean temperatures. Also, precipitation has increased by approximately 1% in the last century, further contributing to the rise in sea levels.

11. (A) Evolution is the descent from a common ancestor. Divergent evolution is a branching out of species from a single source. Adaptive radiation is one example of divergent evolution. The red fox *(Vulpes vulpes)* and the kit fox *(Vulpes macrotis)* provide an example of two species that have undergone divergent evolution. The red fox lives in mixed farmlands and forests, where its red color helps it blend in with surrounding trees. The kit fox lives on the plains and in the deserts, where its sandy color helps conceal it from prey and predators. The ears of the kit fox are larger than those of the red fox. The kit fox's large ears are an adaptation to its desert environment. The enlarged surface area of its ears helps the fox get rid of excess body heat. Similarities in structure indicate that the red fox and the kit fox had a common ancestor. As they adapted to different environments, the appearance of the two species diverged.

12. (D) *Primary economic activities* (III) are at the beginning of the production cycle where humans are in closest contact with resources and the environment. Primary economic activities are located at the site of the natural resources being exploited. In many developing nations approximately three-fourths of the labor force engages in subsistence farming or herding. By contrast, in highly-developed countries only a small fraction of the labor force is directly employed in agriculture (less than 3% in the United States). *Secondary economic activities* (I) use raw materials to produce or manufacture something new and more valuable. Examples of secondary activities include manufacturing, processing, producing power, and construction. Secondary economic activities are located either at the resource site or in close proximity to the market for the manufactured/processed good. Location depends upon whether the raw material or finished product costs more to ship. The major share of global secondary manufacturing activity is found within a relatively small number of major industrial concentrations. Rather than manufacturing/producing within its own country, richer nations often "set up shop" in developing nations because costs are significantly cheaper. *Tertiary economic activities* (IV) include all activities that amount to doing service for others. Doctors, teachers, and secretaries provide personal and professional services. Restaurant staff, store clerks, and hotel personnel provide retail and wholesale services. *Quaternary economic activities* (II) are not connected to resources, access to a market, or the environment but rather include professions that process, administer, and disseminate information. Computer engineers, professors, and lawyers serve as examples of "white collar" professionals who specialize in the collection and manipulation of information. With vast advancement in technology, quaternary economic activities potentially could exist anywhere. However, typically they exist in nations that have access to research centers, universities, efficient transportation and communication networks, and a pool of highly trained, skilled, flexible workers. Quaternary economic activities loom large in highly advanced, developed societies.

13. (D) Heterotrophs are dependent feeders. This means that they cannot synthesize nutritive material from inorganic compounds but must obtain nutrients from preexisting organic material. The original heterotrophs must have obtained energy from organic compounds that were present in the early environment. The first photosynthesizing autotrophs probably arose in response to organic nutrient depletion—evidence that is found in Pre-Cambrian rocks that formed in the presence of free oxygen. These early autotrophs were probably bacterial in nature (cyanobacteria or blue-green algae) with no cell nucleus or specialized organelles with diffuse genetic material and which reproduced through simple body division.

14. (D) Hexavalent chromium (chromium 6) is a byproduct of metal plating and other industrial work. High amounts of the chemical were found in the drinking water of Hinkley, California—a case dramatized in the 1999 movie *Erin Brokovitch* starring Julia Roberts.

15. (C) Coral reefs are among the most ancient of ecosystem types, dating back to the Mesozoic era some 225 million years ago. Coral reefs release carbon dioxide to the atmosphere due to the chemistry of calcium carbonate precipitation. The release of carbon dioxide from coral reefs is very small (probably less than 100 million tons of carbon per year) relative to emissions due to fossil fuel combustion (about 5.7 billion tons of carbon per year). Coral reefs store very little organic carbon and are not very effective "sinks" for carbon dioxide from the atmosphere. Forests are more effective sinks for atmospheric carbon. Although tropical rainforests contain more species than coral reefs, reefs contain more phyla than rainforests. Covering less than 0.2% of the ocean floor, coral reefs contain perhaps one-fourth of all marine species. Despite their limited area, coral reefs may be home to up to 25% of the fish catch of developing countries or 10% of the total amount of fish caught globally for human consumption as food. Coral reefs in 93 of the 109 countries containing them have been damaged or destroyed by human activities. In addition, human impacts may have directly or indirectly caused the death of 5 to 10% of the world's living reefs, and if the pace of destruction is maintained, another 60% could be lost in the next 20 to 40 years. The most important short-term threats to coral reefs are sedimentation (from poor land use such as clear cutting on steep slopes and other activities such as dredging, eutrophication (overfertilization and sewage pollution), and destructive fishing methods (dynamiting and overharvesting).

16. (C) Green taxes are taxes levied by the government on industries for each unit of pollution and are a source of governmental income. Government subsidies for reducing pollution are limited when industries surpass a break-even point or optimum level of pollution—the point at which cleanup costs exceed the harmful costs of pollution. Regulation is a command-and-control governmental approach that incurs costs to: (1) enact and enforce laws; (2) set standards; (3) regulate and monitor potentially harmful activities; and (4) prosecute violators. Furthermore, regulation often focuses on cleanup instead of prevention, discourages innovation by mandating prescribed pollution control strategies, and is often unrealistic with the realities of a competitive global business environment. Charging a "user fee" provides income to the government by charging industries to utilize a natural resource. Trading pollution or resource-use rights occurs between companies. Permits are allocated by the government for certain levels of pollution and companies are free to trade "unused" pollution allocations. Once permits are sold or auctioned by the government, further financial impacts occur between the companies, not between the companies and the government. Permits often allow the largest and most financially secure companies to pollute the most; may concentrate pollutants at the most-polluting sites; and may not create economic incentives to reduce pollution, since pollution levels are simply moved, not reduced.

17. (D) A gene pool contains all the genes of a given population. All the genes in a gene pool are capable of being passed from one generation to the next.

18. (D) According to a 1992 study by the National Oceanographic and Atmospheric Administration, 50% of the spilled oil underwent biodegradation and photolysis (chemical decomposition by the action of radiant electromagnetic energy, especially light). Cleanup crews recovered about 14% of the oil and approximately 13% sunk to the sea floor.

About 2% (some 216,000 gallons) remained on the beaches.

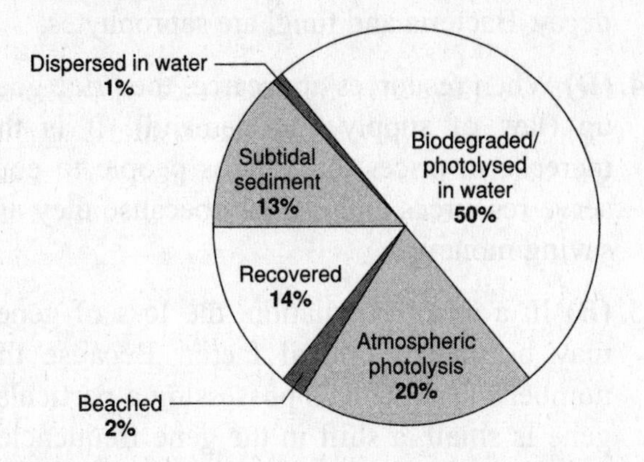

19. (B) Asexual reproduction produces organisms that are genetically identical to the parent. Speciation is the result of genetic variability. There are several types of asexual reproduction. Fission is the simplest form of asexual reproduction and involves the division of a single organism into two complete organisms, each identical to the other and to the parent. A similar form of asexual reproduction is regeneration, in which an entire organism may be generated from a part of its parent. Budding occurs when a group of self-supportive cells sprouts from and then detaches from the parent organism.

20. (A) The Resource Conservation and Recovery Act (RCRA) of 1976 addresses comprehensive management of nonhazardous and hazardous solid waste, proposes minimal standards for all waste disposal facilities and for hazardous wastes, and regulates treatment, storage, and transport. FIFRA is the Federal Insecticide, Fungicide, and Rodenticide Control Act, CERCLA is the Comprehensive Environmental Response, Compensation, and Liability (Superfund) Act; OSHA stands for the Occupational, Safety, and Health Administration, and FEMA stands for Federal Emergency Management Administration.

21. (B) The United States has about 5% of the world's population. However, it produces about one-third of all wastes in the world.

22. (C) The environmental justice movement began in the early 1980s in several locations in the United States. Community groups have been formed all across the country to find solutions to environmental problems that have negative impacts on their residents. While these groups originally formed in neighborhoods in response to a singular event such as the construction of a new incinerator, studies were initiated and found to show statistical relationships between racial/socioeconomic status and environmental risks.

23. (A) Trophic levels describe the routes of energy flow in ecosystems. Autotrophs (green plants) constitute the first trophic level where the energy potential in the food chain and food web is the greatest. Energy flows from the first to the second trophic level that is constituted by plant-eating animals. The flow of energy is routed to the third trophic level, composed of carnivores. Energy is lost at each successive trophic level.

24. (D) The duckweed (*Lemnaceae*) covers the surface of lakes. These plants grow floating in still or slow-moving freshwater except in the coldest regions. The growth of these high-protein plants can be extremely rapid. The blockage of light results in the death and decay of subsurface plants. The lake gradually becomes filled with dead and decaying plants and organisms. Lichens grow on rocks. Elodea is found in shallow water near the edges of lakes and ponds. Maple trees might eventually be the climax community where a lake existed and mushrooms are the decomposers of a terrestrial environment.

25. (A) The ability of daphnia to produce offspring is affected by population density. The reproductive rate decreases as the population increases.

26. (D) Green plants are the producers that can store enough energy to provide the basis upon which a community is built. The least amount of energy is found at the level of the hawks. In (B) and (C), energy is lost at each successive level in a food pyramid. In (E), decomposers recycle minerals into the environment from dead matter.

27. (C) Snakes feed on rodents. At each trophic level, energy is lost; therefore, the snakes must consume more than their biomass to survive. In (A), snakes do not feed on plants. In (B), the snakes must consume more than their biomass. In (D), rodents are the source of food for the snakes. In (E), hawks feed on snakes.

28. (A) The food pyramid reveals the feeding order in a community. The green plants would be producers. Primary consumers would be rodents, which are fed on by secondary consumers (the snakes), which are fed on by tertiary consumers (hawks).

29. (E) The average American gets only 1.5 personal letters each week, compared to 10.8 pieces of junk mail: *4.5 million tons* of junk mail is produced each year, which works out to each person receiving almost 560 pieces. Of all junk mail, 44% is thrown in the trash, unopened and unread. Approximately 40% of the solid mass that makes up our landfills is paper and paperboard waste, and by the year 2010, it is predicted to make up about 48%. Each year, 100 million trees are ground up to produce junk mail. Lists of names and addresses used in bulk mailings are in mass data-collection networks, compiled from phone books, warranty cards, and charity donations, and each name is worth 3 to 20 cents each time it is sold.

30. (D) Less energy is available at each trophic level because energy is lost by organisms through respiration and incomplete digestion of food sources. Therefore, fewer herbivores can be supported by the vegetative material.

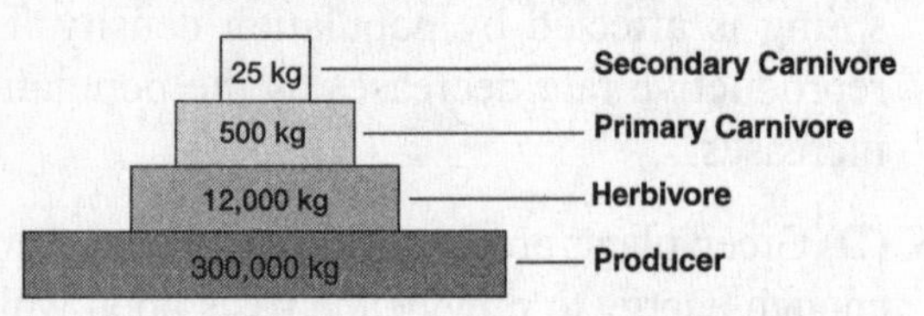

31. (A) Secondary consumers are carnivores that feed upon the flesh of other animals. Snakes eat lizards and arachnids eat grasshoppers.

32. (B) Both lizards and arachnids eat grasshoppers. Competition is the struggle between different species for the same resources.

33. (C) Saprophytic agents gain nourishment from dead organic matter. They are the organisms of decay. Bacteria and fungi are saprophytes.

34. (D) When resources are scarce, the price goes up (law of supply and demand). It is the increase in prices that causes people to conserve resources and use less because they are saving money.

35. (B) In a small population, the loss of genes may become a critical factor. Because the number of individuals possessing a particular gene is small, a shift in the gene frequencies may eventually result in the loss of that gene.

36. (B) Refer to "Federal Legislative Steps" in Chapter 16.

37. (A) Sympatric speciation refers to the formation of two or more descendant species from a single ancestral species all occupying the same geographic location. Mechanisms of sympatric speciation include: (1) anatomical and behavioral barriers (sexual selection), (2) ecological segregation (changes in host, habitat, phenology), (3) hybridization (parapatric speciation), and (4) changes in sexuality (e.g., appearance of asexual lineages).

38. (E) Recent research at the Midwestern Climate Center has pointed to significant reductions in total winter snowfall in the Midwest and Northeast due to a high-pressure ridge over northwestern North America and a low-pressure trough in the southeastern United States. If you did not know the answer (E), you could have gotten the correct answer through the process of elimination, knowing that answers (A) through (D) were true. When answering questions with the word "NOT" in it, focus on looking for the one false statement.

39. (C) This represents a planetary management philosophy.

40. (A) The practice is known as dragging in Canada. Besides the effects already listed, additional effects include:

- Flattens the ocean bottom, filling in holes, leveling "humps," and knocking down protruding organisms, all of which reduce biodiversity;

- Initiates erosion-causing long-term sedimentation problems;
- Changes the microhabitat affecting primary productivity;
- Damages or kills organisms of both commercially harvested species and noncommercial species; and
- Unlike clear cutting in forests, the effects are not visible, thereby not getting much attention.

41. (C) Pressure-treated wood (sometimes called CCA), which usually has a greenish tint to it, contains arsenic, a highly toxic element, as well as chromium and copper. Dog and cat feces may carry diseases that can infect humans. Morning glory, sheep sorrel, ivy, several kinds of grasses, and some other plants can resprout from their roots and/or stems in the compost pile. Bones (and the fat and marrow) are very attractive to pests such as rats. In addition, fatty food wastes can be very slow to break down, because the fat can exclude the air that composting microbes need to do their work. Manures typically contain large amounts of nitrogen (the fresher the manure, the more nitrogen it contains) and are considered a "green" ingredient. Fresh manures can heat a compost pile quickly and will accelerate the decomposition of woody materials, autumn leaves, and other "browns."

42. (A) Sustainability is a concept long recognized and utilized by many cultures. It results from recognition of the need for a harmonious existence between the environment, society, and economy. Sustainability focuses on improving the quality of life for all people without increasing the use of natural resources beyond the capacity of the environment to supply them indefinitely. For hydrogen to be a true, long-term, renewable alternative to fossil fuels (it runs more efficiently than gasoline, but it costs twice to three times as much to produce), it will need to be produced with a clean, low-cost electricity source such as hydroelectric power. Most oil-reliant countries simply don't have access to vast amounts of clean electricity, but Iceland and Norway do.

43. (B) The organism that can adapt to the environment survives. Organism Y has continued to exist since it first appeared. Therefore, organism Y has been the most successful in the environment over time.

44. (E) Greenbelt advocacy groups would encourage locally-owned retail stores located where people live and work. Wal-Marts and Costcos only become magnets drawing in more people to a city than can be effectively accommodated and add to traffic congestion, requiring more parking lots and more roads and decreasing open space (greenbelts).

45. (E) The threshold for pain is at about 120 to 140 dB, but sound begins to damage hearing when it is above 85 dB. Some of the hearing loss after exposure to excessive noise will be permanent. Indication of damage is ringing and noise in the ears (called tinnitus) after sound exposure. Research shows that cumulative exposure to loud sounds, not age, is the major cause of hearing loss. Loud sound can change your heart rate, vision, and reaction time. It may make you more aggressive and, in general, negatively affect you. Hearing loss accumulates. More exposure to loud sounds leads to more hearing loss.

46. (B) Mexico City is the home of 19 million people and 3.5 million vehicles. Emissions from vehicles cause some 70% of pollution in the urban area, which spans 1500 square miles. Almost half of the 1.2 million cars are more than 16 years old and emit an average of 80 times more pollutants than newer models. Infectious diseases like salmonella and hepatitis can be contracted simply by inhaling bacteria suspended in the air.

47. (B) Between 1979 and 1989, the worldwide demand for ivory caused elephant populations to decline to dangerously low levels. During this time period, poaching fueled by ivory sales cut Africa's elephant population in half (from 1.3 million elephants to only 600,000 that still remain). Recently, that number has stabilized, due in large part to the 1990 Convention on International Trade in

Endangered Species (CITES) ban on international ivory sales.

48. (C) Albino voles stand out against a brown background. The increased visibility would have made the voles easy prey for hawks. Predation decreases population.

49. (D) Adaptations encourage the survival of the species in a given environment.

50. (E) The effects of protective coloration, as related to predation, can be evaluated by this experiment because two varieties of voles were subjected to contrasting backgrounds.

51. (B) Both organisms are benefited by the association.

52. (C) The fish is harmed through the loss of its nutritive material, while the clam derives all the benefits from the association.

53. (A) *Natural Capitalism* by Paul Hawken, Amory Lovins, and L. Hunter Lovins introduced four central strategies that are a means to enable countries, companies, and communities to operate by behaving as if all forms of capital were valued: (1) radical resource productivity; (2) biomimicry; (3) service and flow economy; and (4) investing in natural capital.

54. (E) In 1964, Congress established the National Wilderness Preservation System, under the Wilderness Act. The legislation set aside certain federal lands as wilderness areas. These areas, generally 5000 acres or larger, are wild lands largely in their natural state. The act says that they are areas "where the Earth and its community of life are untrammeled by man, where man himself is a visitor who does not remain." Four federal agencies of the U.S. government administer the National Wilderness Preservation System, which includes 644 areas and more than 105 million acres.

1. The National Park Service (Interior) was established to protect the nation's natural, historical, and cultural resources and to provide places for recreation. The Park Service manages 51 national parks. It also oversees more than 300 national monuments, historic sites, memorials, seashores, and battlefields. It manages 13% of Federal lands and 42% of the National Wilderness Preservation System.
2. The U.S. Forest Service (Agriculture) manages national forests and grasslands. It conducts forestry research and works with forest managers on state and private lands. The Forest Service oversees close to 200 million acres of national forest and other lands. It manages 30% federal lands and 33% of the National Wilderness Preservation System.
3. The Bureau of Land Management (Interior) manages nearly 270 million acres. Among other activities, the bureau conserves these lands and their historical and cultural resources for the public's use and enjoyment. It manages 42% of federal lands and 5% of the National Wilderness Preservation System.
4. The U.S. Fish and Wildlife Service (Interior) conserves the nation's wild animals and their habitats by managing a system of more than 500 national wildlife refuges and other areas, totaling more than 91 million acres of land and water. It manages 15% of federal lands and 20% of the National Wilderness Preservation System.

55. (C) The shaded area, which represents the territory shared by the two pairs, is the area where intraspecific competition is likely to occur. Competition arises when organisms utilize similar resources that are limited and often involves food or space. The two most common forms of competition are interference and exploitative. Interference includes aggressive interactions in which one individual actively attempts to exclude another. Exploitative competition is indirect competition in which one species uses more of the limited resource or uses the resource more efficiently than another species.

56. (D) The most likely reason why the range of pair B is larger than that of pair A is that pair B has less food available near its nest than pair A. Pair B has extended its range to increase its chance of capturing food.

57. (B) The most probable explanation for increase in the rabbit population after 1950 was that the coyote population declined drastically. The relationship indicated by the graph is one of predator–prey. The coyotes prey upon rabbits, keeping the population in check.

58. (E) The most probable explanation for the decline of the coyotes is removal by humans. Because the decline in coyotes occurred with cattle introduction and because cattle and coyotes do not compete for the same food, the individuals that introduced the cattle must have interfered with the coyote population. Because coyotes are predators, they were perceived as dangerous to the cattle and were hunted and poisoned by humans. Attempts to control coyotes by poisoning may deplete the numbers of their natural prey and lead to increasing attacks by coyotes on farm animals.

59. (A) Savannas are warm year-round, with prolonged dry season and scattered trees. The environment is intermediate between grassland and forest. An extended dry season followed by a rainy season occurs in Australia; Central, Eastern, and South Africa; India; Madagascar; central South America; Southeast Asia; and Thailand. Savannas consist of grasslands with stands of deciduous shrubs and trees that do not grow more than 30 meters high. Trees and shrubs generally shed leaves during dry season, which reduces need for water. Food is limited during dry season so that many animals migrate during this season. Soils are rich in nutrients. Contain large herds of grazing animals and browsing animals that provide resources for predators.

60. (D) Tundra is located 60° north latitude and above. The weather is influenced by the polar cell. Alpine tundra is located in mountainous areas, above the tree line with well-drained soil; the dominant animals are small rodents and insects. Arctic tundra is frozen treeless plain with low rainfall, low average temperatures (summers average <10°C), and many bogs and ponds. Frozen ground prevents drainage. The growing season lasts 50 to 60 days. Tundra is found in Alaska, Canada, Europe, Greenland, and Russia. Dominant vegetation includes flowering dwarf shrubs, grasses, lichens, mosses, and sedges. Soil has few nutrients due to low vegetation and little decomposition. There are between 60 to 100 frost-free days per year. Arctic tundra is higher latitude than alpine tundra.

61. (D) Biosphere reserves are part of an international network. They are nominated by national governments and must meet a minimal set of criteria and adhere to a minimal set of conditions before being admitted into the network. Each biosphere reserve is intended to fulfill three basic functions, which are complementary and mutually reinforcing:

1. A conservation function—to contribute to the conservation of landscapes, ecosystems, species, and genetic variation;
2. A development function—to foster economic and human development that is socioculturally and ecologically sustainable; and
3. A logistic function—to provide support for research, monitoring, education, and information exchange related to local, national, and global issues of conservation and development.

62. (E) Refer to the table entitled "Net Primary Productivity of Biomes" on page 165.

63. (A) The U.S. District Courts are the trial courts of the federal court system. Within limits set by Congress and the Constitution, the District Courts have jurisdiction to hear nearly all categories of federal cases, including both civil and criminal matters. There are 94 federal judicial districts.

64. (C) At the point of discharge, there is a sudden increase in the amount of toxins, suspended solids, and dissolved organic compounds. The water becomes unsuitable for any human use. Aerobic bacteria increase rapidly, reducing the organic pollutants but using up dissolved oxygen. As the oxygen level falls, anaerobic bacteria multiply resulting in an unpleasant smell or stench. Eventually, as their food supply is used up, the decomposers reduce and algae thrive, re-oxygenating the water. Fish and other animal life reappear and the stream returns to its previous condition.

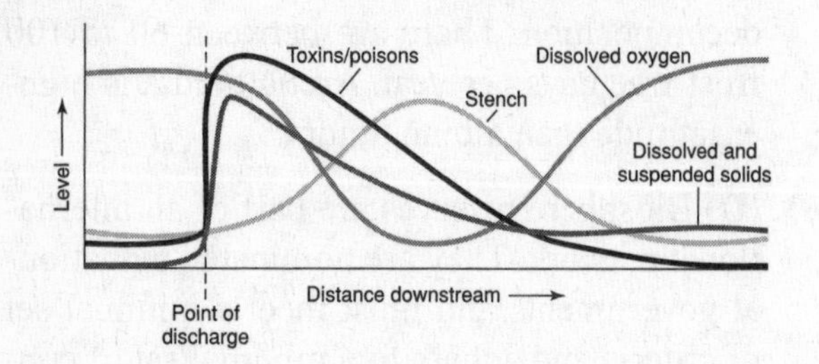

65. **(D)** Thomas Malthus recognized that once the carrying capacity of an area was exceeded, organisms would die of starvation. Darwin's theory of evolution based on natural selection may have been influenced by Malthusian concept.

66. **(E)** Note that only 10% of the population of young oysters reaches the 20th unit of life span.

67. **(A)** Note that almost all of the fruit flies reach the 90th unit of life span before the entire population dies.

68. **(C)** On a cool, inactive day, the average man loses about 12 eight-ounce cups of water, but only consumes about nine cups of water (about half of that from the water in fruits, vegetables, and other solid foods). To avoid even mild dehydration, the following is a simple test to calculate the minimum daily amount of water you should be getting from all sources:

1. Enter your weight: ______ pounds
2. Multiply your weight by 0.04: ______ pounds of water lost
3. Multiply line 2 by 2: ______ cups of water (one cup = 8 oz) needed daily from beverages and food

69. **(C)** There are more than 55,000 community water systems in the United States, processing nearly 34 billion gallons of water per day. There are about 23 million private water wells in the United States: 47% of the U.S. population uses surface water; 53% of the U.S. population uses groundwater. Americans drink more than 1 billion glasses of tap water per day. On average, 50 to 70% of home water is used outdoors for watering lawns and gardens. Daily indoor per capita water use in the typical single family home with no water-conserving fixtures is 74 gallons. If all U.S. households installed water-saving features, water use would decrease by 30%, saving an estimated 5.4 billion gallons per day. This would result in dollar-volume savings of $11.3 million per day or more than $4 billion per year.

Use	Gallons Per Capita	Percentage of Total Daily Use
Showers	12.6	17.3
Clothes washers	15.1	20.9
Dishwashers	1.0	1.3
Toilets	20.1	27.7
Baths	1.2	2.1
Leaks	10.0	13.8
Faucets	11.1	15.3
Other domestic uses	1.5	2.1

70. **(A)** To calculate doubling time, divide the country's growth rate into the number 70 (actually 69.3 for better accuracy). Thus, a growth rate of 2% will double a population in only 35 years, 1% in 70 years, and so forth.

71. **(D)** The age structure diagram for Rwanda in 2000 follows.

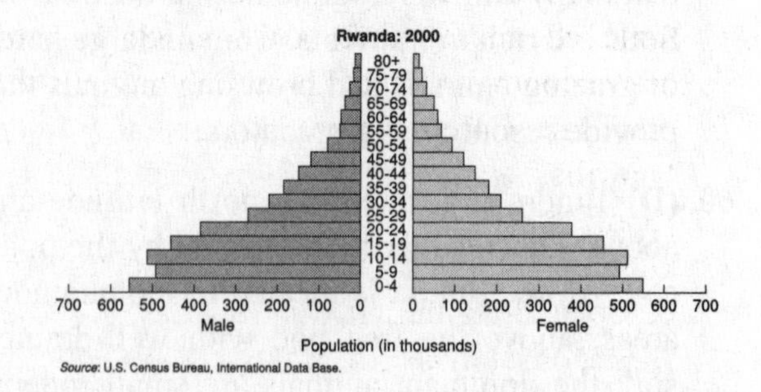

Life expectancy is approximately 41 years. Infant mortality rate (under 5) is 170 per 1000 live births. The growth rate in Rwanda is 1.16% lower than in most least-developed countries due to AIDS and emigration due to civil war.

72. (C) When a project is proposed that will affect the environment, laws state that studies be conducted to evaluate the impact that the project will have on land, water, air, plants, animals, humans, and the total environment and balance of nature. These laws are in place for the protection of the environment and species. The reports generated from these studies are called Environmental Impact Reports (EIR) and Environmental Impact Statements (EIS). In some cases, it is concluded there will be little or no harmful effects, and the project will be approved. In other cases where they conclude some harm will come from the project, recommendations will be made to mitigate (soften and make less severe) the harmful aspects of the project so that less damage will occur, or agreement will be reached to correct the damage that will occur because of the project. In other cases, projects will not be allowed to go forward because the scientific studies have shown that species and/or the environment will be endangered from the project. Statements of financial liability are not currently required in an EIS or EIR.

73. (D) A pure capitalistic economy, in which financial gain is the primary societal motivator, leads to always taking "just one more." Be careful of the double negative in the question.

74. (C) Industrial wastes of unknown content are often commingled with domestic wastes in sanitary landfills. Groundwater infiltration and contamination of water supplies with toxic chemicals have recently led to more active control of landfills and industrial waste disposal. Careful management of sanitary landfills, such as providing for leachate and runoff treatment as well as daily coverage with topsoil, has stopped most of the problems of open dumping. In many areas, however, space for landfills is running out, and alternatives must be found.

75. (D) Bringing in sand from other locations does not solve the problem. While it's best to build away from the beach, sea walls and breakwaters are only built to protect oceanfront property. When the tide is high, waves hit the sea walls and breakwaters. Often their energy is deflected onto sand in front, contributing to erosion. Over time, this can loosen the wall's footing, causing the sea wall to tilt and break up. Restricting access to the beach is draconian and reduces quality of life.

76. (D) About *one-third* of the United States—*737 million acres*—is forested.

77. (D) The depth at which plankton can exist depends on the penetration of sunlight. Sunlight provides the energy for photosynthesis. Since the phytoplankton is abundant near the surface, the region of sunlight penetration, the consumers are also found in this region. The ocean can be divided into three zones based on the amount of light received: (1) the euphotic zone which is the top layer, nearest the surface down to 600 feet where there is enough light penetrating the water to support photosynthesis, and more than 90% of all marine life lives in this zone; (2) the disphotic zone that begins at about 600 feet and extends to 3000 feet, where only animals that have adapted to little light survive (lantern fish, rattalk fish, hatchet fish, viperfish, and midwater jellyfish); (3) aphotic zone, which constitutes 90% of the ocean. It is entirely dark, the water pressure is extreme, and the temperature is near freezing, and the living things found here live close to fissures or vents in the Earth's crust that give off mineral-rich materials and hydrogen sulfide, which are utilized in specialized metabolic pathways by certain species of bacteria.

78. (A) Nonpoint source (NPS) pollution, unlike pollution from industrial and sewage treatment plants, comes from many diffuse sources. NPS pollution is caused by rainfall or snowmelt moving over and through the ground. As the runoff moves, it picks up and carries away natural and human-made pollutants, finally depositing them into lakes, rivers, wetlands, coastal waters, and even our underground sources of drinking water.

79. (B) Soil erosion can be divided into two very general categories: geological erosion and accelerated erosion. Geological erosion occurs where soil is in its natural environment surrounded by its natural vegetation. This has been taking place naturally for millions of years and has helped create balance in uncultivated soil that enables plant growth. A classical example of the results of geological erosion is the Grand Canyon. Accelerated erosion includes such problems as wind erosion and water erosion caused by rain and poor drainage. Three types of accelerated erosion are

1. Sheet erosion—the uniform removal of soil in thin layers from sloping land;
2. Rill erosion—the most common form of erosion, which occurs when soil is removed by water from little streamlets that run through land with poor surface draining (Rills can often be found in between crop rows.);
3. Gully erosion—gullies are larger than rills and cannot be fixed by tillage. Gully erosion is an advanced stage of rill erosion, just as rills are often the result of sheet erosion.

80. (D) John Muir was America's most famous and influential naturalist and conservationist. As a wilderness explorer, he is renowned for his lone excursions in California's Sierra Nevada, among Alaska's glaciers, and worldwide travels in search of nature's beauty. As a writer, he taught the people of his time *(and ours)* the importance of experiencing and protecting our natural heritage. John Muir was the first president of the Sierra Club. John Muir was instrumental in urging President Theodore Roosevelt to protect America's treasures under the authority of the Antiquities Act of 1906.

81. (C) President Roosevelt recommended that the CCC (Civilian Conservation Corps) operate in cooperation with and under the technical supervision of the War Department, the Department of the Interior, the Department of Agriculture, and the Department of Labor. Other agencies such as the Office of Education and the U.S. Veterans Administration also played a role. Many CCC projects centered on forestry, flood control, prevention of soil erosion, and fighting forest fires.

82. (A) In April 1977, President Jimmy Carter gave the first of a series of major addresses to the nation on energy, which was to become one of the dominant concerns of his administration. Congress approved several of Carter's energy proposals, including the deregulation of natural gas prices, and incentives for such conservation measures as conversion to coal in industry and fuel-saving improvements in the home. In a second major energy program, announced in April 1979, Carter ordered the gradual decontrol of domestic oil prices, but a court later struck down his order. Congress approved Carter's tax on the so-called windfall profits of oil companies but rejected his request for standby authority for gasoline rationing. A third major set of energy measures included government underwriting of the development of synthetic fuels, which Congress also approved. President Jimmy Carter, during the last days of his administration, signed the Alaska Lands Act, which set aside nearly 103 million acres as national parks, national monuments, wild and scenic rivers, wilderness areas, and wildlife refuges. The Act put off the issue of drilling in coastal areas of the Arctic Refuge, however. It left the coastal plain without wilderness designation, but said that Congress would have to vote to approve future oil drilling.

83. (E) The Sagebrush Rebellion is known as organized resistance in the West to federal public land policies. The term can be applied to four different movements that have occurred since the 1880s, but it wasn't actually coined until the 1970s and generally applies to the Wilderness Lands Sagebrush Rebellion. One-third of all national land is administered by the federal government, most of it in the West. When states west of the Mississippi River were admitted to the union, each state had to agree to give up all claims to unsettled lands within its borders. In the late 1970s a movement was started to try to gain the land back from the federal government. The eleven states involved felt the land was rightfully theirs and that they could better utilize the land through extrapolation of resources. Prior to his election, Ronald Reagan espoused support for the rebellion. However, after Reagan won the presidential candidacy, he failed to push the cause. The main cause of the defeat of the Sagebrush Rebellion was the states' inability to establish the basic legal claim that the public domain truly belongs to the states. Therefore, the lands remain federal property for use by citizens.

84. (B) The greatest concentration of endangered and threatened species in the continental United States is in the arid Southwest, especially in California, Nevada, and Arizona. The Hawaii State Department of Land and Natural Resources found that 70 of Hawaii's 140 native birds have already become extinct, and 30 of those still surviving are endangered. Fully 18% of the plants on the federal endangered list, and 40% of the birds, are Hawaiian species. Other "hot spots" of endangered and threatened species in the United States include the Florida peninsula, the eastern Gulf coast, and the southern Appalachians.

85. (D) The Endangered Species Act, signed into law by President Nixon in 1973, is the nation's principal protection for animal and plant species that are on the brink of extinction. Currently about 950 U.S. species are listed as endangered or threatened under the act. The act has many parts, but its three main elements are:

1. Listing a species as endangered or threatened;
2. Developing a recovery plan;
3. Designating critical habitat.

86. (C) From a population of tens of thousands in the 19th century, the bald eagle had declined to only 417 pairs in the lower 48 states by 1963. By 1994, after 20 years of protection through the Endangered Species Act, the population had grown to more than 4000 breeding pairs. As a result of this recovery, the bald eagle's status has now been upgraded from endangered to threatened.

87. (D) Couplings are *one-way* linkages in which one component of a system affects another component. In positive coupling between components A and B, a change in A causes a change of the same sign in B. In negative coupling, a change in A causes a change of the opposing sign in B. Two or more couplings acting in sequence form a closed loop. In a positive feedback loop, a positive or negative change in one of the components of the system is amplified by the sequence of couplings, whereas in a negative feedback loop the change is reduced in amplitude.

88. (C) Twenty years after the first clinical evidence of acquired immunodeficiency syndrome was reported, AIDS has become the most devastating disease humankind has ever faced. Since the epidemic began, more than 60 million people have been infected with the virus. HIV/AIDS is now the leading cause of death in Sub-Saharan Africa. Worldwide, it is the fourth biggest killer. In many parts of the developing world, the majority of new infections occurred in young adults, with young women especially vulnerable. About one-third of those currently living with HIV/AIDS are aged 15 to 24. Most of them do not know they carry the virus. Many millions more know nothing or too little about HIV to protect themselves against it.

89. (A) In 2000, beach pollution prompted at least 11,270 closings and swimming advisories in the United States, twice the number that occurred in 1999. The increase was due largely to improved and increased monitoring and reporting. The most frequent pollution sources are polluted runoff and storm water, which led to more than 4102 closings in 2000. Sewage spills and overflows accounted for more than 2208 closings and advisories.

90. (B) Atmospheric CO_2 levels are rising rapidly—currently they are 25% above where they stood before the Industrial Revolution—and Earth's atmosphere now contains some 200 gigatons (GT) more carbon than it did two centuries ago. Hundreds of billions of tons of carbon as CO_2 is absorbed from or emitted to the atmosphere annually through natural processes. These flows include plant photosynthesis, respiration, and decay as well as the oceanic absorption and release of CO_2. These annual carbon flows, although large, are dwarfed by the various carbon sinks. The atmosphere, for example, contains about 750 billion tons of carbon. An additional 800 billion tons are dissolved in the surface layers of the world's oceans. Some 1300 billion tons of carbon are believed to have accumulated in ground litter and soils. Terrestrial organisms, primarily plants, account for an estimated 550 billion tons of carbon. By far the largest carbon reservoirs (sinks) are the deep oceans and fossil· fuel deposits, which account for some 34,000 and 10,000 billion tons of carbon, respectively.

91. (E)

$$\frac{550 \text{ parts}}{\cancel{\text{million}}} \times \frac{1000 \cancel{\text{million}}}{1 \text{ billion}} = \frac{550{,}000 \text{ parts}}{\text{billion}}$$

92. (A) Marine sediments contain an estimated 4×10^9 Tg (1 Tg = 10^{12} grams with an estimated turnover time of 2×10^8 years). Turnover time is the time it takes for a complete cycle to occur within a system or the ratio of the mass of a reservoir to the rate of its removal from that reservoir. In the context of the climate this can be seen as the total amount of carbon dioxide in the atmosphere and its rate of removal via land and ocean processes. Phosphorus sinks in lesser quantities are atmosphere (0.028 Tg with turnover time of 53 hours); land plants and animals combined (3000 Tg with turnover time of approximately 50 years); ocean plants and animals combined (140 Tg with turnover time of 48 days); ocean surface waters (2700 Tg with turnover time of 2.6 years); and mineable phosphorus from land sources (10,000 Tg).

93. (D) Shield volcanoes are built almost entirely of fluid lava flows. Flow after flow pours out in all directions from a central summit vent, or group of vents, building a broad, gently sloping cone of flat, domed shape. They are built up slowly by the accumulation of thousands of flows of highly fluid basaltic lava that spread widely over great distances, and then cool as thin, gently dipping sheets. Lavas also commonly erupt from vents along fractures (rift zones) that develop on the flanks of the cone. Some of the largest volcanoes in the world are shield volcanoes. In northern California and Oregon, many shield volcanoes have diameters of 3 or 4 miles and heights of 1500 to 2000 feet. The Hawaiian Islands are composed of linear chains of these volcanoes, including Kilauea and Mauna Loa on the island of Hawaii—two of the world's most active volcanoes. The floor of the ocean is more than 15,000 feet deep at the bases of the islands. Mauna Loa, the largest of the shield volcanoes (and also the world's largest active volcano), projects 13,677 feet above sea level and its top is over 28,000 feet above the ocean floor.

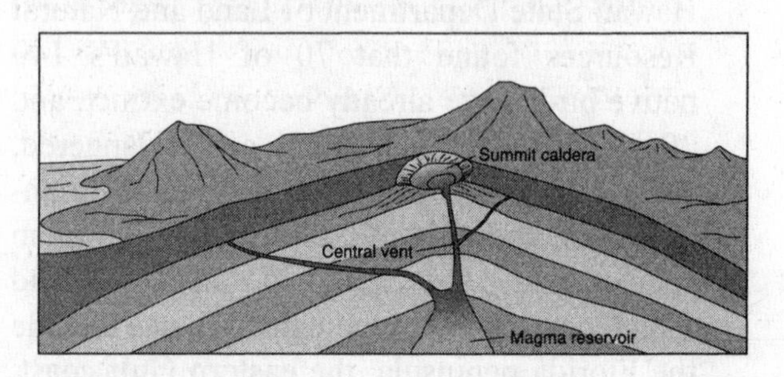

94. (D) In (A), activity is low, and it occurs at very shallow depths. The point is that the lithosphere is very thin and weak at these boundaries, so the strain cannot build up enough to cause large earthquakes. The San Andreas fault is a good example of (B) in which two mature plates scrape by one another. The friction between the plates can be so great that very large strains can build up before they are periodically relieved by large earthquakes. In (C), one plate is thrust or subducted under the other plate so that a deep ocean trench is produced. In (E), shallow earthquakes are associated with high mountain ranges where intense compression is taking place. In (D), the interiors of the plates themselves are largely free of large earthquakes; that is, they are aseismic. There are notable exceptions to this: the 1811 earthquake in New Madrid, Missouri, the largest earthquake ever recorded in American history, is estimated to have had a magnitude of 8.0 or higher and changed the course of the Mississippi River.

95. (C) During the Caboniferous, one of the great die-offs or mass extinctions occurred on Earth. The plant and animal life was killed off in such massive quantities that perhaps 90% of all living beings died roughly simultaneously. Even bacteria must have been affected because the remains were not dispersed by (micro) life but accumulated as asphalt, coal, gas, and oil deposits.

96. (B) Differences in air pressure help cause winds and affect air masses. They are also factors in the formation of storms such as thunderstorms, tornadoes and hurricanes. Differences in air pressure are shown on a weather map with lines called isobars. The weather map illustrates isobars marking areas of high and low pressure. High-pressure areas generally have dry, good weather and areas of low pressure have precipitation. On the weather map, the only area of low pressure is centered in the Midwest. Storms move eastward across the United States.

97. (E) If the soil were unable to form a ball, it would have been sand or gravel. Had they been able to form a ribbon, the soil would have contained clay.

98. (A) SBS (sick building syndrome) and BRI (building-related illness) are associated with acute or immediate health problems. Radon and asbestos cause long-term diseases that occur years after exposure; therefore, they are not considered to be among the causes of sick buildings. Indicators of SBS are: (1) building occupants complain of symptoms associated with acute discomfort (e.g., headache; eye, nose, or throat irritation; dry cough; dry or itchy skin; dizziness and nausea; difficulty in concentrating; fatigue; and sensitivity to odors); (2) the cause of the symptoms is not known; and (3) most of the complainants report relief soon after leaving the building.

99. (D) Tropospheric ozone would be ozone found in the air that we breathe and is generally produced by burning fossil fuels. No relationship has been found that skin cancer is increased by decreases in smog. Stratospheric ozone, otherwise known as the ozone layer, is where dangerous ultraviolet radiation is filtered. The stratosphere extends from about 9 to 31 miles above the Earth's surface. In the stratosphere, temperature increases with altitude, due to the absorption of UV light by oxygen and ozone. This creates a global "inversion layer" that impedes vertical motion into and within the stratosphere—since warmer air lies above colder air, convection is inhibited. The word "stratosphere" is related to the word "stratification," or layering.

100. (B) Frontier ethics would be consistent with "humans are apart from nature." In many developed nations, human values have traditionally expressed a kind of cavalier frontierism that causes humans to pursue their own interests at the expense of the environment, which is further fueled by a capitalistic-based economy that sees national and individual wealth as the purpose of life. Those who subscribe to the greater value assumption believe that humans are superior to all other life forms and that other life forms exist for the use of man.

Explanations for Free-Response Questions

Question 1

(a) 4 points maximum (5 points possible)

Restatement: Total amount of energy (in BTUs) to wash the clothes using the washing machine.

Mass of water: $\frac{20 \text{ gallons of } H_2O}{1} \times \frac{8 \text{ pounds}}{1 \text{ gallon } H_2O} = 160 \text{ pounds}$ (1 point)

Energy to heat water: $\frac{160 \text{ pounds } H_2O}{1} \times \frac{1 \text{ BTU/°F}}{1 \text{ pound } H_2O} \times \frac{(130°F - 60°F)}{1} = 11{,}200 \text{ BTU}$ (1 point)

Energy to run washing machine: $\frac{1500 \text{ watts}}{30 \text{ minutes}} \times \frac{1 \text{ kilowatt}}{1000 \text{ watts}} \times \frac{60 \text{ minutes}}{1 \text{ hour}} = 3 \text{ kWh}$ (1 point)

Energy to run washing machine (BTUs): $\frac{3 \text{ kWh}}{1} \times \frac{3400 \text{ BTU}}{1 \text{ kWh}} = 10{,}200 \text{ BTU}$ (1 point)

Total Energy for Washing Machine: 11,200 BTU + 10,200 BTU = **21,400 BTU** (1 point)

(b) 2 points maximum

Restatement: Total amount of energy (in BTUs) to wash clothes by hand.

$\frac{35 \text{ gallons } H_2O}{1} \times \frac{8 \text{ pounds}}{1 \text{ gallon } H_2O} = 280 \text{ pounds}$ (1 point)

Energy to heat water: $\frac{280 \text{ pounds } H_2O}{1} \times \frac{1 \text{ BTU/°F}}{1 \text{ pound } H_2O} \times \frac{(130°F - 60°F)}{1} = \mathbf{19{,}600 \text{ BTU}}$ (1 point)

Notes:

1. If you do *not* show calculations, no points are awarded.
2. No penalty is assessed if you do not show units. However, you risk setting up the problem incorrectly if you do not show units so that they cancel properly.
3. If your setup is correct but you make an arithmetic error, no penalty is assessed.

(c) 4 points maximum (13 points possible)

Restatement: Environmental impact of washing clothes.

Note: There are no points for just listing ideas. Each idea *must* be explained. Following are ideas that you could use to write your paragraph(s). Take these ideas and create an outline of the order in which you wish to answer them. You do *not* need to use all ideas. Before you begin, decide on the format of how you wish to answer your question (pros vs. cons, chart format with explanations within the chart, compare and contrast, etc.). Each numbered point when explained is worth 1 point.

1. Manufacturing washing machines provides jobs—mining, smelting, design, engineering, manufacturing, transportation, administrative, advertising, sales, etc.
2. Advantages of repairing washing machines rather than discarding them (e.g., discarding washing machine and landfill issues, repairing machines provides jobs)
3. Energy efficiency of different models—labels
4. Using washing machine during "off peak" hours
5. Use of natural resources in manufacturing washing machine
6. Type of energy used to produce electricity—renewable energy sources vs. nonrenewable (e.g., coal, oil, natural gas, solar, hydroelectric)
7. Washing clothes by hand saves (21,400 BTU – 19,600 BTU =) 1800 BTU
8. Washing clothes by machine saves hot water (35 gallons – 20 gallons =) 15 gallons
9. Washing clothes by machine saves time
10. Pollution caused by wastewater. (e.g., phosphates, groundwater contamination)
11. Air pollution caused by heating water.
12. Water treatment options (e.g., using recycled water, using gray water for landscaping use)
13. Choosing a machine that uses the fewest gallons of water per pound of clothes, one that has high, medium, and low water-level controls and an automatic cold rinse cycle. When using the washing machine, following these suggestions to save energy:
    - Presoak heavily soiled clothes.
    - Fill the washer according to load level.
    - Follow the manufacturer's instructions for adding detergent, choosing a biodegradable detergent.
    - Use as low a water temperature for the washing cycle as will give satisfactory cleaning. Cold and medium water temperatures may vary with the season.
    - Always use a cold rinse cycle.
    - Follow maintenance instructions found in the owner's manual.

Question 2

(a)(i) 2 points

Restatement: Biomass as a function of population density.

To determine the optimum population density that produced the greatest biomass would require *dividing the total biomass for each planting by the total amount of seeds planted* per pot; this would provide the average biomass as shown here.

Seeds per Pot	Average Biomass
1	5.0
10	7.0
20	3.8

From the data table, it appears that *10 radish seeds per pot produced the largest average biomass.* When graphed, the results look like this.

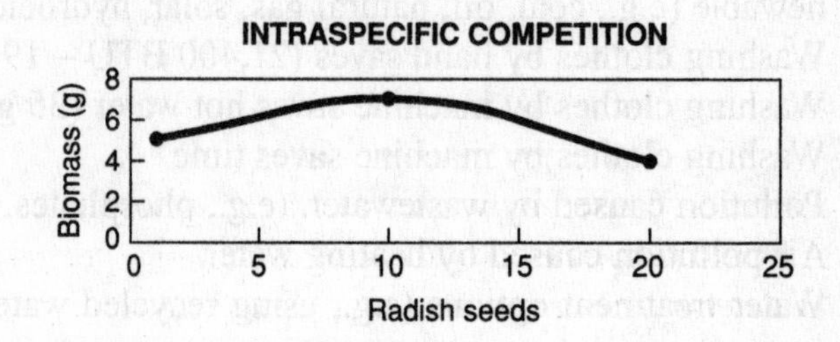

(a) (ii) 2 points

Restatement: Limited resources

The maximum biomass was achieved at 10 seeds per pot. After that, as population density increased, biomass decreased. Factors that may have limited biomass might have included:

1. *Competition for soil nutrients.* As root density increased, less nutrients would have been available for growth.
2. *Competition for light.* As the density increased, less light would have been available for smaller seedlings.
3. *Competition for water.* Given a fixed amount of water provided, as plant density increased, less water would have been available for additional seedlings.

(a)(iii) 2 points

Restatement: Results obtained from the class for Part I in terms of biological laws or principles.

The effect of intraspecific competition in plant populations is usually examined by planting the species over a range of densities. The most common result is that a point is reached where the mean weight per plant decreases as density increases, so that the maximum yield approaches some constant value. This result is called the *law of constant yield.* The maximum plant productivity of a particular environment is called the *carrying capacity.* In this case, the carrying capacity was reached when the biomass reached 7.0 grams for the size of the pot. Environmental variables (nutrients, space, water, and light) that restrict the realized niche are called limiting factors. *Liebig's law of the minimum* states that an organism is most limited by the essential factor that is in least supply. In terms of *survivorship,* the radish seeds represent Type III—characteristics that include large amounts of seeds being produced with few surviving, small size, invaders of disturbed environments, and rapid growth.

(b)(i) 2 points

Restatement: Plant species most affected by competition. 2 POINTS

To determine which plant species was most negatively affected by interspecific competition would require *dividing the total biomass per pot by the total number of seeds per pot to determine average biomass.* The results are provided here.

Seeds per Pot	Average Biomass
1 radish	5.0
1 wheat	3.0
10 radish	5.0
10 wheat	2.5
20 radish	3.8
20 wheat	2.0

From the table, it appears that the *wheat seedlings were negatively impacted by the presence of the radish seeds.* When graphed, the results are as shown here.

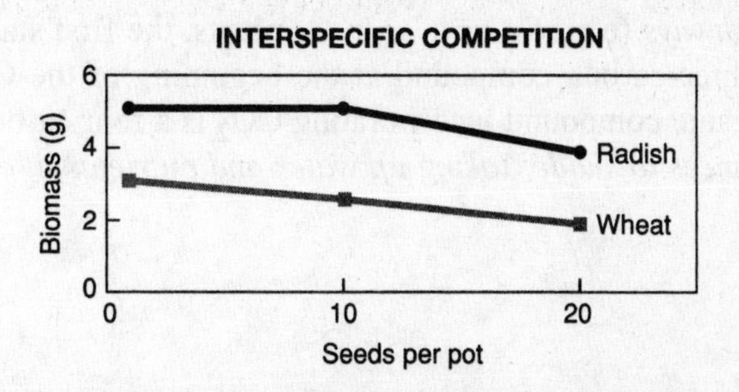

(b)(ii) 2 points maximum

Restatement: Possible reasons why the plant chosen in (b)(i) may have been more successful.

Two different species of plants were grown in the *same* medium. *Nutrient requirements may have been different* between the two species. For example, the amount and ratios of minerals contained in the potting soil may have met the metabolic requirements of the radish plants more than the wheat seedlings. *Water and light availability* (either too much or too little) may also have affected outcome. *Differences in growth patterns* (radish may sprout earlier than wheat) may have existed; in this case, the radish seedlings may have *sprouted earlier* and established a population that was *taking up a fixed amount of nutrients at a faster rate,* or had *established a canopy that decreased the light availability for the wheat seedlings.* Simultaneous to the interspecific competition was intraspecific competition (not only were radish seedlings in competition with wheat seedlings, they were also in competition with other radish seedlings for limited nutrients). More advanced studies and analysis would have employed *Lotka-Volterra equations* and the construction of *zero-isoclines* that would have indicated when growth would have been zero for each species.

(c)(iii) 2 points maximum

Restatement: Results obtained from the class for Part II in terms of biological laws or principles.

In Part II, plant density was held constant in a substitutive or replacement experimental design in which the total density of both species was kept constant while the relative densities were varied (1, 20, and 20). This approach allows investigation of the effects of interspecific competition from the effects of increasing overall plant density on growth or production of a species but requires first quantifying the background effect of intraspecific competition on growth of each species individually. Such an experiment has three possible outcomes: (1) one species prospers at the expense of the other *(competitive exclusion);* (2) one species outperforms the other but only when in higher proportion *(coexistence);* or (3) the two species have no measurable effect on each other *(no competition).* The latter would be the *null hypothesis* (H_o) by which the class would judge whether or not one of the other two scenarios happened. Another scenario, which was not tested for, could have been *instability*—that is, two species are in a constant state of dynamic tension where at one time one species dominates and at other times the other species dominates. Different species of plants are able to coexist within the same biome through *resource partitioning,* which functions through evolution of different *metabolic pathways* (e.g., C_4 *vs.* C_3. In C_3 plants, the first stable compound formed from CO_2 is a three-carbon compound at the beginning of the Calvin cycle. In C_4 plants, the first organic compound incorporating CO_2 is a four carbon compound); having *different tolerances to shade; taking up water and nutrients at different depths;* and so on.

Question 3

(a) 4 points maximum (13 points possible)
List any two characteristics from the column labeled "Sweden" and any two characteristics from the column labeled "Kenya."

Restatement: Given two age structure diagrams, Sweden and Kenya for 2000, compare and contrast the diagrams in terms of population dynamics.

Age structure diagrams are basically divided into three major age categories:

- Prereproductive (0–15 years old)
- Reproductive (16–45 years old)
- Postreproductive (46 years old to death)

Population Characteristic	Affected By	Sweden	Kenya
Birth rate	• Importance of children as a part of the labor force • Urbanization • Cost of raising and educating children • Educational and employment opportunities for women • Infant mortality rate • Average age at marriage • Pensions • Abortions • Birth control • Religious beliefs	• Population has nearly equal proportions of prereproductive and reproductive individuals • Little growth over a long period of time will produce a population with about equal numbers of people in all age groups • Children not required or necessary to support parents • Availability and acceptance of birth control	• Population has pyramid-shaped age structures, with large numbers of prereproductive individuals • Population momentum results from large numbers of prereproductive children becoming reproductive within short period of time • High population rate due to high birth mortality rates • Children viewed as status symbol • Resistance to birth control
Death rate	• Increased food supply • Better nutrition • Improved medical and public health technology • Improvements in sanitation and personal hygiene • Safer water supplies	• Elderly survive longer due to advances in medical technology and availability • Social welfare programs ensure that elderly are taken care of	• Elderly do not survive due to lack of available medical technology • Disease (e.g., malaria, AIDS) and lack of nutritious food decreases life span

(b) 4 points maxium (13 points possible)
List any four characteristics from the column labeled "Affected By."

Restatement: Factors that affect both birth and death rates.

See preceding chart.

(c) 2 points maximum (6 points possible)
List any two of the six methods mentioned.

Restatement: Methods that have been employed by another country to curb population growth.

China: Between 1958 and 1962, an estimated 30 million people died from famine in China. Since then, China has made good progress in trying to feed its people and bring its population growth under control. Much of this reduced population growth was brought about by a drop in the birth rate from 32 to 18 per 1000 between 1972 and 1985. China instituted one of the most rigorous population control programs in the world at an estimated cost of about $1 per person. Some features of the program included:

1. Strongly encouraging couples to postpone marriage;
2. Providing married couples with free access to sterilization, contraceptives, and abortion;
3. Giving couples who sign pledges to have no more than one child economic rewards such as salary bonuses, extra food, larger pensions, better housing, free medical care and school tuition for their child, and preferential treatment in employment when the child grows up;
4. Requiring those who break the pledge to return all benefits;
5. Exerting pressure on women pregnant with a third child to have abortions;
6. Requiring one of the parents in a two-child family to be sterilized.

Question 4

(a) 3 points

Restatement: Definition and three major categories of pesticides.

A pesticide is any substance or mixture of substances intended for preventing, destroying, repelling, or mitigating any pest. Pests can be insects, mice and other animals, unwanted plants (weeds), fungi, or microorganisms like bacteria and viruses. The term "pesticide" also applies to herbicides, fungicides, and various other substances used to control pests. A pesticide is also any substance or mixture of substances intended for use as a plant regulator, defoliant, or desiccant. Types of pesticides include (choose any three):

Algicides Control algae in lakes, canals, swimming pools, water tanks, and other sites.

Antifouling agents Kill or repel organisms that attach to underwater surfaces, such as boat bottoms.

Antimicrobials Kill microorganisms (such as bacteria and viruses).

Attractants Attract pests (e.g., to lure an insect or rodent to a trap); food is not considered a pesticide when used as an attractant.

Biocides Kill microorganisms.

Defoliants Cause leaves or other foliage to drop from a plant, usually to facilitate harvest.

Desiccants Promote drying of living tissues, such as unwanted plant tops.

Disinfectants and sanitizers Kill or inactivate disease-producing microorganisms on inanimate objects.

Fungicides Kill fungi (including blights, mildews, molds, and rusts).

Fumigants Produce gas or vapor intended to destroy pests in buildings or soil.

Herbicides Kill weeds and other plants that grow where they are not wanted.

Insect growth regulators Disrupt the molting, maturity from pupal stage to adult, or other life processes of insects.

Insecticides Kill insects and other arthropods.

Miticides (acaricides) Kill mites that feed on plants and animals.

Microbial pesticides Use microorganisms that kill, inhibit, or out-compete pests, including insects or other microorganisms.

Molluscicides Kill snails and slugs.

Nematicides Kill nematodes (microscopic, wormlike organisms that feed on plant roots).

Ovicides Kill eggs of insects and mites.

Pheromones Use biochemicals to disrupt the mating behavior of insects.

Plant growth regulators Use substances (excluding fertilizers or other plant nutrients) that alter the expected growth, flowering, or reproduction rate of plants.

Repellents Repel pests, including insects (such as mosquitoes) and birds.

Rodenticides Control rats, mice, and other rodents.

(b) 2 points

Restatement: Two positive and two negative effects of pesticide use.

POSITIVE

1. Plants are directly and indirectly mankind's main source of food. They are attacked by tens of thousands of diseases caused by viruses, bacteria, fungi, and other organisms. There are over 30 thousand kinds of weeds competing with crops worldwide; thousands of nematode species reduce crop vigor; and some 10 thousand species of insects devour crops. It is estimated that one-third of the world's food crop is destroyed by these pests annually. *Pesticides increase the world food supply.*
2. There are an estimated 300 million to 500 million cases of malaria per year. The majority of these occur in Africa, while the vast majority of the estimated 1 million annual deaths from the disease occur among children, and mainly among poor African children. Malaria is above all a disease of the poor, impacting at least three times more

greatly on the poor than any other disease. Africa's GDP would be up to $100 billion greater if malaria had been eliminated years ago. Mosquitoes have been estimated to be responsible for half of all human deaths due to transmission of disease. *Pesticides improve human health by destroying disease-carrying organisms* (e.g., West Nile, encephalitis, African sleeping sickness, malaria, Yellow fever, plague).

NEGATIVE

1. If a pesticide is continually applied to a population of the pest species, most susceptible individuals will be killed, leaving only resistant individuals. These resistant individuals breed and multiply, so that eventually a high proportion of the individuals from that pest species are now resistant to the pesticide. We have simply *caused pest populations with a higher tolerance for poisons to survive and breed.* More toxic pesticides in turn are developed and utilized.
2. *Pesticides can accumulate in living organisms.* An example of accumulation is the uptake of a water-insoluble pesticide, such as chlordane, by a creature living in water. Since this pesticide is stored in the organism, the pesticide accumulates and levels increase over time. If this organism is eaten by an organism higher in the food chain that can also store this pesticide, levels can reach higher values in the higher organism than is present in the water in which it lives. Levels in fish, for example, can be tens to hundreds of thousands of times greater than ambient water levels of the same pesticide. This type of accumulation is called bioaccumulation. In this regard, it should be remembered that humans are at the top of the food chain and so may be the most vulnerable to bioaccumulation.

(c) 3 points maximum

Restatement: Three alternatives to the use of pesticides.

Growing pest-resistant crops When landscaping a yard or planning a garden, choose plant varieties that are native to the region and climate. Hearty, native plants resist disease and infestation, and often use less water.

Crop rotation Plant rotation and inter-planting prevent the buildup and spread of pests in one area or among specific plant types.

Beneficial insects and animals Protect and encourage the presence of insect-feeding birds, bats, spiders, praying mantises, ladybugs, predatory mites, parasitic flies, and wasps. Beneficial insect species can often be purchased in volume.

Pheromones Pheromones are chemical signals produced by animals to communicate with others of the same species. In insects, they consist of highly specific perfumes, generally derivatives of natural fatty acids closely related to the aromas of fruit. They are nontoxic. Pheromones may be used to attract insects to traps or to deter insects from laying eggs. However, the most widespread and effective application of pheromones is for mating disruption.

(d) 2 points

Restatement: Describe one U.S. federal law or one international treaty that focuses on the use of pesticides.

The U.S. federal government first regulated pesticides when Congress passed the Insecticide Act of 1910. This law was intended to protect farmers from adulterated or misbranded products. Congress broadened the federal government's control of pesticides by passing the original Federal Insecticide, Fungicide, and Rodenticide Control Act (FIFRA) of 1947. FIFRA required the Department of Agriculture to register all pesticides prior to their introduction in interstate commerce. A 1964 amendment authorized the Secretary of Agriculture to refuse registration to pesticides that were unsafe or ineffective and to remove them from the market. In 1970, Congress transferred the administration of FIFRA to the newly created Environmental Protection Agency (EPA). This was the initiation of a shift in the focus of federal policy from the control of pesticides for reasonably safe use in agricultural production to control of pesticides for reduction of unreasonable risks to man and the environment. This new policy focus was expanded by the passage of the Federal Environmental Pesticide Control Act (FEPCA) of 1972, which amended FIFRA by specifying methods and standards of control in greater detail. In general, there has been a shift toward greater emphasis on minimizing risks associated with toxicity and environmental degradation, and away from pesticide efficacy issues. Under FIFRA, no one may sell, distribute, or use a pesticide unless it is registered by the EPA. Registration includes approval by the EPA of the pesticide's label, which must give detailed instructions for its safe use. The EPA must classify each pesticide as either "general use," "restricted use," or both. "General use" pesticides may be applied by anyone, but "restricted use" pesticides may only be applied by certified applicators or persons working under the direct supervision of a certified applicator. Because there are only limited data for new chemicals, most pesticides are initially classified as restricted use. Applicators are certified by a state if the state operates a certification program approved by the EPA.

Practice Exam 2

Section I—Multiple-Choice Questions

Time: 90 minutes
100 questions
60% of total grade
No calculators allowed.
This section consists of 100 multiple-choice questions. Mark your answers carefully on the answer sheet.

General Instructions

Do not open this booklet until you are told to do so by the proctor.

Be sure to write your answers for Section I on the separate answer sheet. Use the test booklet for your scratch work or notes, but remember that no credit will be given for work, notes, or answers written only in the test booklet. Once you have selected an answer, blacken thoroughly the corresponding circle on the answer sheet. To change an answer, erase your previous mark completely, and then record your new answer. Mark only one answer for each question.

EXAMPLE

The Pacific is

(A) a river.
(B) a lake.
(C) an ocean.
(D) a sea.
(E) a gulf.

SAMPLE ANSWER

To discourage haphazard guessing on this section of the exam, a quarter point is subtracted for every wrong answer, but no points are subtracted if you leave the answer blank. Even so, if you can eliminate one or more of the choices for a question, it may be to your advantage to guess.

Because it is not expected that all test takers will complete this section, do not spend too much time on difficult questions. Answer first the questions you can answer readily, and then, if you have time, return to the difficult questions. Don't get stuck on one question. Work quickly but accurately. Use your time effectively.

Answer Sheet: Practice Exam 2

1. Ⓐ Ⓑ Ⓒ Ⓓ Ⓔ
2. Ⓐ Ⓑ Ⓒ Ⓓ Ⓔ
3. Ⓐ Ⓑ Ⓒ Ⓓ Ⓔ
4. Ⓐ Ⓑ Ⓒ Ⓓ Ⓔ
5. Ⓐ Ⓑ Ⓒ Ⓓ Ⓔ
6. Ⓐ Ⓑ Ⓒ Ⓓ Ⓔ
7. Ⓐ Ⓑ Ⓒ Ⓓ Ⓔ
8. Ⓐ Ⓑ Ⓒ Ⓓ Ⓔ
9. Ⓐ Ⓑ Ⓒ Ⓓ Ⓔ
10. Ⓐ Ⓑ Ⓒ Ⓓ Ⓔ
11. Ⓐ Ⓑ Ⓒ Ⓓ Ⓔ
12. Ⓐ Ⓑ Ⓒ Ⓓ Ⓔ
13. Ⓐ Ⓑ Ⓒ Ⓓ Ⓔ
14. Ⓐ Ⓑ Ⓒ Ⓓ Ⓔ
15. Ⓐ Ⓑ Ⓒ Ⓓ Ⓔ
16. Ⓐ Ⓑ Ⓒ Ⓓ Ⓔ
17. Ⓐ Ⓑ Ⓒ Ⓓ Ⓔ
18. Ⓐ Ⓑ Ⓒ Ⓓ Ⓔ
19. Ⓐ Ⓑ Ⓒ Ⓓ Ⓔ
20. Ⓐ Ⓑ Ⓒ Ⓓ Ⓔ
21. Ⓐ Ⓑ Ⓒ Ⓓ Ⓔ
22. Ⓐ Ⓑ Ⓒ Ⓓ Ⓔ
23. Ⓐ Ⓑ Ⓒ Ⓓ Ⓔ
24. Ⓐ Ⓑ Ⓒ Ⓓ Ⓔ
25. Ⓐ Ⓑ Ⓒ Ⓓ Ⓔ
26. Ⓐ Ⓑ Ⓒ Ⓓ Ⓔ
27. Ⓐ Ⓑ Ⓒ Ⓓ Ⓔ
28. Ⓐ Ⓑ Ⓒ Ⓓ Ⓔ
29. Ⓐ Ⓑ Ⓒ Ⓓ Ⓔ
30. Ⓐ Ⓑ Ⓒ Ⓓ Ⓔ
31. Ⓐ Ⓑ Ⓒ Ⓓ Ⓔ
32. Ⓐ Ⓑ Ⓒ Ⓓ Ⓔ
33. Ⓐ Ⓑ Ⓒ Ⓓ Ⓔ
34. Ⓐ Ⓑ Ⓒ Ⓓ Ⓔ
35. Ⓐ Ⓑ Ⓒ Ⓓ Ⓔ
36. Ⓐ Ⓑ Ⓒ Ⓓ Ⓔ
37. Ⓐ Ⓑ Ⓒ Ⓓ Ⓔ
38. Ⓐ Ⓑ Ⓒ Ⓓ Ⓔ
39. Ⓐ Ⓑ Ⓒ Ⓓ Ⓔ
40. Ⓐ Ⓑ Ⓒ Ⓓ Ⓔ
41. Ⓐ Ⓑ Ⓒ Ⓓ Ⓔ
42. Ⓐ Ⓑ Ⓒ Ⓓ Ⓔ
43. Ⓐ Ⓑ Ⓒ Ⓓ Ⓔ
44. Ⓐ Ⓑ Ⓒ Ⓓ Ⓔ
45. Ⓐ Ⓑ Ⓒ Ⓓ Ⓔ
46. Ⓐ Ⓑ Ⓒ Ⓓ Ⓔ
47. Ⓐ Ⓑ Ⓒ Ⓓ Ⓔ
48. Ⓐ Ⓑ Ⓒ Ⓓ Ⓔ
49. Ⓐ Ⓑ Ⓒ Ⓓ Ⓔ
50. Ⓐ Ⓑ Ⓒ Ⓓ Ⓔ
51. Ⓐ Ⓑ Ⓒ Ⓓ Ⓔ
52. Ⓐ Ⓑ Ⓒ Ⓓ Ⓔ
53. Ⓐ Ⓑ Ⓒ Ⓓ Ⓔ
54. Ⓐ Ⓑ Ⓒ Ⓓ Ⓔ
55. Ⓐ Ⓑ Ⓒ Ⓓ Ⓔ
56. Ⓐ Ⓑ Ⓒ Ⓓ Ⓔ
57. Ⓐ Ⓑ Ⓒ Ⓓ Ⓔ
58. Ⓐ Ⓑ Ⓒ Ⓓ Ⓔ
59. Ⓐ Ⓑ Ⓒ Ⓓ Ⓔ
60. Ⓐ Ⓑ Ⓒ Ⓓ Ⓔ
61. Ⓐ Ⓑ Ⓒ Ⓓ Ⓔ
62. Ⓐ Ⓑ Ⓒ Ⓓ Ⓔ
63. Ⓐ Ⓑ Ⓒ Ⓓ Ⓔ
64. Ⓐ Ⓑ Ⓒ Ⓓ Ⓔ
65. Ⓐ Ⓑ Ⓒ Ⓓ Ⓔ
66. Ⓐ Ⓑ Ⓒ Ⓓ Ⓔ
67. Ⓐ Ⓑ Ⓒ Ⓓ Ⓔ
68. Ⓐ Ⓑ Ⓒ Ⓓ Ⓔ
69. Ⓐ Ⓑ Ⓒ Ⓓ Ⓔ
70. Ⓐ Ⓑ Ⓒ Ⓓ Ⓔ
71. Ⓐ Ⓑ Ⓒ Ⓓ Ⓔ
72. Ⓐ Ⓑ Ⓒ Ⓓ Ⓔ
73. Ⓐ Ⓑ Ⓒ Ⓓ Ⓔ
74. Ⓐ Ⓑ Ⓒ Ⓓ Ⓔ
75. Ⓐ Ⓑ Ⓒ Ⓓ Ⓔ
76. Ⓐ Ⓑ Ⓒ Ⓓ Ⓔ
77. Ⓐ Ⓑ Ⓒ Ⓓ Ⓔ
78. Ⓐ Ⓑ Ⓒ Ⓓ Ⓔ
79. Ⓐ Ⓑ Ⓒ Ⓓ Ⓔ
80. Ⓐ Ⓑ Ⓒ Ⓓ Ⓔ
81. Ⓐ Ⓑ Ⓒ Ⓓ Ⓔ
82. Ⓐ Ⓑ Ⓒ Ⓓ Ⓔ
83. Ⓐ Ⓑ Ⓒ Ⓓ Ⓔ
84. Ⓐ Ⓑ Ⓒ Ⓓ Ⓔ
85. Ⓐ Ⓑ Ⓒ Ⓓ Ⓔ
86. Ⓐ Ⓑ Ⓒ Ⓓ Ⓔ
87. Ⓐ Ⓑ Ⓒ Ⓓ Ⓔ
88. Ⓐ Ⓑ Ⓒ Ⓓ Ⓔ
89. Ⓐ Ⓑ Ⓒ Ⓓ Ⓔ
90. Ⓐ Ⓑ Ⓒ Ⓓ Ⓔ
91. Ⓐ Ⓑ Ⓒ Ⓓ Ⓔ
92. Ⓐ Ⓑ Ⓒ Ⓓ Ⓔ
93. Ⓐ Ⓑ Ⓒ Ⓓ Ⓔ
94. Ⓐ Ⓑ Ⓒ Ⓓ Ⓔ
95. Ⓐ Ⓑ Ⓒ Ⓓ Ⓔ
96. Ⓐ Ⓑ Ⓒ Ⓓ Ⓔ
97. Ⓐ Ⓑ Ⓒ Ⓓ Ⓔ
98. Ⓐ Ⓑ Ⓒ Ⓓ Ⓔ
99. Ⓐ Ⓑ Ⓒ Ⓓ Ⓔ
100. Ⓐ Ⓑ Ⓒ Ⓓ Ⓔ

Directions: Each group of lettered answer choices refers to the numbered statements of questions that immediately follow. For each question or statement, select the one lettered choice that is the best answer and fill in the corresponding circle on the answer sheet.

1. The diagram represents a phylogenetic tree of the evolution of even-toed ungulates.

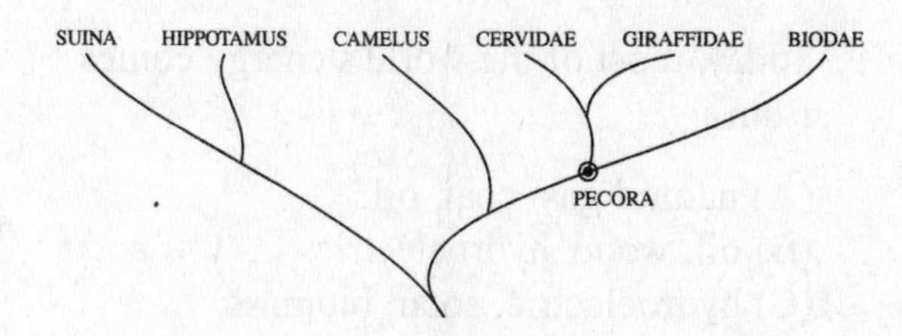

The most likely explanation for the branching pattern seen in the circled region is that

(A) environmental changes caused extinction.
(B) inbreeding led to speciation.
(C) no speciation occurred.
(D) speciation was influenced by environmental change.
(E) only the best-adapted organisms survive from generation to generation.

2. The reason that it might get slightly warmer in the United States before a cold front moves in is that

(A) warm winds move northward along the front.
(B) warm air is pulled from the upper atmosphere.
(C) cold air is sucked backward into the front.
(D) the sun heats the air ahead of the front.
(E) uplifting winds bring warmer air from the surface of the Earth.

3. Droughts, in whatever form, are associated with debilitating negative shocks on national economies. When drought conditions prevail, it is inevitable to see all the following EXCEPT

(A) a decrease in food prices.
(B) a contraction of the GDP.
(C) a decrease in food security.
(D) food imports.
(E) a decrease in the country's balance of trade.

4. The following graphs describe the fates of a hypothetical population of organisms in which there is variation in color. The arrows represent selective pressures. Which graph represents a stabilizing mode of selection?

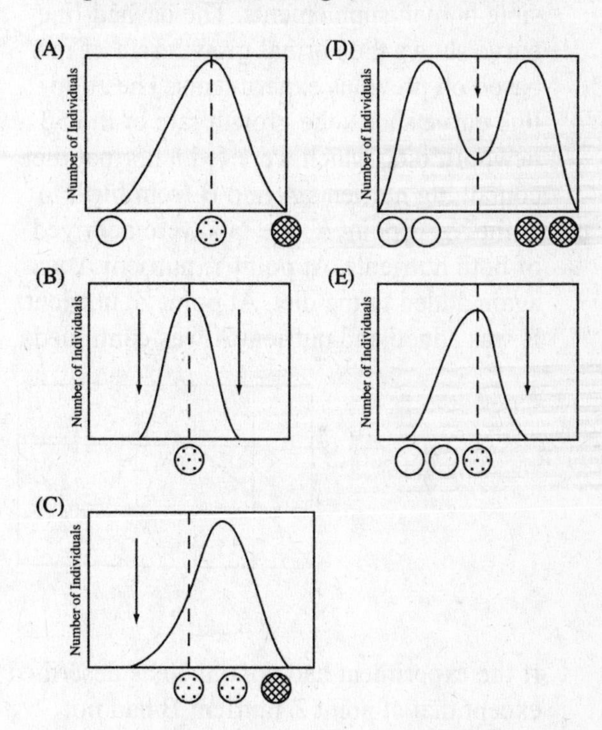

5. On the outskirts of a municipality lies a forest on public property. A person applying the precautionary principle might suggest

(A) clear cutting the forest to provide taxes for the town.
(B) converting the natural woods to tree farms.
(C) harvesting trees at their estimated sustainable yield.
(D) harvesting trees below their estimated sustainable yield.
(E) converting common property to private ownership.

6. The ecological efficiency at each trophic level of a particular ecosystem is 20%. If the green plants of the ecosystem capture 100 units of energy, about _____ units of energy will be available to support herbivores, and about _____ units of energy will be available to support primary carnivores.

(A) 120 . . . 140
(B) 120 . . . 240
(C) 20 . . . 2
(D) 20 . . . 4
(E) 20 . . . 1

7. An experiment with 50 newborn rats was conducted to determine the importance of two nutrients, A and B, in their diets as possible human supplements. The dashed-line curve shows the normal growth rate of rats based on previous experiments. The solid-line curve shows the growth rate of the 50 newborn rats, which were fed a normal diet containing nutrients A and B from birth to point X. At point X, the rats were deprived of both nutrients. At point Y, nutrient A was again added to the diet. At point Z, nutrient B was added and nutrient A was continued.

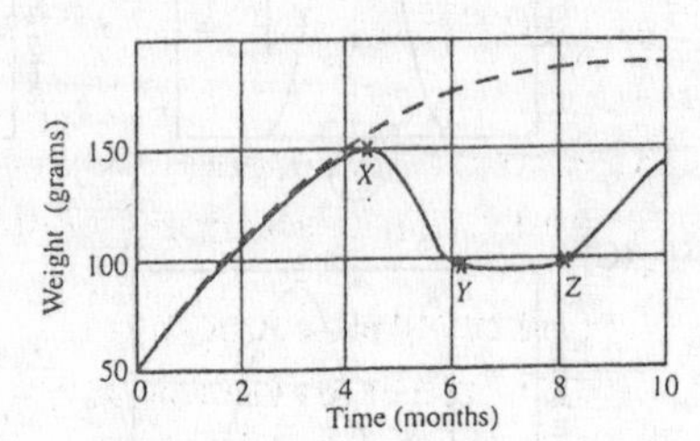

If the experiment had continued as described except that at point Z nutrient B had not been returned to the diet of the 50 rats, it is reasonable to conclude that these rats would most likely have

(A) lived for 4 months and then died.
(B) remained about half the size of normally developed rats.
(C) continued to gain weight, but at a slower rate than the normal rats.
(D) become sexually immature adults.
(E) continuously lost weight.

8. A soil test report recommends 8 lb of "8-0-24" per 1000 square feet. How much phosphorus does it recommend if the area is equal to 10,000 square feet?

(A) 0 pounds
(B) 8 pounds
(C) 24 pounds
(D) 80 pounds
(E) 240 pounds

9. Today, most of the world's energy comes from

(A) natural gas, coal, oil.
(B) oil, wood, hydroelectric.
(C) hydroelectric, solar, biomass.
(D) coal, oil, nuclear.
(E) natural gas, hydroelectric, oil.

10. An AP environmental science class conducted an experiment to illustrate the principles of Thomas Malthus. On day 1, three male and three female fruit flies were placed in a flatbottom flask that contained a cornmeal/banana medium. No other flies were added or removed during the course of this experiment. The students counted the number of flies in the flask each week. The graph shows the results that the class obtained after 55 days.

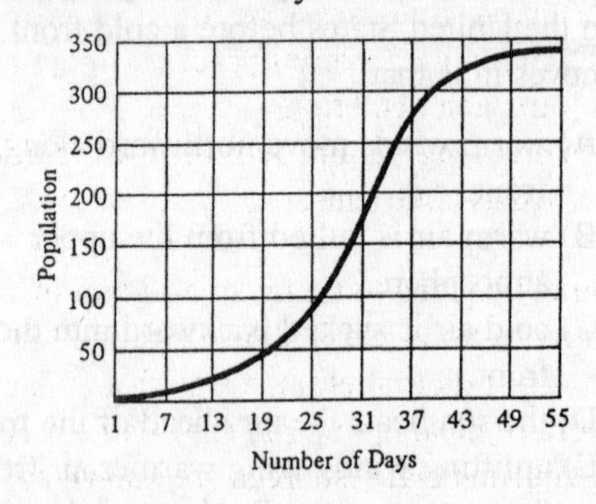

The rate of reproduction is equal to the rate of death on day

(A) 1.
(B) 7.
(C) 25.
(D) 37.
(E) 49.

11. The concept of net primary productivity

(A) is the rate at which producers manufacture chemical energy through photosynthesis.
(B) is the rate at which producers use chemical energy through respiration.
(C) is the rate of photosynthesis plus the rate of respiration.
(D) can be thought of as the basic food source for decomposers in an ecosystem.
(E) is usually reported as the energy output of an area of producers over a given time period.

12. Which of the following is NOT a unit of energy?

(A) Joule
(B) Calorie
(C) Watt
(D) Kilowatt-hour
(E) BTU

13. All of the following are "clean-up" methods for controlling cultural eutrophication except for:

(A) using tertiary waste treatment methods.
(B) treating undesirable plant growth with herbicides and algicides.
(C) harvesting excess weeds.
(D) pumping air through reservoirs to avoid oxygen depletion.
(E) dredging bottom sediments to remove excess nutrients.

14. Suits intended to tie up small nonprofit organizations in frivolous litigation for years to divert their attention away from their real work of cleaning up the environment and to drain financial resources are known as

(A) injunctions.
(B) SLAPPs.
(C) restraining orders.
(D) litigous friviolus.
(E) torts.

15. In the nitrogen cycle, the bacteria that replenish the atmosphere with N_2 are

(A) *Rhizobium.*
(B) nitrifying bacteria.
(C) denitrifying bacteria.
(D) nitrogen-fixing bacteria.
(E) *E. coli.*

16. The interface where plates move apart in opposite directions is known as a:

(A) transform plate boundary
(B) convergent plate boundary
(C) divergent plate boundary
(D) oceanic ridge
(E) trench

17. Which biome, found primarily in the eastern United States, central Europe, and eastern Asia, is home to some of the world's largest cities and has probably endured the impact of humans more than any other biome.

(A) Desert
(B) Coniferous forest
(C) Temperate deciduous forest
(D) Grassland
(E) Chaparral

18. This type of economy exists when supplies and natural resources seem unlimited.

(A) Frontier economy
(B) Free-market economy
(C) Communal resource management system
(D) Command economic system
(E) Natural resource economy

19. A country in Sub-Saharan Africa decided to massively spray the countryside over several months with DDT to rid the country of mosquitoes that were causing large numbers of citizens to contract malaria. Biologists sampled various quadrats for mosquito numbers after the spraying; the results are presented here.

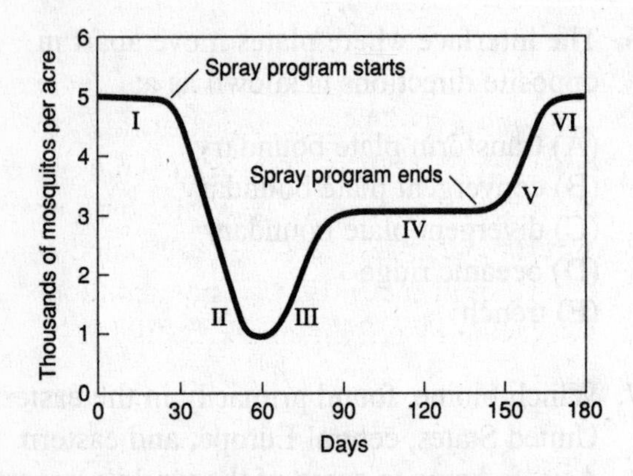

Natural selection is chiefly responsible for the section of the graph labeled
(A) I.
(B) II.
(C) III.
(D) IV.
(E) VI.

20. Cars, trucks, and buses account for approximately ______ of U.S. greenhouse gas emissions.

(A) less than 10%
(B) between 10 and 20%
(C) between 20 and 33%
(D) between 33 and 50%
(E) more than 50%

Questions 21 and 22

Choose the appropriate era to answer Questions 21 and 22.

(A) Cenozoic
(B) Mesozoic
(C) Paleozoic
(D) Precambrian
(E) Archean

21. Humans being evolved.

22. Land plants appeared.

23. It takes on the order of ______ years for adaptive radiations to rebuild biological diversity after a mass extinction.

(A) 100
(B) 100 thousand
(C) 1 million
(D) 10 million
(E) 1 billion

24. On the leeward side of a coastal mountain range, below 4000 feet, which of the following types of trees and/or plants would be most likely to occur?

(A) Epiphytes, lianas, bromeliads
(B) Mangrove, mahogany, cedar
(C) Prickly pear, manzanita, scrub oak
(D) Douglas fir, redwood
(E) Ferns, ivy, rhododendron

25. Primary succession on a sand dune would follow which order?

(A) Grass–shrubs–beech and maple–cottonwoods–pine and black oak
(B) Beech and maple–pine and black oak–cottonwoods–shrubs–grass
(C) Grass–cottonwood–shrubs–beech and maple–pine and black oak
(D) Grass–shrubs–cottonwoods–pine and black oak–beech and maple
(E) Mixture of all species listed would occur simultaneously, with stronger species replacing weaker species.

26. Volcanoes that (1) have a bowl-shaped crater at the summit; (2) only grow to about a thousand feet; (3) are usually made of piles of lava, not ash; (4) blow blobs of lava into the air during the eruption; (5) have small fragments of lava that have fallen around the opening to the volcano; and (6) are typified by Paricutin in Mexico and the middle of Crater Lake in Oregon would be

(A) cinder cones.
(B) shield volcanoes.
(C) composite volcanoes.
(D) mud volcanoes.
(E) spatter cones.

27. Which one of the following best describes what causes sediments to become lithified (turned into rock)?

(A) They are compacted and cemented.
(B) They are subjected to heat and pressure.
(C) They are covered with other sediments.
(D) They come into contact with magma.
(E) Over time, the calcium forms compounds that turn into rock.

28. Which continent has the highest deforestation rate?

(A) Africa
(B) Asia
(C) Europe
(D) South America
(E) Australia/Oceania

29. Excavating and hauling soil offsite to an approved soil disposal/treatment facility would be an example of

(A) sustainability.
(B) remediation.
(C) conservation.
(D) preservation.
(E) mitigation.

30. Which part of the 1982 Law of the Sea Treaty did the United States find contrary to national interest?

(A) Fishing rights
(B) Whaling
(C) Pollution responsibility
(D) Mineral rights
(E) Territorial limits

31. In 1953, Stanley Miller, a graduate student working in the laboratory of Harold Urey, built an apparatus to demonstrate the feasibility of abiotic synthesis. Miller built an apparatus that simulated the presumed conditions of primeval Earth. The conditions included all the following EXCEPT

(A) a gaseous phase containing methane, ammonia, water, and hydrogen gas.
(B) electrical energy provided by spark discharge.
(C) ambient temperature between 0 and 100°C.
(D) sterile conditions (abiotic environment).
(E) primitive nitrifying bacteria.

Questions 32–36

Select from the following locations to answer Questions 32 through 36.

(A) Bhopal, India
(B) Chernobyl, Ukraine
(C) Love Canal, New York
(D) Minamata, Japan
(E) Three Mile Island, Pennsylvania

32. Site of a hazardous chemical dumping ground over which homes and a school were built.

33. Site of mercury poisoning.

34 The most serious commercial nuclear accident in U.S. history.

35. Leakage of poisonous gases from a pesticide manufacturing plant.

36. Nuclear power plant accident that released 30 to 40 times the radiation of the atomic bombs dropped on Hiroshima and Nagasaki.

Questions 37–39

Choose the political party that matches the platform most closely.

(A) Democratic Party
(B) Green Party
(C) Libertarian Party
(D) Natural Law Party
(E) Republican Party

37. "Encourage market-based solutions to environmental problems."

38. "We do not have to choose between economy and environment. Invest in technology and transportation friendly to Earth. We support grants to Amtrak and the states for improving rail routes. We believe in protecting the coasts and the Arctic National Wildlife Refuge from oil and gas drilling."

39. "We believe in cushioning farmers from the instability unique to agriculture and enabling farmers to better pursue financial profitability. We believe that reducing global warming will help the economy. We believe in creating new jobs in energy conservation."

40. The layer of water in a thermally-stratified lake that lies below the thermocline, is noncirculating, and remains perpetually cold is called the

(A) epilimnion.
(B) hyperlimnion.
(C) hypolimnion.
(D) euphotic zone.
(E) benthic zone.

41. India's family planning program has yielded disappointing results for all the following reasons EXCEPT

(A) poor planning and bureaucratic inefficiency.
(B) failure to employ sterilization.
(C) extreme poverty.
(D) a cultural preference for female children.
(E) too little administrative and financial support.

42. Issues of air and water pollution, noise, pesticides, solid waste management, radiation, and hazardous wastes would be the domain of what executive branch office?

(A) Department of the Interior
(B) Department of Health and Human Services
(C) Council on Environmental Quality
(D) Environmental Protection Agency
(E) Office of Management and Budget

43. Which of these threats is NOT one of those that must be decreased to help the survival of the approximately 600 mountain gorillas left in the wild?

(A) Habitat loss
(B) Poaching
(C) War
(D) Exotic species intrusions
(E) Disease

44. The largest user of freshwater worldwide is

(A) mining.
(B) irrigation.
(C) industry.
(D) home use.
(E) production of electrical power.

45. Choose the statement that is FALSE.

(A) Domestic fruits and vegetables are more likely to have pesticide residues than imported ones.
(B) Cancer is not the primary risk from chronic, long-term exposure to pesticides.
(C) When the EPA looks at a pesticide to decide whether to register it for use in the United States, its primary concern is to ensure that there are no significant human health or environmental risks presented by the chemical.
(D) The federal government does not prohibit the use of pesticides known to cause cancer.
(E) Washing and peeling fruits and vegetables does not remove all or most pesticide residues.

46. What happens in the market for airline travel when the price of traveling by rail decreases?

(A) The demand curve shifts left.
(B) The demand curve shifts right.
(C) The supply curve shifts left.
(D) The supply curve shifts right.
(E) The supply curve intersects with the demand curve at the equilibrium price.

47. Taking into account only price, supply, and demand, if plotted on a graph, supply and demand curves

(A) are parallel lines.
(B) are parallel lines running horizontally.
(C) never intersect.
(D) can run in any direction.
(E) intersect at a point called market equilibrium.

48. The power in the wind increases as the cube of the wind speed. About how much more power is produced by a typical wind turbine at 15 mph than at 12 mph?

(A) About the same
(B) About 10% more
(C) About 25% more
(D) About twice as much
(E) About 9 times as much

Questions 49 and 50

For Questions 49 and 50, choose the letter of the item that is MOST closely related to the numbered statement.

(A) Allopatric speciation
(B) Sympatric speciation
(C) Punctuated equilibrium
(D) Parapatric speciation
(E) Divergent evolution

49. Darwin finches found in the Galapagos Islands.

50. The Mexican ground squirrel (*Spermophilus mexicanus*) is found east of a rocky slope in well-drained, generally nonrocky soils especially in open terrace habitats without significant wood vegetation. West of the slope, the rock squirrel (*Spermophilus variegatus*) is found in canyons and rocky uplands.

51. Which one of the following proposals would NOT increase the sustainability of ocean fisheries management?

(A) Establish fishing quotas based on past harvests
(B) Setting quotas for fisheries well below their estimated maximum sustainable yields
(C) Sharply reducing fishing subsidies
(D) Shifting the burden of proof to the fishing industry to show that their operations are sustainable
(E) Strengthening integrated coastal management programs

52. The amount of cultivated land of the world's land resources used to produce over 95% of the world's food is

(A) about 75%.
(B) between 50 and 75%.
(C) about 50%.
(D) between 15 and 50%.
(E) less than 15%.

53. The type of succession that begins in an area where the natural community has been disturbed, removed, or destroyed, but in which the bottom soil or sediment remains is known as

(A) allogenic.
(B) autogenic.
(C) primary.
(D) secondary.
(E) progressive.

54. In general, parasites tend to

(A) become more virulent as they live within the host.
(B) completely destroy the host.
(C) become deactivated as they live within the host.
(D) be only mildly pathogenic.
(E) require large amounts of oxygen.

55. Within a planetary management worldview, which of the following would NOT belong?

(A) "No-Problem" School
(B) Free-Market School
(C) Spaceship-Earth worldview
(D) "Biocentric worldview"
(E) Stewardship School

56. Humans having a finite capacity to manage nature would be consistent with what principle?

(A) Precautionary principle
(B) Integrative principle
(C) Ecological design principle
(D) Humility principle
(E) Environmental justice principle

57. The circulation of air in Hadley cells results in

(A) low pressure and rainfall at the equator.
(B) high pressure and rainfall at the equator.
(C) low pressure and dry conditions at about 30 degrees north and south of the equator.
(D) high pressure and wet conditions at about 30 degrees north and south of the equator.
(E) both A and C.

58. All the following are characteristics of K-strategists EXCEPT

(A) mature slowly.
(B) low juvenile mortality rate.
(C) niche generalists.
(D) Type I or II survivorship curve.
(E) intraspecific competition due to density-dependent limiting factors.

59. First levels of defensive behaviors, used by both predators and prey, to avoid detection would include all of the following EXCEPT

(A) camouflage.
(B) predator swamping.
(C) counter-shading.
(D) Batesian mimicry.
(E) masquerading.

60. In the general pattern of the ocean's currents, the direction of the currents nearest the equator move from

(A) east to west.
(B) west to east.
(C) north to south.
(D) south to north.
(E) southwest to northeast.

61. Which of the following is NOT an example of a chronic condition?

(A) Asthma
(B) Measles
(C) Diabetes
(D) Cancer
(E) Malnutrition

62. Which mobile source pollutant cannot be currently controlled by emission control technology?

(A) Ozone-forming hydrocarbons
(B) Carbon monoxide
(C) Carbon dioxide
(D) Air toxins
(E) Particulate matter

63. Which of the following are examples of trace elements necessary in the human diet?

(A) Ca, Mg, and Na
(B) Al and Fe
(C) I, Cu, and Zn
(D) S, N, and P
(E) C, H, and O

64. Which of the following is the best example of an r-selected species?

(A) Dog
(B) Whale
(C) Condor
(D) Tree
(E) Mouse

65. Which of the following activities causes the most severe impact to backcountry wilderness?

(A) Fishing
(B) Hiking off trails
(C) Littering
(D) Building a fire
(E) Hunting

66. Your are going to buy a soda. You see a vending machine that has sodas in aluminum cans, steel cans, plastic bottles, and glass bottles. Which container has the least environmental impact when recycled?

(A) Aluminum cans
(B) Steel cans
(C) Plastic bottles
(D) Glass bottles
(E) All have the same negative environmental impact.

67. The principle stating that, in general, other things being equal, the higher the price of a good, the greater the quantity of that good sellers will offer for sale over a given period is known as

(A) law of supply.
(B) law of demand.
(C) law of supply and demand.
(D) open access.
(E) neoclassical economics.

68. You are a member of a grassroots environmental organization that has successfully lobbied your U.S. congressperson to co-sponsor a bill to create a small wildlife sanctuary for migratory birds on federal land. After your bill was introduced to the Senate, which committee would it likely be referred to for hearings?

(A) Committee on Agriculture
(B) Committee on Energy and Commerce
(C) Committee on Resources
(D) Committee on Wildlife Conservation
(E) Committee on Preservation

69. In 1995, the population of a small island in Malaysia was 40,000. The birth rate was measured at 35 per 1000 population per year; the death rate was measured at 10 per 1000 population per year. Immigration was measured at 100 per year while emigration was measured at 50 per year. How many people would be on the island after one year?

(A) 39,100
(B) 40,000
(C) 41,050
(D) 42,150
(E) 44,500

70. The best time to catch a predator fish, like a pike or walleye, would be

(A) at dusk, when schools of smaller fish start to break apart.
(B) during midday, when prey moves into shallow, protected waters.
(C) anytime, since lures may look like prey to the predator.
(D) dependant on the amount of vegetation available.
(E) early in the morning, when schools of smaller fish are active.

71. To evaluate the total impact of a disease by combining premature deaths and disability into data rather than to just quantify the effects or mortality is to use a technique called a

(A) risk extrapolation model.
(B) comparative risk analysis.
(C) epidemiology.
(D) risk-based targeting.
(E) disability-adjusted life year.

72. Clumped spacing patterns of plants are most often associated with

(A) pockets of resources within the population's range.
(B) shading and competition for water and minerals.
(C) the random distribution of seeds.
(D) antagonistic chemicals secreted by plants that inhibit germination and growth of nearby individuals.
(E) coincidence.

73. Most of the freshwater found on the Earth is in the form of

(A) ice (glaciers, polar ice caps, etc.).
(B) rivers and lakes.
(C) groundwater.
(D) atmosphere (rain, fog, clouds, vapor, etc.).
(E) oceans.

74. Among adults, which of the following represents the most preventable cause of death?

(A) Cardiovascular disease
(B) AIDS
(C) Alcoholism
(D) Use of tobacco
(E) Traffic accidents

75. Which minerals are not mined in the United States?

(A) Coal and lead
(B) Oil and iron ore
(C) Tin and asbestos
(D) Molybdenum and gold
(E) Manganese and silver

76. Which U.S. president is responsible for creating the National Park System?

(A) Woodrow Wilson
(B) Theodore Roosevelt
(C) Franklin D. Roosevelt
(D) Calvin Coolidge
(E) Herbert Hoover

77. An AP Environmental Science class visited a stream to study biodiversity. They spent the day prior to visiting the stream learning to identify macroinvertebrates by using a dichotomous key. On the day they visited the stream, one group collected the following macroinvertebrates in the following order:

backswimmer–backswimmer–backswimmer–damselfly–damselfly–midge–mosquito larvae–mosquito larvae–mayfly–mayfly–mayfly–mayfly–damselfly–backswimmer

What is the sequential comparison index?

(A) 0.1
(B) 0.25
(C) 0.35
(D) 0.50
(E) 1.0

78. "We can burn coal to produce electricity to operate a refrigerator" is an example of the __________ and "If we burn coal to produce electricity to operate a refrigerator, we lose a great deal of energy in the form of heat" is an example of the __________.

(A) first law of thermodynamics; first law of thermodynamics
(B) second law of thermodynamics; first law of thermodynamics
(C) first law of thermodynamics; second law of thermodynamics
(D) first law of thermodynamics; third law of thermodynamics
(E) third law of thermodynamics; first law of thermodynamics

79. Refer to the following age-structure diagram.

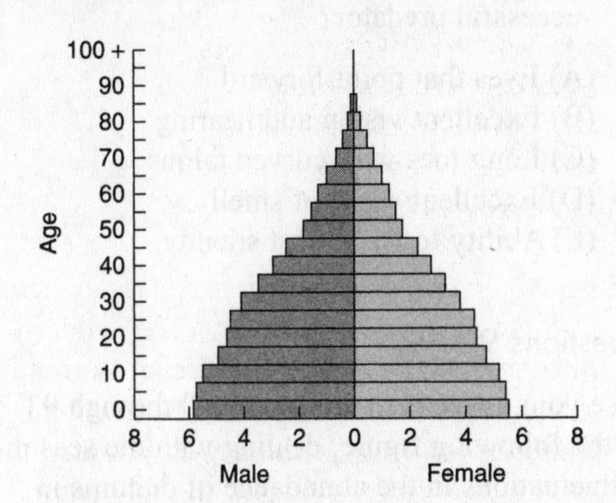

This age-structure diagram would be typical for what country?

(A) United States
(B) Canada
(C) Mexico
(D) Germany
(E) Japan

80. The largest earthquake in the 20th century occurred

(A) off the coast of Alaska.
(B) off the coast of Japan.
(C) off the coast of South America.
(D) off the coast of California.
(E) in Missouri.

81. Butterflies and moths both feed on flowers. Butterflies feed during the day, and moths feed at night. This is an example of

(A) r strategy.
(B) K strategy.
(C) resource partitioning.
(D) commensalism.
(E) mutualism.

82. A point beyond which the application of additional resources yields less than proportional increases in output is known as

(A) external cost.
(B) economic disincentive.
(C) diminishing return.
(D) discount rate.
(E) indirect cost.

83. An AP Environmental Science class was investigating the estimation of population size through the Lincoln-Peterson capture–recapture technique. To measure the population density of monarch butterflies occupying a particular park, 100 butterflies were captured, marked with a small dot on a wing, and then released. The next day, another 100 butterflies were captured, including the recapture of 20 marked butterflies. One would correctly estimate the population to be

(A) 100.
(B) 200.
(C) 500.
(D) 2000.
(E) 20,000.

84. In the eastern United States, water use law is based on the legal principle of

(A) prior appropriation.
(B) private property rights.
(C) common property rights.
(D) public property rights.
(E) riparian rights.

85. Approximately how much of the Earth's land surface is considered "seriously eroded"?

(A) Less than 10%
(B) About 25%
(C) Between 25 and 50%
(D) Between 50 and 75%
(E) More than 75%

86. The concentration of which gas can be reduced by preventing forest depletion?

(A) Carbon dioxide
(B) Nitrous oxide
(C) Oxygen
(D) Methane
(E) CFCs

87. Environmental lawsuits are limited in results for all the following reasons EXCEPT

(A) the plaintiff must be directly harmed by an action, or lack thereof.
(B) corporations may deduct their legal expenses from their federal taxes, whereas public-interest lawyers can usually not recover any fees or tax write-offs.
(C) it is usually not difficult to prove that the defendant is liable and responsible for harmful environmental action.
(D) the courts may take years to come to a decision and appeals may be submitted to higher courts.
(E) bringing a suit is expensive.

88. After a recent storm, an AP Environmental Science class took a field trip to a storm drain outlet entering the Pacific Ocean at Ballona Creek, in southern California. They carefully collected water samples and brought the samples back to the lab. Alpha group took 100.00 mL of the collected water and filtered it through a 1.2-μm Millipore glass fiber filter. The filter was carefully transferred to a premassed stainless steel crucible and placed in a 105°C oven for 24 hours. The following is a hypothetical set of data from Alpha group:

Weight of crucible + Filter + Residue after 24 hr at 105°C = 100.00 g

Weight of crucible + Filter = 80.00 g

What is the *total suspended solid* amount measured in $mg \cdot L^{-1}$?

(A) (80.00 / 100.00) × 100%

(B) 100.00 – 80.00

(C) $\dfrac{(100.00 - 80.00) \times 1000 \times 1000}{100.00}$

(D) $\dfrac{(100.00 - 80.00) \times 1000}{100}$

(E) $\dfrac{(100.00 + 80.00) \times 100}{1000}$

89. Which of the following is NOT an adaptation of the barn owl that allows it to be a successful predator?

(A) Eyes that point forward
(B) Excellent vision and hearing
(C) Long toes with curved talons
(D) Excellent sense of smell
(E) Ability to fly almost silently

Questions 90–93

Base your answers to Questions 90 through 93 on the following figure, dealing with the seasonal fluctuations in the abundance of diatoms in the North Atlantic (solid black line—diatoms; dotted line—light intensity; dashed line—nitrates and phosphates):

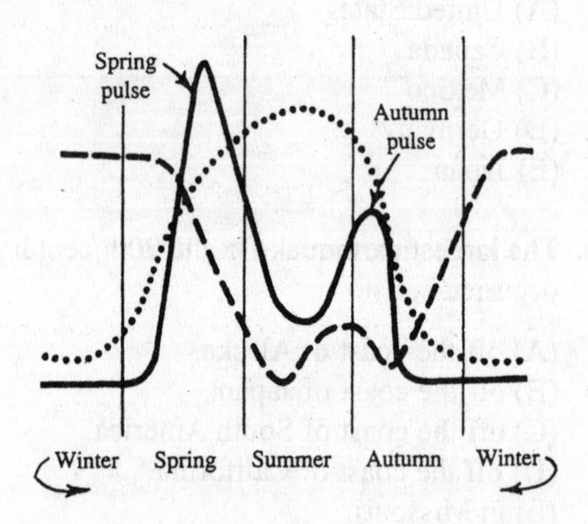

90. Which factor probably contributes to the summer decline of diatoms?

(A) Decreased concentration of nitrates and phosphates
(B) Increased concentration of nitrates and phosphates
(C) Increased intensity of light
(D) Decreased intensity of light
(E) Increase of temperature

91. Which is probably the principal source of nitrates and phosphates?

(A) The water cycle
(B) Nitrogen fixation of lighting
(C) Bacterial decay on the ocean bottom
(D) Changes in environmental temperature
(E) Changes in light intensity

92. A probable reason why the autumn pulse is not as great as the spring pulse is that

(A) diatoms undergo metamorphosis in the autumn.
(B) temperature and light intensity decrease in the autumn.
(C) carnivorous animals increase in the autumn.
(D) the diatoms of the spring have used most of the proteins from the environment.
(E) bacteria of decay increase in the autumn.

93. The low level of the diatom population in winter results partially from the fact that

(A) low temperatures slow down metabolism.
(B) photosynthesis occurs only in summer.
(C) diatoms live in water.
(D) diatoms migrate to warm climates during the winter.
(E) there is a decrease in available nitrates and phosphates in winter.

94. What pulls warm Atlantic water north in the summer?

(A) La Niña
(B) Convection
(C) El Niño
(D) Coriolis effect
(E) Gulf Stream

95. Most marine organisms are concentrated in the

(A) intertidal zone.
(B) pelagic zone.
(C) profundal zone.
(D) neretic zone.
(E) benthic zone.

96. According to the International Union for the Conservation of Nature and Natural Resources (IUCN), the protected area that should be allowed the greatest degree of human impact would be

(A) Lava Beds National Monument.
(B) Upper Newport Bay Ecological Reserve and Regional Park.
(C) Soldier Creek Wilderness Preserve in the Nebraska National Forest.
(D) scenic landscape along the northern California coast.
(E) Joshua Creek Ecological Preserve on the Monterey Peninsula in California.

97. Which of the following indoor pollutants would NOT be contributed by carpeting?

(A) Formaldehyde
(B) Styrene
(C) Mold
(D) Methylene chloride
(E) Mites

98. A woman reading a book of poems turned a page causing a tiny air current to slip out the open window, nudging a passing breeze. That breeze in turn nudged another breeze and eventually was the cause of a tornado in the next state. This series of "conceptual" events is referred to as

(A) the Gaia hypothesis.
(B) the "Butterfly effect."
(C) chaos.
(D) law of common cause.
(E) natural sequencing.

99. All the following are correct statements about the regulation of populations EXCEPT

(A) a logistic equation reflects the effect of density-dependent factors, which can ultimately stabilize population around the carrying capacity.
(B) density-independent factors have a greater effect as a population's density increases.
(C) high densities in a population may cause physiological changes that inhibit reproduction.
(D) because of the overlapping nature of population-regulating factors, it is often difficult to determine their cause-and-effect relationships precisely.
(E) occurrence of population cycles in some populations may be the result of crowding or lag times in response to density-dependent factors.

100. Which of the following energy sources has the lowest quality?

(A) High-velocity water flow
(B) Fuelwood
(C) Food
(D) Dispersed geothermal energy
(E) Saudi Arabian oil deposits

Section II—Free-Response Questions

Time: 90 minutes
Four questions
40% of total grade
No calculators allowed.

Directions: Answer all four questions, which are weighted equally; the suggested time is about 22 minutes for answering each question. Write all your answers on the pages following the questions in the pink booklet. Where calculations are required, clearly show how you arrived at your answer. Where explanation or discussion is required, support your answers with relevant information and/or specific examples.

1. A family is building a new home in Buffalo, New York. Buffalo experiences severe winters. Assume the following.

- The house has 4000 square feet.
- 100,000 BTUs of heat per square foot are required to heat the house for the winter.
- Natural gas sells for $5.00 per thousand cubic feet.
- One cubic foot of natural gas supplies 1000 BTUs of heat energy.
- 1 kilowatt-hour of electricity supplies 10,000 BTUs of heat energy.
- Electricity costs $50 per 500 kWh.

(a) Calculate the following, showing all the steps of your calculations, including units.

(i) The number of cubic feet of natural gas required to heat the house for the winter.
(ii) The cost of heating the house using natural gas.
(iii) The cost of heating the house using electricity.

(b) The homeowners are discussing with their architect the possibility of using either active or passive solar design to reduce their heating and/or cooling costs. Compare these two techniques.

(c) The homeowners wish to incorporate "green design." Discuss five techniques that the homeowners could adopt to make their home "green." ("Green" does not refer to the color of the house!)

2. Charts showing global temperature changes and carbon dioxide emissions follow.

GLOBAL TEMPERATURE CHANGES (1880-2000)

Departure from long-term mean (°F)

1.4
1.2
1.0
0.8
0.6
0.4
0.2
0
−0.2
−0.4
−0.6
−0.8

1880 1890 1900 1910 1920 1930 1940 1950 1960 1970 1980 1990 2000

Year

Source: U.S. National Climatic Data Center, 2001

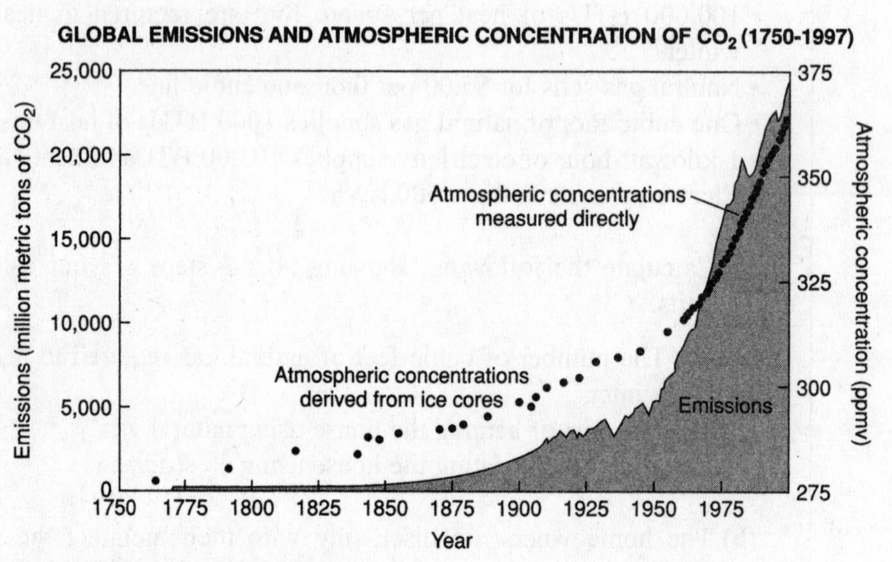

Source: Carbon Dioxide Information Analysis Center, 2001

(a) Describe global warming?
(b) What factors may be increasing global temperature?
(c) What are environmental consequences for global warming?

3. The oxygen content of water is a limiting abiotic environmental factor that determines the biodiversity of an aquatic environment. An AP Environmental Science class visited a stream near their high school and measured several indicators of water quality.

(a) Describe a technique of determining oxygen content of water and what it means.
(b) Briefly describe three other tests that the students could have conducted to determine the water quality of the stream.
(c) Diagram a typical food web based on a freshwater stream ecosystem.

4. In 1900, the average American could expect to live just 47 years. Today, the average American lives 75 years—an increase of nearly three decades of life in just a century's time.

(a) Explain two possible reasons that have occurred within the last 100 years that can account for the increase in human lifespan.
(b) What are the greatest current and projected risks to human health?
(c) Explain the concept and components of "risk analysis." How would information gained from studies of risk analysis increase the human life span?

Answers and Explanations for Practice Exam 2

Answer Key for Multiple-Choice Questions

1. **D**
2. **A**
3. **A**
4. **B**
5. **D**
6. **D**
7. **B**
8. **A**
9. **A**
10. **E**
11. **E**
12. **C**
13. **A**
14. **B**
15. **C**
16. **C**
17. **C**
18. **A**
19. **C**
20. **C**
21. **A**
22. **C**
23. **D**
24. **C**
25. **D**
26. **A**
27. **A**
28. **A**
29. **B**
30. **E**
31. **E**
32. **C**
33. **D**
34. **E**
35. **A**
36. **B**
37. **E**
38. **A**
39. **D**
40. **C**
41. **D**
42. **D**
43. **D**
44. **B**
45. **C**
46. **A**
47. **E**
48. **D**
49. **A**
50. **D**
51. **A**
52. **E**
53. **D**
54. **D**
55. **D**
56. **D**
57. **A**
58. **C**
59. **B**
60. **A**
61. **B**
62. **C**
63. **C**
64. **E**
65. **D**
66. **A**
67. **A**
68. **C**
69. **C**
70. **A**
71. **E**
72. **A**
73. **C**
74. **D**
75. **C**
76. **A**
77. **D**
78. **C**
79. **C**
80. **C**
81. **C**
82. **C**
83. **C**
84. **E**
85. **D**
86. **A**
87. **C**
88. **C**
89. **D**
90. **A**
91. **C**
92. **B**
93. **A**
94. **E**
95. **B**
96. **D**
97. **B**
98. **B**
99. **B**
100. **D**

Explanations for Multiple-Choice Questions

1. (D) Speciation occurs when environmental factors change the composition of the gene pool.

2. (A) Some cold fronts are preceded by warm winds from the south. These hot breezes move from the tropics in a northern direction. If there are few clouds and strong sun, these breezes can send temperatures up several degrees higher than normal.

3. (A) Food prices would increase during a drought due to scarcity.

4. (B) Stabilizing selection operates on the extremes. Stabilizing selection reduces phenotypic variations but maintains the status quo in the gene pool. An example of stabilizing selection found in the human population is phenylketonuria, a genetic disorder that inhibits the proper processing of dietary protein and, without proper dietary control, causes brain damage. In directional selection, a population may find itself in circumstances where individuals occupying one extreme in the range of phenotypes are favored over the others. In other circumstances, individuals at both extremes of a range of phenotypes are favored over those in the middle. This is called disruptive selection.

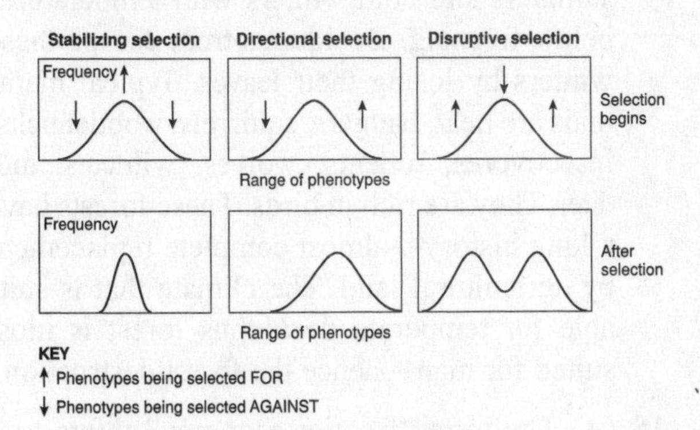

5. (D) The precautionary principle emanates from the wish to protect man and nature, even if there is no certain scientific evidence of the extent and cause of the environmental problem. The precautionary principle focuses on the degree of certainty of knowledge needed before politicians and authorities can decide to initiate action toward possible environmental problems. Scientific uncertainty may prevail for a number of years, whether the problem(s) carries a significant risk of environmental damage, or whether it is of a limited nature. The question is therefore do we act immediately upon suspicion and indication of risks, or do we wait until we have 100% scientific certainty, which may be too late.

6. (D) 100 units of energy are found in the green plants. $100 \times 0.20 = 20$ units available for herbivores. Of the 20 that is now available, $20 \times 0.20 = 4$ that is left for the primary carnivores.

7. (B) The rats would most likely have remained about half the size of normally developed rats if nutrient B had not been returned at point Z. The rats would continue to weigh less than 100 grams.

8. (A) The three numbers on the side of the fertilizer bag refer to the amount of nitrogen, phosphorus, and potassium that the fertilizer contains. The rest of the bag contains minor nutrients and filler material. The most accurate method of determining crop nutrient needs is through tissue analysis. Tissue analysis indicates what the plant is taking up, and along with a soil test can help determine what nutrients need to be added to the soil. Since the number representing phosphorus is 0, no phosphorus is required.

9. (A) Notice that all three forms of energy are nonrenewable sources derived from fossil fuels. Oil is the primary energy source for the world today with the United States at 40% dependency on this single source. Coal and natural gas are each around 22% dependency in the United States and the world.

10. (E) After the 49th day, there is no increase or decrease in the population. The graph levels off.

11. (E) Primary productivity is the amount of biomass produced through photosynthesis per unit area per time by plants and is expressed in units of energy (e.g., joules $\cdot$ m^{-2} $\cdot$ day^{-1})

or in units of dry organic matter (e.g., kg · m^{-2} · year $^{-1}$). Primary production amounts to over 240 billion metric tons of dry plant biomass per year. Gross primary productivity is the total energy fixed by plants through photosynthesis. Because all the energy fixed by the plant is converted into sugar, it is theoretically possible to determine a plant's energy uptake by measuring the amount of sugar produced. A portion of the energy of gross primary productivity is used by plants for respiration. Respiration provides a plant with the energy needed for various activities. Subtracting respiration from gross primary production gives net primary productivity, which represents the rate of production of biomass that is available for consumption by heterotrophic organisms (bacteria, fungi, and animals).

12. (C) A watt is a unit of power.

13. (A) Most of the eutrophication occurring today is caused by humans. Nitrates and phosphates come from several sources: human wastes, animal wastes, industrial wastes, and human disturbance of the land and its vegetation. Sewage from wastewater treatment plants and septic tanks is one source of phosphorus in rivers. Animal waste containing phosphorus sometimes finds its way into rivers and lakes in the runoff from feedlots and barnyards. Soil erosion can also contribute phosphorus to rivers. The removal of natural vegetation for farming or construction exposes soil to the eroding action of rain and melting snow. Soil particles washed into waterways contribute more phosphorus and fertilizers used for crops, lawns, and home gardens usually contain phosphorus. Draining swamps and marshes for farmland releases phosphorus that has remained dormant in years of accumulated organic deposits. Drained wetlands no longer function as filters of silt and phosphorus, allowing more runoff and phosphorus to enter waterways. The simplest, and least effective, method of treatment is primary sewage treatment that allows undissolved solids in raw sewage to settle out of suspension, forming sludge. Such primary treatment removes only one-third of the BOD and virtually none of the dissolved minerals. In secondary treatment, the effluent is brought in contact with oxygen and aerobic microorganisms that break down much of the organic matter to harmless substances such as carbon dioxide. Tertiary waste treatment, using activated carbon filters, is a final step in sewage treatment that targets further removal of fine particles, dissolved organics, and dissolved inorganic materials, especially algal nutrients such as phosphorus.

14. (B) SLAPP = Strategic Lawsuits Against Public Participation. Typical causes of SLAPP suits are trespass, nuisance, harassment, slander, libel, defamation, conspiracy, interference with prospective economic advantage, and interference with contract.

15. (C) Denitrifying bacteria are any strain of bacteria able to utilize nitrate or nitrite in an energy-yielding metabolic sequence that eventually produces nitrogen gas. When colonies of these bacteria occur on croplands, they may deplete the soil nutrients, and make it difficult for crops to grow.

16. (C) A divergent plate boundary is the interface where plates move apart in opposite directions. Rift valleys form between them. Also known as geologic spreading center.

17. (C) Temperate deciduous forests have warm summers and cold winters with temperatures below freezing. Deciduous trees escape these winters by losing their leaves. Typical mammals are bear, badgers, squirrels, woodchucks, insectivores, rodents, wolves, wildcats, and deer. They are rich in birds. These forests have a long history—almost complete replacement by agricultural land. The climate that is suitable for temperate deciduous forest is most suited for man—hence the forest destruction.

18. (A) The term "frontier economy" refers to a time in U.S. history when resources were almost limitless and when the challenge was to harvest them. Today, what used to be limitless has become finite. Our challenge is to conserve what remains and to embrace opportunities that the new economy has to offer.

19. (C) Prior to the development of DDT, it is hypothesized that some mosquitoes had an allele of a gene that could break down DDT and render it harmless—the allele was just a random mutation of a gene for an enzyme. Mosquitoes with that allele may also have had somewhat lower reproductive rates and, because there was no DDT in the environment, these mosquitoes did not receive an advantage from the mutation. Consequently, they were selected against and the allele was maintained at a very low frequency in the environment. However, when DDT was introduced into the environment, the mosquitoes with that allele were selectively favored because they survived and reproduced at a much higher rate than the mosquitoes without the allele. The result was the increase in the frequency of the DDT-resistant allele. Remember, the allele in question was already present in the population; the addition of DDT did NOT cause the allele to arise. Selection acts only on preexisting variation; it does not create adaptive variation. Furthermore, the allele in question may have been actually selected against in one set of environmental circumstances and favored in another. Over the last 50 years, 400 species of insect, 50 species of fungus, and several species of weed have become resistant to pesticides that previously had killed them. Unfortunately, in areas that have been sprayed to kill malaria mosquitoes, up to 43 species are now immune to pesticides. In section III of the graph, the population of mosquitoes that were naturally immune to the effects of DDT are beginning to reproduce at a disproportionately higher rate than those that were susceptible or weakened by DDT.

20. (C) Automobiles and other forms of transportation are responsible for approximately one-third of man-made nitrogen oxide and volatile organic compound emissions, one-fifth of particulate emissions, two-thirds of carbon monoxide emissions, and less than 5% of sulfur dioxide emissions. Government policies on emissions have drastically reduced greenhouse gas emission from automobiles in the last 20 years: Cars built in 2000 emit 97% less hydrocarbons, 96% less carbon monoxide, and 90% less nitrogen oxide than those built in 1980. Between 1970 and 1991, total highway vehicle emissions of hydrocarbons dropped 66%, carbon monoxide emissions decreased by 59%, and nitrogen oxide emissions were reduced by 21% despite the doubling of vehicle miles traveled.

21. (A) Around one million years ago, the ancestors of *Homo sapiens* became dominant. Early humans are thought to have evolved in Africa during the Miocene and were widespread and accomplished toolmakers by the beginning of the Pleistocene. Theory states that *Homo sapiens* evolved in Asia and moved to Europe around 50,000 years ago, developed a form of agriculture in the Middle East about 10,000 years ago, and started writing and building the first cities approximately 4000 years ago.

22. (C) The Paleozoic era lasted from about 570 to 250 million years ago. The 320 million years of the Paleozoic era saw many important events, including the development of most invertebrate groups, life's conquest of land, the evolution of fish, reptiles, insects, and vascular plants, the formation of the supercontinent of Pangea, and no less than two distinct ice ages. The Earth rotated faster than it does today so days were shorter, and the nearer moon meant stronger tides.

23. (D) A mass extinction is an opportunity for adaptive radiation. Perhaps the most dramatic example is the rise of the mammals. Ancestral mammals were small, undifferentiated scavengers. After the demise of the dinosaurs, within ten million years, all of the major orders of mammals (and of birds as well) had differentiated.

24. (C) The leeward side of a coastal mountain range is characteristically dry due to the rain shadow effect. Water is the limiting factor. You would expect to find plants that require little water, those that would be classified as xeric and typically found in chaparral or desert biomes. Choice A refers to tropical plants. Lianas are climbing woody vines that festoon rainforest trees. They have adapted to life in the rainforest by having their roots in the ground and climbing high into the tree canopy to reach available sunlight. Epiphytes are plants that live on the surface of other plants, especially the trunk and branches. They grow on trees to take advantage of the sunlight in the canopy. Most are orchids, bromeliads, ferns, and *Philodendron* relatives. Tiny plants called epiphylls, mostly mosses, liverworts, and lichens, live on the surface of leaves. Some bromeliads grow in the ground, like pineapple, but most species grow on the branches of trees. Their leaves form a vase or tank that holds water. Small roots anchor plants to supporting branches, and their broad leaf bases form a water-holding tank or cup. The tank's capacity ranges from half a pint to 12 gallons or more. The tanks support a thriving ecosystem of bacteria, protozoa, tiny crustaceans, mosquito and dragonfly larvae, tadpoles, birds, salamanders, and frogs. Choice (B), mangrove trees, are found in tropical deltas and along ocean edges and river estuaries. They have adapted to living in wet, marshy conditions and have wide-spreading stilt roots that support the trees in the tidal mud and trap nutritious organic matter. Choice (D) would be found on the windward side because this is the side of the mountain that receives the greatest amount of rainfall. Choice (E) are plants that require a humid environment.

25. (D) After grasses become established, which hold down loose sand, winds bring seeds of more complex plants (shrubs, alders, and willows) to the area. In this open canopy, sun-requiring trees such as cottonwoods and some spruce become established. Next, larger trees like pines, black oak, aspen, and birch begin to grow and dominate. After shade develops from the first large trees, other types of trees, which are more shade-resistant, begin to grow in the area. What has now formed is called a broadleaf forest. The pines, aspens, and birches have grown rapidly, and sun required for their saplings to grow has been shut out. New trees cannot grow in this shade. As a result, the forest's understory is conducive to shade-tolerant oak, tulip trees, beech, maple, ash, hemlock, and others. Slowly, the pines, aspens, and birches die out, leaving the climax forest.

26. (A) Cinder cones are one of the most common types of volcanoes. A steep, conical hill of volcanic fragments called cinders accumulates around a vent, being formed from Stombolian eruptions. The rock fragments, often called cinders or scoria, are glassy and contain numerous gas bubbles "frozen" into place as magma explodes into the air and then cools quickly. Cinder cones range in size from tens to hundreds of meters tall and usually occur in groups.

27. (A) Choices (B) and (D) describe metamorphic rock. Choice (C) is wrong because it takes more than being covered by other sediments to turn into rock.

28. (A) The global rate of net forest loss is approximately 9 million hectares per year, according to the latest global forest assessment by the United Nations Food and Agriculture Organization (FAO). Forests are disappearing most rapidly in Africa and Latin America, whereas in Asia, the reduction of natural forests is largely compensated for by new plantation forests (which reduce biodiversity). In Europe and North America, the forest area is increasing. Overall, the world contains around 6000 square meters of forest for each person, which is reducing by 12 square meters every year. During the 1990s, the world forest cover decreased by an annual 0.2%. In Africa, on the other hand, the annual decrease was 0.8%, making it the continent with the highest deforestation rate. Second to Africa came South America, with a deforestation of 0.4%. The country suffering the highest deforestation rate is Burundi (Africa), which has an incredible annual deforestation rate of 9.0%.

29. (B) Remediation technologies are those that render harmful or hazardous substances harmless after they enter the environment.

30. (E) The major part of the 1982 Law of the Sea Treaty had been supported by U.S. Administrations, beginning with President Reagan, as fulfilling U.S. interests in having a comprehensive legal framework relating to competing uses of the world's oceans. However, the United States and many industrialized countries found some of the provisions relating to deep seabed mining contrary to their interests and would not sign or act to ratify the treaty.

31. (E) The observed products were common amino acids, fatty acids, and other organic molecules. The contemporary conclusion from this effort is that life originated through spontaneous, inanimate processes and that they took place under the conditions that existed on a primitive Earth.

32. (C) During the 1940s and 1950s, the Hooker Chemical Company dumped approximately 21,000 tons of organic solvents, acids, and pesticides as well as their by-products, many of them carcinogenic (causing cancer) or teratogenic (creating birth defects) into an abandoned canal in New York State (near Niagara Falls). A school and homes were built over the site. Chemicals began to leak from the ground, causing illness. Since the disaster, various levels of government have spent around $250 million and 20 years cleaning up the site, but all the waste is still buried there. New York State has since rebuilt homes in the area at reduced prices to attract new residents.

33. (D) From 1932 to 1968, Chisso Corporation (a petrochemical and plastics manufacturer) dumped an estimated 27 tons of mercury compounds into Minamata Bay, Japan. Thousands of people whose normal diet included fish from the bay unexpectedly developed symptoms of methyl mercury poisoning. The illness became known as Minamata Disease. Victims were diagnosed with degeneration of their nervous systems, numbness in their limbs and lips, slurred speech, and constricted vision. Some people had serious brain damage, while others lapsed into unconsciousness or suffered from involuntary movements. To date, 12,615 people have been officially recognized as patients affected by mercury, with estimates that the number of victims could be significantly higher.

34. (E) On March 28, 1979, a minor malfunction occurred in the system which fed water to the steam generators at the Three Mile Island Unit 2 Nuclear Generating Station near Harrisburg, Pennsylvania. This event led eventually to the most serious commercial nuclear accident in U.S. history and caused fundamental changes in the way nuclear power plants were operated and regulated. The accident itself progressed to the point where over 90% of the reactor core was damaged. Despite the severity of the damage, no injuries due to radiation occurred. Eleven days after the events of Three Mile Island, the movie *The China Syndrome,* a film about a nuclear accident, was released.

35. (A) On the night of December 2–3, 1984, 40 tons of methyl isocyanate, hydrogen cyanide, mono-methyl amine, and other lethal gases began spewing from Union Carbide Corporation's pesticide factory in Bhopal, India. Nobody outside the factory was warned because the safety siren was turned off. Over half a million people were exposed to the deadly gases. The gases burned the tissues of the eyes and lungs, crossed into the bloodstream, and damaged almost every system in the body. With an estimated 10 to 15 people continuing to die each month, the number of deaths to date is put at close to 20,000. And today more than 120,000 people are still in need of urgent medical attention. Of the women who were pregnant at the time of the disaster, 43% aborted. Study of growth and development of children whose mothers were exposed to the gases during pregnancy revealed that the majority of children had delayed gross motor and language sector development. Studies have also presented evidence of chromosomal damage.

36. (B) On April 26, 1986, a reactor exploded and released 30 to 40 times the radioactivity of the atomic bombs dropped on Hiroshima and Nagasaki. The world first learned of history's worst nuclear accident from Sweden, where abnormal radiation levels were registered at one of its nuclear facilities. Thirty-one lives were lost immediately. Hundreds of thousands of Ukrainians, Russians, and Belorussians had to abandon entire cities and settlements within a 20-mile zone of extreme contamination. Estimates vary, but it is likely that some 3 million people, more than 2 million in Belarus alone, are still living in contaminated areas. The city of Chernobyl is still inhabited by almost 10,000 people. Billions of rubles have been spent, and billions more will be needed to relocate communities and decontaminate the rich farmland.

37. (E) Refer to the chart entitled "Key American Organization Platforms on the Environment" starting on page 499.

38. (A) Refer to the same chart as above.

39. (D) Refer to the same chart as above.

40. (C) The epilimnion is the upper layer of a lake. The term "hyperlimnion" does not exist. The euphotic zone is the upper layer of water that is penetrated by sunlight and contains waters rich in mineral and organic nutrients that often promotes a proliferation of plant life, especially algae, which may reduce the dissolved oxygen content and often causes the death of other organisms. The benthic zone would be the deepest layer of ocean water.

41. (D) The risk of dying between ages one and five is 43% higher for girls than boys in India. India has less than 93 women for every 100 men against the world average of 105. That accounts to nearly 1.4 million "missing girls" in the age group of 0–6 years based on the assumption that one would typically expect 96 girls for every 100 boys in this age group. The main reason for the widespread female infanticide in parts of India is the dowry system, which, although long prohibited by law, continues to play a significant role in Indian society. Dowries and wedding expenses regularly run to more than a million rupees ($35,000) in a country where the average civil servant earns about 100,000 rupees ($3,500) a year. Added to this is the low status of women in rural India, where they perform the menial tasks of the family such as carrying water and firewood and seeing to feeding the animals. India is estimated to have some 432 million illiterate people. Sixty-four percent of Indian men are literate, but fewer than 40% of women can read and write. About 41% of Indian girls under the age of 14 do not attend school.

42. (D) Refer to the chart on page 503 in Chapter 16.

43. (D) Mountain gorillas are found in Rwanda. The population of Rwanda has more than doubled since the early 1970s. With a continued growth rate of about 3% per year, the population is projected to double approximately every 25 years. The rarest and largest of the great apes, mountain gorillas are among our closest relatives, yet it is one of the most endangered mammals on Earth. They are threatened by poaching, loss of habitat, disease, and war. The gorilla's only known enemies are leopards and humans. Gorillas are commonly hunted for meat or in retaliation for crop raiding, and have been the victims of snares and traps set for antelope and other animals and are also subject to human disease. Poachers have also destroyed entire family groups in their attempts to capture infant gorillas for zoos, while others are killed to sell their heads and hands as trophies. Consider helping these beautiful animals by having your AP Environmental Science class adopt a gorilla. Visit the mountain gorilla conservation fund at www.mgcf.net.

44. (B) Almost 60% of all the world's freshwater withdrawals go toward irrigation purposes. Rice production uses the most water; soybeans and oats use the least. Producing electrical power is also a major use of water in the United States. In 1995, 189,700 million gallons of water each day were used to produce electricity—to cool the power-producing equipment.

45. (C) For (A), a 1999 study by Consumer Reports found that, surprisingly, domestic produce had more toxic pesticide residues than imported in two-thirds of the cases studied. For (B), risks to the human immune, reproductive, and endocrine systems, as well as neurotoxicity, may be equally or even more significant than cancer. Of the 45 environmental contaminants or agents that have been reported to cause changes in mammalian reproductive and hormone systems, 8 are herbicides, 8 are fungicides, and 17 are insecticides. Nine of the 26 most commonly used pesticides have been associated in laboratory tests with sperm abnormalities, reduced sperm production, disrupting male hormones, or damaging male reproductive organs. Use of these pesticides totals over 300 million pounds per year. The legal standard for registration set down by the Federal Insecticide, Fungicide, and Rodenticide Control Act (FIFRA) is a "risk–benefit" standard. EPA must register pesticides if they do not pose "unreasonable risk to man or the environment, taking into account the economic, social, and environmental costs and benefits of the use of any pesticide." (7 USC secs. 136(bb) and 136a(c)(5)(C)). This means that if a pesticide presents substantial benefits to farmers in terms of increased yields or decreased labor costs, those benefits are weighed against health and environmental risks. Even if there are substantial health risks, the EPA may decide the economic benefits outweigh the risks. In (D), 12 of the 26 most widely used pesticides in the United States are classified as possible or probable carcinogens by the EPA based on studies of laboratory animals, with an annual use that totals over 380 million pounds. In (E), a 1994 analysis by the Environmental Working Group, using USDA data, found 12 different carcinogens, 17 neurotoxins, and 11 pesticides that disrupt the endocrine or reproductive system in 12 fruits and vegetables that had been washed, peeled, and prepared for consumption. The foods most likely to be contaminated were (in declining order) peaches, apples, celery, potatoes, grapes, and oranges.

46. (A) The law of demand holds that, other things being equal, as the price of a good or service rises, its quantity demanded falls. The reverse is also true: as the price of a good or service falls, its quantity demanded increases. Think of your trips to the grocery store. When the price of beef rises, you buy less of it.

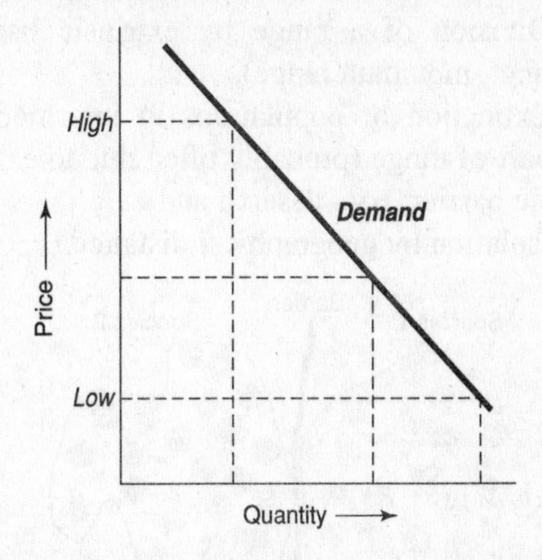

47. (E) See Figure 14.2.

48. (D) If the wind speed is 15 mph, the relative power produced would be 3375 (15^3). For a wind turbine in a 12-mph wind, the relative power would be 1728 (12^3). 3375 is about twice 1728. Wind speeds in this range are classified as "Class 3" wind sites, with an energy density of 300–400 watts per square meter. Characteristic of much of the Midwestern United States, Class 3 sites cover 13% of the total land area of the United States.

49. (A) Allopatric speciation occurs when there is a complete geographic separation between parts of the species range. Gene flow from the other parts of the range stops, and the separated populations evolve to suit their new, restricted environment. This encourages the genetic divergence of the separated populations and might become so great that, if the two populations were rejoined, they would no longer be successful at interbreeding. (Even if limited interbreeding were still possible, the local adaptations might have become so advantageous that there would be strong selection for individuals to discriminate to mate preferentially with members of their local population, rather than with immigrant individuals.) The two sets of organisms have become new species. This concept, in which physical and genetic separation of populations leads to speciation, describes allopatric speciation. Conditions for allopatric speciation include

1. Isolation of a colony (e.g., island, mainland),
2. Division of a range by extrinsic barrier (e.g., mountain range),
3. Extinction of populations in intermediate part of range (probably often due to extrinsic barrier; e.g., desert), and
4. Isolation by geographical distance.

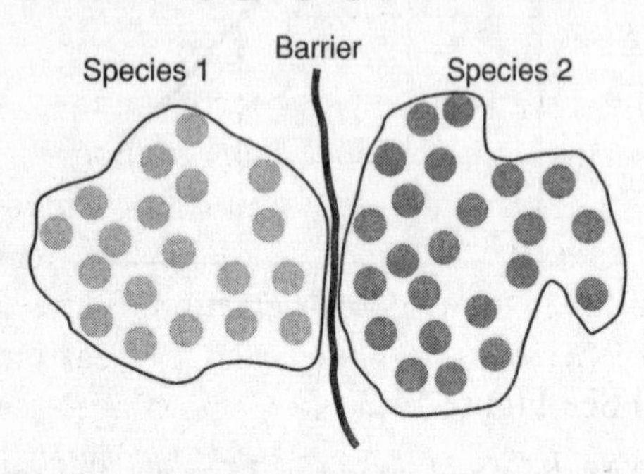

50. (D) Parapatric speciation is speciation involving geographical isolation and divergence of two or more populations from a parent species. The speciation process is completed with the evolution of species-isolating mechanisms that prevent mistaken interbreeding. Parapatric speciation is different than allopatric speciation in that the completion of speciation involves the evolution of postzygotic species-isolating mechanisms. Conditions for parapatric speciation follow: a very strong environmental change → disruptive selection → hybrids less fit than pure parental types.

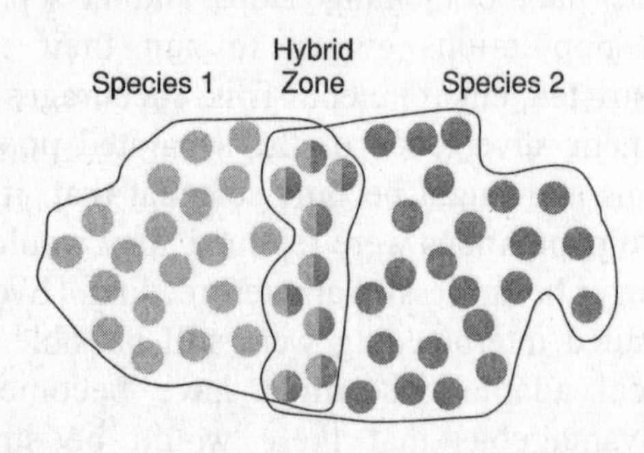

51. (A) Today three million fishing boats operate on the world's oceans and are greatly depleting the supply of fish and other aquatic life forms. Over the past 40 years, fishing quotas have more than tripled. In 1950, 20 million tons of fish and marine products were harvested. By 1990, this amount had increased to 100 million tons per year. The depletion of fish stocks has led to overfishing in all oceans and other bodies of water. Past catches were generally higher than those of today due to overfishing; therefore, to base today's harvest limits on harvest numbers that were probably higher in the past would only accelerate depletion of current fishing stocks.

52. (E) Cultivated land occupies only 11% of the world's land resources but produces about 95% of the world's food.

53. (D) Secondary succession begins in habitats where communities were entirely or partially destroyed by some kind of damaging event. For example, secondary succession begins in habitats damaged by fire, floods, insect devastations, overgrazing, and forest clear-cutting, and in disturbed areas such as abandoned agricultural fields, vacant lots, roadsides, and construction sites. Because these habitats previously supported life, secondary succession, unlike primary succession, begins on substrates that already bear soil. In addition, the soil contains a native seed bank.

54. (D) Virulence is the harm that parasites and diseases cause to their host (e.g., parasite-induced host mortality or reduced fecundity). Parasite virulence is, in general, proportional to the degree that a parasite exploits the host. Parasite offspring are produced by exploiting the host; therefore, some virulence is inevitable. However, too strong a host exploitation leads to high virulence that jeopardizes survival of the host and the parasite itself. Thus, there should be an optimal level of host exploitation, and virulence, by the parasite.

55. (D) In a "No-Problem" School, there are no environmental, resource, or economic problems that cannot be solved by more economic growth, better management, and better technology. In a Free-Market School, the best way to manage the planet is to create a truly free-market global economy with minimal governmental interference and regulations; to convert all public property resources to private property resources; and to let the global marketplace, governed by pure capitalism, decide essentially everything. In a Spaceship-Earth worldview, the Earth is essentially a spaceship that we can understand, dominate, change, and manage to prevent overload and to maintain a satisfactory standard of life. And the last component of a planetary management worldview is the Stewardship School, which believes that we have an ethical responsibility to be caring and responsible managers or stewards who tend the Earth as a garden and that we can and should make the Earth a better place for ourselves and other species through love, care, knowledge, and technology. A "biocentric worldview" believes that the inherent value of all forms of life must be recognized; that species have a hierarchy of values, animal species ranking higher than plant species; that pests and disease-carrying organisms have low value; and that all life forms have inherent rights to struggle to exist.

56. (D) The precautionary principle says to be cautious when making decisions about something that we do not understand and that could have potentially serious side effects. The integrative policy says to make decisions that involve integrated solutions to environmental and other problems. The ecological design principle says to incorporate concepts of good ecological design into decisions and laws. The environmental justice principle says to develop policies so that no one group bears a disproportionate share of harmful environmental risks. The answer (D), the humility principle, reminds us of the limits of human knowledge and, by extension, the limits of our capacity to manage and control the Earth.

57. (A) Large-scale circulations develop in the Earth's atmosphere due to uneven heating of its surface by the sun's rays. Daytime solar heating is greatest near the Earth's equator, where incoming sunlight is nearly vertical to the ground, and least near both poles, where sunlight arrives nearly horizontal to the ground. Near the poles, heat lost to space by radiation exceeds the heat gained from sunlight, so air near the poles is losing heat. Conversely, heat gained from sunlight near the equator exceeds heat losses, so air near the equator is gaining heat. The heated air near the equator expands and rises, while the cooled air near the poles contracts and sinks. Rising air creates low pressure at the equator. Air cools as it rises causing water vapor to condense (rain) as the air cools with increasing altitude. As air mass cools, it increases in density and descends back to the surface in the subtropics (30° N and S), creating high pressure there.

58. (C) See the table in the section entitled "Difference Between r Strategists and K Strategists."

59. (B) Predator swamping occurs among some organisms that produce huge numbers of offspring—predators are simply swamped and can't eat it all. It is not a defensive behavior that relies on avoiding detection. Some squid, wildebeests, springboks, and 17-year cicadas use predator swamping. Camouflage or crypsis occurs when animals match backgrounds for color, shape, and size. Masquerading is a type of crypsis in which the organism resembles something inedible. Countershading occurs when animals are darker on the top and lighter on the bottom, which counteracts the effect of sunlight striking the top surface (some squid have luminescent organs on the ventral, or bottom, side to prevent fish from seeing them from below). Batesian mimicry (common in snakes and butterflies) involves three species—predator, model, and mimic. The model species is noxious or dangerous, so predators avoid it. The mimic species has evolved a resemblance to the model, but it isn't itself noxious. Thus, it engages in "false advertising" but often gains because the predator is fooled and avoids the mimic. In Müellerian mimicry, both the model and the mimic are distasteful.

60. (A) The winds that most affect the oceans' currents are: (1) the westerlies (40–50 degree latitudes) that blow west to east and (2) the trade winds (20 degree latitudes) which are closest to the equator and which blow from east to west. Both winds are a result of warm air from the tropics moving to the poles and incorporating the rotation of Earth into their movement. In the northern hemisphere, they move clockwise. In the southern hemisphere, they move counterclockwise. The currents that are closest to the equator are the north and south equatorial currents, both of which flow west. The equatorial countercurrent, which flows between these currents, flows eastward. Warm western boundary currents flow from the equator to the poles (i.e., Gulf Stream), and cold eastern boundary currents flow from the poles to the equator (i.e., Canary, California).

61. (B) Acute means having a rapid onset, severe symptoms, and a short course (less than six weeks) duration. Chronic refers to an illness that has been or is expected to be a condition that affects the individual for an extended period of time and that historically is an illness that is not expected to be resolved through regular medical treatment or the passage of time. About one million children die worldwide from measles each year.

62. (C) Carbon dioxide is the ultimate result of perfect combustion of any carbon-based fuel. The only ways to reduce CO_2 emissions are to make vehicles more fuel-efficient and/or to drive less, to use a noncarbon fuel such as hydrogen, or to use a "green" fuel such as ethanol that is produced from crops that absorb CO_2 as they grow.

63. (C) Trace elements are minerals that the body requires in amounts of 100 mg or less, per day. For some, including iodine, proper dosage may be as small as one tenth of 1 mg. Minuscule as these amounts are, insufficient intake of trace elements can seriously impair health. Iodine is used by the thyroid gland to produce hormones essential for growth, reproduction, nerve and bone formation, and mental health. Natural sources are fish, shellfish, and iodized salt. Copper is necessary for the formation of blood cells and connective tissue. It is also involved in producing the skin pigment melanin. Natural sources are beef, chicken liver, crab, chocolate, seeds, nuts, fruit, and beans. Zinc is involved in the structure and function of all cell membranes as well as the production of more than 200 enzymes. It also is essential for proper wound healing. Natural sources include oysters, beef, pork liver, beef liver, lamb, crab, and wheat germ.

64. (E) r-selected species have high biotic potential and high mortality, are short-lived, usually are pioneer species, and have small build. A weed would be an r-selected species. As conditions such as soil, water, and light change, r-selected plant species are gradually replaced by K-selected plant species. These more stable species include perennial grasses, herbs, shrubs, and trees. The K-selected species live longer; therefore, their environmental effects slow down the rate of succession.

65. (D) According to the U.S. Department of Agriculture, fire building has the biggest impact, especially if not done with care. Blackened roots, charred wood, limbs broken from trees, and garbage left in fire pits are impacts commonly found. Hunting and fishing are regulated and monitored by both state and federal agencies.

66. (A) Aluminum cans take a lot of energy to be made from scratch. By recycling soda cans, the energy used and air pollution is 95% less than that used for the production of cans from pure bauxite ore. Each aluminum soda can recycled saves energy equal to half a can of gasoline! Aluminum containers can contain 100% recycled content and their light weight conserves energy during transportation. Steel cans, just in U.S. landfills, weighed about 2.5 million tons in 1986. These cans have an outer coating of tin, which is very expensive and is imported into the United States. This tin can be recovered and resold or used to make new cans. Glass bottles comprise 8% of our garbage. Two types of glass bottles are manufactured—refillable and nonrefillable. The reusable type is heavier and sturdier and just needs a thorough cleaning for it to be recycled up to 30 times. The other type, which is non-reusable, can be melted with raw materials to produce new glass bottles. However, this option requires a lot of energy, so the refillable bottles are preferable. The downside to glass containers is that they weigh a lot, so shipping expenses per unit of volume may be more than that of products packaged in lighter materials. Plastics make up 7% of our waste by weight, but 32% by volume. Plastics take hundreds or thousands of years to decompose, so they are essentially permanent in the landfills. Plastic is currently made from petroleum and natural gas, but future plastics are most likely to come from plant and animal matter. In all but a few cases, plastic has not been approved to be recycled. This barrier significantly reduces the recyclability of these products.

67. (A) Price is an important determinant of the quantity of a good supplied. The "Law of Supply" states that the amount offered for sale rises as the price is higher.

68. (C) Any Member in the House of Representatives may introduce a bill at any time while the House is in session by simply placing it in the "hopper." A public bill may have an unlimited number of co-sponsoring members. The bill is referred to the appropriate committee by the speaker. An important phase of the legislative process is the action taken by committees. It is during committee action that the most intense consideration and fact-finding are given to the proposed measures; this is also the time when the people are given their opportunity to be heard. Each piece of legislation is referred to the committee that has jurisdiction over the area affected by the measure. The Committee on Resources has jurisdiction in the following areas:

1. Fisheries and wildlife, including research, restoration, refuges, and conservation;
2. Forest reserves and national parks created from the public domain;
3. Forfeiture of land grants and alien ownership, including alien ownership of mineral lands;
4. Geological Survey;
5. International fishing agreements;
6. Interstate compacts relating to apportionment of waters for irrigation purposes;
7. Irrigation and reclamation, including water supply for reclamation projects and easements of public lands for irrigation projects, and acquisition of private lands when necessary to complete irrigation projects;
8. Native Americans generally, including the care and allotment of Native American lands and general and special measures relating to claims that are paid out of Native American funds;
9. Insular possessions of the United States generally (except those affecting the revenue and appropriations);
10. Military parks and battlefields, national cemeteries administered by the Secretary of the Interior, parks within the District of Columbia, and the erection of monuments to the memory of individuals;
11. Mineral land laws and claims and entries thereunder;
12. Mineral resources of public lands;
13. Mining interests generally;
14. Mining schools and experimental stations;
15. Marine affairs, including coastal zone management (except for measures relating to oil and other pollution of navigable waters);
16. Oceanography;
17. Petroleum conservation on public lands and conservation of the radium supply in the United States;
18. Preservation of prehistoric ruins and objects of interest on the public domain;
19. Public lands generally, including entry, easements, and grazing thereon;
20. Relations of the United States with Native Americans and Native American tribes;
21. Trans-Alaska Oil Pipeline (except rate making).

69. (C) $N_1 = N_0 + B - D + I - E$

$N_1 = 40{,}000 + 35\ (40) - 10\ (40) + 100 - 50$

$\mathbf{N_1 = 41{,}050}$

70. (A) Certain times of the day are more popular for feeding than other times. Smaller fish, occupying shallow water, may feed during daytime when the organisms that make up their diet are most active. The small fish school and search for food, very alert to any approaching predators. Midday is perhaps the poorest time to fish because many larger fish avoid the sunlight and moving waters; however, most fish will feed at any time if prey is available. By dusk, the smaller fish may be less active. Schooling functions, in part, as a means of camouflage and protection from predators. As light decreases, these schools break down. Predator fish become more active. The window of opportunity to catch a walleye or pike lasts as long as dusk is long. As night sets in, even predator fish seek a resting place.

71. (E) The disability-adjusted life year or DALY has emerged as a measure of the burden of disease and it reflects the total amount of healthy life lost, to all causes, whether from premature mortality or from some degree of disability during a period of time. The intended use of the DALY is to assist in: (1) setting health service priorities; (2) identifying disadvantaged groups and targeting of health interventions; and (3) providing a comparable measure of output for intervention, program and sector evaluation, and planning. The number of DALYs estimated at any moment reflects the amount of health care already being provided to the population, as well as the effects of all other actions which protect or damage health.

72. (A) There are basically three patterns of plant distribution—random, regular, and clumped. Random distribution implies neutral interrelations among individuals and no limiting resources. Regular distribution implies negative interactions among individuals and competition for some limiting resource. Clumped distribution implies attraction among individuals or to a common resource.

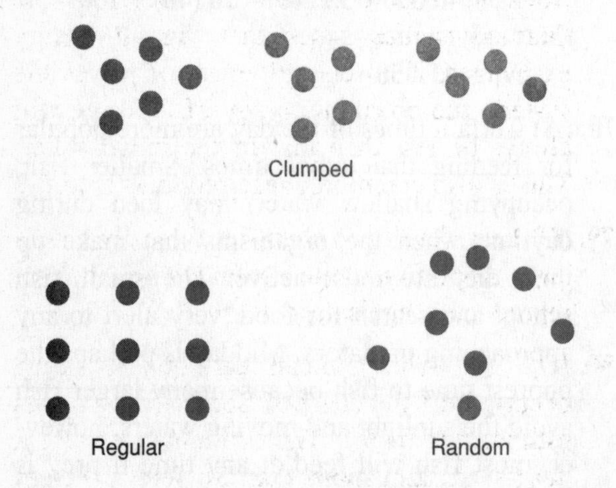

73. (C) Most of the freshwater found on Earth is in the form of groundwater. The estimated and actual relative amounts of freshwater are groundwater (60,000,000 km^3, 66.6%), ice (30,000,000 km^3, 33.3%), rivers/lakes (140,000 km^3, less than 0.01%), and atmosphere (10,000 km^3, less than 0.001%).

74. (D) Every day 3000 Americans under the age of 18 become regular smokers, and a third of them will eventually die of nicotine-related causes. Two out of three 12 to 17 year olds who smoked nicotine in the last year show signs of addiction. Every day over 1000 Americans die as a result of nicotine addiction. In 1900, 4 billion cigarettes were produced in the United States; in 2000, the number had increased to 720 billion.

75. (C) The most important tin mineral is cassiterite, a naturally occurring oxide of tin with the chemical formula SnO_2. In its purest form, cassiterite contains 78.6% tin. The cassiterite occurs either in rocks, often very irregular veins or lodes, or in debris that has built up from the gradual wearing down of tin-bearing rocks to form alluvial deposits that are found in river beds and valleys or on the ocean floor close inshore. Pure tin is a silvery white metal that is soft, ductile, and malleable. Tin readily forms alloys with other metals to create useful materials such as solders, bronzes, pewter, bearing alloys, and fusible alloys. Tin mining is very much a global industry with tin mining operations located worldwide. Major tin deposits are confined to a comparatively small number of areas. More than half of the world's tin ore is mined in the far East (Thailand, Malaysia, Indonesia, and China) and South America (Bolivia and Brazil). About a third of all tin produced today goes to make tinplated steel (tinplate) for food and beverage cans and other packaging. More than 300 million tinplated cans of food are eaten daily worldwide. The United States is the primary consumer of tin, using 35,000 metric tons annually of the total 188,900 metric tons mined annually.

Resistant to wear and heat, asbestos is ideal for friction materials such as clutches and brake linings. As an insulator it was used in the manufacture of pipes and pipe insulation, and in other products such as heat-shielding panels and protective clothing and equipment. Because of its durability and adaptability as a strong bonding agent it was also incorporated into floor tiles and sheet rock. The large-scale use of asbestos began in the early 1900s; however, worldwide production of the mineral, until recently, amounted to approximately 5.5 million tons per year. United States consumption of asbestos reached a peak of 800,000 tons per year in the early 1970s. Since then, consumption has dropped more than 70%. However, much of the asbestos originally installed in buildings is still present. Asbestos fibers are milled from raw asbestos mined from ore deposits, and it is still mined in Australia, Canada, South Africa, and the former Soviet Union. As the incidence of lung cancer and asbestosis in American asbestos workers became apparent, it was declared a hazardous substance and regulated. Legislation was enacted in 1986 by the signing into law of the Asbestos Hazard Emergency Response Act (AHERA). This directed the EPA to directly deal with asbestos exposure in public and private schools.

76. (A) President Woodrow Wilson created the National Park System with the Organic Act of 1916. It was designed to "conserve the scenery and the natural and historic objects and leave them unimpaired for the enjoyment of future generations." There are now 77.5 million acres of land preserved in the park system. Teddy Roosevelt is often wrongly cited as the father of this system. Teddy Roosevelt created the National Wildlife Refuge System in 1903. During his tenure as president, he placed over 230 million acres under federal protection.

77. (D) A sequential comparison index (SCI) is a relative number (no units) that can be used to assess species diversity. A new run is started when a different species appears in the identification sequence. The SCI is computed as follows:

$$\text{SCI} = \frac{\text{Number of runs}}{\text{Number of individuals}}$$

For this sample, let backswimmers = A, damselflies = B, midge = C, mosquito larvae = D, and mayfly = E.

$$\text{SCI} = \frac{\text{run 1}}{\text{AAA}}\ \frac{\text{run 2}}{\text{BB}}\ \frac{\text{run 3}}{\text{C}}\ \frac{\text{run 4}}{\text{DD}}\ \frac{\text{run 5}}{\text{EEEE}}\ \frac{\text{run 6}}{\text{B}}\ \frac{\text{run 7}}{\text{A}}$$

$$\text{SCI} = \frac{\text{Number of runs}}{\text{Number of individuals}} = \frac{7}{14} = 0.5$$

SCI values range from 0.1 to 1.0. An SCI of 1.0 indicates highest biodiversity, dominance of pollution-intolerant species, and outstanding water quality. An SCI value of 0.5 indicates average or moderate biodiversity, and average or moderate water quality. An SCI value of 0.1 indicates lowest biodiversity, dominance of pollution-tolerant species, and degraded water quality. Even though this method is fairly simple, care must be taken in sampling that the samples are representative of the area of the stream in which you are interested. For example, vegetated areas at stream margins tend to have higher indices than the middle stream. Another factor influencing the SCI index is season. For example, the diversity of a stream may be lower in the early summer than in the winter owing to the emergence of many aquatic insects.

78. (C) The first law of thermodynamics states that energy can be changed from one form to another, but it cannot be created or destroyed. The total amount of energy and matter in the universe remains constant, merely changing from one form to another. The second law of thermodynamics states that "in all energy exchanges, if no energy enters or leaves the system, the potential energy of the state will always be less than that of the initial state." This is also commonly referred to as entropy.

79. (C) Less-developed countries typically have a large proportion of their population in the prereproductive age category.

80. (C) The 1960 Chilean earthquake, which occurred off the coast of South America was estimated to be 9.5. The earthquake created a deadly tsunami more than 30 feet in height along the coast of Chile, eliminating entire villages. The tsunami continued across the Pacific, striking Hawaii where it killed 61 people despite a warning that had been issued five hours earlier. Some hours later, the tsunami killed hundreds more in Japan, more than 8000 miles from the epicenter. The New Madrid, Missouri, earthquake in 1812 was the largest earthquake in the continental United States, which was estimated to be higher than 8.0 and which changed the course of the Mississippi River. The 1964 Prince William Sound, Alaska, earthquake, at 9.2 on the Richter scale, is the largest earthquake to occur in the United States so far. *(Author's note: I live 2 miles from the epicenter of the Northridge earthquake that occurred in 1994. I'll never forget it; neither will my house!)*

81. (C) Although communities are variable and dynamic in both space and time, the interactions between species lead to patterns of community structure that are characteristic of particular community types. For example, North American forests have vertical layers—canopy, midstory, and understory—that are composed of different life forms (e.g., trees, shrubs, herbs). Midwestern prairies have species that are active in the spring and fall (cool season species) and others that are active in midsummer (warm season species). In the Northeast, many communities have some species that bloom in spring, some in summer, and some in fall. These other structures are thought to be a result of resource partitioning, an evolutionary strategy that allows species that are potential competitors for a resource (space, light, pollinators) to coexist by specializing in different aspects of the resource.

82. (C) Selling pollution permits is essentially the same as charging each firm for every ton that they emit. Essentially it is charging them in advance. If the pollution permit costs more than abatement, then they will abate. If it costs less, they will buy the permit and continue to pollute. In this case, a permit that costs $151 creates an incentive for the electric utility to reduce its emissions by 2 tons, at a total cost of $250. This is cheaper than having both the smelter and the utility reduce by one ton each, which would cost $200 plus $100, or $300. However, the benefits to society are the same.

83. (C) The marked butterflies from the first sample *(M)* are released where they were captured. The next day, a second sample from the same population is taken *(p)*. Some individuals will be recaptured ($m = 20$) from the first sampling, while others will be captured for the first time, and thus will be unmarked. The ratio of marked animals ($m = 20$) to the total number of animals in the second sample ($p = 100$) is assumed to be the same as the total number of marked animals ($M = 100$) is to the total population size *(P)*. Mathematically $m/p = M/P$ or $20/100 = 100/P$. Therefore, $P = (100 \times 100)/20$; $P = 500$.

84. (E) The doctrine of riparian rights define the rights relating to the bank of a watercourse which says that a landowner adjacent to a stream has the right to the water in that stream. This places the responsibility on the upstream users and protects private rights in streams and lakes. The concept of appropriation as a right to use water came out of the western United States where public land was parceled out to individuals without control over streams. This left water to be treated as though it belonged to no one and could be appropriated in a manner similar to that of a gold claim. In the absence of public control, men took water from streams and used it; that is, they "appropriated" it. When water laws were enacted, this appropriation practice was legalized and the basis of such laws became known as the doctrine of prior appropriation. The doctrine was simply that the first user on a stream has a better right to the supply in times of shortage. Population growth and increased demand for water have produced many limitations and modifications to the doctrine.

85. (D) Erosion is a fundamental and complex natural process that is generally increased by human activities such as land clearance, agriculture (ploughing, irrigation, grazing), forestry, construction, surface mining, and urbanization. It is estimated that human activities have degraded some 15% (2 billion hectares) of the Earth's land surface between 72°N and 57°S latitudes. Slightly over half of this is a result of human-induced water erosion and about a third is due to wind erosion with most of the balance being the result of chemical and physical deterioration. In the United States, soil has recently been eroded at about 17 times the rate at which it forms—about 90% of the U.S. cropland is currently losing soil above the sustainable rate. Soil erosion rates in Asia, Africa, and South America are estimated to be about twice as high as in the United States. The Food and Agricultural Organization of the United Nations (FAO) estimates that 140 million hectares of high-quality soil, mostly in Africa and Asia, will be degraded by 2010 unless better methods of land management are adopted.

86. (A) CO_2 levels are also linked to industry and automobile usage. Similarly, nitrous oxide concentrations are affected by automobile exhaust. N_2O emissions can also be reduced by decreasing the amount of nitrogen-based fertilizers used and by eliminating the use of fossil fuel power plants for electricity and auto exhaust.

87. (C) The burden of proof that the plaintiff caused harmful action is with the defendant.

88. (C) Solids refer to matter suspended or dissolved in water or wastewater. Solids may affect water or effluent quality in a number of ways. Water with highly-dissolved solids is generally of inferior palatability and may cause health problems. Highly-mineralized waters are unsuitable for many industrial applications. Water with high-suspended-solids content can also be detrimental to aquatic plants and animals by limiting light and deteriorating habitat. Total dissolved solids are the amount of filterable solids in a water sample. To calculate total suspended solids:

Let A = Weight of crucible + Filter + Residue after 24 hr at 105°C (mg)

B = Weight of crucible + Filter (mg)

$$\text{Total suspended solids} = \frac{(A - B) \times 1000 \text{ mL/L}}{\text{sample volume (mL)}} \times \frac{1000 \text{ mg}}{\text{g}}$$

$$= \frac{(100.00 \cancel{\text{g}} - 80.00 \cancel{\text{g}}) \times 1000 \cancel{\text{mL}}/\text{L}}{100.00 \cancel{\text{mL}}} \times \frac{1000 \text{ mg}}{1 \cancel{\text{g}}}$$

89. (D) Owls are raptors, or birds of prey, which means they hunt other living things for their food. The forward-facing aspect of the eyes give them a wide range of "binocular" vision (seeing an object with both eyes at the same time). This means that owls can see objects in three dimensions (height, width, and depth) and can judge distances in a similar way to humans. Owls are able to tell direction through hearing because of the minute time difference in which the sound is perceived in the left and right ear. This is important at night when light is low or nonexistent. Owls can detect a left/right time difference of about 0.00003 seconds (30 millionths of a second). Owls can also tell if the sound is higher or lower than the position of the owl by using the asymmetrical or uneven ear openings—the left ear opening is higher than the right, so a sound coming from below the owl's line of site will reach the right ear first. This left-right and up-down positioning system gives the owl a perfect "fix" on the location of prey. When a target is located, the owl will fly toward it, keeping its head in line with the prey until the last moment. This is when the owl pulls its head back and thrusts its feet forward with the talons spread wide—two pointing backward and two forward. The force of the impact is usually enough to stun the prey, which is then dispatched with a snap of the beak. The comblike leading edge of their primary wing feathers effectively muffles the sound of the air rushing over the wing surface, allowing the owl to fly silently.

90. (A) According to the graph, nitrates and phosphates are depleted by the summer growth.

91. (C) The recycling of materials to produce nitrates and phosphates is the function of organisms of decay.

92. (B) According to the graph, light intensity is a factor in autumn growth. It is to be assumed that temperature is also an important factor because the data indicate the seasonal aspect of the cycle.

93. (A) There is an optimum temperature for metabolic activities. Low temperatures slow down the speed of chemical reactions.

94. (E) In summer months, the Gulf Stream pulls warm water up from the tropics. Hurricanes often follow this same path because they need warm water to form. Off the West Coast, the current is pulling cooler water down from the north, suppressing most hurricanes and pushing them further out to sea.

95. (B) Oceans cover about 70% of the Earth's surface, much of this area is relatively unproductive biologically. The ocean zones are marked by differences in depth and light penetration. The intertidal zone is inhabited by plants and animals that are adapted to being alternately exposed to water and air as the tide moves in and out. The neritic zone covers the continental shelf and water above it and extends from the mean low water or edge of the intertidal to the end of the shelf. The maximum depth is 200 meters, but it varies with oceans because of variations in width of shelves. This region makes up only 8% of the total area of the oceans, but it is so fertile that most marine life is concentrated in these nutrient-rich waters. Characteristics of the neritic zone are: water color is frequently green or brown due to plankton; high turbidity is caused by sediment and plankton that are stirred up by waves and tides; water is fed by estuarine and continental runoff; thermoclines are less stable because of winds and strong currents; it is nutrient-rich compared to the pelagic zone; there is a high level of pollutants because it is continuous with estuaries; organisms are diverse and numerous; plankton flourish in the well-aerated photic zone; there is much zooplankton because of rich phytoplankton—large schools of fish grazing on plankton, and large schools of predators eating fish. The pelagic zone is the open ocean and is divided into three subzones based on depth:

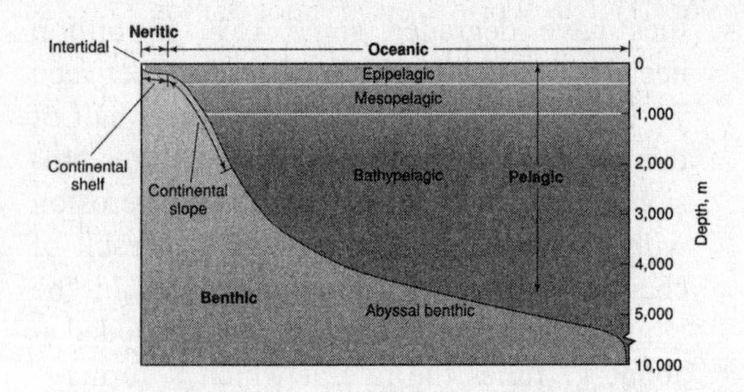

1. Epipelagic (photic) zone—the upper pelagic zone of the ocean. The zonation is based on the daytime distribution of animals. The depth of the zones depends on water clarity, which varies between regions. In clear waters, the lower limit of this zone is about 350 to 400 meters in depth. Above this depth, sufficient light exists to make concealment difficult especially when viewed from below. Characteristics of the epipelagic are fairly constant salinity; good mixing and high productivity; and low nutrients, which are removed rapidly by high-surface productivity.
2. Mesopelagic zone—the middle pelagic zone of the open ocean. In clear waters, the mesopelagic zone extends from about 400 to 1200 meters in depth. The upper limit is defined by light levels that are low enough to allow animals to conceal themselves against the downwelling daylight with bioluminescence (counterillumination). The lower limit is defined as the depth limit for diel vertical migration. Characteristics of the mesopelagic are: light begins to disappear until it cannot be detected at 100 meters; below layer of wind mixing; affected by thermohaline circulation; temperature changes rapidly—from 10 to 4°C, vertically no seasonal change in temperature, light, or salinity; and high in nutrients since they are not utilized because productivity is very low due to not enough light.
3. Bathypelagic zone—the deep pelagic zone of the open ocean; the zone beneath the mesopelagic zone. Characteristics of the bathypelagic zone are constant temperature at 2–4°C, no light, high nutrients, and mostly soft-bodied organisms and large predators with bizarre devices for feeding and reproduction.

96. (D) The World Conservation Union (WCU) was founded in 1948 and brings together 78 states, 112 government agencies, and 10,000 scientists and experts from 181 countries in a unique worldwide partnership. Its mission is to influence, encourage, and assist societies throughout the world to conserve the integrity and diversity of nature and to ensure that any use of natural resources is equitable and ecologically sustainable. Within the framework of global conventions, it has helped over 75 countries to prepare and implement national conservation and biodiversity strategies. According to the WCU, the areas that should be allowed the greatest degree of human impact are cultural and scenic landscapes. National monuments, ecological preserves, and wilderness areas usually have narrowly focused "tourist attractions," which are sensitive to human impact. Scenic and cultural landscapes are usually more diffuse, thereby reducing the impact of large numbers of people within limited areas.

97. (D) Most indoor air pollution experts agree that carpeting should be avoided whenever feasible. The reason for this is that carpeting is made up of some 120 different chemicals, many of which can cause health problems, and that, once installed, carpeting can collect dust and even lead (tracked in from shoes), and grow mold and dust mites. Methylene chloride is given off by paint thinners and paint stippers. It has a fairly long atmospheric half-life, 3–4 months, indicating the fairly long persistence typical of transported pollutants.

98. (B) The "Butterfly effect," based on theories of chaos, is the propensity of a system to be sensitive to initial conditions. Such systems over time become unpredictable. This idea came from a 1979 paper entitled "Does the Flap of a Butterfly's Wings in Brazil Set Off a Tornado in Texas?" which proposed that small differences and changes in the environment can make a massive difference further down the line. Its technical name is sensitive dependence on initial conditions. It affects any complex, dynamic system.

99. (B) A density-independent factor is one where the effect of the factor on the size of the population is independent of and does not depend upon the original density or size of the population. The effect of weather is an example of a density-independent factor. A severe storm and flood coming through an area can just as easily wipe out a large population as a small one. Another example would be a harmful pollutant put into the environment (e.g., a stream). The probability of that harmful substance at some concentration killing an individual would not change depending on the size of the population. Many populations controlled by density-independent factors have resource-limited growth forms. There is much less biological control, and the control is a more haphazard, physical control. The population size often goes over the carrying capacity before some other physical factor decreases the population size. Unlike the case for density-dependent factors, in populations being controlled by density-independent factors, growth rates do not seem to show any trend at all relative to population density. This type of regulation will usually occur in ecosystems where the communities have few species (i.e., where fewer biological interactions are taking place) and in ecosystems that are usually stressed periodically by physical factors (e.g., periodic flooding through a flood plain).

100. (D) High-quality energy is capable of performing a large amount of work, while low-quality energy is capable of performing less work. Energy always changes from high to low quality when work is performed. During the change, some energy is lost in the form of heat, which cannot do work (second law of thermodynamics). The amount of energy lost as heat is often as high as 90% of the total energy involved. The reason that (D) is the answer is because of the word "dispersed." Dispersed energy is not concentrated; therefore, it is low quality. An analogy would be ore deposits—a "rich" gold mine has a lot of gold in concentrated form. A stream might also have a lot of gold (maybe even more than the mine), but if there is only a few small nuggets every few hundred feet, so it is not as "rich."

Explanations for Free-Response Questions

Question 1

(a) 3 points maximum

(i) 1 point

Restatement: Number of cubic feet of natural gas required to heat the house for the winter.

$$\frac{4000 \cancel{\text{square feet}}}{1} \times \frac{100{,}000 \text{ BTU}}{1 \cancel{\text{square foot}}} = \mathbf{400{,}000{,}000 \text{ BTU}}$$

(ii) 1 point

Restatement: Cost of heating the home for one winter using natural gas.

$$\frac{400{,}000{,}000 \cancel{\text{BTU}}}{1} \times \frac{1 \cancel{\text{cubic foot natural gas}}}{1000 \cancel{\text{BTU}}} \times \frac{\$5.00}{1000 \cancel{\text{cubic feet}}} = \mathbf{\$2000}$$

(iii) 1 point

Restatement: Cost of heating the home for one winter using electricity.

$$\frac{400{,}000{,}000 \cancel{\text{BTU}}}{1} \times \frac{1 \cancel{\text{kWh}}}{10{,}000 \cancel{\text{BTU}}} \times \frac{\$50}{500 \cancel{\text{kWh}}} = \mathbf{\$4000}$$

Notes:

1. If you do *not* show calculations, no points are awarded.
2. No penalty is assessed if you do not show units. However, you risk setting up the problem incorrectly if you do not show units so that they cancel properly.
3. If your setup is correct but you make an arithmetic error, no penalty is assessed.

Note: For Parts (b) and (c), there are no points for just listing ideas. Each idea *must* be explained. Following are ideas that you could use to write your paragraph(s). Take these ideas and create an outline of the order in which you wish to answer them. You do NOT need to use all ideas. Before you begin, decide on the format of how you wish to answer your question (e.g., pros vs. cons, chart format with explanations within the chart, compare and contrast). Each numbered point when explained is worth 1 point.

(b) 4 points maximum

Restatement: Active versus passive solar design.

Active Solar System (2 points)

- Collectors that collect and absorb solar radiation. Optimum collector orientation is true south (the highest apparent point in the sky that the sun reaches during the day—not necessarily magnetic south). Collector orientation may deviate up to 20° from true south without significantly reducing the performance of the system. Collectors should be tilted at an angle equal to latitude plus 15° for optimum performance. A collector receives the most solar radiation between 9:00 A.M. and 3:00 P.M. Trees, buildings, hills, or other obstructions that shade collectors reduce their ability to collect solar radiation. Even partial shading will reduce heat output.
- Electric fans or pumps to transfer and distribute the solar heat in a fluid (liquid or air) from the collectors.
- Storage system to provide heat when the sun is not shining.
- May be more flexible than a passive system in terms of location and installation.
- Usually more economical to design an active system to provide 40 to 80% of the home's heating needs.
- "Liquid systems" heat water or an antifreeze solution in a "hydronic" collector, whereas "air systems" heat air in an "air" collector.
- Liquid solar collectors are most appropriate for central heating. They are the same as those used in solar domestic water heating systems. Flat-plate collectors are the most common, but evacuated tube and concentrating collectors are also available. In the collector, a heat transfer or "working" fluid such as water, antifreeze (usually nontoxic propylene glycol), or other type of liquid absorbs the solar heat. At the appropriate time, a controller operates a circulating pump to move the fluid through the collector. The liquid flows rapidly through the collectors, so its temperature only increases 10–20°F (5.6–11°C) as it moves through the collector. The liquid flows to either a storage tank or a heat exchanger for immediate use. Other system components include piping, pumps, valves, an expansion tank, a heat exchanger, a storage tank, and controls.
- Air collectors produce heat earlier and later in the day than liquid systems. Therefore, air systems may produce more usable energy over a heating season than a liquid system of the same size. Also, unlike liquid systems, air systems do not freeze, and minor leaks in the collector or distribution ducts will not cause problems. Air collectors can be installed on a roof or an exterior (south-facing) wall for heating one or more rooms. These systems are easier and less expensive to install than a central heating system. They do not have a dedicated storage system or extensive ductwork. The floors, walls, and furniture will absorb some of the solar heat, which will help keep the room warm for a few hours after sunset. Masonry walls and tile floors will provide more thermal mass, and thus provide heat for longer periods. A well-insulated house will make a solar room air heater more effective. Factory-built collectors and "do-it yourself" for on-site installation are available. The collector has an airtight and insulated metal or wood frame and a black metal plate for absorbing heat with glazing in front of it. Solar radiation heats the plate, which in turn heats the air in the collector. An electrically powered fan or blower pulls cooler air from the home through the collector and distributes it into various rooms in the home. Roof-mounted collectors require ducts for supplying air from the room(s) to the collector and for distribution of the warm air into the room(s). Wall-mounted collectors are placed directly on a south-facing wall. Holes

are cut through the wall for the collector air inlet and outlets. Simple "window box collectors" fit in an existing window opening. They can be active (using a fan) or passive. A baffle or damper keeps the room air from flowing back into the panel (reverse thermosiphoning) when the sun is not shining. These systems only provide a small amount of heat because the collector area is relatively small.
- Local covenants may restrict options; for example, homeowner associations may not allow installation of solar collectors on certain parts of the house.

Passive Solar System (2 points)
- Basic idea of passive solar design is to allow daylight, heat, and airflow into a building only when beneficial. The objectives are to control the entrance of sunlight and air flows into the building at appropriate times and to store and distribute the heat and cool air so it is available when needed.
- Four basic approaches to passive systems: (1) direct gain—solar energy is transmitted through south-facing glazing, and works best when the south window area is double glazed and the building has considerable thermal mass in the form of concrete floors and masonry walls insulated on the outside; (2) indirect gain—a storage mass collects and stores heat directly from the sun and then transfers heat to the living space, but the sun rays do not travel through the occupied space to reach the storage mass; (3) isolated gain—passive solar concept, solar collection and storage are thermally isolated from the areas of the building, allowing collector and storage to function independently of the building; and (4) remote collection—includes a collector space, which intercedes between the direct sun and the living space and is distinct from the building structure.
- Collectors usually receive the most sunlight when placed on the roof. South-facing walls may also work.
- Passive system does not use a mechanical device to distribute solar heat from a collector.
- Example of a passive system for space heating is a sunspace or solar greenhouse on the south side of the house.
- Passive system is simpler in design and less expensive to build.
- May not be possible depending upon location of site (e.g., building an effective sunspace may not be possible due to trees or other buildings in the way).
- Local climate, the type and efficiency of the collector(s), and the collector area determine how much heat a solar heating system can provide.
- Most building codes and mortgage lenders require a backup heating system. Supplementary or backup systems supply heat when the solar system cannot meet heating requirements. They range from a wood stove to a conventional central heating system.
- Passive solar buildings use 47% less energy than conventional new buildings and 60% less than comparable older buildings.

(c) 3 points total

Restatement: Five techniques that the homeowners could adopt to make their home into a "green building."

Definition (1 point): A "green" building focuses on a whole-system perspective, including energy conservation, resource-efficient building techniques and materials, indoor air quality, water conservation, and designs that minimize waste while utilizing recycled materials. Green buildings are a product of good design that minimizes a building's energy needs while reducing construction and maintenance costs over the life cycle of a building.

Techniques (2 points based on choosing any five and a discussion of each one):

- Solar collectors for space heating.
- Solar collectors for water heating.
- Photovoltaics to supply electrical energy.
- Hybrid systems that incorporate more than one power source (e.g., wind or microhydro).
- Energy-efficient appliances (Energy Star—http://www.energystar.gov).

- Products made from environmentally-attractive materials, including products that reduce materials use. Examples: drywall clips that eliminate the need for corner studs, salvaged materials, recycled materials (sprayed cellulose, a recycled material used for insulation), products made from agricultural waste, and certified wood products that carry an FSC stamp, indicating they meet the high standards set by the Forestry Stewardship Council.
- Products that don't contain toxins. Examples: products that substitute for polyvinyl chloride (PVC), ozone-depleting chemicals, and conventional pressure-treated lumber.
- Products that reduce the environmental impacts of construction, renovation, and demolition. Example: erosion-control products and exterior stains with low emissions of volatile organic compounds (VOCs).
- Products that reduce the environmental impacts of building operation, including products that reduce energy and water use, reduce the need for pesticide treatments (e.g., physical termite barriers), or have unusual durability or require little maintenance.
- Products that contribute to a safe, healthy indoor environment, including products that don't release significant pollutants (e.g., low-VOC paints, caulks, and adhesives), products that remove indoor pollutants (e.g., certain ventilation products), and products that warn inhabitants of health hazards (e.g., carbon monoxide detectors and lead paint testing kits).
- Superinsulated houses: (1) are constructed to be air tight; (2) have a higher level of insulation compared to conventional houses; and (3) have a ventilation system to control air quality.
- "Green landscaping"—landscaping that provides shade during summer and allows sunlight to warm house during winter (deciduous trees). Large deciduous trees planted at least 15 to 25 feet away from a house on the south and west sides provide afternoon summer shade (while still allowing cooling winds to pass through the tree canopy) yet permit sun to pass through bare branches for solar gain in winter. In addition, it can reduce cooling costs by shading air conditioning units. And finally, using plants that do not require a lot of water.

Question 2

(a) 3 points

Restatement: What is global warming?

Energy from the sun drives the Earth's weather and climate and heats the Earth's surface; in turn, the Earth radiates energy back into space. Atmospheric greenhouse gases (water vapor, carbon dioxide, and other gases) trap some of the outgoing energy, retaining heat somewhat like the glass panels of a greenhouse. Without the "greenhouse effect," much of the heat trapped by the atmosphere would escape into space and life, as we know it, could not exist. The Earth's climate is predicted to change because human activities are altering the chemical composition of the atmosphere through the buildup of greenhouse gases—primarily carbon dioxide, methane, and nitrous oxide. According to the graph, since the beginning of the industrial revolution, atmospheric concentrations of carbon dioxide have increased nearly 30%. In addition to CO_2, methane concentrations have more than doubled, and nitrous oxide concentrations have risen by about 15%. These increases have enhanced the heat-trapping capability of the Earth's atmosphere. Sulfate aerosols, a common air pollutant, cool the atmosphere by reflecting light back into space; however, sulfates are short-lived in the atmosphere and vary regionally.

(b) 3 points

Restatement: What factors may be increasing global temperature?

Scientists generally believe that the combustion of fossil fuels and other human activities are the primary reason for the increased concentration of carbon dioxide. Plant respiration and the decomposition of organic matter release more than ten times the CO_2 released by human activities, but these releases have generally been in balance during the centuries leading up to the Industrial Revolution with carbon dioxide being absorbed by terrestrial vegetation and the oceans. What has changed in the last few hundred years is the additional release of carbon dioxide by human activities. Fossil fuels burned to run cars and trucks, heat homes and businesses, and power factories are responsible for about 98% of U.S. carbon dioxide emissions, 24% of methane emissions, and 18% of nitrous oxide emissions. Increased agriculture, deforestation, landfills, industrial production, and mining also contribute a significant share of emissions. In 1997, the United States emitted about one-fifth of total global greenhouse gases. By 2100, in the absence of emissions control policies, carbon dioxide concentrations are projected to be 30–150% higher than today's levels. Water vapor is the most abundant greenhouse gas; it occurs naturally and makes up about two-thirds of the natural greenhouse effect. Other gases that accelerate the greenhouse effect include nitrous oxide (N_2O), hydrofluorocarbons (HFCs), perfluorocarbons (PFCs), and sulfur hexafluoride (SF_6). Since preindustrial times atmospheric concentrations of N_2O have increased by 17%, CO_2 by 31%, and CH_4 by 151%. Scientists have confirmed that this is primarily due to human activity. Burning coal, oil, and gas and cutting down forests are largely responsible.

(c) 4 points maximum (1 point for each consequence)

Restatement: Environmental consequences for global warming.

According to the graph, global mean surface temperatures have increased 0.5–1.0°F since the late 1800s. The 20th century's ten warmest years all occurred in the last 15 years. The snow cover in the northern hemisphere and floating ice in the Arctic Ocean has decreased. Globally, sea level has risen 4–8 inches in the last 100 years. Worldwide precipitation over land has increased by about 1% (warmer temperatures cause greater amounts of water to evaporate). The frequency of extreme rainfall events has increased throughout much of the United States. Increasing concentrations of greenhouse gases are likely to accelerate the rate of climate change. Scientists expect that the average global surface temperature could rise 1–5°F in the next 50 years, and 2–10°F in the next century, with significant regional variation. Evaporation will increase as the climate warms, which will increase average global precipitation. Soil moisture is likely to decline in many regions, and intense rainstorms are likely to become more frequent. Sea level is likely to rise 2 feet along most of the U.S. coast. A few degrees of warming will increase the chances of more frequent and severe heat waves, which can cause more heat-related death and illness. Greater heat can also mean worsened air pollution, as well as damaged crops and depleted water resources. Warming is likely to allow tropical diseases, such as malaria, to spread northward in some areas of the world. It will also intensify the Earth's hydrological cycle. This means that both evaporation and precipitation will increase. Some areas will receive more rain, while other areas will be drier. At the same time, extreme events like floods and droughts are likely to become more frequent. Warming will cause glaciers to melt and oceans to expand. Projections are that sea level will rise between 4 inches and 3 feet over the next century, in addition to the local sea level changes caused by other factors such as land subsidence and plate tectonics. This threatens low-lying coastal areas. Scientists are also concerned that warming could lead to more intense storms.

Question 3

(a) 3 points

Restatement: Technique(s) of determining oxygen content of water.

Biological Oxygen Demand (BOD) is a measure of the oxygen used by microorganisms to decompose organic wastes. If there is a large quantity of organic waste in the water supply, bacterial counts will be high; therefore, the demand for oxygen will be high, resulting in a high BOD level. As the waste is consumed or dispersed through the water, BOD levels will begin to decline. Nitrates and phosphates in a body of water can also contribute to high BOD levels. Nitrates and phosphates are plant nutrients and can cause plant life and algae to grow quickly. When plants grow quickly, they also die quickly. This contributes to the organic waste in the water, which is then decomposed by bacteria resulting in a high BOD level. When BOD levels are high, dissolved oxygen (DO) levels decrease because the oxygen that is available in the water is being consumed by the bacteria. Because less dissolved oxygen is available in the water, fish and other aquatic organisms may not survive. The BOD test takes five days to complete and is performed using a dissolved oxygen test kit. The BOD level is determined by comparing the DO level of a water sample taken immediately with the DO level of a water sample that has been incubated in a dark location for five days. The difference between the two DO levels represents the amount of oxygen required for the decomposition of any organic material in the sample and is a good approximation of the BOD level. BOD levels of 1–2 ppm are indicative of good water quality. There will not be much organic waste present in the water supply. A water supply with a BOD level of 3–5 ppm is considered moderately clean. In water with a BOD level of 6–9 ppm, the water is considered somewhat polluted because there is usually organic matter present, and bacteria are decomposing this waste. At BOD levels of 10 ppm or greater, the water supply is considered very polluted with organic waste. At these BOD levels, organisms that are more tolerant of lower dissolved oxygen (i.e., leeches and sludge worms) may appear and become numerous. Organisms that need higher oxygen levels (i.e., caddisfly larvae and mayfly nymphs) will not survive.

(b) 3 points

Restatement: Three other tests that could determine the water quality of the stream.

1. *Turbidity.* Turbidity is a measure of suspended and colloidal particles including clay, silt, organic and inorganic matter, algae, and microorganisms. It is caused by soil erosion, excess nutrients, various wastes and pollutants, and the action of bottom-feeding organisms that stir up sediments in the water. Turbidity is measured by electronically quantifying the degree to which light traveling through a water column is scattered by the suspended organic (including algae) and inorganic particles. The scattering of light increases with a greater suspended load. Turbidity is commonly measured in nephelometric turbidity units (NTU). When turbidity is high, water loses its ability to support a diversity of aquatic organisms. Oxygen levels decrease in turbid water as they become warmer as the result of heat absorption from the sunlight by the suspended particles and with decreased light penetration resulting in decreased photosynthesis. Suspended solids can clog fish gills, reduce growth rates and disease resistance, and prevent egg and larval development. Settled particles can accumulate and smother fish eggs and aquatic insects on the river bottom, suffocate newly-hatched insect larvae, and make river-bottom microhabitats unsuitable for mayfly nymphs, stonefly nymphs, caddisfly larvae, and other aquatic insects.
2. *Nitrate level.* Nitrate (NO_3^-) and nitrite (NO_2^-) are inorganic forms of nitrogen in the aquatic environment. Nitrate along with ammonia are the forms of nitrogen used by plants. Nitrates and nitrites are formed through the oxidation of ammonia by nitrifying bacteria, a process known as nitrification. In turn, these ions are converted to other nitrogen forms by denitrification and plant uptake. Nitrogen in its various forms is usually more abundant than phosphorus in the aquatic environment; therefore, nitrogen rarely limits plant growth, as does phosphorus. Aquatic plants are not usually as sensitive to increases in ammonia and nitrate levels. Sources of nitrates are the atmosphere, inadequately treated wastewater from sewage treatment plants, agricultural runoff, storm drains, and poorly functioning septic systems. Nitrates become toxic to fish (and plants) at levels of 50–300 ppm, depending on the fish species. For fry, however, much lower concentrations become toxic. Nitrate cannot be measured directly. It is first reduced to nitrite by chemical reduction and then measured colorimetrically using a spectrophotometer.

3. *pH.* pH value is measured on a scale of 0 to 14, and measures the concentration of hydrogen ions (H^+), or more accurately, the hydronium ion (H_3O^+). Pure distilled water is considered neutral, with a pH reading of 7. Water is basic if the pH is greater than 7, with 14 being the most basic; and water with pH of less than 7 is considered acidic, with 0 being the most acidic. For every one unit change in pH, there is approximately a ten-fold change in how acidic or basic the sample is. U.S. natural water falls between 6.5 and 8.5 on this scale with 7.0 being neutral. The optimum pH for river water is around 7.4. Water's acidity can be increased by acid rain but is kept in check by limestone, which acts as a buffer. Most valuable species, such as brook trout, are sensitive to changes in pH; immature stages of aquatic insects and immature fish are extremely sensitive to low pH values. Very acidic lakes and streams cause leaching of heavy metals into the water. pH measurements are determined by using either electronic pH meters, which determine the pH through the difference in potential between two electrodes, or pH paper, which has various organic dye indicators sensitive to narrow pH ranges.

(c) 4 points

Restatement: Food web based on a stream ecosystem.

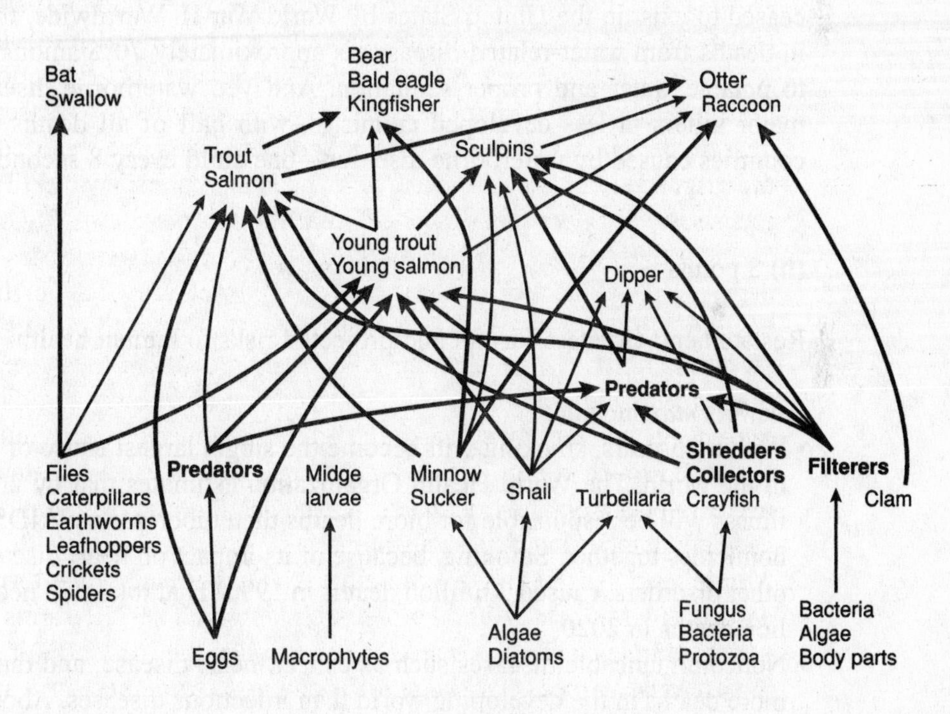

Question 4

(a) 3 points

Restatement: Two possible reasons that human life span has increased dramatically within the last 100 years.

During the 20th century, the health and life expectancy of persons residing in the United States has improved dramatically. Since 1900, the average life span of persons in the United States has increased by more than 30 years; 25 years of this gain are attributable to advances in public health. The two public health interventions that have had the greatest impact within the last 100 years on world health and human life span have been (1) the availability of clean water and sanitation and (2) the development of vaccines. Nearly a third to half of all babies born in the United States in 1900 did not survive past 5 years old. Whooping cough, diphtheria, and rheumatic and scarlet fever were major killers at that time. Major diseases that have been controlled to a greater or lesser extent through the use of vaccines within the last 100 years include plague, smallpox, diphtheria, tuberculosis, polio, measles, and hepatitis B. In terms of the single most infectious carrier, many experts believe that the *Anopheles* mosquito has been responsible for half of all human deaths since the Stone Age!

In 1900, it was common practice to dispose of garbage, industrial wastes, and raw sewage by dumping it into waterways. Few municipalities treated wastewater because it was widely believed that running water purified itself. Typhoid alone killed more than 150 per 100,000 people annually, including Wilbur Wright. Dysentery and diarrhea—the most common waterborne diseases—were the third largest cause of death in the nation 100 years ago. Infections such as typhoid and cholera transmitted by contaminated water, a major cause of illness and death in America early in the 20th century, have been reduced dramatically by improved sanitation. As a result of disinfection of public drinking water and improved sanitation methods, the major waterborne diseases all but ceased to exist in the United States by World War II. Worldwide, the median reduction in deaths from water-related diseases is approximately 70% among people with access to potable water and proper sanitation. And yet, waterborne diseases continue to be major killers in less-developed countries, with half of all deaths of children in poor countries caused by waterborne diseases—one child every 8 seconds.

(b) 3 points

Restatement: Greatest current and projected risks to human health.

Answers *may* include:

- Within 25 years, smoking will become the single largest cause of death and disability in the world. The World Health Organization estimates that by 2020, tobacco-related illness will be responsible for more deaths than tuberculosis, AIDS, car accidents, and homicides together. Smoking, because of its impact on heart disease, lung cancer, and other disorders, caused 3 million deaths in 1990. That total will nearly triple to 8.4 million deaths in 2020.
- Noncommunicable diseases such as cancer, heart disease, and diabetes already cause more deaths in the developing world than infectious diseases. About 56% of all deaths are from noncommunicable diseases, a proportion that is expected to jump to 73% by 2020. More people die of heart disease than any other cause. Of the 6.3 million who died of heart disease in 1990, about one-third were from developed countries. In the same year, strokes killed 1.5 million people in developed countries and a total of 4.5 million worldwide. Worldwide, pneumonia killed 4.5 million people and diarrheal disease 3 million, nearly all of them in the developing countries.
- Depression, thought to be largely associated with affluence, accounts for a full 10% of productive years lost throughout the world. By 2020, depression will account for 15% of the total disease burden.

- Largely as a result of smoking, alcoholism, and accidents, men in the former Soviet Union face a 28% risk of death between the ages of 15 and 60, the highest risk anywhere outside Sub-Saharan Africa.
- Worldwide, one of every three people dies from communicable diseases (50,000 each day), childbirth, or malnutrition. Virtually all those deaths are in developing regions. One of ten deaths result from injuries caused by accidents, wars, suicides, and homicides.
- By 2020, car accidents will be the world's fifth leading cause of death and disability as developing nations build more roads and the number of young adults—those most often killed in traffic mishaps—increases.
- Tuberculosis, pneumonia, and diarrheal diseases currently account for nearly 20% of the global disease burden. Annually, diarrheal diseases, including cholera, typhoid, and dysentery, spread chiefly by contaminated water or food, kill 3 million, most of them children. Tuberculosis kills almost 3 million, mostly adults.
- About 1.5 billion people are infected with intestinal worms.
- In 1990, the major diseases and causes of disability were ranked as

1. Lower respiratory infections
2. Diarrheal diseases
3. Conditions arising during prenatal period
4. Unipolar major depression
5. Ischemic heart disease
6. Cerebrovascular disease
7. Tuberculosis
8. Measles
9. Road traffic accidents
10. Congenital anomalies

In 2020, the ranking worldwide is projected to change to

1. Ischemic heart disease
2. Unipolar major depression
3. Road traffic accidents
4. Cerebrovascular disease
5. Chronic obstructive pulmonary disease
6. Lower respiratory infections
7. Tuberculosis
8. War
9. Diarrheal diseases
10. HIV

(c) 4 points maximum (6 points possible)

Restatement: Concept of "risk analysis." How would information gained from studies of risk analysis increase the human lifespan?

Definition (1 point): Risk, in the context of human health, is the probability of injury, disease, or death from exposure to environmental hazards. In quantitative terms, risk is expressed in values ranging from zero (representing the certainty that harm will not occur) to one (representing the certainty that harm will occur). The process of risk assessment and analysis involves four major steps: (1) hazard identification, (2) dose-response assessment, (3) exposure assessment, and (4) risk characterization.

Describe each component (1 point each):

1. *Hazard identification* is the collection and evaluation of data on the types of health injury or disease that may be produced by a chemical and on the conditions of exposure under which injury or disease is produced. It may also involve characterization of the behavior of a chemical within the body and the interactions it undergoes with organs, cells, or even parts of cells. Data of the latter types may be of value in answering the ultimate question of whether the forms of toxicity known to be produced by a substance in one population group or in experimental settings are also likely to be produced in humans.
2. *Dose-response assessment* involves describing the quantitative relationship between the amount of exposure to a substance and the extent of toxic injury or disease. Data are derived from animal studies or, less frequently, from studies in exposed human populations. There may be many different dose-response relationships for a substance if it produces different toxic effects under different conditions of exposure.
3. *Exposure assessment* involves describing the nature and size of the population exposed to a substance and the magnitude and duration of their exposure. The evaluation could concern past or current exposures or exposures anticipated in the future.
4. *Risk characterization* generally involves the integration of the data and analysis of the first three components of the risk assessment process (hazard identification, dose-response assessment, and exposure assessment) to determine the likelihood that humans will experience any of the various forms of toxicity associated with a substance.

Apply concept of risk analysis to real-world examples (1 point): Risk analysis information is used in the risk management process in deciding how to protect public health and extend human life. Examples of risk management actions include deciding how much of a chemical a company may discharge into a river; deciding which substances may be stored at a hazardous waste disposal facility; deciding to what extent a hazardous waste site must be cleaned up; setting permit levels for discharge, storage, or transport; establishing levels for air emissions; and determining allowable levels of contamination in drinking water. Essentially, risk assessment provides information on the health risk, and risk management is the action taken based on that information.

Appendix I: Units and Measurement

AREA

1 acre (ac)	= 43,560 ft^2 = 0.405 ha
1 hectare (ha)	= 2.471 ac = 1.076×10^6 ft^2 = 10^4 m^2 = 0.01 km^2
1 square centimeter (cm^2)	= 0.155 $in.^2$
1 square foot (ft^2)	= 144 $in.^2$ = 929 cm^2 = 0.0929 m^2
1 square inch ($in.^2$)	= 6.4516 cm^2
1 square kilometer (km^2)	= 10^6 m^2 = 100 ha = 247.1 ac = 0.3861 mi^2
1 square meter (m^2)	= 10^4 cm^2 = 10.76 ft^2 = 1.196 yds^2 = 10^6 mm^2 = 1550 $in.^2$
1 square mile (mi^2)	= 640 ac = 2.59 km^2 = 27,878,400 ft^2
1 square yard (yd^2)	= 9 ft^2 = 0.836 m^2

ENERGY, FORCE, AND POWER

1 British Thermal Unit (BTU)	= 0.293 Wh = 0.000293 kWh = 252 cal = 1055 J
1 calorie (cal)	= 4.184 J
1 erg	= 1 dyne/cm^2
1 foot-pound (ft-lb)	= 1.356 J
1 gigawatt (gW)	= 10^9 W
1 horsepower (hp)	= 640 kcal = 7.457×10^2 W = 550 ft-lb/sec = 33,000 ft-lb/min
1 joule (J)	= 1 watt-second = 10^7 ergs = 9.481×10^{-4} BTU = 0.239 cal = 2.778×10^{-7} kWh = 1 kg · m^2/sec^2 = 1 Newton · meter = 0.7376 ft-lb = 6.24×10^{18} eV
1 kilocalorie (kcal)	= 1000 cal = 0.00116 kWh = 3.97 BTU
1 kilojoule (kJ)	= 1000 J = 0.000278 kWh = 0.9484 BTU = 737.6 ft-lb
1 kilowatt (kW)	= 1000 W = 1 kJ/sec = 1.341 hp
1 kilowatt-hour (kWh)	= 1,000 Wh = 860.421 kcal = 860,421 cal = 3600 kJ = 3,413 BTU
1 langley (ly)	= 1 cal/cm^2
1 langley/minute	= 0.6974 kW/m^2 = 3.687 BTU/ft^2/min
1 megajoule (MJ)	= 1×10^6 J
1 Newton	= 1 kg · m/sec^2
1 quad (Q)	= 10^{15} BTU = 1.054×10^{15} kJ = energy in 1.72×10^8 barrels of oil = 2.93×10^{12} kWh
1 watt (W)	= 3.413 BTU/hr = 14.34 cal/min

FLOW

1 cubic foot/second (cf/s)	= 448.9 gal/min = 7.48 gal/sec = 8.931 × 10^5 m^3/yr = 724.0 acre-feet/yr = 0.0283 m^3/sec = 28.32 L/sec
1 cubic meter/second (m^3/sec)	= 35.32 ft^3/sec
1 ft/sec	= 0.682 mi/hr = 1.097 km/hr
1 ft^3/sec/day	= 1.98 acre-feet
1 m/sec	= 3.6 km/hr = 2.24 mi/hr

LENGTH

1 centimeter (cm)	10^{-5} km = 10^{-2} m = 10 mm = 0.394 in. = 0.0328 ft = 6.2 × 10^{-6} mi
1 chain	4 rods = 66 ft = 20 m
1 fathom	6 ft = 1.83 m
1 foot (ft)	0.305 m = 12 in. = 0.33 yd = 1.89 × 10^{-4} mi = 30.48 cm
1 furlong	10 chains = 40 rods = 660 ft = 200 m
1 inch (in.)	2.54 cm = 0.0254 m = 2.54 × 10^{-5} km = 0.083 ft = 1.58 × 10^{-5} mi
1 kilometer (km)	10^3 m = 10^5 cm = 0.6214 mi = 0.54 nautical miles = 39,370 in. = 3,281 ft
1 meter (m)	10^{-3} km = 10^2 cm = 10^3 mm = 10^6 μm = 39.4 in. = 3.28 ft = 1.09 yd = 6.2 × 10^{-4} mi
1 mile (mi)	5,280 ft = 1760 yd = 8 furlongs = 1.61 km = 63,360 in. = 160,934 cm = 1609 m
1 millimeter (mm)	0.001 m = 0.01 cm = 0.039 in.
1 nautical mile	1.15 mi = 1.85 km
1 rod	16.5 ft = 5 m
1 yard (yd)	0.914 m = 36 in. = 3 ft

MASS AND WEIGHT

1 gram (g)	10^3 mg = 10^6 μg = 0.0353 oz = 0.0022 lb
1 kilogram (kg)	10^3 g = 2.205 lb
1 long ton (lt)	2240 pounds = 1008 kg
1 metric ton (mt)	10^3 kg = 2205 lb = 1.102 t
1 microgram (μg)	10^{-3} mg = 10^{-6} g
1 milligram (mg)	10^{-3} g
1 ounce (oz)	28.3 g
1 pound (lb)	16 oz = 453.6 g = 0.4536 kg = 7000 grains
1 short ton (t)	2000 lb = 907.2 kg

PRESSURE

1 atmosphere (atm)	14.7 lb/in.^{2} = 2116 lb/ft^2 = 1.013 × 10^5 N/m^2 = 101,325 Pa = 101.33 kPa = 760 mm Hg = 29.92 in. Hg = 406.8 in. H_2O = 33.9 ft H_2O = 1013 millibars = 1.013 bars = 760 torrs
1 bar	1 × 10^5 Pa = 0.9869 atm
1 pascal (Pa)	1 N/m^2 = 1.45 × 10^{-4} lb/in.^{2} = 9.87 × 10^{-6} atm
1 pound per square inch (lb/in.^{2})	0.068 atm = 6895 Pa = 51.7 mm Hg = 27.68 in. H_2O
1 torr	133.32 N/m^2 = 1/760 atm

QUANTITATIVE PREFIXES

quintillion	10^{18}	= exa-	(E)	10^{-18}	= atto-	(a)
quadrillion	10^{15}	= peta-	(P)	10^{-15}	= femto-	(f)
trillion	10^{12}	= tera-	(T)	10^{-12}	= pico-	(p)
billion	10^{9}	= giga-	(G)	10^{-9}	= nano-	(n)
million	10^{6}	= mega-	(M)	10^{-6}	= micro-	(μ)
thousand	10^{3}	= kilo-	(k)	10^{-3}	= milli-	(m)
hundred	10^{2}	= hecto-	(h)	10^{-2}	= centi-	(c)
ten	10^{1}	= deca-	(da)	10^{-1}	= deci-	(d)

RADIATION

1 becquerel	1 nuclear disintegration per second
1 curie	3.7 × 10^{10} nuclear disintegrations per second
1 gray (Gy)	absorbed dose of 1 J of energy per kg body tissue = 100 rads = 1 J/kg
1 rad	100 ergs absorbed radiation/g absorbing medium = 0.01 J/kg of medium
1 rem	an equivalent radiation dose = rads. Qualifying Factor (QF) for gamma and X-rays = 1; QF for fast neutrons and protons = 10; QF for alpha particles = 20
1 roentgen	exposure to X-rays or gamma rays resulting in an electric charge of 2.58 × 10^{-4} coulomb/kg dry air = ~ 10 × 10^{-3} sievert dose
1 sievert	100 rems

SPEED

1 meter per second (m/sec)	3.281 ft/sec = 2.237 mph
1 mile per hour (mph)	0.447 m/sec = 1.609 km/hr = 1.467 ft/sec = 88 ft/min

TEMPERATURE

Centigrade (°C)	(°F – 32.0) / 1.80
Fahrenheit (°F)	(°C × 1.80) + 32.0
Kelvin (K)	°C + 273.15

VISCOSITY

1 poise	1.0 g/cm · sec
1 centipoise	0.01 g/cm · sec

VOLUME

1 acre foot	325,851 gal = 1,234,975 l = 1234 m^3 = 43,560 ft^3
1 barrel (oil) (bbl)	42 gal = 159.6 l = 168 qt
1 cubic centimeter (cm^3)	1 mL = 0.001 L = 0.061 $in.^3$
1 cubic ft (ft^3)	1,728 $in.^3$ = 0.0283 m^3 = 28.3 l = 7.480 gal
1 cubic inch ($in.^3$)	0.02 l
1 cubic kilometer (km^3)	10^9 m^3 = 0.2399 mi^3 = 8.106×10^5 acre-feet = 2.64×10^{11} gal
1 cubic meter (m^3)	10^3 L = 10^6 cm^3 = 35.315 ft^3 = 1.307 yd^3 = 61,020 $in.^3$ = 264.2 gal = 6.290 bbl = 1 stere
1 cubic mile (mi^3)	4.166 km^3
1 cubic yard (yd^3)	27 ft^3 = 0.765 m^3 = 202 gal
1 cup	0.24 L
1 fluid ounce (fl oz)	30 mL
1 liter (L)	1000 (cm^3) = 1000 mL = 33.81 fl oz = 1.057 qt = 0.2642 gal = 0.035 ft^3 = 61.02 $in.^3$
1 milliliter (mL)	0.001 L = 1 cm^3
1 pint (pt)	0.47 l
1 quart (qt)	2 pt = 4 cup = 0.95 L = 0.25 gal
1 tablespoon (tbsp)	15 mL
1 teaspoon (tsp)	5 mL
1 U.S. gallon (gal)	4 qt = 3.785 L = 231 $in.^3$ = 0.13 ft^3 = 0.83 imperial (British) gallons = 0.02 bbl

GREEK ALPHABET

Α	α	alpha	Ν	ν	nu
Β	β	beta	Ξ	ξ	xi
Γ	γ	gamma	Ο	ο	omicron
Δ	δ	delta	Π	π	pi
Ε	ε	epsilon	Ρ	ρ	rho
Ζ	ζ	zeta	Σ	σ	sigma
Η	η	eta	Τ	τ	tau
Θ	θ	theta	Υ	υ	upsilon
Ι	ι	iota	Φ	φ	phi
Κ	κ	kappa	Χ	χ	chi
Λ	λ	lambda	Ψ	ψ	psi
Μ	μ	mu	Ω	ω	omega

Appendix II: Comparative Statistics

I. Vital Statistics and Growth

VITAL STATISTICS BY SELECTED COUNTRY FOR 2000 AND PROJECTED FOR 2010

Country	Crude Birth Rate 2000[1]	Crude Birth Rate 2010[1]	Crude Death Rate 2000[2]	Crude Death Rate 2010[2]	Expected Life span 2000	Expected Life span 2010	Infant Mortality 2000[3]	Infant Mortality 2010	Fertility Rate 2000[4]	Fertility Rate 2010
Niger	51.5	45.5	23.2	19.1	41.3	44.6	124.9	111.3	7.16	6.28
Angola	46.9	42.9	25.0	21.6	38.8	41.3	195.8	174.7	6.52	5.94
Mozambique	38.0	31.1	23.3	29.9	37.5	31.4	139.9	131.8	4.93	3.85
Mexico	23.2	19.3	5.1	5.0	71.5	74.1	26.2	18.5	2.67	2.27
United States	14.2	14.3	8.7	8.6	77.1	78.5	6.8	6.2	2.06	2.12
Japan	10.0	8.9	8.2	10.2	80.7	81.7	3.9	3.6	1.41	1.47
Russian Federation	9.0	12.1	13.8	14.5	67.2	68.8	20.3	17.7	1.25	1.56

Source: U.S. Census Bureau, International Database, www.census.gov/ipc/www/idbnew.html.
[1]Births during 1 year per 1000 persons.
[2]Deaths during 1 year per 1000 persons.
[3]Deaths of children under 1 year of age per 1000 live births in 1 year.
[4]Average number of children that would be born if all women lived to end of their childbearing years.

RATE OF POPULATION GROWTH BY COUNTRY, 1990–2000

Country	Annual Rate of Growth (1990–2000)	Population per Square Mile
Gaza Strip	5.7	7,696
Afghanistan	5.6	104
Jordan	4.3	137
Singapore	3.2	16,637
Israel	2.6	732
India	1.8	883
Mexico	1.7	133
World	1.4	120
Australia	1.2	7
United States	1.0	78
Monaco	0.6	40,844
Japan	0.2	829
Russian Federation	–0.1	22

Source: U.S. Census Bureau, International Database, www.census.gov/ipc/www/idbnew.html.

DEATH RATES PER 100,000 POPULATION FROM INJURIES BY MECHANISM AND COUNTRY

Country	Total	Auto	Firearm	Poison	Fall	Suffocation	Drowning	Unspecified	Other
Norway	57.4	7.2	4.3	6.1	6.4	5.3	4.7	16.4	7.0
United States	56.3	16.2	13.7	6.2	4.3	3.9	1.9	3.0	7.1
Canada	44.7	10.5	3.9	6.7	5.0	6.1	2.1	4.9	5.5
Australia	39.7	11.0	2.9	6.8	2.9	4.4	2.2	3.5	6.0
Israel	32.9	10.3	2.8	0.7	2.6	3.1	1.2	8.7	3.5

Source: Organization for Economic Cooperation and Development, Paris, France, *OECD Health Data 99*. Data comes from the 1990s but may be collected over different time periods.

AGE DISTRIBUTION BY COUNTRY: 2000 AND PROJECTED FOR 2010

Country	2000, Under 15 years	2010, Under 15 years	2000, 65 years and over	2010, 65 years and over
Uganda	51.1	66.0	2.2	2.6
Niger	48.0	60.0	2.3	3.0
India	33.6	33.6	4.6	6.1
Mexico	33.8	33.2	4.3	6.3
World Average	29.8	30.0	6.9	8.6
China	25.4	22.5	7.0	8.9
United States	21.2	21.6	12.6	14.4
Russian Federation	18.1	14.9	12.6	12.7
Japan	14.8	14.5	17.0	21.9

Source: U.S. Census Bureau International Database, www.census.gov/ipc/www/idbnew.html.

II. Labor—Education

HIGHEST EDUCATIONAL ATTAINMENT BY COUNTRY, 1998

Country	Preprimary & Primary Education (%)	Secondary Education (%)	College Education (%)
Denmark	21	54	5
Japan	20	50	18
Mexico	79	8	12
Switzerland	19	58	14
Turkey	74	11	6
United States	14	52	15

Source: Organization for Economic Cooperation and Development, Paris, France.

FOREIGN STUDENTS AS PERCENT OF
TOTAL U.S. UNIVERSITY ENROLLMENT, 1998

Region of Origin	Africa	Asia	Europe	Oceania	South America	Japan	South Korea
	0.16	2.08	0.48	0.03	0.17	0.32	0.29

Source: Organization for Economic Cooperation and Development, Paris, France, *Education at a Glance.*

INCOME TAX AND SOCIAL SECURITY CONTRIBUTIONS
AS A % OF LABOR COSTS, 1998

Country	Income Tax (%)	Employee Social Security Contribution (%)	Employer Social Security Contribution (%)
Denmark	34	10	1
Korea (South)	20	—	—
United States	17	7	7
Poland	11	—	33
Japan	6	7	7
Mexico	—	2	20

Source: Organization for Economic Cooperation and Development, Paris, France, *Taxing and Wages, 1998–1999, 2000.*

FOREIGN OR FOREIGN-BORN POPULATION AND LABOR FORCE
IN SELECTED COUNTRIES, 1997

Country	Number of Foreign Population (1000s)	Foreign Population (%)	Number of Foreign Labor Force (%)	Foreign Labor Force (%)
United States	24,600	9.3	14,300	10.8
Australia	3,908	21.1	2,239	24.6
Japan	1,483	1.2	660	1.0
Luxembourg	148	34.9	125	55.1
Switzerland	131	19.0	693	17.5

Source: Organization for Economic Cooperation and Development, Paris, France, *Trends in International Migration.*

III. Food

PER CAPITA CONSUMPTION OF MEAT (IN KILOGRAMS)

Country	Beef and Veal	Pork	Poultry
Argentina	70.2 (1)	NA	NA
Hong Kong	NA	54.3 (6)	67.2 (1)
Denmark	NA	73.7 (1)	NA
United States	45.3 (3)	31.7 (17)	49.4 (2)

( ) = world rank; NA = not available.
Source: U.S. Department of Agriculture, Foreign Agricultural Service, *Livestock and Poultry, World Markets and Trade.*

WORLD FOOD PRODUCTION, 1999

Commodity	Millions of Metric Tons (1,857.7 = 1,857,700,000)
Grains, total	1857.7
Wheat	587.0
Coarse grains	871.0
Corn	600.8
Rice (milled)	400.7
Oils	85.7
Soybeans	154.7
Rapeseed	42.5
Vegetables and melons	628.7
Fruits	444.7
Nuts	6.4
Red meat	137.2
Poultry	57.8
Milk	387.2

Source: U.S. Department of Agriculture, Economic Research Service, *Agricultural Outlook*, monthly.

WORLD FISH CATCH, 1997*

Country	Total Catch (thousands of metric tons)	Catch Per Capita (metric tons per capita)
World	122,138	0.020
China	35,038	0.026
Peru	7,877	0.25
Japan	6,689	0.053
Chile	6,084	0.40
United States	5,448	0.020
Mexico	1,529	0.015

*Includes fish, crustaceans, mollusks, and weight of shells. Does not include mammals or plants.
Source: Food and Agriculture Organization of the United Nations, Rome, Italy.

IV. Energy

NET ELECTRICITY GENERATION BY COUNTRY, 1998

Country	Total (billion kWh)*	% Thermal	% Hydro	% Nuclear
World Total	13,615	62.7	18.8	17.0
China	1,098	80.3	18.5	1.2
Japan	996	56.7	9.0	31.9
Mexico	176	78.1	13.8	5.0
United States	3,619.6	70.3	9.0	18.6
Russian Federation	771.9	67.8	19.5	12.7
India	446.1	80.3	17.1	2.4
France	481.0	10.8	12.5	76.2

*13,615 = 13,615,500,000,000
Source: U.S. Energy Information Administration, *International Energy Annual.*

PRIMARY ENERGY CONSUMPTION AND PRODUCTION BY REGION, 1998

Region	Production (BTU)*	Consumption (BTU)*
United States	99.3	94.8
Far East and Oceania	75.4	99.3
Eastern Europe	58.3	49.0
Middle East	54.5	15.9
Western Europe	43.6	69.5
Africa	26.2	11.8
Central and South America	24.9	19.7
World	382.2	377.7

* in quadrillion BTU (99.3 = 99,300,000,000,000,000)
Source: U.S. Energy Information Administration, *International Energy Annual.*

ENERGY CONSUMPTION AND PRODUCTION BY COUNTRY, 1998

Country	Energy Consumed Per Capita (million BTU)	Natural Gas Production (trillion ft^3)	Crude Petroleum Produced Per Day (1000 barrels)	Coal Production (million short tons)
United Arab Emirates	669	1.3	2,345	0
Norway	420	1.6	3,017	<1
United States	351	18.9	6,252	1,119
Canada	391	6.0	1,981	83
Kuwait	345	0.3	2,085	0
Russian Federation	177	20.9	5,854	272
Japan	168	0.1	9	4
World	64	83.0	66,962	5,043
Mexico	59	1.3	3,070	11
China	27	0.8	3,198	1,351
India	13	0.8	661	359

Source: U.S. Energy Information Administration, *International Energy Annual.*

ENERGY CONSUMPTION AND PRODUCTION BY COMMODITY, 1998

Commodity	Production (BTU)*	Consumption (BTU)*
Oil	143.2	149.7
Coal	88.6	87.5
Natural gas	85.5	84.4
Hydroelectric	26.6	26.8
Nuclear electric power	24.5	24.5
Geothermal, solar, and wind	2.5	2.5

*in quadrillion BTU (143.2 = 143,200,000,000,000,000)
Source: U.S. Energy Information Administration, *International Energy Annual.*

V. Health

HEALTH EXPENDITURES AS A PERCENT OF GDP, 1997

Country	Total Health Expenditures (%)	Public Health Expenditures (%)
United States	13.9	6.5
Germany	10.7	8.3
Norway	7.5	6.2
Japan	7.2	5.7
United Kingdom	6.8	5.8
Mexico	4.7	2.8
Turkey	4.0	2.9

Source: Organization for Economic Cooperation and Development, Paris, France, *OECD Health Data 99.*

MEDICAL DOCTORS AND INPATIENT CARE BY COUNTRY, 1997

Country	Medical Doctors Per 1000 Population	Beds Per 1000 Population	Average Length of Stay in Hospital (days)
Italy	5.8	6.5	9
United States	2.7	4.0	8
Japan	1.8	16.4	43
Mexico	1.3	1.1	4
Turkey	1.1	2.5	6

Source: Organization for Economic Cooperation and Development, Paris, France, *OECD Health Data 99.*

CIGARETTE PRODUCTION, EXPORTS, AND IMPORTS BY COUNTRY, 1999

Country	Cigarettes (billions)
Production	
China	1675
United States	646
Indonesia	219
Russian Federation	190
Japan	188
World	5485
Exports	
United States	151
United Kingdom	106
Netherlands	104
World (total)	951
Imports	
Japan	81
France	61
Russia	45
World (total)	650

Source: U.S. Department of Agriculture, Foreign Agricultural Service, *Tobacco: World Markets and Trade, 2000.*

VI. Pollution

POLLUTION GENERATED PER COUNTRY PER CAPITA

Country	Sulfur Oxides (kg per capita)	Nitrogen Oxides (kg per capita)	Carbon Dioxide (tons per capita)	MSW (kg per capita)	Nuclear Wastes (kg per capita)
Australia	100.7	118.5	16.6	690	—
Canada	88.9	67.1	15.8	490	44.2
Japan	7.3	11.3	9.3	400	7.6
Mexico	23.2	16.4	3.5	300	0.4
Switzerland	4.5	18.0	6.3	600	8.9
United States	69.0	79.9	20.4	720	7.8

Source: Organization for Economic Cooperation and Development, Paris, France, *OCD Environmental Data Compendium, 1999.*

CARBON DIOXIDE EMISSION PRODUCED FROM FOSSIL FUELS BY COUNTRY, 1997

Country	Metric Tons (millions)
World (total)	6,163
United States	1,489
China	785
Russia	411
Japan	296
Canada	135
Mexico	94

Source: U.S. Energy Information Administration, *International Energy Annual, 1998,* and *International Energy Outlook, 2000.*

VII. Transportation—Tourism

MOTOR VEHICLE TRANSPORTATION INDICATORS, 1997

	United States	Japan	Mexico
Number (per 1000 persons)			
Automobiles	484.8	372.2	88.2
Buses	2.6	242	1.4
Roads			
Kilometers per 1000 persons	27.35	9.11	2.58
Kilometers per square kilometer	0.69	3.06	0.13
Vehicle kilometers of travel			
Automobiles (billion)	2,363	464	—
Buses (billion)	11	7	—
Average vehicle kilometers per vehicle			
Automobiles	18,216	9,903	—
Buses	15,149	27,683	—

Source: U.S. Federal Highway Administration, *Highway Statistics, 1998.*

INTERNATIONAL TOURISM INCOME, 1999

Country	Total Income (in millions). % Value in ( ) indicates change from 1998.	World Total (%)
United States	74,448 (+4.5%)	16.4
Spain	32,913 (+10.7%)	7.2
France	31,699 (+5.9%)	7.0
China	14,098 (+11.9%)	3.1
Greece	8,765 (+41.6%)	1.9
Russia	7,771 (+19.4%)	1.7
Mexico	7,587 (–3.9%)	1.7
World (total)	454,553	100

Source: World Tourism Organization, Madrid, Spain, *Tourism Highlights, 2000.*

Appendix III: Less-Developed and More-Developed Countries

Less-Developed Countries: GDP $10,000 and under

Afghanistan, Albania, Algeria, American Samoa, Angola, Anguilla, Antigua, Argentina, Armenia, Azerbaijan, Bangladesh, Barbuda, Belarus, Belize, Benin, Bhutan, Bolivia, Bosnia, Botswana, Brazil, Bulgaria, Burkina Faso, Burma, Burundi, Cambodia, Cameroon, Cape Verde, Central African Republic, Chad, Colombia, Comoros, Cook Islands, Costa Rica, Côte d'Ivoire, Croatia, Cuba, Democratic Republic of the Congo, Djibouti, Dominica, Dominican Republic, Ecuador, Egypt, El Salvador, Equatorial Guinea, Eritrea, Estonia, Ethiopia, Fiji, French Guiana, French Polynesia, Futuna, Gabon, Gambia, Gaza Strip, Georgia, Ghana, Grenada, Grenadines, Guadeloupe, Guatemala, Guinea, Guinea-Bissau, Guyana, Haiti, Herzegovina, Honduras, Hungary, India, Indonesia, Iran, Iraq, Jamaica, Jordan, Kazakhstan, Kenya, Kiribati, Kyrgyzstan, Laos, Latvia, Lebanon, Lesotho, Liberia, Libya, Lithuania, Macedonia, Madagascar, Malawi, Maldives, Mali, Marshall Islands, Martinique, Mauritania, Mayotte, Mexico, Micronesia, Moldova, Mongolia, Montenegro, Montserrat, Morocco, Mozambique, Namibia, Nauru, Nepal, Nevis, New Caledonia, Nicaragua, Niger, Nigeria, Nive, North Korea, Oman, Pakistan, Palau, Panama, Papua New Guinea, Paraguay, Peru, Philippines, Poland, Príncipe, Puerto Rico, Romania, Russia, Rwanda, Samoa, São Tomé, Senegal, Serbia, Seychelles, Sierra Leone, Slovakia, Slovenia, Solomon Islands, Somalia, South Africa, Sri Lanka, St. Kitts, St. Lucia, St. Vincent, Sudan, Suriname, Swaziland, Syria, Tajikistan, Tanzania, Thailand, Togo, Tokelau, Tonga, Tunisia, Turkey, Turkmenistan, Turks and Caicos Islands, Tuvalu, Uganda, Ukraine, Uruguay, Uzbekistan, Vanuatu, Venezuela, Vietnam, Wallis, West Bank, Yemen, Zambia, Zimbabwe

More-Developed Countries: GDP $10,001 and over

Andorra, Aruba, Australia, Austria, Bahamas, Bahrain, Barbados, Belgium, Bermuda, British Virgin Islands, Brunei, Canada, Cayman Islands, Chile, Cyprus, Czech Republic, Denmark, Faroe Islands, Finland, France, Germany, Gibraltar, Greece, Greenland, Guam, Hong Kong, Iceland, Ireland, Israel, Italy, Japan, Kuwait, Liechtenstein, Luxembourg, Macao, Malaysia, Malta, Man, Mauritius, Monaco, Netherlands, Netherlands Antilles, New Zealand, Northern Mariana Islands, Norway, Portugal, Qatar, San Marino, Saudi Arabia, Singapore, South Korea, Spain, St. Pierre and Miquelon, Sweden, Switzerland, Taiwan, Trinidad and Tobago, United Arab Emirates, United Kingdom, United States, Virgin Islands (U.S.)

Appendix IV: Comparison of Environmental Indicators for Selected Countries

Environmental Indicator	United States	China	India	Russia	Mexico
Population (millions), 1999	278.2	1,253.6	997.5	146.2	96.6
Urban population (% of total), 1999	77.0	31.6	28.1	77.3	74.2
GDP ($ billions), 1999	9,152.1	989.5	447.3	401.4	483.7
Agriculture					
Land area (thousand km^2)	9,159	9,327	2,973	16,889	1,909
Agricultural land (% of land area)	45.7	57.4	60.7	12.9	56.2
Irrigated land (% of crop land)	12.0	38.8	34.8	3.6	23.8
Fertilizer consumption (100 g/ha of arable land)	1,117	2,826	1,040	86	677
Food production index (1989–1991=100)	121.8	169.1	126.2	58.4	128.4
Population density, rural (people/km^2 of arable land)	36	689	438	27	98
Forests					
Forest area (thousand km^2)	2,260	1,635	641	8,514	552
Forest area (% of total land area)	24.7	17.5	21.6	50.4	28.9
Annual deforestation (% change, 1990–2000)	–0.2	–1.2	–0.1	0.0	1.1
Biodiversity					
Mammal species, total known	428	394	316	269	450
Mammal species, threatened	37	76	86	42	69
Bird species, total known	650	1,100	923	628	769
Bird species, threatened	55	73	70	38	39
Nationally protected area (% of land area)	13.4	6.4	4.8	3.1	3.5
Energy					
GDP per unit of energy use (PPP [purchasing power parity] $ per kg of oil equivalent)	3.8	4.0	4.3	1.7	5.2
Commercial energy use per capita (kg of oil equivalent)	7,937	830	486	3,963	1,552
Traditional fuel use (% of total energy use)	4	6	21	1	4
Energy imports, net (% of commercial energy use)	22	1	13	–60	–54

Environmental Indicator	United States	China	India	Russia	Mexico
Energy (continued)					
Electric power consumption per capita (kWh)	11,832.4	746.4	383.6	3,937.2	1,513.3
Share of electricity generated by coal (%)	52.7	75.9	75.4	19.4	9.8
Emissions and pollution					
CO_2 emissions per unit of GDP (kg per PPP $ of GDP)	0.7	0.9	0.5	1.4	0.5
Total CO_2 emissions, industrial (1000 kt)	5,467.1	3,593.5	1,065.4	1,444.5	379.7
CO_2 emissions per capita (mt)	20.1	2.9	1.1	9.8	4.0
Passenger cars (per 1000 people)	478	3	5	120	102
Water and sanitation					
Access to improved water source (% of total population)	100	75	88	99	86
Access to improved water source (% of rural population)	100	66	86	96	63
Access to improved water source (% of urban population)	100	94	92	100	94
Freshwater resources per capita (cubic meters)	8,906	2,257	1,913	30,767	4,742
Total freshwater withdrawal (% of total water resources)	18.1	18.6	26.2	1.7	17.0
Withdrawal for agriculture (% of total freshwater withdrawal)	27	77	92	20	78
Access to sanitation in urban areas (% of urban population)	100	68	73	—	87
Access to sanitation in rural areas (% of rural population)	100	24	14	—	32
Under-5 mortality rate (per 1000 live births)	8	37	90	20	36
National accounting aggregates—1999					
Gross domestic savings (% of GDP)	18.4	40.1	20.0	33.0	21.9
Consumption of fixed capital (% of GDP)	12.8	8.3	9.4	9.6	10.5
Net domestic savings (% of GDP)	4.7	31.8	10.5	23.4	11.4
Education expenditure (% of GDP)	4.7	2.0	3.3	3.7	4.4
Energy depletion (% of GDP)	0.7	1.4	1.3	12.8	4.0
Mineral depletion (% of GDP)	0.0	0.3	0.3	0.0	0.1
Net forest depletion (% of GDP)	0.0	0.3	1.8	0.0	0.0
CO_2 damage (% of GDP)	0.4	2.5	1.5	2.0	0.5
Genuine domestic savings (% of GDP)	8.3	29.4	9.0	12.2	11.3

Glossary

abatement costs the costs of reducing emissions.

abiotic referring to things that are not living.

abyssal floor deep-ocean basin approximately 3 miles below ocean surface. It consists of both abyssal hills and abyssal plains—flat, deep ocean floor whose depth may be 2 to 3 miles or more. Thick accumulations of sediment bury topography of oceanic crust.

acceptable risk the probability of suffering disease or injury that is considered to be negligible.

acid deposition acidic pollutants deposited from the atmosphere to the Earth's surface in wet and dry forms.

acid mine drainage drainage produced when rainwater seeps through a mine or the tailings. The solution is acidic due to the presence of sulfuric acid (H_2SO_4) that is produced when aerobic bacteria act on iron sulfide.

acid rain precipitation with increased acidity that can be in the form of rain, sleet, snow, fog, and cloud vapor. The effects were first noticed in 1872 but became a major concern in the 1960s when fishermen noticed declines in fish numbers and diversity in many lakes throughout North America and Europe. Also known as acid precipitation.

Carbonic acid (H_2CO_3) results from carbon dioxide (CO_2) mixing with water vapor. Natural rainfall has a pH of 5.6 due to natural CO_2 in atmosphere. Nitric acid (HNO_3) results from nitrogen dioxide (NO_2) mixing with water vapor. Sulfuric acid (H_2SO_4) results from sulfur dioxide (SO_2) mixing with water vapor. The concentrations of both nitrogen oxides and sulfur dioxides are much lower than atmospheric carbon dioxide that is mainly responsible for making natural rainwater slightly acidic. However, these gases are much more soluble than carbon dioxide and therefore have a much greater effect on the pH of the precipitation. Acid rain is more common in the eastern United States, western Europe, and populated areas of China due to coal burning. It affects human health (primarily respiratory system), can leach toxic metals from metal pipes into drinking supplies, damages building materials, decreases visibility, and affects wildlife, agriculture and fisheries.

active solar heating system heating system that uses solar collectors and additional electricity to power pumps or fans to distribute the Sun's energy. The heart of a solar collector is a black absorber, which converts the Sun's energy into heat. The heat is then transferred to another location for immediate heating or for storage for use later. The heat is transferred by circulating water, antifreeze, or sometimes air. Applications for active solar energy include heating swimming pools, domestic hot water use, heating and ventilation, and water for commercial facilities such as laundries and car washes.

acute toxicity immediate and adverse effects within a short time of exposure.

adaptation a change in morphology (structure) or habits, usually hereditary, by which a species or individual improves its condition in relationship to its environment or a genetic change in the morphology (structure), behavioral pattern, or physiology of an organism that enhances the individual's ability to survive in an environment.

adaptive radiation diversification of a species or single ancestral type into several forms that are each adaptively specialized to a specific environmental niche. Examples: Hawaiian honeycreepers (a finch-like ancestor) evolved into several species- seedeaters, insect eaters, and nectar eaters—each with a beak adapted for a specific food type.

administrative courts courts that hear challenges to agency rules and regulations. Grounds for overturning an agency ruling include (1) the act is vague or unconstitutional; (2) procedures were not followed according to law; and (3) the agency overstepped its authority in the matter. There is no jury in an administrative court.

administrative law the body of rules, regulations, orders, and decisions created by administrative agencies of government. The legal process through which statutory law is implemented.

aerosols small suspended particles in a gas, ranging in size from 1-nm molecules up to 100-μm pollen grains. Aerosols are emitted naturally (e.g., in volcanic eruptions) and as the result of human activities (e.g., by burning fossil fuels). In the stratosphere, aerosols are often crystal groupings of water molecules around a sulfate or nitrate molecule. Stratospheric nitrate and sulfate concentration is a major factor in the formation of polar stratospheric clouds that are involved in the formation of the Antarctic ozone hole.

age structure the proportion of the population or of each sex at each age category (prereproductive, reproductive, and postreproductive). Many species of plants and animals produce a large number of offspring, but the offspring have a high mortality rate because of high predation and other factors that keep the population in check and stable.

agribusiness large-scale farming operation including the production, processing, and distribution of agricultural products and the manufacture of farm machinery, equipment, and supplies.

A horizon layer of soil that contains organic matter (humus), living organisms, and inorganic minerals. It occurs just below the O horizon or surface litter. Also known as topsoil.

algae blooms algae that grows very fast or "blooms" and accumulates into dense, visible patches near the surface of the water. Factors that affect blooms are water temperatures, oxygen content of water, and nutrient levels. Red tide is a naturally-occurring, higher-than-normal concentration of the microscopic algae *Karenia brevis*, which produces a toxin that affects the central nervous system of fish so that they're paralyzed and can't breathe. As a result, red tide blooms often result in dead fish washing up on shores. Also known as HAB (harmful algae bloom).

allelopathy the prevention of growth or establishment of one species of plants by chemicals produced by another species.

alley cropping a system of land use in which harvestable trees or shrubs are grown among or around crops or on pastureland, as a means of preserving or enhancing the productivity of the land. Also known as agroforestry.

allogenic succession a predictable change that occurs within a community due to activity not related to the activities of the community. Examples: fire and flood.

allopatric speciation the evolutionary process through which two geographically separated (and therefore noninterbreeding) populations of the same species become less and less similar to each other over time (via mutation or the success of different traits in each environment) and eventually become distinctly different species.

alpha particle (α) a helium nucleus (two neutrons, two protons).

ammonia (NH_3) component of the nitrogen cycle that is released by organic matter when it decomposes.

ammonification the production of ammonia (NH_3) or ammonium compounds (containing the cation NH_4^+) that is produced in the decomposition of organic matter, especially through bacterial action.

ammonium ion whose formula is NH_4^+. Component of the nitrogen cycle that is produced when organic matter decays.

anthracite a dense form of coal that is high in carbon content and burns with a clean flame. Also known as hard coal.

antienvironmental groups there are six types of antienvironmental groups:

- Public relations firms,
- Corporate front groups,
- Think tanks,
- Legal foundations,
- Endowments and charities, and
- Wise use and share groups, which are timber, mining, ranching, chemical, and recreation companies that band together to fight the environmental movement.

aquifer an underground bed or layer of Earth, gravel, or porous stone that yields water.

arbitration the process by which the parties to a dispute submit their differences to the judgment of an impartial person or group appointed by mutual consent or statutory provision.

assimilation a stage of the nitrogen cycle in which plant roots absorb ammonia, ammonium ions and nitrate ions to manufacture DNA, amino acids and proteins.

$$\text{Plant Roots} + NH_3 + NH_4^+ + NO_3^- \xrightarrow{\text{bacteria}} \text{DNA} + \text{Amino Acids} + \text{Proteins}$$

atmospheric pressure the amount of force exerted over a surface area, caused by the weight of air molecules above it. As elevation increases, fewer air molecules are present; therefore, atmospheric pressure always decreases with increasing height. A column of air in cross section, measured from sea level to the top of the atmosphere, would weigh approximately 14.7 pounds per square inch (psi). The standard value for atmospheric pressure at sea level is: 29.92 inches or 760 mm of mercury; 1013.25 millibars (Mb) or 1013.25 hectopascals (hPa).

atomic mass number the sum of the number of neutrons and protons in an atomic nucleus and represented by the symbol A. Also called atomic weight or nucleon number.

atomic number the number of protons in an atomic nucleus and represented by the symbol Z.

autogenic succession a type of succession where the plant community changes the environment; changes that occur within a community caused by the actions of the community members (e.g., overpopulation and migration).

autotroph an organism capable of synthesizing its own food from inorganic substances, using light or chemical energy. Green plants, algae, and certain bacteria are autotrophs. Also known as primary producers.

backyard composting composting in one's personal backyard, typically in a fenced off area or bin. Backyard composting provides a convenient way to reduce the volume of trash a household produces. It also provides a valuable product that can enhance the soil and increase the growth and health of the yard. It involves converting vegetable matter to compost (a mixture of decaying organic matter, as from leaves and manure) that is used to improve soil structure and provide nutrients. Benefits of composting include

- Keeping organic wastes out of landfills,
- Providing nutrients to the soil,
- Increasing beneficial soil organisms (e.g., worms and centipedes),
- Suppressing certain plant diseases,
- Reducing the need for fertilizers and pesticides,
- Protecting soils from erosion.

balance of trade that part of a nation's balance of payments dealing with imports and exports, that is trade in goods and services, over a given period. If exports of goods exceed imports, the trade balance is said to be "favorable"; if imports exceed exports, the trade balance if said to be "unfavorable."

B horizon layer of soil characterized by its accumulation of iron, aluminum, calcium carbonate, calcium sulfate, and other salts, humic compounds, and clay particles that have been leached into it from O, A, and E horizons that lie above it. Also known as subsoil.

biodegradable plastics plastics that incorporate cornstarch and can be broken down by microorganisms.

biodiversity the total diversity and variability of living things and of the systems of which they are a part. This includes the total range of variation in and variability among systems and organisms at all levels. It also covers the complex sets of structural and functional relationships within and between different levels of organization, including human action, and their origins and evolution in space and time. The biological disciplines comprising biodiversity include evolutionary biology, taxonomy (and the related biosystematics), ecology, genetics and population biology.

biogas a fuel in gas form that is produced through anaerobic respiration and that contains mostly methane, but also smaller amounts of carbon dioxide, water vapor, nitrogen, hydrogen, and other gases.

biological oxygen demand (BOD) the amount of oxygen required for the decomposition of organic compounds by microorganisms in a specific amount of water and usually measured in milligrams of oxygen consumed per liter (mg O_2/L). BOD is used by regulatory agencies for monitoring wastewater treatment facilities and monitoring surface water quality. BOD is the biochemical oxygen demand of the water and is related to the concentration of the bacterial facilitated decomposable organic material in the water. A sample with a 5-day BOD between 1 and 2 mg O/L indicates a very clean water, 3.0 to 5.0 mg O_2/L indicates a moderately clean water and > 5 mg O_2/L indicates a nearby pollution source. BOD is a laboratory test that requires an oxygen-sensing meter, incubator, nitrifying inhibitors, and a source of bacteria.

biomagnification a continued increase in the concentration of pollutants in higher levels of a food chain (e.g., plankton consume mercury). Fish who consume the plankton have a higher concentration of mercury in their bodies than the plankton.

biomass the amount of living material in a habitat measured in terms of either mass, volume, or caloric energy potential.

biome a major regional or global biotic community, such as a grassland or desert, characterized chiefly by the dominant forms of plant life and the prevailing climate. Temperature and precipitation are the most important determinants.

bituminous coal coal with a high percentage of volatile matter that burns with a smoky, yellow flame. Also known as "soft coal."

breeder nuclear fission reactor a type of nuclear reactor that actually creates more nuclear fuel than it consumes, converting nonfissionable U-238 into Pu-239.

British Thermal Unit (BTU) a common unit for heat equivalent to 252 calories or 1055 joules. It represents the amount of energy required to raise the temperature of one pound of water from 60° to 61°F at a pressure of one atmosphere.

caloric intake the number of calories that a person takes in per day. The basic minimum requirement is 2700 calories per day for men and 2000 calories per day for women. In 1950, the *average* worldwide caloric intake was 2000 calories per day. Fifty years later, in 2000, the *average* increased to 2500 calories per day, despite the world's population more than doubling. Problems in food distribution, however, prevent many people from getting proper nutrition—where there are shortages in some areas, there are surpluses in others. The average daily caloric intake in North America and Europe is 3500 calories per day.

calorie a small calorie is the quantity of heat required to raise the temperature of one gram of water by 1°C. A large calorie or dietary calorie is equivalent to 1000 small calories.

capital wealth in the form of money or property used or accumulated in a business by a person, partnership, or corporation.

- Natural capital—goods and services provided by nature
- Human or cultural capital—experience and knowledge
- Manufactured or built capital—technology
- Social capital—shared values, cooperation

capitalist market economy a real-world economic system in which the means of production and distribution are privately or corporately owned and development is proportionate to the accumulation and reinvestment of profits gained in a free market. It is generally found in developed and moderately developed countries. Examples include North and South America, Europe, and Australia. Approximately half of the world's population lives in a capitalist market economy.

carbon dioxide (CO_2) 0.036% of the atmosphere by volume. CO_2 is a natural component of the atmosphere, produced from the decay of vegetation, volcanic eruptions, exhalations of animals, the burning of fossil fuels and deforestation. Human activity introduces 5500 million tons per year of CO_2 into the atmosphere. A CO_2 molecule remains in the atmosphere about 100 years. The concentration of CO_2 in the atmosphere in 1900 was about 290 parts per million. It is projected at current rates, that by 2030, it will reach 550 ppm.

carbon monoxide (CO) a natural component of the Earth's atmosphere. CO forms due to incomplete combustion of fossil fuels and the burning of biomass. Human activities introduce up to 700 million tons of CO into the atmosphere each year as compared to 1300 tons from natural processes. The average time that a CO molecule spends in the atmosphere is several months. CO can cause headaches, fatigue, drowsiness, and even death in humans and is a major air pollutant.

carcinogen substance that increases the rate of neoplasms in humans or animals. Carcinogens include both genotoxic chemicals (affect DNA directly) and nongenotoxic chemicals that induce neoplasms by other mechanisms. Four types of responses are generally accepted as evidence of induction of neoplasms: (1) an increase in incidence of the tumor types that occur in controls; (2) the development of tumors earlier than controls; (3) the occurrence of tumor types not observed in controls; and (4) an increased multiplicity of tumors.

carrying capacity (K) the maximum number of organisms that an ecosystem can support over time without degradation of the ecosystem. Factors that determine the carrying capacity are (1) availability of nutrients, (2) availability of energy, (3) methods of dealing with wastes, and (4) biotic relationships.

case law law that is derived or sets precedence from a civil or criminal court case.

chaparral a biome existing in the midlatitudes that is characterized by hot, dry summers and cool, moist winters and dominated by a dense growth of mostly small-leaved evergreen shrubs.

chlorinated hydrocarbon a hydrocarbon (a compound containing only carbon and hydrogen) that also contains chlorine atoms. Examples: DDT, aldrin, lindane, and kepone. Chlorinated hydrocarbons persist in the environment and can accumulate within living organisms. Many of these products are still manufactured in more-developed countries and sold in less-developed countries, where they have not been banned.

chlorofluorocarbons (CFCs) a series of hydrocarbons containing both chlorine and fluorine first produced by General Motors Corporation in 1928. CFCs were created as a replacement to the toxic refrigerant, ammonia. CFCs have also been used as a propellant in spray cans, cleaner for electronics, sterilizing hospital equipment, fire extinguishing agents, and to produce the bubbles in Styrofoam. CFCs are inexpensive to manufacture and are very stable compounds, lasting up to 200 years in the atmosphere. They have been shown to cause stratospheric ozone depletion.

C horizon layer of soil located just above bedrock. It contains weathered bedrock and inorganic minerals.

Clean Air Act (Title VI) legislation that directs EPA to protect the ozone layer through several regulatory and voluntary programs. Title VI covers production of ozone-depleting substances (ODS), the recycling and handling of ODS, the evaluation of substitutes, and efforts to educate the public.

clear cutting removing all the trees in a tract of timber at one time. Clear cutting increases soil erosion and eliminates habitats for wildlife.

climate the meteorological conditions—including temperature, precipitation, and wind—that characteristically prevail in a particular region of the Earth.

climax community a community of plants that is stable in the type of plant species that inhabits the area. Not changing in terms of species replacing other species.

cogeneration the capture and use of waste heat (e.g., using waste heat from a nuclear power plant to heat water).

cold front the leading portion of a cold atmospheric air mass moving against and eventually replacing a warm air mass.

competitive exclusion principle a rule, derived by G. F. Gause in 1934, stating that two species that occupy the same habitat cannot also occupy the same ecological niche. Any two species that occupy the same niche will compete with each other to the detriment of one of the species, which will thus be excluded.

continental crust land above sea level up to 40 miles thick. It has a lower density than the oceanic crust and "floats" higher than oceanic crust on top of the mantle.

continental drift the movement, formation, or re-formation of continents described by the theory of plate tectonics.

continental margin the area that extends from the shoreline of a continent to the beginning of the ocean floor. It includes the continental rise, continental shelf, and continental slope.

continental shelf submerged part of the continent. It is gently sloping (less than one tenth of a degree) and up to 1500 kilometers wide with an average width of 80 kilometers. Water depth at seaward edge averages 130 meters (about 400 feet).

continental slope the seaward edge of the continental shelf. The boundary between the continental and oceanic crust. It is steeply sloping compared to shelf (averages about 5 degree slope, up to 25 degrees). May be about 20 kilometers wide.

contour farming farming with row patterns nearly level around a hill—not up and down. Crop row ridges built by tilling and/or planting on the contour create hundreds of small dams. These ridges or dams slow water flow and increase water infiltration that reduces erosion. Contour farming can reduce soil erosion by as much as 50% from up-and-down hill farming. By reducing sediment and runoff, and increasing water infiltration, contour farming promotes better water quality.

contour mining a type of surface mining in which resources are removed near the surface. The mining follows the natural contours of the land.

convective lifting air movement that occurs when air heated at the Earth's surface rises in the form of thermal currents.

convergent evolution the adaptive evolution of superficially similar structures, such as the wings of birds and insects, in unrelated species subjected to similar environments. Also called adaptive evolution.

convergent plate boundary an interface formed where the top of convection cells flow toward each other. At the boundary, the oceanic lithosphere moves downward (subduction) forming a subduction zone, where a trench is found.

core Earth's innermost zone. The inner core is a solid ball with a radius of about 700 miles, directly in the center of the Earth. It is made up of mostly iron and some nickel. Temperature of the inner core is approximately 8000°C. The outer core surrounds the inner core and is liquid in nature. The temperature of the outer core is around 7000°C.

cost-benefit analysis an attempt to assign valuation to resources (benefits) and impacts to the environment (costs) to make more informed decisions. It requires identification of what will be affected, what will be the expected outcome, and whether there are alternatives, and then it assigns a monetary value to each item. The items are tabulated and a decision is made.

criteria air pollutants air pollutants that the EPA has regulated by first developing health-based criteria (science-based guidelines) as the basis for setting permissible levels. One set of limits (primary standards) protects health; another set of limits (secondary standards) is intended to prevent environmental and property damage. A geographic area that meets or does better than the primary standard is called an *attainment* area; areas that don't meet the primary standard are called *nonattainment* areas. About 90 million Americans live in nonattainment areas. The consequences of exposure to criteria pollutants are

- Particulates—potent respiratory irritants that can impair lung function by damaging tissue.
- SO_2—respiratory irritant. Ecological damage; damage to materials.
- Ozone—attacks cells and breaks down tissues, particularly lung tissue; also a toxin to plants.
- NO_x—constriction of airway; reduced resistance to infection; increases sensitivity of asthmatics.
- CO—binds to hemoglobin in blood, preventing adequate oxygen transport to cells.
- Lead—circulatory, reproductive, nervous, and kidney damage in adults; impaired mental development in children.

crop rotation crops are changed year by year in a planned sequence. Crop rotation is a common practice on sloping soils because of its potential for soil saving. Rotation also reduces fertilizer needs because alfalfa and other legumes replace some of the nitrogen that corn and other grain crops remove. Pesticide costs may be reduced by naturally breaking the cycles of weeds, insects, and diseases. Grass and legumes in a rotation protect water quality by preventing excess nutrients or chemicals from entering water supplies. Short grasses or small grains cut soil erosion dramatically. Crop rotations add diversity.

crust top layer of the Earth. Its thickness varies from 6 to 40 miles. Eight elements make up 99% of the weight of the crust (in decreasing amounts—O, Si, Al, Fe, Ca, Na, K, Mg). It is divided into continental (30%) and oceanic (70%) sections.

cyanobacteria photosynthetic bacteria of the class *Coccogoneae* or *Homogoneae*. Cyanobacteria are autotrophs—that is they manufacture their own food through photosynthesis. Another group of autotrophs are known as chemotrophs or chemosynthetic bacteria. Cyanobacteria are generally blue-green in color and some species are capable of nitrogen fixation. They were once thought to be algae. Also known as blue-green algae.

decomposer organisms such as earthworms, snails, bacteria, and fungi that consume energy-rich organic molecules (heterotrophs) and that feed upon waste products from other organisms. Also known as a detritivore.

deep ecology system that believes that animals have legal rights. All living things have a basic right to survive.

demand the amount of goods or services people want. Factors that influence demand are

- The price of the good;
- The income of consumers;
- The demand for alternative goods that could be used (*substitutes*);
- The demand for goods used at the same time (*complements*); and
- Whether people like the good (*consumer taste*).

demographic transition in general, how a group of people change over time or a theory that suggests that declines in death rates and birth rates follow the industrialization of a nation. It generally can be demonstrated in the pattern of population growth that occurred during the 19th and 20th centuries in the United States, which saw initial high birth and death rates (preindustrial stage). The result was a net low population growth. The next phase (transitional) was characterized by high economic productivity and growth, which resulted in a higher standard of living. This economic boom resulted in a decline in mortality, but natality rates remained high. This middle phase was characterized by a rapid population growth (about 3%). Countries that are in the transitional stage today include countries in Southeast Asia, Africa, and Latin America. The next phase (industrial stage), as a result of a higher standard of living, produced a decline in birth rate matching the decline in death rates. Access to birth control, decline in infant mortality, increased job opportunities for women, and high cost of education also contributed. This returned population growth to a controlled, steady state. Most developed countries, including the United States, are in this phase. The last stage is called the postindustrial state. This stage is characterized by zero population growth and the death rate exceeding the birth rate. Thirteen percent of the world population in 30 countries, mostly Europe and Japan, are in this stage. In summary, as countries become industrialized, death rates decline first followed by declining birth rates.

denitrification a stage in the nitrogen cycle whereby specialized bacteria convert NH_3 and NH_4^+ into NO_2^- and NO_3^- and then into N_2 and N_2O (nitrous oxide), which are then released into the atmosphere.

$$NH_3 + NH_4^+ \xrightarrow{\text{bacteria}} NO_2^- + NO_3^- \xrightarrow{\text{bacteria}} N_2 + N_2O$$

density-dependent factor a factor that influences population growth and that increases in magnitude with an increase in the size or density of the population. Example: competition, parasitism, and predation. Predation is linked to the size of a population.

density-independent factor a factor that influences population growth and that does not depend on the size or density of the population. Example: the effect of a seasonal drought on a plant population. A seasonal drought is not influenced by the size of a population. Other factors include fires, extreme cold, and floods.

deregulation the lifting of government regulations to allow the market to function more freely.

developing country low- and middle-income countries in which most people have a lower standard of living with access to fewer goods and services than do most people in high-income countries. There are currently about 125 developing countries with populations over 1 million; in 1998, their total population was more than 5.0 billion. Also referred to as less-developed country.

dioxins any of several carcinogenic or teratogenic heterocyclic hydrocarbons that occur as impurities in petroleum-derived herbicides. A family of 75 chlorinated hydrocarbon compounds that are produced when chlorine and hydrocarbons are heated to high temperatures. Estimates are that up to 25% of all cancers may be due to exposure to dioxins that affect the endocrine, reproductive, and immune systems.

direct costs the total costs incurred when a company is held responsible for cleaning up environmental pollution or is required to install pollution control equipment.

divergent evolution change that occurs when a population is separated, usually by geographic barriers, and the separated subpopulations evolve separately but retain some common characteristics. Example: ostrich, rhea, and emu.

dose-response curve a graph that compares a population to its response to varying concentrations of a chemical or physical agent.

durable goods goods that provide a service over a number of years (e.g., cars, major appliances, or furniture).

Dust Bowl an ecological and human disaster that took place in the southwestern Great Plains region of the United States in the 1930s. It was caused by misuse of land and years of sustained drought. Millions of hectares of farmland became useless, and hundreds of thousands of people were forced to leave their homes.

ecofeminism movement led by such feminists as Karen Warren, Vandana Shiva, Carolyn Merchant, Rosemary Ruether, and Ynestra King that focuses on relationship of women to the Earth and to male-dominated societies, which they view as responsible for environmental imbalance and social oppression. They also believe that the environment should be nurtured, not conquered and that women should have the same rights as men (economic, social, and political). The role of humans is to be caretakers. The intrinsic values are represented by relationships whereas the instrumental values are represented by roles.

ecological niche the function or position of an organism or population within an ecological community or the particular area within a habitat occupied by an organism or the biological, chemical, and physical components required for a particular species to survive, grow, and reproduce. The collection of all biological, chemical, and physical factors necessary for a species to grow, survive, and reproduce.

economic resources materials obtained from the environment to meet human needs. Natural resources (also known as Earth capital), human resources (mental and physical talents of people), and financial and manufactured resources (made with Earth capital and human resources) are included.

effluent fees taxes or fines levied on polluters for every unit of pollution discharged into a stream, landfill, or air.

El Niño—Southern Oscillation (ENSO) a warming of the ocean surface off the western coast of South America that occurs every 4 to 12 years when an upwelling of cold, nutrient-rich water does not occur. It causes die-offs of plankton and fish and affects Pacific jet stream winds, altering storm tracks and creating unusual weather patterns in various parts of the world.

emigration to leave one country or region to settle in another. Approximately 25 million people per year emigrate due to environmental reasons—drought, desertification, deforestation, soil erosion, and natural resource shortages. The number could expand to 150 million per year due to global warming and its effects. The overall world average due to emigration to developed countries is about 1% of the total growth in these countries.

emission taxes taxes that are assessed on emissions of pollution, usually on a per ton basis. They generally provide an incentive to reduce pollution.

energy efficiency the percentage of total energy input that does useful work and that is not converted to low-quality, useless heat. Approximately 84% of all energy produced in the United States is wasted due to the second law of thermodynamics. Remaining waste is due to inefficient technology, poor insulation, and poor design. The United States wastes as much energy as two-thirds of the world consumes. Most wastes come from transportation (wastes up to 90% of available energy), lighting (incandescent light bulbs waste 95% of their energy as heat), heating and air-conditioning, and appliances. For more information, go to http://www.eren.doe.gov.

environmental ethics system that focuses on the moral relationship between humans and the world around them.

environmental impact report (EIR) a study of all the factors which a land development or construction project would have on the environment in the area, including population, traffic, schools, fire protection, endangered species, archeological artifacts, and community beauty. Many states require such reports be submitted to local governments before the development or project can be approved, unless the governmental body finds that there is no possible impact, which finding is called a "negative declaration." An EIR contains (1) purpose and need for the project; (2) alternatives to the proposed action; (3) a statement of positive and negative environmental impacts of the proposed activities; (4) the relationship between short-term resources and long-term productivity; (5) identification of any irreversible commitment of resources resulting from project implementation.

environmental indicators a tool for state-of-the-environment reporting, measuring environmental performance, and reporting on progress toward sustainable development. It can be used at both the international and the national level. At the national level they can also be used for clarifying objectives and setting priorities. The Little Green Data book (believe me, it is not little!), published by The World Bank lists the following key environmental indicators for every country in the world. It is a goldmine for class research projects using spreadsheets and databases. See Appendix IV for a country comparison of environmental indicators.

ENVIRONMENTAL INDICATORS

Demographic Population, Urban Population, GDP.

Agriculture Land area, agricultural land %, irrigated land %, fertilizer consumption, food production, population density (rural)

Forests Forest area, forest area (% of total land), and annual deforestation %.

Biodiversity Mammal species known, mammal species threatened, bird species known, bird species threatened, nationally protected area—% of total land area.

Energy GDP per unit of energy used, commercial energy use, energy imports, electric power consumption, share of electricity generated by coal.

- **Emissions and Pollution** CO_2 emission per unit of GDP, total CO_2 industrial emissions, CO_2 emissions per capita, suspended particulates in capital city, passenger cars.
- **Water and Sanitation** Access to improved water source (% of total, % urban, and % of rural population), freshwater resources, freshwater withdrawal, freshwater withdrawal for agriculture, access to sanitation in urban and rural areas, and under-5 mortality rate.

- **National Accounting Aggregates** Gross domestic savings, consumption of fixed capital, net domestic savings, education expenditure, energy depletion, mineral depletion, net forest depletion, CO_2 damage, genuine domestic savings.

environmental justice principle a principle that states that all people, regardless of race, color, creed, economic status, or national origin deserve a safe environment in which to live. This principle is in response to documentation that shows minority communities are affected disproportionately by health risks.

Environmental Protection Agency (EPA) agency of the U.S. government, with headquarters in Washington, D.C. It was established in 1970 to reduce and control air and water pollution, noise pollution, and radiation and to ensure the safe handling and disposal of toxic substances. The EPA engages in research, monitoring, setting, and enforcing national standards.

environmental racism practice wherein minority communities have higher percentages of municipal landfills, incinerators, and hazardous-waste treatment, storage, and disposal facilities.

epicenter location on the surface directly above the focus. An earthquake caused by a fault that offsets features on the Earth's surface may have an epicenter that does not lie on the trace of that fault on the surface. This occurs if the fault plane is not vertical and the earthquake occurs below the Earth's surface.

epilimnion the upper layer of a lake.

epoch a geological unit of time that is shorter than a period. Epochs include the Holocene, Pleistocene, Pliocene, Miocene, Oligocene, Eocene, and Paleocene.

era a unit of geological time that is longer than a period but shorter than an eon. Eras include the Cenozoic, Mesozoic, and Paleozoic.

erosion wearing away of the lands by running water, glaciers, winds, and waves.

estuary the wide part of a river where it nears the sea; it is where fresh and salt water mix.

exponential growth phase population growth that is characterized by a constant percentage of growth over time. Example: if a population was experiencing a 100% growth rate and you started with 10, the growth would be 10, 20, 40, 80, 160, etc. When plotted against time, exponential growth rates have a characteristic J-shape.

external costs harmful effects of a product not included in its market price and paid for by those who do not necessarily use the resource. Example: the cost to society of dumping of wastes generated by the manufacturing of a product into streams.

fault a break in the Earth along which movement occurs where one side moves with respect to the other. Sudden movement along a fault produces earthquakes. Slow movement produces aseismic creep.

Ferrel cell a circulating body of air that occurs in the midlatitudes between 30° and 60° north and south latitude.

first green revolution international agricultural project that began in Mexico in 1944. The Rockefeller Foundation and the Mexican government established a plant-breeding station in northwestern Mexico, with a goal of boosting grain yields. The project was headed by Norman Borlaug, a plant breeder from the University of Minnesota who developed a high-yielding wheat plant and later won the Nobel Prize for his work in this area. The effort was tremendously successful:

1944—Mexico was importing half its wheat
1956—Mexico was self sufficient in wheat production
1964—Mexico exports ½ million tons of wheat

This wheat was also successful when grown in some areas of Asia and Africa. In India, wheat production increased four times in 20 years (from 12 million tons in 1966 to 47 million tons in 1986). At present, over three-quarters of the wheat acreage in India is planted to these new high-yielding varieties. By doubling and tripling yields, it bought time for developing countries to start dealing with rapid population growth. It raised the productivity of the three main staple food crops—rice, wheat, and corn. Between 1950 and 1990, grain yields increased by nearly two and a half times, from 1.06 metric tons per hectare to 2.52 tons. It increased crop yields through:

- Introduction of high-yield genetically engineered varieties of grains;
- Use of pesticides and fertilizers; and
- Improved management techniques involving increased intensity and frequency of cropping.

first-law efficiency the ratio (expressed as a percentage) of the actual amount of energy delivered where it is needed to the amount of energy supplied to meet that need.

first law of energy or thermodynamics a law that states that energy can neither be created nor destroyed, but it may be converted from one form to another. Also known as the law of conservation of energy.

fission nuclei of certain isotopes with large mass numbers are split apart into lighter nuclei. Critical mass required to sustain a chain reaction. Much radioactive wastes are produced.

floodplain a plain bordering a river and subject to flooding.

food chain a succession of organisms in an ecological community that makes up a continuation of food energy from one organism to another as each organism consumes a member lower in the chain and in turn is preyed upon by a higher member.

forest succession a forest community is composed of a wide variety of tree species and ages, and organisms, along a progression known as succession. In primary succession, pioneering species first colonize previously uninhabited landforms facilitating the establishment of later successional forms by developing soil and other conditions favorable to the succeeding organisms leading to the presence of a final self-maintaining community. Secondary succession may also occur following the abandonment of cultivated land or after major or minor environmental changes such as forest fires or tree-falls that leave some trace of previous organic activity.

fusion two isotopes of light elements are forced together at extremely high temperatures. It is technologically not available now. For more information, go to http://fusioned.gat.com/.

General Agreement on Tariffs and Trade (GATT) international trade agreement, that seeks to lower tariffs in order to increase trade and to create a forum to resolve trade and disputes. Criticism is that it prevents a country from using trade barriers to enforce environmental issues and concerns.

genetic engineering (gene splicing) the use of molecular biology techniques to create desired characteristics in plants and animals. Techniques involve gene splicing or recombinant DNA, in which the DNA of a desired gene is inserted into the DNA of a bacterium, which then reproduces itself, yielding more of the desired gene. Genetically engineered products include bacteria designed to break down oil slicks and industrial waste products, drugs (human and bovine growth hormones, human insulin, and interferon), and plants that are resistant to diseases, insects, and herbicides or that yield fruits or vegetables with desired qualities. Genetic engineering techniques have also been used in the direct genetic alteration of livestock and laboratory animals. Also known as "pharming." Examples of genetically modified organisms include:

- Super rice. The International Rice Research Institute (IRRI) in the Philippines has developed a new strain of super rice capable of boosting yields by 25%, amounting to an extra 100 million metric tons a year—enough to feed an additional 450 million people.
- Improved corn. The International Center for the Improvement of Maize and Wheat in Mexico has engineered several improved varieties of corn that could increase yields by up to 40%. These varieties could be grown on marginal land under difficult growing conditions and thus could be raised by poor farmers. If widely used, the new varieties could feed an additional 50 million people a year.
- A new potato. The International Potato Center in Peru is working to produce a new potato that would be resistant to a virulent form of potato blight that has reached every continent except Australia.

geothermal energy heat energy that is extracted and utilized from the Earth's interior.

glaciers flowing bodies of ice formed in regions where snowfall exceeded melting. Due to gravity, glaciers move downhill causing erosion through abrasion. Over the past 800,000 years, there have been several great ice ages during which 30% of the continental surface of the Earth was covered by glacial ice several kilometers thick. Each glacial period lasted about 100,000 years and was followed by a warmer interglacial period lasting between 10,000 to 12,500 years.

Global Warming Potential (GWP) refers to the amount of global warming caused by a substance. The GWP is the ratio of the warming caused by a substance to the warming caused by a similar mass of carbon dioxide. GWP of CO_2 is defined to be 1.0, CFC-12 has a GWP of 8500, and water has a GWP of 0.

greenhouse effect the phenomenon whereby the Earth's atmosphere traps solar radiation, caused by the presence in the atmosphere of gases such as carbon dioxide, water vapor, and methane that allow incoming sunlight to pass through but that absorb heat radiated back from the Earth's surface.

gross domestic product (GDP) the total market value of all final goods and services produced in a country in a given year. GDP equals total consumer investment and government spending, plus the value of exports minus the value of imports. GDP does not distinguish between beneficial and harmful or sustainable and unsustainable goods and services. Calculation of the GDP (and GNP) does not consider the cost of depleting natural resources, encourages and promotes manufacturing rather than reusing and recycling techniques, measures only money spent (not the value received), and does not address issues of income distribution.

gross national product (GNP) a measurement of a nation's output. It can be measured by either (1) measuring the money spent on goods and services or (2) adding up the total cost of producing goods and services.

Hadley cell a circulating body of air that occurs in, at, and near the equator, between 0° and 30° north and south latitude. It consists of a rising mass of air near the equator known as the intertropical convergence zone (ITCZ) and the descending air that occurs near 30° north and south latitude known as the subtropical high.

heavy metals metals with a large atomic number. They include such elements as copper, cadmium, lead, selenium, arsenic, mercury, and chromium. These substances are very dangerous to humans and other life forms, even in small quantities. Lead contamination comes from old paint, solder, lead pipes, lead-based glazes on ceramics, and batteries. Approximately 16,000 children under 9 years old in the United States are treated each year for lead poisoning which results in mental impairment (mental retardation, cerebral palsy, blindness, shortened attention spans, and hyperactivity). In the United States, 200 children die annually from lead poisoning.

high-quality energy energy that is concentrated and capable of performing useful work (e.g., electricity, natural gas, coal).

high-throughput society societies that attempt to sustain ever-increasing economic growth by increasing the throughput of matter and energy in their economic society. Eventually the capacity of the environment to sustain this cycling is exceeded and becomes unsustainable. Also known as a high-waste society.

hurricanes a severe tropical cyclone originating in the equatorial regions of the Atlantic Ocean or Caribbean Sea or eastern regions of the Pacific Ocean, traveling north, northwest, or northeast from its point of origin and usually involving heavy rains and winds with speeds greater than 74 miles (119 kilometers) per hour, according to the Beaufort scale. In the North Atlantic, the hurricane season is from May to November, but the majority of storms occur in August, September, and October.

hypolimnion the layer of water in a thermally stratified lake that lies below the thermocline, is noncirculating, and remains perpetually cold.

immigration the passing or coming into a country for the purpose of permanent residence. In the United States, in the year 2000, legal immigration accounted for about 800,000 people; illegal immigration accounted for about 300,000 people. Canada, Australia, and the United States allow large annual increases to their populations from immigration.

incineration to burn completely. Burning municipal solid waste (MSW) can generate energy while reducing the amount of waste by up to 90% in volume and 75% in weight. 16% of all MSW and 7% of all reported hazardous wastes are incinerated. EPA's Office of Air and Radiation is primarily responsible for regulating combustion of MSW because air emissions from combustion pose the greatest environmental concern. During the process, some of the fly ash may float out with the hot air, depending on the design of the incinerator. Both the fly ash and the ash that is left in the furnace after burning have high concentrations of dangerous toxins such as dioxins and heavy metals. Disposing of this ash can pose a problem. The ash that is buried at the landfills may leach into the groundwater and cause severe contamination. A variety of pollution control technologies have been developed that can reduce the toxic materials emitted in combustion smoke: (1) scrubbers, devices that uses a liquid spray to neutralize acidic gases in smoke; and (2) filters, devices that remove tiny ash particles from the smoke. Burning waste at extremely high temperatures also destroys harmful chemical compounds and disease-causing bacteria. Regular testing ensures that residual ash is nonhazardous before being placed in a landfill. Incineration is usually kept as the last resort and is used mainly for treating infectious waste. In 1999, there were 102 incinerators in the United States with energy-recovery technology with the capacity to burn up to 96,000 tons of MSW per day.

industrial waste material that is left over from the manufacturing process that is recycled outside of the primary manufacturing facility.

infectious disease any disease caused by the entrance, growth, and multiplication of bacteria or protozoans in the body; a germ disease. It may not be contagious. Infectious diseases are responsible for 43% of all disease deaths. In adults, infections of the respiratory system are the leading cause of infectious disease.

integrated pest management (IPM) an ecologically based pest-control strategy that relies on natural mortality factors such as natural enemies, weather, and crop management. It involves the coordinated use of pest and environmental information to prevent unacceptable levels of pest damage by the most economical means, resulting in the least possible hazard to people, property, and the environment. IPM can be promoted through: (1) tax on pesticides to promote research; (2) demonstration projects; (3) economic incentives for farmers that utilize IPM; and (4) reduces federal and state subsidies to farmers that utilize traditional pesticide management practices when IPM methods are available.

intercropping the cultivation of two or more crops simultaneously on the same field or the growing of two or more crops on the same field with the planting of the second crop after the first one has already completed development.

Intergovernmental Panel on Climate Change (IPCC) a United Nations-sponsored organization made up of 2500 scientists from around the world. It is the world's leading authority on global warming.

internal costs direct cost paid by those that use a resource. It usually involves the direct cost of accessing the resource and converting it to a useable product or service.

intertidal zone land at the coastline that is covered by water at high tide and exposed at low tide.

jet stream a high-speed, meandering wind current, generally moving in a west to east direction at speeds often exceeding 250 miles per hour at altitudes of 10 to 15 miles.

joule the unit of force required to accelerate a mass of one kilogram one meter per second. It is equivalent to 100,000 dynes.

kilocalorie (Kcal) energy that is equivalent to 1000 calories.

kilowatt (kW) the amount of power equivalent to 1000 watts.

K strategists large organisms that have long life spans, produce few offspring, and provide care for their offspring.

Kyoto Protocol began in 1992 at the Rio Earth Summit when the United States and 173 other countries signed the United Nations Convention on Climate Change that called for: "a stabilization of greenhouse gas concentrations in the atmosphere at a level that would prevent dangerous anthropogenic interference with the climate system." In 1997, nations met again in Kyoto, Japan, to pass a binding treaty on greenhouse gas emissions. This is when the "arguments" began. The European Union wanted major cuts in emissions (20% below 1900 levels by 2010). The United States, which contributes about 40% of greenhouse gas emissions, wanted to reduce emissions to 1990 levels by 2012 (in the meantime, U.S. emissions of CO_2 had increased 8% since the Rio Earth Summit). The United States also wanted less-developed countries to likewise reduce emissions, even though less-developed countries contribute very little greenhouse gas emissions. Tying the U.S. reductions to less-developed countries would hinder economic growth in less-developed countries. The effect was to penalize less-developed countries. Arguments also developed as to whether emissions should be lowered on a per capita basis or by a country basis. A reduction on total emissions ignores the "polluter pays" principle. Current status is creation of Annex I countries (industrialized) and Annex II countries (less-developed). Emission trading is allowed, credit for forest sinks is allowed, and emission credit for Annex I countries to develop and promote nonpolluting energy sources in Annex II countries is allowed. Details and verification issues have not been worked out.

La Niña a cooling of the ocean surface off the western coast of South America, occurring periodically every 4 to 12 years and causing more Atlantic hurricanes, colder winters in Canada and the northeast United States, warmer and drier winters in the southwest and southeast United States, wetter winters in the Pacific Northwest, and torrential rains in Southeast Asia.

landfills generally located in urban areas where a large amount of waste is generated and has to be dumped in a common place. Unlike an open dump, it is a pit that is dug in the ground. The garbage is dumped and the pit is covered thus preventing the breeding of flies and rats. At the end of each day, a layer of soil is scattered on top of it and is compressed. After the landfill is full, the area is covered with a thick layer of mud and the site can thereafter be developed as a parking lot or a park. Landfills have many problems. All types of waste are dumped in landfills and when water seeps through them it gets contaminated and in turn pollutes the surrounding area. This contamination of groundwater and soil through landfills is known as leaching. The number of landfills in the United States is steadily decreasing—from 8000 in 1988 to 2300 in 1999. The capacity, however, has remained relatively constant. New landfills are much larger than in the past.

law of conservation of energy a law that states that the total amount of energy in the universe is constant and cannot be created nor destroyed.

law of demand the principle that consumers are willing to purchase more of a commodity at lower prices than at higher prices, other things remaining equal. The law of demand holds true because of the widespread availability of substitutes. Demand for a particular product or service represents how much people are willing to purchase at various prices. Thus, demand is a relationship between price and quantity, with all other factors remaining constant. Demand is represented graphically as a downward sloping curve with price on the vertical axis and quantity on the horizontal axis.

Generally, the relationship between price and quantity is negative. This means that the higher the price level is, the lower the quantity demanded will be. Conversely, the lower the price, the higher the quantity demanded will be. Market demand is the sum of the demands of all individuals within the marketplace. Market demand will be affected by other variables in addition to price, such as various value-added services including handling, packaging, location, quality control, and financing.

law of supply the principle stating that, in general, other things being equal, the higher the price of a good, the greater the quantity of that good sellers will offer for sale over a given period. Supply characteristics relate to the behavior of firms in producing and selling a product or service. Market supply is represented by an upward sloping curve with price on the vertical axis and quantity on the horizontal axis.

An increase in price results in producers wanting to increase the quantity of a given product they will bring to the market; therefore, the relationship between the price and supply is positive. With higher prices, the producers of goods and services will receive greater profits. Greater profits will result in the means to expand production increasing the supply. This increased supply will ultimately satisfy the existing demand such that any additional production must be met with new demand for the price increases to be sustained. Other factors that have been identified as important in determining supply behavior include: the number of firms producing the product, technology, the price of inputs, the price of other commodities that could be produced, and the weather.

LD_{50} median lethal dose. The amount of a toxic material administered to a test population that kills half the population in a certain time. Some LD_{50} values for various chemicals and agents follow:

LD_{50} VALUES FOR VARIOUS CHEMICALS

Chemical	LD_{50} (mg/kg body weight)	Route
Table sugar	29,700	Rat, oral
Baking soda	4,220	Rat, oral
Table salt	3,000	Rat, oral
Drinking alcohol	2,080	Rat, oral
Caffeine	192	Rat, oral
Sodium cyanide	6.4	Rat, oral
Sarin nerve gas	24	Human, skin contact
VX nerve gas	0.14	Human, skin contact

legislative riders a clause, usually having little relevance to the main issue that is added to a legislative bill or an amendment or addition to a document or record. Often used to roll back environmental protection laws. Example: Omnibus spending bill in 1999 contained 41 riders that were anti-environment. Other anti-environmental riders have been attached to bills offering aid to victims in Kosovo and victims of hurricanes. Also known as "allonge."

legumes plants that belong to the pea or bean family. Legumes form mutualistic associations with certain species of bacteria *(rhizobium)* that allow the plant to absorb nitrogen in a usable form and allow the bacteria to obtain necessary nutrients and a suitable habitat.

less-developed country a country having an annual gross national product per capita equivalent to $760 or less in 1998. The standard of living is lower in these countries; there are few goods and services; and many people cannot meet their basic needs. There are currently about 58 low-income countries with populations of 1 million or more. Their combined population is almost 3.5 billion. Also known as a low-income country. See Appendix III.

Liebig's law of the minimum the concept that the growth or survival of a population is directly related to the life requirement that is in the least supply.

lignite type of coal that is of more recent origin than anthracite or bituminous coal. Also known as brown coal.

limiting factor an environmental factor that limits the growth or activities of an organism or that restricts the size of a population or its geographical range. The amount of food available would be a limiting factor. Other limiting factors would include lack of dissolved oxygen, competition with other species, competition for sunlight, and disease. Essential nutrients can also be limiting factors.

lobbying to try to influence public officials on behalf of or against proposed legislation. Environmental lobbying groups (both pro and con environment) can take the form of

- Public relations firms,
- Corporate front groups,
- Think tanks,
- Legal foundations,
- Endowments and charities, and
- Wise use and share groups.

logistic growth curve a graph of population growth that is the shape of a sigmoid curve (S on its side). Growth rate is initially high due to abundance of natural resources, it then reaches a maximum or peak when natural resources are maximized, and then it begins to decline as natural resources become limited.

low-throughput society matter and energy efficient society accomplished by: (1) reusing and recycling nonrenewable matter resources, (2) using potentially renewable resources no faster than they are replenished, (3) using matter and energy resources efficiently, (4) reducing unnecessary consumption, (5) emphasizing pollution prevention and waste reduction, and (6) controlling population growth. Also known as a low-waste society or Earth-Wisdom society.

low-waste societies or low-throughput societies (Earth-Wisdom Society) a most efficient society. Accomplished by: (1) reusing and recycling nonrenewable matter resources, (2) using potentially renewable resources no faster than they are replenished, (3) using matter and energy resources efficiently, (4) reducing unnecessary consumption, (5) emphasizing pollution prevention and waste reduction, and (6) controlling population growth. In addition, low-waste societies believe that up to 80% of the solids and hazardous wastes could be eliminated through reduction, reuse, and recycling.

magma molten rock. Magma typically consists of (1) a liquid portion (often referred to as the melt), (2) a solid portion made of minerals that crystallized directly from the melt, (3) solid rocks incorporated into the magma from along the conduit or reservoir, called xenoliths or inclusions; and (4) dissolved gases.

Malthusian system of thought named after Thomas Malthus who believed that human population would soon exceed food supplies. The result would be catastrophic famine. Today proponents believe in a pessimistic view that limits on natural resources will result in catastrophic consequences in the world economy.

market economy a decentralized economic system wherein buyers and sellers interact to set prices and demand.

maximum sustained yield (MSY) a population size grows over time to its carrying capacity, which is when the rates of births and deaths exist in a dynamic equilibrium and the population grows no further. As a population grows, its rate of change in size increases, but at a decelerating rate. When the population size is half as large as the carrying capacity, the population's rate of change is at its maximum and declines thereafter, reaching zero when the population reaches carrying capacity. In theory, such a population could be continuously harvested to one-half of its carrying capacity, thereby producing a perpetual, maximum yield without compromising the ability of the population to be replenished.

mediation an attempt to bring about a peaceful settlement or compromise between disputants through the objective intervention of a neutral party.

methane a hydrocarbon with the formula CH_4. A major greenhouse gas. Since 1750, the concentration of methane in the atmosphere has increased 140%. Major sources of additional methane are oil and gas extraction, coal mining, landfills, termites, domestic grazing animals, and rice cultivation.

midlatitude cyclone storm that occurs primarily in the middle latitudes (between 30° and 60° north and south latitudes) and within the boundary of a Ferrel cell, which is caused by a drop in pressure in a large body of warmer air with colder air moving counterclockwise (northern hemisphere) into that area.

monsoon a wind from the southwest or south that brings heavy rainfall to southern Asia in the summer. Created by temperature gradients that exist between ocean and land surfaces. Monsoons occur over very large areas and are seasonal. In summer, humid wind blows from cooler ocean areas (higher pressure) to warmer land masses (lower pressure). As air rises over land masses, it cools and is unable to retain water, producing great amounts of rain. In winter, the ocean is now warmer and the cycle reverses and the drier air travels from the continent out to the ocean. Monsoon winds exist in Australia, Africa, and North and South America.

morals belief system that reflects predominant feelings of a culture about ethical issues. It emphasizes the distinction between right and wrong.

more-developed country a country having an annual gross national product per capita equivalent to $9361 or greater in 1998. Most high-income countries have an industrial economy. There are currently about 28 high-income countries in the world with populations of one million people or more. Also known as a high-income country. See Appendix III.

municipal solid waste (MSW) garbage or trash generated from residential, commercial, institutional, and industrial sources that falls into six basic categories: (1) durable goods, (2) non-durable goods, (3) containers and packaging, (4) food wastes, (5) yard trimmings, and (6) miscellaneous organic and inorganic wastes. It consists of everyday items such as product packaging, grass clippings, furniture, clothing, bottles, food scraps, newspapers, appliances, paint, and batteries. In 1999, U.S. residents, businesses, and institutions produced more than 230 million tons of MSW, which is approximately 4.6 pounds of waste per person per day, up from 2.7 pounds per person per day in 1960. MSW management practices include: (1) source reduction; (2) recycling; (3) composting, and (4) preventing or diverting materials from the wastestream.

mutualism an association between organisms of two different species in which each member benefits.

National Oceanic and Atmospheric Administration (NOAA) a branch of the U.S. Department of Commerce whose mission is to promote global environmental stewardship to conserve and wisely manage the nation's marine and coastal resources, and to describe, monitor, and predict changes in the Earth's environment to ensure and enhance sustainable economic opportunities. For more information, go to http://www.noaa.gov./

natural selection a process whereby organisms that possess inheritable adaptive traits that offer an advantage to an organism's survival achieve a higher rate of reproduction, thereby influencing the makeup of the gene pool. Natural selection is the mechanism of evolution.

naturalism system of thought based on the separation of nature from society and the related proposition that social organization should be modeled on the laws of nature, and that we should live by the laws of nature. At its most radical, naturalism is reflected in what has been called the deep ecology movement, providing the philosophical and tactical underpinning of the most oppositional positions in recent environmental conflict and politics. This has been the case in wilderness and native forest conflicts in America, Canada, and Australia. Naturalism is the common thread in ecocentric moral and social philosophy. Naturalism claims that if nature's authority and laws formed the basis of a new social formation, an ecological society, environmental problems would be solved.

neo-Luddites a group of people who believed that all large-scale human endeavors eventually fail, and that science and technology cause more problems than they solve. They believe in going back to a low-tech pastoral or hunter-gatherer society. An early champion of this philosophy was Ned Ludd.

niche the physical and functional location of an organism within an ecosystem; where a living thing is found and what it does there.

nihilism a philosophy that states that the world makes no sense at all and that everything is arbitrary. The only principle or truth is the instinct to survive. Morals do not exist. It is seen in the works of Schopenhauer.

nitrification a step in the nitrogen cycle in which ammonia is converted by aerobic bacteria to nitrite ions (NO_2^-), which are then converted to nitrate ions (NO_3^-) through bacterial action.

$$NH_3 \xrightarrow{\text{bacteria}} NO_2^- \xrightarrow{\text{bacteria}} NO_3^-$$

nitrogen fixation the process by which nitrogen gas (N_2) which cannot be utilized by plants, is converted to ammonium ion (NH_4^+). Nitrogen fixation is accomplished naturally through microbial action and lightning.

North American Free Trade Agreement (NAFTA) an agreement signed in 1993 between the United States, Mexico, and Canada reducing trade barriers between the three countries. Opposition to NAFTA included concerns of relocation of American companies to Mexico where environmental regulation is not as stringent and labor is cheaper. A subsequent treaty, the North American Agreement for Environmental Cooperation, addressed many of the environmental issues.

NO_x generally refers to nitric oxide (NO) and nitrogen dioxide (NO_2). Both gases are natural components of the Earth's atmosphere. Both gases are involved in acid precipitation and the production of photochemical smog. NO_x are released from burning fossil fuels and biomass. Up to 50 million tons per year are released into the environment from human activity as compared to up to 20 million tons per year from natural processes. Average length of time that a NO_x molecule spends in the atmosphere is two or three days.

oceanic plate a rigid section of the lithosphere beneath the ocean, consisting primarily of granite, that due to density, rides on top of the asthenosphere and is able to move slowly over the Earth's surface. Formed less than several hundred million years ago at one of the Earth's mid-oceanic ridges, oceanic plates have an average thickness of 50 miles (~75 kilometers).

open-pit mining large open pits carved out of the Earth to extract mineral resources.

organophosphates broad and narrow spectrum agents that kill insects by deactivating the chemical in their nervous system that transmits nerve impulses. Examples: TEPP, methyl parathion, and diazinon. Some of them are also used as fungicides and herbicides. Their persistence is low to moderate, lasting anywhere from weeks to several years. Because most of them break down quickly, they can be applied in higher volume and frequency than would otherwise be advisable. One very dangerous aspect of organophosphates is their polarity: they are water soluble and can easily and quickly contaminate water supplies. Although organophosphates do not bioaccumulate or magnify, they are highly toxic to humans and other wildlife. These account for most cases of pesticide poisoning and death.

ozone depletion potential (ODP) a number that ranges from 0.01 to 1.0 and that refers to the amount of ozone depletion caused by a substance. It is the ratio of the impact on ozone of a chemical compared to the impact of a similar mass of CFC-11. Thus, the ODP of CFC-11 is defined to be 1.0. Halons have ODPs ranging up to 10. HFCs have an ODP of 0 because they do not contain chlorine.

ozone layer an atmospheric layer that occurs 11 to 16 miles above the surface of the Earth, in the stratosphere. The concentration is ~12 ppm. The forward rate normally equals the reverse rate. The ozone layer blocks out most high-energy UV-C; half of the next most energetic radiation, UV-B; and only a small portion of lowest energy UV-A. It prevents 95% of UV from reaching Earth. The ozone layer also creates a thermal cap, which prevents warm gases in the troposphere from entering the stratosphere.

passive solar heating system a heating system that uses direct solar radiation for space heating.

plate tectonics a theory that states that the surface of the Earth is divided into massive sections known as plates. The plates move slowly over time, sinking in areas of volcanic island chains, folded mountain belts, and usually trenches and rising up from ridges and rift valleys.

polar jet stream a stream of fast-moving air in the upper atmosphere existing in the midlatitudes (between 30° and 60° north and south latitudes). This stream of air flows from west to east in the northern hemisphere.

polar stratospheric clouds (PSC) clouds that form in cold, polar stratospheric winters where, despite the dryness of the stratosphere, the temperature drops low enough for condensation to occur. PSCs are made of nitric acid and ice. PSCs provide the surfaces upon which chemical reactions involved in ozone destruction take place. These reactions lead to the production of free chlorine and bromine, released from CFCs and other ozone depleting chemicals (ODCs), which directly destroy ozone molecules. Also known as mother-of-pearl or nacreous clouds.

policy life cycle steps that occur for public policies to be enacted:

1. Identify the problem.
2. A group decides on the strategies that will be used to implement a policy.
3. Solutions to the problem are developed and weighed.
4. Build public support for the proposal—special interest groups, lobbyists, etc.
5. Bring the proposal to the proper agency for consideration.
6. If after study, public hearings, modification, and compromise, the majority of the members of the agency or body agree with the policy, it is enacted as law.
7. The effectiveness of the policy and other ramifications are evaluated.
8. Modifications or amendments may be made to the policy.

pollution costs pollution can be seen as good for the economy. Creating pollution adds to the gross national product by: (1) encouraging a "throw away economy" where more material needs to be manufactured rather than being reused; (2) creating jobs to clean up pollution; and (3) spending money to treat those affected by pollution. However, the gross national product does not factor the depletion and degradation of Earth capital. In calculating GNP, all monetary transactions benefit the economy—whether they have positive environmental effects or not. From an economic standpoint, energy saving devices reduce the amount of money people spend, and are therefore economically undesirable.

polychlorinated biphenyls (PCB) mixtures of synthetic organic chemicals that are non-flammable and chemically stable and that have high boiling points and electrical insulating properties. PCBs were used in hundreds of industrial and commercial applications including electrical, heat transfer, and hydraulic equipment; as plasticizers in paints, plastics and rubber products; and in pigments, dyes, and carbonless copy paper. More than 1.5 billion pounds of PCBs were manufactured in the United States prior to cessation of production in 1977. Inuit people of Broughton Island, above the Arctic Circle, have the highest level of PCBs in their blood of any group on Earth.

positivists analytical thinkers including Rene Descartes, Francis Bacon, and Isaac Newton who believed that universal laws of morality could be derived from science. Understanding science would allow understanding of nature and mankind.

poststructuralist or postmodernists philosophers including Jacques Derrida, Jean-Francois Lyotard, and Michael Foucault, who believed there is no universal philosophy. Nobody can ever know anything for sure. There is no universal truth, reality, or meaning. They focus on marginalized, disempowered groups and local, contingent knowledge. No one viewpoint has more validity than another. It can lead to social paralysis if everybody is correct.

precautionary principle the principle that states that when an activity raises threats of harm to human health or the environment, safety should prevail and control measures should be initiated even if cause-and-effect relationships are fully established.

P waves the fastest seismic waves produced by an earthquake. They oscillate the ground back and forth along the direction of wave travel. Also known as a primary wave, compressional wave, or longitudinal wave.

rain shadow effect condition that occurs on leeward (side of mountain away from ocean) side. Moist air from ocean rises when it hits mountains. As air rises, it cools and loses its moisture as rain and snow on windward side. On leeward side, air is dry, and semiarid to arid conditions exist. Examples: eastern side of Sierra Nevada mountain range in California; Himalayas and Karakorum ranges of south Asia, and Mount Waialeale on the island of Kauai.

relativists the Sophists who were proponents of a philosophy that states that moral principles are relative to a particular person, society, or situation. No transcendent or absolute principles apply, regardless of circumstances. The quotation, "There are no facts, only interpretations," by Friedrich Nietzsche, sums up the viewpoint of a relativist.

renewable water supplies freshwater surface water runoff including infiltration of water into underground aquifers that is usable and accessible for human use.

reservoir a man-made facility for the storage, regulation, and controlled release of water. Types of reservoirs include flood control, water supply, and power generation.

- A reservoir gains water through inflows such as rivers and rain.
- Outflows are ways reservoirs lose water such as by evaporation, river outflow, and human usage.
- If the inflow and outflow balance, the reservoir remains the same size.
- Because water is constantly entering and leaving the system, a given quantity of water only stays in the reservoir for a certain amount of time, called the residence time.

Richter scale developed in 1935 by Charles F. Richter of the California Institute of Technology as a mathematical device to compare the size of earthquakes. The magnitude of an earthquake is determined from the logarithm of the amplitude of waves recorded by seismographs. A magnitude 7.0 is 100 times greater than a 5.0. Insignificant (Less than 4.0); Minor (4.0–4.9); Damaging (5.0–5.9); Destructive (6.0–6.9); Major (7.0–7.9); and Massive (over 8.0). As an estimate of energy, each whole number step in the magnitude scale corresponds to the release of about 31 times more energy than the amount associated with the preceding whole number value.

riparian pertaining to the banks of a stream (e.g., riparian vegetation).

risk assessment the qualitative or quantitative estimation of the likelihood of adverse effects that may result from exposure to specified health hazards. The four steps of risk assessment are

1. **Hazard identification**—the determination of whether a particular chemical is or is not causally linked to particular health effect(s);
2. **Dose-response assessment**—the determination of the relation between the magnitude of exposure and the probability of occurrence of the health effects;
3. **Exposure assessment**—the determination of the extent of human exposure; and
4. **Risk characterization**—the description of the nature and often the magnitude of human risk.

risk management the process of minimizing risk that entails consideration of political, social, economic, and engineering information with risk-assessment information by developing systems to identify and analyze potential hazards to prevent accidents, injuries, and other adverse occurrences.

r-strategists small organisms that reproduce many offspring, have a short life span, and generally do not reach carrying capacity. Examples: rats, mice, cockroaches, and grasshoppers.

sanitary landfill rehabilitated land in which garbage and trash have been buried. Of all MSW in the United States, 54% is buried in sanitary landfills. This compares to 90% in the United Kingdom and 80% in Canada. Sanitary landfills solve the problem of leaching to some extent. They are lined with materials that are impermeable such as plastics and clay and are usually built over impermeable soil. Constructing sanitary landfills is very costly. Some authorities claim that the plastic liner develops cracks as it reacts with various chemical solvents present in the waste. The rate of decomposition in sanitary landfills is variable due to the fact that less oxygen is available as the garbage is compressed very tightly. It has also been observed that some biodegradable materials do not decompose in a landfill. Methane gas also develops since little oxygen is present (i.e., anaerobic decomposition). In some countries, the methane being produced from sanitary landfills is tapped and sold as fuel.

second green revolution international agricultural project that began in 1967 with the introduction of fast-growing dwarf varieties of rice and wheat that were bred for tropical and subtropical climates. Output of new genetically-modified plants that can produce up to three crops per year can increase agricultural output up to five times over traditional varieties. By producing more output per acre, the effect is to decrease the effects of deforestation and the conversion of wetlands and grasslands for agricultural purposes. The debate is, however, whether the advantages of the second green revolution really reduce the need for new land given that:

- Food output is tied to economic development;
- The increase in the world population will place even more demand on more agricultural output requiring more land utilization; and
- Techniques employed in the second green revolution require even more use of fertilizer, freshwater, and pesticides and their impact on the environment.

second-law efficiency the ratio of the minimum available work needed to perform a particular task compared to the actual work used to perform the task, expressed as a percentage.

second law of energy or thermodynamics law that states that when energy is changed from one form to another, some of the useful energy is always degraded to lower-quality, more-dispersed (higher-entropy), and a less-useful form of energy.

secondary consumer an animal that feeds on smaller plant-eating animals in a food chain. Also known as a primary carnivore.

sedimentary rock rock that forms from the accumulation of smaller rocks under pressure or from compacted shells and/or remains of organisms. Examples are dolomite, limestone, diatomaceous earth, rock salt, gypsum, lignite and bituminous coal. Sedimentary rock covers nearly three-quarters of the Earth's surface.

shield volcano the largest volcanoes on Earth that have broad, gentle slopes and that were built by the eruption of fluid basalt lava.

sink storage deposit. Example: tropical rainforests and the oceans are important carbon sinks.

solid waste any garbage, or refuse, sludge from a wastewater treatment plant, water supply treatment plant, or air pollution control facility and any other discarded material, including solid, liquid, semisolid, or contained gaseous material resulting from industrial, commercial, mining, and agricultural operations, and from household or community activities. It does not include solids or dissolved material in domestic sewage or other significant pollutants in water resources, such as silt, dissolved or suspended solids in industrial wastewater effluents, dissolved materials in irrigation return flows or other common water pollutants. Examples: paper, biomass or agricultural leftovers, glass, plastic, metals, textiles.

statute of limitations a law that sets the maximum period that one can wait before filing a lawsuit, depending on the type of case or claim.

stewardship responsibility to manage and care for a particular place. The role of humans is to be a caretaker. Intrinsic value is represented by humans and nature whereas the instrumental values are tools. Rene Dubos and Wendell Berry were two of the many supporters of stewardship.

subduction zone an area where oceanic lithosphere is carried downward under an island arc or continent at a convergent plate boundary. The place where two lithosphere plates come together, one riding over the other. Most volcanoes on land occur parallel to and inland from the boundary between the two plates.

subsidy monetary assistance granted by a government to a person or group in support of an enterprise regarded as being in the public interest. Oil subsidies have created a national energy policy by default—a policy that is actually the reverse of stated national priorities by increasing the dependence on foreign oil supplies and burdening taxpayers with unacceptable costs to human health, the environment, and the economy. Federal corporate income tax credits and deductions have resulted in an effective income tax rate of 11% for the oil industry, compared to the non-oil industry average of 18%. If the oil industry paid the industrywide average federal tax rate (including oil) of 17%, the industry would have paid an additional $2.0 billion in 1991. Making matters worse are the losses on a state and local level. Alliance to Save Energy (www.ase.org) found that state and local governments also taxed gasoline at about half the rate as other goods—approximately 3% versus 6% resulting in an estimated $2.7 billion revenue loss from gasoline sales alone in 1991. When home, industry, and office petroleum products are included, the total state and local revenue loss sums to $4.1 billion. When federal and state losses are added, subsidies to the oil companies amounts to almost $7 billion dollars a year (~$25 for each American). Oil imports are equal to almost half of U.S. oil consumption and half of the trade deficit. The situation is likely to worsen with U.S. refineries running at full capacity and the remaining inexpensive oil reserves lying outside U.S. borders, which impacts national military security. This de facto energy policy also discourages private investments in new, cleaner technologies such as electric vehicles. Finally, hidden subsidies waste taxpayer dollars by undermining government programs to promote fuel efficiency, alternative fuels, and environmental protection.

succession the gradual and orderly process of ecosystem development.

sustainable agriculture agricultural practices and procedures that attempt to maintain the productivity of the land for the future.

sustainable yield catch that can be removed over an indefinite period without causing the stock to be depleted. This could be either a constant yield from year to year or a yield that is allowed to fluctuate in response to changes in abundance.

symbiosis a close, prolonged association between two or more different organisms of different species that may, but does not necessarily, benefit each member. There are three common types of symbiosis:

1. *antagonistic or antipathetic symbiosis* is an association that is destructive to one of the symbionts or partners involved in the association. Example: parasitism; one organism of an association benefits at the expense of the other.
2. *conjunctive symbiosis* occurs when there is bodily union (in extreme cases so close that the two form practically a single body, as in the union of algae and fungi to form lichens and in the inclusion of algae in radiolarians).
3. *disjunctive symbiosis* occurs if there is no actual union of the organisms (as in the association of ants with myrmecophytes).

S waves seismic waves that oscillate the ground perpendicular to the direction of wave travel. They travel about 1.7 times slower than P waves. Because liquids will not sustain shear stresses, S waves will not travel through liquids like water, molten rock, or the Earth's outer core. Also known as secondary waves.

temperature inversion an atmospheric condition in which the air temperature rises with increasing altitude, holding surface air down and preventing dispersion of pollutants. There are two types of inversions: (1) radiation temperature inversion—generally occurring at night, a condition where a layer of warm air lies on top of a layer of cooler air. As the sun warms the surface of the Earth, the inversion generally dissipates during the day—dispersing pollutants, and (2) subsidence temperature inversion—large mass of warm air at high altitude moves in and traps colder air near ground, preventing mixing of air and dispersion of pollutants.

third law of thermodynamics law that states that if all molecular motion within a sample ceased, wherein the kinetic energy would be equal to zero, then a state known as absolute zero would result.

total fertility rate (TFR) the number of children an average women would have assuming that she lives her full reproductive lifetime. Replacement TFR is 2.1. Global TFR averages 2.9, developing countries have an average 3.2 TFR, and developed countries have an average 1.3 TFR. Within the United States (average 2.1), Hispanic TFR is ~2.9; Asian TFR, ~1.9; African American TFR, ~2.2, and white TFR, ~1.8. Worldwide TFRs are Italy, 1.2; Europe, 1.6; Israel, 2.6; United States, 2.1; Uganda, 7.3; Gaza Strip, 8.8. For more information on TFRs, visit www.un.org/popin.

toxicity the quality of being poisonous, especially the degree of virulence of a toxic microbe or of a poison. More chemicals are produced than there is time to test their toxicity or interaction effects. Toxicity depends upon

- Bioaccumulation—increase in concentration of a pollutant or toxin from the environment to the first organism in a food chain;
- Biomagnification—increase in concentration of a pollutant from one link in a food chain to another;
- Response an individual has to the toxin, depending on health status, genetic factors, and so on;
- Synergistic or antagonistic interactions;
- Persistence;
- Chemical characteristics of the toxin (solubility, phase, etc.);
- How long the individual was exposed;

- How much of the toxin the individual was exposed to (was the toxin cumulative or non-cumulative?);
- Amount of toxin the individual received.

Toxicity is determined through: (1) epidemiology—detailed studies of people who have been exposed to the toxin; (2) case reports—written records and data of people who have been exposed to the toxin; and (3) laboratory investigations—results from work with laboratory animals extrapolated to human populations.

traditional subsistence agriculture small parcels of land that families use to produce food. The agricultural output is often barely sufficient to feed a family and the purchase of seed and fertilizer is often too expensive. Even if a surplus can be produced, it is difficult to transport it to markets.

universalists the philosophy that states that the fundamental principles of ethics are universal, unchanging, and eternal. The basic rules of right and wrong are true regardless of human interest, attitudes, feelings, or preferences. Some believe universal truth is revealed by God; others believe it is derived through human knowledge and reason. Plato and Kant were universalists.

urban renewal the planned upgrading of a deteriorating urban area, involving rebuilding, renovation, or restoration. It frequently refers to programs of major demolition and rebuilding of blighted areas.

utilitarianism a philosophy that claims that something is right if it produces the greatest good for the greatest number of people as seen in the works of Jeremy Bentham, Socrates, and Aristotle. It equates goodness with happiness and happiness with pleasure. The good life is the one that gives the greatest pleasure; it can be totally hedonistic or modified as by John Stuart Mill. Mill believed that intellectual pleasure was superior to physical pleasure. Therefore the greatest pleasure is to be educated. This viewpoint is called "enlightened utilitarianism" and inspired Gifford Pinchot and early conservationist to believe that conservation should be "for the greatest good for the greatest number for the longest time." Utilitariansism can be used to justify or rationalize human behavior (e.g., bullfighting).

values a principle, standard, or quality considered worthwhile or desirable; the ultimate worth of actions or things.

watershed a ridge of high land dividing two areas that are drained by different river systems or the region draining into a river, river system, or other body of water. Timber cutting on the Chang Jiang (Yangtze) River watershed in 1998 killed 30,000 people due to flooding. Also known as catchment or drainage basin.

watt a unit of power equal to one joule per second.

weather the state of the atmosphere at a given time and place, with respect to variables such as temperature, moisture, wind velocity, and barometric pressure.

Index